# Animal Behavior

## NINTH EDITION

# Animal Behavior
## *An Evolutionary Approach*
### NINTH EDITION

### John Alcock
#### ARIZONA STATE UNIVERSITY

Sinauer Associates, Inc. • Publishers
Sunderland, Massachusetts  U.S.A.

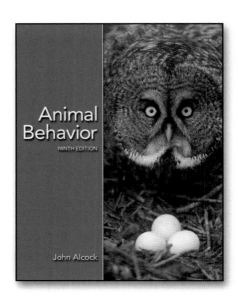

## About the Cover

The great grey owl makes itself look larger, and thus more dangerous, by erecting its feathers when confronting an intruder near its nest. Photograph © Michael Quinton/Minden Pictures.

Animal Behavior: An Evolutionary Approach, Ninth Edition
Copyright © 2009

For information, address:
Sinauer Associates, Inc., 23 Plumtree Road, Sunderland, MA 01375 U.S.A.
Fax: 413-549-1118
Email: publish@sinauer.com
Internet: www.sinauer.com

Library of Congress Cataloging-in-Publication Data

Alcock, John, 1942-
  Animal behavior : an evolutionary approach / John Alcock. -- 9th ed.
     p. cm.
  Includes bibliographical references and index.
  ISBN 978-0-87893-225-2 (pbk. : alk. paper)
  1. Animal behavior--Evolution.  I. Title.
  QL751.A58 2009
  591.5--dc22

                                        2008050144

Printed in China

10  9  8  7  6  5  4  3  2  1

*In memory of my parents,*
*John P. and Mariana C. Alcock*

# Contents in Brief

# Contents

# Preface

I am completing this ninth edition of my textbook soon after having retired from the School of Life Sciences at Arizona State University. I began my long and happy tenure as a university professor in classrooms dominated by chalkboards, which were then supplanted by pull-down screens and overhead projectors, before PowerPoint presentations became standard operating procedure. Concurrently, I saw manual typewriters give way to electric ones, which disappeared in the competition with the first personal computers running on MS-DOS, whatever that was, to the latest version of Windows, wireless connections, thumbdrives and the like. The standard 50-minute uninterrupted lecture has largely gone the way of the dodo, and now most of my still-employed colleagues use versions of the Socratic method or divide their classes into teams for cooperative learning exercises or rely on electronic personal response systems so that students can click their way into an interactive, 21st century education.

The field of animal behavior has changed mightily as well during my professorial career, which began in 1969, the year of my first job, to 2008, the year when I said good-bye to teaching at ASU. When I started out, I wanted to write a textbook in order to bring news to my students of a revolution in the field that was not covered by the textbooks of the time. This revolution was stimulated by the work of W. D. Hamilton, who showed us how to use gene thinking to identify interesting theoretical challenges, like how could self-sacrificing altruism evolve in animal species, and George C. Williams, who knew why for-the-good-of-the-group explanations for behavior were almost certainly wrong. I had barely absorbed these messages myself 40 years ago but I knew that they were hugely important and well worth presenting to students who were then only a little younger than I was.

After the first edition came out in 1975, I continued to keep tabs on developments in animal behavior in order to revise the textbook every four or five years. I could and did rationalize revision on the grounds that every year brought a host of excellent new papers on behavior. The nature of science is such that the earth is constantly changing under our feet as old ideas are revisited and changed or thrown out altogether while their replacements are presented, challenged, and modified in turn. Therefore, it was both possible and desirable to update my textbook regularly. Of the 502 reference citations in the first edition, only a handful have survived the multiple revisions to date and are still used in this, the 9th edition, which features over 1500 references. Science is not static.

In this edition, I have followed the rules of revision used in many of my preceding editions. My main goal has been to add information from recent papers that change or amend conclusions based on what is now older material as well as to provide readers with newer, fresher examples that illustrate particular points of theoretical importance. With each revision, I have been impressed with both the quantity and quality of published research in animal behavior available to a textbook writer. Experimental procedures are more sophisticated, statistical tests more complex, reliance on the hypothetico-deductive method more pervasive than in the past.

So these days I have an embarrassment of riches to consider when it comes to updating and revising my text. In selecting what to use, I have been helped by one of my early Ph.D. students, David Skryja, who sends me (and a number of other colleagues) news releases on behavioral research that has caught his eye. I save these items in order to identify promising research reports as I begin the revision process. I track down the original articles via the Web of Science (http://www.webofscience.com), a wonderful search engine that every undergraduate with an interest in science should learn about and use. Highly recommended.

Subsequently, the revised chapters go out for review to persons who not only catch errors in what I have written but also direct my attention to fine papers that I did not know about. As always, their usually gentle criticisms help me avoid making a hash of things and steer me toward an improved version. This time around, I have had the benefit of help from Steve Phelps (Chapter 1), Sarah Woolley (Chapter 2), Maydianne Andrade (Chapter 3), Ken Catania (Chapter 4), Brian Trainor (Chapter 5), Don Owings (Chapter 6), Todd Blackledge (Chapter 7), Anthony Zera (Chapter 8), Stephanie Dloniak (Chapter 9), Gerry Borgia (Chapters 9 and 10), Dustin Rubenstein (Chapter 11), Jeff Hoover (Chapter 12), David Queller (Chapter 13), and Joan Silk (Chapter 14). Taewon Kim and Juergen Leibig also alerted me to errors in the eighth edition, so that I could correct them this time around. I haven't accepted every bit of advice I received, although I probably should have, but the changes I have made have been all to the good.

Many colleagues have generously provided illustrations for this edition. I have acknowledged the collegial suppliers of photographs and drawings in the caption to the relevant figures in the text but would like to especially thank Bob Montgomerie for working with me on the box on microsatellite analysis (see page 389). Doug Emlen was also exceptionally helpful; he spent much time constructing an entirely new figure for Chapter 10 (see page 343). Acknowledgements to publishers who have kindly granted permission to use their copyrighted material appear in the Illustration Credits.

My editor at Sinauer Associates, Graig Donini, has been hugely helpful over the years. Among his many contributions has been to organize the chapter reviews that make the revision process better. I have also received much assistance from other members of the Sinauer Associates team, including Joanne Delphia, Laura Green, David McIntyre, Elizabeth Morales, and Chris Small, all of whom have been extraordinarily helpful and competent at their respective publishing specialties. My copy editor, Lou Doucette, did what good copy editors do, which is to clean up prose that needs help, a task that requires immense patience and a meticulous sense of detail. Thanks for that.

Even though I am retired, my chair at the School of Life Sciences, Rob Page, has not dragged me into the street, but has encouraged me to remain on campus and to this end he has permitted me to retain my office, for which I thank him. I have been spending most weekdays in my office at ASU, the better to have lunch with my non-retired colleagues, the same group who has been with me for decades now (except for Jim Collins who moved on to an administrative post at the National Science Foundation). With Dave Brown, Stuart Fisher, Tony Lawson, Dave Pearson, and Ron Rutowski still alive and kicking, we continue to have a quorum most weekdays for a robust discussion of character defects in higher administrators and assorted companions (provided they are not present at the table), our views on a full spectrum of political issues and personalities, how to prepare financially and psychologically for retirement, and what decent movies are in town or available through Netflix, all matters of much import to us.

One aspect of life has remained constant over my academic lifetime: I remain married to Sue Alcock, a tolerant and accommodating wife who still talks to me, as do our two grown sons, Joe and Nick, one whom (Joe) has during the last four years provided us with a daughter-in-law (Satkirin) while the other (Nick) has in collaboration with his wife Sara come up with a granddaughter for us named Abby, who calls me Yoyo, if she is in a good mood. Abby and company make my existence richer and more amusing than it would be otherwise, for which I am grateful.

# Media & Supplements

*to accompany* Animal Behavior,
*Ninth Edition*

### NEW! eBOOK (ISBN 978-0-87893-344-0; www.coursesmart.com)

New for the Ninth Edition, *Animal Behavior* is available as an eBook via CourseSmart, at a substantial discount off the price of the printed textbook. The eBook reproduces the look of the printed book exactly, and is available either online or as a download. Features include convenient tools for searching the text, highlighting passages of text, and adding notes.

*For more information, please visit www.coursesmart.com*

### INSTRUCTOR'S RESOURCE LIBRARY (ISBN 978-0-87893-285-6)

Available to qualified adopters of *Animal Behavior,* the Ninth Edition Instructor's Resource Library contains a variety of teaching and laboratory resources. The IRL includes the following elements:

#### Textbook Figures and Tables

All of the textbook's figures, tables, and photographs are provided in both JPEG (high- and low-resolution) and PowerPoint® formats. All of the images have been formatted and optimized for excellent projection quality.

#### NEW! Animal Behavior Video Collection

New for the Ninth Edition, this collection of video segments brings to life many of the specific behaviors discussed in the textbook. A great many of these high-quality segments were selected by the author from the collection of the Cornell Lab of Ornithology's Macaulay Library. Great for use in class, the segments are short and easy to incorporate into lectures. All segments are provided both as MPEG movies and in ready-to-use PowerPoint® presentations.

*For a sample of the new video collection, please visit www.sinauer.com/alcock9e/sample*

#### Teaching Animal Behavior:
#### An Instructor's Manual to accompany Animal Behavior, *Ninth Edition*

*Teaching Animal Behavior* provides instructors with several resources to facilitate the preparation of lectures, quizzes, and exams. Contents include:

- Answers to the discussion questions presented in the textbook
- Sample quiz questions and answers
- Sample exam questions and answers
- Descriptions to accompany the new collection of animal behavior videos
- A listing of films on animal behavior for use in the classroom

### Learning the Skills of Research: Animal Behavior Exercises in the Laboratory and Field

Edited by Elizabeth M. Jakob and Margaret Hodge

Students learn best about the process of science by carrying out projects from start to finish. Animal behavior laboratory classes are particularly well-suited for independent student research, as high-quality projects can be conducted with simple materials and in a variety of environments. The exercises in this electronic lab manual are geared to helping students learn about all stages of the scientific process: hypothesis development, observing and quantifying animal behavior, statistical analysis, and data presentation. Additional exercises allow the students to practice these skills, with topics ranging from habitat selection in isopods to human navigation. Both student and instructor documentation is provided. Data sheets and other supplementary material are offered in editable formats that instructors can modify as desired.

# Animal Behavior

## NINTH EDITION

# 1

# An Evolutionary Approach to Animal Behavior

For hundreds of thousands of years, our ancestors observed animals hungrily, learning the fine details of their behavior in order to put the next meal on the table. Even today, the subject of animal behavior still has great practical significance. If you'd like to maximize the production of wood ducks for duck hunters, then you had best determine the consequences of putting out lots of conspicuous nest boxes in the ducks' breeding areas.[1305] If you wish to protect your cotton crop against the pestiferous pink bollworm, it might be good to know how adult females of this moth attract mates with special scents, the better to interfere with these signals.[228] Likewise, if you think it desirable to reduce the incidence of date rape, then perhaps you should learn something about the biological basis of human sexual behavior.[1438] Or if your goal is to minimize the ever-increasing environmental destruction caused by our species, then why not get to the evolutionary roots of the problem?[1115, 1116]

Although some people study animal behavior to help solve one of the many problems facing human societies today, others become behavioral biologists simply because they find the subject intrinsically fascinating. Among these researchers are those who want to find out why males of some, but not all, species of dragonflies fly around holding onto their mates long after they have copulated with them. Others would like to know how night-flying moths avoid getting nailed by bats, which maneuver so much faster than their prey. And why do some seabird parents sit back and let their two babies fight until one has killed the other? And how can male red-sided garter snakes copulate in spring, when they have almost no testosterone circulating through their bodies?

In the pages ahead, you will learn something about dragonfly mate guarding, moth evasive tactics, avian siblicide, and the relationship between testosterone and sexual activity in red-sided garter snakes. The practical applications of these particular findings are modest, but there is something fundamentally satisfying about the study of animal behavior, a satisfaction regularly experienced by the thousands of curious naturalists who try to figure out why animals do what they do. Those researchers have provided the material that appears in my textbook. I hope that you will find their discoveries worth learning about, even entertaining on occasion. But in addition, I would like you to understand how scientists reach conclusions of the sort that are presented here. I believe that the process of doing science is every bit as interesting as the findings that are its end product. Therefore, the focus throughout this textbook will be on how scientific logic promotes effective thinking and leads to convincing conclusions. So let's get started by examining what some scientists have learned about the behavior of the prairie vole.

## Understanding Monogamy

The prairie vole (*Microtus ochrogaster*) is a mouselike mammal that lives in burrows in grasslands in the central parts of the United States and southern Canada. In almost every respect, the drab little prairie vole is nothing to write home about. But one feature of its behavior does stand out: the prairie vole is often **monogamous** (Figure 1.1). In other words, many male prairie voles content themselves with a single sexual partner, with whom they stay for an entire reproductive cycle, sometimes for a lifetime.[532] In contrast, males of many other voles (of which there are dozens of species), and most other mammals, are generally polygynous. Unlike the more or less monogamous (one male–one female) prairie vole, males of polygynous species, such as the meadow vole (*Microtus pennsylvanicus*), roam from one female to another and another, and some persuade several females to copulate with them in a single breeding season.

So why should male prairie voles be capable of monogamy when most other mammalian males are polygynous? One research team, led by Larry Young, has come up with an answer.[865] The team found that cells in certain parts of male prairie vole brains are loaded with protein receptors that bind chemically with a hormone called vasopressin. Vasopressin is produced and released into the bloodstream by other brain cells when a vole copulates a number of times with a given female. Molecules of vasopressin are carried to the *ventral pallidum*, the label given a particular structure near the brain's base in mammals and other vertebrates that plays an important role providing rewarding sensations associated with certain behaviors. When receptor proteins in the ventral pallidum, called V1a receptors, are stimulated by vasopressin, they trigger activity in the receptor-rich cells (Figure 1.2). This activity in turn affects neural pathways in the brain that provide the vole with positive feedback.

Researchers believe that these rewards encourage the male vole to remain in the company of his mate, forming a long-term social bond with her. In con-

**FIGURE 1.1   The monogamous prairie vole.** In this species, males in at least some populations form long-term relationships with females, pairing off as couples that live together and coordinate their parental care activities. Photograph by Lowell Getz.

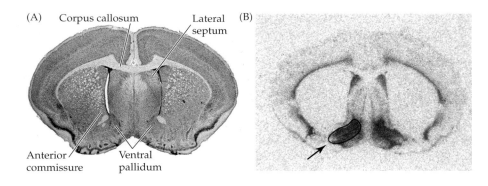

(A) Corpus callosum    Lateral septum

Anterior commissure    Ventral pallidum

(B)

**FIGURE 1.2**   **The brain of the prairie vole** is a complex, highly organized machine. (A) A cross section of the brain with just a few of its anatomically distinct regions labeled. The ventral pallidum contains many cells with receptor proteins that bind to the hormone vasopressin. (B) A brain section that has been treated in such a way that regions with large numbers of vasopressin receptors appear black. The ventral pallidum occurs in both the left and right halves of the brain; the left-hand portion of the ventral pallidum is outlined in black (arrow). After Lim et al.[865]

trast, the reward system in the brains of males of polygynous vole species is different, in part because V1a receptors are less numerous in the ventral pallidum in these species. As a result, when these males copulate, their brains do not provide the same kind of feedback that leads to the formation of a durable social attachment between a male and his mate. Therefore, these males move on after copulating, instead of staying put.

In addition, Young and his colleagues have also provided a somewhat different explanation for why some prairie vole males live with one female for life, an answer that focuses on a possible genetic basis for the monogamous mating system rather than on the fine details of brain cell functioning.[1149] Young's group knew that the V1a receptor protein, which is so important in the prairie vole's vasopressin-based system of social bonding, is encoded by a specific gene, the *avpr1a* gene. The prairie vole's *avpr1a* gene has a specific chunk of DNA that is lacking in the polygynous montane vole's version of the same gene. The extra DNA of the prairie vole's gene might increase the abundance of vasopressin receptors in the ventral pallidum.

Suspecting that the *avpr1a* gene had something to do with the mating systems of voles, Young and his team reasoned that if they could transfer extra copies of the prairie vole's form of the *avpr1a* gene into the right cells of the right part of the brain of male prairie voles, then they should be able to make these voles bond even more eagerly with a female than they would naturally. Using a safe viral vector, the researchers inserted extra copies of the gene in question directly into cells in the ventral pallidum in their prairie vole subjects. Once in place, the "extra" genes enabled these particular cells to make even more V1a receptor proteins than they would have otherwise. The genetically modified, receptor-rich males did indeed form especially strong social bonds with female companions, even if they had not mated with them (Figure 1.3). Apparently, by increasing the number of brain cells with an active form of one particular gene, the research team was able to boost the tendency of male prairie voles to remain close to a social partner. They concluded, therefore, that the *avpr1a* gene contributes to the monogamous behavior of male prairie voles in nature.[1632]

Other researchers have come up with still another explanation for why prairie voles are monogamous. Jerry Wolff and his coworkers argue that monogamy occurs in this species because prairie vole males that formed close attachments to a mate in the past left more descendants than males with a tendency toward polygyny.[1611] Wolff and company think that a male that bonds with one mate is reproductively successful because he can keep his female partner from

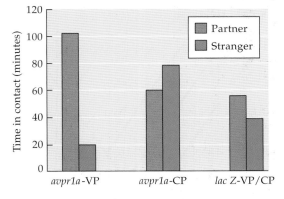

**FIGURE 1.3**   **A gene that affects male pairing behavior in the prairie vole.** Males that received added copies of the *avpr1a* gene (also known as *V1aR*) in the ventral pallidum (*avpr1a*-VP males), when given a choice between spending time with a familiar female or with a strange one, spent significantly more time in contact with the familiar female over a 3-hour trial period. In contrast, familiar partners were not significantly preferred by males that received the *avpr1a* gene in the caudate putamen of the brain (*avpr1a*-CP males) or by males that received a different gene, the *lacZ* gene, in the ventral pallidum or the caudate putamen. After Pitkow et al.[1149]

copulating with other males. They found that when they experimentally prevented male prairie voles from associating with their mates in a laboratory setting, 55 percent of the females copulated with more than one male. (The females in this experiment could choose among three males that had been tethered so that they could not interfere with one another, nor could they prevent the females from leaving them and moving on to another male.) In nature, males that left their mates might well lose them to other males, which would translate into reduced paternity for the less monogamous males.

So here is a very different answer to the monogamy puzzle, one that revolves around the possible reproductive benefits to males that prevent their mates from mating with other males. According to this view, prairie vole monogamy exists because in the past, males that lived with their mates, and thereby kept them under surveillance, sired most or all of the offspring of their monogamous partners. This reproductive tactic apparently has resulted in more descendants for males than if they had employed the alternative tactic of mating and moving on. This outcome would be especially likely if prairie voles were sparsely distributed in the past, as in fact they often are today.[533] In low-density populations, a male that left one female would have difficulty finding another available partner, particularly if other males guarded their mates.

Thus, under some conditions, monogamy can actually enhance male reproductive success, even though monogamous males, by definition, forgo having offspring with more than one female. If mate-guarding males in the past tended to have more surviving descendants than males that behaved in other ways, then those monogamous individuals would have shaped the evolutionary history of their species. According to this argument, when we see monogamy in today's prairie voles, we are looking at the historical result of reproductive competition between males that differed in their mating tactics.

But in addition to explaining monogamy in terms of the reasons why this mating system spread through the vole species some time ago, we can also provide a different sort of historical explanation for the behavior. This other angle requires that we trace the sequence of events that took place during evolution as monogamy originated and spread among some vole lineages (see Box 1.1 on how this can be done). There was surely a time when the prairie vole, or, more likely, some now extinct species that eventually gave rise to the modern prairie vole, was not monogamous. As noted above, in the overwhelming majority of living mammalian species, males attempt to mate with more than one female, which suggests that this pattern probably also prevailed long, long ago in the lineage that eventually evolved into the first vole. Indeed, not all voles are monogamous; the red-backed vole, for example, appears to exhibit the ancestral pattern of polygyny even today (Figure 1.4). If polygyny was the original mating system, then at some time in the past, polygyny gave way to monogamy in a population ancestral to modern prairie voles. Judging from the data presented in Figure 1.4, the shift toward monogamy may have occurred in an ancestral species that gave rise to two modern genera, *Lemmiscus* and *Microtus*. A number of the species in these genera combine tendencies toward male parental care and monogamy. Within this group is the prairie vole.

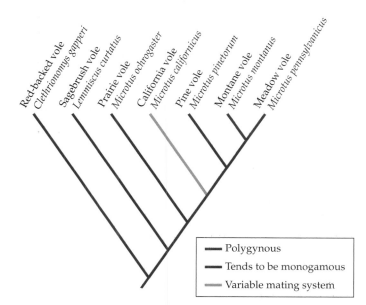

**FIGURE 1.4   The evolutionary relationships of the prairie vole and six of its relatives.** The diagram suggests that a tendency toward monogamy may have originated early in the history of the genus *Microtus* and been retained in the monogamous living species within this group, which includes the prairie and pine voles. Polygyny as the male mating system was almost certainly the ancestral state that preceded the evolution of paternal care and mate guarding in *Microtus*, but a few modern species in the genus (e.g., the montane and meadow voles) have evidently shifted from monogamy back to polygyny. Data from Conroy and Cook[293] and McGuire and Bemis.[968]

## BOX 1.1   *How are phylogenetic trees constructed and what do they mean?*

The diagram in Figure 1.4 is intended to represent the evolutionary history of seven of the modern species of voles living today. To create a **phylogeny** of this sort, it is necessary to determine which species are most closely related to each other, and thus which are descended from a more recent common ancestor. Phylogenetic trees can be drawn up on the basis of anatomical, physiological, or behavioral comparisons among species, but more and more often, molecular comparisons are used for their construction. The molecule DNA, for example, is very useful for this purpose because it contains so many "characters" on which comparisons can be based, namely, the specific sequences of nucleotide bases that are linked together to form an immensely long chain. Each of the two strands of that chain has a *base sequence* that can now be read automatically by the appropriate machine. Therefore, one can, in theory and in practice, compare a cluster of species by extracting a specific segment of DNA from either the nuclei or mitochondria in cells from each species and identifying the base sequence of that particular segment.

For the purposes of illustration, here are three made-up base sequences from one of the two strands of DNA that constitute part of a particular gene found in three made-up species of animals:

| Species X | A T T G C A T A T G T T A A A |
| Species Y | A T T G C A T A T G G T A A A |
| Species Z | G T T G T A C A T G T T A A T |

These data could be used to claim that species X and Y are more closely related to each other than either is to species Z. The basis for the claim is the fact that the base sequences of species X and Y are nearly identical

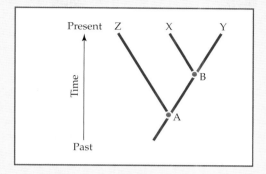

(differing by a single change in position 11 of the chain), whereas species Z differs from the other two by four and five base changes, respectively.

The shared genetic similarity between X and Y can be explained in terms of their history, which must have featured a very recent common ancestor (B in the phylogenetic tree shown here). Species B must have split so recently into the two lineages leading to living species X and Y that there has not been sufficient time for multiple mutations to become incorporated in this segment of DNA. The lesser, but still substantial, similarities among all three species can be explained in terms of their more ancient common ancestor, species A. The interval between the time when A split into two lineages and the present was long enough for several genetic changes to accumulate in the different lineages, with the result that species Z differs considerably from both species X and species Y.

## Discussion Question

**1.1**   Use the data in Figure 1.4 and information in Box 1.1 to outline the history of the polygynous mating system of the montane vole. Why does the meadow vole also exhibit polygyny? How similar genetically are the montane and meadow voles? The red-backed vole and the montane vole?

The change from polygyny to monogamy in an ancestor of the prairie vole may have originated in an ancestral species in which low-ranking males developed a novel tactic, infanticide, to combat the ability of a few dominant males to control many females. By killing the pups sired by other males, a subordinate male might have been able to mate with the pups' mother and produce pups of his own (a topic covered in more detail later in this chapter). Females that then mated with a number of males might have benefited if these males were loath to kill the pups of their mates. In response to female promiscuity, males that acquired a mate might have employed a countermeasure, guarding the female to protect her against such competitors. But **mate guarding** restricts a male's mobility, reducing his chances to acquire several mates in short order. If under these circumstances a mate-guarding male also

**FIGURE 1.5    The possible history behind monogamy in the prairie vole.** Evolution consists of one change layered on another. Monogamy in prairie voles may be the product of a series of behavioral shifts, with polygyny leading to infanticide, which may have favored females that tended to mate with several males to keep the males confused about the paternity of their young. But female promiscuity could have then favored any male that guarded a single mate and even cared for the offspring she produced, leading to the kind of monogamy exhibited by today's prairie voles. After Wolff and Macdonald.[1612]

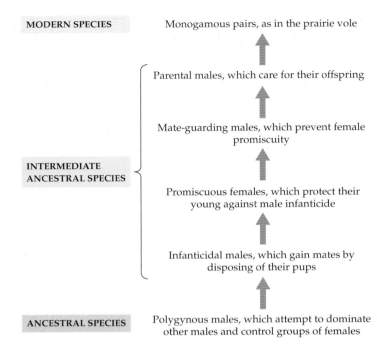

MODERN SPECIES — Monogamous pairs, as in the prairie vole

Parental males, which care for their offspring

INTERMEDIATE ANCESTRAL SPECIES

Mate-guarding males, which prevent female promiscuity

Promiscuous females, which protect their young against male infanticide

Infanticidal males, which gain mates by disposing of their pups

ANCESTRAL SPECIES — Polygynous males, which attempt to dominate other males and control groups of females

produced more surviving young by protecting his partner's pups, male parental behavior would have provided a further impetus for monogamy. Once the trait appeared in one ancestral species, it could have been retained in some or all of the descendant species that trace their history back to that ancestor. In other words, the unusual monogamous mating system of today's prairie voles may be the result of a series of changes in populations that preceded them, including the change from polygyny to monogamy (Figure 1.5).

### Levels of Analysis

So which answer is correct? Is prairie vole monogamy the product of brain physiology, or is it due to a special form of a gene that is present in the prairie vole's brain cells, or is it the product of the reproductive benefits gained by mate-guarding males, or does it stem from a series of modifications that gradually converted a polygynous species into a monogamous one? Here we come to a critical point, central to everything that follows in this book: all four of these answers could be right, because none of these explanations for prairie vole monogamy precludes any of the others. In the jargon of behavioral biology, these four different explanations for monogamy represent different *levels of analysis* (Table 1.1). Each level potentially contributes one element that could be integrated into a satisfyingly complete explanation for the behavior.[1316]

Let's illustrate the connections between the different levels of analysis of prairie vole monogamy in the following way. Imagine that at some point in evolutionary time, a male vole, perhaps a member of a species that no longer exists, happened to have a mutant (altered) gene that somehow changed the way he behaved toward his mate (perhaps by altering the number of vasopressin receptors in his brain). Instead of loving her and leaving her, this male remained with the female, and in so doing, he prevented her from mating with other males and sired all her offspring. Imagine also that because of his behavior, this novel mate-guarding male was somewhat more successful at producing descendants than the other males of his species, which tended to seek out many copulatory partners instead of keeping an eye on just one. Because of the differences in the reproductive success of the two genetically

| TABLE 1.1  *Levels of analysis in the study of animal behavior* | |
| --- | --- |
| **Proximate Causes** | **Ultimate Causes** |
| 1. Genetic–developmental mechanisms<br>    Effects of heredity on behavior<br>    Development of sensory–motor systems<br>      via gene–environment interactions<br>2. Sensory–motor mechanisms<br>    Nervous systems for the detection of environmental<br>      stimuli<br>    Hormone systems for adjusting responsiveness to<br>      environmental stimuli<br>    Skeletal–muscular systems for carrying out responses | 1. Historical pathways leading to a current behavioral trait<br>    Events occurring over evolution from the origin of the<br>      trait to the present<br><br>2. Selective processes shaping the history of a behavioral trait<br>    Past and current usefulness of the behavior in promoting a<br>      lifetime of reproductive success |

*Sources*: Holekamp and Sherman,[666] Sherman,[1316] and Tinbergen[1449]

different kinds of males, the genetic makeup of the next generation changed, with the special form of the gene associated with mate-guarding monogamy becoming somewhat more common.

If this pattern were repeated generation after generation, the ancestors of today's prairie voles would have shifted from polygyny to monogamy as the frequencies of the different forms of various genes changed. Imagine that a particular form of the *avpr1a* gene had a developmental effect that contributed to the kind of monogamy associated with higher reproductive success. Male parents with this gene would pass it on to their relatively numerous offspring, where it would be used during the development and operation of vole brains in the next generation.

Voles with the modern form of the *avpr1a* gene have cells in the ventral pallidum that acquire a certain amount of the protein encoded by this gene. The protein in question acts as a receptor for vasopressin, which is released when male prairie voles copulate with their mates. The chemical interaction between vasopressin and the V1a receptors in cells of the ventral pallidum causes a cascade of neural activity that rewards a male for being close to a female. Thus, to secure additional positive reinforcement, a male forms a social attachment to one copulatory partner and remains with her over the long haul, guarding her against other males. If monogamous males in the current generation leave more descendants on average than any novel types that happen to have genetically different mating tendencies, then most males of the next generation will continue to exhibit the mating behavior that leads to monogamy, the mating system that has been more successful in the past than any other option that has arisen. But if in the future, the environment of the prairie vole were to change in certain ways, any males that happened to have a hereditary tendency to be polygynous might well begin to leave more descendants than their more monogamous conspecifics; if so, the population would evolve away from monogamy and toward polygyny. This capacity to change with changing environmental circumstances demonstrates that species have no preset endpoint, no evolutionary goal.

## Proximate and Ultimate Explanations in Biology

In sketching this compound explanation for prairie vole monogamy, we have integrated the different levels of analysis pursued by different teams of behavioral researchers. These teams have discovered something about (1) how a gene contributes to the development of the behavior in male voles, (2) the physiological foundation for the behavior in terms of the operation of the male vole's brain, (3) the **adaptive value** of the behavior in terms of its contribu-

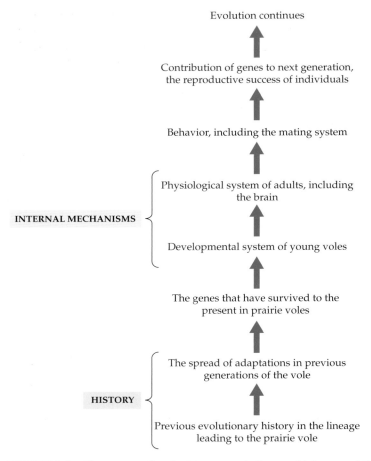

Evolution continues

Contribution of genes to next generation, the reproductive success of individuals

Behavior, including the mating system

**INTERNAL MECHANISMS** {

Physiological system of adults, including the brain

Developmental system of young voles

The genes that have survived to the present in prairie voles

**HISTORY** {

The spread of adaptations in previous generations of the vole

Previous evolutionary history in the lineage leading to the prairie vole

**FIGURE 1.6   The connection between evolutionary history and the mechanisms of behavior.** Evolutionary processes determine which genes survive over time. The genes that animals possess influence the development of the mechanisms that make behavior possible. Behavior affects the genetic success of individuals in the current generation. Evolutionary change is ongoing.

tion to male reproductive success, and (4) the transformation of a polygynous ancestor into the monogamous modern prairie vole (Figure 1.6). But we can consolidate these four levels of analysis into two larger categories, which are labeled *proximate* and *ultimate*.[957, 1071]

Proximate explanations about behavior deal with what is responsible for the building and operation of an animal that enables it to behave in a particular way. We can, for example, view the prairie vole as a machine with internal devices that make it capable of being monogamous. These devices include the sensory mechanisms that come into play during copulation, the brain cells that release vasopressin after mating takes place, the system that transports excreted vasopressin to other brain centers, cells in the ventral pallidum with their receptor proteins that bind with vasopressin, the visual and olfactory systems that provide inputs that enable males to recognize particular females with whom they have mated, and on and on.

All of these proteins, cells, and systems of connected brain regions emerge during the course of genetically guided development. As a male vole grows from a fertilized egg to an adult male, simple chemical reactions shape which genes are transcribed into mRNA and then translated into functional proteins.

These hereditary proteins interact with other chemicals within the growing vole to create a diverse pool of specialized cells. These cells constitute the complex battery of internal devices that can form memories about a copulation, release vasopressin under special circumstances, and motivate a male to stay close to a female.

Therefore, we can explain how an animal is able to behave in particular ways if we understand, first, how that animal developed and, second, how its internal mechanisms work after they are constructed. In effect, by looking inside the creature, we may be able to figure out what causes it to behave one way instead of another. Because these internal developmental and physiological causes occur during one animal's lifetime, they are said to be the immediate, or **proximate**, causes of its behavior.

In contrast, another set of causes of an animal's behavior can be traced to events that took place over many generations. Prairie vole behavior is the product of a long history, during which time some forms of certain genes have survived while other forms have not. Voles that reproduced successfully in the past transmitted their special genes to future generations, with the result that these chunks of DNA are still around to influence the development of each living vole's nervous system, muscles, hormones, and skeleton. In contrast, voles that failed to reproduce took any different forms of genes they may have had to the grave with them. If we can determine why some voles were able to reproduce better than others in the past, then we gain understanding about why some genes have survived over time. This, in turn, provides a long-term explanation for the existence of particular proximate causes of behavior.

Furthermore, if we can secure information on the precise sequence of events that took place over the long term, we provide an extra historical dimension to the explanation for monogamy. We could in theory learn what mating system was replaced when monogamy first appeared in the prairie vole, when that original form of monogamy arose and what it was like, and how many modifications subsequently took place before the modern mating system of today's prairie voles was fully in place.

Therefore, if we want to know why prairie voles are monogamous, we need to know more than the proximate, or immediate, causes of male behavior. We need to know things about the history of the species, the long-term processes that gradually shaped vole attributes over time. Because these historical causes involve events that happened in previous generations, they are said to be the evolutionary, or **ultimate**, causes of behavior.

## Discussion Questions

**1.2**  The four main questions for behavioral researchers according to the great Niko Tinbergen[1445] can be paraphrased as follows:

1. How does the behavior promote an animal's ability to survive and reproduce?
2. How does an animal use its sensory and motor abilities to activate and modify its behavior patterns?
3. How does an animal's behavior change during its growth, especially in response to the experiences that it has while maturing?
4. How does an animal's behavior compare with that of other closely related species, and what does this tell us about the origins of its behavior and the changes that have occurred during the history of the species?

Place these questions within the four levels of analysis framework and then assign each to the proximate or ultimate category. If you heard that because evolutionary questions are "ultimate" ones, they are therefore more important than questions about proximate causes, you would respectfully disagree. Why?

**1.3**   When a female baboon copulates, she vocalizes loudly, but her cries are longer and louder if her partner happens to be a high-ranking, "alpha" male.[1306] A primate researcher has suggested that females cry out more vigorously when copulating with top males because this warns low-ranking baboons to stay clear. (Subordinate males sometimes harass mating pairs to such an extent that the copulation ends prematurely, but if that happens, they may be attacked by the dominant male whose mating they so rudely interrupted.) The same researcher also says, however, that the more vigorous cries may simply reflect the fact that females are more strongly stimulated by the larger, more energetic, alpha males. Are the two explanations in competition with each other? Explain why they could both be right.

## How to Discover the Causes of Behavior—Scientifically

In describing the findings of various research teams on what causes prairie voles to be monogamous, I presented the conclusions that the different groups reached, emphasizing that these explanations were different, but complementary. But why did these researchers think that their discoveries were valid? And why should we take their claims seriously? The answer to these questions requires that we take a look at the logical foundations of scientific inquiry.[1154]

Let's take the proximate explanation for prairie vole monogamy that focuses on the presence of large numbers of V1a receptors in the ventral pallidum of the vole's brain (see Figure 1.2). Before the researchers reached the conclusion that brain cells with this protein had a major role to play in the formation of social attachments between a male and a female vole, they compared the brains of prairie voles with those of polygynous montane voles. The comparison revealed that pallidum receptors for vasopressin were numerous in the prairie vole but much less so in the montane vole. This receptor difference suggested a possible explanation for the behavioral difference between the two kinds of voles, a working **hypothesis** that required verification before it could be accepted.

To determine whether vasopressin receptors in the ventral pallidum really are essential to the formation of monogamous bonds in prairie voles, the researchers had to develop a way to test their idea. They did so by employing what can be called "if–then logic": they reasoned that if the vasopressin receptor explanation was correct, then a given experiment should generate specific results. For example, if the researchers were able to increase the number of V1a receptors in the brains of living prairie voles, then these rodents should be even more inclined to monogamy than unaltered males. In other words, the researchers predicted what would happen if the receptor hypothesis was correct, namely, that the experimental males would be especially likely to form social partnerships with females.

Once the research team had a logical expectation or prediction to work with, they did the experiment to secure the data needed to test the prediction.[865] They found that the receptor-rich male voles did indeed attach themselves more strongly to females, so much so that they would bond with females without mating with them (see Figure 1.3). In this case, the actual results matched the researchers' expectations, providing solid evidence in support of the vasopressin receptor hypothesis. Had the experimental voles been no more monogamous than control groups with unaltered numbers of vasopressin receptors, then the researchers' confidence in the vasopressin receptor hypothesis would have fallen.

The vasopressin receptor hypothesis can be used to make many other predictions, one of which is that if you could transfer the prairie vole's vasopressin receptor gene to male meadow voles, you should be able to make meadow voles monogamous. This experiment has been done by using a safe viral vector to transfer the *avpr1a* gene of prairie voles to the ventral pallidum in male meadow voles. The genetically altered males formed much stronger attachments to their particular mating partners than did unaltered control animals (Figure 1.7).[866] This result provides another line of support for the vasopressin receptor explanation for monogamy in prairie voles.

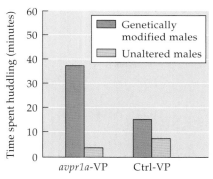

**FIGURE 1.7   Testing the hypothesis** that male monogamy is influenced by a single gene. A male meadow vole made to express high levels of *avpr1a* in the ventral pallidum (*avpr1a*-VP) huddles more closely with a female sexual partner than do males with normal levels of *avpr1a*. Males that express elevated levels of an unrelated gene (*lacZ*; Ctrl-VP genetically modified males) are not significantly more likely to huddle with their mates than genetically unaltered meadow voles. After Lim et al.[866]

## Discussion Question

**1.4** As it turns out, not all prairie voles are monogamous.[1070] Some can be classified as "wanderers" because these males travel widely in search of receptive females that are paired with other males. Wanderers are eager to mate with any female that they can reach, although consummating this goal is difficult because resident monogamous males attack male intruders vigorously. A research team studied the two kinds of males, promiscuous wanderers and monogamous residents, and found that there was no difference between the two types in the number of V1a receptors in the ventral pallidum. What prediction (based on studies reported above) led to this research finding? Why were the researchers surprised by their finding? What prediction can you make about the amount of vasopressin available to interact with the V1a receptors in the ventral pallidum in monogamous pair-bonded males versus wandering single males?

To test an ultimate hypothesis, researchers use the same procedures as they do for proximate explanations. Take the hypothesis that prairie voles are usually monogamous because over evolutionary time, monogamous males have been able to force their mates to be "faithful," which has meant that the offspring produced almost always carry the genes of the mate-guarding male. If this hypothesis is true, and if we were to devise an experiment in which we prevented male prairie voles from guarding their mates, then we would expect to find at least some female voles willing to mate with more than one male. Jerry Wolff and his coresearchers conducted the necessary experiment, as described above. They found that, yes, many female prairie voles did indeed copulate with more than one partner when they had the opportunity to do so.[1612] The fact that this prediction proved to be correct provides support for one particular adaptive explanation for prairie vole monogamy.

One test is good; more tests are better, which is why most hypotheses are subjected to multiple tests. So, for example, monogamous male prairie voles can be predicted to have higher reproductive success than those males unable (or unwilling) to settle down with one female. Indeed, in one study, resident males sired more than twice as many pups as nonmonogamous wanderer males.[1070] Because the adaptive monogamy hypothesis leads to multiple predictions that have subsequently been found to be correct, we can accept it as probably true, whereas hypotheses that regularly fail their tests are discarded as probably false.

**FIGURE 1.8   Charles Darwin**, shortly after returning from his around-the-world voyage on the *Beagle*, before he wrote *The Origin of Species*.

## Darwinian Theory and Ultimate Hypotheses

When Jerry Wolff and his colleagues became interested in explaining prairie vole behavior from an evolutionary perspective, they used Charles Darwin's theory of evolution by natural selection as presented in *The Origin of Species* (Figure 1.8).[348] From 1859 onward, biologists have used Darwin's theory to guide them whenever they have wanted to explain something in ultimate terms. Because of the importance of this theory to the study of animal behavior, and because it is so often misunderstood, we will now review what the theory is—and is not.

Darwinian theory is based on the premise that evolutionary change is inevitable if just three conditions are met:

1. **Variation**, such that members of a species differ in some of their characteristics (Figure 1.9)
2. **Heredity**, with parents able to pass on some of their distinctive characteristics to their offspring
3. **Differences in reproductive success**, such that some individuals have more surviving offspring than others in their population, thanks to their distinctive characteristics

**FIGURE 1.9   A variable species.** The ladybird beetle *Harmonia axyridis* exhibits hereditary variation in its color pattern. Photographs by Mike Majerus.

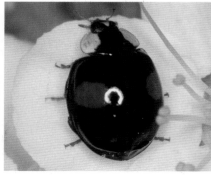

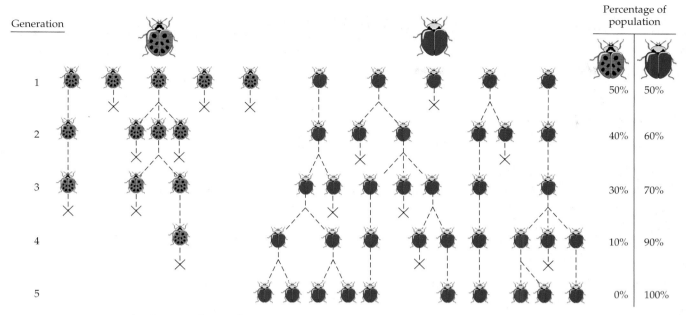

**FIGURE 1.10    Natural selection.** If the differences in the color patterns of ladybird beetles are hereditary and if one type of beetle leaves more surviving offspring on average than another, then the population will evolve, becoming increasingly dominated by the reproductively successful type.

If there is hereditary variation within a species (and there almost always is) and if some hereditary variants consistently reproduce more successfully than others, then the increased abundance of living descendants of the more successful types will change the makeup of the species (Figure 1.10). The species will evolve as it becomes dominated by individuals that possess the traits associated with successful reproduction in the past. (You may ask, why, then, is there still so much genetic variation related to color pattern within the ladybird beetle *Harmonia axyridis*—and most other species? For some possible answers to this question, see Mayr.[958])

Because the process that causes evolutionary change is natural, Darwin called it **natural selection**. Darwin not only laid out the logic of natural selection theory clearly, but also provided abundant evidence that hereditary variation is common within species and that high rates of mortality are also the rule. Thus, alternative types within a species are forced into an unconscious competition to be among the relatively few survivors. In other words, the conditions necessary and sufficient for evolutionary change are standard in living things.

When Darwin developed his theory of natural selection, he and his fellow scientists knew very little about heredity. Now, however, biologists know that considerable variation among individuals within a species arises because of differences in their **genes**, the segments of DNA that faithfully encode the information needed for the synthesis of proteins, such as the stretch of DNA that codes for the vasopressin receptor protein present in the prairie vole. Since genes can be copied and transmitted to offspring, parents can pass on the hereditary information needed for the development of critically important attributes, such as those that affect the propensity for monogamy among male prairie voles.

Genetic variation within a species occurs when a given gene exists in two or more forms, or **alleles**, within the species' gene pool. The different alleles sometimes affect the nature or abundance of the protein coded for by the gene such that genetically different individuals transmit different instructions for protein manufacture to their offspring. If some alleles are superior to others in their ability to help make individuals reproductively successful, then those alleles will get themselves passed on from generation to generation and will become more common over time, gradually displacing their "competitors" over the course of evolution. We can summarize evolution at the genetic level with a simple equation:

$$\text{Genetic variation} + \text{differential reproduction} = \text{evolutionary change at the genetic level}$$

The logical conclusion of this way of thinking about selection at the level of the gene is that alleles will spread in proportion to how well they help build bodies that are unusually good at reproducing. As E. O. Wilson puts it, a chicken is really the way that chicken genes make more copies of themselves.[1588] Given the way in which selection acts, we can assume that chickens (and all other organisms) are probably very good at reproducing and passing on their special genes.

### Discussion Questions

**1.8**  Imagine that a mutation occurs this year in the gene that codes for the V1a receptor protein in prairie voles. The altered protein increases the tendency of a male to form social bonds with a mate, in effect making him somewhat more monogamous than the typical male prairie vole. What is required for enhanced monogamy to become more common in this species over time? If monogamous males have 3.7 pups on average during their lifetime while enhanced monogamous males have 4.1 on average, will the allele associated with enhanced monogamy necessarily make up an increasing proportion of the population? ("No" is the correct answer. Why? Hint: How does natural selection "measure" individual reproductive success?) Now imagine two types of prairie voles—one type that usually lives for 1.5 years and another for 0.8 years. Would it ever be possible for the one that died sooner to replace the longer-lived type over time?

**1.9**  If you wanted to invent the term "genetic success" for use in evolutionary studies based on natural selection theory, what would your definition be?

### Darwinian Theory and the Study of Behavior

No matter how the concept of natural selection is presented, it is a blockbuster of an idea. Not only is the logic of the theory ironclad, but the conditions required for natural selection apply to nearly every organism, which means that almost every species has probably been shaped by natural selection in the past. This claim can be and has been tested. Darwin engineered one such test himself by demonstrating that evolution would indeed happen when people created the conditions needed for selection to occur as they domesticated certain useful animals and plants. As Darwin pointed out, all the various breeds of domesticated pigeons were derived in remarkably short order from one species, the rock dove, as pigeon fanciers selected hereditarily distinctive individuals and permitted them to reproduce, while disposing of less desired members of the species. In nature, of course, people do not control selection as they do during the process of domestication, but still, the fact that different breeds of pigeons, dogs, and cats emerged from populations in which (1) there was variation (2) that was hereditary and (3) that affected the reproductive success of the variants gave Darwin confidence that his logic was correct.

We can test Darwinian theory in formal experiments by attempting to generate evolution in the laboratory, starting with a population that exhibits hereditary variation in attributes that affect the reproductive success of individuals (because the researcher controls which individuals get to leave offspring). Consider the **artificial selection** experiment done by Carol Lynch with house mice, which build nests of soft grasses and other plant materials in nature but will happily accept cotton as nesting material in the laboratory.[902] The amount of cotton a mouse collects can be quantified as the number of grams it pulls into a nest cage over a 4-day period. In the starting generation of Lynch's experiment, individual mice moved between 13 and 18 grams of cotton into their cages from an external cotton supply.

Lynch worked with this variation on the assumption that some of the differences among individuals were hereditary. She attempted to evolve a "high line," by interbreeding males and females that collected a relatively large amount of cotton, as well as a "low line" (by crossing males and females that gathered relatively little cotton) and a "control line" (by crossing males and females chosen at random from each generation). The offspring produced by these crosses were reared under the same conditions their parents were, eliminating environmental variation as a possible cause for differences in their behavior. When the youngsters became adults, the amount of nest material they collected in 4 days was measured. The most avid cotton collectors in the high line were permitted to breed, creating a second selected generation, as were the least eager collectors in the low line.

Lynch repeated these procedures over 15 generations, with the eventual result that the high-line mice gathered about 50 grams of cotton for their nests on average, while the low-line mice brought in about 5 grams on average. The 15th-generation control mice brought in about 20 grams, roughly the same amount as their ancestors (Figure 1.11). Evolution had occurred in the lab

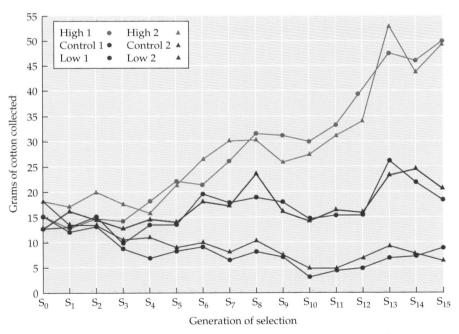

**FIGURE 1.11  Artificial selection causes evolutionary change** as predicted by the theory of natural selection. Here the experimenter only permitted mice that collected large amounts of cotton for their nests to breed with each other, which led to the evolution of populations (the high lines) whose members collected much more cotton on average than the control lines, whose behavior was not selected and so did not evolve. Likewise, selective breeding of mice that gathered relatively little cotton resulted in the evolution of low lines, whose members made very small nests. The symbols indicate the average amounts of cotton collected over 4 days by each generation. After Lynch.[902]

under the conditions that should have resulted in evolutionary change if the theory of natural selection is correct.

Still more confidence in the reality of natural selection comes from the many demonstrations that natural, as opposed to human-controlled, selection can in fact cause evolutionary change in nature, and it can do so with great speed. For example, people began rearing milk cows successfully only about 5000 years ago in northern Europe, probably because this area had few deadly contagious diseases of cattle.[132] Once dairying originated, a mutant gene spread naturally through this human population, a gene that contributed to the ability of adults to digest milk sugar.[663] Although children everywhere can digest lactose, most lose this ability when they are no longer being nursed by their mothers, after which time they become lactose intolerant. But natural selection acting on variation in the ability of adults to absorb milk sugar resulted in the rapid spread of lactose tolerance in adults within the northern European cattle-herding culture and also in a number of African groups where dairy cattle provide milk for children and adults alike.[1450]

### Discussion Question

**1.10** Figure 1.12 shows the results of an experiment in which domestic dogs and hand-reared wolves, the closest relatives of dogs, were tested with respect to their ability to locate food hidden in one of two covered bowls. When the test subjects were allowed to watch humans intently gazing at or pointing at the food-containing bowl, the domestic dogs went directly to the correct bowl significantly more often than the wolves. Even chimpanzees cannot match dogs at this task. Moreover, puppies that have been reared in a kennel with minimal human contact also use the cues provided by people to find food, and they do as well as puppies that have been reared in a household setting.[622] How is this case similar to the case of the evolution of lactose tolerance in humans?

Even more rapid evolutionary change has been documented for Darwin's finches in the Galápagos Islands. There Peter and Rosemary Grant discovered, for example, that in drought years, when the smaller seeds eaten by the medium ground finch (*Geospiza fortis*) were scarce relative to the large seeds of the plant *Tribulus cistoides*, selection "favored" relatively large-billed individuals. Which is to say that birds with the hereditary tendency to develop larger-

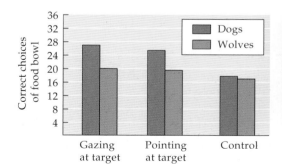

**FIGURE 1.12  A test of whether dogs are more sensitive to signals from human beings than are hand-reared wolves.** Note that hungry dogs, but not hungry wolves, use the information provided by humans pointing at or looking at a bowl with hidden food to find the food. The results are the mean number of correct choices made over a series of trials by seven dogs and seven wolves. The control represents a situation in which a human did not gaze at or point at the correct bowl. After Hare et al.[622]

than-average beaks (and bill size is passed on from parents to offspring)[754] survived and reproduced because of their ability to crack open big tough seeds; meanwhile their hereditarily smaller-beaked cousins were starving to death rather than reproducing.[567] In rainy years, the situation was reversed inasmuch as birds with smaller bills handled the then abundant smaller seeds better than large-beaked finches. The temporary feeding advantage of smaller-billed finches translated into a reproductive advantage, leading to the evolution of a somewhat smaller-billed population. The annual changes in selection caused the average bill size of the finches to fluctuate *annually* in response to changes in the resources available to the birds.

Many other examples exist of the evolutionary effects of nonrandom variation in reproduction, including the evolution of bacterial resistance to antibiotics,[286] the change in overwintering behavior of some songbirds in the last few decades (see Chapter 3), the spread and decline of melanic wing coloration in the moth *Biston betularia* over the last 150 years (see Chapter 6), and the evolution in Australian black snakes of an aversion to a highly poisonous, very recently introduced toad.[1132]

Given the abundant evidence that evolutionary change occurs when the conditions required by natural selection are met, let us assume that individuals of most species are endowed with alleles that have spread because they were better than any alternative at getting their bearers to reproduce successfully. If that is true, almost every hereditary attribute of almost every one of the millions of species of plants, animals, fungi, protozoans, and bacteria probably has something to do with reproductive success. Therefore, when biologists wish to understand the ultimate reasons why an animal does something, they almost always try to come up with a working hypothesis that is consistent with natural selection theory.

Admittedly, some traits do not look especially likely to increase an individual's reproductive success. For example, male Hanuman langurs expend much energy and time trying to kill the infants of the females that they live with in bands in northern India.[687] These graceful, long-limbed primates form groups composed of one or more large adult males and several smaller females and their offspring (Figure 1.13). Young langurs are especially likely to be attacked when a new adult male becomes dominant, generally as a result

**FIGURE 1.13    Hanuman langur females and offspring.** Males fight to monopolize sexual access to the females in groups like this one.

(A)

(B)

**FIGURE 1.14** **Male langurs commit infanticide.** (A) A nursing baby langur that has been paralyzed by a bite to the spine (note the open wound) by a male langur. This infant was attacked repeatedly over a period of weeks, losing an eye and finally its life at age 18 months. (B) An infant-killing male langur flees from a female belonging to the band he is attempting to join. A, photograph by Carola Borries; B, photograph by Volker Sommer, from Sommer.[1368]

of having chased off the previous top male. The new male on the block, which may have come from another band, may then try to separate infants from their mothers, harassing the mothers and biting their babies viciously when he gets a chance (Figure 1.14A).

But killing baby langurs is not easy. Infanticidal males have to deal with female langurs, which join forces to fight back in defense of their infants (Figure 1.14B).[1369] A male risks injury whenever he tries to destroy a youngster. Why should he take the risk, especially since his behavior would hardly seem to endear him to the very females that must become his mates if he is to ever produce offspring of his own? Given the apparent disadvantages to would-be infanticidal males, perhaps infanticide is not the product of natural selection, but rather an abnormal, pathological response to overcrowding. In fact, this nonevolutionary hypothesis for infanticidal behavior was advocated by some langur watchers,[330] who noted that langurs are often fed by Indian villagers and so perhaps now live at densities much higher than in the past. Under these novel conditions, male behavior could be thrown out of sync, leading to maladaptive, hyperaggressive reactions to youngsters.

But another langur observer, Sarah Hrdy, felt that Darwinian theory provided a legitimate solution to the puzzle of infanticide.[687] She thought that killer males could boost their reproductive chances by leaving the mothers of dead infants with no other adaptive option except to mate with the killers. With her nursing infant eliminated, the female might resume her reproductive cycling sooner than otherwise and so might be impregnated sooner by the infant killer, who would leave more descendants as a result.

For the moment, let's put aside the question of whether Hrdy's hypothesis is right or wrong. The point here is that her answer to, why infanticide? derives from Darwinian theory by focusing on how males could conceivably increase their reproductive success by practicing infanticide. If they did, then the proximate bases for infanticide could have spread in past generations of Hanuman langurs living under completely natural conditions.

So we have two potential explanations for infanticide by male langurs: the social pathology hypothesis and Hrdy's quicker reproduction hypothesis. But here's another idea. Perhaps male langurs commit infanticide as a means of population regulation. Langurs living at high densities may very well stress their food supplies, which could favor a mechanism for the prevention of overpopulation. Infanticide, although brutal and vicious, could nevertheless help keep langur bands from destroying the resources vital to their long-term survival.

## The Problem with Group Selection

This population regulation hypothesis is certainly evolutionary because it explains infanticide in historical terms. It proposes that the trait evolved because in the past, groups (even entire species) that lacked the means to keep their populations in reasonable harmony with the available food supplies went extinct, while those that had some infanticidal males were more likely to keep their numbers in line with resources and so persist over the long haul. But note that although the hypothesis claims that infanticide spread because of its beneficial consequences, the beneficiary of infanticide is not the male who does the killing, but rather the entire group to which he belongs. Thus, the evolutionary process that causes infanticide to spread is not Darwinian selection, which is based on differences among individuals in their reproductive success, but rather a form of **group selection**, which is based on differences among groups in their ability to survive.

Group selection theory in its original form was laid out in detail in a book written in 1962 by V. C. Wynne-Edwards.[1622] According to Wynne-Edwards, species have such great potential to destroy the very things on which they depend that only those that happened to acquire the capacity for population regulation would have been able to avoid self-destruction. In contrast, species that were fortunate enough to have some self-sacrificing individuals would have been able to keep their numbers down and their essential resource bases intact. Wynne-Edwards specifically identified infanticide as one of the mechanisms of "social mortality" that contributed to population stabilization in the animal kingdom (although in 1962 no one knew of its occurrence in langurs).

Group selection theory was challenged in 1966 by George C. Williams in *Adaptation and Natural Selection*, probably the most important book on evolutionary theory written since *The Origin of Species*. Williams showed that the survival of alternative alleles was much more likely to be determined by differences in the reproductive success of genetically different individuals than by survival differences among genetically different groups.[1578] We can illustrate his point with reference to langurs. Imagine that in the past there really were male langurs prepared to risk serious injury, even death, by killing infants in order to regulate their band's size for the survival benefit of the group. In such a case, group selection would be said to favor the allele(s) for male infanticide because the group as a whole would benefit from the removal of excess infants.

But in this species, Darwinian natural selection would also be at work, provided that at any time there were two genetically different kinds of males, one that practiced infanticide for the good of the group and the other that permitted other males to pay the price for population reduction. If the non-killers lived longer and reproduced more, which of these two types would become more common in the next generation? Whose hereditary material would increase in frequency over time? Would infanticide long persist in such a population of langurs?

This kind of thought experiment convinced Williams and his readers that Darwinian selection acting on differences among individuals within a population or species will usually have a stronger evolutionary effect than group selection acting on differences among entire groups. Yes, group selection can occur, provided that groups can retain their integrity for long periods and differ genetically in ways that influence their chances of survival. But if group selection favors a trait that involves reproductive self-sacrifice while natural selection acts against it, natural selection seems likely to trump group selection, as we have just seen in our hypothetical langur example. Although some other forms of group selection theory have acquired strong advocates,[1360, 1586, 1590] almost all behavioral biologists have been persuaded by Williams to distinguish between naïve group selection a la Wynne-Edwards and individual (or gene) selection hypotheses. Most researchers exploring ultimate questions about behavior look first to Darwinian theory when producing their hypotheses.

## Discussion Questions

**1.11**   Lemmings are small rodents that live in the Arctic tundra. Their populations fluctuate wildly. At high population densities, large numbers leave their homes and travel long distances, during which time many die, some by drowning as they attempt to cross rivers and lakes. One popular explanation for their behavior is that the travelers are actually committing suicide to relieve overpopulation. By heading off to die, the suicidal lemmings leave shelter and food for those who stay behind. These surviving individuals will perpetuate their species, saving it from extinction. What theory was used to produce this hypothesis? How would George C. Williams use Gary Larson's cartoon (Figure 1.15) to evaluate the hypothesis critically?

**1.12**   As noted above, some researchers believe that group selection has a role to play in evolutionary change. For example, David S. Wilson and Edward O. Wilson promote what they call multilevel selection, which they say incorporates a form of group selection.[1586] Read their article and then read a paper by Kern Reeve.[1205] What is Reeve's view of Wilsonian group selection? Are the Wilsons arguing for a revival of Wynne-Edwardsian group selection, or do they have something else in mind? See also Shavit and Millstein.[1310]

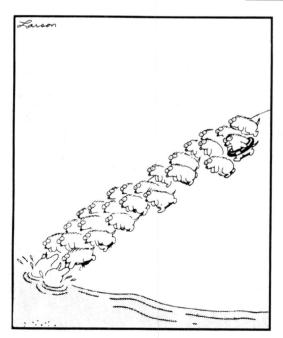

**FIGURE 1.15**   Variation in suicidal tendencies in a make-believe lemming-like species. Courtesy of Gary Larson.

## Testing Alternative Hypotheses

We can argue that the group benefit hypothesis for infanticide has some logical deficiencies that reduce its plausibility, while Hrdy's quicker reproduction hypothesis has the virtue of resting on a logical and extremely powerful theory. But, as we have already noted, scientists do not call it a day after coming up with a hypothesis, no matter how reasonable it may seem. Hypotheses have to be tested. In the case of langur infanticide, we have three alternatives to consider: the nonevolutionary social pathology hypothesis and the two evolutionary possibilities outlined above. To test these alternatives, we first need to derive testable predictions from them.

If high population density really does cause male langurs to behave abnormally (the social pathology hypothesis) or if it truly threatens the survival of langur groups and so activates infanticide by self-sacrificing males (the population regulation hypothesis), then we would expect to see infanticide by males only in areas where Hanuman langurs live in exceptionally high densities. However, contrary to this prediction, infanticide regularly occurs in troops living at moderate or even low densities in areas where they are not fed by people.[148, 1043] This finding weakens our confidence in both the social pathology and population regulation hypotheses for infanticide by langur males.

What about Hrdy's quicker reproduction hypothesis? It is not the only Darwinian explanation for infanticide. Another possibility is that males kill infants after taking over a band of females in order to eat the dead youngsters, thereby replenishing energy reserves that have been depleted in the takeover struggle. Cannibalism is widely practiced in the animal kingdom, even by some primates,[974] so we really ought to consider it a possibility for langurs as well. As it happens, both the (1) cannibalism and (2) quicker reproduction hypotheses generate one prediction that is identical; namely, infanticide should occur soon after takeovers, when the males (1) are presumably energetically stressed and (2) need to reproduce quickly if they are to have as many descendants as possible. The fact that infanticide is associated with the period soon after a takeover does not enable us to discriminate between the two hypotheses. Nonetheless, if the cannibalism hypothesis is true, we can also predict that male langurs will consume the infants they kill. Since no one has observed them doing so, it is unlikely that infanticide has evolved in Hanuman langurs because of its nutritional benefits for infanticidal males.

The quicker reproduction hypothesis leads to some additional predictions. For one thing, we would not expect to see males killing their own babies, which carry their genes, because that would obviously damage a male's reproductive success. In one group of langurs that contained several adult males, not just one dominant male, 16 cases of infanticide were recorded in which it was possible to compare the presumptive killer's DNA with that of the dead infant. In every instance, as predicted, the killer was not the father of the victim.[150]

The quicker reproduction hypothesis also yields the prediction that females who have lost an infant to a killer male will promptly resume sexual cycling and become receptive to that very same male. In fact, langur females who have lost offspring are quick to regain their sexual receptivity, which is suppressed when females are nursing young. The infant-deprived females soon become pregnant and give birth to babies whose DNA matches that of the infanticidal males.[687, 1368] Thus, tests of the quicker reproduction hypothesis have been positive.

But the more predictions and tests, the better. For example, if infanticide is adaptive for male langurs because it speeds up their chances of impregnating females who would otherwise be tied up in the care of some other male's progeny, then infanticide should also be adaptive in other species whose social systems resemble that of the Hanuman langur. We can check this prediction by observing what happens when male lions compete to control a pride, a group of females that live together. On occasion, new males oust the previous pride masters. Under these circumstances, Darwinian selection should result in the spread of infanticidal behavior by newcomers, who, by eliminating the cubs of the previous residents, leave the lionesses with little option except to come into heat at once in order to replace their dead offspring. As predicted, incoming males hunt down youngsters less than 9 months old and try to kill them, often succeeding (Figure 1.16) despite resistance from the cubs' mothers.[1180] Once a lioness has lost her cubs, she will resume sexual cycling and mate with her infanticidal companion. Had she retained her infant, she would not have become sexually receptive until her progeny were 2 years old. Since the average tenure of an adult male with a pride of females is only 2 years, the reproduc-

**FIGURE 1.16  Infanticide by a male lion.** This male carries a cub he has killed after displacing the adult males that once lived with his pride. Photograph by George Schaller.

**FIGURE 1.17 An evolved response to the risk of infanticide.** This male water bug guards a clutch of eggs against infanticidal females that may destroy his current clutch in order to replace these eggs with their own. Photograph by Bob Smith.

tive benefits of infanticide for the male after a takeover are obvious. Indeed, in some populations, male lions may kill a quarter of all the cubs that die in their first year of life.[1180]

The existence of male infanticide in both lions and langurs, despite the fact that the two species are not closely related, provides strong evidence that the trait permits reproductive gains for killer males because it speeds up their access to fertile females. Infanticide by males has now been recorded in over 50 species of mammals,[1612] as well as in birds[1488] and spiders.[1288] Each case helps test the hypothesis that infanticide has evolved in langurs because it increases reproductive opportunities for successful infant killers.

Other kinds of infanticide can also provide relevant evidence for testing the quicker reproduction hypothesis. For example, in a species in which females compete for sexual access to males, we can predict that females will have evolved the capacity for infanticide if it leads males to accept them sooner than otherwise. This prediction has been confirmed for a giant water bug whose males take care of large masses of eggs deposited by their mates (Figure 1.17).[1354] The male will not mate with additional females if the clutch of eggs under his care is sufficiently large. Therefore, a female that finds a brooding male sometimes attacks and destroys all his eggs. Under these circumstances, the now eggless male may mate with the egg-destroying female and care for her clutch after she lays them on a stick or branch.

Much the same sort of thing also occurs in the wattled jacana, a waterbird in which males incubate the eggs of a partner and guard the chicks when they hatch. Here, too, females sometimes commit infanticide by killing the chicks of other females before mating with the males that have lost their broods. The killer female deposits her eggs in the care of the male, gaining a caretaker for her offspring sooner than if she had waited for the male to finish rearing his current brood.[440]

## Discussion Questions

**1.13** Male Hanuman langurs apparently use a rule of thumb when deciding which infants to kill: they attack youngsters of females that they did not copulate with prior to the birth of those offspring. In light of this finding, some observers have an explanation for why it is that after a takeover, pregnant females may mate with newcomers. These researchers have suggested that the females' behavior may create confusion about the paternity of their babies, the better to limit male infanticidal behavior. If females also engage in paternity confusion when their band contains several adult males, what predictions can you make about (1) the duration of female receptivity over each reproductive cycle, (2) the relationship between the period when females are in heat and the timing of ovulation, and (3) the occurrence of copulations by subordinate males as well as the dominant male langur, who attempts to monopolize sexual access to the female? In light of the paternity confusion hypothesis, what significance do you attach to the finding that male Hanuman langurs copulate more frequently with fertile females as compared with pregnant females?[1083]

**1.14** All scientific puzzles, not just the issue of infanticide, can be tackled by testing alternative hypotheses. Consider, for example, the fact that burrowing owls collect dried mammal dung and bring it back to scatter the stuff around the entrances to their underground burrows. One hypothesis to account for this odd, time-consuming behavior is that the scents coming from the dung make the nests safer for the adult owls and their youngsters by masking any owlish odors that might help a predator track down its prey. What other hypotheses can you present for dung-collecting behavior? What predictions can you derive from your alternative explanations for the behavior? *After* completing this part of the problem, compare your hypotheses and predictions with those developed by Matthew Smith and Courtney Conway.[1352]

## Certainty and Science

You must have deduced from my summary of the research on infanticide that I think the quicker reproduction hypothesis applies to langurs and lions, water bugs and jacanas. I do—but I could be wrong, and indeed, some other researchers surely think that I am wrong.[78, 135, 333] These differences of opinion remind us that all scientific conclusions must be considered tentative, to some degree. In the past, conclusions have shifted dramatically when new evidence has thrown a once widely accepted hypothesis into doubt. For example, when I was a student at Amherst College, my paleontology professor convinced me that the Earth's continents have always been where they are now located. However, as more and more data to the contrary came in, the old view was abandoned, and now all professors of paleontology instruct their classes on the virtues of plate tectonics theory and the idea that today's continents have drifted far from their original positions.

The rejection of established wisdom happens all the time in science. Scientists tend to be a skeptical lot, perhaps because special rewards go to those who can show that published conclusions are erroneous. Researchers constantly reexamine their colleagues' findings, in good humor or otherwise, sometimes causing their fellow scientists to reevaluate a published hypothesis. Take the *avpr1a* gene's proposed causal connection to monogamy in the prairie vole. Data from the gene transplant experiment described earlier in the chapter clearly support the claim that this gene causes its owners to behave monogamously. Additional work done after the transplant experiments led the vole research team to the conclusion in 2004 that the key genetic difference between monogamous and nonmonogamous voles lay in a segment of DNA (the promoter region) adjacent to the protein-coding component of *avpr1a*.[614] But a Swiss research team led by Gerald Heckel looked again at the arginine vasopressin 1a receptor gene in 2006.[465] When they examined the DNA of 25 rodent species, not just four vole species, they found that almost all these rodents, whether they were monogamous or not, had very similar *avpr1a* genes right down to the supposedly critical promoter region of the gene that was said to be the basis for the different mating systems of the monogamous prairie vole and the polygynous meadow vole. The title of the paper where they present these results says it all: "Mammalian monogamy is not controlled by a single gene." True, the meadow vole and a close relative do not have the special sequence of bases found in the prairie vole's *avpr1a* promoter region. But many other thoroughly polygynous voles and other polygynous mammals do have the key piece of DNA in the promoter region of their *avpr1a* gene. This finding is at odds with the prediction that monogamous and polygynous voles should consistently differ with respect to this one component of one gene. Given the conflicting data, it is appropriate to be cautious about the one-gene hypothesis for vole monogamy.

The uncertainty about Truth that scientists accept, at least when talking about someone else's ideas, often makes nonscientists nervous, in part because scientific results are usually presented to the public as if they were written in stone. But anyone who has taken a look at the history of science will learn that new ideas continually surface and old ones are regularly replaced or modified.[817] Indeed, a great strength of science lies in the willingness of at least some scientists to consider new ideas and to test old hypotheses repeatedly, even when some of their colleagues may think that doing so is a waste of time.

Please keep this point in mind as we review some current scientific conclusions in the chapters ahead. We will first examine the proximate and ultimate aspects of bird song (in Chapter 2) before looking more closely at different proximate analyses of behavior (in Chapters 3 through 5). Then we will turn to ultimate questions about evolutionary history and adaptation (in Chapters 6 through 13). The book concludes with a chapter on the evolution of human behavior. Thanks to the many behavioral researchers who have explored these issues, there is much to write about, so let's get started.

## Summary

1. The causes of any behavior can potentially be understood in terms of four different levels of analysis, which deal with (1) how the behavior develops, (2) how physiological mechanisms work to make the behavior possible, (3) how the behavior promotes the animal's reproductive success, and (4) how the behavior originated and has been changed over evolutionary time.

2. These four levels of analysis can be condensed into two: (1) those that deal with the proximate, or immediate, causes of behavior, which are linked to the operation of internal developmental and physiological systems, and (2) those that have to do with the ultimate, or long-term evolutionary, causes of behavior, which are linked to questions of adaptive value and historical modification.

3. Proximate and ultimate causes are interrelated. The genes present in today's animals have survived a historical process dominated by the effects of past differences among individuals in their reproductive success. A living animal's naturally selected genotype influences an individual's development, affecting the nature of the proximate mechanisms that the animal comes to possess, which in turn enable it to do certain things.

4. Both proximate and ultimate questions about behavior can be investigated scientifically in much the same way:

   a. We begin with a question about what causes an animal to do something.

   b. We devise a working hypothesis, a possible answer to the question.

   c. We use this potential explanation to make a prediction about what we expect to observe in an experiment or in nature if the hypothesis is true.

   d. We then collect the necessary data to determine whether the prediction is correct or incorrect.

   e. If the actual results do not correspond to the expected ones, we conclude that the underlying hypothesis is probably incorrect; if the evidence does match the predicted results, we conclude that the hypothesis can be tentatively accepted as correct.

5. The nature of our theories affects the kinds of working hypotheses that we generate. Darwinian evolutionary theory leads us to consider how a given behavior might advance the reproductive success of individuals; if the behavior has been shaped over time by Darwinian natural selection, then it must have been better than all the others that have appeared over the generations in terms of promoting individual reproductive success.

6. An alternative theory, the group selection theory of V. C. Wynne-Edwards, generates working hypotheses that focus on how a given behavior helps groups survive; if the behavior in question has been shaped over time by group selection, then it must be superior to all others in terms of helping entire groups avoid extinction.

7. Almost all behavioral biologists today use Darwinian theory rather than Wynne-Edwardsian group selection theory as the foundation for their hypotheses because selection at the level of the individual is likely to be

more powerful than group selection in causing evolutionary change. The logic of this conclusion can be seen if one imagines what would happen over time to a group-benefiting behavior that caused those who behaved in this manner to leave fewer descendants than others who did not sacrifice themselves (and their distinctive genes) for the benefit of their group.

8. The beauty of science lies in the ability of scientists to use logic and evidence to evaluate the validity of competing theories and alternative hypotheses. Persons who use the scientific approach can make progress by eliminating explanations that fail their tests while accepting others that pass their exams.

## Suggested Reading

Great books written by scientists who have studied the proximate basis of animal behavior include Vincent Dethier's *To Know a Fly*[384] and Kenneth Roeder's *Nerve Cells and Insect Behavior.*[1234] Niko Tinbergen's *Curious Naturalists*[1446] and Konrad Lorenz's *King Solomon's Ring*[886] bridge the gap between proximate and ultimate approaches. See also Howard Evans's *Life on a Little Known Planet*[447] and Michael Ryan's *The Túngara Frog.*[1260]

For other books by scientists that capture the delight of field research, consider Evans's *Wasp Farm,*[448] Bernd Heinrich's *In a Patch of Fireweed*[644] and *The Geese of Beaver Bog,*[647] George Schaller's *The Year of the Gorilla,*[1282] and Tinbergen's *The Herring Gull's World,*[1448] as well as Jane Goodall's *In the Shadow of Man,*[551] Shirley Strum's *Almost Human,*[1398] Cynthia Moss's *Elephant Memories,*[1014] and *Journey to the Ants* by Bert Hölldobler and E. O. Wilson.[670] Craig Packer's *Into Africa* tells us what it is like to work on lions in the wild,[1093] while Sarah Hrdy's *Langurs of Abu* provides an account of her study of infanticide.[687] Bernd Heinrich's superb *Ravens in Winter* offers an unusually clear picture of how scientists test alternative hypotheses.[646] The complex story of monogamy in prairie voles with its proximate and ultimate components has been recently reviewed by Steve Phelps and Alexander Ophir.[1131] For a provocative essay on the nature of science itself, read Woodward and Goodstein's "Conduct, misconduct, and the structure of science."[1617]

Charles Darwin had something useful to say about the logic of natural selection in *On the Origin of Species,*[348] and so do Daniel Dennett[382] and Richard Dawkins.[364, 368] G. C. Williams's classic *Adaptation and Natural Selection*[1578] demolishes naïve "for the good of the species" arguments. D. S. Wilson and E. O. Wilson attempt to explain why we should now adopt a new form of group selection theory.[1586]

# 2

# Understanding the Proximate and Ultimate Causes of Bird Song

I began bird-watching avidly at age 6. Since then, my "life list" has grown to well over a thousand species, ranging from the tiny calliope hummingbird to the giant Andean condor. Although most birders, myself included, enjoy the challenge of identifying bird species by their appearance, we also get pleasure from learning to recognize species by their songs. So it was that long ago my father taught me that yellow warblers sing "sweet-sweet-sweet-sweeter-than-sweet," whereas the common yellowthroat (another warbler) goes "witchety-witchety-witchety," while the ovenbird (yet another warbler) belts out a loud "teacher-teacher-teacher-teacher-teach."

In fact, almost every bird species has its own distinctive vocalizations, a phenomenon that has long been explained in ultimate terms as an adaptation that helps a male attract a female of the "right" species. Thus, when a female ovenbird returns from her wintering ground in southern Mexico to the Ohio woodland where she will breed, she surely can hear male ovenbirds singing well before she ever lays eyes on a potential mate. By singing "teacher-teacher-teacher," the male ovenbird enables females to find him efficiently, which is in his reproductive interest. Females benefit as well from this system, in part because it helps them track down ovenbird males while avoiding males of other species, which decreases their risk of producing infertile or otherwise disadvantaged hybrid offspring.

But if males of different species sing different songs in order to attract females of their species, why should different individuals of the same species produce somewhat different versions of their species' song, as they often do? Yes, every yellow warbler sings something like "sweet-sweet-sweet-sweeter-than-sweet," but all sorts of variations on this theme exist within the species. So what's going on here? What happens during a bird's development that causes it to sing a little differently from some other members of its species? How does the operation of an individual male's brain differ from that of other males in ways that affect his song? Do the song differences among individuals influence how many offspring they have? Are these behavioral differences rooted in their species' evolutionary history? This chapter attempts to answer these questions as a way to reinforce the main point of Chapter 1—if we want to understand what causes animals to behave a particular way, we have to tackle the problem from both proximate and ultimate angles.

## Different Songs: Proximate Causes

In the short film *Why Do Birds Sing?* Peter Marler describes how he became interested in the proximate causes of song variation. In the 1950s, when he was a fledgling limnologist studying the mud in British lakes, he also casually kept an eye and ear open for the birds in the vicinity of his research sites. An experienced birder, he knew the chaffinch's song by ear, a song that has been described as a rattling series of chips terminated by a descending flourish. Marler realized that the chaffinches living near one lake sang songs that were definitely different from the chaffinch songs he heard around other lakes not very far away. He came to believe that the birds lived in geographically discrete populations, each endowed with its own special version of their species' basic song, its own dialect so to speak.

When Marler moved to the University of California in the 1960s, he and his students investigated the same phenomenon in white-crowned sparrows, the males of which sing a complex whistled vocalization during the breeding season. Marler's team found that different populations of this bird often have distinctively different dialects. Thus, white-crowns living in Marin, north of San Francisco Bay, sing a song type that is easily distinguished from that produced by white-crowns that live in Berkeley, even though the two cities are only 50 miles apart (Figure 2.1).[938] Although white-crowned sparrow dialects sometimes change gradually over time,[1035] in at least some populations, the local dialect has persisted for decades with only modest changes,[618] showing that white-crown dialects can be relatively stable (just like human ones).

We now know that dialects are a common phenomenon in songbirds. In fact, in some species, the birds occupy small neighborhoods only a mile or so in diameter, with each population singing its own distinctive version of the species' song.[789] So what developmental factors are responsible for avian dialects? Marler knew that one proximate explanation for the dialect differences was that Marin white-crowns might differ genetically from Berkeley birds in ways that affected the construction of their nervous systems, with the result that birds in the two populations came to sing different songs. One way to test the genetic differences hypothesis is to check the prediction that groups of birds singing different dialects will be genetically distinct from one another. In one such study, however, researchers found little genetic differentiation among six different dialect groups of white-crowned sparrows.[1364]

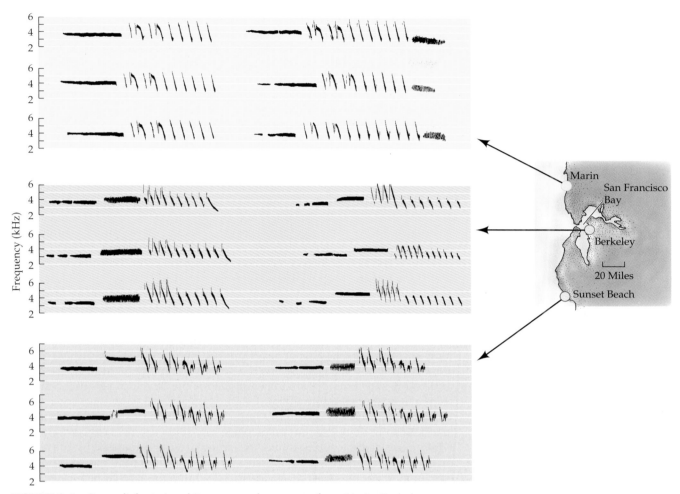

**FIGURE 2.1    Song dialects in white-crowned sparrows** from Marin, Berkeley, and Sunset Beach, California. Males in each location have their own distinctive song dialect, as revealed in these sonograms of the songs of six birds from each location. Sonograms shown in the same color are the same dialect. Sonograms courtesy of Peter Marler.

So let us consider an alternative hypothesis for dialect differences, namely that these differences might not be hereditary, but rather caused by differences in the birds' environments. Perhaps young males in Marin learn to sing the dialect of that region by listening to what adult Marin males are singing, while farther south in Berkeley, young male white-crowns have different formative experiences as a result of hearing males singing the Berkeley dialect. After all, a young person growing up in Mobile, Alabama, acquires a dialect different from someone in Bangor, Maine, simply because the Alabaman child hears a different brand of English than the youngster reared in downeast Maine.

Marler and his colleagues explored the development of white-crown singing behavior by taking eggs from the nests of white-crowned sparrows and hatching them in the laboratory, where the baby birds were hand reared. Even when these young birds were kept isolated from the sounds made by singing birds, they still started singing when they were about 150 days old, but the best they could ever do was a twittering vocalization that never took on the rich character of the full song of a wild male white-crowned sparrow from Marin or Berkeley or anywhere else.[939]

Zebra finch

Father

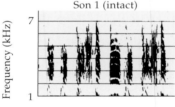

Son 1 (intact)

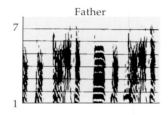

Son 2 (deaf)

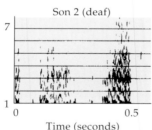

Time (seconds)

**FIGURE 2.2 Hearing is critically important for song learning** in the zebra finch (and many other songbirds). A sonogram of a male zebra finch's song is shown with those of two of his male offspring. The first son's hearing was intact, and he was able to copy his father's song accurately. The second son was experimentally deafened early in life, and as a consequence, he never sang a typical zebra finch song, let alone one that copied his father's song precisely. Photograph courtesy of Atsuko Takahashi; sonograms from Wilbrecht, Crionas, and Nottebohm.[1574]

This result suggested that something critical was missing from the hand-reared birds' environment, perhaps the opportunity to hear the songs of adult male white-crowned sparrows. If this was the key factor and if a young male were isolated in a soundproof chamber but exposed to tapes of white-crowned sparrow song, then the experimental subject ought to be able to sing a complete white-crown song in due course. And that is exactly what happened when 10- to 50-day-old white-crowns were allowed to listen to tapes of white-crowned sparrow song. These birds also started singing on schedule when they were about 150 days old. At first their songs were incomplete, but by the age of 200 days, the isolated birds not only sang the species-typical form of their song, they closely mimicked the exact version that they had heard on tape. Play a Berkeley song to an isolated young male white-crown, and that male will come to sing the Berkeley dialect. Play a Marin song to another male, and he will eventually sing the Marin dialect.

These results offer powerful support for the environmental differences hypothesis for white-crown dialects. Young birds that grow up near Marin hear only the Marin dialect as sung by older males in their neighborhood. They evidently store the acoustical information they acquire from their tutors and later match their own initially incomplete song versions against their memories of tutor song, gradually coming to duplicate a particular dialect. Along these lines, if a young hand-reared white-crown is unable to hear itself sing (as a result of being deafened after hearing others sing but before beginning to vocalize itself), then it never produces anything like a normal song, let alone a duplicate copy of the one it heard earlier in life.[798] Indeed, the ability to hear oneself sing appears to be critical for the development of a complete song in a host of songbirds (Figure 2.2).

Marler and others did many more experiments designed to determine just how song development takes place in white-crowns. For example, they wondered whether young males were more easily influenced by the stimuli provided by singing adults of their own species than by those of other species. In fact, young, isolated, hand-reared white-crowns that hear only songs of another species almost never come to sing that kind of song (although they may incorporate notes from the other species' song into their vocalizations). If 10- to 50-day-old birds listen only to tapes of song sparrows instead of white-crowns, they develop aberrant songs somewhat similar to the "songs" produced by males that never hear any bird song at all. But if an experimental bird has the chance to listen to tapes of the white-crowned sparrow along with songs of another sparrow species, then by 200 days of age it will sing the white-crowned sparrow dialect that it heard earlier.[799] The young bird's developmental system is such that listening to the other sparrow's song has no apparent effect on its later singing behavior.

## Discussion Question

**2.1** Work on the development of singing behavior in male white-crowned sparrows has demonstrated that the birds must learn to sing a particular dialect of the full song of their species. Would we be right, therefore, to conclude that the genetic information present in the cells of white-crowned sparrows is irrelevant for the development of the bird's singing behavior? In this regard, what importance do you attach to the finding that white-crown males apparently can learn their species' song far more easily than the song of other sparrows? What about the finding that white-crown males that hear white-crown song only during a 40-day period early in life can nevertheless generate a complete song, although they do not start singing themselves for several months after their early exposure to a tutor's song?

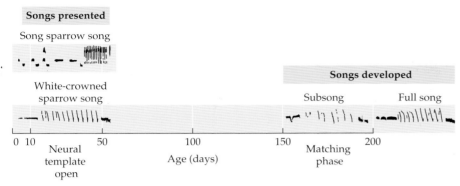

**FIGURE 2.3** **Song learning hypothesis based on laboratory experiments** with white-crowned sparrows. According to this hypothesis, young white-crowns have a critical period 10 to 50 days after hatching, when their neural systems can acquire information from listening to white-crown song, but not to any other species' song. Later in life, the bird matches his own subsong with his memory of the tutor's song and eventually imitates it perfectly—unless he is deafened. Based on a diagram by Peter Marler.

## Social Experience and Song Development

The experiments with isolated white-crowns exposed to taped songs in laboratory cages led Marler to conclude that song development in this species follows a particular course (Figure 2.3). At a very early age, the white-crown's still immature brain is able to selectively store information about the sounds made by singing white-crowns while ignoring other species' songs. At this stage of life, it is as if the brain possesses a restricted computer file capable of recording only one kind of sound input. Then, months later, when the bird begins to sing, it accesses the file. By listening to its own subsongs (incomplete versions of the more complex full song that it will eventually sing) and comparing those sounds against its memories of the full songs it has heard, the maturing bird is able to shape its own songs to match its memory of the song it has on file. When it gets a good match, it then repeatedly practices this "right" song, and in so doing crystallizes a full song of its own, which it can then sing for the rest of its life.

The ability of male white-crowned sparrows to learn the songs of other male white-crowns in their birthplace just by listening to them sing provides a plausible proximate explanation for how males come to sing a particular dialect of their species' full song. However, very occasionally, observers have heard wild white-crowns singing songs like those of other species, including the song sparrow. These rare exceptions led Luis Baptista to wonder whether some other factor, in addition to acoustical experience, might influence song development in white-crowns. One such factor might be social experience, a variable excluded from Marler's famous experiments with isolated, hand-reared birds whose environments offered acoustical stimuli but not the opportunity to interact with living, breathing companions.

To test whether social stimuli can influence song learning in white-crowns, Baptista and his colleague Lewis Petrinovich placed young hand-reared white-crowns in cages where they could see and hear living adult song sparrows or strawberry finches.[73] Under these circumstances, white-crowns learned their social tutor's song, even when they could hear, but not see, adult male white-crowned sparrows (Figure 2.4). In fact, social experience with another species can trump early tape tutor experience with white-crown song even after the young males are more than 50 days old, the end of the window of opportunity for song learning from tape tutors.[72] Social experience has extremely powerful developmental effects on white-crowned sparrow singing behavior.

**FIGURE 2.4   Social experience influences song development.** A white-crowned sparrow that has been caged next to a strawberry finch will learn the song of its social tutor. (A) The song of a tutor strawberry finch. (B) The song of a white-crowned sparrow caged nearby. The letters beneath the sonograms label the syllables of the finch song and their counterparts in the song learned by the sparrow. Sonograms courtesy of Luis Baptista.

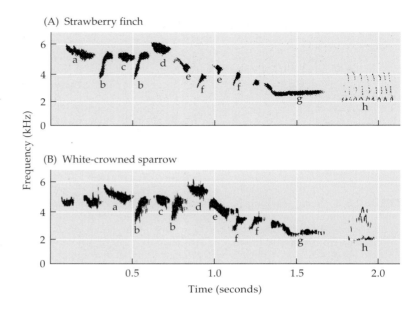

We now know that social factors influence song acquisition in many other birds as well.[93] Captive starlings, for example, are adept at mimicking human speech, producing such phrases as "see you soon baboon" or "basic research" as well as the sounds of laughter or kissing or coughing. But they will do so only if they are hand-reared in a human household where they literally become part of the family (Figure 2.5).[1544] In nature, young starlings and white-crowned sparrows are evidently strongly stimulated by interactions with vocalizing companions of their own species, and these experiences then influence song development. The same holds true for young song sparrows, which also typically learn their songs from adults of their own species. Interestingly, the social effect is stronger when the birds are 8 months old, as opposed to a month or two in age; moreover, an 8-month-old male is more likely to learn particular song types from a tutor when he overhears the adult

**FIGURE 2.5   Social effects on song learning.** Kuro the starling learned to include words in his vocalizations because he had a close relationship with the family of Keigo Iizuka. Photograph by Birgitte Nielsen, courtesy of Keigo Iizuka.

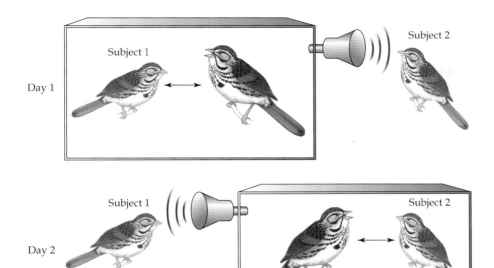

**FIGURE 2.6  The design of an experiment** to test whether young song sparrow tutees learn their songs by interacting with a singing adult male or by overhearing an adult male interacting with another inexperienced young male. The tutees were paired so that on one day the young male was confronted by a singing male and on another day the subject "eavesdropped" on an adult singing at another male. After Beecher et al.[95]

interacting with another bird rather than when he is directly interacting with a singing adult himself (Figure 2.6).[95]

Whether white-crowned sparrows are also more prone to pick up a dialect from a male that is singing aggressively at another individual is not known, but social experience clearly plays a key role in the song-learning process of this species. Young males respond to their social tutors by acquiring song memories that they will draw on later when they begin singing. The birds use their stored song memories as models to imitate, on the road to producing their own full songs. Males with different model song memories will therefore develop different songs as adults.

## Discussion Question

**2.2**  A natural experiment sometimes occurs in Australian woodlands when galahs (a species of parrot) lay eggs in tree hole nests that are then stolen from them by pink cockatoos (another species of parrot). Thus, the cockatoos inadvertently become foster parents for baby galahs. The young foster-reared galahs produce begging calls and alarm calls that are identical to those produced by galahs cared for by their genetic parents. However, the adopted galahs eventually give contact calls very much like those of their adoptive cockatoo parents, as you can see from Figure 2.7. (The birds produce these signals to help maintain contact with others when traveling in flocks.[1244]) Someone claims that these observations show that galah begging and alarm calls are genetically determined, whereas contact calls are environmentally determined. Explain why this claim is wrong (you may find it helpful to read ahead in Chapter 3). Then defend the superficially similar statement that the differences between the alarm calls given by adopted galahs and by their cockatoo foster parents are the result of genetic differences between them. What other behavioral differences are the result of differences between the social environment of the adopted galahs and that of certain other individuals?

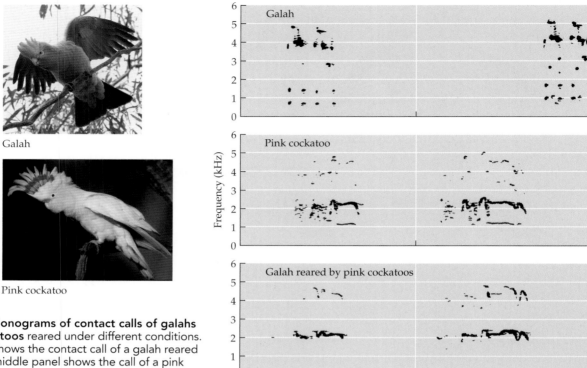

Galah

Pink cockatoo

**FIGURE 2.7** **Sonograms of contact calls of galahs and pink cockatoos** reared under different conditions. The top panel shows the contact call of a galah reared by galahs; the middle panel shows the call of a pink cockatoo reared by its genetic parents; the bottom panel shows the call of a galah reared by pink cockatoo foster parents. After Rowley and Chapman.[1244]

## The Development of the Underlying Mechanisms of Singing Behavior

If we want to understand more about the development of a white-crowned sparrow's dialect, we have to go beyond identifying those elements of the young bird's social and acoustical environment that affect its later behavior. For one thing, where are a month-old white-crown's song memories stored? And what part of the brain controls the sounds that it produces at 5 months of age? And how does the young male know how to match his song memories with his own initially simpler songs? These questions require us to consider the internal devices that the young male possesses that are capable of using social and acoustical inputs to steer his singing behavior along a particular developmental pathway. The mechanisms that make song learning possible are located in the brain of the young male. How did the male's brain acquire the hardware for song acquisition?

We can start by exploring why male white-crowns sing a full and complex song, whereas female white-crowns do not. If singing behavior is dependent on the way in which the bird's brain is built, then it follows that male and female brains must differ in white-crowns. These brain differences could in theory arise because of genetic or environmental differences (or both) between the two sexes that influence how the birds' nervous systems develop.

No one doubts that male and female birds differ genetically. Male birds have two Z chromosomes, whereas females have a Z and a W chromosome. (This sex determination system differs from that of mammals, in which males have two different chromosomes, an X and a Y, while females have two X chromosomes.) Because chromosomes are where the genes are located, and because the avian W chromosome has many fewer genes than the Z chromosome, it follows that male and female birds are genetically distinct.[433] These genetic differences are known to have large effects on development. An embryonic female white-crown, with her W and Z chromosomes, develops gonadal cells that will give rise to her ovaries. An embryonic male, with his different genes, develops differently, such that he eventually acquires sperm-producing testes, not egg-producing ovaries.

Thus, from a very early stage, male and female gonadal cells follow different developmental paths as their different genes interact with the compounds available for the construction of more cells and the assembly of the testes or ovaries. Moreover, gonadal cells in the two sexes also differ in the kinds of chemicals they produce for transport to other cells. In particular, cells in the young testes, but not in the pre-ovaries, manufacture the hormone testosterone. This hormone, as it travels through the male white-crown's developing body, becomes part of the chemical environment of other cells, including those in the male's developing brain. Testosterone can be picked up by cells that have the appropriate protein receptors in their nuclei, setting in motion a chain of chemical events that alters gene activity within the testosterone-sensitive cells. The result is the growth of special neural circuits that will eventually be used by singing males.

Note that the differences between the brains of male and female sparrows are the product of both genetic and environmental differences. A chromosomal (genetic) difference translates into a hormonal (environmental) difference, which leads to differences in the genetic activity of certain cells in the brains of males and females, which produce additional differences in the proteins manufactured by those cells, with follow-on (environmental) effects on other cells, and so on. When developmental biologists speak of development as an interactive process, they envision a cascade of alterations in genetic activity regulated by changes in the chemical environment of cells. (Remember that the genetic contribution to development is the information encoded in the organism's DNA; everything else constitutes the environmental contribution to development, including the cellular chemical products of preceding gene–environment interactions.)

As it turns out, the gonads are not the only, or even the most important, source of chemical signals needed for male brain development. Another key substance is the hormone estrogen. Although we usually think of estrogen as a female hormone (because it is produced by the ovaries), cells in male bird brains convert testosterone into estrogen for transport to other cells. There, this self-manufactured environmental signal activates the development of specific neural pathways found in the male's brain. These circuits link the elaborate network of neural structures, known as the song control system, whose integrated operation is required if the male is to sing his songs correctly.[672]

The importance of estrogen for development of a male song control system has been demonstrated by testing the prediction that insertion of small estrogen pellets under the skin of a nestling female should result in the masculinization of her song control system. This experiment has been done with zebra finches, whose brains are anatomically similar to those of white-crowned sparrows. Normally, certain special clusters of brain cells that grow rapidly in size in young male zebra finches shrink over time in young females (Figure 2.8).[389, 777] In immature females subjected to estrogen treatment, however, these song control units increase in size, as predicted, providing support for the hypothesis that a criti-

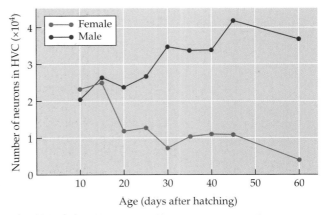

**FIGURE 2.8  Changes in the song system of young male and female zebra finches.** Between 10 and 40 days after hatching, the number of neurons in the female's HVC (higher vocal center), a component of the song system, declines rapidly, whereas the number of HVC cells in males increases greatly. After Kirn and DeVoogd.[777]

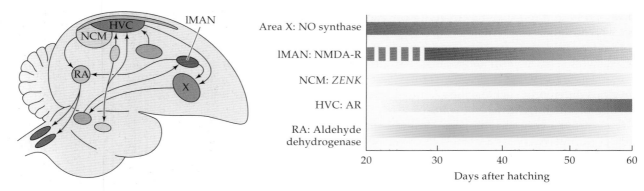

**FIGURE 2.9** **The timing of gene activity** in different components of the avian song control system in males. Different genes follow different schedules of activity over the course of development of the young zebra finch. The more intense the color, the more active the gene. Thus, for example, the gene that encodes the enzyme nitric oxide (NO) synthase is moderately active in area X of the nestling's brain early in life, but production of that protein soon declines at this site. See Figure 2.13 for abbreviations. After Clayton.[270]

cal chromosomal–hormonal connection sets off a series of gene–environment interactions underlying the development of a male song control system in a zebra finch's brain.[594]

Although in zebra finches both sexes listen to and memorize their species' song in preference to the songs of other species,[156] only the male's brain eventually acquires the structures needed if the finch is to sing. To do so requires a whole series of highly coordinated changes in gene activity in the brain structures that make song learning and song production possible. And indeed, when geneticists have examined when particular genes are turned on and then off during the early stages of brain development in the zebra finch, they have found a special pattern that almost all males share (Figure 2.9).

Nor do changes in gene activity cease when the brain achieves its mature form. If a male zebra finch's or white-crowned sparrow's song control system is to serve as a song-learning mechanism, certain cells within the system must in effect anticipate certain events. Thus, when a young bird is bombarded with sounds produced by singing adults of its own species, these sounds activate special sensory signals that get relayed to particular parts of the brain. In response to these distinctive inputs, some cells in these locations must then alter their biochemistry in order to change the bird's behavior. Biochemical changes typically require changes in gene expression (in which the information encoded in a gene is actually used to produce a product, such as an enzyme). So, for example, when a young white-crown hears its species' song, certain patterns of sensory signals generated by acoustical receptors in the bird's ears are relayed to song control centers in the brain where learning occurs. These sensory inputs are believed to alter the activity of certain genes in the set of responding cells, leading to new patterns of protein production that reshape those cells biochemically. Once the cells have been altered, the modified song control system can do things that it could not do before the bird was exposed to the song of other males of its species.

This theory of learning has been tested by examining cells in song control centers for changes in gene expression after a bird has heard a relevant song. You will recall that one acoustical stimulus that apparently affects the acquisition of song by white-crowned sparrows is the young bird's own songs. The sparrow's ability to hear itself sing when it is 150 to 200 days old is essential for the crystallization of a normal full song from the variable sub-songs it initially sings. Likewise, male zebra finches go through a period early in life in which they appear to be matching elements in their initial subsongs against stored memories of the full songs they heard others singing previously. During this process, certain neurons in the finch's anterior forebrain become more and more responsive to the bird's own song as opposed to tutor songs; by the time the finch is an adult, with a fully crystallized song, certain of its auditory neurons have become highly selective, responding much more strongly to this song than to any other.[409] Presumably this developmental process involves a biochemical restructuring of the neurons in question, which in turn demands genetic information that can be turned on or off by specific environmental events.

We now know that as zebra finches attempt to match a tutor's song, the activity of a gene called *ZENK* rapidly increases in certain song control neurons, resulting in corresponding increases in cellular amounts of the protein encoded by that gene.[973] In other words, when a zebra finch listens to itself sing, it generates sensory feedback that activates a particular gene in certain cells. This gene's activity translates into the production of a specific protein, which is believed to have something to do with subsequent alterations in the neural circuits that control the finch's song (Figure 2.10). This hypothesis has received support from the observation that as the zebra finch gets closer and closer to singing an accurate copy of a tutor song, *ZENK* gene activity falls in a particular part of the brain.[726] As the finch gains a crystallized full song, further changes in cell architecture or biochemistry in its song control system cannot improve its ability to sing that song, and therefore the *ZENK* gene shuts down in the appropriate set of cells.

*ZENK* is not the only zebra finch gene whose information is known to contribute to song learning and production.[1489] For example, the gene *FoxP2* is also a player, as demonstrated by researchers who injected a chemical inhibitor of the gene into a particular part of the brains of 23-day-old male finches. (Interestingly, the same gene appears to play several roles in language acquisition and use in humans.) The treated birds were held in cages with social tutors of their own species. When the young males began to sing, their songs were recorded and compared with those produced by control finches of the same age whose brains had been injected with a chemical that targeted a different segment of DNA, not *FoxP2*. The experimental group failed to copy their social tutors' songs as well as the control males (Figure 2.11), and in addition, their vocal output was more variable.[603] Because the birds lacked normal levels of the protein coded for by *FoxP2*, they appear to have been handicapped in their use of their auditory memories of tutor songs when attempting to imitate these songs accurately.

Studies like this one illustrate the intimate connection between the developmental and physiological levels of analysis of bird song. Changes in genetic activity in response to key environmental stimuli translate into changes in neurophysiological mechanisms that control the learning process. An understanding of both developmental and neurophysiological systems is required to give us a full picture of the proximate causes of behavior.

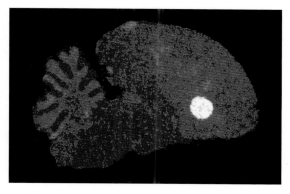

**FIGURE 2.10    Gene expression in a component of the zebra finch song system.** The white-and-yellow area corresponds to area X. The brightness of this area is related to the high level of activity of the *ZENK* gene, which has resulted in the production of relatively large amounts of the protein coded for by this gene in a finch about 40 days old. The absence of white and yellow elsewhere in this brain image indicates that the *ZENK* gene was not active in those areas. Photograph courtesy of David Clayton.

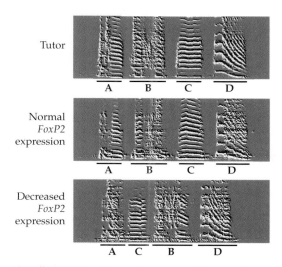

**FIGURE 2.11    A gene important in song learning by male zebra finches.** The experimental subjects (bottom) were treated in such a way that the *FoxP2* gene was unable to express its information at normal levels within certain brain cells. These males failed to imitate the songs they heard from a tutor finch (top) nearly as well as control males (middle) whose *FoxP2* had not been altered and so could produce normal amounts of the protein coded for by the gene. A, B, C, and D are different syllables of the song. After Haesler et al.[603]

**FIGURE 2.12** **The song preferences of female starlings,** as measured by their willingness to perch near a nest box from which a long starling song is played versus one from which a shorter song is played. After Gentner and Hulse.[528]

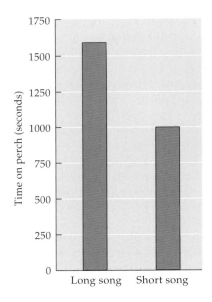

European starling

### Discussion Question

**2.3** Naturally, female starlings possess the same *ZENK* gene males do. In the female brain, the ventral caudomedial neostriatum, or NCMv, responds to signals sent to it from auditory neurons that fire when the bird is exposed to sounds, such as those made by singing male starlings. When captive female starlings are given a choice between perching next to a nest box where they can hear a long song versus perching next to another nest box where a shorter song is played, they spend more time at the long song site (Figure 2.12). What proximate hypothesis could account for the song preferences of female starlings? What prediction can you make about the activity of the *ZENK* gene in the NCMv of female starlings exposed to long versus short songs? How might you check your prediction? What would be the scientific point of collecting the data necessary to evaluate your prediction?

### *How the Avian Song Control System Works*

Having identified just a few of the many genetic and environmental factors involved in the development of avian singing behavior, let's move to another proximate level of analysis, one that focuses explicitly on the operating rules that characterize the brain of the bird. White-crowned sparrows, and other songbirds, have a brain with many anatomically distinct clusters of neurons, or **nuclei**, such as the just-mentioned NCMv, as well as neural connections that link one nucleus to another. The various components of the brain are made up of cells (neurons) that communicate with one another via bioelectric messages (action potentials) that travel from one neuron to another via elongate extensions of the neurons (axons) (see Figure 4.10). Some components of the brain are deeply involved in the memory of songs, while others are necessary for the imitative production of memorized song patterns.[548] Figuring out which anatomical unit does what is a task for neurophysiologists more interested in the operational mechanics of the nervous system than in its development.

These researchers have long been especially interested in the higher vocal center (HVC) in the brains of white-crowned sparrows and other songbirds. This dense collection of neurons connects to the robust nucleus of the arcopallium (mercifully shortened to RA by anatomists), which in turn is linked with the tracheosyringeal portion of the hypoglossal nucleus (whose less successful acronym is nXIIts). This bit of brain anatomy sends messages to the syrinx, the sound-producing structure of birds that is analogous to the larynx in humans. The fact that the HVC and RA can communicate with the nXIIts, which connects to the syrinx, immediately suggests that these brain elements exert control over singing behavior (Figure 2.13).

This hypothesis about the neural control of bird song has been tested. For example, if neural messages from the RA cause songs to be produced, then the experimental destruction of this center, or surgical cuts through the neural pathway leading from the RA to the nXIIts, should have devastating effects on a bird's ability to sing. Experiments designed to test these predictions have been done with a variety of songbirds,[250] with the result that we can now say with confidence that the RA does indeed play a critical role in song production. If this is true, then in bird species like the white-crowned sparrow, in which males sing and females do not, the RA should be larger in male brains than in female brains, and it is (Figure 2.14).[61, 1031, 1057]

Other brain nuclei appear to be essential for song learning, rather than for the production of vocal signals. Destruction of the lateral magnocellular nucleus of the anterior nidopallium (lMAN), for example, does not strongly interfere with an adult zebra finch's ability to sing the song it has learned earlier in life, but if the operation is performed on a juvenile bird before it has acquired a mature song, then the bird will fail to sing a normal song in adulthood. Further evidence of the importance of this component of the control system for song learning came from checking the prediction that the lMAN should be much reduced or absent in bird species that sing but do not learn their songs. Sure enough, many species of this type lack the well-defined lMANs (and certain other forebrain nuclei) of white-crowned sparrows and other vocal learners.[250]

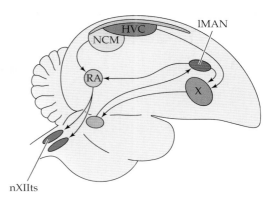

**FIGURE 2.13   The song system of a typical songbird.** The major components, or nuclei, involved in song production include the robust nucleus of the arcopallium (RA), the higher vocal center (HVC), the lateral portion of the magnocellular nucleus of the anterior nidopallium (lMAN), the caudomedial neostriatum (NCM), and area X (X). Neural pathways carry signals from the HVC to the tracheosyringeal portion of the hypoglossal nucleus (nXIIts) to the muscles of the song-producing syrinx. Other pathways connect the nuclei, such as lMAN and area X, that are involved in song learning rather than song production. After Brenowitz, Margoliash, and Nordeen.[170]

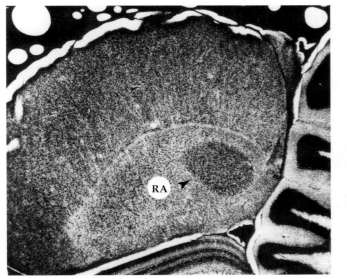
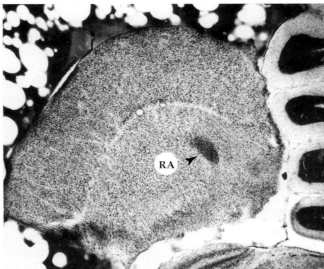

**FIGURE 2.14   Differences in the size of one nucleus of the song system,** the robust nucleus of the arcopallium (RA), in (left) the male and (right) the female zebra finch. Photographs courtesy of Art Arnold; from Nottebohm and Arnold.[1057]

Evidence that the HVC, as well as the lMAN, plays a role in song learning comes from a study of a group of closely related European warbler species that differ in the extent to which males learn a number of different songs. Among these species, the larger the song repertoire, the larger the HVC. In addition, individuals of the same species can differ in the size of their song repertoire. An analysis of all currently available studies on the relation between the volume of the HVC and the size of the male song repertoire within a species shows that overall, the two variables are significantly correlated: a male of a given species with a relatively large repertoire usually has a larger HVC than other, less gifted members of his species.[516]

In the studies just outlined, however, an assumption has been made that large amounts of neural tissue are required if the male is to learn a complex song or songs. But one could argue that the experience of singing a complex song or acquiring a large repertoire causes the HVC to expand in response to stimulation of this region of the brain. One way to see whether the HVC must be large if the bird is to learn or whether learning makes the HVC grow would be to create two groups of male songbirds, one that is permitted to shape its song repertoire through learning and the other that is raised in acoustical isolation. If it is the learning experience that causes certain parts of brains to grow, then isolated males should have smaller HVCs than those that learn songs. This experiment has been done with sedge warblers; isolated males had brains that were in no way different from those of males that had learned their songs by listening to the songs of other males.[850] Prior to this work, similar results came from a study in which some male marsh wrens were given a chance to learn a mere handful of songs while another group listened to and learned up to 45 songs.[169] Thus, in both the warbler and the wren, the male brain develops largely independently of the learning experiences of its owner, suggesting that the production of a large HVC is required for learning, rather than the other way around.

Although the neurophysiology of song learning has been explored at the level of entire brain nuclei, neuroscientists could in theory find out how a given neuron contributes to communication between birds. In fact, this kind of research has been done by Richard Mooney and his coworkers in a study of song perception by swamp sparrows.[1005] Swamp sparrows sing two to five song types, each type consisting of a "syllable" of sound that is repeated over and over in a trill that lasts for a couple of seconds. If a young male swamp sparrow is to learn a set of song types, he must be able to discriminate between the types being sung by males around him, and then later he must be able to tell the difference between his own song types as he listens to himself sing.

One mechanism that could help a young male control his song type output would be a set of specialized neurons in the HVC that respond selectively to a specific song type. Activity in these cells could contribute to his ability to monitor what he is singing so that he could adjust his repertoire in a strategic manner (by, for example, selecting a song type that would be particularly effective in communicating with a neighboring male, as we will see below). Of course, the contribution that any one cell makes to a behavioral decision is dependent on a host of other cells with which that neuron communicates. But the existence of song type specialists would help us understand how the component parts of various neural mechanisms contribute to song discrimination.

The technology exists to permit researchers to record the responses of single cells in a swamp sparrow's HVC to playbacks of that bird's own songs. In so doing, Mooney and his associates discovered a number of HVC relay neurons that generated intense volleys of action potentials when receiving neural signals from other cells upon exposure to one song type only.[1005] Thus, one relay neuron, whose responses to three different song types are shown in Figure

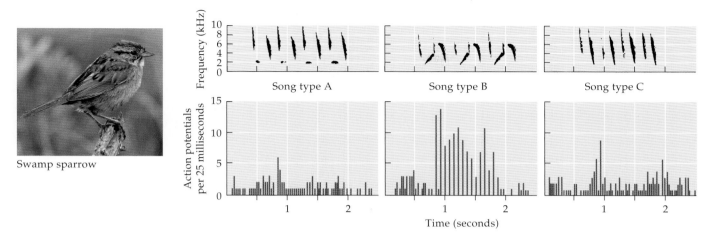

**FIGURE 2.15  Single cells and song learning in the swamp sparrow.**
(Top) Sonograms of three song types: A, B, and C. (Bottom) When the three songs
are presented to the sparrow, one of its HVC relay neurons reacts substantially only to
song type B. Other cells not shown here respond strongly only to song type A, while
still others fire rapidly only when song type C is the stimulus. After Mooney, Hoese,
and Nowicki.[1005]

2.15, produces large numbers of action potentials in a short period when the
song stimulus is song type B. This same cell, however, is relatively unresponsive when the stimulus is song type A or C. So here we have a special kind
of cell that could help the sparrow identify which song type it is hearing, the
better to select the best response for a particular situation.

## Different Songs: Ultimate Causes

Although a great deal has been learned about how the song control system of
songbirds develops and operates, we still do not know everything about the
underlying proximate mechanisms of singing behavior. But even if we had
this information in hand, our understanding of singing by white-crowned
sparrows would still be incomplete until we dealt with the ultimate causes of
the behavior. Because the proximate mechanisms underlying bird song did
not emerge out of thin air, we can ask questions such as, when in the distant past did an ancestral bird species start learning its species-specific song,
thereby setting in motion the events that led to dialects in birds like the white-
crowned sparrow? Evidence relevant to this question includes the finding that
song learning occurs in members of just 3 of the 23 avian orders: the parrots,
the hummingbirds, and the "songbirds," which belong to that portion of the
Passeriformes that includes the sparrows and warblers, among others (Figure 2.16).[168] Members of the remaining 20 orders of birds produce complex
vocalizations, but they do not have to learn how to do so, as shown in some
cases by experiments in which young birds that were never permitted to hear
a song tutor, or were deafened early in life prior to the onset of song practice,
nevertheless came to sing normally (Figure 2.17).[809]

One evolutionary question is, did the song learning that occurs within each
of the three orders evolve independently of the other groups? The fact that the
nearest relatives of each of the three song-learning orders do not learn their
songs suggests one of two things: either song learning originated three different times in the approximately 65 million years since the first truly modern

**FIGURE 2.16** **The phylogeny of song learning in birds.** If we assume that the long-extinct bird that gave rise to all modern species did not learn elements of its songs but instead produced vocalizations instinctively, as do many modern bird groups, then song learning must have evolved independently in three different lineages of modern birds. On the other hand, song learning may have originated in a common ancestor (see arrow) of parrots (Psittaciformes), hummingbirds (Trochiliformes), and passerine songbirds (Passeriformes) and may have been retained in these three lineages while being lost in other descendants of that ancestral song-learning species. After Brenowitz.[168]

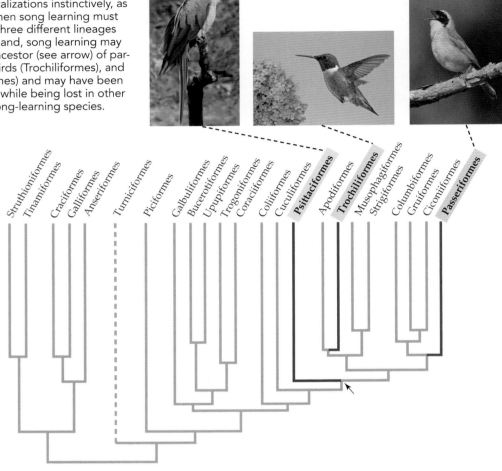

bird appeared, or song learning originated in the ancestor of all the lineages from Psittaciformes to Passeriformes but then was lost three times: (1) at the base of the evolutionary line leading to the Apodiformes, (2) at the base of the combined Musophagiformes and Strigiformes lineages, and (3) at the base of the lines constituting the Columbiformes, Gruiformes, and Ciconiiformes. In other words, either there have been three independent origins of song learning or one origin followed by three separate losses of the trait (see Figure 2.16).

The first scenario is more parsimonious since it requires just three evolutionary innovations, not four, and is therefore generally considered somewhat more likely to have occurred. If the three-independent-origins scenario is true, we would expect to find major differences among the song control systems of the three vocal learning groups. If we are to check this prediction, we need to identify the brain structures that endow parrots, hummingbirds, and songbirds with their singing abilities. One way to do so is to capture birds that have just been singing (or listening to others sing) and kill them immediately in order to examine their brains for regions in which a product of the *ZENK* gene can be found.[712] As noted above, this gene's activity is stimulated in particular parts of the brain when birds produce songs and in different parts of the brain when birds hear songs. Other stimuli and activities do not turn the gene on, so its

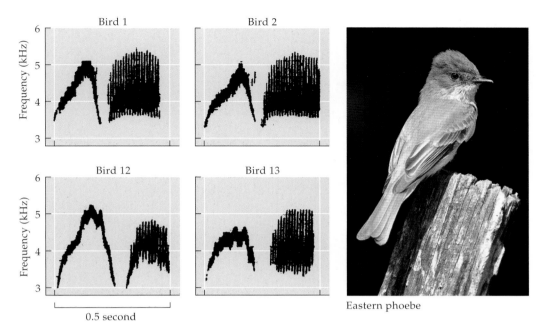

Eastern phoebe

**FIGURE 2.17   The song of a vocal non-learner, the eastern phoebe,** a member of part of the Passeriformes that does not need to hear others sing in order to produce a complete song of its own. Birds 1 and 2 were normal males, capable of hearing; their songs were recorded in the field. Birds 12 and 13 were males that were deafened early in life and so had not heard other phoebes calling, nor were they able to hear themselves sing. Learning apparently does not figure in the development of song in this species. After Kroodsma and Konishi.[809]

product is absent or very scarce when the birds are not communicating. By comparing how much ZENK protein appears in different parts of the brains of birds that were engaged in singing or listening to songs immediately before their deaths, one can effectively map those brain regions that are associated with bird song.

The *ZENK* activity maps of the brains of selected parrots, hummingbirds, and songbirds reveal strong similarities in the number and organization of discrete centers devoted to song production and processing. In all three groups, for example, cells that form the caudomedial neostriatum (NCM) activate their *ZENK* genes when individuals are exposed to the songs of others. The NCM is located in roughly the same part of the brain in all three groups, where it constitutes part of a larger aggregation of anatomically distinct elements that contribute to the processing of song stimuli. When parrots, hummingbirds, and songbirds vocalize, other brain centers, many in the anterior part of the forebrain, respond with heightened *ZENK* gene activity. Once again, there is considerable (but not perfect) correspondence in the locations of these distinctive song production centers in parrots, hummingbirds, and sparrows (Figure 2.18). The many similarities in brain anatomy among the three groups of vocal learners would seem to rule against the hypothesis that song-learning abilities evolved independently in the three groups of birds. Thus, we have to take seriously the possibility that the ancestor of all those birds in the orders sandwiched between parrots and songbirds (see Figure 2.16) was a song learner and that the mechanism for vocal learning possessed by that bird was then retained in some lineages while dropping out early on in others—a hypothesis that continues to generate debate.[457, 712]

**FIGURE 2.18  The song control systems of parrots, hummingbirds, and oscine songbirds** are distributed throughout the brain in remarkably similar patterns. On the left is a diagram of the evolutionary relationships among some major groups of birds, including the three orders of vocal learners. On the right are diagrams of the brains of these groups, with the various equivalent components of the song control system labeled (e.g., HVC, higher vocal center; NCM, caudomedial neostriatum). After Jarvis et al.[712]

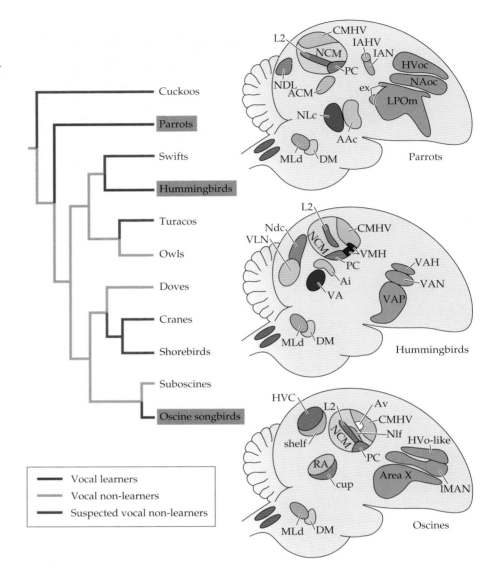

## The Reproductive Benefits of Song Learning

Let us accept the possibility that in three bird lineages, neural mechanisms that promote song learning have been retained over evolutionary history because of their reproductive value to individuals. What might these fitness benefits be in species of parrots, hummingbirds, and songbirds? We can begin to tackle this question by asking another: What do birds gain by having a distinctive vocal message, whether it is learned or unlearned? One hypothesis is that a distinctive vocalization conveys information about species membership, as we noted at the outset of this chapter. After all, if humans can tell an ovenbird from a yellow warbler by their songs, it is not unreasonable to assume that members of the two species of warblers can also make this acoustical discrimination. One benefit that might flow to males that sang differently from members of other species would be enhanced deterrence of competitors of their own species for territories and mates. Rival males of the same species pose the greatest threat to a singer's reproductive success, because they may want to oust him from good breeding habitat or to lure his mate away from him. By broadcasting a clear message that a particular site or a particular female is

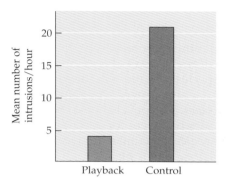

**FIGURE 2.19  Does bird song repel territorial intruders?** Territories from which resident male white-throated sparrows were experimentally removed attracted fewer intruders when the taped song of the removed male was broadcast from his vacant territory. After Falls.[453]

White-throated sparrow

defended by a physiologically fit singer, a male might encourage male conspecifics (members of the same species) to move on rather than engage in time-consuming conflicts with the songster.

Evidence in support of the species identity announcement hypothesis comes from experiments in which resident male white-throated sparrows (a relative of white-crowns) were removed from their territories, after which speakers were placed in the vacant territories. In half of the territories, the taped songs of the former territory holders were broadcast from the speakers. New males were slower to move into territories with broadcast songs than into those in which the speakers were silent (Figure 2.19).[453] Likewise, in paired experiments in which two song sparrow males were simultaneously removed from their territories but only one was replaced with a taped song broadcast, the silent territory was always the first to be invaded by an intruder.[1059] These results support the hypothesis that singers can gain by repelling other males of their species.

In much the same way, male singers able to communicate their species identity to females might attract mates more readily than those whose songs are less distinctive, and so less recognizable. Females that rapidly locate males of their species can begin reproducing sooner with a member of their own species, avoiding the risks of mating with members of another species. The mate attraction hypothesis for song distinctiveness correctly predicts that females will respond to taped songs of their own species more strongly than to taped songs of another species (Figure 2.20).[1036] If in the past female songbirds have been particularly strongly drawn to easily identified members of their species, then natural selection driven by mate choice would have resulted in the spread of the favored songs because of the clarity with which they announced the species membership of the singer.

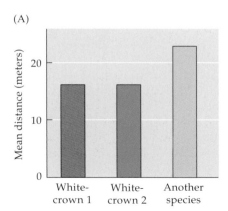

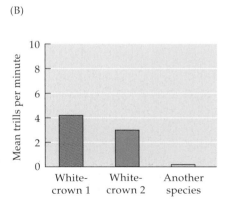

**FIGURE 2.20  White-crowned sparrow females** are attracted to the songs of male white-crowned sparrows but not to the songs of other species. In an experiment with female white-crowns, researchers temporarily removed their mates from their territories and then played two digital versions of white-crowned sparrow song (composed of elements taken from two different males) and vocalizations of song sparrows and dark-eyed juncos. The females (A) came closer to the speakers and (B) trilled more in response to the playbacks when white-crown songs were played. After Nelson and Soha.[1036]

### The Benefits of Learning a Dialect

Although much more could be said about the reasons why so many bird species can be recognized on the basis of their songs alone, this chapter revolves around why males of the same species may differ in the songs they sing. So let's shift our focus to why male white-crowns sing dialects that they have learned from others of their species. The puzzle here is that song learning has some obvious reproductive disadvantages for individual sparrows. Song learning takes time, energy, and special neural mechanisms, all of which could be devoted to other reproduction-enhancing activities. The fact that many—indeed, most—birds do not make these costly investments, but instead produce perfectly good songs instinctively, tells us that song learning is not essential for the acquisition of a species-distinctive signal. So we must identify some reproductive advantages for learning that are large enough to outweigh its costs if we are to account for the spread of the trait in a given songbird species.

One possible benefit of song learning would be the ability of a young male to fine-tune his song so that it resembled the songs of one or more other individuals in a particular region. Fine-tuning could conceivably help the young male in several ways. First, imagine that males in a particular location have acquired a dialect that can be transmitted unusually effectively in that habitat. A young male that learns his song from his elders might then generate calls that travel farther and with less degradation than if he sang another dialect better suited to a different acoustical environment.[250] And in fact, males of the satin bowerbird that live in dense forest habitats sing songs characterized by relatively low frequencies, whereas those that live in more open, less cluttered forests tend to employ higher frequencies in their versions of bowerbird song.[1044] High-frequency sounds become less degraded in open habitats than in places with thick foliage,[137] which helps explain why males of the great tit, like satin bowerbird males, sing different versions of their species' song in different woodlands with different foliage densities (Figure 2.21).[699] The

Great tit

**FIGURE 2.21    Songs match habitats.**
Great tits from dense forests produce pure whistles of relatively low frequency, whereas males of the same species that live in more open woodlands use more and higher sound frequencies in their more complex songs. After Hunter and Krebs.[699]

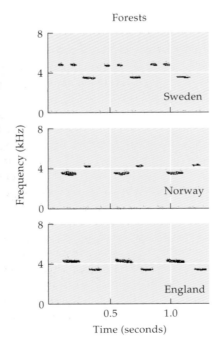

Forests

Sweden

Norway

England

Frequency (kHz)

Time (seconds)

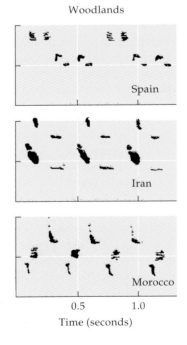

Woodlands

Spain

Iran

Morocco

Time (seconds)

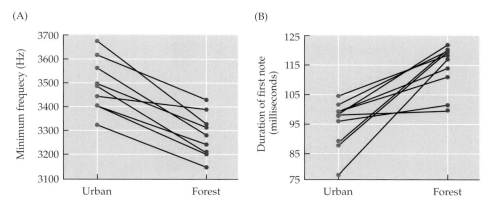

**FIGURE 2.22** **The songs sung by great tits differ in cities versus forests.** (A) City males sing higher-pitched songs (based on a higher minimum frequency), and (B) they sing faster as well (based on the shorter duration of the first note in the song). After Slabbekoorn and den Boer-Visser.[1342]

great tit also lives in many European cities, enabling the Dutch ecologist Hans Slabbekoorn and his coworker to compare the dialects of city birds with the songs of their country cousins. The vocalizations of the two groups differed substantially, with city birds having shorter, higher-pitched songs while the birds in nearby forests produced longer and lower-pitched songs (Figure 2.22). The Dutch ecologists linked these differences to the fact that the traffic noise in cities is primarily made up of low-frequency sounds, which are largely absent in woodlands. When urban great tits sing in the high-frequency range, they occupy a channel not obscured by cars and trucks rumbling through city streets. In contrast, great tits living in forests employ low-frequency signals because that channel is open and effective in their environment.[1342] Young birds in both types of environments are likely to acquire the song types that they can hear most easily, and as a result, they too can communicate over greater distances when they sing as adults.

A second hypothesis on the benefits of song learning centers on the advantages of matching songs to the singer's *social* environment.[171] The idea is that males able to learn the local version of their species-specific song can communicate better with rivals that will also be singing that particular learned song variant. If this is so, then we can predict that young males should learn directly from their territorial neighbors, using those individuals as social tutors. By adopting the song of a neighbor, a new boy on the block could signal his recognition of that male as an individual and demonstrate his capacity to learn a new song, which could be based on his health, body condition, or other indicators of competitive ability. Males that are evenly matched competitors have something to gain by accepting each other's presence rather than engaging in fruitless and expensive combat. Thus, a mutual nonaggression pact based on learned song signals could benefit both an established resident and a rival newcomer by helping them conserve time and energy.

Note that this hypothesis leads to the prediction that males should be able to fine-tune their songs even after settling on initial versions relatively early in life. This prediction has been checked by examining the songs sung by young male white-crowns before and after they acquired their first territory. As predicted, males in at least some populations did change what had appeared to be their crystallized full songs in order to more closely match the songs of neighbors[74, 101] (but see Nelson[1033]).

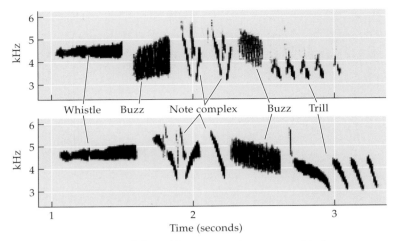

**FIGURE 2.23**   **Two songs of the white crowned sparrow** with the component parts labeled, including the "note complex" and the "trill" at the end of the song. After Nelson, Hallberg, and Soha.[1035]

---

### Discussion Questions

**2.4**   Note that the song of the white-crowned sparrow is composed of several parts or phrases, one of which is the "note complex" and another the terminal "trill" (Figure 2.23). When the responsiveness of male birds was tested in playback experiments to songs that had been modified in various ways, the researchers found that changes to the trill were more likely to reduce the male's aggressive reaction to the taped song than were changes to the note complex component.[1035] With this background, what do you predict about the frequency of improvisation as opposed to accurate mimicry of the two parts of the song by young captive hand-reared birds exposed to social tutors in the lab? What is the basis for your prediction?

**2.5**   At the proximate level, two hypotheses exist for the ability of some yearling white-crowned sparrows to match the dialect of neighboring males. The late acquisition hypothesis states that yearlings have a developmental window similar to that of very young birds, which enables them to listen to and learn directly from their immediate neighbors, overriding (if need be) any dialect learning that took place earlier in life. In contrast, the selective attrition hypothesis argues that fledglings can memorize a number of dialect versions of their species' song early in life; then, after settling next to some older males, these birds gradually discard certain of their learned variants until they are left with the one that best matches the dialect of their neighbors. In light of these two alternative hypotheses, what significance do you attach to Figure 2.24? Outline the entire scientific process here, from the question that motivated the work to the scientific conclusion.

---

The ability of males to produce the songs, or at least song elements, of rival neighbors is also characteristic of the song sparrow, a bird that has a repertoire of several different, distinctive song types rather than a single vocalization of a particular dialect. Young song sparrows usually learn their songs from tutors that are their neighbors in their first breeding season, with their final repertoire tending to be similar to that of an immediate neighbor.[1055] In fact, Michael Beecher and his colleagues found through playback experiments that when a

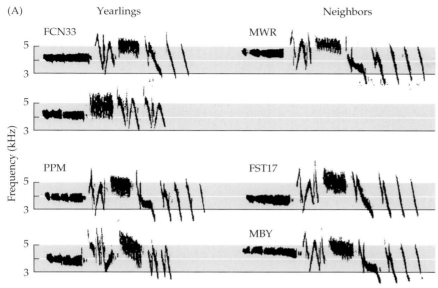

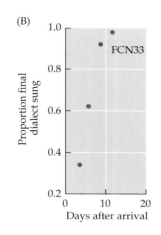

**FIGURE 2.24   Dialect selection by male white-crowned sparrows.** (A) After arriving on their first territories, yearling males FCN33 and PPM initially each sang two different dialects, but after a time they each settled on one song type (the upper of the two song types shown for each male in the left-hand part of the figure). The song type each came to use matched that of his neighbor(s) (MWR for FCN33; FST17 and MBY for PPM). (B) Initially, male FCN33 mostly sang one of his two song dialects, but soon after his arrival on a territory near MWR, he stopped singing that song in favor of the one that matched his neighbor's song. After Nelson.[1034]

male heard a tape of a neighbor's song coming from the neighbor's territory, he tended to reply to that tape by singing a song from his own repertoire that matched one in the repertoire of that particular neighbor. This kind of type matching occurred when, for example, male BGMG heard the taped version of male MBGB's song type A (Figure 2.25), which led him to answer with his own song type A.

Another kind of response involves repertoire matching, in which a bird exposed to a song type from a neighbor responds not with the exact same type, but with a song type drawn from their shared repertoire. Repertoire matching would occur if male BGMG answered back with song type B or C when he heard song type A from male MBGB's territory (see Figure 2.25).

A third option for a male song sparrow is to reply to a song from a neighbor with an unshared song. This kind of mismatched response would be illustrated by male BGMG singing song type D, F, or H (see Figure 2.25) upon hearing male MGBG's song type A.

The fact that song sparrows type-match and repertoire-match so often indicates that they recognize their neighbors and know what songs they sing and that they use this information to shape their replies.[92] But how do male sparrows benefit from their selection of a song type to reply to a neighbor? Perhaps their choice enables them to send graded threat signals to their neighbors, with type matches telling the targeted receiver that the singer is highly aggressive, whereas replying with an unshared song could signal a desire to back off, while a repertoire match might signal an intermediate level of aggressiveness. If so, then playback experiments should reveal differences in the responses of territorial birds to taped songs that contain a type match, a repertoire match, or neither. When Michael Beecher and various colleagues checked this prediction, they found that a tape containing a type match did indeed elicit the most

Song sparrow

**FIGURE 2.25 Song type matching in the song sparrow.** Males BGMG and MBGB occupy neighboring territories and share three song types (A, B, and C: the top three rows of sonograms); six unshared song types (D, E, F, G, H, and I) appear on the bottom three rows. After Beecher et al.[90]

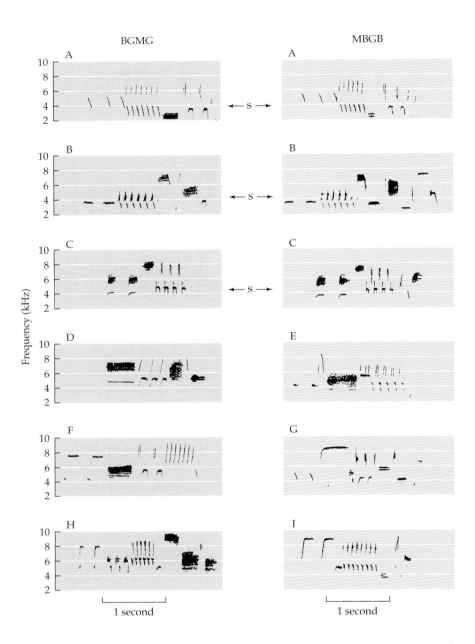

aggressive response from a listening neighbor, while a repertoire match song generated an intermediate reaction, and nonshared song types were treated less aggressively still (Figure 2.26).[94, 217]

In another experiment, Michael Beecher and Elizabeth Campbell waited by a song sparrow's territory until the male sang a song shared by a neighbor. They then played a tape of this song type. When the subject approached the speaker, they switched on a new song, either a song type that both birds had in their repertoire or a song that was not shared by the aggressive territory defender. The responding song sparrow left more quickly when it heard a nonshared song than when it heard a song in its own repertoire.[94] These results support the hypothesis that male song sparrows that are able to learn songs from a neighbor can convey information about just how strongly they are prepared to challenge that individual. Males that are seriously aggressive match the song types of their opponent; those that wish to avoid conflict can signal their desire to do so by selecting nonshared song types from their repertoires to sing when a neighbor is listening.

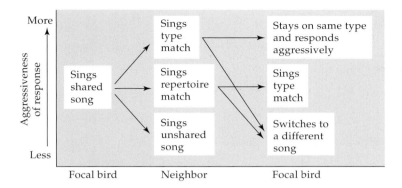

**FIGURE 2.26  Song matching and communication of aggressive intent in the song sparrow.** Song sparrow males can control the level of conflict with a neighbor by their selection of songs. When a focal male sings a shared song at a rival, the neighbor has three options: one that will escalate the contest, one that keeps it at the same level, and another that de-escalates the interaction. Likewise, the initiator of the contest can use his ability to select a matching song type, a repertoire match, or an unshared song type. The three different kinds of songs convey information about the readiness of the singer to escalate or defuse an aggressive encounter. After Beecher and Campbell.[94]

If this ability to modulate song challenges is truly adaptive, then we can predict that male territorial success, as measured by the duration of territorial tenure, should be a function of the number of song types that a male shares with his neighbors. Song sparrows can hold onto their territories for up to 8 years, which tends to create a relatively stable community of neighbors. And, in fact, the number of years that an individual male song sparrow holds his territory increases about fourfold as the number of song types shared with his neighbors goes from fewer than 5 to more than 20.[91] This finding suggests strongly that the potential for neighbors to form long-term associations in song sparrow communities selects for males able to advertise exactly what their aggressive intentions are toward particular individuals. If so, then in the fluid social settings of certain other species in which territorial males come and go quickly, song learning and sharing of the sort exhibited by song sparrows should not occur. Instead, under these different ecological circumstances, males should acquire songs characteristic of their species as a whole to facilitate communication with any and all conspecifics, rather than developing a dialect or a set of shared song types characteristic of a small, stable population.

Donald Kroodsma and his colleagues have tested this proposition by taking advantage of the existence of two very different populations of the sedge wren. In the Great Plains of North America, sedge wren males are highly nomadic, moving from breeding site to breeding site throughout the summer. In this population, song matching and dialects are absent; instead, the birds improvise variations on songs heard early in life as well as inventing entirely novel songs of their own, albeit employing the general pattern characteristic of their species.[810] In contrast, sedge wren males living in Costa Rica and Brazil remain on their territories year-round. In these areas, dialects and song matching are predicted to occur, and these learning-proficient males do have dialects and do sing like their neighbors.[811] The differences in song learning that occur within this one species indicate that learning and dialects do evolve when males gain by communicating specifically with other males that are their long-standing neighbors.

**FIGURE 2.27 Evidence that male Cassin's finches direct their songs at females.** When a male is paired off with a female on day 1, he sings relatively little. But when the female is removed on day 2, the male invests considerable time in singing, presumably to call her back. If the male is caged without a female on day 1, however, he allocates little time to singing, and he sings even less when a female is introduced into his cage, indicating that it is the loss of a potential mate that stimulates the male to sing. After Sockman et al.[1361]

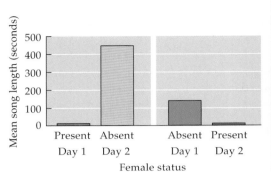

Cassin's finch

## Female Preferences and Song Learning

Still other ultimate hypotheses on song learning take aim at the social environment provided by females. There can be little doubt that females are the intended recipients of at least some bird songs. When, for example, a male Cassin's finch has been close to a female for a while and she then disappears, the number of songs he sings and the time he spends singing increase dramatically, almost certainly in an attempt to get his partner back (Figure 2.27).

But could the fact that male Cassin's finches learn their songs contribute to a male's ability to attract a female back to him? Consider a species that is divided into stable subpopulations. In such a species, males in each of these groups are likely to have genes that have been passed down for generations by those males' successful ancestors. Therefore, by learning to sing the dialect associated with their place of birth, males could announce their possession of traits (and underlying genes) well adapted for that particular area. Females hatched in that area might gain by having a preference for males that sang the local dialect, because they would endow their offspring with genetic information that promotes the development of locally adapted characteristics.[62]

This hypothesis received some support from the discovery that male white-crowned sparrows that sang the local dialect around Tioga Pass, California, were less infected with a blood parasite than non-dialect singers. This finding means that female white-crowns in this population could potentially use song information to secure healthier mates. And, in fact, males with songs not matched to the local dialect fathered fewer offspring than local dialect singers, a result consistent with the hypothesis that females prefer to mate with males that sing the local dialect.[909]

On the other hand, if female preferences are designed to get them to mate with males produced in their natal area, then female white-crowns should prefer males with the dialect that they heard while they were nestlings—namely, their father's dialect—and they do not, at least in one Canadian population.[260] Furthermore, young male white-crowns are not locked into their natal dialects, but are able to change them[390] if they happen to move from one dialect zone to another; therefore, female white-crowns cannot rely on a male's dialect to identify his birthplace with complete certainty. All of this casts doubt on the proposition that female song preferences enable them to endow their offspring with locally adapted gene complexes from locally hatched partners.

Another version of the female choice hypothesis for male song learning is based on the possibility that females look to the learned details of a potential mate's song for information about his developmental history. If a female could tell just by listening to a male that he was unusually healthy, she could acquire a mate whose genes had worked well during his development and so should be worth passing on to her offspring. For example, in zebra finches, males with more complex songs have a larger-than-average HVC;[4, 5] in song sparrows, males with larger song repertoires also have a larger-than-average HVC and moreover are in better condition as judged by their relatively large fat reserves and by their apparently more robust immune systems.[1127] Male song sparrows with larger repertoires have more offspring and grandoffspring,[1210] suggesting that females that mated with males of this sort might be able to endow their male offspring with genes that could give them a competitive edge (assuming that there is a hereditary component to HVC size and body condition).

Alternatively, a healthier partner might be able to provide his offspring with above-average parental care. In the great reed warbler, well-fed nestlings have larger learned song repertoires as first-year adults than individuals that have been nutritionally stressed in the nest.[1060] Thus, superior songsters may be in better shape and therefore capable of offering their offspring better paternal care than the average male. This pattern is also seen in the sedge warbler, a reed warbler relative, whose males learn their song repertoires and whose females prefer males with larger repertoires. Katherine Buchanan and Clive Catchpole have shown that these song-rich males bring more food to their offspring, which grow bigger—a result that almost certainly raises the reproductive success of females that find large song repertoires sexually appealing, which in turn selects for males that are able to sing in the favored manner.[203]

But the question still remains, why might female choice favor males that *learn* their complex songs, rather than simply manufacturing an innately complex song with no learned component? One hypothesis is that the quality of vocal learning could give female songbirds a valuable clue about the quality of the singers as potential mates. The key point here is that song learning occurs when males are very young and growing rapidly. If rapid growth is difficult to sustain, then young males that are handicapped by genetic defects or nutritional stress should be unable to keep up, resulting in suboptimal brain development. According to this hypothesis, individuals with even slightly deficient brains may be less able to meet the demands of learning a complex species-specific song.[1058]

These predictions have been tested in experiments with swamp sparrows. By bringing very young nestling males into the laboratory, Steve Nowicki and his coworkers were able to control what they were fed. One group of nine males (the controls) received as much food as they could eat, while another group of seven (the experimentals) received 70 percent of the food volume consumed by the control males. During the first 2 weeks or so, when the sparrows were totally dependent on their handlers for their meals, the controls came to weigh about a third more than the experimentals, a difference that was then gradually eliminated over the next 2 weeks as the sparrows came to feed themselves on abundant seeds and mealworms. Even though the period of nutritional stress was brief, this handicap had large effects on both brain development and song learning (Figure 2.28). Components of the song system in the food-deprived experimental group were significantly smaller than the equivalent regions in the control sparrows.[1061] Likewise, in the related song sparrow, nestlings that were subject to food shortages soon after hatching developed smaller HVCs than those fed as much as they wanted. The effect manifested itself by the time of fledging, even before the young birds had begun to learn their songs.[908] In the case of swamp sparrows, the experimentally deprived birds came to sing poorer copies, compared with the controls, of the taped song that both groups listened to during their early weeks of captivity.[1061] These results, and similar ones from other studies,[1375] have not been repeated by some other researchers.[538] If, however, we accept the results showing a relationship between nutritional shortfalls, stress, development of the song nuclei, and song learning, then we can accept the possibility that adult female songbirds might learn something about the developmental history, and thus the quality, of potential mates just by listening to their learned songs.

(A)

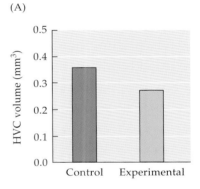

(B)

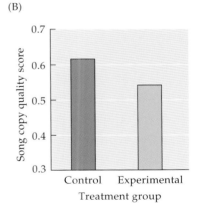

**FIGURE 2.28  Nutritional stress early in life has large effects** on both (A) brain development, as expressed in the volume of the HVC, and (B) song learning, as measured by the match between learned songs and tutor tapes. Nestling swamp sparrows were hand reared and exposed to tutor tapes. The control group was fed all they could eat, whereas the experimental group received only 70 percent of that amount for about 2 weeks. After Nowicki, Searcy, and Peters.[1061]

(A) Tape tutor song

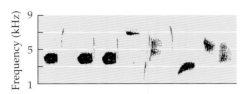

(B) Good copy of (A)

(E)

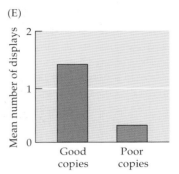

(C) Tape tutor song

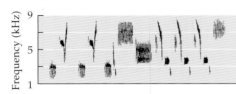

(D) Poor copy of (C)

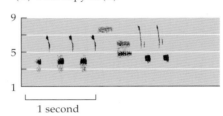

1 second

**FIGURE 2.29 The mean number of precopulatory displays given by female song sparrows** in response to playback of the songs of males that had been able to copy their tutor songs very accurately and of males of lower copying skill. The upper panel shows (A) a tape tutor song and (B) a copy sung by a male able to copy a high proportion of the notes in the taped song it listened to; the bottom panel shows (C) a tape tutor song that was (D) poorly copied by another individual. (E) The females' response to good and poor song copies. After Nowicki, Searcy, and Peters.[1062]

But do female birds actually pay attention to the information about mate quality potentially encoded in male songs? Nowicki's team again supplied supporting evidence by playing male song sparrow songs to female sparrows in the laboratory.[1062] Some of the songs had been more accurately copied from tape tutors than others by males of diverse learning abilities. Female sparrows can be hormonally primed to respond with a tail-up precopulatory **display** to male songs they find sexually stimulating. The more accurately copied songs elicited significantly more precopulatory displays from the females than those that were copied less perfectly (Figure 2.29).

Another study of this sort with similar results involved zebra finch females that were given an opportunity to fly to perches near speakers from which were broadcast songs made by either stressed or non-stressed males that had learned their songs from the same social tutor.[1376] The stressed males had been food-deprived or had had their corticosterone levels elevated to induce the physiological state of a stressed bird. The songs of these handicapped males were shorter and less complex than those of the control males (Figure 2.30). As predicted from the hypothesis that females use a male's learned song performance to evaluate his quality as a mate, the female zebra finches preferred to fly to and land upon perches near speakers from which were broadcast the songs of control males. This result supports the notion that males able to learn their songs fully and well will be rewarded sexually by potential mates.

Along these lines, the fact that normally reared male zebra finches change their songs when in the company of females suggests that their potential mates prefer songs that are faster and slightly more stereotyped. This hypothesis was tested by giving female zebra finches a choice of entering a cage compartment near a speaker that played a male song of the apparently preferred type (a "directed" song) versus a compartment near a speaker that played the kind of slower, more variable ("undirected") song that males sing in the absence of a female (Figure 2.31).[1619] As predicted, females spent more time next to the speaker playing the directed song of either an unfamiliar male or her mate,

(A)

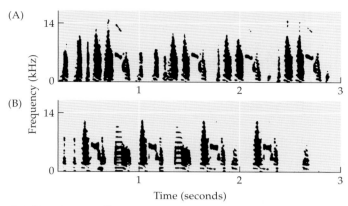

(B)

**FIGURE 2.30** **Stress influences song structure in the zebra finch.** (A) The upper sonogram is representative of songs produced by males that were reared under good conditions. (B) The lower sonogram is representative of the songs of males that had been developmentally stressed either by injecting them with corticosterone or by rearing them under food shortage conditions. After Spencer et al.[1376]

compared with the speaker playing her mate's undirected song. Sarah Woolley and Allison Doupe linked this preference to the properties of cells in a particular part of the auditory cortex of the zebra finch brain, the caudomedial mesopallium, or CMM. When the female finch hears directed songs, many cells within her CMM express the *ZENK* gene, which, as we have seen, plays an important role in various song-related behaviors in many songbirds. Thus, the female zebra finch is programmed to respond sexually to songs that are presumably more challenging for the male to produce.

Other female songbirds definitely prefer songs that are relatively difficult to produce. Male swamp sparrows sing a trilling song (see Figure 2.15). The faster the trill, the harder it is to sing, especially if the components of the trill cover a relatively wide range of sound frequencies. Females are attracted to males that sing right up to the limits of performance.[67] Likewise, female serins (a small finch) prefer males that sing at relatively high frequencies, another physiological challenge for males.[229] By basing their mate preferences on male performance, females are in effect favoring males likely to be healthy and in good condition, attributes that could make the male a superior donor of genes or a superior caregiver to offspring.

(A)

(B)

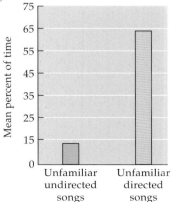

**FIGURE 2.31** **Measuring female preferences for different kinds of zebra finch songs.** (A) The test apparatus in which females placed in a central compartment could choose to spend time close to either of two speakers. (B) Females consistently went to the speaker from which was being broadcast a recording made of a male singing to a female, as opposed to a tape of a male singing in the absence of a female. After Woolley and Doupe.[1619]

---

**Discussion Questions**

**2.6** William Searcy and a team of researchers played taped songs to captive female song sparrows that had been given hormone implants shortly after being taken to the laboratory.[1298] The recorded songs came from male song sparrows that lived in the population from which the females had been taken, as well as from males living various distances (18, 34, 68, 135, and 540 kilometers) from the female subjects. Songs from males living 34 or more kilometers from the populations from which the females came were not nearly as effective in eliciting the precopulatory display as songs from local males; in contrast, songs from males living only 18 kilometers away were about as sexually stimulating as local songs. These data have relevance for more than one ultimate hypothesis on song learning by male sparrows. What are the hypotheses, and what importance do these findings have for them?

**2.7** Parasites are often microscopic in size but have large negative effects on their hosts. If this is true for the parasites of songbirds, what predictions follow about their effects on male song performance, and how should females respond to the song of infected males as opposed to uninfected individuals? Check your predictions by reading Garamszegi.[517]

---

## Proximate and Ultimate Causes Are Complementary

We have reviewed only a small portion of what is known about learning and dialects in bird song, but we have covered enough to illustrate the links between the proximate and ultimate causes of behavior. If we want to understand why male white-crowns from Marin and Berkeley sing somewhat different songs, we must understand how the song control mechanisms develop. Male white-crowns differ genetically from females, a fact that guarantees that different gene–environment interactions will take place within embryonic male and female birds. The cascading effects of differences in gene activity and in the protein products that result from gene–environment interactions affect the assembly of all parts of the bird's body, including its brain and nervous system. The way in which a male white-crown's brain, with its special subsystems, can respond to experience is a function of the way in which the male has developed. By the time a male sparrow has become an adult, he possesses a large, highly organized song control system whose physiological properties permit him to produce a very particular kind of song. The learned dialect he sings is a manifestation of both his developmental history and the operating rules of his brain.

The proximate differences among white-crowns that generate differences in their dialects have an evolutionary basis. In the past, in the lineage leading to today's white-crowns, males have differed in their song control systems and thus in their behavior. Some of these differences have been hereditary, and some males have surely been hereditarily better than others at announcing their desirability as mates or their capacity to cope with rival males. These males' superior behavioral skills have translated into greater genetic contributions to the next generation, where those genes have been available to participate in interactive developmental processes within members of that generation. The hereditary attributes of those animals will in turn be measured against one another today in terms of their ability to promote genetic success in the current round of selection. This selective process links the proximate bases of behavior with their ultimate causes in a never-ending spiral through time (Figure 2.32).

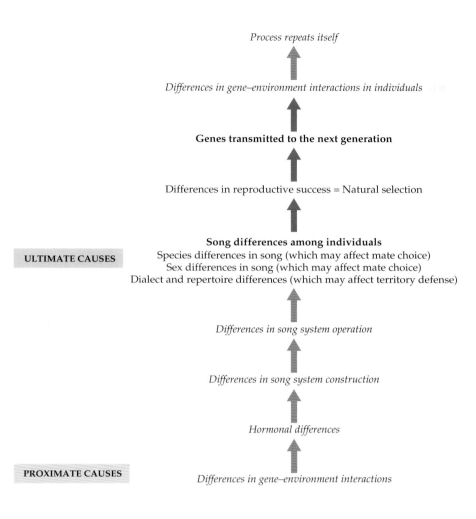

*Process repeats itself*

*Differences in gene–environment interactions in individuals*

**Genes transmitted to the next generation**

Differences in reproductive success = Natural selection

**Song differences among individuals**
Species differences in song (which may affect mate choice)
Sex differences in song (which may affect mate choice)
Dialect and repertoire differences (which may affect territory defense)

ULTIMATE CAUSES

*Differences in song system operation*

*Differences in song system construction*

*Hormonal differences*

PROXIMATE CAUSES     *Differences in gene–environment interactions*

**FIGURE 2.32   What causes differences among individuals?** The development of an individual's traits depends on the interaction between the genetic information inherited from its parents and the nutrients, chemical substances, and experiences derived from its environment. Therefore, differences among individuals can be caused by differences in either their genes or their environments, or both. The individual differences resulting from these proximate causes have the potential to affect the evolution of a species.

## Discussion Question

**2.8**  What features of language learning in humans are similar to song learning in birds? Do these similarities suggest certain hypotheses on the proximate bases of human language learning, especially the genetic and developmental components? Do comparisons with birds also suggest some interesting hypotheses on the adaptive value of learned language for members of our species? After you have attempted to answer these questions, go to the ISI Web of Science, if it is available at your college or university library, and try to find references on the shared proximate and ultimate causes of vocal learning by humans and birds. You may find it useful to know that those who have written knowledgeably on this subject include Peter Marler and Fernando Nottebohm.

## Summary

1. Different species of birds sing different songs, and even within a single species, individuals may vary in how they sing their species' special vocalization. These features of bird song can be subjected to both proximate and ultimate analyses.

2. Some birds learn their species' song, which may lead to geographic differences in the songs sung by members of the same species. The underlying developmental processes of song learning and dialect formation are dependent on both genetic information and environmental inputs, which include the bird's acoustical and social experiences as well as the proteins and other chemical constituents of its brain and other body parts.

3. In addition to focusing on how the mechanisms of behavior are assembled during development, proximate analyses include attempts to understand the operating rules of these mechanisms. Birds that learn their songs possess an elaborate song control system whose component parts contribute to early memorization of a song or songs, which are then copied when the learners begin to practice singing.

4. The ultimate causes of behavior can also be explored on two levels. Thus, understanding why some birds learn a regional dialect can lead us to explore the historical changes that occurred during the evolution of vocal learning in hummingbirds, parrots, and some passeriform songbirds. Another layer of evolutionary analysis of learned bird song involves studying why this means of song acquisition was favored by natural selection. Much evidence now exists that learned songs may enable individual males of some species to target signals to conspecific rival males as well as to potential mates, which may acquire information on the developmental history of the singer by listening to how well he sings.

5. In the past, any heritable variant mechanism of song learning that helped individuals leave more copies of their genes would have become more common over time, thanks to the action of natural selection. As certain forms of genes became more common, certain developmental outcomes would likewise have predominated, leading to the transformation of the song control systems in the brains of a songbird species under selection. Changes in the song control mechanisms would have consequences for how a songbird acquired its song, which in turn could affect its ability to reproduce successfully. Proximate and ultimate causes of behavior are thus linked together in a never-ending cycle.

## Suggested Reading

*Bird Song: Biological Themes and Variations*[250] is a very readable review of the proximate and ultimate causes of bird song. *Neuroscience of Birdsong* focuses on the proximate issues surrounding this behavior.[1646] Eliot Brenowitz and Mike Beecher have written a concise review that focuses on the differences in what birds learn and how they learn their songs, while highlighting how much we still have to learn about these differences.[171] Donald Kroodsma's *The Singing Life of Birds* has been written for the general reader, who will learn how exciting it can be to listen to and understand bird song.[812] Highly recommended.

# 3

# The Development
# of Behavior

When you and I listen to a white-crowned sparrow's song, we are hearing the product of an extraordinarily long and complex developmental process. As the previous chapter indicated, behavioral development is influenced both by the genetic information the bird possesses in its DNA and by a host of environmental influences, ranging from the nongenetic materials in the egg yolk, to the hormones that certain of the bird's cells manufacture and transport to other cells, to the sensory signals generated when a baby sparrow hears its species' song, to say nothing of the neural activity that occurs when a young adult male interacts with a neighboring territorial male.

The sparrow example highlights an extremely important concept, namely, that development is an interactive process in which genetic information interacts with changing internal and external environments in ways that assemble an organism with special properties and abilities. The process occurs because some of the genes in the organism's cell nuclei can be turned on or off by the appropriate signals, which are ultimately derived from the external environment. As genetic activity changes within an organism, the chemical reactions within its cells change, building (or modifying) the proximate mechanisms that underlie an organism's characteristics and capacities.

◀ *One leafcutter ant worker carries a leaf back to the colony while a much smaller worker rides on the leaf to protect her sister against parasitic flies. How did these ants develop their different behavioral specifications? Photograph by Gavriel Jecan.*

As mentioned in the preceding chapter, when a young zebra finch listens to other males sing or when he listens to his own song, the expression (protein-producing activity) of certain genes in his brain increases. As the protein products of those genes are produced in greater quantities, they in turn modify the structure and function of some of the finch's brain cells. These changes underlie the bird's ability to remember what it has heard and later to mimic the zebra finch songs it has listened to.[1489]

The fact that the development of any attribute of any living multicellular organism is dependent on both genes and environment means that no trait—not one—is "genetic" as opposed to "environmental," nor is any attribute environmentally determined in the sense of developing without genetic input. This claim is counterintuitive to many who want to divide the features of living things into those that are caused by "nature," the so-called genetically determined traits, and those that are caused by "nurture," the so-called environmentally determined traits. The nature-versus-nurture approach is strongly embedded in the popular view of animal behavior, in which instincts are often said to be genetic, or more strongly affected by hereditary factors, than is learning, which is commonly believed to be largely or entirely due to the animal's environment. A main goal of this chapter is to address this misconception head-on. Having stressed the importance of gene–environment interactions for development, we will then examine evidence that both genetic and environmental *differences* among individuals can lead to *differences* in development, which in turn can produce *differences* in how individuals behave. This point has great significance for an understanding of behavioral evolution, a subject that will be covered here with examples of how aspects of the developmental process promote the reproductive success of individuals.

## The Interactive Theory of Development

Although most people think of insects as fairly boring automatons with only a limited set of basic instincts, in fact many insect species possess extremely sophisticated behavioral abilities. Take worker honey bees, for example. These little creatures spend their lives helping their fellow hivemates. Although workers leave the egg laying to their mother queen bee (because workers are largely sterile), they are responsible for care of the larvae that hatch from their mother's eggs, the construction of honeycomb, the regulation of the hive's temperature, the defense of the colony against parasites and predators, and, of course, the collection of the pollen and nectar that they and their colonymates will need to survive.

One of the many fascinating things about bee behavior is that a worker changes its occupational role over the course of her lifetime. When a worker emerges from a brood cell in the waxy comb tended by other workers, her first job is a humble one, the cleaning of comb cells. She then becomes a nurse bee that feeds honey to larvae in the brood comb before making the transition to a distributor of food to her fellow workers. The last phase of her life, which begins when she is about 3 weeks old, is spent foraging for pollen and nectar outside the hive (Figure 3.1).[871]

So what causes a worker to go through these different developmental stages? According to the interactionist approach that we outlined in the preceding chapter, the information in some of the bee's many thousands of genes (the bee's **genotype**) must respond to the environment in ways that influence the development of her measurable characteristics (the bee's **phenotype**), which include the proximate mechanisms underlying her behavior, such as her nervous system, and her behavioral traits as well. One result of these genotype–environment interactions is the standard shift in the behavioral phenotype of

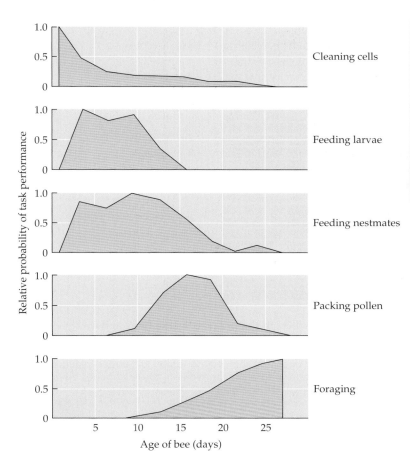

**FIGURE 3.1** **Development of worker behavior** in honey bees. The tasks, such as foraging for pollen (seen here), adopted by worker bees are linked to their age, as demonstrated by following marked individuals over their lifetimes. After Seeley.[1300]

workers, with 3-week-old adults no longer feeding stored honey and pollen to larval bees, but instead flying from the hive to gather nectar and pollen from surrounding fields and woods. A developmental biologist might ask, does the active component of a worker bee's genotype change between her nurse phase and her forager phase, just as gene activity changes in a structured fashion as a young nestling zebra finch develops into an adult (see Figure 2.9)?

This question can now be answered, thanks to the ability of researchers to use microarray technology, a procedure that makes it possible to analyze the activity of a large set of genes by detecting certain products (messenger RNAs) made when those genes have been "turned on." In order to see which gene products are abundant and which are not, molecular biologists scan a sheet on which minute amounts of tissue have been spread before being subjected to a fluorescent dye that reacts with nucleic acids. When Charles Whitfield and his coworkers ran brain extracts from nurses and foragers through a variety of microarrays, they were able to compare the activity of about 5500 genes (of the roughly 14,000 in the bee genome) for these two kinds of individuals.[1552] Some of the genes that were turned on in nurse bees' brains differed substantially and consistently from those active in foragers' brains, and vice versa. Indeed, about 2000 of the surveyed genes showed different levels of product output in the two kinds of bees. These changes in genetic activity are correlated with the transformation of a nurse into a forager (Figure 3.2).

Additional microarray analyses have shown that nearly 2000 genes change their activity during the first 4 days of an adult bee's life, while another 600 or so alter their expression in the next 4 days. These findings were made by comparing gene activity in recently emerged adults with 4-day-old individuals and by comparing 4-day-olds with 8-day-old bees.[1553] The genetic changes

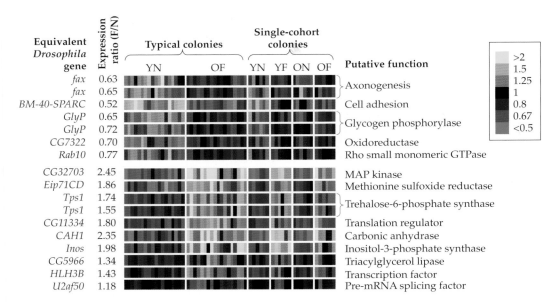

| Equivalent *Drosophila* gene | Expression ratio (F/N) | Typical colonies | | Single-cohort colonies | | | | Putative function |
|---|---|---|---|---|---|---|---|---|
| | | YN | OF | YN | YF | ON | OF | |
| *fax* | 0.63 | | | | | | | Axonogenesis |
| *fax* | 0.65 | | | | | | | |
| *BM-40-SPARC* | 0.52 | | | | | | | Cell adhesion |
| *GlyP* | 0.65 | | | | | | | Glycogen phosphorylase |
| *GlyP* | 0.72 | | | | | | | |
| *CG7322* | 0.70 | | | | | | | Oxidoreductase |
| *Rab10* | 0.77 | | | | | | | Rho small monomeric GTPase |
| *CG32703* | 2.45 | | | | | | | MAP kinase |
| *Eip71CD* | 1.86 | | | | | | | Methionine sulfoxide reductase |
| *Tps1* | 1.74 | | | | | | | Trehalose-6-phosphate synthase |
| *Tps1* | 1.55 | | | | | | | |
| *CG11334* | 1.80 | | | | | | | Translation regulator |
| *CAH1* | 2.35 | | | | | | | Carbonic anhydrase |
| *Inos* | 1.98 | | | | | | | Inositol-3-phosphate synthase |
| *CG5966* | 1.34 | | | | | | | Triacylglycerol lipase |
| *HLH3B* | 1.43 | | | | | | | Transcription factor |
| *U2af50* | 1.18 | | | | | | | Pre-mRNA splicing factor |

Reference code: >2, 1.5, 1.25, 1, 0.8, 0.67, <0.5

**FIGURE 3.2   Gene activity varies in the brains of nurse bees and foragers.** Shown here are individual records (each bar represents one bee's gene expression record for a particular gene) of bees from typical and manipulated colonies. The 17 genes surveyed here were the ones that showed the largest difference in activity between the brains of nurses and of foragers. In addition, these 17 genes are highly similar to genes found in fruit flies (*Drosophila*); the function of the corresponding gene in fruit flies has been established and is noted here. The activity level of a gene in forager, relative to nurse, bee brains is color coded from high (>2) to low (<0.5); see the reference code at the upper right. The left-hand portion of the figure shows gene activity for young nurses (YN) and old foragers (OF) from unmanipulated colonies; the right-hand portion shows those for young foragers (YF) paired with nurses of the same age and old nurses (ON) paired with foragers of the same age taken from colonies manipulated to contain either all young or all old workers. The expression ratio (F/N) is calculated by dividing the activity score of a given gene in foragers by that for nurses. After Whitfield, Cziko, and Robinson.[1552]

that are taking place during this time are believed to contribute to the developmental changes occurring in the brains of the young bees, changes that take place well before the bees leave the hive in search of flowers but that are necessary if the insects are to acquire the capabilities needed to forage.

Microarray studies do *not* demonstrate that the environment is irrelevant to behavioral development. Consider, for example, that when the queen bee produces queen mandibular **pheromone** (pheromones are chemicals used by animals to communicate with one another), this volatile compound causes changes in the expression of many genes in the brain cells of workers.[591] Likewise, workers themselves produce certain pheromone signals that can change gene expression in the brains of other workers when these individuals are exposed to these special chemicals.[7] Indeed, environmental factors are critical for *every* element of gene expression within organisms, especially because it is the environment that supplies the molecular building blocks that are essential if the information in DNA is to be used to make messenger RNAs and proteins. The cellular environment must contain the precursors of these constituents of living things if they are to be produced. These chemicals ultimately come from substances consumed by the queen prior to making her eggs, as well as from the honey and pollen eaten by the larvae and adults that develop from those eggs. Some of the resultant gene–environment products may play a special role in changing the activity of one or more key genes in an individual, initiating a cascading series of gene–environment outputs that eventually alters the development of the brain and, thus, the behavior of the bee.

One particularly important developmental product appears to be a substance called juvenile hormone, which is found in low concentrations in the blood of young nurse workers but in much higher concentrations in older foragers. As one might predict, if young bees are treated with juvenile hormone, they become precocious foragers,[1229] but if one removes a bee's corpora allata (the glands that produce juvenile hormone), the bee delays its transition to foraging. Moreover, bees without corpora allata that receive hormone treatment regain the normal timing of the switch to foraging.[1402]

So it appears that changes in juvenile hormone production have something to do with getting the ball rolling in ways that lead to a shift in behavior. But what causes the gene that codes for juvenile hormone to boost its output when the worker is about 3 weeks old? Actually, a whole series of genetic changes occurs about the time the bee makes its transition from being confined to the hive to launching its career as a forager.[1553] Part of the basis for these changes surely lies in the preceding sequence of gene–environment interactions that have taken place since the worker bee metamorphosed into an adult. But some gene changes in expression are responsive to the current social environment of the worker, as shown by research in which experimental colonies were formed with a worker force consisting of uniformly young bees of the same age. Under these unusual conditions, a division of labor still manifested itself, with some individuals remaining nurses much longer than usual, while others began foraging as much as 2 weeks sooner than average.

What enabled the bees to make these developmental adjustments? One hypothesis is that a deficit in social encounters with older foragers may have stimulated an early developmental transition from nurse to forager behavior. This possibility has been tested by adding groups of older foragers to experimental colonies made up of only young workers. The higher the proportion of added older bees, the lower the proportion of young nurse bees that undergo an early transformation into foragers (Figure 3.3).[690] As it turns out, the behavioral interactions between the young residents and the older transplants inhibit the development of foraging behavior, because transplants of young bees have no such effect on young resident bees. The inhibiting agent has been traced to a fatty acid compound called ethyl oleate, which only foragers manufacture and store in a chamber (the crop) off the digestive tract.[856] When returning foragers pass nectar contained in the crop to nurses back at the hive, ethyl oleate is probably transferred as well. The more foragers in a hive, the more likely nurses are to receive quantities of this chemical, which slows their transition to foraging status.

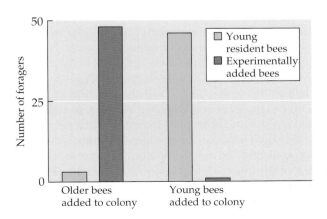

**FIGURE 3.3  Social environment and task specialization** by worker honey bees. In experimental colonies composed exclusively of young workers (residents), the young bees do not forage if older forager bees are added to their hive. But if young bees are added instead, the young residents develop into foragers very rapidly. After Huang and Robinson.[690]

**FIGURE 3.4** Levels of the messenger RNA produced when the *for* gene is expressed in the brains of nurses and of foragers in three typical honey bee colonies. After Ben-Shahar et al.[103]

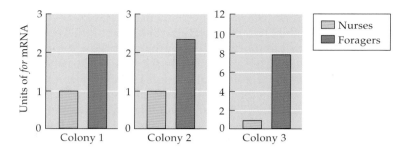

## Discussion Questions

**3.1** In the context of discussing the causes of developmental differences between individuals, what significance do you attach to the fact that the queen bee behaves very differently from her workers even though she has essentially the same genome as her worker sisters and daughters? Develop at least one hypothesis on why the two categories of bees behave so differently after you find out about how worker bee and queen larvae are reared. The Internet will be helpful in this regard.

**3.2** Honey bees possess a gene (labeled *for*) that contains information for the production of a particular enzyme called PKG. If this genetic information is important for foraging activity in worker honey bees, what prediction follows about the levels of PKG enzyme present in the heads of foragers versus nonforagers taken from a typical bee colony composed of workers of many different ages? Figure 3.4 presents data on the quantity of the messenger RNA coded by *for* that is present in the brains of nurses and foragers in three typical honey bee colonies; this RNA is required for the production of PKG. Are these results consistent with your prediction? But since foragers are older than nurses in typical colonies, perhaps the greater activity of *for* is simply an age-related change that has nothing to do with foraging. What additional prediction and experiment are required to reach a solid conclusion about the causal role of *for* in regulating foraging in the honey bee? Suggestion: Take advantage of the ability to create experimental colonies of same-age workers. If the *for* gene contributes to the switch to foraging by worker bees,[103] how can this change be influenced by the workers' social environment as well? Can you produce a hypothesis that integrates the genetic and environmental contributions to the switch in bee behavior?

## The Nature-or-Nurture Fallacy

The honey bee example clearly illustrates why it would be a mistake to say that some behavioral phenotypes are more genetic than others. Worker foraging behavior cannot be purely "genetically determined" because the behavior is the product of literally thousands of gene–environment interactions, all of which are required to construct the bee's brain and the rest of its body. Indeed, the information in the DNA that makes up a gene is expressed only when the gene is in the appropriate environment. As Gene Robinson puts it, "DNA is both inherited and environmentally responsive."[1230] Environmental signals, such as those provided by juvenile hormone and ethyl oleate, influence gene activity. When a gene is turned on or off by changes in the environment, changes in protein output can directly or indirectly alter the activity of other genes within affected cells. A multitude of precisely timed and well-integrated changes in gene–environment interactions are responsible for the construction of every trait, and therefore no trait can be purely "genetic."

By the same token, it would be flat wrong to say that a given phenotype is "environmentally determined." The development of every attribute of every living thing requires the information contained in large numbers of genes and expressed in a multitude of gene–environment interactions. Neither genotype nor environment can be said to be more important than the other, just as no one would say that a chocolate cake owed more to the recipe used by the cook than to the ingredients that actually went into the finished product.

### Discussion Question

**3.3** The nature–nurture controversy involves those who believe that our nature (essentially our genes) dominates our behavioral development and others who argue just as forcefully that our nurture (especially our upbringing as children) is what shapes our personalities. Some have dismissed the controversy by saying that the two sides might as well be fighting about whether a rectangle's area is primarily a matter of its height or mostly a function of its width. What's the point of the rectangle analogy? Does the analogy have any weaknesses?

## Behavioral Development Requires Both Genes and Environment

The contribution that DNA makes to behavioral development can be demonstrated by examining the development of the ability to learn, since **learning** (a change in an animal's behavior linked to a particular experience it has had) is so often said to be environmentally determined, even though it is not. Of course, the environment is involved when an animal learns something, but since learning takes place within a brain whose properties have been shaped by *gene*–environment interactions, the genetic influence on development simply cannot be ignored. This point is evident when we consider just how circumscribed and focused learned behaviors actually are. These constraints on learning are a consequence of specialized features of the brain, which in turn arise through the interplay between information-rich genes and the environment.

A classic example of the circumscribed nature of learning is provided by **imprinting**, in which a young animal's early social interactions, usually with its parents, lead to its learning such things as what constitutes an appropriate sexual partner. Thus, a group of greylag goslings, having imprinted on behavioral biologist Konrad Lorenz (Figure 3.5) rather than a mother goose, formed both a learned attachment to Lorenz[886] and, in the case of the male greylags when they reached adulthood, a preference for humans as mates. The experience of following a particular individual early in life must have somehow altered those regions of the male goose's nervous system responsible for sexual recognition and courtship. The special effects of imprinting could not have occurred without a "prepared" brain, one whose genetically influenced development enabled it to respond to the special kinds of information available from its social environment.

The fact that different species exhibit different imprinting tendencies provides further circumstantial evidence for a genetic contribution to

**FIGURE 3.5  Imprinting in greylag geese.** These goslings have imprinted on the behavioral biologist Konrad Lorenz and are following him wherever he goes. Photograph by Nina Leen.

(A)

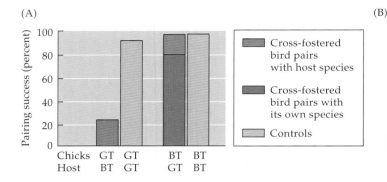

(B)

**FIGURE 3.6 Cross-fostering has different imprinting effects in two related songbirds.** (A) Males of the great tit (GT) that have been reared by blue tit (BT) foster parents try to pair with blue tit females, but only a fraction succeed. In contrast, cross-fostered blue tits always find mates, generally of their own species. Control birds, which were reared by their own species, always find mates and pair with members of their own species. (B) When blue tit females pair with great tit males, they also copulate with male blue tits. Here a female blue tit (far left) paired with a male great tit (upper left) rear a brood together that consists entirely of blue tit nestlings. After Slagsvold;[1343] photograph by Tore Slagsvold.

learning. A group of Norwegian researchers provided this kind of evidence when they switched broods from blue tit nests into the nests of breeding great tits, and vice versa. Some of the cross-fostered youngsters grew up and survived to court and form pair bonds with members of the opposite sex (Figure 3.6A). Of the surviving fostered great tits, only 3 of 11 found mates—all of which were blue tit females that had been fostered by great tits. Of the surviving fostered blue tits, all 17 found mates, although 3 of these were females that socially mated with cross-fostered male great tits.[1345]

However, although some individuals of both species became imprinted on another species as a result of their foster-care experiences, the degree to which individuals imprinted on their foster parents differed between the two species of songbirds. None of the cross-fostered great tits mated with a member of its own species, whereas most of the cross-fostered blue tits did. Moreover, each blue tit female that had a great tit as a social partner must have mated with a blue tit male on the side, because all 33 offspring produced by those females were blue tits, not hybrids (Figure 3.6B). Thus, although misimprinting occurred in both species, the developmental effect of being reared by members of another species was far greater for great tits than for blue tits, an indication that the hereditary basis of the imprinting mechanism was not the same for these two species.

In addition to imprinting, bird species possess other specialized learning abilities, including the ability to remember where they have hidden food. The black-capped chickadee is especially good at this task. This bird's spatial memory enables it to relocate large numbers of seeds or small insects that it has hidden in bark crevices or patches of moss scattered throughout its environment. To establish just how good chickadees are at relocating their food caches, David Sherry provided captive chickadees with a chance to store food in holes drilled in small trees placed in an aviary. After the chickadees had placed sunflower seeds in 4 or 5 of 72 possible storage sites, they were shooed into a holding cage for 24 hours. During this time, Sherry removed

(A)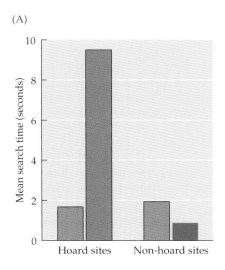

(B)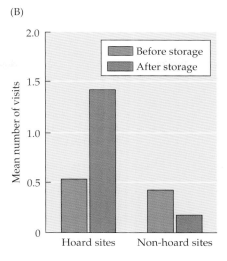

Black-capped chickadee

**FIGURE 3.7** **Spatial learning by chickadees.** (A) Black-capped chickadees spent much more time at sites in an aviary where they had stored food 24 hours previously (hoard sites) than they had spent during their initial exposure to those sites, even though experimenters had removed the stored food. (B) The chickadees also made many more visits to hoard sites than to other sites, evidently because they remembered having stored food there. After Sherry.[1319]

the seeds and closed each of the 72 storage sites with a Velcro cover. When the birds were released back into the aviary, they spent much more time inspecting and pulling at the covers at their hoard sites than at sites where they had not stored food 24 hours before (Figure 3.7). Because the storage sites were all empty and covered, there were no olfactory or visual cues provided by stored food to guide the birds in their search; they had to rely solely on their memories of where they had hidden seeds.[1319] In nature, these birds store only one food item per hiding spot and never use the same location twice, yet they can relocate their caches as much as 28 days later.[655]

Clark's nutcrackers may have an even more impressive memory, for they scatter as many as 33,000 seeds in up to 5000 caches that may be as many as 25 kilometers from the harvest site (Figure 3.8). The bird digs a little hole in the earth for each store of seeds, then completely covers the cache. A nutcracker does this work in the fall and then relies on its stores through the winter and into the spring, recovering an estimated two-thirds of the caches, often months after making them.[65, 982]

It could be that nutcrackers do not really remember where each and every seed cache is, but instead rely on a simple rule of thumb, such as "look near little tufts of grass." Or they might remember only the general location where food was stored and, once there, look around until they see disturbed soil or some other indicator of a cache. But experiments similar to those performed with chickadees show that the birds do remember exactly where they hid their food. In

**FIGURE 3.8** **A Clark's nutcracker holding a seed** in its bill that the bird is about to cache underground. Photograph by Russ Balda.

one such test, a nutcracker was given a chance to store seeds in a large outdoor aviary, after which it was moved to another cage. The observer, Russ Balda, mapped the location of each cache and then removed the buried seeds and swept the cage floor, removing any signs of cache making. No visual or olfactory cues were available to the bird when it was permitted to go back to the aviary a week later and hunt for the food. Balda mapped the locations where the nutcracker probed with its bill, searching for the nonexistent caches. The bird's spatial memory served it well, for it dug into as many as 80 percent of its ex-cache sites, while only very rarely digging in other places.[65] Other long-term experiments on nutcracker memory have demonstrated that nutcrackers can remember where they have hidden food for at least 6, and perhaps as long as 9, months.[66] Indeed, when Balda tested one of his graduate students as if he were a food-storing bird, the student did only about half as well as a typical nutcracker when tested a month after making his caches.[982] The birds can even remember the size of the seeds they have hidden, as demonstrated by their tendency to spread their bills farther apart when probing the earth for large cached seeds as opposed to smaller ones. Because nutcrackers retrieve one seed at a time from their underground stores, they can secure and process them more efficiently by opening their beaks just the right distance to grasp and pluck a seed of a given size out of a cache.[998]

The extraordinary ability of nutcrackers and chickadees to store spatial information in their brains is surely related to the ability of certain brain mechanisms to change biochemically and structurally in response to the kinds of sensory stimulation associated with hiding food. These changes could not occur without the genes needed to construct the learning system and the genes that are responsive to key sensory stimuli relevant to the learning task. The more general point is that even learned behaviors, which are obviously environment-dependent, are gene-dependent as well.

## What Causes Individuals to Develop Differently?

One of the facts of development is that members of the same species frequently differ in their behavior. Thus, for example, members of an Alaskan population of black-capped chickadees store food more often, and retrieve caches more efficiently, than black-caps that live in the lowlands of Colorado (Figure 3.9).[1164] The superior food-storing ability of Alaskan chickadees may be related to their larger hippocampus[272] (but see MacDougall-Shackleton et al.[910]). If the behavioral differences observed among these chickadees are truly caused by differences in the size of the hippocampus, the differences among individual birds in either their genetic information or environmental inputs could be responsible (because the construction of the hippocampus is dependent on both genes and environment).

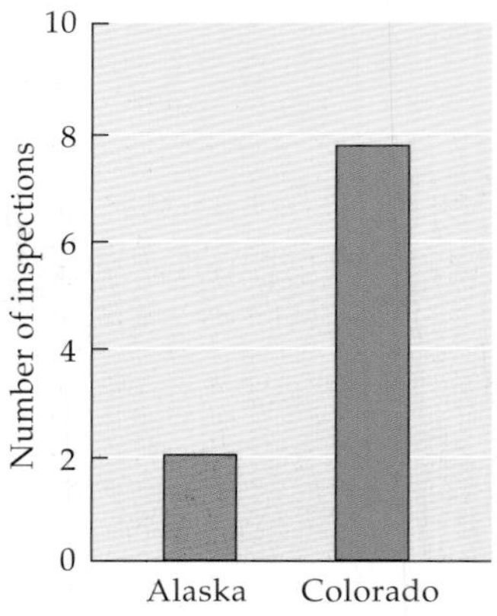

**FIGURE 3.9 Differences within a species in learned behavior.** Black-capped chickadees that live in the severe cold of Alaska not only store more food, but remember better where they have put their caches than chickadees of the same species that live in Colorado, where the climate is less demanding. The mean number of site inspections per food item found is much less for Alaskan chickadees than for their Coloradan cousins. After Pravosudov and Clayton.[1164]

---

**Discussion Question**

**3.4** Return to the chocolate cake analogy (see page 69), and use it to illustrate how a change in either genes or environment could lead to developmental differences between individuals.

---

Although, in theory, the differences between an Alaskan chickadee and a member of the same species from Colorado could be due to genetic differences between them, or environmental differences, or both, no direct evidence is available to resolve this case. However, it is known for some other species that an environmental difference can generate differences in the size of the hippocampus. When Nicky Clayton and John Krebs hand-reared some marsh tits—close relatives of the black-capped chickadee—in the laboratory, they gave some of the young birds opportunities to store whole sunflower seeds, while others were always fed powdered sunflower seeds, which the birds ate but never stored.[271] The individuals that had opportunities to store food gained more cells in the hippocampus than those birds lacking food-storing experience, a reflection of the fact that genes in hippocampal cells responded in different ways to different kinds of experiential inputs.

## Environmental Differences and Behavioral Differences

The marsh tit case is merely one of thousands in which certain phenotypic differences among individuals have been linked to differences in their environment, rather than genetic differences. You will remember that the dialect differences among male white-crowned sparrows are also a product of differences in the birds' acoustical and social environments, which affect what young sparrows learn when they listen to males singing around them. Environmental differences are important whenever members of a species differ in a learned behavior. Consider why one spiny mouse huddles with a companion while another refuses to do so. The difference between these mice could stem entirely from differences in the experiences they had as pups; baby spiny mice learn who their littermates are, and later in life, they prefer to cluster together with familiar individuals rather than unfamiliar ones. Typically, members of this species huddle preferentially with their siblings with whom they have grown up, but if one experimentally creates a litter composed of nonsiblings that are cared for by the same female, these unrelated littermates will treat one another as if they were siblings.[1160] The spiny mice offer one example of many in which individuals differ in their response to others based on whether they have had an association with them earlier in life.[676]

While living with one another in their mother's nest, baby spiny mice might learn the distinctive odors of their companions, which could be the basis for their later discrimination in favor of these individuals. The use of variable odors as recognition cues is widespread in the animal kingdom,[1427] including insects like *Polistes* paper wasps. Some female paper wasps emerge from the paper nest and stay on as nonbreeding workers. These individuals learn to recognize one another as nestmates in large part because they have acquired the special odor of the nest (Figure 3.10) in which they were reared by other adult members of the colony.[512] If this hypothesis is correct, then it should be possible to fool paper wasps into tolerating wasps they would otherwise attack by transferring newly emerged females for a few hours to a portion of another nest where they can acquire the odor of that nest. As predicted, the transferred females are less likely to fight with other females reared in the host nest than with their sisters that emerged from and remained on their original nest. It is as if paper wasps inherit a behavioral command that reads, "Learn what your nest smells like soon after you emerge as an adult, and then respond nonaggressively to those individuals that share this odor." Differences among nests in their odor can therefore cause emerging wasps to differ in their social responses to others.

Not only do females of the paper wasp *Polistes fuscatus* use olfactory cues to discriminate among potential companions, they can even learn to recognize individuals on the basis of the distinctive appearance of a face (Figure 3.11).[1440] When Elizabeth Tibbetts altered the facial features of some members of a colony of female

**FIGURE 3.10 Nests of *Polistes* paper wasps** contain odors that adhere to the bodies of the wasps reared in them, providing a proximate cue for the learned recognition of nestmates in these insects. Note the larvae with their bodies pressed up against the cell walls. Photograph by the author.

**FIGURE 3.11 Paper wasps are capable of using the variation in the color patterns** of their nestmates' faces to recognize them as individuals. (A) The faces of a set of paper wasps, *Polistes dominulus.* (B) Nestmates of a related wasp (*Polistes fuscatus*), which also exhibits individual variation in facial patterns, behaved more aggressively toward females that had had their faces altered than to females whose faces were painted without changing the facial color pattern. A, photographs courtesy of Elizabeth Tibbetts. B, after Tibbetts.[1440]

(A)

(B)

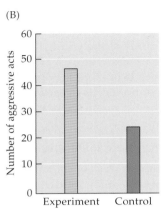

paper wasps, the paint-altered individuals were subject to more aggression from their fellow females than were those individuals whose faces had been painted in such a way as to maintain their original color pattern.[1442]

Note that the ability to record information about the odor or appearance of one's nestmates requires genetic information, which is needed for the construction of a nervous system with the capacity for this kind of learning. The same point can be made with reference to a study of Belding's ground squirrels in which the newborn offspring of captive females were switched around at birth, creating four classes of individuals: (1) siblings reared apart, (2) siblings reared together, (3) nonsiblings reared apart, and (4) nonsiblings reared together. After having been reared and weaned, the juvenile ground squirrels were placed in an arena in pairs and given a chance to interact. In most cases, animals that were reared together, whether actual siblings or not, tolerated each other, whereas animals that had been reared apart tended to react aggressively to each other.[673] Here, the young squirrels learned something about their nestmates thanks to a nervous system primed to record certain information about the olfactory cues associated with individuals.

But in addition, biological sisters *reared apart* engaged in fewer aggressive interactions than nonsiblings reared apart (Figure 3.12). In other words, the squirrels had some way of recognizing their sibs that was not dependent on living with them as youngsters.[673, 675] Instead, a different kind of learning was probably involved, one that goes informally by the indelicate label of the "armpit effect."[948] That is, if individuals can learn what they themselves smell like, then they can use this information as a reference against which to compare the odors of other individuals.[674]

This hypothesis has been examined by Jill Mateo,[949] who noted that Belding's ground squirrels possess several scent-producing glands,[950] including one around the mouth and another on the animal's back. Moreover, these

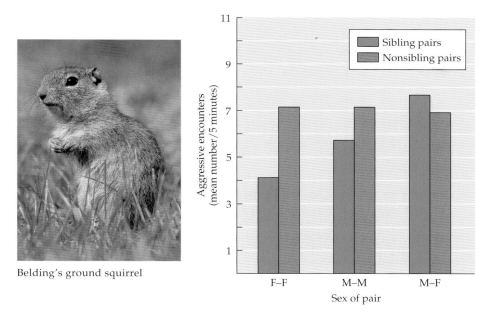

Belding's ground squirrel

**FIGURE 3.12    Kin discrimination in Belding's ground squirrels.** Sisters reared apart display significantly less aggression toward each other than other combinations of siblings reared apart, which are as aggressive to each other when they meet in an experimental arena as nonsiblings reared apart. After Holmes and Sherman.[673]

squirrels regularly sniff the oral glands of other individuals, as if they were acquiring odor information that could conceivably be compared with the sniffers' own scents. By capturing pregnant ground squirrels and moving them to laboratory enclosures, Mateo was able to observe their juvenile offspring investigating objects (plastic cubes) that had been rubbed on the dorsal glands of other squirrels of varying degrees of relatedness to the youngsters. Because the captive squirrels had been separated from some of their relatives, they had never met them and so had no prior experience with their odors. If, however, the test animals had learned what they themselves smelled like, and if close relatives produced scents more similar to their own than do distant relatives, an inexperienced youngster could in theory discriminate between unfamiliar relatives and nonrelatives on the basis of odor cues alone. As it turns out, the length of time a Belding's ground squirrel sniffs an object is an indicator of its interest in that object, which depends on the object's odor similarity to the squirrel's own scent. Thus, cubes that have been rubbed on a fairly close relative, genetically speaking, receive only a cursory inspection. Items smeared with odors of a more distant relative are given a significantly longer sniff, and the inspection time increases again for cubes daubed with a nonrelative's odor. Belding's ground squirrels are odor analyzers, spending less time with scents similar to their own and increasingly more time with odors less like their own (Figure 3.13). They therefore have the capacity to treat individuals differently on the basis of this learned label of relatedness.[949]

Belding's ground squirrels possess a highly specific form of learning that enables them to remember what they themselves smell like in order to use this information to make decisions about which individuals to mate with and which individuals to ignore or reject. Differences among individuals in their odor "environment" translate into learned differences in their behavior—one example among many of how environmental differences can lead to the development of behavioral differences within a species.

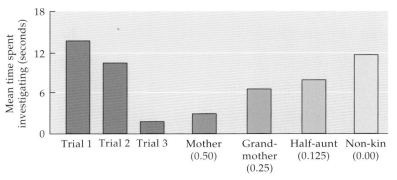

**FIGURE 3.13   Evidence for the ability of Belding's ground squirrels to learn their own odor,** which they may match against that of other individuals. Juvenile squirrels were first given three trials during which they could investigate their own odors applied to plastic cubes. Note the squirrels' decline in responsiveness to their own dorsal gland scents over these initial trials. Then the squirrels were provided with plastic cubes daubed with dorsal gland odors from four categories of individuals. Cubes with scents from close relatives received less attention than those with scents from distant relatives or non-kin. Numbers in parentheses show the genetic similarity of the individual to the tested ground squirrel. After Mateo.[949]

## Discussion Question

**3.5**   A good predictor of a young person's vocabulary is the amount of time parents spent talking to their child when he or she was very young. Some have concluded that family environment is therefore the essential factor in determining a person's language skills. What's a logical problem with this conclusion?

### Genetic Differences and Behavioral Differences

Although a great many differences in behavioral phenotypes have been traced to differences in the environment, others have been linked to genetic differences among individuals, which makes sense given the interactive, dual-factor nature of development. To see if a genetic difference underlies why some blackcaps, a type of warbler, spend the winter in southern Great Britain while most other members of this species migrate to Africa (Figure 3.14), Peter Berthold checked whether the offspring of "winter in Britain" birds would inherit their parents' behavior. To conduct this research, he and his colleagues captured some wild blackcaps in Britain during the winter and took them to a laboratory in Germany, where the birds spent the rest of the winter indoors. Then, with the advent of spring, pairs of warblers were released into outdoor aviaries, where they bred, providing Berthold with a crop of youngsters that had never migrated.[115]

Once the young birds were several months old, Berthold's team placed some in special cages that had been electronically wired to record the number of times a bird hopped from one perch to another. The electronic data revealed that when fall arrived, the young warblers became increasingly restless at night, exhibiting the kind of heightened activity characteristic of songbirds preparing to migrate. The immature blackcaps' parents also became nocturnally restless when placed in the same kind of cages in the fall. These observations showed that the British wintering population is not composed of birds that have simply lost their ability to migrate. Instead, the birds wintering in Britain must be migrants that flew to Britain from somewhere else.

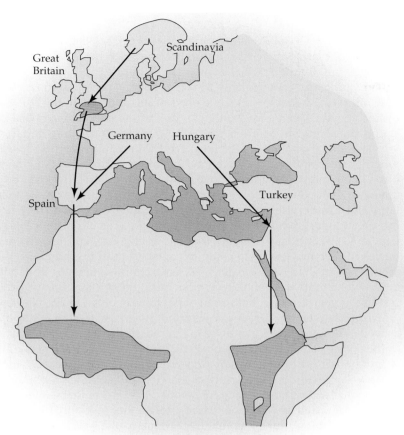

FIGURE 3.14 Different wintering sites of blackcaps. Blackcaps living in southern Germany and Scandinavia first go southwest to Spain before turning south to western Africa. Blackcaps living in eastern Europe migrate southeast before turning south to fly to eastern Africa. Where do the birds that winter in Great Britain come from?

Blackcap warbler

But just where does the British wintering population come from? To answer this question, the researchers put some about-to-migrate warblers in cages shaped like funnels and lined with typewriter correction paper. Whenever the bird leaped up from the base of the funnel in an attempt to take off, it landed on the paper and left scratch marks, which indicated the direction in which the bird was trying to go (Figure 3.15). Berthold's subjects, experienced adults and young novices alike, oriented due west, jumping up in that direction over and over, judging from the footmarks left on the paper. These data showed that the adults, which had been captured in wintery Britain, must have traveled there by flying west from Belgium or central Germany, a point eventually confirmed by the discovery of some blackcaps in Britain that had been banded earlier in Germany.

If the differences in migratory behavior are hereditary and therefore subject to selection, it should be possible to do an artificial selection experiment that leads to behavioral evolution in the laboratory. Although no one seems to have done an experiment of this sort on the destination preferences of blackcaps, one research group was able to exert selection on the timing of migratory behavior in this species. By breeding males that started their fall journey south late with females that shared a similar tendency, the blackcap research team quickly produced a late-departing

FIGURE 3.15 Funnel cage for recording the migratory orientation of captive birds. The bird can see the night sky through the wire mesh ceiling of the cage. As the bird jumps up onto the surface of the funnel, it leaves marks that show the direction in which it is intending to fly. Photograph by Jonathan Blair.

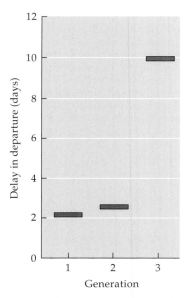

**FIGURE 3.16** **Response to artificial selection** on the fall migration departure date of blackcaps. After two generations of selection, the onset of migratory activity in captive birds had been shifted by nearly 8 days. After Pulido et al.[1175]

line of birds.[1175] Just two generations of artificial selection were sufficient to create a population that began autumnal migration more than a week later on average than did the original population (Figure 3.16).

The blackcap is not the only bird that has the potential to rapidly evolve new migratory behaviors in some parts of its range.[1176] In the 1990s, German ornithologists became aware that some European blackbirds had begun to spend the winter in Munich while those in nearby forests departed when winter came, as is the rule for northern European blackbirds. To determine whether the difference in the winter behavior of urban versus forest blackbirds in Germany had a hereditary component, Jesko Partecke and Eberhard Gwinner took young nestlings from the two locations and reared them in the laboratory under identical, seasonally changing conditions that resembled those in the city environment. They then checked the birds' migratory restlessness at night, finding that males that had been removed from city nests were far less active in their first spring in captivity than their male counterparts whose parents had been forest dwellers. In contrast, females from both urban and wild habitats were equally restless during both migratory seasons. These results suggest strongly that urban males have evolved sedentary tendencies, which cause them to stay put year-round, whereas forest males and females, as well as urban females, continue to migrate south in the fall before returning in the spring. The female blackbirds that spend the summer in German cities would probably die if they attempted to overwinter there, because male blackbirds are larger and more aggressive than females and therefore more likely to monopolize scarce food during the winter.[1107] Thus, only the males have been selected for sedentary behavior during winter, because only they can successfully exploit the resources available in German cities during this season.

## Discussion Questions

**3.6** A few blackcaps live year-round in southern France, although 75 percent of the breeding population migrates from this area in winter. Perhaps the difference between the two behavioral phenotypes is environmentally induced and not hereditary. Make a prediction about the outcome of an artificial selection experiment in which the experimenter tries to select for both nonmigratory and migratory behavior in this species. Describe the procedure and present your predicted results graphically. Check your predictions against the actual results (see Berthold[113]).

**3.7** The black redstart is a bird species that migrates a relatively short distance from Germany to the Mediterranean region of Europe, whereas the common redstart travels as much as 5000 kilometers from Germany to central Africa. The scale in Figure 3.17 shows the duration of migratory restlessness in three groups of captive birds all hand raised under identical conditions: black redstarts, hybrids created by crossing black and common redstarts, and common redstarts. Why do black redstarts exhibit migratory restlessness at night for fewer days than common redstarts? What does the behavior of the hybrids tell us about the genetic differences hypothesis for the difference in the duration of migratory restlessness in the two parental species?

**3.8** Robert Plomin and his colleagues have compared the cognitive abilities of children with those of their parents (genetic or adoptive) and twin siblings.[1156] What significance do you attach to these data (Figure 3.18) in the context of determining whether genetic or environmental differences are responsible for the differences between humans in their spatial and

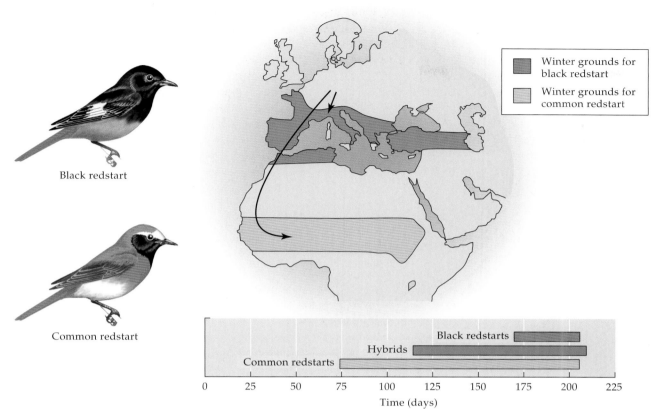

**FIGURE 3.17** **Differences in the migratory behavior of two closely related birds,** the black redstart and the common redstart. The scale at the bottom shows the periods after hatching (day 0) when the young birds exhibit migratory restlessness at night. After Berthold and Querner.[116]

verbal abilities? If environmental differences are the key to understanding differences in these human phenotypes, what is the predicted relationship between the number of years a child has spent in an adoptive home and the degree of difference between the child's spatial and verbal attributes and those of his or her genetic parents?

## Hereditary Differences in the Food Preferences of Garter Snakes

The garter snake *Thamnophis elegans* occupies much of dry inland western North America as well as foggy coastal California, a region that it almost certainly invaded relatively recently.[45] The diets of snakes in the two areas, referred to hereafter as coastal and inland snakes, differ markedly. Whereas inland snakes feed primarily on the fish and frogs found in lakes and streams in the arid West, coastal snakes regularly eat the banana slugs that thrive in the wet forests of coastal California (Figure 3.19). You can watch a brief video of a snake swallowing a slug at http://www.birdsamore.com/videos/snake-eatingslug.htm. I can only marvel at the ability of the snakes to consume these prey. When I once made the mistake of picking up a banana slug, the creature promptly covered my hand with massive amounts of an exceedingly sticky and repulsive mucus that greatly reduced my desire to touch these animals ever again.

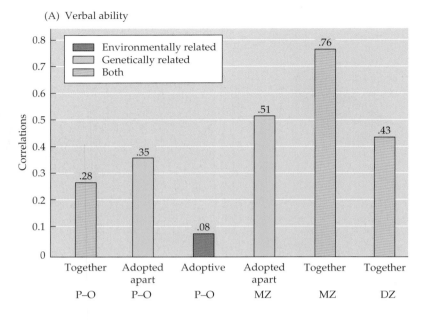

(A) Verbal ability

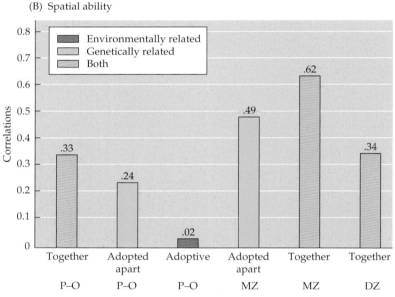

(B) Spatial ability

**FIGURE 3.18** **Why do people differ in their test scores?** The graphs show correlations in (A) verbal ability scores and (B) spatial ability scores for parents and offspring (P–O), monozygotic (identical) twins (MZ), and dizygotic (fraternal) twins (DZ) either living together or apart. The data are composite measures based on a number of different studies. After Plomin et al.[1156]

If the preference for banana slugs exhibited by coastal garter snakes has a hereditary basis, then these snakes should differ genetically from inland snakes. To check this prediction, Steve Arnold took pregnant female snakes from the two populations into the laboratory, where they were held under identical conditions. When the females gave birth to a litter (garter snakes produce live young rather than laying eggs), each baby snake was placed in a separate cage, away from its littermates and its mother, to remove these possible environmental influences on its behavior. Some days later Arnold offered each baby snake a chance to eat a small chunk of freshly thawed banana slug by placing it on the floor of the young snake's cage. Most naïve young coastal snakes ate all the slug hors d'oeuvres they received; most of the inland snakes did not (Figure 3.20). In both populations, slug-refusing snakes ignored the slug cube completely.

Arnold took another group of isolated newborn snakes that had never fed on anything and offered them a chance to respond to the odors of different prey items. He took advantage of the readiness of newborn snakes to flick their tongues at, and even attack, cotton swabs that have been dipped in fluids from some species of prey (Figure 3.21). Chemical scents are carried by the tongue to the vomeronasal organ in the roof of the snake's mouth, where the odor molecules are analyzed as part of the process of detecting prey. By counting the number of tongue flicks that hit the swab during a 1-minute trial, Arnold measured the relative responsiveness of inexperienced baby snakes to different odors.

Populations of inland and coastal snakes reacted about the same to swabs dipped in toad tadpole solution (a prey of both groups), but they behaved very differently toward swabs daubed with slug scent. Almost all inland snakes ignored the slug odor, whereas almost all coastal snakes rapidly flicked their tongues at it. Because all the young snakes had been reared in the same environment, the differences in their willingness to eat slugs and to tongue-flick in reaction to slug odor appear to have been caused by genetic differences among them.

**FIGURE 3.19   A coastal Californian garter snake about to consume a banana slug,** a favorite food of snakes in this region. Photograph by Steve Arnold.

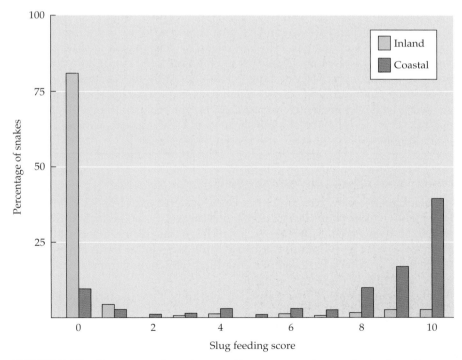

**FIGURE 3.20   Response of newborn, naïve garter snakes to slug cubes.** Young snakes from coastal populations tended to have high slug feeding scores (e.g., a score of 10 indicates that the snake ate a slug cube on each of the 10 days of the experiment). Inland garter snakes were much less likely than coastal snakes to eat even one slug cube (which would yield a score of 1). After Arnold.[45]

**FIGURE 3.21   A tongue-flicking newborn garter snake** senses odors from a cotton swab that has been dipped in slug extract. Photograph by Steve Arnold.

If the feeding differences between the two populations arise because most coastal snakes have a different allele or alleles than most inland snakes, then crossing adults from the two populations should generate a great deal of genetic and phenotypic variation in the resulting group of hybrid offspring. Arnold conducted the appropriate experiment and found the expected result, confirming again that the behavioral differences between populations have a strong genetic component.

If among the early occupants of coastal habitat were a very few garter snakes with a rare allele or two for slug acceptance, these slug-eating snakes would have been able to exploit an abundant, if mucus-covered, food in their new habitat. Currently, coastal snakes are much more efficient at digesting and assimilating the nutrients in slugs than are inland snakes.[175] If, as a result of securing useful energy from slugs, the reproductive success of slug-eating individuals in the past was even 1 percent higher than that of their slug-ignoring fellows, the coastal population could have reached its present state of divergence from the inland population in less than 10,000 years.[45] This study illustrates once again that if there are genetic differences among individuals that affect their reproductive success, natural selection can be a powerful agent for evolutionary change.

## Discussion Question

**3.9**   Debi Fadool at Florida State University headed a research team that studied a strain of genetically modified mice that lacked the ability to make a protein called Kv1.3.[452] In unaltered mice, this protein is found in regions of the brain that process olfactory information, leading Fadool and her team to predict that the two kinds of mice should differ in their ability to smell things. In fact, the genetically modified mice were able to smell scents at much lower concentrations than mice that possessed the protein; the mutant mice found odorous foods, such as peanut butter crackers, much faster than their wild-type cousins. What evolutionary question is raised by these findings? What ultimate explanation do you have for the fact that mice with Kv1.3 protein are actually less sensitive to food odors than mice without that protein?

### Single-Gene Effects on Development

The blackcap breeding experiments and the garter snake crosses do not tell us how many genetic differences are responsible for the behavioral differences present in these species. In theory, a single genetic difference could be the starting point for a series of downstream differences in the gene–environment interactions occurring in different individuals, which may translate into large behavioral differences between them.

Single-gene effects of this sort exist, and they have been documented in several different ways, perhaps most dramatically via gene knockout experiments. Researchers are now able to inactivate a given gene in an animal's genome in order to determine how that particular gene contributes to development in a particular environment. Sometimes the developmental effect of knocking out a gene is spectacular, as demonstrated by the effects of scrambling the genetic code of the *fosB* gene of laboratory mice. Females with the experimental "mutation" are normal in most respects but are totally indif-

**FIGURE 3.22** **A single genetic difference between females** has a large effect on their maternal behavior. Wild-type female mice gather their pups together and crouch over them (above), but females with inactivated *fosB* genes (below) do not exhibit these behaviors (the pups can be seen scattered in the foreground). Photographs courtesy of Michael Greenberg; from Brown et al.[196]

ferent to their newborn pups, which they fail to retrieve should they wriggle away from the nest. In contrast, normal females with two copies of the active *fosB* gene invariably gather displaced pups together and crouch over them, keeping them warm and permitting them to nurse (Figure 3.22).[196]

## Discussion Question

**3.10** Does the *fosB* knockout experiment demonstrate that a single gene determines the maternal behavior of a female mouse? You should know by now that the answer is no. Use this example to illustrate the difference between claiming that maternal behavior is genetically determined and claiming that certain differences among individuals in their maternal behavior phenotypes are genetically determined. How is the idea that genes are responsive to particular kinds of environmental inputs illustrated by the fact that when a female mouse inspect her pups after birth, she receives olfactory stimulation, which affects the mouse's brain and triggers *fosB* gene activity in a mouse with the typical genotype? How might this gene's activity initiate additional changes in other genes, leading to a specific pattern of biochemical events?

Other knockout mutations also have highly damaging, specific developmental consequences for mice. Males whose *Oxt* gene has been knocked out cannot produce oxytocin, an important brain hormone,[460] with the correlated effect that these males cannot remember females with whom they have recently interacted. Each time a given female is removed from and then returned to

(A)

(B)

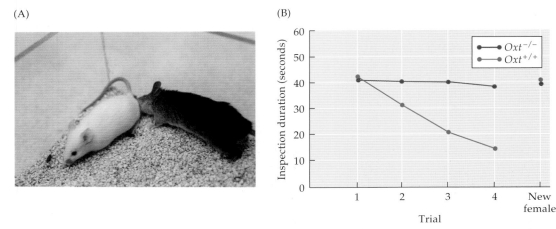

**FIGURE 3.23** **Social amnesia is related to the loss of a single gene.**
(A) Male mouse inspecting a female. (B) A knockout male mouse that lacks a functional *Oxt* gene carefully inspects the same female every time she is reintroduced into his cage, whereas a male with the typical genotype shows less and less interest in a female that he has inspected previously. A, photograph by Larry J. Young; B, after Ferguson et al.[460]

the cage she shares with an *Oxt* mutant male, the male gives her a thorough and lengthy sniffing that is no different from his response the very first time they met (Figure 3.23). In contrast, if a female is placed in the cage of a normal male with a functional *Oxt* gene, he remembers what she smells like, so if she is taken from his cage but then later returned to it, he spends less time sniffing her on the second occasion than he did the first time they met. The presence of a functional *Oxt* gene seems critical, therefore, if the male is to remember that he has interacted with a familiar female.

In yet another experiment, the *Trpc2* gene was knocked out. When a male mouse without this gene encounters another male in his laboratory cage, he attempts to mate with the intruder, who rarely responds enthusiastically.[1394] In contrast, a male with functional copies of the *Trpc2* gene makes war, not love, when he finds another male in his cage. Apparently a knockout male cannot identify a fellow male by his distinctive scents and so treats every mouse as a potential copulatory partner, perhaps because of an alteration in his olfactory apparatus. The vomeronasal organ, an olfactory device in the mouse nose, contains a cluster of neurons that responds to sex-identifying mouse scents. However, if these cells lack the *Trpc2* gene, they are incapable of reacting to male odors, so the knockout male never receives signals from the vomeronasal organ that a male is in the vicinity. Because the genetically altered mouse does not detect the key cue of maleness, he responds to the other mouse as if it were a female.

What about the behavior of *female* mice with two inoperative copies of *Trpc2*? Researchers in Catherine Dulac's lab created these females, which exhibited *male* sexual behavior. The mutant females closely inspected any males they encountered and attempted to copulate with them to the best of their ability, although their "partners" were, not surprisingly, uncooperative in this regard.[773] Apparently both females and males have the same neural circuits in the vomeronasal organ and allied olfactory apparatus in the brain. In nonmutant females, however, olfactory signals received from other mice and processed by the vomeronasal organ prevent females from acting like males. Those mutant females lacking just one gene have vomeronasal organs that cannot respond to male pheromonal signals in the "proper" manner.

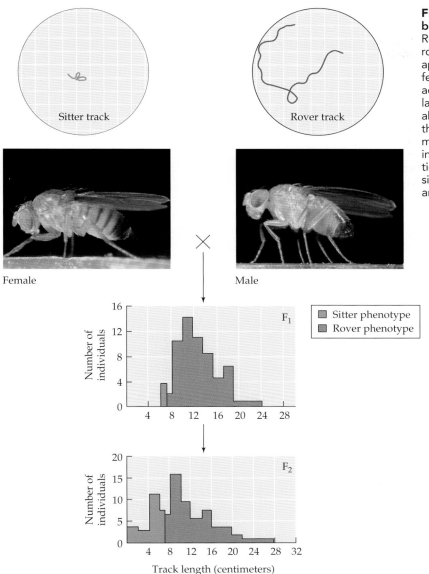

**FIGURE 3.24  Genetic differences cause behavioral differences in fruit fly larvae.** Representative tracks made by sitter and rover phenotypes feeding in a petri dish appear at the top of the figure. When adult female flies of the sitter strain mate with adult male flies of the rover strain, their larval offspring (the $F_1$ generation) almost all exhibit the rover phenotype (that is, they move more than 7.6 centimeters in 5 minutes). When flies from the $F_1$ generation interbreed, their offspring (the $F_2$ generation) are composed of rovers (blue) and sitters (red) in the ratio of 3:1. After de Belle and Sokolowski.[372]

Knockout experiments are not the only way to test the hypothesis that even a single genetic difference can translate into a behavioral difference between individuals. For example, persons studying the wormlike larvae of *Drosophila melanogaster*, the humble fruit fly, have found two naturally occurring, distinctively different phenotypes. Some larvae (called rovers) travel about four times farther when feeding on a yeast-coated petri dish during a 5-minute period than larvae of the other type (which are labeled sitters).[372] When adults reared from rover larvae are bred with adults reared from sitter larvae, these pairs of flies produce larval offspring (the $F_1$ generation) that are all rovers. When these larvae mature and interbreed, they produce an $F_2$ generation with three times as many rovers as sitters (Figure 3.24). Persons familiar with Mendelian genetics will recognize that rovers are likely to have at least one copy of the dominant allele of a gene affecting larval foraging behavior, whereas sitters are likely to have two copies of the recessive allele. If this analysis is correct, and if one could transfer the dominant allele associated with rover behavior to an individual of the sitter genotype, then the genetically altered larva should exhibit rover behavior. This experiment has been done, with positive results.[1078] Thus, the difference in foraging behavior between rovers and sitters stems from a difference in the information contained in a single gene—just one of the 13,061 genes[1117] located on the four chromosomes of *Drosophila melanogaster*.[373]

The techniques now available to molecular biologists have enabled researchers to identify the gene in question. This gene, which has been given the label *for*, is the same one mentioned earlier in the context of honey bee behavior. In both species, the gene codes for a cGMP-dependent protein kinase, more of which is produced by larvae carrying the rover form of the allele (*forR*) than by those endowed with the sitter allele (*fors*).[1078] This enzyme is produced in certain cells of the larval fruit fly's brain, where they presumably affect neuronal activity and thereby shape the larva's behavior. One study has shown that individuals with *forR* exhibit better short-term memory for olfactory stimuli, whereas those with *fors* do better at a simple long-term memory task involving an odor.[976] These different learning abilities may be linked to the different foraging tactics of the two kinds of fruit fly larvae. In any case, individuals that differ in their *for* alleles produce different forms of a key protein, with different efficacies in promoting a particular chemical reaction in particular parts of the fly's brain, with behavioral consequences of several sorts.

Humans have a nervous system and a genome too, of course, and so we can expect that some of the behavioral differences among us have a genetic component. The search for genetic variation that affects human behavioral development has led some researchers to one particular part of the brain, the lateral frontal cortex, which is known to contribute to human intelligence.[420] Cells in this part of the brain become especially active when people try to solve either spatial or verbal problems. (Brain cell activity can be visualized by positron-emission tomography [PET], a technology that measures the extent of blood flow to particular parts of the brain.) It is possible, therefore, that a portion of the differences in intelligence among people in some populations will eventually be traced to differences among them in the gene–environment interplay that contributes to the development or activity of the lateral frontal cortex.

The lateral frontal cortex is such a complex chunk of brain tissue that we can be certain that literally thousands of genes are necessary for its complete development and effective operation.[287] In fact, about half of the human genome, perhaps 10,000 genes or thereabouts, is active at some time in some part of the brain.[880] Thus, variation in any one of these thousands of genes could conceivably contribute to variation in brain phenotypes and, thus, variation in the cognitive ability or behavior of human beings. Indeed, one variable gene (*COMT*) that codes for an enzyme called catechol-O-methyltransferase has been shown to affect performance on at least one intelligence test.[1604] The difference between two common alleles of *COMT* translates into a single difference in the long chain of amino acids that make up the enzyme in question. One variant form of the enzyme is four times as active at body temperature as the other type, which means that persons possessing the "fast" enzyme carry out one particular biochemical reaction at a relatively high rate. The reaction that is mediated by *COMT* breaks down a substance called dopamine, which is an important chemical communicator between certain brain cells. The rate at which dopamine is removed therefore affects signal transmission between cells in the prefrontal cortex, which in turn evidently affects a person's ability to perform certain cognitive tasks.

Another example of a relationship between variation in a gene, a neurotransmitter, and human behavior involves a segment of DNA found in a particular portion of chromosome 17 (humans have 23 pairs of chromosomes in all). The gene in question produces a protein that regulates the uptake of serotonin—another chemical, like dopamine, that relays messages between neurons in certain parts of the human brain. The activity of this serotonin transporter gene (labeled *5-HTT*) is controlled by a segment of DNA some distance from *5-HTT*. This regulatory chunk of DNA comes in two forms, one longer than the other; the shorter form causes the *5-HTT* gene to produce about a third less protein per unit of time than the longer form. As a result, a person's genotype affects how much protein is available to remove serotonin from the spaces between certain brain neurons, thus affecting the nature of neural activity in these cells, which depend on serotonin to communicate with one another. The regions of the brain that rely heavily on serotonin as a neurotransmitter include structures thought to play major roles in controlling our emotions, mood, and anxiety levels. Indeed, a small part of the difference among people in just how anxious they are has been linked to variation in the *5-HTT* regulator genotype.[857]

### Discussion Question

**3.11** Infant humans learn languages by listening to the speech of other persons. Given the obvious importance of this environmental factor on language acquisition, what do you make of the finding that certain alleles of two genes (*ASPM* and *microcephalin*) are much more likely to be found in people who speak a so-called tonal language (like Mandarin Chinese) than in those of us who speak a nontonal language (like English)?[378] (In tonal languages, the meaning of a word depends not just on its consonants and vowels but also on the tone or pitch, higher or lower, that the speaker imparts to a given syllable.) Explain this genetic finding in the context of the interactive theory of development, and link it to the evolution of language learning in our species.

## Evolution and Behavioral Development

The developmental features of living things have a history, which can be explored in two quite different ways. First, there is the question of the sequence of evolutionary events that resulted in the modification of an ancestral pattern and its reconfiguration into a modern attribute. This kind of question is at the heart of what has been called the field of evolutionary development or "evo-devo."[238, 1456] A spectacular product of this approach has been the discovery that creatures as different as fruit flies and humans share a set of homeobox (or *Hox*) genes whose operation is critical for the developmental organization of their bodies. These genes, which originated in a distant ancestor, have been retained in flies, humans, and many other organisms because of their importance and utility in regulating the development of functional body structures. The base sequence of the genes has, of course, been altered somewhat from species to species, and the way in which their products influence the gene–environment interactions can differ markedly, leading to dramatically different developmental outcomes, but the imprint of history on the process can still be seen in the information contained within this particular set of genes, or "toolkit."

One example of the phenomenon that relates specifically to animal behavior involves the *for* gene in *Drosophila* fruit flies (see page 86), a gene that also occurs in very similar form in the honey bee.[1456] As we have noted, in fruit flies this gene codes for a protein that, when produced, leads to chemical changes that eventually affect the operation of the brain of larval fruit flies. Depending on the allele, the larval flies either engage in little movement (the sitter phenotype) or move about over much greater distances (the rover phenotype). The honey bee has inherited this same gene from a common ancestor of flies and bees. But over evolutionary time, the now modified gene has taken on a different but allied function in the honey bee, where it plays a role in regulating the transition from being a sedentary young adult that stays within the hive to becoming a long-distance forager worker that collects food for the colony outside the hive. This transition is linked to an increase in the expression of the allele in the brains of the older workers.[103]

The other kind of evolutionary approach to development focuses on the possible adaptive significance of a developmental trait, rather than on its origin and historical modifications. This approach examines the possible role of natural selection in the evolution of the attribute. Persons interested in this possibility know that in organisms living today, a single genetic difference can sometimes lead to developmental differences between individuals. If there is genetic variation that leads to behavioral variation in animal populations today, then surely the same applied to populations in the past. If so, natural selection could have operated on previous generations, leading to the spread of advantageous developmental characteristics that have a reproductive payoff. Has this happened with respect to the underlying molecular mechanisms that guide behavioral development in animals?

## Adaptive Features of Behavioral Development

Given the fact that most organisms have thousands of genes and are subject to thousands of variable environmental factors, development errors must often occur. The genomes of most individuals do have some damaging mutant alleles, and few organisms grow up in ideal environments. Yet, despite the potential for developmental problems, most animals look and behave reasonably normally. In fact, although gene knockout experiments sometimes do have dramatic phenotypic effects, in many cases, blocking the activity of a particular gene has little or no developmental effect. These findings have led some geneticists to conclude that genomes exhibit considerable information redundancy, which would explain why the loss of one gene–environment product is not fatal to the acquisition of one or more traits of importance to the individual.[766, 1134]

We also know that many animals overcome what you might think would constitute considerable environmental obstacles to normal development. For example, some young birds lack the opportunity to interact with their parents and so cannot acquire the information that in other species is essential for normal social and sexual development (as discussed earlier in this chapter). When chicks of the Australian brush turkey hatch from eggs placed deep within an immense compost heap of a nest, they dig their way out and walk away, often without ever seeing a parent or sibling. So how do they manage to recognize other members of their species? Ann Göth and Christopher Evans studied captive young brush turkeys in an aviary in which they were exposed to feathered robots that looked like other youngsters. All that was required to elicit an approach from a naïve youngster was a peck or two at the ground by the robot. Thus, young brush turkeys do not require extensive social experience in order for rudimentary social behavior to develop,[553] and as adults, the birds are completely capable of normal sexual behavior despite having lived primarily by themselves beforehand.

Other experimenters have created genuinely abnormal rearing environments, only to find that various forms of sensory deprivation have little or no effect on the development of normal behavior. Bring up baby Belding's ground squirrels without their mothers, and they still stop what they are doing to look around when they hear a tape of the alarm call of their species.[947] Male crickets that live in complete isolation sing a normal species-specific song despite their severely restricted social and acoustical environment.[106] Captive hand-reared female cowbirds that have never heard a male cowbird sing nevertheless adopt the appropriate precopulatory pose when they hear cowbird song for the first time, if they have mature eggs to be fertilized.[774]

## Developmental Homeostasis: Protecting Development against Disruption

The ability of many animals to develop more or less normally, despite defective genes and deficient environments, has been attributed to a process called **developmental homeostasis**. This property of developmental systems reduces the variation around a mean value for a phenotype (see Figure 3.31B) and reflects the ability of developmental processes to suppress some outcomes in order to generate an adaptive phenotype more reliably. A clear demonstration of this ability comes from a classic experiment on the development of social behavior in young rhesus monkeys deprived of contact with others of their species by Margaret and Harry Harlow.[623, 624] (The Harlows' experiments were conducted nearly 4 decades ago, when animal rights were not the issue they are today; readers can decide for themselves whether the Harlows' harsh treatment of infant monkeys was justified.) In one such study, the Harlows separated a young rhesus from its mother shortly after birth. The baby was placed in a cage with an artificial surrogate mother (Figure 3.25), which might be a wire cylinder or a terry cloth figure with a nursing bottle. The baby rhesus gained weight normally and developed physically in the same way that non-isolated rhesus infants do. However, it soon began to spend its days crouched in a corner, rocking back and forth, biting itself. If confronted with a strange object or another monkey, the isolated baby withdrew in apparent terror.

The isolation experiment demonstrated that a young rhesus needs social experience to develop normal social behavior. But what kind of social experience—and how much—is necessary? Interactions with a mother are insufficient for full social development of rhesus monkeys, since infants reared alone with their mothers fail to develop truly normal sexual, play, and aggressive behavior. Perhaps normal social development in rhesus monkeys requires the young animals to interact with one another. To test this hypothesis, the Harlows isolated some infants from their mothers but gave these infants a chance to be with three other such infants for just 15 minutes each day.[624] At first, the young rhesus monkeys simply clung to one another (Figure 3.26), but later they began to play. In their natural habitat, rhesus babies start to play when they are about 1 month old, and by 6 months they spend practically every waking moment in the company of their peers. Even so, the 15-minute play group developed nearly normal social behavior. As adolescents and adults, they were capable of interacting sexually and socially with other rhesus monkeys without exhibiting the intense aggression or withdrawal of individuals that had been completely isolated as infants.

Naturally one wonders about the relevance of these studies for another primate species, *Homo sapiens*, whose intellectual development is often said to be dependent on the early experiences that children have with their parents and peers. But is this true? We cannot, of course, do social isolation experiments with human babies, but we can examine evidence of another sort regarding the resilience of intellectual development in the face of nutritional deprivation. Consider, for example, the results of a study of young Dutch men who

**FIGURE 3.25  Surrogate mothers used in social deprivation experiments.** This isolated rhesus infant was reared with wire cylinder and terry cloth dummies as substitutes for its mother. Photograph by Nina Leen.

**FIGURE 3.26** **Socially isolated rhesus infants** that are permitted to interact with one another for short periods each day at first cling to each other during the contact period. Photograph by Nina Leen.

were born or conceived during the Nazi transport embargo during the winter of 1944–1945, which caused many deaths from starvation in the larger Dutch cities.[1384] For most of the winter, the average caloric intake of city people was about 750 calories per day. As a result, urban women living under famine conditions produced babies of very low birth weights. In contrast, rural women were less dependent on food transported to them, and the babies they had that were conceived at the same time were born at more or less normal birth weights.

One would think that full brain development depends on adequate nutrition during pregnancy, when much of brain growth occurs. However, Dutch boys who were born in urban famine areas did not exhibit a higher incidence of mental retardation at age 19 than rural boys whose early nutrition was far superior (Figure 3.27A). Nor did those boys born to food-deprived mothers score more poorly than their relatively well-nourished rural counterparts when they took the Dutch intelligence test administered to draft-aged men (Figure 3.27B).[1407] These results are buttressed by the discovery that Finnish adults who experienced severe nutritional shortfalls in utero (during a 19th-century famine) lived just as long on average as those who were born after the famine was over.[746]

No one believes that pregnant women or young children should be deprived of food,[1009] and some continue to argue that the nutritional state of the fetus is critical for a person's health later in life (see review in Rasmussen[1197]). But the survival of a fetus, the intellectual development of a young person, and the later health of the individual are not necessarily harmed even by highly adverse early-life conditions,[331] perhaps because our developmental systems

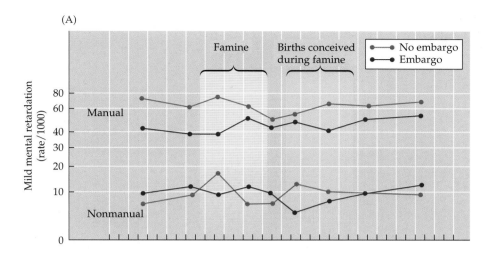

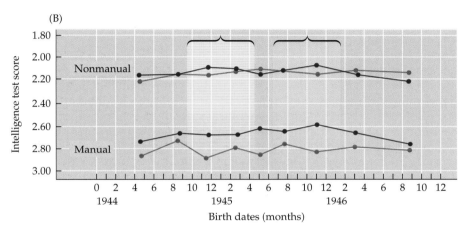

**FIGURE 3.27   Developmental homeostasis in humans.** Maternal starvation has surprisingly few effects on intellectual development in humans, judging from (A) rates of mild mental retardation and (B) intelligence test scores among 19-year-old Dutch men whose mothers lived under Nazi occupation while pregnant. (In this case, the lower the intelligence test score, the greater the intelligence of the subject.) The subjects were grouped according to the occupations of their fathers (manual or nonmanual) and whether their mothers lived or gave birth to them in a city subjected to food embargo by the Nazis (embargo) or in a rural area unaffected by the embargo (no embargo). Those who were conceived or born under famine conditions exhibited the same rates of mental retardation and the same levels of intelligence test scores as men conceived by or born to unstarved rural women. After Stein et al.[1384]

evolved in past environments in which episodes of nutritional deprivation, even starvation, were not uncommon. In this light, it is relevant that fetal malnutrition has less effect on fetal brain development than on the growth of other organ systems. Developmental mechanisms that buffer the growing brain against nutritional trauma testify to the adaptively guided, structured nature of development, which can be thrown off course only by extremely unusual environmental shortfalls or severe genetic deficits.

Developmental homeostasis probably contributes to the development of symmetrical bodies, an adaptive outcome in species in which symmetrical individuals are more likely to acquire mates than their less symmetrical competitors. For example, in the damselfly *Lestes viridis*, mated males tend to have more symmetrical hindwings than their unmated rivals (Figure 3.28).[374] Individuals with symmetrical wings might be better at maneuvering in flight and thus better able to engage in the aerial duels that determine territorial winners and successful reproducers in this damselfly.

Another way in which developmentally buffered, symmetrical males of some species can gain a reproductive advantage is through female mate choice. In the barn swallow, for example, females have been reported to prefer males whose long outer tail feathers are the same length on each side,[1000] while females of the Iberian rock lizard associate preferentially with males endowed with a symmetrical distribution of pheromone-releasing pores on

**FIGURE 3.28** Mating males of the damselfly *Lestes viridis* (red bars) have more symmetrical wings than unmated males (orange bars) at two dates during the breeding season. After De Block and Stoks.[374]

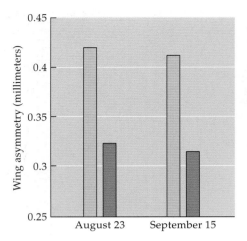

*Lestes viridis*

their thighs.[942] As for humans, some researchers have reported that both men and women find symmetry in facial features appealing (Figure 3.29).[693, 1218] Perhaps prospective mates in these and other species respond positively to body or facial symmetry because these attributes announce the individual's capacity to overcome challenges to normal development.[1001] Disruptions to development caused by mutations or by an inability to secure critical material resources early in life could generate asymmetries in appearance. If body asymmetry reflects suboptimal development of the brain or other important organs, then a preference for symmetrical traits (or attributes closely allied with them) could enable the selective individual to acquire a partner with "good genes" to be transferred to their offspring. Alternatively, the benefit to the choosy individual could be the acquisition of a mate in excellent physiological condition, someone who would be more fertile or better at the delivery of parental care to offspring. In keeping with this prediction, symmetrical young women have significantly higher levels of estradiol and, thus, are presumably more likely to conceive than less symmetrical women, all other things being equal.[715] As for young men, symmetrical individuals are more capable dancers than their less symmetrical counterparts.[198] You can rate the dancers yourself at http://people.brunel.ac.uk/~hsstwmb/.

Debate exists on all aspects of this scenario, however.[1412] Although, as noted, some researchers report that asymmetrical individuals have indeed experienced developmental deficits,[58, 1002] other researchers disagree.[127, 419] Moreover, although symmetrical individuals apparently enjoy a mating advantage in some species, no such advantage has been observed in other species.[1159, 1454] Finally, in some species in which a mate preference for symmetry has been reported, the differences between preferred symmetrical and rejected asymmetrical individuals are often so slight that it seems unlikely that the degree of symmetry per se provides the basis for making the choice. Starlings, for example, have been shown to be simply incapable of perceiving the kinds of very small differences that characterize most naturally occurring body asymmetries in their species.[1410] Furthermore, in our species, when women are asked to rate photographs of men's faces in terms of their attractiveness, their rankings do correlate with male facial symmetry, but the same rankings emerge when the women are provided with photographs of only the left or right side of the face, thereby eliminating information about facial symmetry. These results suggest that facial symmetry correlates with some other feature that women actually use to make their judgments.[1285]

One species that meets all the criteria for visual mate choice based on body symmetry is the brush-legged wolf spider, whose males wave their hairy-

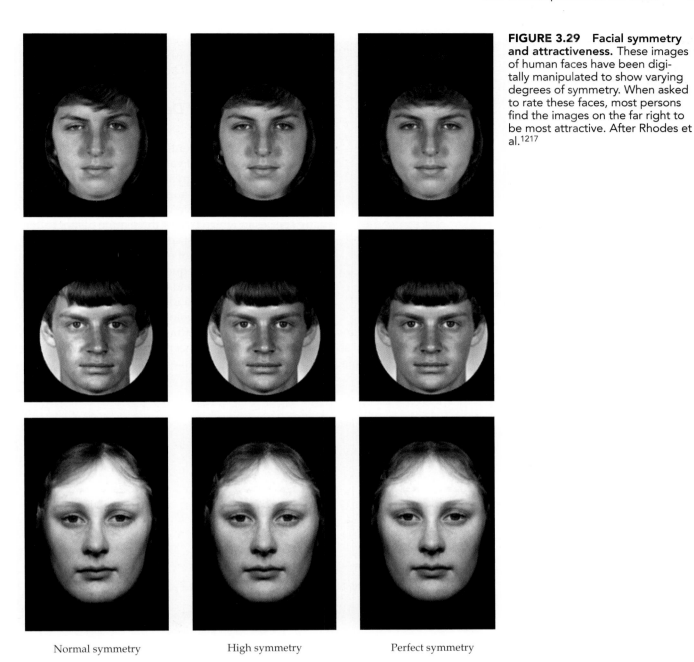

|                    |                    |                    |
| :----------------: | :----------------: | :----------------: |
| Normal symmetry    | High symmetry      | Perfect symmetry   |

**FIGURE 3.29  Facial symmetry and attractiveness.** These images of human faces have been digitally manipulated to show varying degrees of symmetry. When asked to rate these faces, most persons find the images on the far right to be most attractive. After Rhodes et al.[1217]

tufted forelegs at females during courtship. Males that have larger tufts on one leg than the other (asymmetrical males) tend to be smaller and in poorer body condition than those whose tufts are symmetrical. To determine whether tuft symmetry was important, George Uetz and Elizabeth Smith took advantage of the willingness of female wolf spiders to signal their sexual receptivity while watching videos of courting males played on a tiny Sony Watchman micro television (Figure 3.30). Uetz and Smith recorded the reactions of females to digitally manipulated videotapes of a courting spider that were identical in every respect except for the degree to which the male's foreleg tufts were symmetrical. Female spiders signaled their readiness to mate (by raising the abdomen) more often when they saw the symmetrical male, showing that they found this kind of individual more sexually stimulating than the asymmetrical male the researchers had created digitally.[1477] At least in species of this

**FIGURE 3.30 Testing mate choice in a female wolf spider.** The female (on the floor of the arena, to the left) responds to a moving image of a displaying male on the screen of a tiny television (to the right of the female). Photograph by George Uetz.

sort, developmental homeostasis seems highly likely to confer a reproductive advantage by increasing the odds that individuals will be able to attract mates and leave descendants.

### The Adaptive Value of Developmental Switch Mechanisms

The effect of developmental homeostasis is often a restriction in the degree of variation among individuals, which as a result have a greater likelihood of acquiring an adaptive phenotype, such as a symmetrical body, rather than a less effective version of it. But there are many species in which two or three quite distinct alternative phenotypes coexist comfortably, with the differences arising as a result of environmental differences among the individuals in question (Figure 3.31).[1050, 1543] At the proximate level, one challenge associated with such **polyphenisms** is to identify the environmental cues that activate the developmental mechanisms that steer development down one or another pathway (the process of "canalization") so that an individual acquires one or another distinct phenotype, rather than any of a variety of intermediates between the alternative forms.

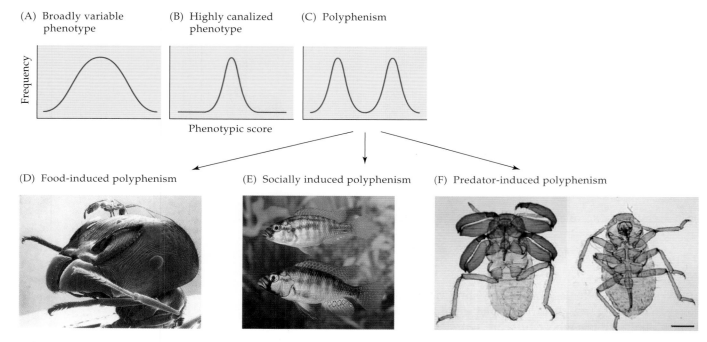

**FIGURE 3.31 Developmental switch mechanisms can produce polyphenisms** within the same species. Different phenotypes can arise when developmental switch mechanisms are activated in response to critical environmental cues. Top panel: Phenotypic variation within a species can range from (A) continuous, broad variation about a single mean value to (B) continuous but narrow variation about a single mean value to (C) discontinuous variation that generates several distinct peaks, each representing a different phenotype. Bottom panel: (D) In some cases, the amount or nature of the food eaten contributes to the production of certain polyphenisms, as in the castes of ants and other social insects. (E) In others, social interactions play a key role in switching phenotypes, as in the territorial and nonterritorial forms of the cichlid *Astatotilapia burtoni*. (F) In still other instances, the presence or activity of predators contributes to the development of an antipredator phenotype, as in the soldier caste (left) of some aphids, which possess more powerful grasping legs and a larger, stabbing proboscis than nonsoldier forms (right). D, photograph by Mark Moffett; E, photograph by Russ Fernald; F, photograph by Takema Fukatsu; from Ijichi et al.[705]

**FIGURE 3.32**  **Tiger salamanders occur in two forms.** The typical form (being eaten by its companion) feeds on small invertebrates and grows more slowly than the cannibal form (which is doing the eating). Cannibals have broader heads and larger teeth than their insect-eating companions. Photograph by David Pfennig, courtesy of James Collins.

A representative example of the phenomenon is provided by a tiger salamander in which there are two immature forms: (1) a typical aquatic larva, which eats small pond invertebrates such as dragonfly nymphs, and (2) a cannibal form, which grows much larger, has more powerful teeth, and feeds on other tiger salamander larvae unfortunate enough to live in its pond (Figure 3.32). The development of the cannibal type, with its distinctive form and behavior, depends on certain factors in the salamanders' social environment. For example, cannibals develop only when many salamander larvae live together.[289] Moreover, they appear more often when the larvae in a pond (or aquarium) differ greatly in size, with the largest individual much more likely to become a cannibal than its smaller companions.[931] In addition, the cannibal form is more likely to develop when the population consists largely of unrelated individuals than when many siblings live together.[1128] If a larger-than-average salamander larva occupies a pond with many other young salamanders that do not smell like its close relatives,[1129] its development may well be switched from the typical track to the one that produces a giant, fierce-toothed cannibal. Thus, at the proximate level, any of several environmental cues can activate the developmental pathway leading to the cannibal form.

What selective advantages do tiger salamanders derive from having two possible developmental pathways and a switch mechanism that enables them to "choose" how to grow and behave? Individuals with some developmental flexibility may do better at coping with an environment with two or three distinct niches than individuals stuck with a one-size-fits-all phenotype. Larval salamanders are faced with two distinctly different sources of potential nutrients: insect prey and their fellow salamanders. If numerous salamander larvae occupy a pond, and if most are smaller than the individual that becomes a cannibal, then shifting to the cannibal phenotype gives that individual access to an abundant food source that is not being exploited by its fellows, so it can grow quickly. But a relatively small individual that was locked into becoming a big-jawed cannibal form would surely starve to death in a pond that lacked numerous potential victims of appropriate size. Because salamanders have no way of knowing in advance which of two food sources will be more available in the place where they happen to be developing and because the two food sources are very different, selection appears to have favored individuals with the ability to develop in either one of two ways depending on the information they receive from their environments.

More generally, any time there are discrete ecological problems to be solved that require different developmental solutions, the stage may be set for the evolution of sophisticated developmental switch mechanisms that enable individuals to develop the phenotype best suited for their particular circumstances. The existence of two nonoverlapping categories of food (large versus small) or two levels of risk (predators present versus predators absent) or an environment in which members of the same species compete for limited food[1130] may select for the kind of developmental mechanism that can produce very different specialist phenotypes, rather than a mechanism that generates a full range of intermediate forms.

So, for example, the fact that males of the cichlid fish *Astatotilapia burtoni* (see Figure 3.31E) are either competitively superior or socially inferior to others helps explain why they have the capacity to shift between two different phenotypes. In this fish, males compete for a mate-attracting territory (a location they defend), with winners holding sites until ousted by a stronger intruder. In such an either/or social environment, it pays to be either aggressively territorial (and to signal that state with bright colors) or nonaggressive (and to signal that state with dull colors).[461, 488] Fish that behave in some intermediate fashion will almost certainly fail to hold a territory against motivated rivals, but they will also fail to conserve their energy, which they can do only by dropping out (at least temporarily) from territorial competition. To this end, the fish respond to changes in their social status with changes in gene activity (Figure 3.33) within specific brain cells;[1550] indeed, when a subordinate male is experimentally given a chance to become dominant (through the removal of a rival), gonadotropin-releasing (GnRH1) nerve cells in the anterior parvocellular preoptic nucleus quickly begin to ramp up the activity of a gene (*egr-1*) that codes for a protein that regulates another gene (*GnRH*). In socially ascendant males, *egr-1* was expressed twice as much in the target cells as in males that were stable subordinates or longtime dominant territorial individuals (Figure 3.34). By a week later, the males had been transformed not only in terms of their appearance and behavior, but also in the size of their GnRH1 neurons and the size of their testes. The initial rapid boost in *egr-1* expression in the brains of previously subordinate males in response to the chance to become dominant appears to act as a trigger for a whole series of genetic and developmental changes. These changes enable an ascendant subordinate to take advantage of his good fortune and become reproductively active while suppressing reproduction in the other males in his neighborhood.[215]

Although polyphenisms are common, they are far from universal, perhaps because many environmental features vary continuously rather than discontinuously. Under these conditions, individuals may not benefit from developmental systems that produce a particular phenotype targeted at a narrow part of the entire range of environmental variation. Instead, selection may favor the ability to shift the phenotype by degrees in such a way as to generate a broad distribution of phenotypes, each one representing an adaptive response to one or more variable factors. So, for example, the final body size of a male redback spider varies considerably in response to variation in available food and the cues associated with virgin female redbacks (potential mates) and male redbacks (rivals for females).[750] When males are reared in the presence of

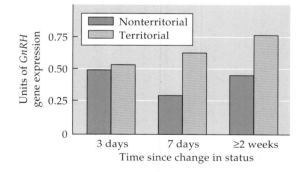

**FIGURE 3.33   Activity of the gene that codes for gonadotropin-releasing hormone** in the cichlid fish *Astatotilapia burtoni*. After males switch from nonterritorial to territorial status, the *GnRH* gene becomes increasingly active over time in certain brain cells. Conversely, those males that switch from territorial to nonterritorial status show reduced activity in the *GnRH* gene. After White, Nguyen, and Fernald.[1550]

(A)

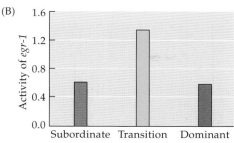

(B)

FIGURE 3.34   **Subordinate males of the fish *Astatotilapia burtoni*** react very quickly to the absence of a dominant rival. (A) Within minutes of the removal of the dominant male, a subordinate may begin to behave more aggressively than before. (B) This change in behavior is correlated with a surge in the activity of a specific gene in the preoptic region of the fish's brain. This gene may initiate a sequence of other genetic changes that provide the male with the physiological foundation for dominance behaviors. Note that the gene *egr-1* ramps up its activity during the transition from subordinate to dominant status but then falls back once the male has become truly dominant. Photograph of two males of *A. burtoni* in an aggressive encounter courtesy of Russell Fernald; data from Burmeister, Jarvis, and Fernald.[215]

virgin females, they develop more quickly and achieve adulthood at a smaller size compared with males reared on the same diet but in the absence of the odor cues associated with virgin females (Figure 3.35). The developmental effects of growing up in the presence of males are exactly the opposite, a fine example of how evolutionary pressures can favor developmental plasticity that produces a wide range of phenotypes within one species.

## Discussion Questions

**3.12**   Identify the probable adaptive basis for the flexible development of body size in the redback spider. Predict what effect large body size must have on female choice in this species versus the effect of large body size on the ability of male redbacks to compete physically with rival males. Check your answer with Kasumovic and Andrade.[750]

**3.13**   Some marine fishes exhibit a spectacular polyphenism in that individuals can, under special circumstances, change their sex from female to male (in other species, the switch goes from male to female). This developmental change involves reproductive organs, hormones, and mating behavior.[1514] A key social cue for the switch in some species is a change in the makeup of the social unit in which the sex-altering individual lives; the removal of a dominant, breeding male from a cluster of females triggers a sex change in the largest female present. Here we have a case of socially induced polyphenism. Identify the apparent restrictions imposed on this system, starting with the most obvious one, namely, the ability to be transformed into a member of the opposite sex rather than some sort of intermediate sex. Speculate on the benefits associated with each restriction.

## The Adaptive Value of Learning

Learning is the adaptive modification of behavior based on experience. As such, it can be considered a polyphenism of sorts because it too confers a highly focused behavioral flexibility that requires developmental modifications in the nervous system. Learning does not produce behavioral change just for the sake of change. Instead, selection favors investment in the mechanisms underlying learning only when there is environmental unpredictability that

(A)

(B)

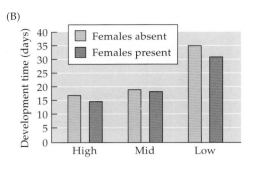

(C)

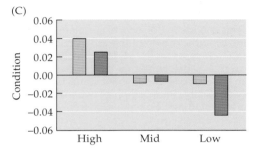

(D)

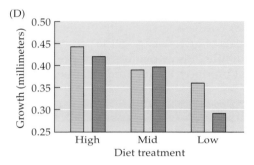

**FIGURE 3.35   Developmental flexibility in redback spiders.** (A) An adult male redback. (B–D) Immature males that grow to adulthood in the presence of females (red bars) develop more rapidly and so reach a smaller adult size in poorer physical condition than immatures that develop in the absence of females (orange bars). Adult males that develop in places without females are likely to encounter male competitors when they reach a web with a female somewhere else. A, photograph by Ken Jones and copyright by M. C. B. Andrade; B–D, after Kasumovic and Andrade.[750]

has reproductive relevance for individuals. What we have here is another cost–benefit argument, which presupposes that any proximate mechanisms that enable individuals to learn come with a price tag. We can check this assumption by predicting, for example, that the brains of male long-billed marsh wrens living in the western United States should be larger than those of their East Coast counterparts because young West Coast wrens learn nearly 100 songs by listening to others, whereas East Coast wrens have much smaller learned repertoires of about 40 songs.[808] When the birds' brains were examined, the song control systems of West Coast wrens weighed on average 25 percent more than the equivalent nuclei of East Coast wrens.

If learning mechanisms are costly, then we can expect learning to evolve only when there is some major counterbalancing benefit. As mentioned at the outset of this chapter, people tend to think of insects as instinct-driven automatons, but honey bees are, for example, quite capable of learning where to search for food; what odors, shapes, and colors are associated with different pollen- or nectar-producing flowers; when during the day a particular plant species will open its flowers; how to get back to the hive after a foraging expedition; and much more.[1500] These abilities are all related to the fact that the conditions a worker bee will encounter cannot be precisely predicted before she goes out to forage. Instead, selection has favored a bee brain that can incorporate information about key variables in the bee's environment—information that alters genetic activity in brain cells, changing the structure of the brain,

(A)

(B)

**FIGURE 3.36** **Male thynnine wasps can be deceived into "mating" with a flower.** (A) A wingless female thynnine wasp that is releasing a sex pheromone to attract males. (B) Some Australian orchids possess flowers with a female decoy petal that can stimulate males into attempting to copulate with it. The orchid may secure a pollinator in this fashion. Note the yellow pollen sacs stuck to the male's back. Photographs by the author.

and ultimately modifying the behavior of the individual so that she can better exploit the particular pattern of food resources in her neighborhood.

Likewise, male thynnine wasps exhibit a special spatial learning ability that comes into play when a mimetic sex pheromone is released by a freshly opened orchid flower. Various orchids possess flowers with female decoy petals that smell and look vaguely like female thynnine wasps. A male can be fooled into rushing to these flowers and attempting to mate with the petal (Figure 3.36);[1392] indeed, in some cases, males are so stimulated by the experience that they ejaculate upon grasping the orchid's decoy petal.[522] When a tricked male comes to a second orchid, he will, if fooled again, transfer pollen from orchid 1 to orchid 2. But having once been deceived by a particular flower, male wasps sometimes learn to avoid the spot where that flower occurs, which explains why, when researchers move an orchid to a new spot, large numbers of males show up initially but then fly away and do not return (Figure 3.37).

Male thynnine wasps evidently store information about the locations of pseudofemales and will avoid responding to the scent coming from those sites.[1112] The reproductive benefits of the male wasp's behavioral flexibility are clear. Male wasps cannot be programmed in advance to know where female wasps and deceptive orchid flowers are on any given day. By using experience to learn where particular orchids are (in order to avoid them) while remaining responsive to novel sources of sex pheromone, the male wasp saves time and energy and improves his chance of encountering a receptive female that has begun to release sex pheromone.

That spatial learning evolves in response to particular ecological pressures can also be seen by comparing the learning abilities of four bird species, all members of the crow family (Corvidae), that vary in their predisposition to store food—a task that puts a premium on spatial memory. As we have seen, Clark's nutcracker is a food-storing specialist, and it has a large pouch for the transport of pine seeds to storage sites. The pinyon jay also has a special anatomical feature, an expandable esophagus, for carrying large quantities of seeds to hiding places. In contrast, the scrub jay and Mexican jay lack special seed transport devices and appear to hide substantially less food than their relatives.

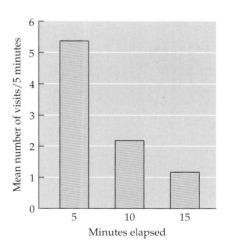

**FIGURE 3.37** **Male thynnine wasps can learn to avoid being deceived by an orchid.** The frequency of visits to a deceptive orchid soon falls after the male wasps in an area have interacted with it and learned that an unrewarding source of sex pheromone is associated with that particular location. After Peakall.[1112]

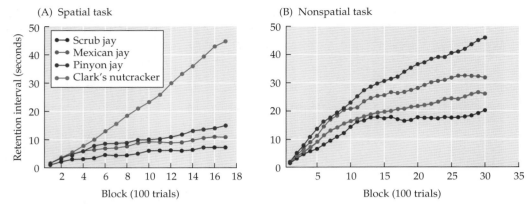

**FIGURE 3.38** **Spatial learning abilities differ among members of the crow family.** (A) Captive Clark's nutcrackers performed much better than three other corvid species in experiments that required the birds to retain information about the location of a circle. (B) But when the birds' ability to remember the color of a circle was tested, the nutcrackers did not excel in this nonspatial learning task. After Olson et al.[1069]

Individuals from the four species were tested on two different learning tasks in which they had to peck a computer screen to receive rewards. One task required the birds to remember the color of a circle on the screen (a nonspatial learning task), and the other required them to remember the location of a circle on the screen (a spatial task). When it came to the nonspatial learning test, pinyon jays and Mexican jays did substantially better than scrub jays and nutcrackers. But in the spatial learning experiment, the nutcracker went to the head of the class, followed by the pinyon jay, then the Mexican jay, and finally the scrub jay (Figure 3.38), but see de Kort and Clayton[375] and Pravosudov and de Kort.[1165] These results suggest that the birds have not evolved all-purpose learning abilities; instead, their learning skills are designed to promote success in solving the special problems that they face in their natural environments.[1069]

The logic of an evolutionary approach to learning leads us to expect that if males and females of the same species differ in the benefits derived from a particular learned task, then a sex difference in learning skills should evolve. The pinyon jay provides a case in point. As just noted, the jay hides large numbers of pinyon seeds, when they are available; it retrieves them up to 5 months later when food is scarce. But males are more likely than females to have to relocate old caches because they provide their mates and young with recovered food while their female partners spend their time instead at the nest, incubating their eggs and young, rather than searching for stored pinyon nuts. As predicted, males appear to have evolved better long-term memory than females. When captive birds of both sexes were tested during what was the nesting season, males made fewer errors than females when trying to find their own and their mates' caches, which they had hidden months earlier (Figure 3.39).[421]

The sex differences hypothesis has also been tested by Steve Gaulin and Randall FitzGerald in their studies of spatial learning in three species of voles, all members of the genus *Microtus*. Males of the polygynous meadow vole move about in areas more than four times as large as those occupied by each of their several mates. In contrast, males and females of the monogamous prairie vole (see Figure 1.1) and the monogamous pine vole share the same living space. When tested in a variety of mazes, which the animals had to solve in order to receive food rewards, males of the wide-ranging meadow vole consistently made fewer errors than females of their species. Given these findings, you should not be surprised to learn that male meadow voles invest more heavily in the hippocampus than do females of their species.[708] In the monogamous prairie

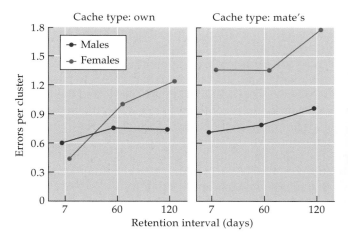

Cache type: own  Cache type: mate's

**FIGURE 3.39  Male pinyon jays make fewer errors** than females do when retrieving seeds from caches they (or their mates) have made, especially after intervals of 2 to 4 months. This result accords with expectation, because females are the incubators of eggs and youngsters while the males provide the female and offspring with seeds relocated in caches made up to several months previously. After Dunlap et al.[421]

Pinyon jay

vole, however, males and females did equally well on these spatial learning tests (Figure 3.40).[524] The same is true for the monogamous pine vole,[523] a species in which the sexes do not differ in hippocampal size.[708]

## Discussion Question

**3.14**  In a study in which men and women were asked to sit at a computer and navigate through a virtual maze (Figure 3.41), the men were able to complete the task more quickly and with fewer errors over five trials than the women.[996] The conclusion that men do better at location learning than women has been supported by other research as well (e.g., Jones and Healy[736]). (Note, however, that in other tests, involving language skills, women score higher on average than men.) What possible proximate developmental mechanisms might be responsible for this sex difference in navigational ability? Keeping in mind the evolutionary explanation for sex differences in spatial learning ability in voles, what prediction can you make about the nature of human mating systems over evolutionary time?

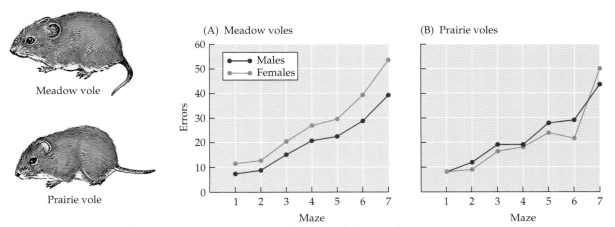

Meadow vole

Prairie vole

**FIGURE 3.40  Sex differences in spatial learning ability are linked to home range size.** Spatial learning in voles was tested by giving individuals opportunities to travel through seven different mazes of increasing complexity in the laboratory and then letting them run through each maze again. (A) Polygynous male meadow voles, which roam over wide areas in nature, consistently made fewer errors (wrong turns) on average than the more sedentary females of their species. (B) In contrast, females matched male performance in the monogamous prairie vole, a species in which males and females live together on the same territory. After Gaulin and FitzGerald.[524]

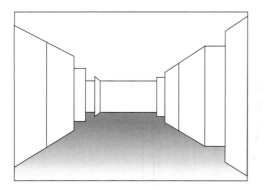

**FIGURE 3.41 A virtual maze used for computer-based studies of navigational skills.** After Moffat, Hampson, and Hatzipantelis.[996]

In a species in which females face greater spatial challenges than males, we would expect females to make larger investments in the expensive neural foundations of spatial learning. The brown-headed cowbird is such a species because cowbirds are **brood parasites** that lay their eggs in other birds' nests. A female must search widely for nests to parasitize, and she must remember where potential victims have started their nests, in order to return to them up to several days later when the time is ripe for her to add one of her eggs to those already laid. In contrast, male cowbirds do not confront such difficult spatial problems. As predicted, the hippocampus (but no other brain structure) is considerably larger in female brown-headed cowbirds than in males. No such difference occurs in some nonparasitic relatives of this species (Figure 3.42).[1320]

Moreover, it is not just spatial learning that bears the clear imprint of natural selection. Consider **operant conditioning**, in which an animal learns to associate a voluntary action with the consequences that follow from that action.[1341] Operant conditioning (or trial-and-error learning) does occur outside psychology laboratories, but it has been studied extensively in Skinner boxes, named after the psychologist B. F. Skinner. After a white rat has been introduced into a Skinner box, it may accidentally press a bar on the wall of the cage (Figure 3.43), perhaps as it reaches up to look for a way out. When the bar is pressed down, a rat chow pellet pops into a food hopper. Some time may pass before the rat happens upon the pellet. After eating it, the rat may continue to explore its rather limited surroundings for a while before again happening to press the bar. Out comes another pellet. The rat may find it quickly this time, and then turn back to the bar and press it repeatedly, having learned to associate this particular activity with food. It is now operantly conditioned to press the bar.

Skinnerian psychologists once argued that one could condition with equal ease almost any operant (defined as any action that an animal could perform). Indeed, the successes of operant conditioning are legion, including the ability to get nutcrackers and jays to become computer users, as mentioned above. White rats also can be conditioned to do all sorts of things in the laboratory, such as avoid novel, distinctively flavored foods or fluids after they are exposed to nausea-inducing X-ray radiation. However, John Garcia and his colleagues found that the ability of these animals to learn to avoid certain punishing foods or liquids had some specifications.[518, 519] The degree to which an irradiated rat rejects a food or fluid is proportional to (1) the intensity of the resulting illness, (2) the intensity of the taste of the substance, (3) the novelty of the substance, and (4) the shortness of the interval between consumption and illness.[519] But even if there is a long delay (up to 7 hours) between eating a distinctively flavored food and exposure to radiation and consequent illness, the rat still links the two events and uses the information to modify its behavior.

In contrast, white rats never learn that a distinctive sound (a click) is a signal that always precedes an event associated with nausea. Nor can rats easily make an association between a particular taste and shock punishment (Figure 3.44). If, after drinking a sweet-tasting fluid, the rat receives a shock on its feet, it often remains as fond of the fluid as it was before, as measured by the amount drunk per unit of time, no matter how often it is shocked after drinking this liquid. These failures surely relate to the fact that in nature, particular sounds are never associated with illness-inducing meals, any more than the consumption of certain fluids causes a rat's feet to hurt.

Understanding the natural environment of the ancestor of white rats, the Norway rat, also helps explain why white rats

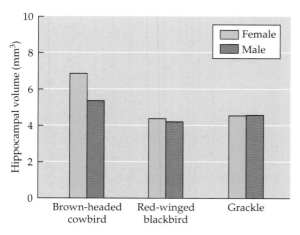

**FIGURE 3.42 Sex differences in the hippocampus.** Female brown-headed cowbirds have a larger hippocampus than males, as would be expected if this brain structure promotes spatial learning and if selection for spatial learning ability is greater on female than on male cowbirds. Red-winged blackbirds and common grackles do not exhibit this sex difference. After Sherry et al.[1320]

**FIGURE 3.43  Operant conditioning exhibited by a rat in a Skinner box.** The rat approaches the bar (top left) and then presses it (top right). The animal awaits the arrival of a pellet of rat chow (bottom left), which it consumes (bottom right), so the bar-pressing behavior is reinforced. Photographs by Larry Stein.

are so adept at learning to avoid novel foods with distinctive tastes that are associated with illness, even hours after ingesting the food. Under natural conditions, a Norway rat becomes completely familiar with the area around its burrow, foraging within that area for a wide variety of foods, plant and animal.[885] Some of these foods are edible and nutritious; others are toxic and potentially lethal. A rat cannot clear its digestive system of toxic foods by vomiting. Therefore, the animal takes only a small bite of anything new. If it gets sick later, it avoids this food or liquid in the future, as it should because eating large amounts might kill it.[519] This case suggests that even what appears to be a general, all-purpose form of learning is actually a specialized response to particular kinds of biologically significant associations that occur in nature.

If this argument is correct, then other mammals that are dietary generalists, which also run the risk of consuming dangerous, toxic items, should behave

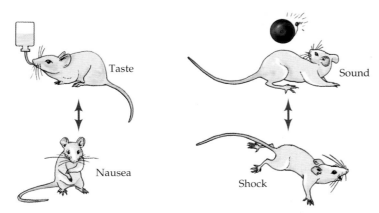

**FIGURE 3.44  Biases in taste aversion learning.** Although white rats can easily learn that certain taste cues will be followed by sensations of nausea and that certain sounds will be followed by skin pain caused by shock, they have great difficulty forming learned associations between taste and consequent skin pain or between sound and subsequent nausea. After Garcia, Hankins, and Rusiniak.[519]

**FIGURE 3.45 Vampire bats cannot form learned taste aversions.** Instead they continued to consume a flavored fluid even if, immediately after accepting this novel substance, they were injected with a toxin that caused gastrointestinal distress. In contrast, three insect-eating bat species completely rejected the novel dietary item when it was combined with injection of the toxin, no matter whether this was done immediately after feeding or after a delay. Two control groups were also used in the experiment, one in which the consumption of the novel food was paired with a harmless injection of saline solution, and another in which the toxin was injected but not in conjunction with feeding on the fluid. After Ratcliffe, Fenton, and Galef.[1198]

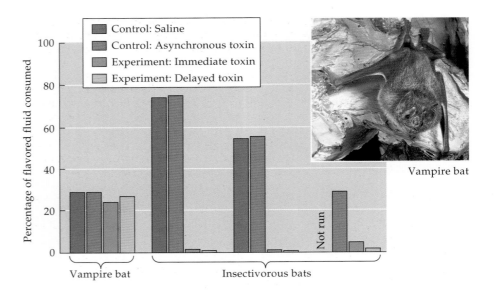

Vampire bat

like the Norway rat, which is to say that they should also quickly form taste aversions to bad-tasting, illness-inducing items. And they do. Three bat species that feed on a range of foods behaved in the predicted manner: they rapidly formed taste aversions when fed a meal laced with an unfamiliar flavor, cinnamon or citric acid, before being injected with a chemical that made them vomit. When later offered a choice between food with and without cinnamon or citric acid, these three generalists avoided foods spiced with the novel additives.[1198]

In contrast, dietary specialists, which concentrate exclusively on one or a very few safe foods, should be unable to acquire taste aversions in this manner. The vampire bat, a blood-feeding specialist, is in fact quite incapable of learning that consumption of an unusual-tasting fluid will lead to gastrointestinal distress (Figure 3.45).[1198] The difference between the specialist vampire bat and its generalist relatives supports the hypothesis that taste aversion learning is an evolved response to the risk of food or fluid poisoning. Just as is true for all aspects of behavioral development, the changes associated with learned behavior are worth the cost only if they confer a net fitness benefit on individuals capable of modifying their behavior in a particular way.

## Summary

1. The development of any trait is the result of an interaction between the genotype of a developing organism and its environment, which consists not only of the food it receives and the metabolic products produced by its cells (the material environment), but also its sensory experiences (the experiential environment). The value of genetic information lies in the ability of genes to respond to signals from the environment by altering their activity, leading to changes in the gene products available to the developing organism.

2. Because development is interactive, no measurable product of development (a phenotype) can be genetically determined. The statement "in garter snakes, there is a gene for eating banana slugs" is shorthand for the following: A particular allele in a garter snake's genotype codes for a distinctive protein; if the protein is actually made, which requires an interaction between the gene and its environment, the protein may influence the development or operation of the physiological mechanisms underlying the snake's ability to recognize slugs as food.

3. By the same token, the interactive nature of development means that no phenotype can be purely environmentally determined. The statement "the rat's learned avoidance of rat poison is caused by its experience with this chemical" is shorthand for the following: A specific experience with rat poison led to chemical changes in the rat's body, which were translated eventually into chemical changes in the rat's brain cells. These changes, in turn, altered the pattern of genetic activity in some parts of its nervous system and thereby modified the rat's response to the poison upon a second encounter with this stimulus.

4. Because development is interactive, changes in either the genetic information or the environmental inputs available to an individual can potentially alter the course of its development by changing the gene–environment interactions that take place within that individual. Therefore, the behavioral differences between two individuals can be genetically determined or environmentally determined or both. Note that this claim is very different from the misconception that a given behavioral phenotype is caused either by an animal's genes or by its environment alone.

5. Because some differences between individuals are caused by genetic differences, populations have the potential to evolve by natural selection, which acts on genetic variation within groups.

6. Because behavior can evolve, we expect to find that behavioral development has adaptive features. One such feature is developmental homeostasis, the capacity of the developmental process to ignore or overcome certain environmental or genetic shortfalls that might conceivably prevent animals from acquiring valuable traits that provide reproductive success. Indeed, normal physiological and behavioral phenotypes often develop in animals growing up in challenging, suboptimal environments and in animals burdened with potentially harmful mutations.

7. Other adaptive features include developmental switch mechanisms, which guide development into one of two or three alternative developmental pathways in response to specific environmental cues. Each of the resulting phenotypes can master the distinctive obstacles to success associated with its particular niche within the larger environment.

8. Another adaptive aspect of development involves learning mechanisms that can respond to particular environmental inputs related to individual experience and that generate functional changes in the behavior of animals. Learning, like other forms of developmental flexibility, reflects past selection for the capacity to make adaptive adjustments in behavior that match the environment of the individual.

## Suggested Reading

Gene Robinson discusses how to integrate molecular biology, developmental biology, neurobiology, and evolutionary biology in the context of the nature–nurture debate.[1229] Environmental effects on development are nicely illustrated by studies of kin and individual discrimination, a topic that is the subject of a collection of papers edited by Philip Starks.[1383] The relation between genes and migratory behavior has been reviewed by Francisco Pulido,[1176] while Ralph Greenspan examines some alternative approaches to behavior genetics in general.[575] See also Fitzpatrick et al.[469] You can gain a sense of the controversy about how to analyze the development of human behavior by reading papers by Thomas Bouchard [151, 152] in conjunction with a counterview from Marla Sokolowski and Doug Wahlsten.[1365] Jeremy Gray and Paul Thompson review a range of approaches to the genetics of human intelligence.[571]

# 4

# The Control of Behavior: Neural Mechanisms

B ecause nervous systems are the foundation for animal behavior, behavioral biologists have been eager to learn how these systems work, a topic that we first considered in the context of bird song and bird brains (see Chapter 2). Although our current understanding of these systems has been greatly helped by the application of sophisticated technologies, even simple observations of animals in action can sometimes provide considerable information about the properties of neural mechanisms. Consider that males of the bee *Centris pallida* will sometimes try to copulate with a person's thumb, as I learned one day after unkindly pulling a male bee from his sexual partner, the better to measure his size with a pair of calipers. In the midst of this exercise, I found more or less by accident that if I perched the male on my upturned thumb (Figure 4.1), he would grasp it firmly and stroke it with his legs and antennae as if he were holding a female of his species. (Incidentally, male bees don't sting, so my actions were neither courageous nor stupid.) Despite the fact that my thumb has only the vaguest similarity to a female *C. pallida*, it is evidently close enough for males of this bee.

(A)

(B)

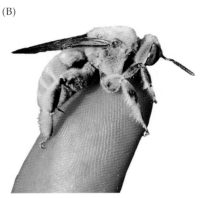

**FIGURE 4.1** A complex response to simple stimuli. (A) A male bee copulating with a female of his species. (B) A male of this species attempts (unsuccessfully) to copulate with the author's thumb. Photographs by the author.

Even though I am not a neurophysiologist, I could see that the bee's nervous system had some special operating rules. Apparently, when a sexually motivated male *C. pallida* grasps an object approximately the size of a female of its species, the sensory signals generated by its touch receptors travel to other parts of its nervous system, where messages are produced that eventually translate into a complex series of muscle commands. The behavioral result is the sequence of movements that passes for courtship in *C. pallida*. That these activities can be stimulated by my thumb instead of a female bee indicates that the nervous system of a male *C. pallida* is not terribly discriminating. Nor is "my" bee at all unusual in this respect, given that males of the ivy bee, *Colletes hederae*, will attempt to mate with a mass of tiny blister beetle larvae (Figure 4.2).[1490]

Cases of this sort show that nervous systems can generate complex responses to very simple stimuli. This phenomenon has attracted the attention of those who wish to figure out how **neurons** (nerve cells) or neural networks acquire information from objects in the environment and then order the nervous system's owner to respond in particular ways. Thanks to research of this sort, we now know a great deal about such things as how moths flying at night can detect and avoid hungry bats rushing in for the kill, how the star-nosed

**FIGURE 4.2** A male of the solitary bee *Colletes hederae* attracted to a cluster of larval blister beetles (some of which are indicated with the arrow). After attempting to mate with the mass of beetles, the bee becomes covered in the tiny larvae, which climb aboard for transport to a female of this bee, should the male succeed in finding a mate of the appropriate species. Photograph by Nicolas Vereecken.

mole uses its amazing nose to locate tasty worms in its underground tunnels, and how birds, butterflies, and sea turtles can accurately navigate over great distances to reach particular destinations. The proximate mechanisms that make these feats possible have adaptive properties that help individuals survive and reproduce in the environments utilized by their species. The interaction between proximate and ultimate analyses of behavioral control systems is the focus of this chapter.

## How Neurons Control Behavior

The study of how nerve cells activate behavior took a step forward when Niko Tinbergen began his work on the link between simple stimuli and complex responses in gulls.[1445] When he waved a small stick with black and white bands on its tip in front of a baby gull, the chick often pecked at the bands in exactly the same way that it would peck at the tip of its parent's bill (Figure 4.3). The adult gull typically reacts to these pecks by regurgitating a half-digested fish or other delicacy, which the chick enthusiastically consumes. You might think that a baby gull would require a living three-dimensional gull in order to run through its bill-pecking routine, but painted sticks and two-dimensional cardboard cutouts of a herring gull head elicit the pecking reaction (Figure 4.4).[1444] Experiments with these and other models have revealed that herring gull chicks, at least very young ones, apparently ignore almost everything except the shape of the "bill" and the red dot at the end of the beak. Tinbergen proposed that when a young gull sees certain key stimuli, sensory signals are relayed by neurons to its brain, where other neurons eventually generate the motor commands that cause the chick to peck at the stimulus—whether it is located on its mother's bill or a piece of cardboard or the end of a stick.

Tinbergen and his friend Konrad Lorenz collaborated on another famous experiment that identified a simple stimulus capable of triggering a complex behavior. They found that if they removed an egg from under an incubating greylag goose and put it a half meter away, the goose would retrieve the egg by stretching its neck forward, tucking the egg under its lower bill, and rolling

**FIGURE 4.3  Begging behavior by a gull chick.** A silver gull chick is being fed regurgitated food by its parent after pecking at the adult's bill. Photograph by the author.

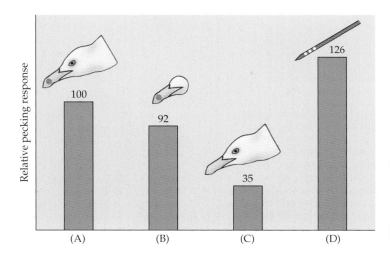

**FIGURE 4.4  Effectiveness of different visual stimuli** in triggering the begging behavior of young herring gull chicks. A two-dimensional cardboard cutout of the head of a gull with a red dot on its bill (A) is not much more effective in eliciting begging behavior in a gull chick than is a model of the bill alone (B), provided the red dot is present. Moreover, a model of a gull head without the red dot (C) is a far less effective stimulus than is an unrealistically long "bill" with contrasting bars at the end (D). After Tinbergen and Perdeck.[1444]

the egg carefully back into its nest. If they replaced the goose egg with almost any roughly egg-shaped object, the goose would invariably run through its egg retrieval routine. And if the researchers removed the object as it was being retrieved, the bird would continue pulling its head back just as if an egg were still balanced against the underside of its bill.[1445] From these results, Tinbergen and Lorenz concluded that the goose must have a special perceptual mechanism that is highly sensitive to certain visual cues provided by eggs (and other objects of similar shape). Moreover, that sensory mechanism must send its information to neurons in the brain that automatically activate a more or less invariant motor program for egg retrieval.

The gull chick's pecking response and the greylag goose's retrieval behavior are only two of many instincts that Tinbergen and Lorenz studied. These founders of **ethology**, the discipline dedicated to the study of both the proximate and ultimate causes of animal behavior, were especially interested in the instincts exhibited by wild animals living under natural conditions. An **instinct** can be defined as a behavior pattern that appears in fully functional form the first time it is performed, even though the animal may have had no previous experience with the cues that elicit the behavior. You may recall an example from Chapter 3: the tongue-flicking response of baby coastal garter snakes to banana slug extract. You may also remember that tongue flicking cannot be "genetically determined," nor can any other instinct, because these behaviors are dependent on the gene–environment interactions that took place during development. In the case of a herring gull chick, these interactions led to the construction of a nervous system that contains a network that enables the little bird to identify the key components of an adult gull's bill and to peck at the red dot on that bill. The neural network responsible for detecting the simple cue (the **sign stimulus** or **releaser**) and activating the instinct, or **fixed action pattern** (FAP), was given the label **innate releasing mechanism** (Figure 4.5).[1445]

The simple relationship between an innate releasing mechanism, sign stimulus, and FAP is highlighted by the ability of some species to exploit the

(A)

Niko Tinbergen

(B)

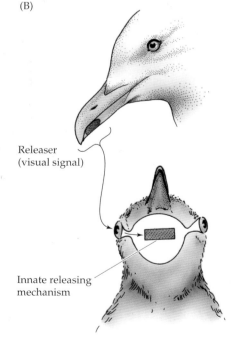

Releaser
(visual signal)

Innate releasing
mechanism

**FIGURE 4.5  Instinct theory** was developed by (A) Niko Tinbergen and Konrad Lorenz (see Figure 3.5). They proposed that simple stimuli, such as (B) the red dot on a parent gull's bill, can activate or release complex behaviors, such as a gull chick's begging behavior. This effect is achieved, according to these ethologists, because certain sensory messages from the releaser are processed by innate releasing mechanisms (neuronal clusters) higher in the nervous system, leading to motor commands that control a fixed action pattern, a preprogrammed series of movements that constitute an adaptive reaction to the releasing stimulus. Photography by B. Tschanz.

FAPs of other species, a tactic known as code breaking.[1562] In the preceding chapter, we discussed an example—orchids whose flower petals may provide the visual, tactile, and olfactory releasers that trigger attempted copulation by certain male wasps, much to the insects' disadvantage (see Figure 3.36). In at least one species of orchid, the plant produces an entire library of volatile chemicals similar or identical to those produced by the females of the pollinating species, and these chemicals attract the males of this species and induce them to "mate" with the orchid.[923] Likewise, the blister beetle larvae mentioned earlier use scents that mimic those released by receptive female ivy bees. When a male ivy bee pounces on a ball of beetle larvae, he gets covered with little parasites that may later be transferred to the female bee, if the male is fortunate enough to find a real sexual partner. After moving onto a female, the larvae will eventually be transported to the underground nest of the bee, where they can drop off and make their way into the food-containing brood cells. There they consume the provisions the mother bee has collected and stored for her own offspring.[1490]

The close similarity between the olfactory cues provided by a deceptive signaler and those of another species has been meticulously documented in the case of the Alcon blue butterfly, a European species whose caterpillar larvae smell very much like the larvae of two ant species (Figure 4.6). As a result of their chemical mimicry, a baby caterpillar that has just hatched from an egg laid on a plant may attract an ant worker of either of the two species. The duped ant will then cart the caterpillar back to its nest, where other worker ants will assiduously feed and protect the young butterfly as if it were an immature ant.[1029]

**FIGURE 4.6   A chemical code breaker.** (A) The larva of the Alcon blue butterfly has attracted an ant (*Myrmica rubra*), which has picked the caterpillar up and is in the process of transporting it back to the ant colony's nest where it will be cared for by other ants. The inset shows an adult Alcon blue butterfly. (B) The larva's success in deceiving ants stems from its chemical mimicry of the scents present on the cuticle of the ant that acts as its host. The top gas chromatogram shows the compounds present in the butterfly cuticle (each spike represents a particular compound); the middle chromatograms come from two species of ants that are fooled into treating the butterfly larvae as larval ants of their kind; the lower chromatogram shows the chemical compounds found in the cuticle of a third species of *Myrmica* ant, which does not respond to Alcon blue butterfly larvae. A, photograph by David Nash. B, after Nash et al.[1029]

**FIGURE 4.7 A visual and acoustic code breaker.** This young cuckoo begs for food from its foster parent, a reed warbler, which provides for the cuckoo at great cost to itself and its own offspring. Photograph by Ian Wyllie.

Code breaking has also been mastered by the offspring of parasitic birds, such as the European cuckoo and North American cowbird, whose adult females deposit their eggs in the nests of other bird species. After the parasite's egg hatches, the baby cuckoo or cowbird exploits its host by supplying it with the acoustical and visual signals that the adult birds usually use to decide which of their own nestlings to feed.[361, 1562] Typically, a parent songbird that has returned to its brood with food favors cheeping youngsters that are able to reach up high, with head moving and mouth gaping. Cuckoo and cowbird nestlings grow rapidly and become larger than their hosts' own offspring and therefore can generate these releasers of parental feeding better than their smaller nestmates.[861] Because the parasite begs for food so effectively, it gets more than its fair share, eventually growing into a demanding youngster far larger than its duped caregivers (Figure 4.7).

## Discussion Questions

**4.1** Suggest how a modern behavioral biologist might explore the effect of a releaser, such as a red dot on a moving gull bill, in terms of changes in gene expression and neural activity in selected portions of the brain of the gull chick. It might be helpful to review the relationships between bird song, neural networks, and the *ZENK* gene as presented in Chapter 2.

**4.2** Males of various beetles have been seen trying to copulate with everything from beer bottles to large yellow signs (Figure 4.8). Apply ethological terminology (see Figure 4.5) to these cases by identifying the releaser, the fixed action pattern, and the innate releasing mechanism. Then develop an ultimate hypothesis to account for what clearly is maladaptive behavior on the part of these obtuse beetles (which sometimes die rather than leave the inanimate and unresponsive copulatory partners that they have chosen).

## *Sensory Receptors and Survival*

You do not need to use ethological jargon in order to study how simple cues trigger generally adaptive, but occasionally exploited, responses in animals of all sorts. Kenneth Roeder, for example, studied how night-flying moths manage to escape from bats without referring to innate releasing mechanisms and the like. His interest in the neural control of escape behavior was piqued when

(A)

(B)

**FIGURE 4.8** **Male beetles trying to mate with objects other than female beetles.** This large Australian beetle will attempt to mate with any object with approximately the same color as a female of his species, such as (A) a beer bottle or (B) a telecommunication sign. A, photograph courtesy of Darryl T. Gwynne; B, photograph by the author.

he spent some time outdoors at night in summer, armed with "a minimum amount of illumination, perhaps a 100-watt bulb with a reflector, and a fair amount of patience and mosquito repellent."[1234] With these items in place, you too might see a moth that has been attracted to the light turn abruptly away right before a bat swoops into view. Or you might see a moth dive straight down just as a bat appears. You can also sometimes get moths approaching a light to turn away abruptly, or even plummet earthward, if you jangle a set of keys. The moths' responses suggest that they can detect an acoustical cue, which provides the trigger for a particular behavior, just as simple visual cues are sufficient to activate begging behavior in a baby gull.

As it turns out, the hypothesis that acoustical stimuli trigger the turning or diving behavior of moths is correct, but the sounds that the moth hears when the keys are rattled together are not the sounds that you and I hear. Instead, the moth detects the very high-frequency sounds produced by the clashing keys, which makes sense when you consider that most night-hunting bats vocalize ultrasonically using sound frequencies between 20 and 80 kilohertz (kHz)—well outside the hearing range of humans, but not moths.

Bats use ultrasonic calls to navigate at night—something that was not suspected until the 1930s, when researchers with ultrasound detectors were able to eavesdrop on the sounds produced by flying bats. At that time, Donald Griffin suggested that night-flying bats use high-frequency cries in order to listen for weak ultrasonic echoes reflected back from objects in their flight paths.[581] Skeptics of the echolocation hypothesis came round after reading about Griffin's experiments with little brown bats, a common North American species. When Griffin placed captive bats in a dark room filled with fruit flies and wires strung from ceiling to floor, his subjects had no trouble catching the insects while negotiating the obstacle course—until Griffin turned on a machine that filled the room with high-frequency sound. As soon as the machine-produced ultrasound bombarded the bats, they began to collide with obstacles and crash to the floor, where they remained until Griffin turned off the jamming device. In contrast, loud sounds of 1 to 15 kHz (which humans can hear) had no effect on the bats because these stimuli did not mask the high-frequency echoes from objects in the room. Griffin rightly concluded that the little brown bat employs a sonar system to avoid obstacles and detect prey at night.

(A)

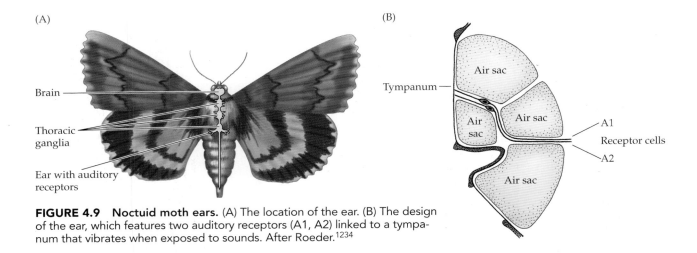

Brain

Thoracic ganglia

Ear with auditory receptors

(B)

Tympanum

Air sac

Air sac

Air sac

A1

Receptor cells

A2

Air sac

**FIGURE 4.9 Noctuid moth ears.** (A) The location of the ear. (B) The design of the ear, which features two auditory receptors (A1, A2) linked to a tympanum that vibrates when exposed to sounds. After Roeder.[1234]

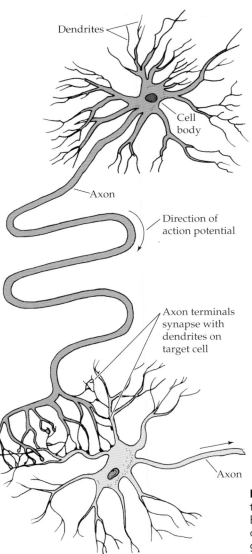

Dendrites

Cell body

Axon

Direction of action potential

Axon terminals synapse with dendrites on target cell

Axon

As Roeder watched moths evading echolocating bats, he felt sure that the insects were able to hear pulses of bat ultrasound. He knew that if he was right, then he should be able to find ears on some moths. Taking advantage of some earlier work on moth hearing organs, he located the structures that noctuid moths use to hear ultrasound (Figure 4.9). As it turns out, a noctuid moth has two ears, one on each side of the thorax. Each ear consists of a thin, flexible sheet of cuticle—the tympanic membrane, or tympanum—lying over a chamber on the side of the thorax. Attached to the tympanum are two neurons, the A1 and A2 auditory receptors. These receptor cells are deformed when the tympanum vibrates, which it does when intense sound pressure waves sweep over the moth's body. Roeder decided to focus his attention on these auditory receptors, using a cellular approach to get at the proximate basis of moth behavior.

The moth's A1 and A2 receptors work in much the same way that most neurons do: they respond to the energy contained in selected stimuli by changing the permeability of their cell membranes to positively charged ions. The effective stimuli for a moth auditory receptor appear to be provided by the movement of the tympanum, which mechanically stimulates the receptor cell, opening stretch-sensitive channels in the cell membrane. As positively charged ions flow in, they change the electrical charge inside the cell relative to the charge on the other side of the membrane. If the inward movement of ions is sufficiently great, a substantial, abrupt, local change in the electrical charge difference across the membrane may occur and spread to neighboring portions of the membrane, sweeping around the cell body and down the axon—the "transmission line" of the cell (Figure 4.10). This brief, all-or-nothing change in electrical charge, called an **action potential**, is the signal that one neuron uses to communicate with another.

**FIGURE 4.10 Neurons and their operation.** This diagram illustrates the structure of a generalized neuron with its dendrites, cell body, axon, and synapses. Electrical activity in a neuron originates with the effects of certain stimuli on the dendrites. Electrical changes in a dendrite's cell membrane can, if sufficiently great, trigger an action potential, which begins near the cell body and travels along the axon toward the next cell in the network.

When an action potential arrives at the end of an axon, it may cause the release of a neurotransmitter at this point. This chemical signal diffuses across the narrow gap, or **synapse**, separating the axon tip of one cell from the surface of the next cell in the network. Neurotransmitters can affect the membrane permeability of the next cell in a chain of cells in ways that increase (or decrease) the probability that this neuron will produce its own action potential(s). If a neuron fires an action potential in response to stimulation provided by the preceding cell in the network, the message may be relayed on to the next cell, and on and on. Volleys of action potentials initiated by distant receptors may have excitatory (or inhibitory) effects that reach deep into the nervous system, eventually resulting in action potential outputs that reach the animal's muscles and cause them to contract.

In the case of the noctuid moth studied by Roeder, the A1 and A2 receptor cells are linked to relay cells called **interneurons**, whose action potentials can change the activity of other cells in one or more of the insect's thoracic ganglia (a ganglion is a neural structure composed of a highly organized mass of neurons) that relay messages on to the moth's brain (Figure 4.11). As messages flow through these parts of the nervous system, certain patterns of action potentials produced by cells in the thoracic ganglia trigger other interneurons, whose action potentials in turn reach motor neurons that are connected to the wing muscles of the moth. When a motor neuron fires, the neurotransmitter it releases at the synapse with a muscle fiber changes the membrane permeability of the muscle cell. These changes initiate the contraction or relaxation of muscles, which drive the wings and thereby affect the moth's movements.

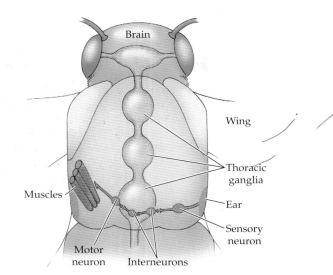

**FIGURE 4.11   Neural network of a moth.** Receptors in the ear relay information to interneurons in the thoracic ganglia, which communicate with motor neurons that control the wing muscles.

Thus, the moth's behavior, like that of any animal, is the product of an integrated series of chemical and biophysical changes in a network of cells. Because these changes occur with remarkable rapidity, a moth can react to certain acoustical stimuli in fractions of a second, which helps the moth avoid bats zooming in for the kill.

Although the neurons of many animals are very much alike at a fundamental level, they differ greatly in their functions. Thus, the auditory receptors of noctuid moths are highly specialized for the detection of ultrasonic stimuli. Roeder demonstrated this point by attaching recording electrodes to the A1 and A2 receptors of living, but restrained, moths.[1234] When he projected a variety of sounds at the moths, the electrical responses of the receptors were relayed to an oscilloscope, which produced a visible record. These recordings revealed the following features (Figure 4.12):

1. The Al receptor is sensitive to ultrasounds of low to moderate intensity, whereas the A2 receptor begins to produce action potentials only when an ultrasound is relatively loud.

2. As a sound increases in intensity, the A1 receptor fires more often and with a shorter delay between the arrival of the stimulus at the tympanum and the onset of the first action potential.

3. The A1 receptor fires much more frequently in response to pulses of sound than to steady, uninterrupted sounds.

4. Neither receptor responds differently to sounds of different frequencies over a broad ultrasonic range. Thus, a burst of 30 kHz sound elicits much the same pattern of firing as an equally intense sound of 50 kHz.

5. The receptor cells do not respond at all to low-frequency sounds, which means that moths are deaf to stimuli that we can easily hear, such as the chirping calls and trills of night-singing crickets. Of course, we are deaf to sounds that moths have no trouble hearing.

**FIGURE 4.12 Properties of the ultrasound-detecting auditory receptors of a noctuid moth.** (A) Sounds of low or moderate intensity do not generate action potentials in the A2 receptor. The A1 receptor fires sooner and more often as sound intensity increases. (B) The A1 receptor initially reacts strongly to pulses of ultrasound but then reduces its rate of firing if the stimulus is a constant sound.

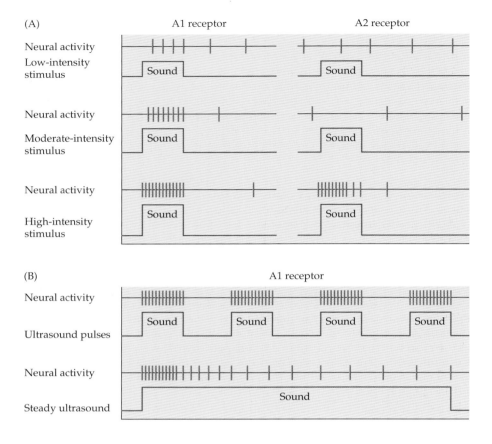

Although the moth's ears have just two receptors each, they could potentially provide an impressive amount of information to the moth's nervous system about echolocating bats. The key property of the A1 receptor is its great sensitivity to pulses of ultrasound, which enables it to begin generating action potentials in response to the faint cries of a little brown bat 30 meters away, long before the bat can detect the moth. In addition, because the rate of firing in the A1 cell is proportional to the loudness of the sound, the insect has a system for determining whether a bat is getting closer or going farther away.

The moth's ears also gather information that could be used to locate the bat in space. If a hunting bat is on the moth's left, for example, the A1 receptor in its left ear will be stimulated a fraction of a second sooner and somewhat more strongly than the A1 receptor in its right ear, which is shielded from the sound by the moth's body. As a result, the left receptor will fire sooner and more often than the right receptor (Figure 4.13A). The moth's nervous system could also detect whether a bat is above it or below it. If the predator is higher than the moth, then with every up-and-down movement of the insect's wings, there will be a corresponding fluctuation in the firing rate of the A1 receptors as they are exposed to, then shielded from, bat cries by the wings (Figure 4.13B). If the bat is directly behind the moth, there will be no such fluctuation in neural activity (Figure 4.13C).

As neural signals initiated by the receptors race through the moth's nervous system, they may ultimately generate motor messages that cause the moth to turn and fly directly away from the source of ultrasonic stimuli.[1233] When a moth is moving away from a bat, it exposes less echo-reflecting area than if it were presenting the full surface of its wings to the bat's vocalizations. If a bat receives no insect-related echoes from its calls, it cannot detect a meal. Bats rarely fly in a straight line for long, and therefore the odds are good that a moth will remain undetected if it can stay out of range for a few seconds. By then the bat will have found something else within its 3-meter moth detection range and will have veered off to pursue it.

In order to employ its antidetection response, a moth must orient itself so as to synchronize the activity of the two A1 receptors. Differences in the rate of firing by the A1 receptors in the two ears are probably monitored by the brain, which relays neural messages to the wing muscles via the thoracic ganglia and allied motor neurons. The resulting changes in muscular action steer the moth away from the side of its body with the ear that is more strongly stimulated. (You can imagine what would happen if the moth turned itself toward the side of its body with the more strongly stimulated ear!) As the moth turns away from the relatively intense ultrasound reaching one side of its body, it will reach a point at which both A1 cells are equally active; at this moment, it will be facing in the opposite direction from the bat and will be heading away from danger (see Figure 4.13C).

Although this reaction is effective if the moth has not been detected, it is useless if a speedy bat has come within 3 meters of the moth. At this point, a moth has at most a second, and probably less, before the bat reaches it.[743] Therefore, moths in this situation do not try to outrun their enemies but instead employ drastic evasive maneuvers, including wild loops and power dives, that make it harder for bats to intercept them. A moth that executes a successful power dive and reaches a bush or grassy spot is safe from further attack because echoes from the leaves or grass at the moth's crash-landing site mask those coming from the moth itself.[1233] Other nocturnal insects have independently evolved the capacity to sense ultrasound, and they also take evasive action when bats attack them (Figure 4.14).[986, 1626, 1627]

Roeder speculated that the physiological basis for this erratic escape flight lies in the neural circuitry leading from the A2 receptors to the brain and then back to the thoracic ganglia.[1235] When a bat is about to collide with a moth, the intensity of the sound waves reaching the insect's ears is high. It is under these conditions that the A2 cells fire. Roeder believed that the A2 signals, once relayed to the brain, might shut down the central steering mechanism that regulates the activity of the flight motor neurons. If the steering mechanism was inhibited, the moth's wings would begin beating out of synchrony, irregularly, or not at all. As a result, the insect might not know where it was going—but neither would the pursuing bat, whose inability to plot the path of its prey could permit the insect to escape.

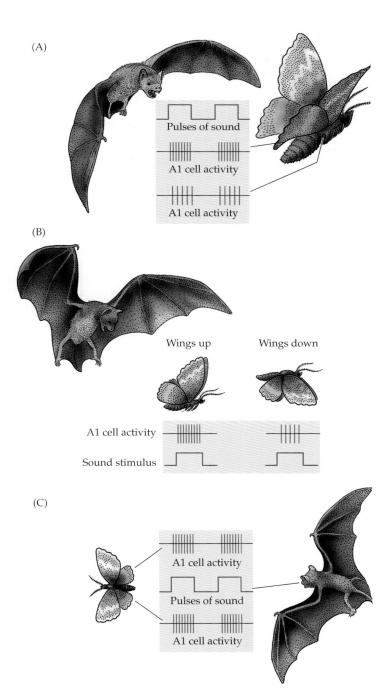

**FIGURE 4.13    How moths might locate bats in space.** (A) When a bat is to one side of the moth, the A1 receptor on the side closer to the predator fires sooner and more often than the shielded A1 receptor in the other ear. (B) When a bat is above the moth, activity in the A1 receptors fluctuates in synchrony with the moth's wingbeats. (C) When a bat is directly behind the moth, both A1 receptors fire at the same rate and time. (Figures not drawn to scale.)

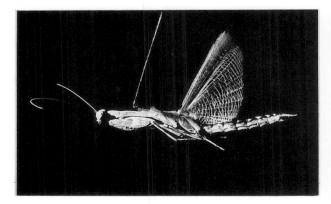

**FIGURE 4.14  Bat ultrasonic cries trigger evasive behavior in a number of insects.** In normal flight, a praying mantis holds its forelegs close its the body (top left), but when it detects ultrasound, it rapidly extends its forelegs (bottom left), which causes the insect to loop and dive erratically downward. Likewise, a lacewing may employ an anti-interception power dive (right) when approached by a hunting bat. The numbers superimposed on this multiple-exposure photograph show the relative positions of a lacewing and a bat over time. (The lacewing survived.) Top left and bottom left from Yager and May,[1627] photographs courtesy of D. D. Yager and M. L. May; right, photograph by Lee Miller.

Although Roeder's hypotheses about the functions of the A1 and A2 cells were plausible and supported by considerable evidence, especially with respect to the A1 receptor, other persons have continued to look at how these cells moderate the interactions between moths and bats. As a result, we now know that notodontid moths, even though they have just one auditory receptor per ear, still appear to exhibit a two-part response to approaching bats: the turning away from distant hunters and then the last-second erratic flight pattern when death is at hand. Thus, two cells may not be necessary for the double-barreled response of moths to their hunters.[1405] Even in moths with two receptors per ear, the A1 cell's activity changes greatly as a bat comes sailing in toward it because the bat's ultrasonic cries speed up and become much more intense (Figure 4.15).[504] Presumably, higher-order neurons up the chain of command could analyze the correlated changes in the activity of the A1 receptor alone and make adaptive adjustments accordingly without involvement of the A2 receptor.

More doubts that the A2 cell is necessary to trigger erratic evasive behavior come from the finding that in some noctuid moths, both the A1 and A2 cells may more or less stop firing during the terminal buzz phase—the last 150

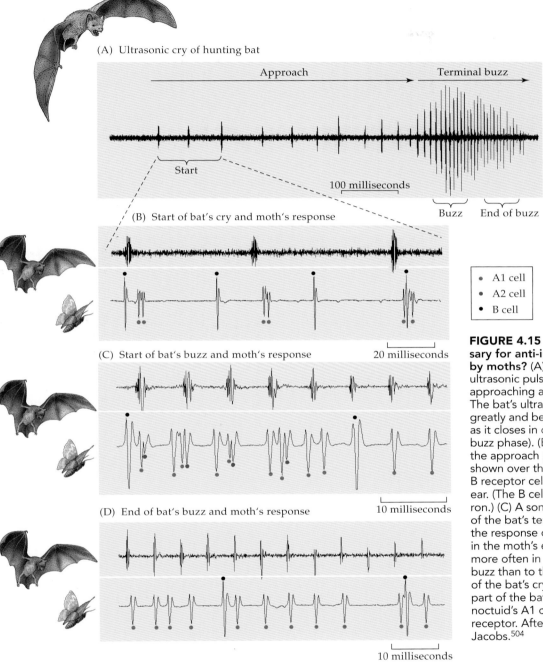

(A) Ultrasonic cry of hunting bat

Approach        Terminal buzz

Start

100 milliseconds

Buzz    End of buzz

(B) Start of bat's cry and moth's response

- A1 cell
- A2 cell
- B cell

(C) Start of bat's buzz and moth's response

20 milliseconds

10 milliseconds

(D) End of bat's buzz and moth's response

10 milliseconds

**FIGURE 4.15 Is the A2 cell necessary for anti-interception behavior by moths?** (A) A sonogram of the ultrasonic pulses produced by a bat approaching and then attacking a prey. The bat's ultrasonic cries speed up greatly and become much more intense as it closes in on the prey (the terminal buzz phase). (B) A sonogram of part of the approach portion of the bat's cry is shown over the response of the A1 and B receptor cells in the noctuid moth's ear. (The B cell is a nonauditory neuron.) (C) A sonogram of the initial part of the bat's terminal buzz is shown over the response of the A1, A2, and B cells in the moth's ear. The A1 receptor fires more often in response to the terminal buzz than to the approach component of the bat's cry. (D) During the latter part of the bat's terminal buzz, only the noctuid's A1 cell is active, not the A2 receptor. After Fullard, Dawson, and Jacobs.[504]

milliseconds—of a bat attack. One would think that these cells would keep signaling if either one was truly important in controlling the last-gasp evasive maneuvers of moths under attack. James Fullard and his coworkers suggest that perhaps these cells fail to signal at this late stage simply because the extremely loud and rapid attack vocalizations of a nearby bat incapacitate the cells. Thus, we have reason to question whether a connection exists between A2 cell activity and the moth's response to onrushing bats.[504]

(A)

(B)

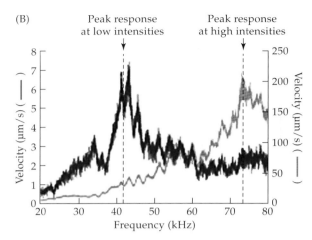

**FIGURE 4.16** **The tympanum of the moth** *Noctua pronuba* (shown in A with the arrow pointing to the attachment site of the mechanosensory neuron) vibrates differently in response to a low-intensity ultrasound stimulus (shown in green) than to a high-intensity ultrasound (shown in orange). The peak response of the tympanum to less intense sounds is at a lower frequency than the peak response to highly intense sounds. After Windmill, Fullard, and Robert.[1596]

## Discussion Questions

**4.3** Figure 4.16A shows the location of the tympanic membrane of a noctuid moth. Figure 4.16B shows how the tympanum vibrates in response to sounds ranging from 20 to 80 kHz when these sounds are at low intensity (the green curve) and at high intensity (the orange curve). What is surprising about these results? Could these properties promote the moth's ability to detect and respond adaptively to calling bats in its environment? Two points to consider: (1) moths do best at hearing relatively low-frequency ultrasound, and (2) bats shift into high-frequency ultrasound during the last phase of their attack as they attempt to capture flying prey.[1595]

**4.4** An American cockroach can begin to turn away from approaching danger, such as a hungry toad lunging toward it or a flyswatter wielded by a cockroach-loathing human, in as little as a hundredth of a second after the air pushed in front of the toad's head or the descending flyswatter reaches the roach's body. A cockroach has wind sensors that react to even slight air movements; these sensors are concentrated on its cerci, two thin projecting appendages at the end of its abdomen. One cercus points slightly to the right, the other to the left. Use what you know about moth orientation to bat cries to suggest how this simple system might provide the information the roach needs to turns away from the toad, rather than toward it. How might you test your hypothesis experimentally?

## Relaying and Responding to Sensory Input

The moth–bat story also demonstrates how knowledge of natural history and the real-world problems confronting an animal can help biologists formulate productive hypotheses about a species' sensory mechanisms. Kenneth Roeder knew that noctuid moths live in a world filled with echolocating moth killers; he searched for, and found, a specialized proximate mechanism that helps some moths cope with these enemies. Many other researchers have followed Roeder in exploring the relationship between the ecology of a species and its neurophysiological mechanisms.

We have focused thus far primarily on how the noctuid moth's auditory receptors collect information about ultrasound. Of course, if the insect is to act on that information, its receptors must be able to forward messages to those parts of its central nervous system that can process acoustical inputs and order appropriate reactions. Just how this is accomplished has not been worked out for noctuid moths, but we know something about the proximate mechanisms of information relay in another insect, the cricket *Teleogryllus oceanicus*, which also flees from ultrasound-producing bats.[997] As in noctuid moths, the cricket's ability to avoid bats begins with the firing of certain ultrasound-sensitive auditory receptors in its ears, which are found on the cricket's forelegs. Sensory messages from these receptors travel to other cells in the cricket's central nervous system. Among the receivers of these messages is a pair of sensory interneurons called int-1, also known as AN2, one of which is located on each side of the insect's body. Ron Hoy and his coworkers established that int-1 plays a key role in the perception of ultrasound by playing sounds of different frequencies to a cricket and recording the resulting neural activity. Their recordings revealed that these cells became highly excited when the cricket's ears were bathed in ultrasound. The more intense a sound in the 40 to 50 kHz range, the more action potentials the cells produced and the shorter the latency between stimulus and response—two properties that match those of the A1 receptor in noctuid moths.

These results suggest that the int-1 cell is part of a neural circuit that helps the cricket respond to ultrasound. If this is true, then it follows that if one could experimentally inactivate int-1, ultrasonic stimulation should not generate the typical reaction of a tethered cricket suspended in midair, which is to turn away from the source of the sound by bending its abdomen (Figure 4.17).

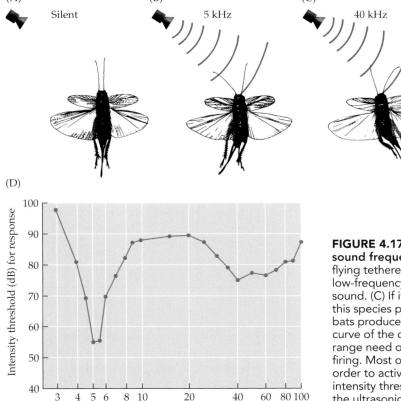

(A) Silent

(B) 5 kHz

(C) 40 kHz

(D)

Intensity threshold (dB) for response — Sound frequency (kHz)

**FIGURE 4.17  Avoidance of and attraction to different sound frequencies by crickets.** (A) In the absence of sound, a flying tethered cricket holds its abdomen straight. (B) If it hears low-frequency sound, the cricket turns toward the source of the sound. (C) If it hears high-frequency sound, it turns away. Males of this species produce sounds in the 5 kHz range; some predatory bats produce high-frequency calls of about 40 kHz. (D) The tuning curve of the cricket's int-1 interneuron. Sounds in the 5 to 6 kHz range need only be about 55 decibels (dB) loud in order to trigger firing. Most other sound frequencies have to be much louder in order to activate a response in this cell, although note that the intensity threshold dips in the neighborhood of 40 kHz, which is in the ultrasonic range commonly produced by bats. After Moiseff, Pollack, and Hoy.[997]

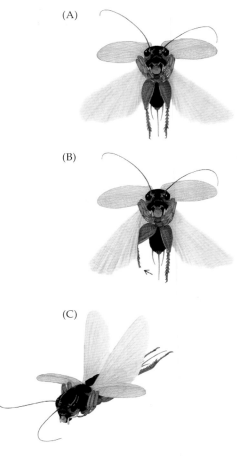

**FIGURE 4.18** How to turn away from a bat—quickly. (A) A flying cricket typically holds its hindlegs so as not to interfere with its beating wings. (B) Ultrasound coming from the cricket's left causes its right hindleg to be lifted up into its right wing. (C) As a result, the beating of the right wing slows, and the cricket turns right, away from the source of ultrasound as it dives to the ground. Drawings by Virge Kask, from May.[955]

As predicted, crickets with temporarily inactivated int-1 cells do not attempt to steer away from ultrasound, even though their auditory receptors are firing. Thus, int-1 is necessary for the steering response.

The corollary prediction is that if one could activate int-1 in a flying, tethered cricket (and one can, with the appropriate stimulating electrode), the cricket should change its body orientation as if it were being exposed to ultrasound, even when it was not. Experimental activation of int-1 is sufficient to cause the cricket to bend its abdomen.[1052] These experiments convincingly establish that int-1 activity is both necessary and sufficient for the apparent bat evasion response of flying crickets; therefore, this interneuron can be considered a critical part of the relay apparatus between the receptors and the central nervous system that enables the cricket to react adaptively to an ultrasonic stimulus.

And just how does a flying cricket carry out orders from its brain to steer away from a source of ultrasound? This problem attracted the attention of Mike May, who began his study not by conducting a carefully designed experiment, but instead by "toying with the ultrasound stimulus and watching the responses of a tethered cricket."[955] As May zapped the cricket with bursts of ultrasound, he noticed that the beating of one hindwing seemed to slow down with each application of the stimulus. Crickets have four wings, but only the two hindwings are directly involved in flight. If the hindwing opposite the source of ultrasound really did slow down, that would reduce power or thrust on one side of the cricket's body, with a corresponding turning (or yawing) of the cricket away from the stimulus.

On the basis of his informal observations, May proposed that the flight path of a cricket was controlled by the position of the insect's hindleg, which when lifted into a hindwing, altered the beat of that wing and thereby changed the cricket's position in space (Figure 4.18). May went on to take a number of high-speed photographs of crickets with and without hindlegs. Without the appropriate hindleg to act as a brake, both hindwings continued beating unimpeded when the cricket was exposed to ultrasound. As a result, crickets without hindlegs required about 140 milliseconds to begin to turn, whereas intact crickets started their turns in about 100 milliseconds. These findings led May to assert that neurons in the ultrasound detection network order the appropriate motor neurons to induce muscle contractions in the opposite-side hindleg of the cricket. As these muscles contract, they lift the leg into the wing, interfering with its beating movement, thereby causing the cricket to veer rapidly away from an ultrasound-producing bat.[955]

### Discussion Question

**4.5** Outline Mike May's research in terms of the question that provoked his study and his hypothesis, prediction(s), evidence, and scientific conclusion. In addition, what contribution to this research could come from learning that locusts, a group of insects not closely related to crickets, also possess a special mechanism for very rapidly altering leg positions and wingbeat patterns in reaction to ultrasound, such that a flying individual banks sharply downward away from the side stimulated by the stimulus?[370]

### Central Pattern Generators

In crickets and moths, ultrasound stimulation turns on auditory receptors that send sensory signals to interneurons, which may relay the message to other neurons within the central nervous system. When these target cells respond, they can generate signals that turn on a set of motor neurons. The end result is that certain muscles contract or relax in ways that cause the moth to stop beating its wings or the cricket to lift one leg upward, braking the activity of one wing. The escape dives of moths and the evasive swerves of flying crickets pursued by bats are effective one-step responses triggered by simple releasing stimuli in these animals' environments. Most behaviors, however, involve a coordinated series of muscular responses, which cannot result from a single command from a neuron or neural network. Consider, for example, the escape behavior of the sea slug *Tritonia diomedea*, which is activated when the slug comes in contact with a predatory sea star (Figure 4.19). Stimuli associated with this event cause the slug to swim in the ungainly fashion of sea slugs, by bending its body up and down.[1583] If all goes well, it will move far enough away from the sea star to live another day.

How does *Tritonia* manage its multistep swimming response, which requires from 2 to 20 alternating bends, each involving the contraction of a sheet of muscles on the slug's back followed by a contraction of the muscles on its belly? As it turns out, the dorsal and ventral muscles are under the control of a small number of motor neurons. The dorsal flexion neurons (DFN) are active when the animal is being bent into a U, and the ventral flexion neuron (VFN) produces

**FIGURE 4.19 Escape behavior by a sea slug.** The slug in this photograph has just begun to swim away from its deadly enemy, a predatory sea star. The slug's dorsal muscles are maximally contracted, drawing the slug's head and tail together. Soon the ventral muscles will contract and the slug will begin to thrash away to safety. Photograph by William Frost.

a pulse of action potentials that turn the slug into an inverted U (Figure 4.20). But what controls the alternating pattern of DFN and VFN activity?

The escape reaction begins when sensory receptor cells (S) in the skin of *Tritonia* detect certain chemicals on the tube feet of its sea star enemy (Figure 4.21). The receptors then relay messages to interneurons, among them the dorsal ramp interneurons (DRI), which, upon receipt of sufficiently strong stimulation, begin to fire steadily. This category of interneurons sends a stream of excitatory signals to several interneurons (the dorsal swim interneurons, or DSI), which in turn are part of an assembly of interconnected cells, among them the ventral swim interneurons (VSI) and cerebral neuron 2 (C2), as well as the flexion neurons mentioned already.[530, 531] A web of excitatory and inhibitory relations exists within this cluster of interneurons such that, for example,

Dorsal flexion

Ventral flexion

DFN

VFN

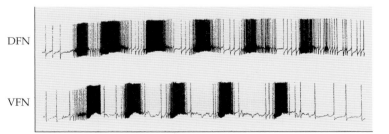

**FIGURE 4.20 Neural control of escape behavior in *Tritonia*.** The dorsal and ventral muscles of the sea slug are under the control of two dorsal flexion neurons (DFN) and a ventral flexion neuron (VFN). The alternating pattern of activity in these two categories of motor neurons translates into alternating bouts of dorsal and ventral bending—the movements that cause this animal to swim. After Willows.[1583]

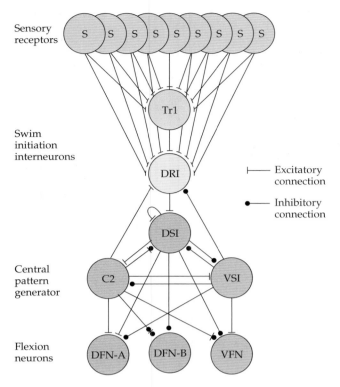

**Sensory receptors**

**Swim initiation interneurons**

Excitatory connection

Inhibitory connection

**Central pattern generator**

**Flexion neurons**

**FIGURE 4.21   The central pattern generator of** *Tritonia* in relation to the dorsal ramp interneurons (DRI) that maintain activity in the cells that generate the sequence of signals necessary for the sea slug to swim to safety. These interneurons receive excitatory input from receptor cells (S) and from another interneuron (Tr1). DRI cells in turn interact with three other categories of neurons (DSI, C2, and VSI), which are the cells that send messages to the flexion neurons. There are two kinds of dorsal flexion neurons (DFN-A and DFN-B) with somewhat different properties and one kind of ventral flexion neuron (VFN). After Frost et al.[495]

activity in DSI turns on C2, which leads to excitation of the DFN and the contraction of the dorsal flexion muscles. After a short period of excitation, however, C2 begins to block the DFN while sending excitatory messages to the VSI, leading to activation of the VFN and contraction of the ventral flexion muscles. The situation then reverses. Alternating bouts of activity in the interneurons regulating the DFN and VFN lead to alternating bouts of DFN and VFN firing and, thus, alternation of dorsal and ventral bending.[495]

The capacity of the simple neural network headed by the DRI to impose order on the activity of the motor neurons that control the dorsal and ventral flexion muscles means that this mechanism qualifies as a **central pattern generator**. Systems of this sort have been particularly well studied in invertebrates, especially with regard to locomotion, because the small number of neurons involved and their relatively large size facilitate their investigation.[268] The neural clusters labeled central pattern generators play a preprogrammed set of messages—a motor tape, if you will—that helps organize the motor output underlying movements of the sort that Tinbergen would have labeled fixed action patterns.

Central pattern generators are also found in vertebrates, as we can illustrate by looking at the plainfin midshipman, a fish that "sings" by contracting and relaxing certain muscles in a highly coordinated fashion.[82] Only the large males of this rather grotesque fish sing "humming" songs, which last more than a minute, and they do so only at night during spring and summer while guarding certain rocks. Their songs are so loud that they can annoy houseboat owners in the Pacific Northwest. The male fish sings to attract females of his species; the fish spawn at the defended rocks, and the male guards the eggs his mates lay in his territory.

How do the male fish produce their songs? When Andrew Bass and his coworkers inspected the anatomy of the fish's abdomen, they found a large, air-filled swim bladder sandwiched between layers of muscles (Figure 4.22). The bladder serves as a drum; rhythmic contractions of the muscles "beat" the drum, generating vibrations that other fish can hear. Muscle contractions require signals from motor neurons, which Bass found connected to the sonic muscles. He applied a cellular dye called biocytin to the cut ends of these motor neurons, which absorbed the material, staining themselves brown. And the stain kept moving along, crossing the synapses between the first cells to receive it and the next ones in the circuit, and so

**FIGURE 4.22   Song-producing apparatus of the male plainfin midshipman fish.** The sonic muscles control the movement of the swim bladder, thereby controlling the fish's ability to sing. After an illustration by Margaret Nelson, in Bass.[82]

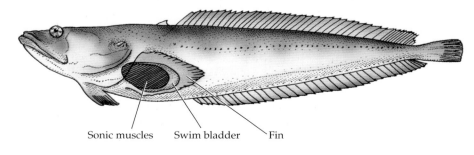

Sonic muscles      Swim bladder      Fin

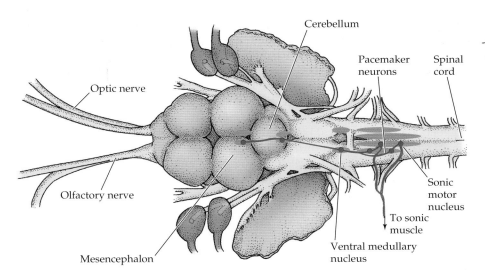

Cerebellum

Pacemaker neurons    Spinal cord

Optic nerve

Sonic motor nucleus

To sonic muscle

Olfactory nerve

Mesencephalon

Ventral medullary nucleus

**FIGURE 4.23    Neural control of the sonic muscles in the plainfin midshipman fish.** Signals from the central region of the brain (the mesencephalon) travel by way of the cerebellum and ventral medullary nuclei to the sonic motor nuclei in the upper part of the spinal cord. The firing of the pacemaker neurons regulates the frequency of firing by the neurons in the sonic motor nuclei; these signals, in turn, set the rate of contraction of the sonic muscles and thus the frequency of the sounds produced by the fish. After an illustration by Margaret Nelson, in Bass.[82]

on, through the whole network of cells connected to the sonic muscles. By cutting the brain into fine sections and searching for cells stained brown by biocytin, Bass and his colleagues mapped the fish's sonic control system. In so doing, they discovered two discrete collections of interrelated neurons that generate the signals controlling the coordinated muscle contractions required for midshipman humming. These two clusters, called the sonic motor nuclei, consist of about 2000 neurons each and are located in the upper part of the spinal cord near the base of the brain (Figure 4.23). The long axons of their neurons travel out from the brain, fusing together to form nerves that reach the sonic muscles.

In addition to these components, two other anatomically distinct elements of the nervous system ally themselves with the sonic motor nuclei. First, next to each nucleus lies a sheet of pacemaker neurons—a multicellular central pattern generator—that adjusts the activity of the sonic motor neurons so that the sequence of muscle contractions will yield a proper song. Second, in front of the pair of sonic motor nuclei are some special neurons that appear to connect the two nuclei, probably to coordinate the firing patterns coming from the left and the right nucleus so that the muscle contractions and relaxations will be synchronous, the better to produce a humming sound.[82]

## The Proximate Basis of Stimulus Filtering

We have now looked at the attributes of individual neurons and neural clusters that are involved in the detection of certain kinds of sensory information, the relaying of messages to other cells in the nervous system, and the control of motor commands that are sent to muscles. The effective performance of these basic functions is promoted by **stimulus filtering**, the ability of neurons and neural networks to ignore—to filter out—vast amounts of potential information in order to focus on biologically relevant elements within the diverse stimuli bombarding an animal.

The noctuid moth's auditory system offers an object lesson on the operation and utility of stimulus filtering. First, the A1 receptors are activated only by acoustical stimuli, not by other forms of stimulation. Moreover, as noted, these cells completely ignore sounds of relatively low frequencies, which means that moths are not sensitive to the stimuli produced by chirping crickets or croaking frogs—sounds they can safely ignore. Finally, even when the A1 receptors do fire in response to ultrasound, they do little to discriminate between different ultrasonic frequencies (whereas human auditory receptors produce distinct messages in response to sounds of different frequencies, which is why we can tell the difference between C and C-sharp). The noctuid moth's sensory apparatus appears to have just one task of paramount importance: the detection of cues associated with its echolocating predators. To this end, its auditory capabilities are tuned to pulsed ultrasound at the expense of all else. Upon detection of these critical inputs, the moth can take effective action.

Noctuid moths are not the only animals with biased auditory systems designed to filter out the irrelevant and focus on the important. The relationship between stimulus filtering and a species' special obstacles to reproductive success is evident in every animal whose sensory systems have been carefully examined. Consider the male midshipman fish that listen to the underwater grunts, growls, and hums produced by others of their species. These signals are dominated by sounds in the 60 to 120 hertz (Hz) range. The auditory receptor cells in the hearing organs of these fish are most sensitive to sounds in exactly that range.[1339] In summer, however, reproducing females listen to the humming "songs" of territorial males. These songs have components that range up to 400 Hz in frequency. As expected, when males are singing lustily for spawning partners, female hearing is much more sensitive to the higher-frequency sounds present in male songs than at other times of the year.[1338] Thus, the auditory system undergoes changes that enable females to detect and respond to acoustical stimuli in the summer that they ignore in the winter.

Female midshipman fish employ stimulus filtering on a seasonal basis as they listen to sounds in their underwater world. The screening of acoustical stimuli also occurs in certain parasitoid flies, which use their hearing to locate singing male crickets, the better to place their larvae on these insects. The little maggots burrow into the unlucky crickets and proceed to devour them from the inside out. Larvae-laden female flies of the species *Ormia ochracea* can find food for their offspring because they have ears tuned to cricket calls, as researchers discovered when they found *Ormia* coming to loudspeakers that were playing tapes of cricket song at night.

The unique ears of the female fly consist of two air-filled structures with tympanic membranes and associated auditory receptors on the front of the thorax. Vibration of the fly's tympanic "eardrums" activates the receptors, just as in noctuid moth ears, and thus provides the fly with information about sound in its environment—but not every sound. As predicted by a trio of evolutionary biologists, Daniel Robert, John Amoroso, and Ronald Hoy, the female fly's auditory system is tuned to (i.e., most sensitive to) the dominant frequencies in cricket songs (Figure 4.24). That is, the female fly can hear sounds of 4 to 5 kHz (the sort produced by crickets) more easily than sounds of 7 to 10 kHz, which have to be much louder if they are to generate any response.[1227] In contrast, male *Ormia* are not especially sensitive to sounds of 4 to 5 kHz. Although males can hear those sounds, they do not depend on finding singing male crickets, and as a result, their auditory system has evolved its own distinctive stimulus-filtering properties.

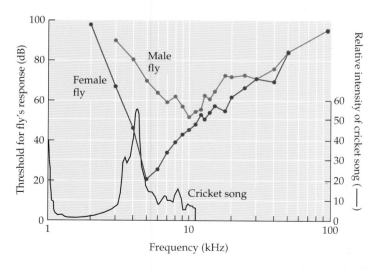

**FIGURE 4.24  Tuning curves of a parasitoid fly.** Females, but not males, of the fly *Ormia ochracea* find their victims by listening for the calls of male crickets, which produce sound with a frequency–intensity spectrum that peaks between 4 and 5 kHz. The female fly, unlike the male fly, is maximally sensitive to sounds around 5 kHz. After Robert, Amoroso, and Hoy.[1227]

## Discussion Question

**4.6** Females of another parasitoid fly related to *Ormia* track down singing male katydids (*Poecilimon veluchianus*), whose ultrasonic mate-attracting calls fall largely in the 20 to 30 kHz range.[1400] What sound frequencies should elicit maximal response in the ears of this katydid-hunting parasitoid if stimulus filtering enables the animal to achieve goals that are biologically relevant? What conclusion can you reach based on the data in Figure 4.25?

### Cortical Magnification in the Tactile Mode

The sensory receptor system of every animal species is designed to screen out some kinds of sensory stimulation while reacting to other kinds. Much the same sort of biased treatment of different kinds of potential information occurs in the central nervous system, as we will illustrate by examining the operating rules of the star-nosed mole's brain. This weird mammal lives in wet, marshy soil, where it burrows about in search of earthworms and other prey. In its dark tunnels, earthworms cannot be seen, and indeed, the mole's eyes are greatly reduced in size, so it largely ignores visual information even when light is available. Instead, the mole relies heavily on touch to find its food, using its wonderfully strange nose to sweep the tunnel walls as it moves forward. Its two nostrils are ringed by 22 fleshy appendages, 11 on each side of the nose (Figure 4.26). These appendages cannot

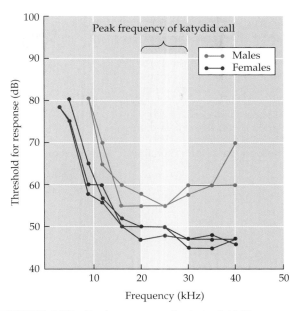

**FIGURE 4.25 Tuning curves of a katydid killer.** Females of the fly *Therobia leonidei* parasitize male katydids, whose stridulatory calls contain most of their energy in the range of 20 to 30 kHz. After Stumpner and Lakes-Harlan.[1400]

**FIGURE 4.26 The star-nosed mole's nose** (top left) differs greatly from that of the eastern mole (top right) and even more from those of its distant relatives, which include the African hedgehog (bottom left) and the masked shrew (bottom right). All four species, however, rely on tactile information to a considerable degree in locating prey, which range from insects to earthworms. Photographs by Ken Catania.

**FIGURE 4.27   A special tactile apparatus.** The 22 appendages of the star-nosed mole's nose are covered with thousands of Eimer's organs. Each organ contains a variety of specialized sensory cells that respond to mechanical deformation of the skin above them. After Catania and Kaas;[245] photographs by Ken Catania.

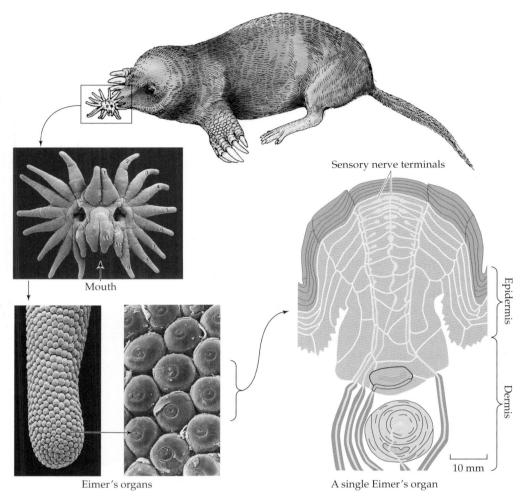

grasp or hold anything but instead are covered with a thousand or so tiny sensory devices called Eimer's organs. Each of these organs contains several different kinds of sensory cells that appear to be dedicated to the detection of objects touching the nose (Figure 4.27).[924] With these mechanoreceptors, the animal can collect extremely complex patterns of information about the things it encounters underground, enabling it to identify prey items in the darkness with extreme rapidity.[244, 248]

Whenever the mole brushes an earthworm with, say, appendage 5, it instantly sweeps its nose over the prey so that the two projections closest to the mouth, labeled appendage 11 (Figure 4.28A), come in contact with the object of interest. The tactile receptors on each appendage 11 generate a volley of signals, which are carried by nerves to the brain of the animal. Although these two nose "fingers" contain only about 7 percent of the Eimer's organs on the star nose, more than 10 percent of all the nerve fibers relaying information from the nose's touch receptors to the brain come from these two appendages. In other words, the mole uses relatively more neurons to relay information from appendage 11 than from any other appendage.

Not only is the relay system biased toward inputs from appendage 11, but the animal's brain also gives extra weight to signals from this part of the nose. The information from the nose travels through nerves to the somatosensory cortex, the part of the brain that receives and decodes sensory signals from

(A)

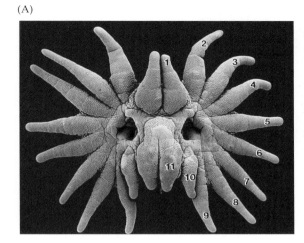

(C)

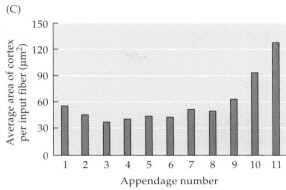

(B)

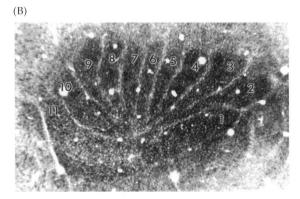

**FIGURE 4.28** **The cortical sensory map** of the star-nosed mole's tactile appendages is disproportionately weighted toward appendage 11. (A) The nose of the mole, with each appendage numbered on one side. (B) A section through the area of the somatosensory cortex that is responsible for analyzing sensory inputs from the nose. The cortical areas that receive information from each appendage are numbered. (C) The amount of somatosensory cortex devoted to the analysis of information from each nerve fiber carrying sensory signals from the different nose appendages. After Catania and Kaas;[245] photographs by Ken Catania.

touch receptors all over the animal's body. Of the portion of the somatosensory cortex that is dedicated to decoding inputs from the 22 nose appendages, about 25 percent deals exclusively with messages from the two appendage 11s (see Figure 4.28).[245] This discovery was made by Kenneth Catania and Jon Kaas when they recorded the responses of cortical neurons as they touched different parts of the anesthetized mole's nose. Perhaps the mole's brain is "more interested in" information from appendage 11 because of its location right above the mouth; should signals from this appendage activate a cortical order to capture a worm and consume it, the animal is in the right position to carry out the action immediately.[244]

In the star-nosed mole, the disproportionate investment in brain tissue to decode tactile signals from one part of the nose is mirrored on a larger scale by the biases evident in the somatosensory cortex as a whole, which focuses on signals from the hands and nose at the expense of other parts of the body. This pattern of cortical magnification makes adaptive sense for this species because of the biological importance of the mole's hands for burrowing and the mole's nose for locating prey (Figure 4.29A).

If this argument is correct, then we can predict that the allocation of cortical tissue to somatosensory inputs will differ from species to species in ways that make sense given the special environmental problems each species has to solve. And it is true that other insectivores, even though they are related to the star-nosed mole, exhibit their own adaptively different patterns of cortical magnification (Figure 4.29B–D).

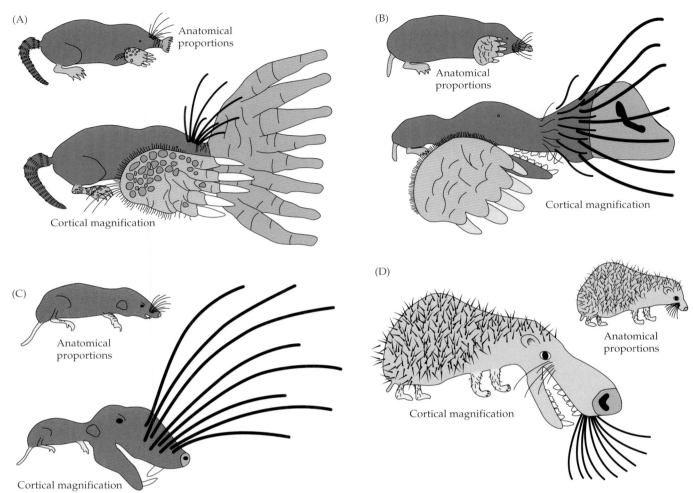

(A)   Anatomical proportions

Cortical magnification

(B)   Anatomical proportions

Cortical magnification

(C)   Anatomical proportions

Cortical magnification

(D)   Anatomical proportions

Cortical magnification

**FIGURE 4.29   Sensory analysis in four insectivores.** In each case, the smaller drawing shows the actual anatomical proportions of the animal; the larger drawing shows how the body is proportionally represented in the somatosensory cortex of the animal's brain. (A) The star-nosed mole devotes much more somatosensory cortex to processing inputs from its nose and forelimbs than it does to processing those coming from receptors in other parts of its body. (B) Cortical magnification in the eastern mole focuses on sensory inputs from the feet, nose, and sensory hairs, or vibrissae, around the nose. (C) Cortical magnification in the masked shrew also reveals the importance of the vibrissae. (D) Sensory signals from the vibrissae are magnified cortically to a lesser degree in the African hedgehog. After Catania and Kaas[244] and Catania.[246]

## Discussion Question

**4.7**   Cortical magnification occurs in all mammals. The two cartoonish drawings in Figure 4.30 are cortical maps based on the amount of brain tissue devoted to the sensory analysis of tactile inputs from different parts of the bodies of human beings and of the naked mole rat (see Figure 13.41), a strange, nearly hairless mammal that uses its large front teeth to dig a vast network of underground tunnels while also excavating and processing tuberous roots for food. In what ways do these two maps support the argument that animal brains exhibit adaptive sensory biases? For an additional comparison of the cortical maps of laboratory rats and the naked mole rat see Catania and Henry.[249]

**FIGURE 4.30   Sensory analysis in humans and naked mole rats.** Brains evolve in response to selection pressures associated with particular physical and social environments. For each species, the drawing on the left shows the actual anatomical proportions; the drawing on the right shows how the body is proportionally represented in the somatosensory cortex. Cortical map of human male is based on data from Kell et al.[752] Cortical map of naked mole rat drawn by Lana Finch, from Catania and Remple.[247]

## Adaptation and the Proximate Mechanisms of Behavior

Our brief survey has revealed that both individual cells and neural networks have special operating rules that filter the information an animal receives, relays, and processes. These rules adaptively shape the ways in which members of different species perceive their environments. Many bees, for example, can see ultraviolet (UV) light (which we humans cannot see), which helps the bees find the nectar source within a UV-reflecting flower quickly (Figure 4.31A). Likewise, the ability of some butterfly species to see UV light helps them respond to the UV-reflecting patterns on the wings of other butterflies, which announce species membership and sex (Figure 4.31B)[1258] and may also signal the quality of a potential mate. Captive females of the butterfly *Eurema hecabe* were more likely to mate with males with bright ultraviolet patches on their wings than with males whose UV-reflecting wing patches had been dulled by about 25 percent.[758]

Ultraviolet light perception is not limited to bees and butterflies[105] but also occurs in some fishes, lizards, and many songbirds. One experimental demonstration of this point comes from a study of three-spined sticklebacks. A gravid female of this fish species selects a male with a nest in which to deposit her eggs, which are then fertilized by the male. To test whether female mate choice was affected by the UV component of a male's coloration, researchers constructed an aquarium with three compartments (Figure 4.32). Gravid females were placed in the end compartment, where they could view a nesting male in each of the two adjacent compartments. Each test female was separated

(A)

**FIGURE 4.31 Ultraviolet-reflecting patterns** have great biological significance for some species. In both sets of photographs, the image on the left shows the organism as it appears to humans, while the image on the right shows the organism's UV-reflecting (pale) surfaces. (A) The ultraviolet pattern on this daisy advertises the central location of food for insect pollinators. (B) Only males (top specimens) of this sulphur butterfly species have UV-reflecting patches on their wings, which helps signal their sex to other individuals of their species. A, photographs by Tom Eisner; B, photographs by Randi Papke and Ron Rutowski.

(B)

from both males by filters, one of which screened out UV radiation while the other did not. Females spent more time closely observing the male whose UV coloration could be seen, as opposed to the male whose UV coloration had been blocked.[153]

Signals containing UV wavelengths can play a role in male–male interactions, as well as influencing female mating decisions. For example, adult male collared lizards aggressively display to one another by opening their mouths wide, and when they do so, whitish UV-reflecting patches become visible at the corners of their mouths (Figure 4.33). Both the width of the gape *and* the size of the pale patches at the mouth corners contain information about how strongly a male can bite an opponent. It seems likely that the UV aspect of the signal plays a role in making a male lizard's threat display more conspicuous and thus more intimidating to a less powerful rival.[842] Research work with another lizard showed that males with experimentally reduced UV-reflecting throat patches were subject to more attacks than control males, an indication that the size of the UV patch is a signal of male fighting ability.[1378]

(A)

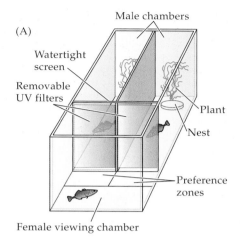
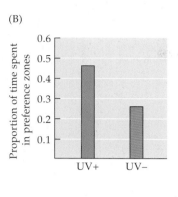

Male chambers

Watertight screen

Removable UV filters

Plant

Nest

Preference zones

Female viewing chamber

(B)

Proportion of time spent in preference zones

UV+    UV−

**FIGURE 4.32   Ultraviolet (UV) reflectance from male stickleback bodies** influences female mate preferences. (A) The experimental setup. A female could view males in adjacent chambers that were separated from her compartment by a screen and filters, which could either permit the passage of UV (UV+) or block UV (UV−). (B) When given a choice of which male to spend time close to, the females preferred the male whose UV signals could reach them via the UV+ filter. After Boulcott, Walton, and Braithwaite.[153]

Similar experimental manipulations have been done with some songbirds, such as the bluethroat of northern Europe and Asia. The bird's common name refers to a patch of blue feathers located, as you may have suspected, on the male's throat (Figure 4.34). These feathers reflect light that we perceive as blue, but they also reflect ultraviolet light that is invisible to us. If female bluethroats evaluate a courting male's suitability as a mate on the basis of his UV-reflecting throat patch, then males whose blue throats have been altered to absorb, rather than reflect, ultraviolet radiation should become less attractive to females. To check this prediction, a Scandinavian research team placed captured male bluethroats of the same age in pairs in an aviary. One member of each pair had sunscreen plus a fatty substance from an avian preening gland rubbed onto its blue throat patch, while the other received an application of preening gland oil only. The chemicals in the sunscreen absorbed the UV wavelengths, whereas the glandular secretions did not. In 13 of 16 paired trials, a female bluethroat approached the UV-reflecting male more often than she approached the male whose ornament had been altered to become UV-absorbing,[36] indicating that UV reflectance affects female mate choice in this species as well as sticklebacks.

Thus, many species have visual systems very different from our own, and they put their extraordinary (from our perspective) abilities to use in activities of reproductive significance.

## Discussion Questions

**4.8**  The blue tit is another songbird with a blue patch of feathers that reflect ultraviolet radiation, which is visible to this species. Some ornithologists have reported that female blue tits prefer to pair with males that have relatively bright UV-reflecting feathers on their crowns,[696] while others have found that females mated with such males supply the eggs fertilized by their attractive mates with more carotenoids, a potentially valuable pigment that may enhance the development of their offspring.[1418] In light of these results,

**FIGURE 4.33   Ultraviolet reflectance can be used as a threat signal.** When a male collared lizard opens his mouth in a threat display, he unfolds and exposes pale patches of skin at the corners of his mouth that reflect ultraviolet light. The size of these patches is proportional to the strength of the male's bite. Photographs by A. Kristopher Lappin.

**FIGURE 4.34   A bird that can sense ultraviolet light.**
The male bluethroat's throat feathers appear purely blue
to us but not to bluethroats themselves, which also see the
ultraviolet light reflecting from the feathers. Photograph by
Bjørn-Aksel Bjerke, courtesy of Jan Lifjeld.

you should be surprised to learn that another team of avian ecologists
found that in one population, male blue tits whose crowns reflected *less* UV
produced *more* offspring than males with more UV-ornamented crowns.
This team suggested that perhaps the males with low-UV crowns were better able to sneak onto neighboring territories and sire "extra-pair" offspring
with their neighbors' mates than were males with high-UV crowns (see page
338). How would you test this hypothesis experimentally? List your predictions and then check the results presented in Delhey et al.[381]

**4.9**   The ocean-dwelling crab *Bythograea thermydron* goes through three
life stages, which live at three different depths and habitats. The minute
larvae float in waters about 1000 meters below the surface, where only a
faint blue light penetrates. The older, larger juvenile forms sink into deeper,
darker waters, where the only sources of light are luminous fish and other
deep-sea creatures that produce their own blue-green light. Finally, the
adults sink to the ocean floor, where they live in the vicinity of deep-sea
hydrothermal vents 2500 meters or thereabouts below the surface. Here
only the faintest flickers of light are given off at the vents themselves. What
would an evolutionary biologist predict about the properties of the eyes of
the different life stages of this species? You can check your predictions with
information in Jinks et al.[727]

## Adaptive Mechanisms of Human Perception

Even though the studies I have just described have not involved direct inspection of the nervous systems of bluethroats or butterflies, they tell us something about the properties of the sensory receptors and information decoders
possessed by these animals. Bluethroats and butterflies have nervous systems with perceptual capabilities that help these animals manage their special environments. If we were to apply this principle to our own species, we
could predict that natural selection should have endowed us with proximate
mechanisms suited to the ecological problems we face. Indeed, much evidence
suggests that our highly specialized auditory skills, for example, have evolved
to match the human social environment, which is dominated by language.[1143]
Moreover, our visual perception complements our acoustical analysis of language in a most interesting way. As it turns out, a listener's understanding of
spoken language is heavily influenced by visual cues provided by the moving
lips of a speaker. When someone sees a video of a person mouthing the non-

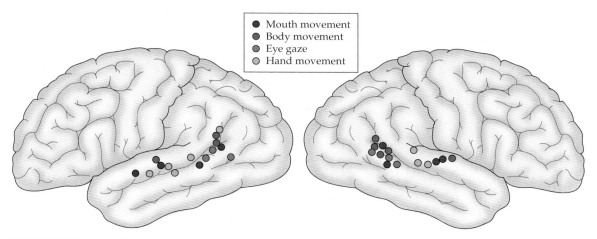

**FIGURE 4.35**  **Socially relevant movements** of the lips, mouth, hands, and body activate neurons in different parts of the superior temporal sulcus in the human brain. The left and right hemispheres are shown on the left and right, respectively. Each circle represents findings by a particular research team. After Allison, Puce, and Mc-Carthy.[23]

sense sentence "my gag kok me koo grive" while an audiotape synchronously plays the nonsense sentence "my bab pop me poo brive," the viewer/listener will hear quite clearly "my dad taught me to drive." This result demonstrates that our brains have circuits that integrate the visual and auditory stimuli associated with speech, using the visual component to alter our perception of the auditory channel.[945] Even though most people are totally unaware of their lipreading skills, the ability increases the odds that they will understand what others say to them.

Lipreading depends on specific neural clusters within the visual cortex. In particular, a region of the brain called the superior temporal sulcus becomes especially active when we view moving mouths, as well as hands and eyes (Figure 4.35).[23] People are remarkably good at detecting even subtle movements of these body parts because this skill enables the viewer not only to read lips, but also to deduce the intentions of other individuals, a useful ability for members of a highly social species. Cells in the superior temporal sulcus are also activated by certain static visual stimuli, especially those associated with faces. They perceive, for example, the direction in which a companion's eyes are gazing, which says something about this person's current focus of interest. Because neurons within the superior temporal sulcus are "tuned" to particular facial stimuli, such as eye position, we can better predict the actions of those around us.[23]

Nor is the superior temporal sulcus the only brain area devoted to the analysis of faces. Indeed, each component of a network of brain areas may be required to provide a person with the sense of recognizing a *familiar* face while also supplying the observer with information on the personality and intentions of this individual, as well as providing an emotional context for an encounter with this other person. Without all parts of the system, recognition may be hampered or absent.[547]

One of the many brain areas that is involved in face recognition is called the fusiform gyrus, which is on the underside of the cerebral cortex. If an experimenter gives someone 5 seconds to inspect each of 50 photographs of unfamiliar faces, odds are that when the subject is tested later, he will be able to pick out about 90 percent of the previously glimpsed faces from a large collection of photographs.[232] We know that this kind of face recognition is

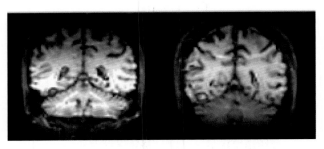

**FIGURE 4.36   A special-purpose module in the human brain: the face recognition center.** Magnetic resonance images of the brains of two persons looking at a photograph of a human face. The most active part of the brain, the facial fusiform area, is shown in red. From Kanwisher et al.[748]

dependent on an intact fusiform gyrus, because persons who have suffered an injury to this part of the brain lose this ability,[434] a phenomenon that gave Oliver Sacks the title for his book *The Man Who Mistook His Wife for a Hat*.[1268] Some persons with a damaged fusiform gyrus can see things perfectly well, name many objects, and identify particular individuals by the sound of the voice or by familiar clothing, but when shown pictures of the faces of their friends, their spouses, even themselves, they are at a complete loss.[376] Interestingly, other slightly brain-damaged persons have just the opposite problem. They cannot identify ordinary objects when they see them but have no difficulty recognizing particular faces, suggesting that the special neural mechanisms dedicated to face recognition are still intact in their brains.[97]

Functional magnetic resonance imaging reveals that neurons in a small part of the posterior fusiform gyrus fire only when a person looks at a face (Figure 4.36). This neural module, called the facial fusiform area, does not respond to pictures of inanimate objects, although another nearby region of the brain does; this provides strong evidence of task specialization by modules that provide biologically relevant perceptions for their owners.[608, 747]

Yet another approach to identifying which part of the brain does what has been to place electrodes directly on the surface of the cerebral cortex and record the electrical activity of neurons. This technique is used to map the brains of epileptic patients prior to operations designed to remove the tissue responsible for their epileptic seizures. Needless to say, the idea is to destroy as little of the brain as possible, which means finding the dysfunctional tissue in order to leave the rest alone when the surgery is performed. In the course of recording electrical activity in different parts of the brains of these patients, who remain conscious, researchers have been able to locate those parts of the brain that react to different kinds of visual stimuli. This approach has confirmed that different sections on the underside of the brain are devoted to different kinds of visual analyses. The sites that respond strongly to images of entire faces are different from those that react primarily to parts of faces, such as eyes, which in turn are different from those that are activated when the subject is shown a picture of an object (Figure 4.37).[1174] Our brains are computers with stimulus-filtering circuits that enable us to perceive some things much more readily than others.

## Discussion Question

**4.10**   Some evolutionary biologists have argued that the human cerebral cortex contains an assortment of regions specialized for the analysis of an assortment of biologically relevant stimuli. Thus, according to this view, our face recognition mechanism evolved because of the reproductive value for individuals of quickly recognizing the identity of others, given the highly social nature of our species. What problem is posed for this argument by the discovery that we also possess a "visual word form area" located in the left fusiform gyrus (Figure 4.38)?[959] You are using this particular region of your brain at this very moment as you read these words. In addition, what significance do you attach to the finding that when we read, we recognize each letter independently by its simple features? We never come to recognize words as wholes on the basis of their complexly distinctive patterns, even though we could read far more efficiently if we worked with entire word forms rather than moving from one letter to the next.[1113]

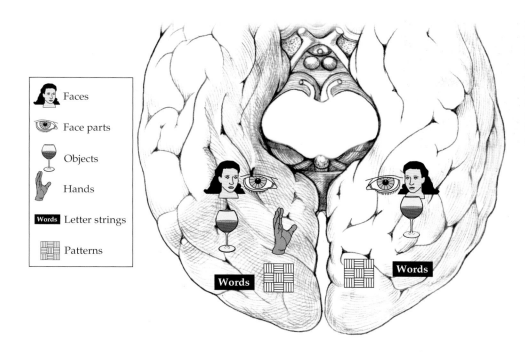

**FIGURE 4.37** **Specialization of function in different parts of the visual cortex of humans.** Different neural circuits perform different analyses of images in our environment. The ability to remember faces, for example, is dependent on a specialized site in the fusiform gyrus. After Puce, Allison, and McCarthy.[1174]

## Adaptive Mechanisms of Navigation

The human visual system is biased toward gathering biologically significant information from our social environment. Persons in the past who were better than average at remembering who was friend and who was foe, and at deducing the intentions of others by examining their faces, have in effect left us a legacy in the form of our brains, with all their adaptive filters, oddities, and quirks. The point that nervous systems have been shaped by natural selection can be reinforced by considering the relationship between the brains of animals and their ability to travel to destinations of their choice. Because our hunter–gatherer ancestors wandered widely from a home base in search of food and other resources, we can predict that our brains should have features that promote efficient movement through space, just as behavioral ecologists predicted that food-storing birds should possess neural mechanisms—notably an enlarged hippocampus—that help them remember where they hid their food (see page 70). Does the human hippocampus play a part in getting us from point A to point B?

In order to answer this question, researchers using magnetic resonance imaging gave their computer game–playing subjects time to explore a virtual town with a complex maze of streets before giving them the task of navigating from point A to point B as quickly as possible.[916] Individuals able to navigate more accurately evidently relied more heavily on the right hippocampus, compared with other, less accurate subjects, judging from the higher levels of activity observed in this part of the brains of especially skillful navigators (Figure 4.39).

Among the best navigators are London taxi drivers. Before they receive their operating license, they undergo a training program that lasts up to 4 years, during which time they have to learn a map containing 25,000 streets in the city. For a sample of London cabbies, the average posterior hippocampal size, as revealed by magnetic resonance imaging, was larger than that in a comparable group of men who did not drive taxis for a living.[917] Navigational experience itself appears to be responsible for the development of a large hippocampus capable of storing a great deal of spatial information: the more years of taxi driving, the larger the posterior hippocampus.

(A)

(B)

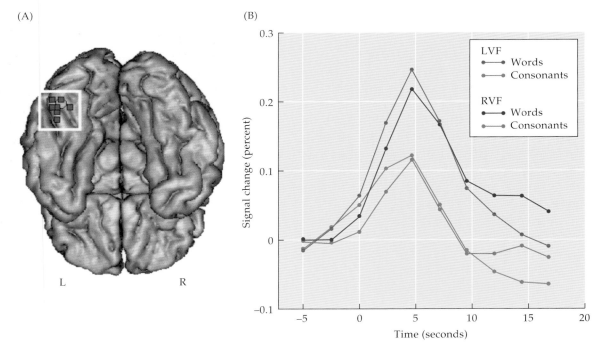

**FIGURE 4.38 A cerebral word analysis center.** (A) The fusiform gyrus, shown by the green squares and yellow circle, responds especially strongly to the visual stimulation provided by words. (B) Shown are the differences in neural activity over time of the fusiform gyrus on the left (LVF) and right (RVF) sides of the brains of human subjects who were permitted to see words or strings of consonants as opposed to checkerboards of black and white squares. Seeing words produces a greater change in activity in this part of the brain than seeing strings of consonants; both kinds of stimuli generate a greater response than seeing a checkerboard stimulus. A from and B after Cohen et al.[285]

But the enlarged hippocampal region in taxi operators might arise from other factors besides map learning and navigational experience. After all, taxi drivers are in motion for more hours per day than the average person, and any special brain features might simply reflect the fact that taxi drivers do much more driving than the typical adult.[918]

One way to test these alternatives is to find persons who drive for as many hours per day as London cabbies do but without the navigational challenges that confront the taxi driver. Other long-day drivers do exist in the form of London bus operators, who are in motion for the same number of hours per day but travel over fixed routes rather than having to find new destinations all the time. If many hours of daily movement or extensive driving experience resulted in a hippocampus with greater volume in certain of its components, then bus drivers should have the same hippocampal volumes as taxi drivers. But they do not.[918]

These and other research results suggest that our ability to learn spatial information may be dependent upon a developmentally flexible hippocampus, although other brain regions are also involved in various ways.[919] But no matter how large the hippocampus of an experienced London taxicab driver, if he were to be dropped off in Detroit with the requirement that he get to 16th and Ash without assistance from others, he would almost certainly be totally lost. Nor would he set off confidently if he were given a compass by a bystander. A compass would help only if he knew where he was and where he was headed, which would require a map of some sort. Although humans

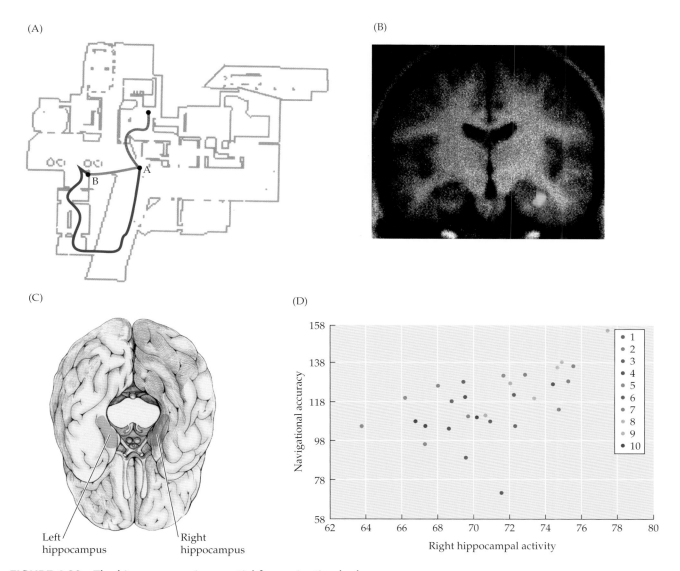

**FIGURE 4.39   The hippocampus is essential for navigation by humans.**
(A) A map (aerial view) of the virtual town through which experimental subjects navigated. Three examples of attempts to get from point A to point B are illustrated. The shortest route (in yellow) between the two points was considered the most accurate. (B) Average magnetic resonance image for ten subjects while performing the navigation task, showing the location of peak neural activity (bright yellow), which lies in the right hippocampus. (C) The location of the right hippocampus in a brain viewed from the underside. (D) The accuracy of the ten navigators (represented by different-colored dots) was a function of the intensity of the activity in the right hippocampus. After Maguire et al.[916] and Carter.[239]

without a map and compass are handicapped when it comes to navigating across unfamiliar areas, many other animals show no such disability, because they possess an internal map sense (knowing the location of home or some other goal) and an internal compass sense (knowing in what direction to move) (Figure 4.40).

Take the honey bee and the homing pigeon, for example. Both species are skilled navigators capable of crossing unfamiliar terrain on the way home, as demonstrated by a honey bee's ability to make a beeline back to its hive after a meandering outward journey in search of food and by a homing pigeon's ability to make a "pigeon line" back to its loft after having been released in a

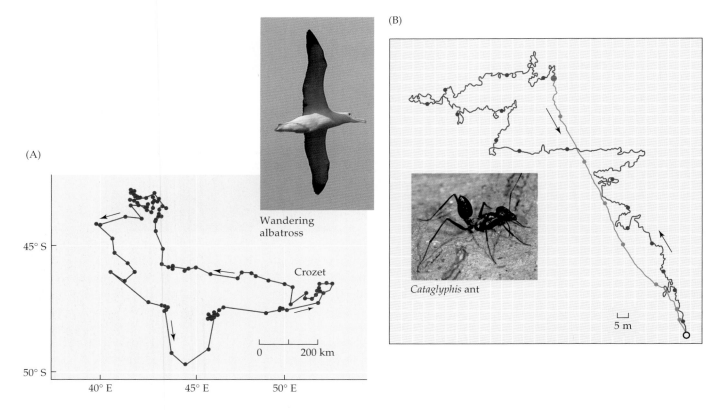

(A)

45° S

50° S

40° E          45° E          50° E

0          200 km

Wandering
albatross

Crozet

(B)

*Cataglyphis* ant

5 m

**FIGURE 4.40** **The ability to navigate over unfamiliar terrain** requires both a compass sense (knowing in what direction to move) and a map sense (knowing the location of a goal). (A) The flight path taken by a wandering albatross on a foraging journey of over 4000 kilometers from its nest in the Crozet Islands in the southern Indian Ocean, north of Antarctica, and back again. (B) A trip of 592 meters by a foraging ant out from its nest (at the large open circle at the bottom) and then directly back home after capturing a prey (at the spot marked with a large red circle, 140 meters away from home). A, after Weimerskirch et al.;[1531] B, after Wehner and Wehner.[1528]

distant and strange location.[1507, 1508] Because honey bees and homing pigeons are both active during the daytime, they both, as one might suspect, use the sun's position in the sky as a directional guide.[257, 1594] Even we can do this to some extent, knowing that the sun rises in the east and sets in the west—provided that we also know approximately what time of day it is. Every hour the sun moves 15 degrees on its circular arc through the sky. Therefore, one has to adjust for the sun's movement if one is to use its position as a compass.

A honey bee leaving its hive notes the position of the sun in the sky relative to the hive as it flies off on a foraging trip. It might spend 15 to 30 minutes on its trip and might move into unfamiliar terrain in search of food. If on the homeward flight, it oriented as if the sun had not shifted, the bee would be lost. In reality, honey bees rarely get lost, in part because they can compensate for the sun's movement, thanks to an internal clock mechanism (see page 157).[871] This ability can be demonstrated by training some marked bees to fly to a sugar-water feeder some distance away from their hive (say, 300 meters due east of the hive). One can then trap the bees inside the hive and move it to a new location. After 3 hours have passed, the hive can be unplugged and the bees can be free to go in search of food. They will not have familiar visual landmarks to guide them, and yet some marked individuals will remember that food is found 300 meters due east. They will fly 300 meters due east of the new hive location, to the spot where the food source "should be." They will have compensated for the 45-degree shift in the position of the sun that has taken place during their 3-hour confinement.

Pigeons can also be tricked into demonstrating how important a clock sense is for their navigation system.[1506] You can reset a pigeon's biological clock by placing the bird in a closed room with artificial lighting that is turned on and off so as to shift the light and dark periods out of phase with sunrise and sunset in the real world. If sunrise is at 6:00 a.m. and sunset at 6:00 p.m., for example, one might set the lights in the room to go on at noon (6 hours later than the actual sunrise) and off at midnight (6 hours later than actual sunset). A pigeon exposed to this routine for several days would become clock-shifted 6 hours out of phase with the natural day. If taken from the room and released at noon at a spot some distance from its loft, the bird would behave as if the sun had just come up (as if it were 6:00 a.m.), which would cause it to orient improperly. For example, let's say that the pigeon is released at a place 30 kilometers due east of its loft. Its map sense somehow tells it this, and it attempts to orient itself to fly west. To fly west at 6:00 a.m., a pigeon will fly away from the sun, but when the clock-shifted birds take this route and fly away from the sun in its actual southerly noontime position, they fly north, at a 90-degree angle away from the correct route (Figure 4.41).

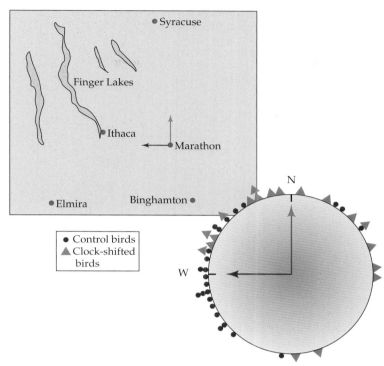

**FIGURE 4.41    Clock shifting and altered navigation in homing pigeons.** The results of an experiment in which pigeons were released at Marathon, New York (about 30 kilometers east of Ithaca, where their home loft was located). On sunny days, birds in the control group generally headed west, back toward the home loft. But pigeons whose biological clocks had been shifted by 6 hours usually misoriented such that, on average, they headed north. After Keeton.[751]

## Discussion Question

**4.11**  We put a homing pigeon on an experimental light–dark cycle in which the lights come on at noon and go off at midnight at a time of the year when actual sunrise is at roughly 6:00 a.m. and sunset is at 6:00 p.m. After several weeks on this schedule, we release the pigeon at noon on a clear day in unfamiliar territory due north of its home loft. First, in what direction will the pigeon fly? Second, are you surprised to learn that on a completely overcast day, the pigeon would fly directly home? What does this finding suggest about the homing mechanism(s) of this bird? In this light, consider the findings of a research team in New Zealand who released nearly 100 pigeons near the Auckland Junction Magnetic Anomaly, a place where the Earth's magnetic field is distorted by unusual underground geological features. The New Zealanders found that nearly 60 percent of the birds initially flew on a track aligned with or perpendicular to the local geomagnetic field, but when they got past the Anomaly, they changed direction and headed home.[383]

## Adaptive Mechanisms of Migration

Although what honey bees and pigeons can do is most impressive, other long-distance migrants are even more amazing navigators. The most familiar of these are the many birds whose migratory journeys cover thousands of kilometers, such as the white-crowned sparrows that move between Mexico and

**FIGURE 4.42    The fall migration route of monarch butterflies** takes some butterflies from Canada to Mexico, where the monarchs cluster in a few forest patches high in the central Mexican mountains. After Brower.[186]

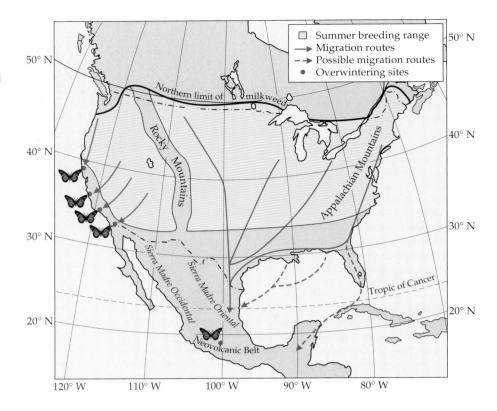

Alaska twice a year (see also pages 261–270). Although much has been learned about the proximate mechanisms of bird migration,[1594] here we will focus on two other animals, the monarch butterfly and the green sea turtle.

In the fall, adult monarch butterflies that had flown to Canada turn around and head south. They do not stop until they reach one of a select group of forest groves high in the mountains of central Mexico, a trip that requires flying as far as 3600 kilometers (Figure 4.42). Once in Mexico, the butterflies congregate by the millions in stands of Oyamel firs, where they mostly sit and wait, day after day, through the cool winter season until spring comes. Then the monarchs rouse themselves and begin a return trip to the Gulf Coast of the United States, arriving in time to lay their eggs on growing milkweed plants, which their caterpillar progeny will consume.

Migrating monarchs fly during the daytime, which suggests that they might use the sun as a compass to guide them in a southwesterly direction during the fall migration (and in a northeasterly direction during the spring return flight). If so, then it ought to be possible to do unto monarchs what has been done unto pigeons, namely, to trick them into misorienting by manipulating their biological clocks.[496] To test this prediction experimentally, a research team captured migrating monarchs in the fall and held them in a laboratory under a regime of 12 hours of light and 12 hours of dark. They placed the butterflies in two groups, one that received its 12 hours of light beginning daily at 7:00 a.m. and the other that had the lights go on 6 hours earlier, at 1:00 a.m. After a number of days in captivity, the monarchs were tethered and released in a flight cage under natural daylight conditions. There they could fly and, in so doing, indicate the direction in which they wanted to go (without actually leaving the test arena, thanks to the tethers).

The results showed that the monarchs attended to the sun's position in the sky. Those that had had their biological clocks set to the natural daily cycle for the fall (lights on at 7:00 a.m.) struggled against their tethers to fly off to

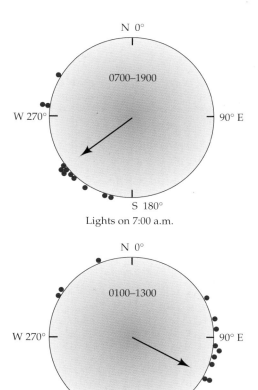

Lights on 7:00 a.m.

Lights on 1:00 a.m.

Monarch butterfly

**FIGURE 4.43 Experimental manipulation of the biological clock** changes the orientation of migrating monarchs. Individuals were tested in an outdoor flight cage after one group had been held indoors under an artificial light–dark cycle with lights on at 7:00 a.m. and lights off at 7:00 p.m. This group tried to fly in a southwesterly direction. A second group of butterflies that had been held with lights on at 1:00 a.m. and lights off at 1:00 p.m. flew in a southeasterly direction, evidence of the importance of a sun compass in helping monarchs stay on course. After Froy et al.[496]

the southwest. But those individuals whose clocks had been shifted 6 hours earlier oriented 90 degrees to the left, which meant that they were heading in a generally southeasterly direction (Figure 4.43).[496]

Sunlight is actually composed of many different wavelengths of light, and so the question arises, what wavelengths are critical for the navigational orientation of monarch butterflies? Perhaps ultraviolet light is a key factor. As all sunscreen users know, the sun produces plenty of ultraviolet radiation, and although you and I cannot see this radiation, many other animals can, as noted earlier in this chapter. To test the hypothesis that monarch navigation is dependent on ultraviolet radiation, experimenters captured migrating monarchs in the fall and then permitted tethered monarchs to start flying in the flight cage where they could orient in a direction of their choosing. The team then covered the flight cage with a UV interference filter, which screened out this component of sunlight. The monarchs quickly became confused, and many stopped flying altogether. Most individuals (11 out of 13) resumed flight, however, as soon as the filter was removed, evidence that the sun's UV radiation is indeed essential for monarch navigation.[496]

Indeed, we now know that ultraviolet light is detected by specialized cells in the eye and relayed to a particular part of the monarch's brain, the pars lateralis, where the cells that constitute the butterfly's biological clock reside. In turn, the clock mechanism apparently communicates with another part of the brain, the central complex, which controls the butterfly's compass sense (Figure 4.44).[1645] This connection between clock and compass could provide the proximate basis for changes in flight orientation during the day with the passage of time.

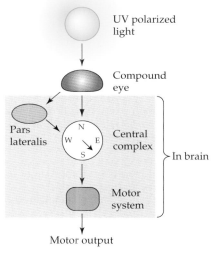

**FIGURE 4.44 Ultraviolet polarized light** affects the circadian clock and sun compass of the monarch butterfly. Polarized light from the sun is detected by special cells in the butterfly's compound eye, which in turn communicate directly with the clock mechanism in the pars lateralis of the butterfly's brain. This system then sends signals to the central complex in the insect's brain, where the sun compass is housed. As a result, the motor system of the butterfly receives messages that enable it to navigate accurately via a clock-guided sun compass. After Zhu et al.[1645]

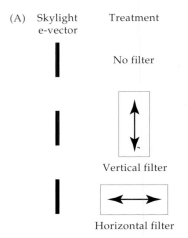

(A) Skylight e-vector Treatment

No filter

Vertical filter

Horizontal filter

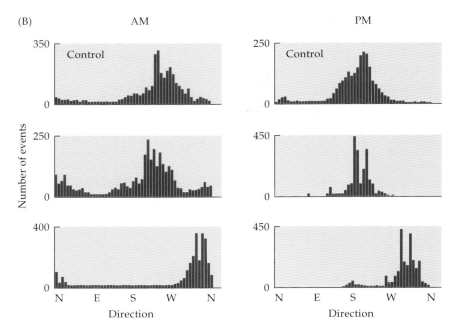

**FIGURE 4.45 Polarized light affects the orientation of monarch butterflies.** (A) Butterflies tethered in a flight cage received three treatments with respect to the angle of polarized light reaching them from the sky (the skylight e-vector): (1) no filter, which meant that they could see the natural pattern of polarized light in the sky; (2) vertical filter, which did not interfere with the pattern of polarized light visible from the flight cage; and (3) horizontal filter, which shifted the angle of polarized light arriving from the sky by 90 degrees. (B) Orientation choices made by two flying monarchs in the flight cage, one tested in the morning (AM) and the other in the afternoon (PM). During the period of flight, a computer automatically recorded the orientation of the butterfly every 200 milliseconds, yielding a record composed of many "events" during the minutes when the monarch was flying. The data show that when the insects could observe polarized light in the natural skylight pattern, they tended to fly to the southwest, but when the filter shifted the angle of polarized light, altering the skylight pattern by 90 degrees, the monarchs shifted their flight orientation accordingly. After Reppert, Zhu, and White.[1211]

Ultraviolet light may help monarchs get started flying in the right direction. But once airborne, the butterflies need to maintain their compass orientation, and as it turns out, they achieve this end with assistance from the cues provided by the polarized skylight pattern. Polarized light, which you and I cannot see, is produced when sunlight enters the Earth's atmosphere and the light scatters so that some light waves are vibrating perpendicular to the direction of actual sunrays. The three-dimensional pattern of polarized light in the sky created in this fashion depends upon the position of the sun relative to the Earth, and so this pattern changes as the sun moves across the sky. Therefore, if an animal on the ground is able to perceive the pattern of polarized light in the sky, this stimulus can serve as a proxy for the position of the sun at any given time. As a result, creatures with a biological clock and the capacity to see polarized light can use the information in skylight as a compass in the same way that the sun's position in the sky can be used as a compass. Being able to make use of the directional information in polarized light has real advantages because this cue is available even when the sun is hidden behind clouds or mountains.

A team of researchers established that monarchs can orient to polarized light information by tethering butterflies captured on their fall migration in a small walled arena where the insects could not see the sun but could look at the sky overhead. Under these conditions, when the butterflies flew, they were able to orient consistently to the southwest. Flying tethered monarchs retained this ability when a light filter was placed over the apparatus and aligned so as to permit the entry of linear polarized light waves in the same plane as occurs in the sky at the zenith (the highest point in the sky). However, when the filter was turned 90 degrees from its original orientation, changing the angle of entry of the polarized light visible to the monarchs at the zenith, they altered their flight orientation by 90 degrees as well, demonstrating their dependence on a polarized-light compass (Figure 4.45).[1211]

Green sea turtles are the monarch butterflies of the turtle world. Like the monarch, these animals travel thousands of kilometers on their migrations, which take some turtles from their nesting beaches on Ascension, a tiny island in the middle of the South Atlantic Ocean, to their feeding grounds just off the coast of Brazil, about 2000 kilometers away (Figure 4.46).[899] Other green

**FIGURE 4.46  Migratory routes taken by five green sea turtles** that nested on Ascension Island and then returned over 2000 kilometers to feeding areas in the South Atlantic Ocean near Brazil. After Luschi et al.;[899] photograph by Ursula Keuper-Bennett and Peter Bennett.

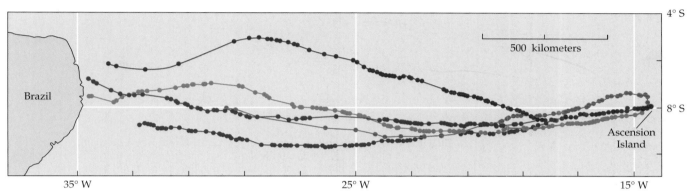

sea turtles shuttle between other feeding and breeding sites separated by vast distances, often spending several years at sea before making their way back to the beach where they nested before. Satellite transmitters attached to green sea turtles revealed that the animals travel long distances at night, suggesting that they do not need to track the sun's position in order to make their remarkable journeys. Therefore, Ken Lohmann and his co-workers hypothesized that the turtles must make use of some other navigational cue, of which there are several potential candidates, among them the Earth's magnetic field.[884]

To test whether lines of magnetic force are used by the turtles to make maps, Lohmann and company captured some young turtles at sea, brought them back to the mainland, tethered them in a cloth harness, and plunked them into an "ocean simulator," namely, a plastic pool of seawater in the back-yard of a house in Melbourne Beach, Florida. The pool was surrounded by a computer-driven magnetic coil system that the researchers used to alter the magnetic field around the pool, thereby simulating the conditions that a magnetic field detector would experience hundreds of kilometers to the north or south. If the turtles were capable of sensing the Earth's magnetic field and using it as a map, then an individual that perceived conditions associated with an area 340 kilometers to the north of the Melbourne Beach area should swim steadily south, rowing along in its tether, getting nowhere but establishing its orientation preference. If the magnetic field was one that a turtle would encounter 340 kilometers to the south, then the turtles should orient in such a way as to head north. The turtles did what was expected of them (Figure 4.47),[884] demonstrating that they are indeed geomagnetic-map navigators.

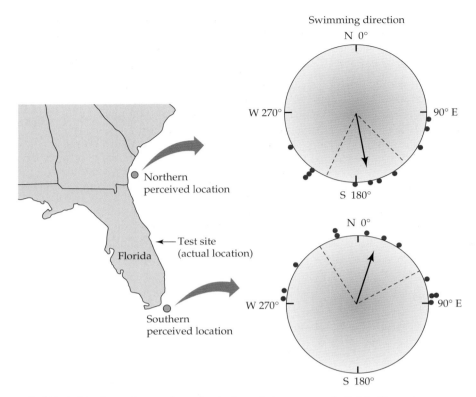

**FIGURE 4.47 Experimental manipulation of the magnetic field** affects the orientation of green sea turtles. Individuals that experience the magnetic field associated with an area to the north of their actual location swim south; turtles that sense the magnetic field of an area to the south of their actual location swim north. After Lohmann et al.[884]

Although monarch butterflies and green sea turtles do not use the same navigational systems,[1015] they each possess perceptual skills absent in our own species. These two species provide additional evidence, as if any more were needed, that the sensory mechanisms and brains of animals have been shaped by natural selection arising from each animal's special requirements for survival and reproduction.

## Summary

1. The operating rules of neural mechanisms constitute a proximate cause of behavior. Neurons acquire sensory information from the environment, relay and process that information, and order appropriate motor responses to environmental events. Different species have different neural mechanisms and therefore perform these tasks differently, providing proximate reasons why species differ in their behavior.

2. The classic ethological approach to the study of nervous systems showed that some animals possess neural elements shaped by natural selection to detect key stimuli and to order appropriate species-specific responses to these biologically relevant cues. Modern neurophysiological studies confirm that animals possess highly specialized sensory receptors whose design facilitates the acquisition of critical information from the environment, as well as interneurons and motor neurons with unique properties that contribute to particular behavioral abilities. Moreover, central pattern generators within the central nervous system can produce a programmed series of messages to selected muscles, facilitating complex patterned responses to certain stimuli.

3. Stimulus filtering, the filtering out of irrelevant information, along with other forms of biased sensory processing are fundamental properties of all animal nervous systems. Sensory receptors ignore some stimuli in favor of others, while interneurons relay some, but not all, of the messages they receive. Within the central nervous system, many cells and circuits are devoted to the analysis of certain categories of information, although this means that other inputs are not as thoroughly analyzed.

4. The proximate mechanisms of stimulus filtering and biased data analysis are the basis for adaptive perception and adaptive responses. By ignoring some potential information, animals are able to focus on the biologically relevant stimuli in their environment, increasing the odds of a prompt and effective reaction to these key stimuli. Because the obstacles to reproductive success differ among species, selection has resulted in the evolution of proximate mechanisms with different functional attributes. That is why some animals have perceptual abilities not possessed by humans, such as the capacity to hear ultrasound, or to see ultraviolet light, or to form geomagnetic maps useful for navigating to a particular destination across unfamiliar terrain.

## Suggested Reading

Hans Kruuk describes Niko Tinbergen's life and science dispassionately and well,[816] while Richard Burkhardt's *Patterns of Behavior* provides a superb history of how Tinbergen and Konrad Lorenz founded the discipline of animal behavior.[212] Kenneth Roeder's *Nerve Cells and Insect Behavior*[1234] is a classic on how to conduct research on the physiology of behavior. Textbooks on the neurophysiology of behavior include those by Peter Simmons and David Young[1332] and Thomas Carew.[230] *Mapping the Brain* by Rita Carter[239] is a beautiful book about the operation of the human brain. *Evolving Brains* by John Allman focuses on the evolution of the human brain but contains a wealth of fascinating information on other species as well.[25] For an advanced text on the neural basis of our cognitive abilities, see *Principles of Cognitive Neuroscience*.[1178]

One of the main findings of neurobiologists is that different species have evolved remarkable abilities in response to natural selection in different environments. Examples include infrasound-detecting elephants,[520, 1064] the electric sense of certain fishes (see Hopkins[681] and Kalmijn[744]), and the infrared perception of some snakes.[1042] Another illustration of this point comes from a comparison of the vomeronasal organs of assorted vertebrates (see Michael Meredith's Web site http://www.neuro.fsu.edu/faculty/meredith/vomer/). This organ, which serves a special olfactory role, has evolved several very different functions, each appropriate to a task of biological significance for the species in question. Likewise, the different mechanisms underlying the feats of animal navigation are fascinating (see Wehner, Lehrer, and Harvey[1529]). If you have a special interest in sea turtle navigation, go to http://www.unc.edu/depts/oceanweb/turtles/ where you will find information on the amazing proximate mechanisms underlying the ability of hatchling loggerhead turtles to get into the sea, away from land, and on great journeys that will take them across the Atlantic Ocean and back again.

# 5

# The Organization of Behavior: Neurons and Hormones

In the previous chapter, we talked about animals as if they were endowed with neural computers designed to detect key stimuli, discriminate among patterns of inputs, and order adaptive reactions. So, for example, when a flying moth is exposed to ultrasound from a predatory bat, its auditory receptors fire, setting in motion a chain of neural events, which help the insect respond adaptively to this acoustical stimulus. The capacity of neural mechanisms to filter out irrelevant information, to perceive some things very reliably, and to order effective reactions makes biological sense.

But nervous systems do more than just turn on response X in the presence of stimulus Y. Imagine a flying male moth hot on the scent trail of a distant pheromone-releasing female. If the male's nervous system operated simply by keeping the moth flying whenever female sex pheromone was in the air, a scent-tracking moth would be in real trouble if bats were around. But the moth's nervous system balances risk against reproductive opportunity[1340] so that if the moth hears very loud ultrasonic pulses, he generally aborts his scent-tracking activity and dives for cover.[2, 1409] Thanks to the way his nervous system works, the male moth can stop searching for females when a hunting bat is heading his way, which means the moth may live to mate another day.

◀ **These male red-sided garter snakes** *emerging from hibernation are ready to mate, despite the fact they have almost no testosterone in their blood. Photograph by François Gohier.*

(A)

(B)

**FIGURE 5.1**   **Different courtship displays of the male ring dove** are under the control of different hormones. (A) The aggressive strutting and bowing display, which is linked to testosterone. (The male is the bird at the left.) (B) The nest-soliciting display, which is linked to estrogen. Photographs by Leonida Fusani.

Hormones often help animals keep their behavioral options straight. To pick just one example, when a male ring dove courts a female, he starts off by aggressively chasing his potential mate around, as well as strutting and bowing to her in a pushy manner. If that were his only courtship behavior, he would not get very far, because females will not mate unless this early aggressive display is followed by a calmer display called nest soliciting, in which the male stands on the nest with his tail raised (Figure 5.1). As it turns out, the hormone testosterone increases the likelihood of chase-strutting, whereas the hormone estrogen facilitates nest soliciting.[509] The male ring dove's brain gets its various sexual activities in the right order by producing an enzyme, aromatase, at the appropriate stage of courtship; aromatase catalyzes the conversion of testosterone to estrogen, thereby enabling the male ring dove to switch from strutting to nest soliciting in the proper sequence.

The ability of nervous and hormonal systems to perform these feats of behavioral organization is what this chapter is all about. The fundamental problem that we will examine is how proximate mechanisms structure an individual's behavior—from moment to moment, over the course of a day, over a few weeks or a breeding season, or even over a whole year. We will examine three classes of mechanisms that carry out these functions: neural command centers that communicate with one another, clocks that schedule the activity of these command centers, and hormonal systems that track changing physical and social environments and adjust the priorities of competing command centers.

## Neural Command Centers Organize Behavior

Because most animals have the capacity to do many different things in response to many different stimuli, at any given moment they face the question of which behavior to activate. At an ultimate level, it is easy to understand why animals rarely try to do two things at once. But at a proximate level, how are animals' nervous systems organized so that maladaptive conflicts do not occur?

One way to establish behavioral priorities is to have a nervous system endowed with **command centers**, which include the innate releasing mechanisms, central pattern generators, song control systems, and the like that we have met in preceding chapters. Let each command center be primarily responsible for activating a particular response, but have the various centers communicate with one another in a hierarchical fashion so that an active center can suppress competing signals from another center (or vice versa). Note that although a "command center" could be a single bundle of neurons found in a particular part of the brain, it might instead consist of any number of interconnected batteries of neurons that are capable of unified decision making.

Kenneth Roeder used command center theory to examine decision making in the praying mantis.[1234] A mantis can do many things: search for mates, sunbathe, copulate, fly, dive away from bats, and so on. Most of the time, however, the typical mantis remains motionless on a leaf until an unsuspecting bug wanders within striking distance. When the mantis's visual receptors alert it to the presence of the prey, the mantis makes very rapid, accurate, and powerful grasping movements with its front pair of legs.

Roeder proposed that the mantis's nervous system sorts out its options thanks to inhibitory relationships among an assortment of command centers within its neural network. The design of

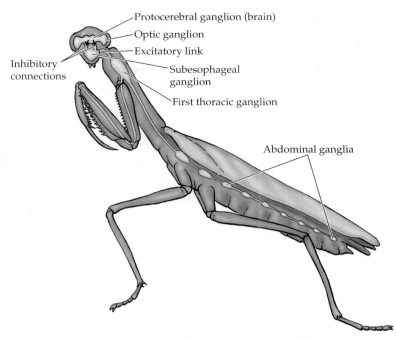

**FIGURE 5.2  Nervous system of a praying mantis.** If the connections between the protocerebral ganglion and the subesophageal ganglion are cut, the subesophageal ganglion sends a stream of excitatory messages to the segmental ganglia in the thorax and abdomen; the mantis then attempts to do several competing activities simultaneously.

the mantis's nervous system (Figure 5.2) suggested that the control of muscles in each of the insect's segments is the responsibility of that segment's ganglion. Roeder tested this possibility by cutting one segmental ganglion's connections with the rest of the nervous system. Not surprisingly, the muscles within the neurally isolated segment subsequently failed to react when the mantis's nervous system became active elsewhere. However, if the isolated ganglion was stimulated electrically, the muscles and any limbs in that segment made vigorous, complete movements.

If the segmental ganglia are indeed responsible for telling the muscles in particular segments to carry out given movements, what is the mantis's brain doing? Roeder suspected that certain brain cells are responsible for inhibiting (blocking) neural activity in the segmental ganglia, keeping cells in a ganglion quiet until they are specifically ordered into action by an excitatory command center in the brain. If so, then cutting the connection between the inhibitory brain cells and the segmental ganglia should have the effect of removing this inhibition and inducing inappropriate, conflicting responses. When Roeder severed the connections between the protocerebral ganglion (the mantis's brain) and the rest of its nervous system, he produced an insect that walked and grasped simultaneously, something that would be disastrous in nature. The protocerebral ganglion apparently makes certain that an intact mantis either walks or grasps but does not do both at the same time.

When Roeder removed the entire head, however—a procedure that eliminates the subesophageal ganglion as well as the protocerebral ganglion—the mantis became immobile. Roeder could induce single, irrelevant movements by poking the creature sharply, but nothing more. These results suggest that the protocerebral ganglion of an intact mantis typically sends out a stream of inhibitory messages to the subesophageal ganglion, preventing these neurons from communicating with the other ganglia. When certain sensory signals

**FIGURE 5.3   A no-brainer.** Losing his head and brain did not extinguish the mating behavior of this male praying mantis mounted on the back of an intact female. Photograph by Mike Maxwell.

reach the protocerebral ganglion, however, neurons there stop inhibiting certain modules in the subesophageal ganglion. Freed from suppression, these subesophageal neurons send excitatory messages to various segmental ganglia, where new signals are generated that order muscles to take specific actions. Depending on what sections of the subesophageal ganglion are no longer inhibited, the mantis walks forward, or strikes out with its forelegs, or flies, or does something else.

Interestingly, mature male mantises do not always obey the "rule" that complete removal of the head eliminates behavior. Instead, a headless adult male performs a series of rotary movements that swing its body sideways in a circle. While this is happening, the mantis's abdomen is twisting around and down, movements that are normally blocked by signals coming from the protocerebral ganglion. This odd response to decapitation begins to make sense when you consider that a male mantis sometimes literally loses his head over a female, when she grabs him and consumes him, head first. Even under these difficult circumstances, the male can still copulate with his cannibalistic partner (Figure 5.3), thanks to the nature of the control system regulating his mating behavior. Headless, his legs carry what is left of him in a circular path until his body touches the female's, at which point he climbs onto her back and twists his abdomen down to copulate competently.

Whether male or female, adult or immature, the mantis, like many other animals, has a nervous system that appears to be functionally organized as a cluster of command centers, each with specific responsibilities. Some centers produce their own output, inhibiting the activities of other centers, which makes it possible for the mantis to do just one thing at a time. The importance of inhibitory relationships within nervous systems is also evident in the feeding blowfly, the subject of Vincent Dethier's classic studies.[385] These humble insects drink various exudates from plants, juices of liquefying animal corpses, and other savory fluids rich in sugars and proteins. During the night, the nutrients collected in the day's meals are metabolized to provide energy for the insect. In the morning, the blowfly flies off to seek additional food, which it locates in part by smell and in part by tasting with its feet when it happens to step into something wet. The appropriate sensory inputs activate feeding commands: the fly extends its proboscis, spreads its labellum, and imbibes.

Dethier showed experimentally that the speed at which a fluid is sucked up and the duration of feeding are proportional to the concentration of sugar in the fluid. If the liquid is not very sugary, the fly's oral receptors cease firing quickly, and sucking stops. If sugar concentrations are high, however, the oral receptors may keep firing for 90 seconds or thereabouts before they quit, ending that bout of feeding. In short order, the sensory cells on the feet may become active again, leading to reextension of the proboscis and more drinking. Sooner or later, however, the fly stops drinking altogether, even when standing in the richest of sugary liquids. Dissection of a sated fly reveals that its crop, a storage sac off the digestive tract, is filled to overflowing, forcing fluid back into the foregut. Dethier hypothesized that distension of the foregut is detected by stretch receptors attached to this part of the digestive tract. He believed that those receptors might send messages to the brain, stimulating cells there to inhibit the feeding response.

As predicted from this hypothesis, receptors similar in design to stretch receptors in other organisms were found in the fly's foregut. As predicted,

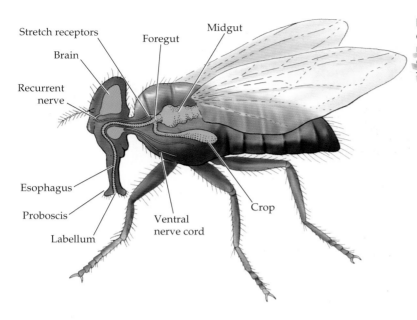

Stretch receptors
Foregut
Midgut
Brain
Recurrent nerve
Esophagus
Proboscis
Labellum
Ventral nerve cord
Crop

**FIGURE 5.4   The blowfly's nervous system and digestive system.** Severing the recurrent nerve eliminates feedback to the brain about the fullness of the crop and eliminates the signals that eventually block feeding in a blowfly whose stomach and crop are full.

these receptors feed their sensory input into a pathway called the recurrent nerve, which runs from the foregut to the brain (Figure 5.4). As predicted, if one experimentally severs the recurrent nerve, the fly will not stop feeding. The insect will continue on with bout after bout of drinking until its body literally explodes. Flies with intact recurrent nerves do not blow themselves up, thanks to inhibitory arrangements between elements of the nervous system.[385]

## Discussion Question

**5.1**   Figure 5.5 shows a record of the activity of a particular neuron in a cricket's brain that appears to act as a command cell controlling the male's chirping call, which is produced when the male rubs his wings together. About 3 seconds into the recording, the cricket was subjected to a sharp puff of air that struck its cerci, a pair of sensory appendages that project out from the rear of the abdomen. What is the relevance of this figure to the topic that we have just discussed? What adaptive value do you attach to the apparent proximate mechanisms that help male crickets make the behavioral decisions shown here?

## Behavioral Schedules

The ability of neural command centers to communicate with and inhibit one another helps set an animal's behavioral priorities. But what if the hierarchical ordering of command centers has to change in order to meet the demands of a changing environment? Some activities are better done at some times of the day than others. Female *Teleogryllus* crickets, for example, usually hide in burrows or under leaf litter during the day and move about only after dusk, when it is relatively safe to search for mates.[883] Not surprisingly, male crickets wait for the evening to start calling for mates.[882] These observations suggest that the inhibitory relationships between the calling center in a male cricket's brain and other neural elements responsible for other behaviors must change cyclically over a 24-hour period. A clock mechanism could be useful for the control of singing behavior, just as the monarch butterfly's biological clock enables it to adjust its sun compass as the hours pass during the daytime (see Figure 4.43).[1645]

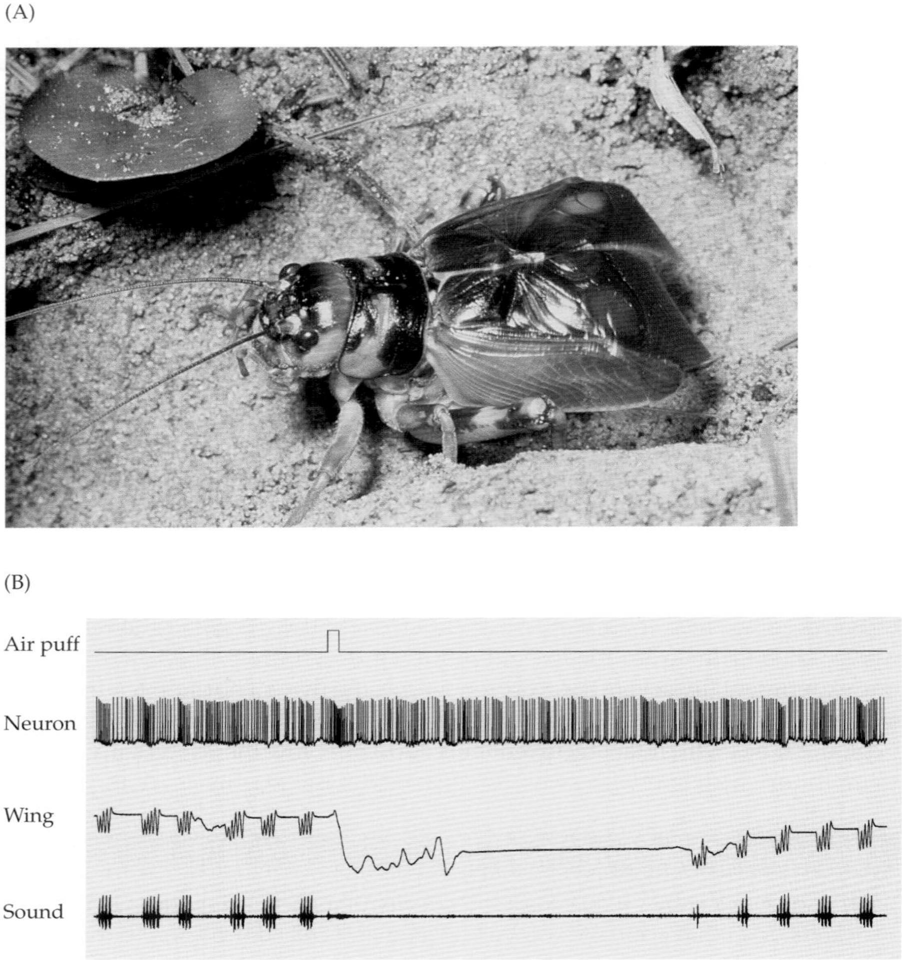

**FIGURE 5.5 Record of neural and behavioral activity of a calling cricket.** (A) Male crickets call by rubbing their forewings together. (B) The record of a command neuron in the brain of a calling male that was subjected to a brief tactile stimulus, a puff of air on the tip of the abdomen. The cricket's wing movements and the sounds of its calls are shown on separate lines. A, photograph by Edward S. Ross; B, from Hedwig.[641]

(A)

(B)

Researchers interested in how animals can change their priorities over time have developed two major competing theories. First, animals might change their priorities in response to a **biological clock**, a timing mechanism with a built-in schedule that acts independently of any cues from the animal's surroundings. That such environment-independent timing mechanisms might exist should be plausible to anyone who has flown across several time zones and then tried to adjust immediately to local conditions. The second theory, however, suggests that animals alter the relationships among command centers in their nervous systems strictly on the basis of feedback information gathered by mechanisms that monitor the surrounding environment. Such devices would enable individuals to modulate their behavior in response to certain changes in the world around them, such as a decrease in light intensity as evening comes on.

Let's consider these two possibilities in the context of the calling cycle of male *Teleogryllus* crickets. Each day's calling bout could begin at much the same time because the crickets possess an internal timer that measures how long it has been since the last bout began; they could use this environment-independent system to activate the onset of a new round of chirping each evening at dusk. Alternatively, the insect's neural mechanisms might be designed to initiate calling when light intensity falls below a particular level. If this second hypothesis is correct, then crickets held under constant bright light should

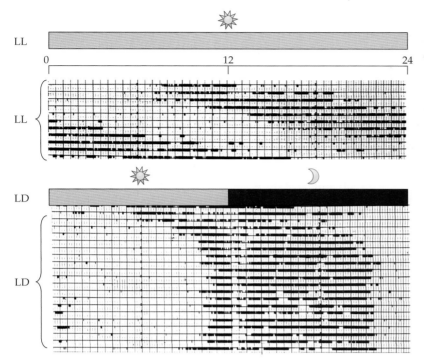

LL

LL

LD

LD

**FIGURE 5.6** **Circadian rhythms in cricket calling behavior.** Each horizontal line on the grid represents one day; each vertical line represents a half hour on a 24-hour time scale. Dark marks indicate periods of activity—in this case, calling. The bars at the top and middle of the figure represent the lighting conditions; thus, for the first 12 days of this experiment, male crickets are kept in constant light (LL), and for the remainder, they are subjected to 12-hour cycles of light and dark (LD). Male crickets held under constant light exhibit a daily cycle of calling and noncalling, but the calling starts later each day. The onset of "nightfall" on day 13 acts as a cue that resets the calling rhythm, which soon stops shifting and eventually begins an hour or two before the lights are turned off each day. After Loher.[882]

not call. But in fact, laboratory crickets held in rooms in which the temperature stays the same and the lights are on 24 hours a day still continue to call regularly for several hours each day. Under conditions of constant light, calling starts about 25 to 26 hours after it did the previous day (Figure 5.6). A cycle of activity that is not matched to environmental cues is called a **free-running cycle.** Because the length, or period, of the free-running cycle of cricket calling deviates from the many 24-hour environmental cycles caused by the Earth's daily rotation about its axis, we can conclude that the cyclical pattern of cricket calling is caused in part by an environment-independent internal **circadian rhythm** (*circadian* means "about a day").

Now let's place our crickets in a regime of 12 hours of light and 12 hours of darkness. The switch from light to darkness offers an external environmental cue that the crickets can use to adjust their timing mechanism. In fact, they do, just as monarch butterflies and pigeons can reset their clocks when moved into a laboratory with artificial lighting (see Chapter 4). After a few days, the males start calling about 2 hours before the lights go off, accurately anticipating "nightfall," and they continue until about 2.5 hours before the lights go on again in the "morning" (see Figure 5.6). This cycle of calling matches the natural one, which is synchronized with dusk; unlike the free-running cycle, it does not drift out of phase with the 24-hour day but is reset, or **entrained**, each day so that it begins at the same time in relation to lights-out.[882] From these results, we can conclude that the complete control system for cricket calling has both environment-independent and environment-dependent components: an environment-independent timer, or biological clock, set on a cycle that is not exactly 24 hours long, and an environment-activated device that synchronizes the clock with local light conditions.

Researchers with an evolutionary perspective would expect that members of the same species should possess different circadian mechanisms, if these individuals gain by having different daily behavioral schedules. One such animal is the sand cricket *Gryllus firmus*, a species that comes in different forms including a long-winged, flight-capable morph and a short-winged, flightless morph. The two forms differ in their wing size, their wing muscles, their hormones, their genetics, their circadian rhythms, and their behaviors.[1236] The form that can fly exercises that ability primarily at night in order to disperse safely from an unsuitable area or to search far and wide for nocturnally singing males. Flight-capable females use their ability to track down distant mates, while flying males may approach calling rivals in order to intercept the females attracted to these other males. In contrast, short-winged individuals lack the ability to fly and so do not have to schedule a time when it is profitable to sail off in search of callers elsewhere. Instead, they stay put and major in producing offspring, inasmuch as they do not have to invest in the structures needed for dispersal.[1236]

To compare the circadian rhythms of the two types of sand crickets, Anthony Zera and colleagues brought recently collected crickets to the laboratory and then took blood samples at intervals over the course of 24-hour periods. When they analyzed the juvenile hormone (JH) concentrations in the blood of short-winged crickets, they found no significant changes in relation to time of day. In contrast, however, the blood of flight-capable crickets revealed a very different story, with JH concentrations rising sharply in the late afternoon or evening from baseline levels comparable to those in the flightless crickets (Figure 5.7).[1644]

Zera and company were surprised to find that the flying form of *G. firmus* actually had higher JH concentrations for a number of hours each day than the flightless form. The hormone is known to cause the breakdown of flight muscles under some situations, which seems highly counterproductive for a cricket intent on aerial dispersal. However, in some cases JH has been reported to facilitate flight. For example, as you may recall, increases in JH concentrations in honey bee workers are associated with the age-related transformation that bees undergo as they shift from sedentary hivebound nurses to wide-ranging foragers that fly out from the hive in search of pollen and nectar. So it seems likely that the circadian surge in JH in fully winged crickets in some way helps prepare these individuals for a round of nocturnal flight, although just how JH exerts its effects remains to be fully determined.[1644]

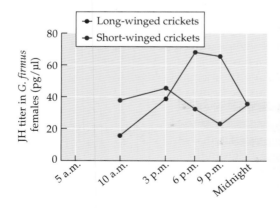

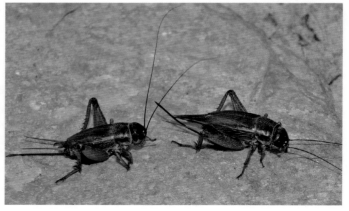

**FIGURE 5.7** **In the early part of the night,** the long-winged, flight-capable form of the cricket *Gryllus firmus* (on the right in the photo) has higher concentrations of juvenile hormone (JH) in its blood than the short-winged form of the cricket (on the left). Photograph of short-winged and long-winged forms of *G. firmus* by Derek Roff; after Zera, Zhao, and Kaliseck.[1644]

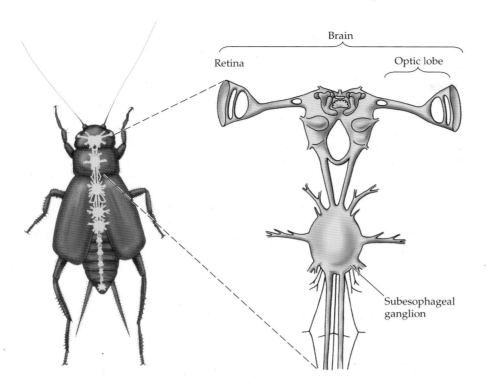

**FIGURE 5.8** **The cricket nervous system.** Visual information from the eyes is relayed to the optic lobes of the cricket's brain. If the optic lobes are surgically disconnected from the rest of the brain, the cricket loses its capacity to maintain a circadian rhythm. Based on diagrams by F. Huber and W. F. Shurmann.

## How Do Circadian Mechanisms Work?

In some crickets, cyclical release of juvenile hormone appears to contribute to the regulation of behavioral activity on a daily schedule. Other elements of their circadian mechanisms include the optic lobes. If one cuts the nerves carrying sensory information from the eyes of a male cricket to the optic lobes of his brain (Figure 5.8), the insect enters a free-running cycle. Visual signals of some sort are evidently needed to entrain the daily rhythm to local conditions, but a rhythm persists in the absence of this information. If, however, one separates both optic lobes from the rest of the brain, the calling cycle breaks down completely; all hours are now equally probable times for cricket calling. These results are consistent with the hypothesis that a master clock mechanism (Figure 5.9) resides within the optic lobes, sending messages to other regions of the nervous system[732, 1096] and almost certainly receiving and integrating hormonal signals generated by the animal's endocrine system as well.

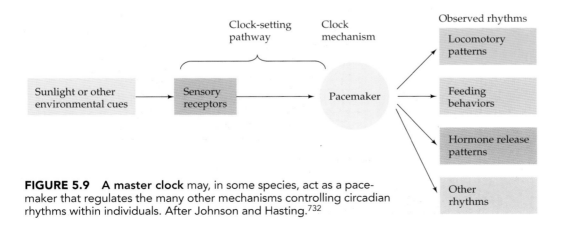

**FIGURE 5.9** **A master clock** may, in some species, act as a pacemaker that regulates the many other mechanisms controlling circadian rhythms within individuals. After Johnson and Hasting.[732]

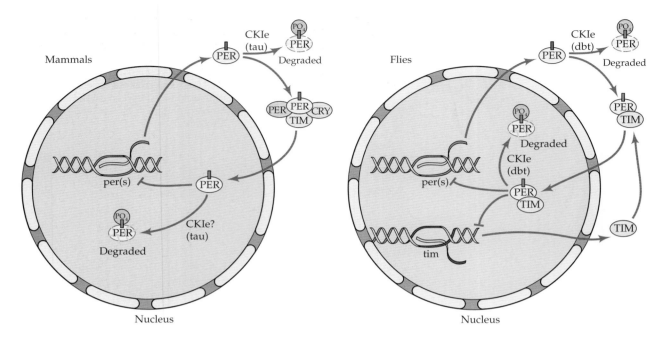

**FIGURE 5.10   The genetics of biological clocks in mammals and fruit flies.** In both groups, a set of three key genes produces proteins that interact with one another to regulate the activity of certain other genes on a cycle lasting approximately 24 hours. One of the genes (*per*) codes for a protein (PER) that gradually builds up inside and outside the cell nucleus over time. Another key gene, called *tau* in mammals and *dbt* in flies, codes for an enzyme, CKIe, that helps break down PER, slowing its rate of accumulation in the cell. But during peak periods of production of PER, more PER is available to bond with another protein (TIM), coded for by a third gene (*tim*). When the PER protein is bound in complexes with TIM (and another protein, CRY, in the case of mammals), it cannot be broken down as quickly by CKIe. Therefore, more intact PER is carried back into the nucleus, where it blocks the activity of the very gene that produces it, though only temporarily. Then a new cycle of *per* gene activity and PER protein production begins. After Young.[1633]

Biologists interested in the control of circadian rhythms in mammals and other vertebrates have focused on one important structure of the brain, the hypothalamus, with special emphasis on the suprachiasmatic nucleus (SCN), a pair of hypothalamic neural clusters that receive inputs from nerves originating in the retina. The SCN is therefore a likely element of the mechanism that secures information about day and night length, information that could be used to adjust a master biological clock.

If the SCN contains a master clock or pacemaker that is critical for maintaining circadian rhythms, then damage to the SCN should cause individuals to lose those rhythms. Such an experiment has been done by selectively destroying SCN neurons in the brains of hamsters and white rats, which subsequently exhibit arrhythmic patterns of hormone secretion, locomotion, and feeding.[1647] If arrhythmic hamsters receive transplants of SCN tissue from fetal hamsters, they sometimes regain their circadian rhythms, but not if the tissue transplants come from other parts of the fetal hamster brain.[377] Moreover, if an arrhythmic hamster gets an SCN transplant from a mutant hamster with a circadian period that is much shorter than the standard one of approximately 24 hours, the experimental subject adopts the circadian rhythm of the donor hamster, further evidence in support of the hypothesis that the SCN controls this aspect of hamster behavior.[1195]

Perhaps the SCN clock of hamsters, rats, and other mammals operates via rhythmic changes in gene activity. A key candidate gene in this regard appears to be the *per* gene, which codes for a protein (PER) whose production varies over a 24-hour schedule in concert with that of the product of another mammalian gene, called *tau* (Figure 5.10). The product of *tau* is an enzyme whose production is turned on when PER is at peak abundance in the cell. The enzyme degrades PER, contributing to a 24-hour cycle in which PER first increases in abundance and then falls.[1633]

A striking feature of this system, whose complexity is daunting, is that the key clock genes regulating cellular circadian rhythms in mammals are also present in insects. *Drosophila* fruit flies and honey bees also have the *per* gene, a chain of DNA composed of somewhat over 3500 base pairs, which provides the information needed to produce the PER protein chain of nearly 1200 amino acids. Alterations in the base sequence involving as little as a single base sub-

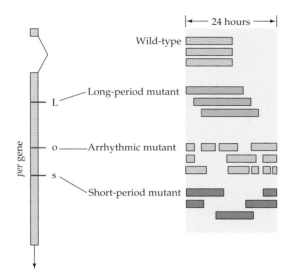

**FIGURE 5.11   Mutations of the *per* gene affect the circadian rhythms of fruit flies.** On the left is a diagram of the DNA sequence that constitutes the *per* gene. The locations of the base substitutions present in three mutant alleles found in fruit flies are indicated on the diagram. The activity patterns of wild-type flies, and those associated with each mutation, are shown on the right. After Baylies et al.[85]

stitution can result in dramatically different circadian rhythms in fruit flies (Figure 5.11), as well as in humans (carriers of one *per* mutation typically fall asleep at 7:30 in the evening and arise at 4:30 in the morning[1452]). These results strongly suggest that the gene's information plays a critical role in enabling circadian rhythms.

If expression of *per* does have this effect, then animals in which the *per* gene is relatively inactive should behave in an arrhythmic fashion. Very little PER protein is manufactured in young honey bees, which generally remain within the hive to care for eggs and larvae. And young honey bees are in fact just as likely to perform these nursing tasks at any time during the day or night over a 24-hour period. In contrast, older honey bees, which forage for food during the daytime only, exhibit well-defined circadian rhythms, leaving the hive to collect pollen and nectar only during that part of the day when the flowers they seek are most likely to be resource-rich. Foragers have almost three times as much PER protein in their brain cells as do young nurse bees, thanks to their heightened *per* gene activity.[1453]

## Discussion Question

**5.2** You may recall that the transition to foraging in honey bees depends on the makeup of the colony, such that if there is a shortage of nurse workers within the hive, older bees will delay their shift to the foraging role. What prediction follows about *per* gene expression in the brains of these socially delayed older nurses relative to foragers of the same age from other colonies with numerous young nurse bees? Provide proximate and ultimate hypotheses for the fact that social interactions can alter circadian rhythms in honey bees—and even fruit flies,[859] which do not live in highly organized societies.

Because the fruit fly, the hamster, the honey bee, and you and I all have the same gene serving much the same clock function, evolutionary biologists believe that we inherited this gene, as well as some others involved in the regulation of activity patterns, from a very ancient animal that lived perhaps 550 million years ago.[1633] In mammals and some other vertebrates, the *per* gene and certain others are expressed in the neurons of the SCN, which is usually viewed as the home of a master clock or circadian pacemaker that regulates many other tissues, thereby keeping many different behaviors on a

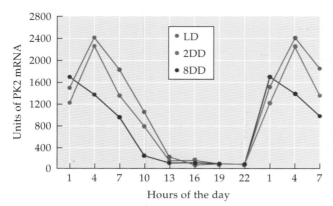

**FIGURE 5.12** **Expression of the gene that codes for PK2 in the SCN.** Mice were exposed to a standard light–dark cycle of 12 hours of light and 12 hours of darkness (LD) prior to the start of the experiment. The graph shows hourly changes in the production of the messenger RNA encoded by the *PK2* gene. The gene was expressed in a circadian rhythm whether the animals were held under the standard light–dark cycle, kept in complete darkness for 2 days (2DD), or held in darkness for 8 days (8DD). After Cheng et al.[258]

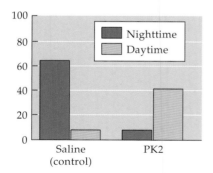

**FIGURE 5.13** **Circadian control of wheel running by white rats** changes when the brains of rats are injected with PK2. These rats are active primarily during the daylight hours, whereas control rats injected with saline exhibit the standard preference for nighttime activity. After Cheng et al.[258]

daily schedule. However, other parts of the brain may also have their own biological clocks. For example, the olfactory bulb in mice exhibits environment-independent cyclical changes in genetic activity, with the result that mice are more sensitive to odors at night than during the day, an adaptive effect of a timing mechanism for a nocturnally active creature.[563]

Nonetheless, the SCN clearly contains a major biological clock, which must send chemical signals to the target systems it controls. If so, we can predict that (1) the molecule that relays the clock's information ought to be secreted by the SCN in a manner regulated by the clock's genes, (2) there should be receptor proteins for that chemical messenger in cells of the target tissues, and (3) experimental administration of the chemical messenger should disrupt the normal timing of an animal's behavior.

The classic candidate chemical of this sort is melatonin, which has been the subject of much research demonstrating its involvement in the regulation of animal circadian rhythms.[1473] But melatonin is not the only such chemical. More recently, researchers have demonstrated that a protein, prokineticin 2 (PK2), has the key properties expected of a clock messenger.[258] So, in normal mice living under a cyclical regime of 12 hours of light and 12 hours of darkness, the PK2 protein is produced in a strongly circadian pattern by the SCN (Figure 5.12). Moreover, mice with certain mutations in key clock genes lack the circadian rhythm of PK2 production. Furthermore, as also predicted, only certain structures within the brain of the mouse produce a receptor protein that bonds with PK2. These regions are linked to the SCN by neural pathways and are believed to contain command centers that control various behavioral activities in a circadian fashion. Finally, if one injects PK2 into the brains of white rats during the night, when the animals are normally active, the behavior of these animals changes dramatically. Instead of running in their running wheel, they sleep, shifting to daytime activity instead (Figure 5.13).[258]

All of these lines of evidence point to PK2 as the chemical messenger that the mammalian SCN uses to communicate with and regulate target centers in the brain. The SCN, in turn, receives information from the retina about the light–dark cycle of the animal's environment so that it can fine-tune its autoregulated pattern of gene expression. The adaptive value of the environment-independent component of the clock system may be that a clock of this sort enables individuals to alter the timing of their behavioral and physiological cycles without having to constantly check the environment to see what time it is. The presence of an environment-dependent element, however, permits individuals to adjust their cycles in keeping with local conditions. As a result, a typical nocturnal mammal automatically becomes active at about the right time each night, while retaining the capacity to shift its activity period gradually to accommodate the changes in day length that occur as spring becomes summer, or summer becomes fall.

One way to test this hypothesis about the adaptive value of circadian rhythms would be to predict that if there were a mammal for which the day–night cycle was biologically irrelevant, then this species should lack a circadian pattern of activity. The naked mole rat (Figure 5.14) is such a species. This animal lives in groups that occupy an extensive network of underground tunnels. Because mole rats almost never come to the surface but instead live in total darkness, feeding on roots and tubers collected from their tunnels, what goes on aboveground is not important to them. As predicted, naked mole rats

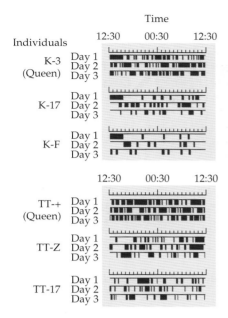

Naked mole rat

**FIGURE 5.14  Naked mole rats lack a circadian rhythm.** Patterns of activity are shown for six individuals from two captive colonies held under constant low light. Dark bars indicate periods when the individual was awake and active. After Davis-Walton and Sherman.[362]

lack a circadian rhythm. Instead, individuals scatter generally brief episodes of activity among longer periods of inactivity, with the pattern changing irregularly from day to day (see Figure 5.14).[362]

## Long-Term Cycles of Behavior

Because of their unusual lifestyle, naked mole rats do not have to deal with cyclically changing environments, and they have apparently lost their circadian rhythm as a result. But almost all other creatures confront not only daily changes in food availability or risk of predation, but also changes that cover periods longer than 24 hours, such as the seasonal changes that occur in many parts of the world. If circadian rhythms enable animals to prepare physiologically and behaviorally for certain predictable daily changes in the environment, might not some animals possess a **circannual rhythm** that runs on an approximately 365-day cycle?[597] A circannual clock mechanism could be similar to the circadian master clock, with an environment-independent timer capable of generating a circannual rhythm in conjunction with a mechanism that keeps the clock entrained to local conditions.

Testing the hypothesis that an animal has a circannual rhythm is technically difficult because individuals must be maintained under constant conditions for at least 2 years after being removed from their natural environments. One successful study of this sort involved the golden-mantled ground squirrel[1114] of North America, which in nature spends the late fall and frigid winter hibernating in an underground chamber. Five members of this species were born in captivity, then blinded and held thereafter in constant darkness and at a constant temperature while supplied with an abundance of food. Year after year, these ground squirrels entered hibernation at about the same time as their fellows living in the wild (Figure 5.15).

In another study, several nestling stonechats were taken from Kenya to Germany to be reared in laboratory chambers in which the temperature and **photoperiod** (the number of hours of light in a 24-hour period) were always the same. Needless to say, these birds, and their offspring, never had a chance to encounter the spring rainy season in Kenya, which heralds a period of insect abundance and is the time when Kenyan stonechats must reproduce if they

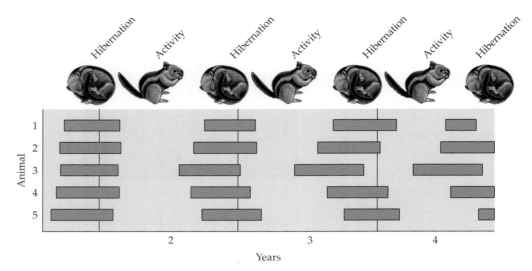

**FIGURE 5.15** **Circannual rhythm of the golden-mantled ground squirrel.** Animals held in constant darkness and at a constant temperature nevertheless entered hibernation (green bars) at certain times year after year. After Pengelley and Asmundson.[1114]

are to find sufficient food for their hungry nestlings. Thus, wild stonechats exhibit an annual cycle of reproductive physiology and behavior. The transplanted stonechats, despite their constant environment, also exhibited an annual reproductive cycle, but one that shifted out of phase with that of their Kenyan compatriots over time (Figure 5.16). One male, for example, went through nine cycles of testicular growth and decline during the 7.5 years of the experiment. Evidently, the stonechat's circannual rhythm is generated in part by an internal, environment-independent mechanism, just as is true for the golden-mantled ground squirrel.[596]

### The Physical Environment Influences Long-Term Cycles

In nature, environmental cues entrain circadian and circannual clocks so as to produce behavioral rhythms that match the particular features of the animal's environment, such as the times of sunrise and sunset, or the onset of the rainy season in a given year, or the increasing day lengths associated with spring. This fine-tuning of behavioral cycles involves mechanisms of great diversity that respond to a full spectrum of environmental influences, which can vary from species to species according to their ecological circumstances.

So, for example, nocturnal animals that are at risk of attack from visually alert predators should vary their activity in accordance with the lunar cycle. This prediction receives support from the behavior of tropical katydids that have to deal with bats that can spot them at night. (Some bats hunt using visual cues rather than relying totally on sonar echolocation.) When moonlight is minimal, the katydids on Barro Colorado Island in Panama are far more likely to be active, judging from the number caught in mist nets on dark nights versus nights when the moon is full.[834]

Likewise, banner-tailed kangaroo rats are more likely to stay in their underground retreats when moonlight is available to aid their nocturnal predators, such as great horned owls.[878, 879] Robert Lockard and Donald Owings reached this conclusion by monitoring the activity of free-living kangaroo rats in a valley in southeastern Arizona. To measure kangaroo rat activity, Lockard invented an ingenious food dispenser/timer that released very small quanti-

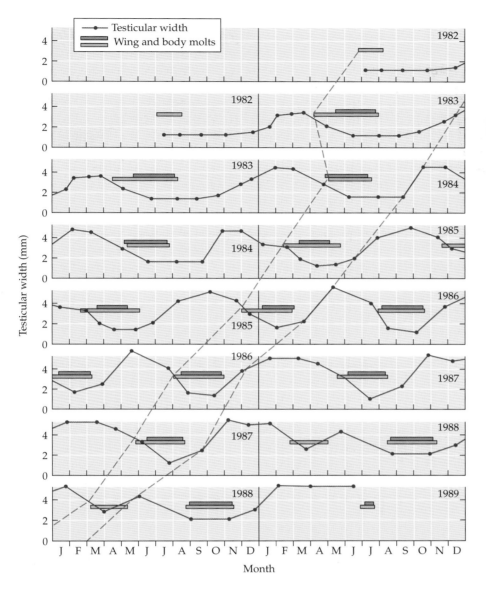

Stonechat

**FIGURE 5.16   Circannual rhythm in a stonechat.** When transferred from Kenya to Germany and held under constant conditions, this male stonechat still underwent a regular long-term cycle of testicular growth and decline (purple lines), as well as regular feather molts (the two bars, which refer to wing and body molts). The cycle was not exactly 12 months long, however, so the timing of molting and testicular growth shifted over the years (see the dashed lines that angle downward from right to left). After Gwinner and Dittami.[596]

ties of millet seed at hourly intervals. To retrieve the seeds, an animal had to walk through the dispenser, depressing a treadle in the process. The moving treadle caused a pen to make a mark on a paper disk that turned slowly throughout the night, driven by a clock mechanism. When the paper disk was collected in the morning, it carried a temporal record of all nocturnal visits to the dispenser.

Data collection was sometimes frustrated by ants that perversely drank all the ink or by Arizonan steers that trampled the recorders. Nevertheless, Lockard's surviving records showed that when the kangaroo rats had accumulated a large cache of seeds in the fall, they were selective about foraging, usually coming out of their underground burrows only at night when the moon was not shining (Figure 5.17). Because the predators of kangaroo rats (coyotes and owls) can see their prey more easily in moonlight, banner-tails are safer when they forage in complete darkness. For this reason, the kangaroo rats apparently possess a mechanism that enables them to shift their foraging schedule in keeping with nightly moonlight conditions.

**FIGURE 5.17 Lunar cycle of banner-tailed kangaroo rats.** Each thin black mark represents a visit made by a banner-tail to a feeding device with a timer. From November to March, the rats were active at night only when the moon was not shining. A shortage of seeds later in the year caused the animals to feed throughout the night, even when the moon was up, and later still, to forage during all hours of the day. After Lockard.[879]

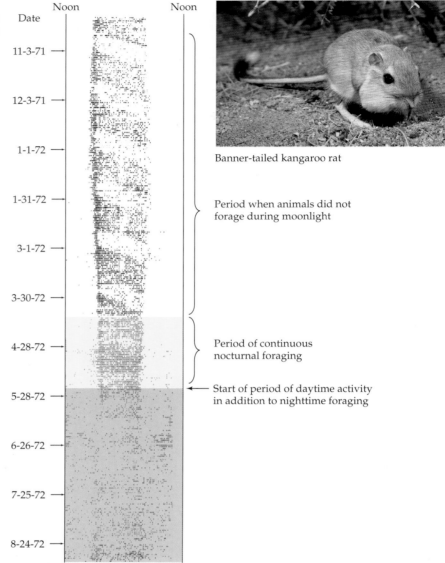

Banner-tailed kangaroo rat

Period when animals did not forage during moonlight

Period of continuous nocturnal foraging

Start of period of daytime activity in addition to nighttime foraging

Whereas desert kangaroo rats match their foraging behavior to the lunar cycle and Kenyan stonechats employ a circannual rhythm to breed at the right time in their tropical environment, temperate-zone birds such as white-crowned sparrows have their own behavioral control system that is well suited for coping with the dramatic seasonal changes that occur in their environments. In the spring, males fly from their wintering grounds in Mexico or the southwestern United States to their distant summer headquarters in the northern United States, Canada, or Alaska. There they establish breeding territories, fight with rivals, and court sexually receptive females. In concert with these striking behavioral changes, the gonads of the birds grow with dramatic rapidity, regaining all the weight lost during the winter, when they shrink to 1 percent of their breeding-season weight. In order to properly time the regrowth of their gonads and the onset of their reproductive activities, the birds must somehow anticipate the spring breeding season. How do they manage this feat?

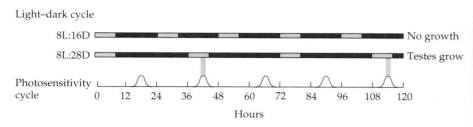

White-crowned sparrow

**FIGURE 5.18    A cycle of photosensitivity.** An experiment with white-crowned sparrows tested the hypothesis that these birds possess a clock mechanism that is especially sensitive to light between hours 17 and 19 of each day. The lower line represents these hypothetical periods of photosensitivity. The yellow and black sections of the two upper horizontal bars show the light and dark periods of two different light–dark regimes. Only sparrows under the 8L:28D experimental regime were exposed to light during the supposed photosensitive phase of the cycle, and only they responded with testicular growth. After Farner.[455]

The sparrows' ability to change their physiology and behavior depends on their capacity to detect changes in the photoperiod, which grows longer as spring advances in temperate North America.[456] One hypothesis on how such a system might work proposes that the clock mechanism of white-crowns exhibits a daily change in sensitivity to light, with a cycle that is reset each morning at dawn. During the initial 12 hours or so after the clock is reset, this mechanism is highly insensitive to light; this insensitivity then steadily gives way to increasing sensitivity, which reaches a peak 16 to 20 hours after the starting point in the cycle. Photosensitivity then fades very rapidly to a low point 24 hours after the starting point, at the start of a new day and a new cycle. Therefore, if the days have fewer than 12 hours of light, the system will never become activated because no light is present during the photosensitive phase of the cycle. However, if the photoperiod is longer than 14 or 15 hours, light will reach the bird's brain during the photosensitive phase, initiating a series of hormonal changes that lead to the development of its reproductive equipment and the drive to reproduce.

If this model of the photoperiod-measuring system is correct, it should be possible to deceive the system. William Hamner, working with house finches,[615] and Donald Farner, in similar studies with white-crowned sparrows,[455] stimulated testicular growth by exposing captive birds to light during the hypothesized photosensitive phase of their circadian rhythms. In Farner's experiment, birds that had been on a regular schedule of 8 hours of light and 16 hours of darkness (8L:16D) were shifted to an 8L:28D schedule. Because the light periods were now out of phase with a 24-hour cycle, these birds sometimes experienced light during the time when their brains were predicted to be highly photosensitive. The male birds' testes grew under these conditions, even though there was a lower ratio of light to dark hours than under the 8L:16D cycle, which did not stimulate testicular growth (Figure 5.18).[455]

## Discussion Question

**5.3**   In another experiment with white-crowned sparrows, groups of males that had been held on an 8L:16D schedule were housed in complete darkness for variable lengths of time (anywhere from 2 to 100 hours) before being exposed to an 8-hour block of light. A few hours later, the researchers measured the concentrations in the birds' blood of luteinizing hormone (a hormone released by the anterior pituitary and carried to the testes, where it stimulates the growth of these tissues). Do the collected data shown in Figure 5.19 provide a test for the photosensitivity hypothesis just described?

Photoperiod changes are useful guides for breeding activity for birds that live in environments where food resources track the seasons very predictably. But there are some birds for which the arrival of spring does not guarantee food resources sufficient for the rearing of a brood. For these species, one might expect the evolution of proximate mechanisms that respond to cues more closely

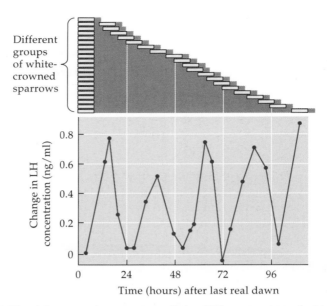

**FIGURE 5.19 A hormonal response to light.** Different groups of white-crowned sparrows were kept in complete darkness for different lengths of time before being exposed to a single 8-hour period of light. The upper diagram illustrates when the 8-hour light exposures occurred. The lower graph shows the changes from the start of the experiment in luteinizing hormone (LH) concentrations in the blood in each group of birds. LH is a hormone known to stimulate the growth of the gonads. After Follett, Mattocks, and Farner.[473]

related to the availability of food itself. This prediction can be checked by studying the rufous-winged sparrow of the Sonoran Desert. In southern Arizona, desert plants and the insects that feed upon them may be in very short supply in springs that follow dry winters. Thus, food for a sparrow's offspring may become available only after summer rains have fallen, and in fact, rufous-winged sparrows often wait to reproduce until after the onset of the monsoon, which can begin anywhere from early July to mid-August.[1346]

Sparrows that skip the spring in favor of summer breeding might do so by waiting for rainfall itself to trigger development of their gonads, rather than using less reliable photoperiodic cues. But when male rufous-winged sparrows are exposed to an increasing photoperiod in March, their testes grow, following the standard pattern for temperate-zone songbirds. However, under drought conditions, these males will not attempt to breed, despite their enlarged testes. Instead, a summer rainstorm is required to initiate the process. Rainfall appears to stimulate the production of luteinizing hormone, which in turn may cause the testes to produce testosterone, priming the male to reproduce. In conjunction with these changes, the song control system (see Chapter 2) in the sparrow's brain can grow even in July, when photoperiod is decreasing, something that does not happen in the brains of other sparrows living in more seasonally predictable regions. The increased volume of the song control system is associated with a dramatic increase in song production compared with the pre-monsoon period (Figure 5.20), an activity associated with reproductive territoriality.[1395]

Rufous-winged sparrows have evolved proximate mechanisms that respond to summer rainfall because in their desert environment, rainfall is a more reliable indicator of future food for offspring than photoperiod changes are. For two seed-eating finch species, the white-winged and red crossbills, it is food intake itself that acts as the primary determinant of breeding.[104] Craig Benkman has found that there are years when conifer seed production is so high

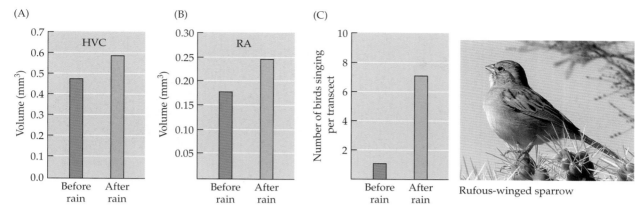

**FIGURE 5.20  Changes in the song control regions** of the rufous-winged sparrow's brain occur after summer rainfall and lead to an increase in singing behavior. The sizes of (A) the higher vocal center (HVC) and (B) the robust nucleus of the archistriatum (RA) in the sparrow increase following monsoon thunderstorms in southern Arizona. (C) The neural changes are linked to an increase in singing after the monsoon has begun. Green bars represent data collected before the first thunderstorms. Orange bars represent data collected after the start of the monsoon. Photograph of a rufous-winged sparrow by Pierre Deviche; A, B, and C after Strand, Small, and Deviche.[1395]

that the birds can find ample food for themselves and a brood in almost any month, and they accordingly take advantage of the good times to breed. This pattern can be linked to a distinctive feature of crossbill hormonal physiology, which is that birds that have been on long photoperiod days (20L:4D) do not completely shut down their gonadotropin-releasing neurons, as do other, related birds such as redpolls and pine siskins.[1118] Therefore, they appear to retain the capacity to stimulate the release of reproduction-regulating hormones should conditions for reproduction become especially favorable.

This flexibility, however, does not mean that crossbills ignore all environmental cues except seed abundance.[607] In studies of crossbills in the wild, Thomas Hahn noticed a break in crossbill breeding in December and January (Figure 5.21), even in environments where food was plentiful. Therefore, although these birds are more flexible and opportunistic than the average songbird, Hahn wondered if crossbills still have an underlying reproductive cycle dependent on photoperiod. When he held red crossbills at a constant temperature with unlimited access to food while letting the birds experience natural photoperiod changes, he found that male testis length fluctuated in a cyclical fashion (Figure 5.22), becoming smaller during October through December of each year, even when the birds had all the seeds a crossbill could hope to eat.[606] In addition, free-living red crossbills undergo declines in the concentrations of sex hormones in their blood at this time of year, even in areas where their food is abundant.[606] Therefore, the reproductive opportunism of these birds is not absolute, but rather superimposed on the photoperiod-driven timing mechanism characteristic of temperate-zone songbirds in general.

The persistence of the standard timing system in the flexibly breeding crossbills could be explained at the evolutionary level in at least two ways: (1) the photoperiod-driven mechanism might be a nonadaptive holdover from the past, or (2) crossbills might derive reproductive benefits from the retention of a physiological system that reduces the likelihood that they will attempt to reproduce at times when factors other than food supply make successful reproduction unlikely (such as the need to molt and replace their feathers in the fall, which requires a large caloric investment).[388]

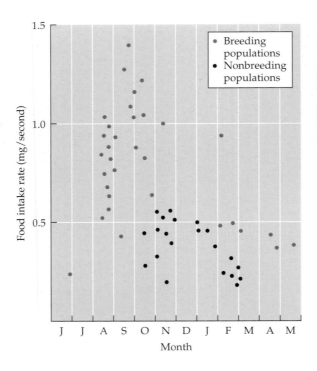

White-winged crossbill

**FIGURE 5.21 Food intake and reproductive timing in the white-winged crossbill.** Breeding populations usually occur in areas with relatively high food availability. Nonbreeding populations generally occur in areas where the birds have low food intake. Note, however, the absence of breeding populations in all locations in December and January. After Benkman.[104]

Red crossbill

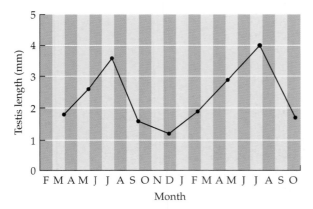

**FIGURE 5.22 Photoperiod affects testis size in the red crossbill.** Six captive birds were held under natural photoperiods, which changed over the seasons, but temperature and food supply were held constant. The data represent the average testis length among these birds at different times during the year. After Hahn.[604]

## Changing Priorities in Changing Social Environments

As we have just seen, different features of the physical environment, such as moonlight, day length, rainfall, or food supply, are used by different species to establish their behavioral priorities. In addition, animals can use changes in their social environments to make adaptive adjustments in their physiology and behavior. Thus, for example, when Hahn and several coworkers performed an experiment on crossbills in which some captive males were caged with their mates while others were forced into bachelorhood but were kept within sight and sound of the paired crossbills in a neighboring aviary, the bachelor males experienced a slower return to reproductive condition after the winter break than did the paired males.[605] Similarly, just one 60-minute exposure to a female resulted in elevated testosterone concentrations in captive male starlings,[1145] which may have contributed to subsequent changes in the males' behavior, such as increased courtship singing.

The social environment also affects the behavioral priorities of female and male house mice. When females are given a chance to show a preference for a potential mate, they can do so by spending time sniffing the male of their choice when placed in the center cubicle of a cage with males in compartments at either end. Females that have had prior experience only with a subordinate male do not exhibit a preference when given a choice between sniffing a dominant and a subordinate male. But those that have had prior experience with the odor of a dominant male strongly favor the dominant individual in the three-compartment cage in which female choice has been measured (Figure 5.23). Exposure to the scent of the dominant male promotes the addition of neurons in two regions of the mouse brain, effectively rewiring her

(A)

(B)

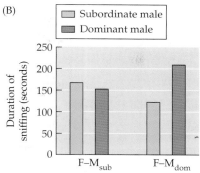

neural machinery such that she can identify a dominant male should her social environment provide her with such a partner.[920]

Dominant male house mice also are programmed to permit social experience to alter their behavioral decisions. For example, after a male house mouse mounts a female and ejaculates, he immediately becomes highly aggressive toward mouse pups, killing any he finds. For almost 3 weeks after mating, he is likely to commit infanticide, but after that time, he becomes more and more likely to protect any young pups he encounters. When about 7 weeks have passed since ejaculation, he becomes infanticidal once again.[1119]

This remarkable cycle has clear adaptive value. After a male transfers sperm to a partner, 3 weeks pass before she gives birth. Attacks on pups during these 3 weeks will invariably be directed against a rival male's offspring, with all the benefits attendant on their elimination (see page 23). After 3 weeks, a male that switches to paternal behavior will almost always care for his own neonatal offspring. After 7 weeks, his weaned pups will have dispersed, so once again he can practice infanticide advantageously.

At the proximate level, what kind of mechanism could enable a male to switch from infanticidal Mr. Hyde to paternal Dr. Jekyll 3 weeks after a mating? One possible explanation involves an internal timing device that records the number of days since the male last copulated. If such a sexually activated timing mechanism exists, then an experimental manipulation that either increases or decreases the length of a "day," as perceived by the mouse, ought to have an effect on the absolute amount of time that passes before the male makes the transition from killer to caregiver.

Glenn Perrigo and his coworkers manipulated day length by placing groups of mice under two different laboratory conditions, one with "fast days," in which 11 hours of light were followed by 11 hours of darkness (11L:11D) to make a 22-hour "day," and another with "slow days" (13.5L:13.5D) that lasted for 27 hours. As predicted, the total number of light–dark cycles, not the number of 24-hour periods, controlled the infanticidal tendencies of males (Figure 5.24). Thus, when male mice in the fast-day group were exposed to mouse pups 20 real days after mating, only a small minority committed infanticide, because these males had experienced 22 light–dark cycles during this period. In contrast, more than 50 percent of the males in the slow-day group attacked newborn pups at 20 real days after mating, because these males had experienced only 18 light–dark cycles during this time. These results demonstrate that a timing device registers the number of light–dark cycles that have occurred since mating and that this information provides the proximate basis for the control of the infanticidal response.[1119]

Perhaps the timing device that controls the male mouse's treatment of mouse pups influences the hormonal state of the male. If so, then you might expect that during their infanticidal phase, male mice would have high concentrations of testosterone in their blood, given the well-established relationship between testosterone and male aggression. An alternative hypothesis

**FIGURE 5.23  Dominant male odors change female mate preferences in the house mouse.** (A) The three-compartment cage in which a female (center chamber) can approach and smell males in either of two end compartments. (B) Females that have only experienced subordinate male pheromones do not exhibit a preference between subordinate and dominant males; females that have been exposed to dominant male mouse odors spend much more time sniffing the more dominant of two males in the experimental cage. A, photograph courtesy of Gloria Mak; B, after Mak et al.[920]

**FIGURE 5.24** **Regulation of infanticide by male house mice.** (A) Male mice were held under artificial "slow-day" and "fast-day" experimental conditions. (B) Most of the males held under fast-day conditions had stopped being infanticidal by 20 real days (22 fast days) after mating; males experiencing slow days did not show the same decline in infanticidal behavior until nearly 25 real days had passed. After Perrigo, Bryant, and vom Saal.[1119]

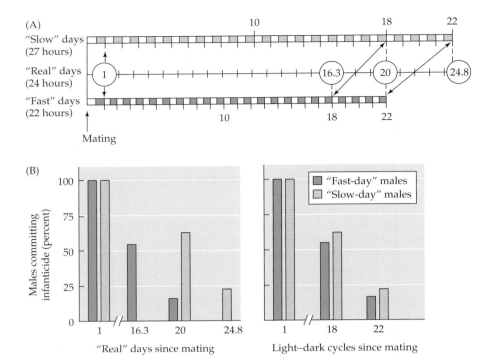

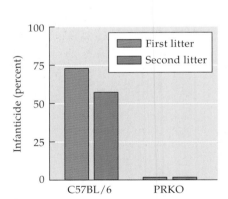

**FIGURE 5.25** **A hormonal effect on infanticidal behavior in laboratory mice.** Males of the strain C57BL/6 are highly likely to attack their own first or second litters, rather than caring for their young. If, however, the progesterone receptor gene is removed from the genome, the resulting progesterone knockout (PRKO) mice cannot detect the progesterone in their bodies, and they do not exhibit infanticidal behavior toward their offspring. After Schneider et al.[1289]

is that extreme male aggression toward mouse pups occurs when the males are under the influence of progesterone, because this hormone is known to suppress parental behavior in female rodents. To test these alternatives, Jon Levine and his coworkers used genetic knockout techniques to create a population of laboratory mice that lacked progesterone receptors, which meant that males of this line could not detect the progesterone in their bodies. When the knockout males were exposed to a pup, they never attacked it, whereas over half of the males from a genetically unmodified strain of laboratory mice assaulted a test infant (Figure 5.25). (This strain is an unusually aggressive one in which untreated males may kill their own pups.) There were no significant differences in testosterone or progesterone levels between the two groups of mice, only a difference in the ability of their brain cells to detect progesterone.[859] Of course, this difference was present in the developing male as well and so may have affected brain development, producing abnormalities that led to the behavior seen in this experiment. Nonetheless, this work suggests that when progesterone is present in certain concentrations within an intact male, the mouse becomes primed to be an infanticidal killer. As progesterone levels slowly fall (or sensitivity to the hormone declines in brain cells) after mating, the paternal capacity of the male may slowly increase in time, keeping him from harming his own brood.

## Discussion Question

**5.4** In the California mouse, *Peromyscus californicus*, males are highly paternal. Explanations for this behavior include the reduced progesterone hypothesis that we have just explored for laboratory mice. Alternatively, it is possible that increased levels of estrogen are responsible for male parental behavior. Make some predictions derived from these two hypotheses, including at least one that has to do with testosterone (if necessary, reread the introduction to this chapter to refresh your memory on the relation between testosterone and estrogen). Then examine Figure 5.26 and evaluate your hypotheses in light of these data.

(A)

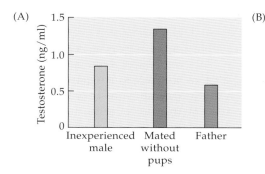

(B)

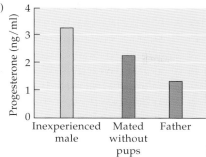

California mouse

**FIGURE 5.26    Testosterone and progesterone concentrations** in three categories of male California mice: inexperienced males with neither a mate nor offspring, mated males without pups, and father mice with a mate and offspring. (A) The testosterone differences between inexperienced males and fathers are not statistically significant. (B) In contrast, progesterone occurs in far higher amounts in the blood of inexperienced males than in males with offspring. Photograph of parental male California mouse by Brian Trainor; A and B, after Trainor et al.[1461]

Mating alters the behavioral priorities of many animals, not just house mice and prairie voles (see Chapter 1). Take the male Japanese quail, for example. When a male copulates with a female, even once, his behavior changes dramatically. Prior to mating, a male housed next to a female in a two-compartment cage will spend relatively little time peering through a window between the compartments in order to look at her. But after mating, a male appears positively fascinated by his partner, to the extent that he will stare at her for hours on end (Figure 5.27A).[71] This behavior, which presumably has the function in nature of bringing the male together with a receptive female for yet another copulation, is heavily influenced by the presence of testosterone in particular regions of the male's brain. This conclusion is based in part on the finding that removal of the male's testes eliminates all sexual behavior—unless the male receives an implant containing testosterone, in which case he regains his motivation to seek out mates.

(A)

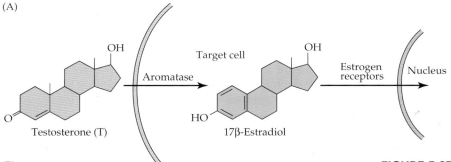

(B)

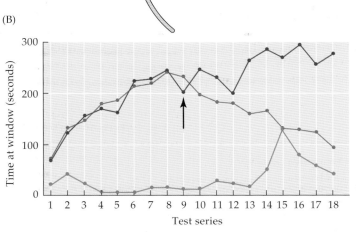

- Testosterone implanted
- Testosterone implant + aromatase inhibitor (at test 9)
- Controls (no testosterone implant)

**FIGURE 5.27    Testosterone and the control of sexual motivation in male Japanese quail.** (A) Hormonal changes contribute to changes in how sexually experienced males respond to the opportunity to view a mature female quail. Testosterone is released from the testes and travels to certain parts of the brain. When testosterone enters target cells, chemical reactions catalyzed by the enzyme aromatase convert it to 17$\beta$-estradiol. That substance bonds with an estrogen receptor to form an estrogen–receptor complex, which is transferred to the target cell nucleus, where it promotes chemical changes that ultimately cause a male quail to stare at a female through a window in their two-compartment cage. (B) Castrated males that have received testosterone implants exhibit the staring-at-female response, unlike controls that have not been given testosterone implants. But when implanted males are injected with an aromatase inhibitor, they lose the response, presumably because testosterone can no longer be converted to 17$\beta$-estradiol, the essential hormone for modulating male sexual motivation. After Balthazart et al.[71]

Interestingly, testosterone itself is not the signal that turns a male Japanese quail into an apparently lovelorn individual when separated from his partner. Instead, testosterone is converted into 17β-estradiol, an estrogen, in target cells within the preoptic area of the brain. The conversion requires an enzyme, aromatase, which is coded for by a gene that becomes much more active after a male quail has mated. The estrogen produced in the presence of aromatase then bonds with an estrogen receptor protein, and the resulting estrogen–receptor complex relays a signal to the nucleus of the cell, leading to further biochemical events. These events ultimately translate into the neural signals that cause a male to stare intently at a currently unreachable sexual partner.

The importance of the enzyme aromatase to this process has been established experimentally by creating three groups of castrated male quail. The controls received no replacement testosterone, whereas males in two other groups received testosterone implants. The implanted males spent much more time at the window than the controls. After eight trials, one of the two groups of implanted males was given daily injections of an aromatase inhibitor; these birds' interest in the female in the next compartment fell steadily as seen in the reduction in time they spent looking through the window over the next 10 trials (Figure 5.27B).

### Hormones and the Organization of Reproductive Behavior

Many of the hormonally mediated adjustments that animals make directly affect their reproductive behavior. In the Japanese quail, 17β-estradiol plays a major role in adjusting the degree to which a male becomes attached to one mate, whereas in the prairie vole, vasopressin is more important in regulating sexual bonding (see page 4).

Fruit fly males and females do not form lasting sexual attachments. Instead, soon after mating, the female becomes completely unreceptive, rejecting not only her previous partner but all other males, should they attempt to court her. Rather than spend time copulating, she goes about for some days laying eggs, an adaptive shift in behavior given that she can fertilize all her mature eggs with stored sperm received from a single mating. As it turns out, the female's dramatic switch from sexual receptivity to sexual refusal is regulated by a hormone—not one that she manufactures herself, but one that she receives from the seminal fluid the male transfers to her during copulation.

The male-donated hormone is called SP (sex peptide). Using a technique called RNA interference, it is possible to block the specific gene in male fruit flies that codes for SP. The blocked males are normal except for the fact that they cannot make SP and, therefore, when they copulate with females, they are unable to transfer SP to their mates. If SP is indeed the critical protein signal that stops females from responding to courting males, then females mated with SP-deficient males should remain receptive following copulation, despite having received sperm and seminal fluid. This prediction was shown to be correct (Figure 5.28).[254]

As noted earlier, a chemical messenger acts by bonding with receptor molecules on the surface of a target cell. The fruit fly hormone SP should therefore bond with a specific receptor protein, and accordingly, the appropriate molecule, unimaginatively called sex peptide receptor (SPR), has been found. Here too it is possible to block the single gene that codes for SPR, creating "mutant" females whose cells are incapable of bonding with the male-donated hormone. When this is done experimentally, SPR-missing females that have mated once copulate again when given a chance, just as if they were virgin females.[1628]

Additional research was needed to identify the target organ(s) of the male's SP. These targets were found by first supplying the tissues of female fruit flies with a compound that bonds specifically to SPR and then staining the tissues with a dye specific to the attached compound. The stain was concentrated in

(A)

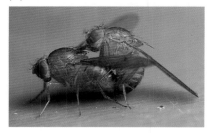

(B)

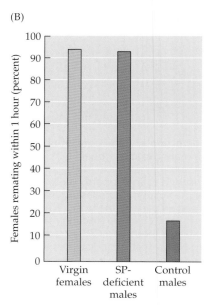

**FIGURE 5.28** **Female fruit flies mated to males unable to supply sex peptide (SP)** are as likely to copulate again within 48 hours as are virgin females in the same period. A, photograph of mating *Drosophila* by Brian Valentine, www.flickr.com/photos/lordv; B, after Chapman et al.[254]

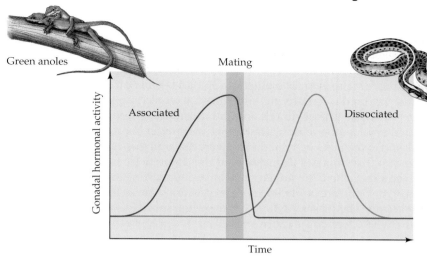

Green anoles

Mating

Associated

Dissociated

Red-sided garter snake

Gonadal hormonal activity

Time

**FIGURE 5.29** **An associated reproductive pattern** is one in which environmental cues trigger internal hormonal changes, which then activate behavioral responses within a relatively short time (as illustrated by the green anole). In a **dissociated reproductive pattern**, mating may be dissociated in time from gonadal (and hormonal) activity (as in the red-sided garter snake). After Crews.[316]

the sperm storage organs of the female and a particular portion of the brain whose neurons express a gene called *fruitless*. This gene's activity in a select set of nerve cells in the male brain is critical for the male's capacity to court females effectively.[1267] In the female fly, the neurons that express the information contained in *fruitless* are the ones that can detect SP. When these neurons are exposed to SP after a mating, their activity is apparently altered after the male-donated hormone bonds with the special receptors in the nerve cells. It seems likely that SP affects *fruitless* gene expression in the target cells, provided that they possess the typical form of the receptor needed to bond with SP.[1628]

Hormonal signals are important not only in the interactions between males and females but also in encounters between males. Thus, when a male cichlid fish is ousted from his mate-attracting territory, a change in gonadotropin-releasing hormone production contributes to reductions in his aggressiveness, the size of his testes, and his efforts to reproduce (see pages 94–96).[1550] In a host of other species, hormones also underlie the integrated shifts in sexual and aggressive behavior that facilitate reproduction at times when external environmental and internal physiological conditions are most favorable.

Hormonal organization of reproduction often involves coordinated changes in gamete production and in sexual activity—the so-called associated reproductive pattern (Figure 5.29). This pattern is, for example, evident in the seasonal reproductive behavior of red deer stags living in Great Britain. Stags that have been living peacefully with one another all summer become aggressive as September approaches prior to the mating season. At this time, their testes generate sperm and testosterone. Adult males that have been castrated prior to the fall rutting season show little aggression and do not try to mate with sexually receptive females. If the behavioral differences between castrated and intact males stem from an absence of circulating testosterone in the castrated stags, then testosterone implants should restore their aggressive and sexual behavior during the rut, and they do.[869]

Likewise, in the green anole when males first become active after a winter dormant phase, their circulating concentrations of the hormone testosterone are very low. As this hormone begins to be produced in greater amounts, however, the males' testes grow in size, and mature sperm are made. At this time some males begin to defend territories and court females.[721] These individuals tend to be larger and more powerful biters, to have much more testosterone in their blood than smaller male anoles, and to have larger dewlaps with which to court females and threaten rivals.[702]

These apparent effects of testosterone are in part dependent upon the season of the year, a conclusion reached by Jennifer Neal and Juli Wade, who implanted two groups of captive male green anoles with testosterone capsules. One group had been exposed to warmer temperatures and long day lengths, mimicking the environmental conditions that the lizards would encounter during their normal summer breeding season. The other group had been kept under cooler conditions and shorter day lengths, the situation that would apply during the time of the year when the lizards do not normally reproduce. Testosterone-implanted males in the "breeding season" (BS) group were much more likely to display to females and to copulate with them than were males in the "nonbreeding season" (NBS) group. The brains of BS males were also different from those of NBS males in that neurons in their amygdalae were larger, suggesting that here, too, specific regions of the brain are under hormonal regulation. The fact that this effect is not independent of the season of the year suggests a mechanism for the associated reproductive pattern of the green anole.[1030]

## Discussion Question

**5.5**  In women, the menstrual cycle involves hormonally mediated changes that regulate the production of a mature egg. Might the link between hormones and ovulatory physiology extend to female reproductive behavior, producing an associated reproductive pattern? What would evolutionary biologists predict about the relationship between the menstrual cycle and sexual desire? Why would they predict a difference in this relationship for married women versus those without a steady partner? Why might evolutionary biologists also predict that at the time of ovulation, women should find males with masculinized facial features especially attractive? (See Figure 14.9, page 521, for an example of variation in male facial features.) Check your responses against Gangestad et al.,[514] Macrae et al.,[912] and Pillsworth, Haselton, and Buss.[1142]

Although many species studied to date appear to possess associated reproductive patterns, the theory that reproductive behavior is under this kind of hormonal control continues to be tested, as it should be.[318] If testosterone is required for sexual behavior in birds, then castration should eliminate male sexual behavior, because removal of the testes eliminates a major source of testosterone. This prediction has been confirmed for some species, including the Japanese quail, which if castrated, stops reproducing but is back in business if given a testosterone implant,[1087] just like the red deer. But the prediction fails for white-crowned sparrows. Even without his testes, a male white-crown will mount females that solicit copulations, provided that he has been exposed to long photoperiods.[1007] Moreover, some populations of white-crowned sparrows raise more than one brood per breeding season. Those male white-crowns that mate with females to produce a second or third clutch of fertilized eggs in the summer have relatively low testosterone concentrations in their blood at that time (Figure 5.30), further evidence that high levels of testosterone are not essential for male sexual behavior in this species.

## Discussion Question

**5.6**  In studying the hormonal control of behavior, it is common to remove an animal's ovaries or testes and then inject the creature with assorted hormones to see what behavioral effects they have. What advantage does this technique have over another approach, which is simply to measure the concentrations of specific hormones in the blood of animal subjects from time to time? The far less invasive direct measurement approach would show, for example, whether testosterone or estrogen concentrations were elevated when mating was occurring.

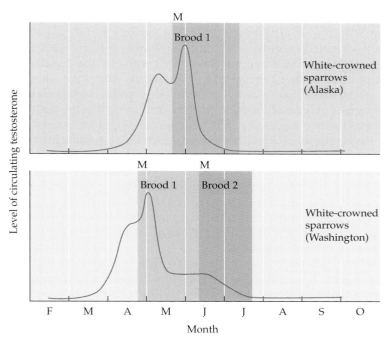

**FIGURE 5.30  Hormonal and behavioral cycles** in single-brood and multiple-brood populations of white-crowned sparrows. Testosterone concentrations in the blood of male white-crowns peak shortly before the time when the males mate with females (M) in their first breeding cycle of the season. In populations that breed twice in one season, however, copulation also occurs during a second breeding cycle, at a time when testosterone concentrations are declining. After Wingfield and Moore.[1598]

## Discussion Question

**5.7**  In the guinea pig, individual males vary in their sex drives, as measured (for example) by the number of times a male ejaculates when given access to receptive females for a standard period of time. One hypothesis for this variation is that male sex drive correlates with circulating testosterone concentrations. What prediction follows from this hypothesis? Figure 5.31 presents data from an experiment with three guinea pigs in which male sex drive was measured. All three males were castrated, after which their sex drives continued to be monitored until finally, after some weeks, the three males were all given the same amount of supplemental testosterone, and their sex drives were measured again at intervals. What is the relevance of these data for the hypothesis in question? What scientific conclusion can you reach based on these results?

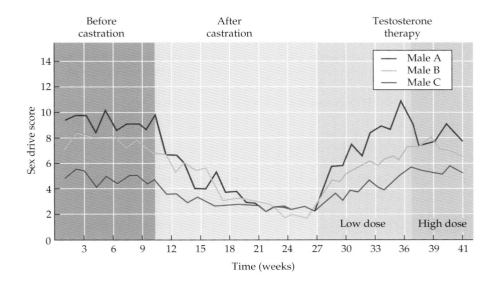

**FIGURE 5.31  The effects of castration followed by testosterone therapy** on three male guinea pigs. "Sex drive score" is a measure of copulatory activity. After Grunt and Young.[592]

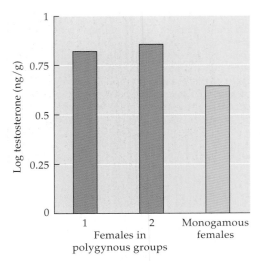

**FIGURE 5.32 Testosterone and female aggression in the dunnock.** Testosterone concentrations were higher in females competing for males in polygynous groups (red bars) than in females in monogamous relationships. After Langmore, Cockrem, and Candy.[836]

## The Variable Role of Testosterone

The white-crowned sparrow is not the only bird in which high testosterone concentrations are not necessary for courtship and mating.[1571] Could testosterone have functions in some species other than stimulating sexual behavior? One possibility is that testosterone regulates singing behavior, as noted earlier in this chapter. Another is that the hormone acts as a facilitator of aggression. If this hypothesis is correct, then we can predict that in seasonally territorial birds, testosterone concentrations should be especially high early in the breeding season, when males are aggressively defending a territory against rivals. This phenomenon has been observed in white-crowned sparrows (see Figure 5.30) and some other birds.[1598] Even territorial spotted antbirds, which live in the tropics and do not exhibit elevated testosterone concentrations in any particular seasonal pattern, respond to recorded songs of their species with a rapid buildup of circulating testosterone.[1571]

The hypothesis that testosterone has a role to play in regulating aggression leads to another prediction, which is that when competition among females is a regular feature of a bird's natural history, then aggressive females should possess relatively high testosterone levels. One bird species with competitive females is the dunnock (see Figure 10.30). Female dunnocks often wind up breeding in small groups that share one or a couple of males. In these situations, a female's reproductive success depends on how much assistance she can secure from the male or males with whom she lives. Therefore, females try to keep other females away from "their" partner(s), chasing them, giving aggressive "tseep" calls, even singing to announce their readiness to fight. When the testosterone levels of females in multi-female groups were compared with those of unchallenged females who were each paired with a single male, the results supported the hypothesis that testosterone is a hormonal facilitator of aggression (Figure 5.32),[836] although it is possible that fighting causes testosterone levels to rise, rather than the other way around.

Even in species in which testosterone definitely promotes adaptive sexual or aggressive behavior, it is the rule that testosterone concentrations fall to almost nothing outside the breeding season or after the likelihood of territorial challenges subsides. Why should this be? Perhaps the hormone comes with a price, reducing fitness at certain times or in certain situations. The hormone does have multiple effects (Figure 5.33), not all of them positive, including its interference with the immune system,[1649] which may explain why males of so many vertebrates are more likely than females to be infected by viruses, bacteria, and parasites.[780] High testosterone concentrations may also contribute to high glucocorticosteroid concentrations, which may contribute to or be reflective of physiological stress,[1600] and thus greater vulnerability to disease-causing organisms.

In addition, the direct behavioral effects of testosterone can be costly, leading individuals under the influence of the hormone to expend energy at a much higher rate than otherwise.[936] For example, when male barn swallows had their chest feathers painted a darker red, they became more attractive to females *and* a target of other aggressive males. As a result, their testosterone concentrations went up—and their body weight went down over time, again almost certainly because the hormone induced the birds to become more physically active at the cost of using up their energy reserves.[1270]

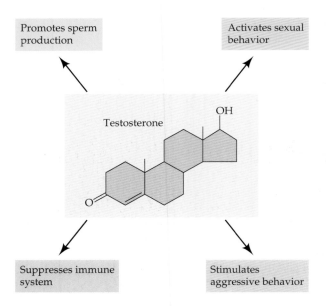

**FIGURE 5.33 The chemical structure of testosterone** and its diverse effects on physiology and behavior. After Wingfield, Jacobs, and Hillgarth.[1599]

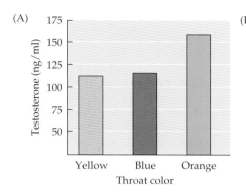

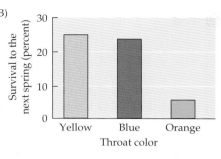

*Uta stansburiana*

**FIGURE 5.34   A survival cost of testosterone.** (A) The three different forms of the lizard *Uta stansburiana* have different throat colors and different amounts of testosterone circulating through their blood on average. (B) The form with the highest testosterone concentrations also has a very high annual mortality rate. Photograph by the author; A and B, after Sinervo et al.[1334]

Testosterone-saturated males can also become so focused on mating with females or fighting with rivals that they become sitting ducks for predators or parasites. In some species of this sort, males with higher testosterone concentrations are less likely to survive than those with more modest amounts of the hormone in their blood (Figure 5.34)[1334] (see also Figure 8.26). Even if a testosterone-saturated male survives, he may neglect his young in favor of fighting with other males. Thus, although male dark-eyed juncos that have received testosterone implants do not experience higher mortality than control birds, these songbirds do feed their young less often,[768] which may be why the offspring of testosterone-enhanced males produce smaller offspring that do not survive as well as those of control males.[1201]

Testosterone implants may keep a male junco's testosterone concentrations abnormally high for long periods, whereas in nature, males can boost their testosterone concentrations as needed in response to certain events, rather than being constantly under the influence of the hormone. To determine the consequences of more natural temporary surges in testosterone, a team of behavioral ecologists led by Ellen Ketterson injected gonadotropin-releasing hormone (GnRH) into captured male juncos. This treatment induced a brief and variable increase in testosterone concentrations in the team's subjects, which were then released back to their nesting territories. Those males whose testosterone concentrations were relatively high after the GnRH challenge behaved more aggressively toward a simulated intruder (a caged male junco placed in the center of the resident's territory). But males that increased their testosterone concentrations the most in response to the GnRH challenge also brought food to their nestlings at the lowest rate.[965] Even a short period of elevated testosterone concentrations apparently has the potential cost of making a male a less helpful parent.

### Discussion Question

**5.8** Here is another question about the unusually paternal California mouse *Peromyscus californicus*, whose males are highly protective of their offspring. If it is true that testosterone interferes with paternal behavior, then castration of males of the California mouse ought to have what effect? Check your prediction against data collected by Brian Trainor and Catherine Marler and presented in Figure 5.35.[1460] What is your conclusion about the trade-off hypothesis as it applies to this species?

Given the potential disadvantages of testosterone, which vary from species to species, it is not surprising that the pattern of hormonal control of aggression and reproduction is not the same for every animal (Figure 5.36). In the song sparrow, for example, males defend territories long after

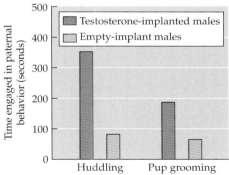

**FIGURE 5.35   The time spent in parental behaviors** (huddling with pups and grooming pups) by castrated male California mice that had received replacement testosterone (red bars) versus males that had been given empty implants (orange bars). After Trainor and Marler.[1460]

**FIGURE 5.36** **Testosterone and territorial behavior.** No one pattern exists for the relationship between testosterone concentrations and the duration of male territoriality. In some bird species (top three panels), a surge of testosterone occurs at the onset of territoriality and breeding, but in other birds (bottom two panels), males are territorial at times when they have little or no circulating testosterone. After Wingfield, Jacobs, and Hillgarth.[1599]

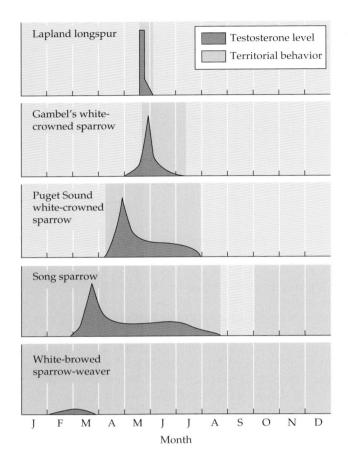

the breeding season is over, when testosterone has essentially disappeared from their blood. But these birds produce nongonadal sex hormones, especially estrogen, which are thought to play a critical role in the post-breeding territorial phase. If estrogen does boost male territoriality, then males treated with fadrozole (FAD), which blocks the manufacture of estrogen, should sing less often and stay farther away from a simulated intruding rival than untreated controls or males that have received both FAD and replacement estrogen. You can evaluate the data for yourself (Figure 5.37).[1367]

The red-sided garter snake offers another demonstration that hormone–behavior relationships differ among species. This reptile lives as far north as southern Canada, where it spends much of the year dormant in a sheltered underground hibernaculum. On warm days in the late spring, the snakes begin to stir, and they soon emerge, sometimes by the thousands, from their hibernaculum (Figure 5.38). Before going their separate ways, they engage in an orgy of sexual activity, with males slithering after females and attempting to copulate with them. At this time, they will ignore food even if it is available, as they focus on females to the exclusion of nearly everything else. Later in the season, when the odds of finding receptive females fall, male snakes become more motivated to exercise the foraging option, a nice example of a species with proximate systems that enable it to resolve competition among behavioral choices adaptively.[1065]

During the mating frenzy, males compete for females by trying to contact receptive partners before their fellow males do, but they do not fight with one another for the privilege of copulation. Examination of the sex hormone concentrations in their blood reveals that these nonaggressive snakes have almost no circulating testosterone or any other equivalent substance. Yet they have

(A)

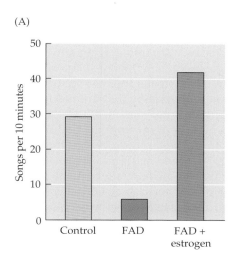

(B)

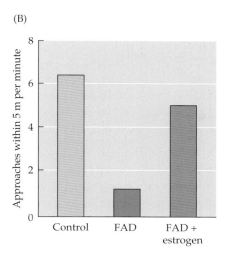

**FIGURE 5.37  Estrogen and territorial behavior.** Male song sparrows treated with fadrozole (FAD), a chemical that blocks the production of estrogen, were significantly less likely (A) to sing or (B) to come within 5 meters of a simulated intruder than were control males or males given both FAD and estrogen. After Soma, Tramontin, and Wingfield.[1367]

no trouble mating, so red-sided garter snakes are animals with a dissociated reproductive pattern (see Figure 5.29).[315] Indeed, various hormonal manipulations have been performed on adult male garter snakes without any effect on their sexual behavior. Removal of the pineal gland prior to hibernation, however, produces male snakes that almost always fail to court females the following spring.[318] The pineal gland provides a critical mechanism for detecting temperature increases following a period of hibernation, and warmer weather suffices to activate sexual behavior in male garter snakes independently of their testosterone concentrations.

This does not mean, however, that testosterone has no role at all to play in the reproductive cycle of the garter snake. In the fall, male snakes have high concentrations of testosterone, which contribute to the production of sperm, which are stored internally over the winter in anticipation of the spring mating frenzy. Furthermore, although temperature increases may be the cue for the activation of sexual activity, testosterone may play an organizational role in the development of the mechanisms underlying reproductive behavior in the red-sided garter snake, as it does in so many other vertebrates. Evidence on this point comes from experiments on adult male snakes that were castrated shortly before the breeding season. Without their testes, these individuals

**FIGURE 5.38  Spring mating aggregation of red-sided garter snakes.** The males search for and avidly court females emerging from hibernation. Photograph by Nic Bishop.

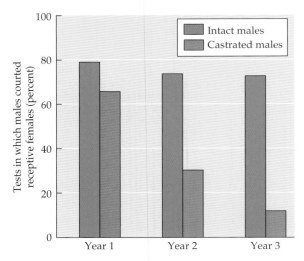

**FIGURE 5.39 Testosterone and the long-term maintenance of mating behavior.** Male garter snakes whose testes were removed shortly before the breeding season in year 1 remained sexually active during that breeding season, despite the absence of testosterone. But in years 2 and 3, these males became less and less likely to court receptive females, compared with males that still possessed their testes. After Crews.[319]

could not produce testosterone, but they still exhibited courtship behavior after a period of hibernation in the laboratory. If, however, the castrated snakes were tested again after a second hibernation period, their sexual activity fell sharply. These results suggest that the surge of testosterone that occurs prior to hibernation primes the neural systems that are needed for sexual behavior the following spring (Figure 5.39).

This hypothesis receives more support from the finding that testosterone implants given in the summer before the male snake's first hibernation turn 1-year-old males into sexually active animals, although snakes of this age are normally sexually immature. Therefore, testosterone production, which normally begins in the snake's second or third year, appears to be necessary for the full development and maintenance of the mechanisms that control garter snake sexual behavior.[319] However, high concentrations of testosterone are not necessary for copulation to occur in the spring—another example of the rule that there is no one hormonal mechanism that regulates the sexual behavior of all animals, or even all vertebrates, in exactly the same way.[3, 320]

Despite the fact that hormonal systems for organizing behavior are highly diverse, you will have noted that the same hormones (especially testosterone and estrogen) have appeared repeatedly in this chapter. Likewise, in many vertebrates, much the same set of neural structures, including specific regions of the hypothalamus, interacts with certain hormones in ways that make behavioral organization possible. Thus, the diversity of neural and hormonal systems consists mainly of variations on a theme, rather than arrays of uniquely different mechanisms. Even the red-sided garter snake, whose reproductive behavior appears to be relatively independent of sex hormones, has neurons with receptors that bond with sex hormones in the neural pathways that control its mating behavior.[807] Evidently, as clusters of species have arisen and diverged from a common ancestor, each descendant species has retained elements of its hormonal past, which have been modified and rearranged over time but not replaced by something totally new. The imprint of evolution is apparent in the proximate mechanisms that organize the behavior of today's animals.

## Discussion Question

**5.9** The endangered Amargosa River pupfish lives in Death Valley, where different populations of the species live in total isolation from one another in tiny permanent pools and short flowing stream segments.[851] Researchers have estimated that some populations have been cut off from others for only 400 to 4000 years. Yet the reproductive behavior of the different populations can be quite different. In some populations, males aggressively defend territories and court females drawn to those sites. But in other populations, males are not aggressive toward one another and do not form territories; instead they pursue and court females as they become receptive. In other words, dramatic shifts in behavior have occurred in this species in just a few hundred to a few thousand years. Imagine that you wish to explain how these changes could occur so rapidly, knowing what you now know about the ability of hormones to organize sexual and aggressive behavior. Note that arginine vasotocin (AVT) is a brain hormone that has been linked to decreased aggressive behavior in some fish. How might you establish experimentally that the hormone had this effect in the pupfish? If the hormone did lower aggression in this species, what predictions could you make about AVT or about AVT receptor protein differences among different populations, as well as between territorial and nonterritorial males of the same population? If your results supported a particular proximate hypothesis about AVT's effect on aggression, describe how selection could lead to rapid changes in territorial behavior in this species.

## Summary

1. Because an animal's environment provides various stimuli that can trigger contradictory responses, and because its physical and social environment often change over time, animals gain by having mechanisms that set priorities for their different behavioral options. One such proximate system includes behavioral command centers with the capacity to inhibit one another so that animals do not try to do several things simultaneously.

2. As the environment changes, the nature of the inhibitory relationships between neural command centers may also change. Devices to achieve this end include the various pacemaker or clock mechanisms that regulate nervous system functioning and hormonal output in cycles that typically last either 24 hours or 365 days. Circadian and circannual clocks have environment-independent components, but they can also adjust their performance by acquiring information from the environment about local conditions, such as the time of sunrise or sunset.

3. Hormones function within those mechanisms that establish behavioral priorities. In many animals, changes in the physical environment (such as seasonal changes in photoperiod) and in the social environment (such as the presence of potential mates) are detected by neural mechanisms and translated into hormonal messages. These chemical signals often set in motion a cascading series of physiological and behavioral changes that make reproductive activity the top priority at times when it is most likely to translate into the production of surviving offspring.

4. The precise roles played by hormones in effecting behavioral change vary from species to species. Male sexual behavior, for example, may or may not be dependent on high testosterone concentrations in the blood. Nevertheless, the proximate mechanisms of behavioral organization of many different species show similarities, such as a reliance of one sort or another on testosterone and certain other widely distributed hormones for organizing male reproductive behavior. This pattern reflects the nature of evolutionary change, in which the attributes of today's species are modified versions of previous ones, not inventions that have arisen out of thin air.

## Suggested Reading

Kenneth Roeder's classic *Nerve Cells and Insect Behavior*[1234] and Vincent Dethier's *The Hungry Fly*[385] discuss how some animals avoid conflicting responses and structure their behavior over the short haul. Randy Nelson has written a fine textbook that covers all the topics in this chapter in much more detail.[1037] In just a few pages, Michael Young explains the exceedingly complex molecular basis of circadian rhythms about as clearly as possible.[1633] The dissociated reproductive pattern of the red-sided garter snake has attracted a great deal of attention, which has resulted in many interesting papers on topics only briefly touched on in this chapter, including Krohmer,[807] Lemaster and Mason,[852] and Mendonca et al.[975] These papers deal with some of the factors that control male and female behaviors in this species; more articles can be tracked down online via the ISI Web of Science. The snake's proximate mechanisms for organizing sexual behavior can be contrasted with those of the green anole,[1504] about which there has been some debate (e.g., see Jenssen, Lovern, and Congdon[721] as well as Crews,[314] and then Winkler and Wade[1603]).

# 6
# Behavioral Adaptations for Survival

It is hard to pass on your genes when you are dead. Not surprisingly, therefore, most animals do their best to stay alive long enough to reproduce. But surviving can be a challenge in most environments, which are swarming with deadly predators. You may remember that bats armed with sophisticated sonar systems roam the night sky hunting for moths, katydids, lacewings, and praying mantises. During the day, these same insects are at grave risk of being found, captured, and eaten by keen-eyed birds. Life is short for most moths, lacewings, and the like.

Because predators are so good at finding food, they place their prey under intense selection pressure, favoring those individuals with attributes that postpone death until they have reproduced at least once. The hereditary life-lengthening traits of these survivors can then spread through their populations by natural selection—an outcome that creates reciprocal pressure on predators, favoring those individuals who manage to overcome their prey's improved defenses. For example, the hearing abilities of certain moths and other night-flying insects may have led to the evolution of bats with unusually high-frequency sonar that their prey cannot detect as easily.[500, 1266] But now some night-flying mantises do hear unusually high-frequency

◄ **Canyon treefrogs rely on camouflage to protect themselves** *from predators, which means they must pick the right rocks to which they cling tightly wthout moving. Photograph by the author.*

sounds, presumably as a counter to the innovative high-frequency sonar used by their hunters.[325] The back and forths between predator and prey constitute evolutionary arms races.

With this chapter, which looks at the results of this ongoing contest between the hunters and the hunted, the focus of the book shifts from the proximate to the ultimate causes of behavior. The main goals of the chapter are to establish what is meant by an adaptation and to show how one can use a cost–benefit approach to produce hypotheses on the possible adaptive value of a behavioral trait while using the comparative method to test those hypotheses.

## Mobbing Behavior and the Evolution of Adaptations

When I was in New Zealand some years ago, I visited a coastal nature reserve rich in wildlife. As I walked along the shore, I came to a place where hundreds of pairs of silver gulls had built their nests on the stony ground. While I was watching the gulls from a distance, a young researcher came down to the coast carrying a scale for weighing gull chicks and a clipboard for recording data. As she walked toward the colony, the gulls took notice, and soon those closest to her began to fly up, calling raucously. By the time she came within a few meters of the first nests, the colony was in an uproar, and many of the adult gulls were swooping about, some diving at the intruder, others yapping loudly (Figure 6.1).

In gull colonies around the world, whenever a human, hawk, crow, or some other potential consumer of eggs or chicks comes close to nesting birds, the gulls usually react strongly. Among the masses of screaming gulls, some may launch what appear to be kamikaze attacks on the unwanted visitor. No one likes being hit on the head by a gull's trailing foot as the bird pulls out of a wing-roaring dive; nor do visitors to gull rookeries enjoy being splattered by liquid excrement released by agitated gulls flying overhead.

**FIGURE 6.1** **Mobbing behavior of colonial, ground-nesting gulls.** Silver gulls reacting to a trespasser in their breeding colony in New Zealand. Photograph by the author.

**FIGURE 6.2**  A nesting colony of black-headed gulls.

We can easily guess why the gulls become upset when potential predators get close to their nests. The parents' assaults probably keep hungry intruders away from their youngsters, helping them survive. If so, then mobbing by gulls could increase their reproductive success, passing on the hereditary basis for joining other gulls in screaming at, defecating on, and hitting those who might eat their eggs and youngsters. Indeed, this hypothesis was the one that quickly came to mind when Hans Kruuk, a student of Niko Tinbergen, decided to investigate **mobbing behavior** in the black-headed gull, another ground-nesting, colony-forming species (Figure 6.2) but one that lives in Europe rather than New Zealand.[814]

Kruuk was interested in studying the evolutionary, ultimate causes of mobbing, not the proximate ones, which would have required examination of the genetic, developmental, hormonal, and neural bases of the behavior—all interesting and valuable subjects but not on Kruuk's agenda. To get at the evolutionary foundation of the behavior, Kruuk employed what is now called the **adaptationist** approach. In effect, he wanted to know whether the mobbing response of black-headed gulls was an adaptive product of natural selection. His working hypothesis was that mobbing behavior distracted certain predators, reducing the chance that they would find the mobbers' offspring, which would boost the fitness of mobbing parent gulls.

Kruuk used his hypothesis to make testable predictions about gull mobbing behavior.[814] He knew that natural selection occurs when individuals vary in their hereditary attributes in ways that affect how many surviving offspring they contribute to the next generation (see Chapter 1). Imagine a population of gulls, some of whose members mob nest predators while others do not. This population will surely evolve if the difference between the two types of individuals is hereditary and if one type consistently leaves more surviving progeny than the other. If, for example, the mobbing phenotype outreproduces the non-mobbing type generation after generation, then mobbing behavior will eventually become the norm, and non-mobbing will disappear (always assuming that the differences between the two types are hereditary). Thereafter, any hereditary change in the nature of the mobbing response that enhances individual success in passing on genes will also spread through the species, given enough time.

---

**TABLE 6.1**  *Constraints on adaptive perfection*

**CONSTRAINT 1: Failure of appropriate mutations to occur**

Evolutionary constraints on adaptive perfection can arise from the failure of appropriate mutations to occur, which will prevent selection from keeping up with environmental change. Thus, maladaptive or nonadaptive traits can persist, especially in environments only recently invaded by a species. So, for example, some arctic moths live in regions where bats are absent, but the moths still cease flying in response to an ultrasonic stimulus.[1265] Likewise, arctic ground squirrels react defensively upon experimental exposure to snakes, even though there are no snakes living in the Arctic.[301]

Man-made changes in the environment are especially likely to lead to inappropriate expression of previously adaptive traits.[1286] Thus, some moths are so strongly attracted to artificial lights that bats visit lights in order to make some easy kills.[503] Likewise, male buprestid beetles (see Figure 4.8) may die while persistently attempting to mate with beer bottles,[599] while sea turtles sometimes expire after consuming plastic bags that they mistake for edible jellyfish.[205, 831] The current obesity epidemic in Western societies may well be caused in part by the once adaptive desire of humans to consume calorie-rich foods in an "unnatural" modern environment where it is entirely possible to eat too much of a good thing.[1010]

**CONSTRAINT 2: Pleiotropy**

Developmental constraints on adaptive perfection can occur as a result of **pleiotropy** (the multiple developmental effects that most genes have). Not all the effects of a given gene are positive. If the negative consequences of a gene outweigh the positive ones, the gene will be selected against. The converse is that because some gene effects are so valuable, the less significant, mildly negative consequences of an otherwise adaptive proximate mechanism may be maintained in the population by selection. For example, misdirected parental care is not uncommon in nature, a result of the intense drive to care for offspring, which usually is adaptive but in relatively rare instances can lead an adult to provide assistance to a youngster of a genetic stranger (see also Figure 14.5).[638]

**CONSTRAINT 3: Coevolution**

Coevolution (the kind of evolution that occurs when different species interact in ways that affect the fitness of each other's members) means that evolutionary stability may never be reached. Instead, each species changes in response to selection pressure imposed by the other, so first one and then the other species gains the upper hand, as in the coevolutionary arms races between predator and prey. The inability of selective processes to generate an immediate effective solution to an environmental problem means that less-than-perfect traits can persist within a species.

---

So, what kind of mobbing behavior did Kruuk expect to observe in black-headed gulls? Absolutely perfect, 100-percent-effective mobbing behavior that always saved the mobber's eggs and chicks? No, for any number of reasons (Table 6.1).[313, 1038] Selection cannot create the best of all possible genes for a particular task but has to wait for mutations to occur by chance; only then can it winnow out the less effective alleles, leaving the one that does better at promoting reproduction in place. If a "better" allele does not appear, there is nothing selection can do about it, because selection is a consequence of the confluence of certain conditions (hereditary variation that causes differences in individual reproductive success) that are not designed to act for the good of the species. Moreover, even if a mutant allele happened along with a particularly positive developmental effect on mobbing, the new allele might

**FIGURE 6.3    An arms race with a winner?** Although this salamander is extremely toxic, the garter snake is able to safely consume even the most poisonous members of the prey species. Photograph by Edmund D. Brodie III.

very well damage the construction of other traits. Almost all genes do have multiple developmental consequences. An allele with a net negative effect on reproductive success would never spread, no matter how useful its effect on the mobbing trait. Finally, gulls have been around for a long time, but so have their predators, and selection works on both sides of the predator–prey equation. As we noted at the outset, any improvement in a gull's ability to thwart a certain predator would tend to select for improvement in that predator's ability to circumvent the bird's defense. As a result of this arms race, we could find that neither prey nor predator has complete control of the situation at any given moment,[444] although it does appear that in one such case, the predator (a garter snake) has won the race with a prey species (a toxic salamander) because the snake can safely consume even the most deadly salamander (Figure 6.3).[617]

  Given the constraints that usually prevent the evolution of adaptive perfection, what does evolve? As just noted, if there are hereditary alternative phenotypes in a population, the "better" one will spread, namely, the one that confers higher **fitness** (defined as higher reproductive success or higher genetic success) on the individuals that happen to have this superior alternative. In other words, an **adaptation** is a hereditary trait that either (1) spread through the population in the past and has been maintained by natural selection to the present or (2) is currently spreading relative to alternative traits because of natural selection. In all such cases, the trait in question has conferred and continues to confer (or is just beginning to confer) higher genetic success, on average, on individuals that have it compared with other individuals with alternative traits. There are other definitions of adaptation,[627, 845] but the one we shall use permits us to test hypotheses about possible adaptations by focusing on the current benefits of a given trait, which has some major practical advantages over trying to test whether certain traits provided genetic benefits to individuals in the past.[1207]

### Discussion Questions

**6.1** Many people think that an adaptation is a trait that improves the survival chances of an organism. Under what circumstances would such a trait be an adaptation? Under what other circumstances would a survival-enhancing attribute actually be selected against?

**6.2** Stephen Jay Gould and Richard Lewontin claimed that adaptationists make the elementary mistake of believing that every characteristic of living things is a perfected product of natural selection,[557] when in reality many attributes of living things are not adaptations (see Table 6.1). Moreover, in their eagerness to explain everything as an adaptation, adaptationists have, according to Gould and Lewontin, invented fables as absurd as the fictional "just-so" stories of Rudyard Kipling, who made up amusingly silly explanations for the leopard's spots and the camel's hump. How might adaptationists defend themselves against these charges? Do adaptationists have the means to discover whether their tentative explanations for a particular trait are wrong?

With our definition of adaptation in place, we can identify a naturally selected adaptation if we can establish that it is better than other alternatives that have appeared in the past. But how do we do that? Adaptationists tackle this problem with a very useful tool borrowed from economics, namely, the **cost–benefit approach**, which enables them to analyze phenotypes in terms of their fitness benefits and fitness costs. (In evolutionary biology, **fitness benefit** refers to the positive effect of a trait on the number of surviving offspring produced by an individual or the number of copies of its alleles that it contributes to the next generation, while **fitness cost** refers to the damaging effects of the trait on these measures of individual genetic success.) So, for example, Kruuk knew that mobbing behavior comes with significant disadvantages for individuals in terms of their reproductive chances. These costs include the time and energy that mobbers expend when they are screaming, diving, and flapping about in response to an unwanted visitor near their nests. In addition, mobbers can lose not only calories, but also their lives. More than one black-headed gull has made a lethal miscalculation when dive-bombing an unusually agile fox that was able to wheel about, leap up, and catch the gull as it passed by. Moreover, all the noise made by a mobber when dealing with one predator may draw others to the spot, and one of these additional enemies may make off with the offspring that the mobbing individual is trying to protect.[801]

Given the very real fitness costs associated with mobbing, Kruuk knew that the trait could not be an adaptation unless there were some equally obvious, substantial benefits for the mobbers. Only when, on average, the benefits of mobbing exceed its costs can a trait possibly be considered an adaptation.

Therefore, Kruuk predicted that mobbing gulls should force nest-robbing predators to expend more searching effort than they would otherwise, a simple prediction that can be tested just by watching gull–predator interactions.[814] Kruuk saw that egg-eating carrion crows have to continually face swooping gulls and so, while being mobbed, they cannot look around comfortably for nests and eggs. Because distracted crows are probably less likely to find their prey, Kruuk established that a probable benefit exists for mobbing. Moreover, the benefit of mobbing crows plausibly exceeds the costs, given that these predators do not attack or injure adult gulls.

However, the predator distraction hypothesis yields much more demanding predictions as well. Because adaptations are better than the traits they replace, we can predict that the benefit experienced by mobbing gulls in pro-

Black-headed gull

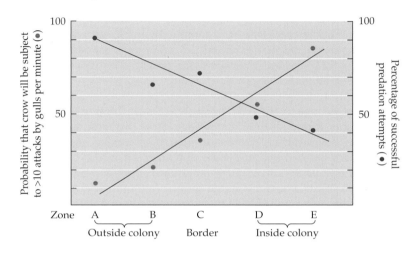

**FIGURE 6.4  Does mobbing protect eggs?** When chicken eggs were placed outside and inside a black-headed gull nesting colony, crows searching for the eggs within the colony were subject to more attacks by mobbing gulls (red circles), and as a result, they discovered fewer hen eggs (blue circles). After Kruuk.[814]

tecting their eggs should be directly proportional to the extent that predators are actually mobbed. Kruuk used an experimental approach to test this prediction.[814] He placed ten chicken eggs, one every 10 meters, along a line running from the outside to the inside of a black-headed gull nesting colony. The eggs placed outside the colony, where mobbing pressure was low, were much more likely to be found and gobbled up by carrion crows and herring gulls than were eggs placed inside the colony, where the predators were communally harassed by the many parents whose offspring were threatened by them (Figure 6.4).

Thus, Kruuk assembled some observational and experimental evidence in support of the hypothesis that mobbing is an adaptation that helps adult black-headed gulls protect their eggs and youngsters. Note that these tests did not involve measuring gull reproductive success by counting the number of surviving offspring produced by individuals in their lifetimes. Kruuk looked instead at the number of hen eggs that were uneaten, on the reasonable assumption that had they been gull eggs in a gull nest, they would have had a chance to become surviving offspring for the gull parent, in which case they would constitute part of the parent's lifetime genetic contribution to the next generation.

Behavioral ecologists often have to settle for an indicator or correlate of reproductive success or genetic success when they attempt to measure fitness. In the chapters that follow, *fitness* or *reproductive success* is often used more or less interchangeably with such things as egg survival (Kruuk's measure), young that survive to fledging, or number of mates inseminated, or even more indirectly, the quantity of food ingested per unit of time, the ability to acquire a breeding territory, and so on. Keep in mind, however, that the bottom line with respect to an individual's reproductive success is the total number of his or her offspring (or grandoffspring) that reach the age of reproduction. Proxies for this measure can only approximate actual reproductive success.

## Discussion Questions

**6.3** For many evolutionary biologists, the term "adaptation" must be reserved for a characteristic that provides "current utility to the organism and [has] been generated historically through the action of natural selection for its current biological role."[84] What could "current utility" mean, and what do you think it should mean? Make use of the terms "fitness benefits" and "fitness costs" in your answer. If a trait originated for function X and later took on a different, but still adaptive, biological role Y, does that mean it is not an adaptation? Track down the evolutionary history of the flight feathers on the wings of modern birds (see, for example, Prum[1172]). Where did these feathers come from, and what function did their predecessor feathers exhibit? If you go back far enough in time, will the ancestral form of any current trait have the same function that it does now?

**6.4** The arctic skua, a close relative of gulls, also nests on the ground and mobs colony intruders, including the great skua, a larger predator that eats many arctic skua eggs and chicks. In one study, hatching success and post-fledging survival were greater for arctic skuas that nested in dense colonies than in low-density groups (Figure 6.5); the number of near neighbors was, however, negatively correlated with the growth rate of their chicks.[1133] Rephrase these findings in terms of the fitness costs and benefits of communal mobbing by the arctic skua. If *adaptation* meant a perfect trait, would communal mobbing by arctic skuas be labeled an "adaptation"?

### The Comparative Method for Testing Adaptationist Hypotheses

Experiments are highly valued in science, so much so that most persons believe that scientific research can be performed only in high-tech laboratories by officious white-coated researchers. As Kruuk's work shows, however, good experimental science can be done in the field. Moreover, the manipulative experiment is only one of several ways in which predictions from hypotheses can be tested. Another powerful technique for testing adaptationist hypotheses is the **comparative method**. This approach involves testing predictions about the evolution of an interesting trait by looking at species other than the

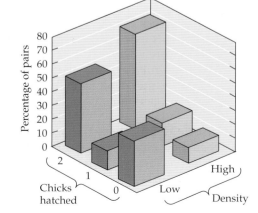

**FIGURE 6.5  Benefit of high nest density for the arctic skua.** In one population studied in 1994, skuas nesting with relatively many nearby neighbors were more likely to rear two chicks than were individuals nesting in areas with a lower density of breeding pairs. After Phillips, Furness, and Stewart.[1133]

Arctic skua

the one whose characteristics are under investigation.[279] We used the comparative method informally earlier when considering the evolution of infanticide (see page 23) and again when dealing with the responses of moths and other insects to bat ultrasonic calls (see page 117). The idea was that if a given trait, such as infanticide, is adaptive for one species, such as Hanuman langurs, then it should also evolve in other species, such as lions, that are subject to the same selective pressures as langurs. Here we shall use the comparative method to test the adaptationist explanation for mobbing by black-headed gulls. This technique, as applied to this case, yields the following prediction: if mobbing by ground-nesting black-headed gulls is an evolved response to predation on gull eggs and chicks, then other gull species whose eggs and young are at low risk of predation should not exhibit mobbing behavior.

The rationale behind this prediction is as follows: the costs of mobbing can be outweighed only by the benefits derived from distracting predators. If predators were not a major problem for a species, then the benefits of mobbing would be reduced, increasing the odds that the overall benefit-to-cost ratio of mobbing would be negative, leading to its eventual loss from a species whose ancestors possessed the trait.

There is good reason to believe that the ancestral gull was a ground-nesting species with a range of nest-hunting predators against which the mobbing defense would have been helpful. If we look at the 50 or so species of gulls living today, we find that the large majority nest on the ground and exhibit communal mobbing behavior against enemies that hunt for their eggs and chicks.[1447] These similarities among gulls, which also have many other features in common, are believed to exist in part because all gulls are the descendants of a relatively recent common ancestor, from which they all inherited the genetic package that predisposes all gulls to develop a similar set of traits. However, a few of today's gull species nest on cliff ledges, rather than on the ground. Perhaps these gull species are the descendants of a more recent cliff-nesting gull that evolved from a ground-nesting ancestor. The alternative possibility, that the original gull was a cliff nester, requires the cliff-nesting trait to have been lost and then regained, which produces an evolutionary scenario that requires more changes than the competing one (Figure 6.6). Many evolutionary biologists, although not all, believe that simpler scenarios involving fewer transitions are more probable than more complicated alternatives. In this, the majority accepts a commonly held philosophical principle, known as Occam's razor or the principle of parsimony, which holds that simpler

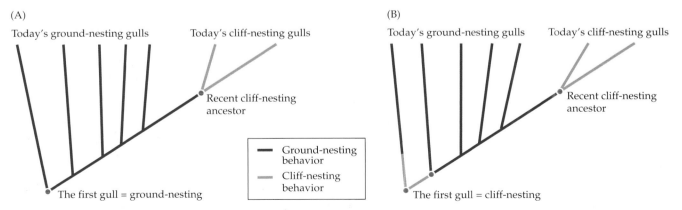

**FIGURE 6.6   Gull phylogeny and two scenarios for the origin of cliff-nesting behavior.** (A) Hypothesis A requires just one behavioral change, from ground nesting to cliff nesting. (B) Hypothesis B requires two behavioral changes, one from the ancestral cliff nester to ground nesting, and then another change back to cliff nesting.

(A)

(B)

**FIGURE 6.7  Not all gulls nest on the ground.** (A) Steep cliffs are utilized for nesting by kittiwake gulls, which appear in the lower half of this photograph. (B) Kittiwakes are able to nest on extremely narrow ledges; their youngsters crouch in the nest, facing away from the cliff edge. A, photograph by the author; B, photograph by Bruce Lyon.

explanations are more likely to be correct than complex ones—all other things being equal. (The contrary view is that, because of the many chance events that influence the course of evolution, historical pathways may often be quite convoluted, with traits lost, then regained, then lost again, violating expectations based on Occam's razor.[1207])

In any event, cliff-nesting gulls currently have relatively few nest predators because it is hard for small mammalian predators to scale cliffs in search of prey, while predatory birds have a difficult time maneuvering near cliffs in turbulent coastal winds. Thus, a change in nesting environment meant a change in predation pressure, which ought to have altered selection on these gulls. The evolutionary result should have been a shift away from the ancestral mobbing behavior pattern. By finding such cases of **divergent evolution** and identifying the selective reason for the change, one can, in principle, establish why the ancestral trait has been retained in some species but has been modified or lost in other species.

The kittiwake nests on nearly vertical coastal cliffs (Figure 6.7), where its eggs are relatively safe from predators. These small, delicate gulls have clawed feet, and they can land and nest on tiny ledges. As a result, predation pressure on kittiwake eggs and young has been greatly reduced[946] compared with that affecting their ground-nesting relatives. The relatively small size of these gulls may also make the adults more vulnerable to personal attack by nest predators, making the benefit-to-cost ratio for mobbing still less favorable. As predicted, groups of nesting adult kittiwakes do not mob their predators, despite sharing many other structural and behavioral features with black-headed gulls and

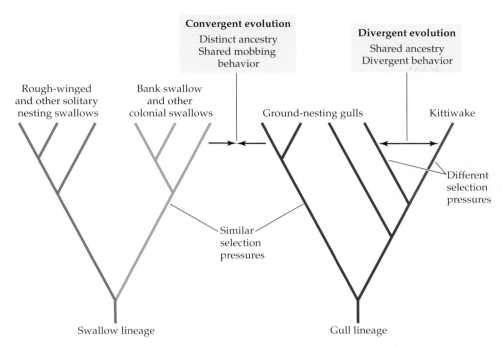

**FIGURE 6.8    The logic of the comparative method.** Members of the same evolutionary lineage (e.g., gull species of the family Laridae) share a common ancestry, and therefore share many of the same genes, and thus tend to have similar traits, such as mobbing behavior. But the effects of shared ancestry can be overridden by a novel selection pressure. A reduction in predation pressure has led to divergent evolution by the cliff-nesting kittiwake, which no longer mobs potential enemies. The various ground-nesting gulls, including the black-headed gull, mob nest predators, as do some colonial swallows, including the bank swallow, even though gulls and swallows are not related, having come from different ancestors long ago. These colonial swallows and gulls have converged on a similar antipredator behavior in response to shared selection pressure from predators that have fairly easy access to their nesting colonies. What kind of evolution is responsible for the difference between bank swallows, which mob their enemies, and rough-winged swallows, which do not? This difference constitutes evidence in support of what evolutionary hypothesis about mobbing?

other ground-nesting species. The kittiwake's behavior has become less like that of its close relatives, providing a case of divergent evolution in support of the hypothesis that mass mobbing by black-headed gulls evolved in response to predation pressure on the eggs and chicks of nesting adults.[324]

The other side of the comparative coin is that species from different evolutionary lineages that live in similar environments, and therefore experience similar selection pressures, can be predicted to evolve similar traits, resulting in cases of **convergent evolution**. If true, these species will adopt the same adaptive solution to a particular environmental obstacle to reproductive success, despite the fact that their different ancestral species had very different genes and attributes (Figure 6.8). We can predict, for example, that if mobbing by colonial, ground-nesting gulls is an adaptation to distract predators from vulnerable offspring, similar behavior should have evolved in other, quite unrelated animals that nest or breed in groups and are often visited by predators intent on eating their offspring.

As predicted, mobbing behavior has evolved convergently in many other bird species only distantly related to gulls (e.g., Sordahl[1370]), including the bank swallow. This species also nests in colonies, which are visited by predators, including snakes and blue jays, which enjoy eating swallow eggs and nestlings.[677] The common ancestor of bank swallows and gulls occurred long,

**FIGURE 6.9   Colonial California ground squirrels mob their snake enemies.** One squirrel kicks sand at a rattlesnake, while others give a variety of alarm signals. Courtesy of R. G. Coss and D. F. Hennessy.

long ago, with the result that the lineages of the two groups have evolved separately for eons—a fact recognized by the taxonomic placement of gulls and swallows in two different families, the Laridae and the Hirundidae. Despite their evolutionary and genetic differences from gulls, bank swallows behave like gulls when they are nesting. As they swirl around and dive at their predators, they sometimes distract hunting jays or snakes that would otherwise destroy their offspring.

Even some colonially nesting mammals have evolved mobbing behavior.[1088] Adult California ground squirrels, which live in groups and dig burrows in the ground, react to a hunting rattlesnake by gathering around it and shaking their tails vigorously, a visual signal to the snake to encourage it to depart before it is physically attacked by the ground squirrels. To enhance communication with rattlers, which can sense the infrared radiation coming from the warm bodies of potential prey, the adult squirrels also heat up their tails, the better to signal to a rattlesnake that the tail-shaking squirrels are about to kick sand in its face.[1255] Rattlers assaulted in this fashion cannot hunt leisurely for nest burrows to enter in search of vulnerable young ground squirrels (Figure 6.9).

Although adult squirrels have a partial antivenin that reduces the costs of being envenomated, squirrels that are struck do not get off scot-free. Given the costs of a rattlesnake bite, one team of researchers led by Ronald Swaisgood predicted that squirrels would be able to adjust their behavior in relation to the level of risk of snakebite, which they could assess by listening to the defensive rattling buzzes made by rattlesnakes harassed by their prey. These sounds vary according to the size and body temperature of the snake. Larger snakes present a greater risk to squirrels, as do warmer snakes, which can move more quickly. Squirrels were, as predicted, less eager to approach speakers playing the rattles of larger and warmer snakes. In nature, this ability to assess the risk of snakebite would enable the squirrels to reduce the costs of their mobbing behavior.[1413] In much the same vein, during the daytime Siberian jays mob models of hawks more cautiously than models of owls, a result that reflects the greater danger posed by diurnal hawks than by sleepy nocturnal predators.[580] These results suggest that when the net benefits of intense mobbing behavior fall, animals are less likely to engage in the activity.

Because mobbing behavior has evolved independently in several unrelated species whose adults can sometimes protect their vulnerable offspring by distracting predators, we can conclude tentatively that mobbing is an antipredator adaptation. But what if I have presented only supportive examples while ignoring other colonial species in which parents do not mob the consumers of their youngsters (and do not have alternative means of protecting their offspring)? If for every group-living species in which mobbing occurred under the expected conditions there were two in which the behavior was absent, you would be skeptical of the predator distraction hypothesis, and rightly so. For this and other reasons, researchers increasingly require that the comparative method be used in a statistically rigorous fashion.[627] For our purposes, however, the point is that one can, in principle, test adaptationist hypotheses by predicting that particular cases of divergent or convergent evolution will have occurred—a prediction about the past that can be checked by making disciplined comparisons among species living today.

## Discussion Questions

**6.5**  The ability to hear ultrasound in one species of noctuid moth is considered an antipredator adaptation because it apparently enables individuals to hear and avoid nocturnal, ultrasound-using bats. Imagine that you wished to test this hypothesis via the comparative method. Identify the utility of each of the following lines of evidence about the hearing abilities of other insect species. Specify whether these cases involve convergent evolution, divergent evolution, or neither.

1. Almost all other species of noctuid moths also have ears that respond to ultrasound.

2. Almost all the species in the evolutionary lineage that includes the noctuid moths and many other moths belonging to several other superfamilies also have ears that respond to ultrasound.[1625]

3. Some diurnal noctuid moths have ears but are largely or totally incapable of hearing ultrasound.[501]

4. Almost all butterflies, which belong to the same large evolutionary grouping as the noctuids but are usually active during the day, lack ears and so cannot hear ultrasound.[502]

5. Six species of noctuid moths found only on the Pacific Ocean islands of Tahiti and Mooréa have ears and can hear ultrasound but do not react to this stimulus with anti-bat responses.[505]

6. Members of one small group of nocturnal butterflies have ears on their wings and can hear ultrasound; they respond to ultrasonic stimulation by engaging in unpredictable dives, loops, and spirals.[1625]

7. Lacewings and praying mantises fly at night and have ears that detect ultrasound and lead to anti-bat defensive behavior (see page 118).

**6.6**  Some persons would say that the fact that most noctuid moths have ultrasound-sensitive ears is "simply" a reflection of their shared ancestry, a holdover from the past, and therefore that ultrasound sensitivity is not an adaptation in these species.[182] Others disagree, arguing that it makes no sense to define adaptations in a way that limits them to just those traits that have diverged from the ancestral pattern.[1203] Who is right?

## The Cost–Benefit Approach to Antipredator Behavior

We have used mobbing behavior to illustrate how adaptationists can employ natural selection theory to analyze the possible **adaptive value** of behavioral traits. All behaviors have fitness costs and fitness benefits, but only those whose benefits exceed their costs can be directly selected for. A demonstration that a trait has substantial reproductive benefits, as mobbing does by virtue of helping parents protect their offspring, constitutes evidence that the trait in question could be the evolutionary product of natural selection.

Many investigators have tried to test whether a putative antipredator adaptation really does confer significant benefits on individuals that employ the behavior of interest. Some persons, for example, have been interested in why butterflies aggregate in large, densely packed groups around mud puddles on tropical riverbanks, where they suck up fluid containing valuable mineral nutrients from the soil (Figure 6.10A). While they are "mud-puddling," the butterflies could be attacked by various birds, with large groups particularly likely to attract a predator. But this probable cost of communal mud-puddling might be offset by a dilution in the chance that any one individual would be the target of a predator.

Imagine that five insect-eating birds inspect mud-puddling areas and that each bird kills two prey per day there. Under these conditions, the risk of death for a member of a group of 1000 butterflies is 1 percent per day, whereas it is ten times higher for members of a group of 100 butterflies. This advantage associated with being part of a large group of mud-puddling butterflies has been confirmed by Joanna Burger and Michael Gochfeld (Figure 6.10B). The observations of these researchers in the field show that any butterfly puddling by itself or with only a few companions would be safer if moved to an even slightly larger group. In fact, a butterfly in a group of 20 would significantly reduce its risk of being captured and eaten by moving to a group of 30 or more,[209] suggesting that the tendency to join many other mud-puddlers would provide benefits in excess of the costs of this trait.

(A)

(B)

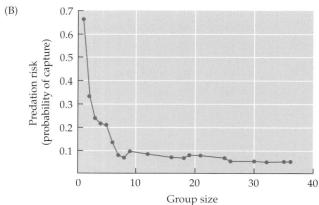

**FIGURE 6.10   The dilution effect in butterfly groups.** (A) Masses of butterflies on a Brazilian riverbank. (B) Individual butterflies that "mud-puddle" in large groups experience a lower risk of predation than those that suck up fluids from the ground by themselves or in small groups. A, photograph by Joanna Burger; B, after Burger and Gochfeld.[209]

**FIGURE 6.11  A recently hatched black-necked stilt.** Its parent has flown off with part of the egg shell from which the chick has just hatched. Is this an adaptive thing to do? Photograph by Tex Sordahl.

## Discussion Questions

**6.7**  Figure 6.11 shows a nest with a recently hatched black-necked stilt chick and three eggs. A parent has removed most, but not all, of the shell of the egg from which the chick emerged. The adult will be back soon to take the remaining fragment far from the nest. Develop at least one antipredator hypothesis to account for this behavior. List the possible benefits and the likely costs of the parent's actions. Under what circumstances would the benefits probably outweigh the costs?

**6.8**  In studies of the effect of the spiny water flea, an introduced predator, on *Daphnia*, a small aquatic crustacean once abundant in the Great Lakes, researchers have documented several evolutionary responses. For example, *Daphnia* exposed to spiny water fleas now tend to have larger defensive spines, and they also tend to stay in deeper waters away from their enemies. But spinier *Daphnia* move more slowly, and so they secure less food per unit time spent foraging, while deep water *Daphnia* reproduce more slowly because the water is colder. As a result, *Daphnia*'s reproductive rate fell sharply in the Great Lakes,[1100] leading some to conclude that the evolutionary responses linked to avoiding the introduced predator had done ten times more damage to the population than would have been done if the prey species had remained unchanged and had simply accepted a higher mortality rate from predation. Is this conclusion based on a cost–benefit analysis of the sort we have just been discussing above? Defend your answer.

Could the dilution effect also contribute to the tendency of some mayflies to synchronize their metamorphosis from aquatic nymphs to flying adults so that most individuals emerge from the water during just a few hours on a few days each year?[1415] If so, then the higher the density of emerging individuals, the lower the risk to any one mayfly of being inhaled by a trout as the mayflies make the risky transition to adulthood. To check this prediction, Bernard

Mayfly

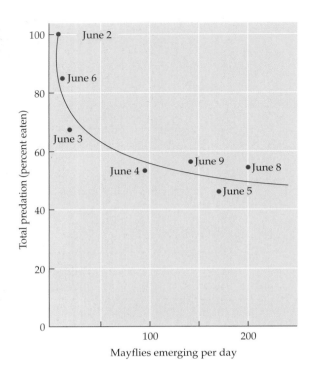

**FIGURE 6.12   The dilution effect in mayflies.** The more female mayflies that emerged together on a June evening, the less likely any individual mayfly was to be eaten by a predator. After Sweeney and Vannote.[1415]

Sweeney and Robin Vannote placed nets in streams to catch the cast-off skins of mayflies, which molt on the water's surface as they change into adults, leaving the discarded cuticle to drift downstream on the current. Counts of the molted cuticles revealed how many adults emerged on a particular evening from a particular segment of stream. The nets also caught the bodies of females that had laid their eggs and then died a natural death; a female's life ends immediately after she drops her clutch of eggs into the water, provided a nighthawk or a whirligig beetle does not consume her first. Sweeney and Vannote measured the difference between the number of molted cuticles of emerging females and the number of intact corpses of spent adult females that washed into their nets on different days. The greater the number of females emerging together on a given day, the stronger the dilution effect, and the better the chance each mayfly had to live long enough to lay her eggs before expiring (Figure 6.12).[1415]

Of course, there are other possible benefits of getting together with others, including the potential for group attack on a shared enemy (Figure 6.13A). The social insects, including termites, ants, wasps, and bees, are famous for the ability to go after a predator together, using their stingers or jaws to injure or even kill intruders (Figure 6.13B and C). The race of the honey bee that lives in Africa, where bee colonies have been relentlessly exploited over the millennia for their honey and larvae by a host of mammalian predators, including human beings, is renowned for the ferocity of its response to creatures that threaten to rob its nest. In fact, more than 80 times as many African honey bees will persistently pursue beekeepers that have disturbed their hive compared with Europeanized honey bees, which have been artificially selected for docility.[595] The willingness of the African race of honey bees to sting potential predators en masse is reflected in the fact that more than a thousand people have died from its attacks as the bee has spread through the Americas following its release by Brazilian beekeepers.[166]

**FIGURE 6.13    Fighting back by terns and wasps.** (A) A group of royal terns confronts a gull interested in stealing a tern egg. (B,C) A colony of *Polybia* wasps (B) just before and (C) just after a nest was tapped by an observer. These wasps will leave their nest to assault any predator foolish enough to persist in disturbing them. A, photograph by Bruce Lyon; B and C, photographs by Bob Jeanne.

Among the insects, organized defensive attack is not limited to stinging bees and biting termites. In southwestern Australia, one frequently comes across sawfly larvae clustered in balls of ten or so individuals. These caterpillar-like insects feed on eucalyptus leaves, which contain highly resinous, toxic oils that deter most herbivores—but not sawflies. The sawfly grubs not only ingest the eucalyptus oils safely, but also store them in special sacs, from which they can be regurgitated onto attacking ants or birds.[1013] Disturb a group of sawfly larvae resting in a defensive circle during the day, and they all spew out large, pale, sticky droplets of resinous fluid (Figure 6.14), which they are prepared to smear communally on any approaching enemy, the better to send it on its way. In this case, and in the others mentioned above, because researchers have been able to identify the antipredator benefits that individuals derive from their behaviors, we now have the minimum evidence needed to label these behaviors evolved responses to predation.

**FIGURE 6.14  Communal defense by sawfly larvae.** These larvae form clusters that rest with their heads facing outward during the day. When threatened, they raise and wave their abdomens above their heads in warning while regurgitating droplets of sticky eucalyptus oils, which they hold in their mouths to apply to an enemy. Photograph by the author.

### Discussion Question

**6.9**  In my front yard, I sometimes find several hundred male native bees clustering in the evening on a few bare plant stems (Figure 6.15). An assassin bug sometimes approaches the cluster and kills some bees as they are settling down for the night. Devise at least three alternative hypotheses on the possible anti-assassin bug value of these sleeping clusters, and list the predictions that follow from each hypothesis.

**FIGURE 6.15  A group of sleeping bees.** In this species, males spend the night clustered together. Photograph by the author.

### The Costs and Benefits of Camouflage

One can also test hypotheses on the possible antipredator benefits of behaviors used by solitary animals. For example, many persons have thought it likely that certain apparently camouflaged animals have evolved the ability to select the kind of resting background where they would be hard to see (Figure 6.16). A classic attempt to test whether preferred resting places really do enhance the camouflage of a prey species involved the peppered moth, *Biston betularia* (Figure 6.17). In some parts of Great Britain and the United States, the melanic (black) form of this moth, once extremely rare, almost completely replaced the once abundant whitish salt-and-pepper form in the period from about 1850 to 1950.[564] Most biology undergraduates have heard the standard explanation for the initial spread of the melanic form (and the special allele associated with its mutant color pattern): as industrial soot darkened the color of forest tree trunks in urban regions, the whitish moths living in these places became more conspicuous to insectivorous birds, which ate the whitish form and thereby removed the genetic basis for this color pattern. Despite some recent claims to the contrary,[308] this story remains largely valid,[565,1250] especially with respect to the significance of H. B. D. Kettlewell's famous experiments.[769] Kettlewell placed the two forms of the peppered moth on dark tree trunks and on pale tree trunks, finding that whichever form was more conspicuous to humans was also taken by birds much more quickly than the other form. Thus, paler individuals were at special risk of attack when they perched on dark backgrounds.

**FIGURE 6.16  Cryptic coloration depends on background selection.** The Australian thorny devil possesses remarkable camouflage, but it works only when the lizard is motionless in areas littered with bark and other varicolored debris, not on roads. Photographs by the author.

In nature, *B. betularia* apparently rarely sits on tree trunks; instead, it tends to settle within the shaded patches just below the junctions of branches with the trunk. If the moth's actual perch site selection is an adaptation, then we can predict that individuals resting underneath tree limbs should be better protected against predators than if they were to choose an alternative location. R. J. Howlett and M. E. N. Majerus attached samples of frozen moths to open trunk areas and to the undersides of limb joints. Their data showed that birds were particularly likely to overlook moths on shaded limb joints (Fig-

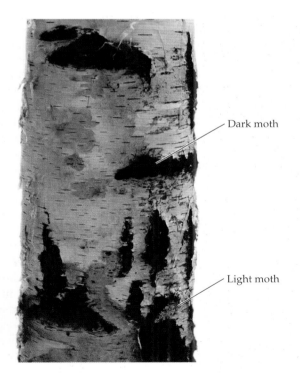

Dark moth

Light moth

**FIGURE 6.17  The camouflaged moth, *Biston betularia*.** One typical (whitish) individual and one melanic (black) individual are shown in each photograph. Left, photographs by Michael Tweedie; right, photograph by Bruce Grant.

**FIGURE 6.18  Predation risk and background selection by moths.** Specimens of typical and melanic forms of the peppered moth were attached to tree trunks or to the undersides of limb joints. Moths of both types were less likely to be found and removed by birds on limb joints than on trunks, but overall, melanic forms were less often discovered by birds in polluted (darkened) woods, while typical forms "survived" better in unpolluted woods. After Howlett and Majerus.[686]

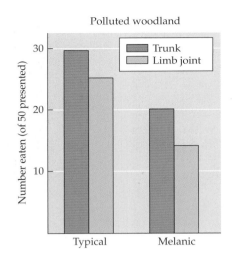
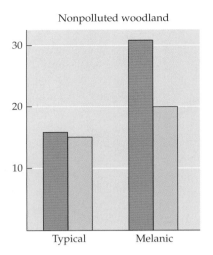

ure 6.18).[686] Here is another demonstration that a putative adaptation almost certainly does provide a survival benefit.

---

### Discussion Question

**6.10**  Consider the following finding: In the years since 1950, pollution controls have reduced the amount of soot deposited on tree trunks, and the melanic form of *B. betularia* has correspondingly become increasingly scarce in Europe,[163, 294] and North America,[566] where the species also occurs. Put this statement in the context of a scientific investigation into whether the typical salt-and-pepper coloration of some members of this species constitutes an adaptation. Begin with a research question and proceed through hypothesis, prediction, test, and conclusion.

---

**FIGURE 6.19  Cryptic coloration and body orientation.** The orientation of a resting *Catocala* moth determines whether the dark lines in its wing pattern match up with the dark lines in birch bark. Photograph by H. J. Vermes, courtesy of Ted Sargent, from Sargent.[1278]

Moths other than *B. betularia* make decisions about where to perch during the day. For example, the whitish moth *Catocala relicta* usually perches head-up, with its whitish forewings over its body, on white birch and other light-barked trees (Figure 6.19). When given a choice of resting sites, the moth selects birch bark over darker backgrounds.[1278] If this behavior is truly adaptive, then birds should overlook moths more often when the insects perch on their favorite substrate. To evaluate this prediction, Alexandra Pietrewicz and Alan Kamil used captive blue jays, photographs of moths on different backgrounds, and operant conditioning techniques (Figure 6.20).[1137] They trained the blue jays to respond to slides of cryptically colored moths positioned on an appropriate background. When a slide flashed on a screen, the jay had only a short time to react. If the jay detected a moth, it pecked at a key, received a food reward, and was quickly shown a new slide. But if the bird pecked incorrectly when shown a slide of a scene without a moth, it not only failed to secure a food reward, but had to wait a minute for the next chance to evaluate a slide and get some food. The caged jays' responses showed that they saw the moth 10 to 20 percent less often when *C. relicta* was pinned to pale birch bark than when it was placed on darker bark. Moreover, the birds were especially likely to overlook a moth when it was oriented head-up on birch bark. Thus, the moth's preference for white birch resting places and its typical perching orientation appear to be anti-detection adaptations that thwart visually hunting predators such as blue jays.

Among insects and other prey species, the many different kinds of potentially protective color patterns appear to work by exploiting certain aspects of the visual processing systems of their predators.[1389, 1390] So, for example, in many vertebrate predators, interneurons relaying messages from the retina will produce a burst of signals when the small area monitored by these relay cells is stimulated by an image that contains a dark band bordered by a much lighter one. These *edge detectors* probably help a hunting bird spot an insect resting on a trunk or leaf by responding to the contrast between the prey's body and the substrate on which it is sitting. However, an insect whose body contains contrasting color patterns that produce "false" edges (Figure 6.21) may live to reproduce another day by drawing the predator's attention away from the outline of its body and onto the distracting features within that outline, which keep the predator from recognizing what it is seeing.

Perhaps because prey items often use visual deceit and distraction, some hunting predators rely instead on prey odors to help them track down their prey. This may be why California ground squirrels chew on cast rattlesnake skins and then lick their fur, applying the scent of snakes over their own body odors. Barbara Clucas and her coworkers tested this hypothesis by giving captive rattlers a chance to investigate pieces of filter paper, some of which had been rubbed on the bodies of ground squirrels and later exposed to segments of rattlesnake skin while others had acquired ground squirrel scents alone. The snake subjects in this experiment spent nearly twice as much time inspecting the pure squirrel-scented targets, as opposed to those that had the combination of squirrel and rattlesnake scents.[275]

The larvae of skipper butterflies have evolved a different technique for reducing their vulnerability to odor-searching predators. These caterpillars

**FIGURE 6.20    Does cryptic behavior work?** Images of moths on different backgrounds and in different resting positions are shown to a captive blue jay, which is rewarded for detecting a moth. Photograph by Alan Kamil.

(A)

Edge and line detection

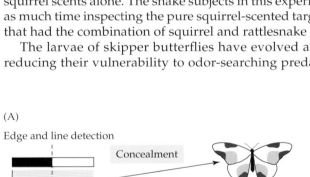

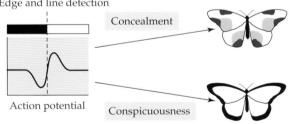

(B)

(C)

**FIGURE 6.21    Safety lies in false edges** for prey that exploit their predator's edge detectors (A). Disruptively colored prey, like the grasshopper (B) with its brown and white patches, create false edges that obscure the animal's actual outline (see also the *Catocala* moth in Figure 6.19). Warningly colored prey, like the largely black horse lubber grasshopper (C), emphasize their outline (note the yellow line on the thorax and the distinctive coloration of the wings), making them more conspicuous. Predators can quickly learn to recognize these prey as inedible and to be avoided (see also the monarch butterfly in Figure 6.23A). Photographs by the author; A, after Stevens.[1390]

**FIGURE 6.22 Personal hygiene by a skipper butterfly larva** may be an antipredator adaptation. (A) A larva of a skipper butterfly inside its partly opened leaf shelter. Butterfly larvae expel pellets of waste from their shelters. (B) When waste pellets are added experimentally to shelters from which the larvae have been removed, wasps focus on those sites in preference to shelters containing an equivalent quantity of odorless glass beads. Shown are the percentage of initial visits by ten wasps to the two kinds of sites and the percentage of time the wasps spent investigating both shelters. A, photograph by the author; B, after Weiss.[1533]

(A)

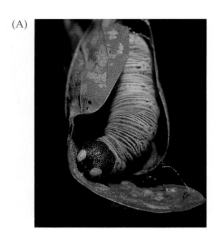

(B)

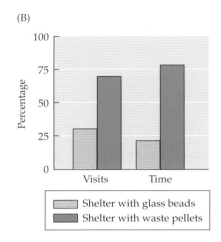

hide within rolled-up leaves (Figure 6.22A) and fire their fecal pellets far away from their shelters.[1533] They do so by using a special device that holds a firm waste pellet in the anus until a rapid change in blood pressure expands the terminal segment of the abdomen, causing the little fecal ball to be discharged explosively into space like a miniature pea from a peashooter. Some skipper caterpillars regularly blast their waste pellets 20 body lengths or more from where they are hiding.

If this form of waste management is truly an antipredator adaptation, then these larvae should be hunted by enemies that could use the odor cues associated with the fecal pellets to locate their prey if those cues were available. Martha Weiss tested this prediction by removing skipper larvae from a set of leaf shelters and then adding fresh waste pellets to some of the shelters, while the remaining shelters received the same number of odorless glass beads about the same size and color as the pellets. The experimental and control shelters were then placed in a cage with a colony of paper wasps, known predators of butterfly larvae. The wasps were far more likely to inspect leaf shelters with caterpillar waste than those with glass beads (Figure 6.22B). Moreover, when Weiss offered a foraging paper wasp a choice between two shelters, each containing a hidden caterpillar but one surrounded with 25 waste pellets and the other accompanied by 25 glass beads, the wasp first found and killed the caterpillar associated with fecal matter on 14 of 17 trials.[1533] These butterfly larvae do well to rid themselves of their waste.

## Discussion Question

**6.11** Weiss also collected information on the growth rates of caterpillars that either were forced to inhabit shelters that she contaminated with their feces or were allowed to mature in clean shelters. She found no difference in weight between the pupae that experienced these two different conditions as larvae; moreover, the days required for the larvae to pupate did not differ between individuals growing up with and without waste pellets in their shelters. Why did Weiss gather these data?

## Some Darwinian Puzzles

Although many insects possess beneficial attributes that reduce their risk of being spotted or smelled, other species seem to go out of their way to make themselves obvious to their predators (Figure 6.23). This is the kind of thing

(A)

(C)

(B)

**FIGURE 6.23   Warning coloration and toxins.** Animals that are chemically defended typically behave in a conspicuous manner. (A) The monarch butterfly's body and wings may contain potentially lethal cardiac glycosides, which it sequesters from its milkweed diet. (B) Blister beetles, which have blood laced with cantharidin, a highly noxious chemical, may mate for hours right out in the open on flowering plants. (C) This conspicuous moth exudes a toxic froth from thoracic glands when disturbed. A and B, photographs by the author; C, photograph by Tom Eisner.

that attracts the attention of adaptationists because of the obvious fitness costs incurred by a prey animal that is conspicuous to its predators. Traits whose costs seem likely to exceed their benefits qualify as **Darwinian puzzles** because they would seem unlikely to evolve by natural selection. Take the monarch butterfly, for example—a species whose orange-and-black wing pattern makes it easy to spot. Bright coloration of this sort is correlated with greater risk of attack in some cases.[1399] How can it be adaptive for monarchs to flaunt themselves in front of butterfly-eating birds? In order to overcome the costs of conspicuous coloration, the trait ought to have some very substantial benefits, and in the case of the monarch, it does. These benefits appear to be linked to the ability of monarch larvae to feed on poisonous milkweeds, from which they sequester an extremely potent plant poison in their tissues.[185] In dealing with this highly toxic species, you would be wise to ignore the recommendation of the lepidopterist E. B. Ford, who wrote, "I personally have made a habit, which I recommend to other naturalists, of eating specimens of every species which I study."[476] Any bird that makes the mistake of accepting the Ford challenge with respect to monarchs usually finds the experience most unpleasant, albeit highly educational (Figure 6.24). After vomiting up a noxious monarch just once, a surviving blue jay will avoid this species religiously thereafter.[184, 185]

If you have absorbed the essence of the cost–benefit approach, you will now be asking how an eaten monarch could gain fitness by inducing vomiting, given that regurgitated monarchs rarely fly off into the sunset. Is this case an exception to the rule that dead animals cannot pass on their genes? Probably not, although the possibility exists that toxicity has evolved via indirect selection (see pages 470–472), provided that the dead individuals educate the potential predators of their close relatives, thereby helping these genetically similar monarchs to pass on shared genes to the next generation. Another

**FIGURE 6.24 Effect of monarch butterfly toxins.** A blue jay that eats a toxic monarch vomits a short while later. Photographs by Lincoln P. Brower.

**FIGURE 6.25 Why behave conspicuously?** This tephritid fly (top) habitually waves its banded wings, which give it the appearance of a leg-waving jumping spider (bottom). When the spiders wave their legs, they do so to threaten one another. The fly mimics this signal to deter attack by the spider. Photographs by Bernie Roitberg; from Mather and Roitberg.[951]

hypothesis, however, is that monarchs recycle poisons from their food plants to make themselves so bad tasting that most birds will release any monarch immediately after grabbing it by the wing.[185] In fact, when caged birds are offered a frozen but thawed monarch, many pick it up by the wing and then drop it at once, evidence that in nature monarchs could benefit personally from being highly unpalatable, a fact advertised by their color pattern.

Bright coloration is not the only means by which a prey species can make itself conspicuous. Take the tephritid fly that habitually waves its banded wings as if trying to catch the attention of its predators. This puzzling behavior attracted two teams of researchers, who noticed that the wing markings of the fly resemble the legs of jumping spiders, important fly predators. The biologists proposed that when the fly waves its wings, it creates a visual effect similar to the aggressive leg-waving displays that the spiders themselves use (Figure 6.25).[573, 951] Thus, the fly could be a code breaker (see Chapter 4) whose appearance and behavior releases escape behavior on the part of the predator.

In order to test the deception hypothesis, one group of researchers became expert fly surgeons. Armed with scissors, Elmer's glue, and steady hands, they exchanged wings between clear-winged houseflies and pattern-winged tephritid flies. After the operation, the tephritid flies behaved normally, waving their now plain wings and even flying about their enclosures. But these modified tephritids with their housefly wings were soon eaten by the jumping spiders in their cages. In contrast, tephritids whose own wings had been removed and then glued back on repelled their enemies in 16 of 20 cases. Houseflies with tephritid wings gained no protection from the spiders, showing that it is the combination of leglike color pattern and wing movement that enables the tephritid fly to deceive its predators into treating it as a dangerous opponent rather than a meal.[573] By comparing the proportions of survivors among those endowed with a presumptive adaptation and those with an alternative, the researchers showed that the tephritid fly's wings and behavior work better than other options.

Some vertebrates also behave in ways that paradoxically make them easier to spot. Take the Thomson's gazelle. When pursued by a cheetah

**FIGURE 6.26** **An advertisement of unprofitability to deter pursuit?** Stotting behavior by a springbok, a small antelope that leaps into the air when threatened by a predator, just as Thomson's gazelles do.

or lion, this antelope may leap several feet into the air while flaring its white rump patch (Figure 6.26). Any number of possible explanations exist for this behavior, which is called stotting. Perhaps a stotting gazelle sacrifices speed in escaping from one detected predator in order to scan ahead for other as-yet-unseen enemies lying in ambush (as lions often do).[1148] The anti-ambush hypothesis predicts that stotting will not occur on short-grass savanna, but will instead be reserved for tall-grass or mixed grass-and-shrub habitats, where predator detection could be improved by jumping into the air. But gazelles feeding in short-grass habitats do stot regularly, so we can reject the anti-ambush hypothesis and turn to some others:[233, 234]

- *Alarm signal hypothesis*: Stotting might warn conspecifics, particularly offspring, that a predator is dangerously near. This signal could increase the survival of the signaler's offspring and relatives, thereby improving the stotter's fitness (see page 473).

- *Social cohesion hypothesis*: Stotting might enable gazelles to form groups and flee in a coordinated manner, making it harder for a predator to cut any one of them out of the herd.

- *Confusion effect hypothesis*: By stotting, individuals in a fleeing herd might confuse and distract a following predator, keeping it from focusing on one animal.

- *Pursuit deterrence hypothesis*: Stotting might announce to a pursuing predator that the individual is in excellent condition and is therefore unlikely to be captured, which, if true, would favor predators that stopped chasing that gazelle.

Table 6.2 lists some predictions that are consistent with these hypotheses. Because the same prediction sometimes follows from two different hypotheses, we must consider multiple predictions from each one if we are to discriminate among them. This study illustrates the value of coming up with multiple hypotheses and then thinking through the predictions derived from each one.

Tim Caro learned that a single gazelle will sometimes stot when a cheetah approaches, an observation that helps eliminate the alarm signal hypothesis (if the idea is to communicate with other gazelles, then lone gazelles should not stot) and the confusion effect hypothesis (because the confusion effect

**TABLE 6.2** *Predictions derived from four alternative hypotheses on the adaptive value of stotting by Thomson's gazelle*

| Prediction | Alternative hypotheses | | | |
|---|---|---|---|---|
| | Alarm signal | Social cohesion | Confusion effect | Signal of unprofitability |
| Solitary gazelle stots | No | Yes | No | Yes |
| Grouped gazelles stot | Yes | No | Yes | No |
| Stotters show rump to predator | No | No | Yes | Yes |
| Stotters show rump to gazelles | Yes | Yes | No | No |

can occur only when a group of animals can flee together). We cannot rule out the social cohesion hypothesis on the grounds that solitary gazelles stot, because there is the possibility that solitary individuals stot in order to attract distant gazelles to join them. But if the goal of stotting is to communicate with fellow gazelles, then stotting individuals, solitary or grouped, should direct their conspicuous white rump patch toward other gazelles. Stotting gazelles, however, orient their rumps toward the predator. Only one hypothesis is still standing: gazelles stot to announce to a predator that they will be hard to capture. Cheetahs get the message, since they are more likely to abandon hunts when the gazelle stots than when the potential victim does not perform the display (Figure 6.27).[233]

Gazelles are not the only species in the order Artiodactyla to exhibit stotting or something similar. Therefore, it is possible to make use of the comparative method to test the hypothesis that stotting behavior acts as a pursuit deterrence signal to predators. A comparative analysis of 200 species belonging to the Artiodactyla showed that there was the predicted association between leaping during pursuit for those species in the family Bovidae that had stalking predators but that deer (family Cervidae) did not behave as predicted.[236] These mixed results provide, at best, mixed support for a predator-signaling function for stotting by gazelles.

Nonetheless, if we accept the possibility that gazelles have evolved a visual signal with which to communicate with cheetahs, we must assume that the behavior honestly advertises their uncatchability to their enemies. If not, cheetahs would gain by ignoring the signal given by stotting gazelles. The prediction of honesty in signaling has not been tested for gazelles, but it has been for an *Anolis* lizard that performs pushup displays when it spots an approaching lizard-eating snake. Snakes tend to break off their hunt in response to seeing the lizard bobbing up and down. Because the number of pushups performed by individual lizards varies, Manuel Leal realized that he had an opportunity to test the prediction that these displays did indeed convey accurate information about a lizard's capacity for escape.

To conduct the test, Leal and an assistant first counted the number of pushups each individual performed in the laboratory when exposed to a model snake. They then took the lizard to a circular runway, where they induced it to keep running by lightly tapping it on the tail. The total running time

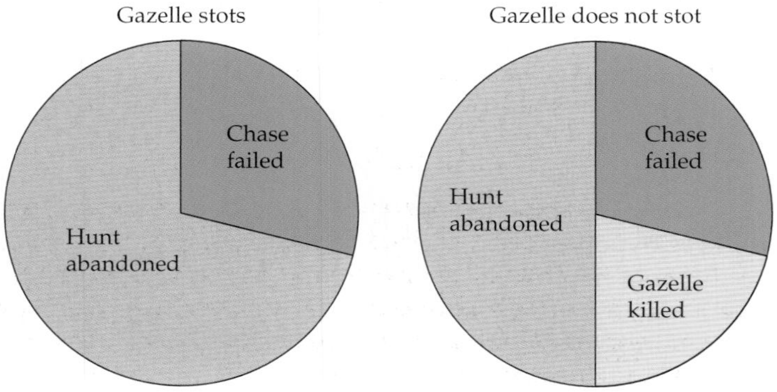

**FIGURE 6.27 Cheetahs abandon hunts more often when gazelles stot** than when they do not, supporting the hypothesis that these predators treat stotting as a signal that the gazelle will be hard to capture. After Caro.[233]

(A)

(B)

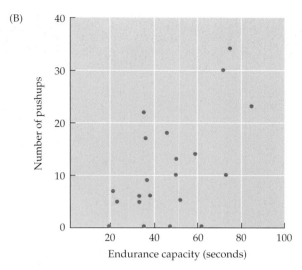

**FIGURE 6.28** ~~Are pushup displays an honest signal of a lizard's physiological condition?~~ (A) The lizard *Anolis cristatellius* performs a pushup display when it
spots an approaching snake. (B) The time a lizard spent running until exhaustion was
positively correlated with the number of pushups performed by that individual under
perceived threat from a model snake. A, photograph by Manuel Leal; B, after Leal.[846]

sustained by tapping was proportional to the number of pushups the lizard
performed in response to the model of its natural predator (Figure 6.28).[846]
Thus, as the attack deterrence hypothesis requires, predators could derive
accurate information about the physiological state of a potential prey by
watching its pushup performance. Since this anole sometimes does escape
when attacked, it could well pay predatory snakes to make foraging decisions
based on the signaling behavior of their would-be victims.

## Discussion Question

**6.12** The Bonaire whiptail lizard (Figure 6.29) runs a short distance from
potential predators and then raises one foreleg, which it waves about
ostentatiously.[296] This arm-waving behavior might be another example of a
pursuit deterrence signal. What predictions follow from this hypothesis with
respect to when the arm-waving behavior should be performed in response
to the approach of a human being (a predator substitute)? That is, should
arm waving occur more often when a person approaches slowly or rapidly?
In response to a direct or a tangential approach? And which arm should be
waved when the lizard is not directly facing the human?

One final Darwinian puzzle comes from the observation that some prey species produce conspicuous sounds. For example, the nocturnally active moth
*Euchaetes egle*, a perfectly edible insect, possesses devices (tymbals) on the sides
of its thorax that can produce a loud ultrasonic click whenever muscles pull
the tymbals in and then again when the structure is released to pop back into
its original shape.[75] Why make noises that make the moth's location all that
more obvious to killer bats, which can hear ultrasound wonderfully well?

As it turns out, this edible moth flies in areas where other related but thoroughly unpalatable moths, like *Cycnia tenera*, cruise the night skies. These
other moths feed as larvae on highly poisonous food plants, including milk-

**FIGURE 6.29** The lizard *Cnemidophorus murinus* often waves a foreleg at humans that disturb it. Why? Photograph by Laurie Vitt.

weeds containing cardiac glycosides, which are also eaten by monarch butterfly larvae. Just as is true for the monarchs, the caterpillars of *C. tenera* store the poisonous compounds they have eaten for their own protection later in life. When they become adults, they click their tymbals whenever they hear a bat approaching. By using an auditory channel that bats can hear, the clicking moths warn off experienced predators that have learned to associate moth-generated ultrasonic clicks with the bad taste of a toxic *C. tenera* and other noisy, poisonous arctiid moths.[689] The arctiid *E. egle* has the best of both worlds because it does not have to invest in the metabolic equipment needed to sequester and store the toxic compounds in milkweeds but can still deceive educated bats into giving it a wide berth through its use of acoustic deception.[75]

Thus, what looks like an especially maladaptive "come get me" signal from a moth turns out to be an adaptive use of deception that confers a survival advantage on the successful mimic. But what about the extraordinarily loud and piercing screams that rabbits and some birds give while in the clutches of a predator? Perhaps there is no benefit at all from screaming, which is simply a nonadaptive product of the prey's ability to feel pain. On the other hand, perhaps a loud scream might so startle a predator that it would release its captured prey.[661] Goran Högstedt studied this phenomenon by noting the responses of birds captured in Sweden upon their removal from a mist net. In his paper, "Adaptation unto Death," Högstedt reported that the fear screams of birds that cried out in this situation were loud and noisy, as required by the startle hypothesis. However, the captured birds kept calling and calling, which should reduce their ability to startle a predator (but see Conover[292]).

Högstedt considered whether screaming might warn conspecifics of danger from the predator or attract others that might distract the predator. If so, then the screamer should be easy to locate so that listeners would know where the predator is. The acoustical features of fear screams do make it easy for other animals to pinpoint the source of the noise, but as Högstedt pointed out, other members of the injured prey's species largely ignore its cries. This reaction makes a certain amount of sense, since a predator with a captive will be occupied with that victim for some time, which means that uncaptured specimens nearby have little to fear for the moment.

In a more recent study on the calls given by mist-netted birds in Costa Rica, Diane Neudorf and Spencer Sealy asked whether flocking species would be more prone to give loud distress calls than species that live alone. They found no difference between the two groups, suggesting that Costa Rican birds are not screaming to call for help, or to warn relatives of deadly dangers.[1040]

Högstedt's evidence pointed to a fourth hypothesis, namely, that the captured animal's screams attract other predators to the scene—predators that may turn the tables on the prey's captor, or at least interfere with it, sometimes enabling the prey to escape in the resulting confusion. This hypothesis requires that predators be attracted to fear screams, and they are, as Högstedt showed by broadcasting the taped screams of a captured starling, which brought hawks, foxes, and cats to the recorder. (On the other hand, in another study, when coyotes attracted to starling distress calls came toward a fellow coyote with a captured starling, the attacker intensified its effort to dispatch the bird, which was not at all helpful for the distressed starling.[1607])

The attract-competing-predators hypothesis also produces the prediction that birds living in dense cover should be more prone to give fear screams than birds of open habitats. In areas where vision is blocked, a captured bird cannot rely on other nearby predators to see it struggling with its captor. Therefore, as a final effort to avoid death, it may attempt to call competing predators to the scene. Högstedt found that mist-netted birds of species that live in dense cover are in fact more likely to give fear screams when handled than are species that occupy open habitats.[661] Under natural conditions, those screamers that attract other predators and survive as a result derive a benefit from their behavior, which could make it adaptive to scream even at death's door.

A Spanish team of ornithologists more recently considered a hypothesis that was not on Högstedt's list of alternative explanations for intense distress calls, namely, the possibility that these sounds convey information to a predator about the health and condition of the caller. Such information could conceivably influence whether a predator would pursue a bird that managed to break free after having been captured (or was on the verge of being captured). If this hypothesis had validity, there ought to be a connection between the "quality" of a distress call and physiological state of the distressed signaler.

To see if this prediction held up, the researchers caught lesser short-toed larks and held them overnight in cotton bags before releasing them the next day. Much earlier, the team had measured the body mass and wing length of each bird, as these measurements permit one to identify which individuals are relatively heavy for their size and thus in relatively good condition. When comparing the quantitative estimate of body condition of each bird with the "harshness" of its distress call (which is related to how broad a range of frequencies appear in the call), the researchers found that birds in better condition did indeed produce harsher calls.[830] Thus, a predator could conceivably evaluate the likelihood of recapturing an escaping lark and use this information to make a decision about whether to pursue the prey or not, making the fear scream of a departing bird another example of a pursuit deterrence signal (see pages 207–209). Whether some predators do make use of what they can learn from a bird's fear screams remains to be determined.

## Optimality Theory and Antipredator Behavior

Thinking about fitness benefits ($B$) and fitness costs ($C$) has drawn the attention of behavioral researchers to a variety of evolutionary puzzles and motivated a search for the reproductive advantages that might have driven the evolution of behavioral traits that initially appear to make individuals more vulnerable to predators, not less. Here I shall introduce two cost–benefit approaches, both derived from natural selection theory, that can potentially yield precise quantitative predictions, rather than the more general qualitative ones that we have been discussing up to this point. Both theories attempt to construct hypotheses that take into account the net benefit ($B - C$) of a trait, rather than merely focusing on whether a trait provides a benefit or not. Both theories reflect the reality that adaptations have to do more than merely confer a benefit if they are to increase in frequency in a population. An adaptation, by definition, is better

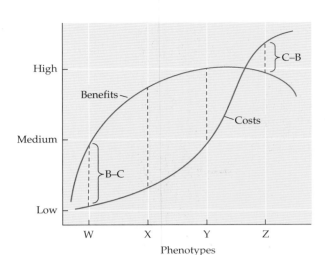

**FIGURE 6.30** **An optimality model.** If one can measure the fitness costs and benefits associated with four alternative behavioral phenotypes in a population, then one can determine which trait confers the greatest net benefit on individuals in that population. Such a trait is an adaptation—an optimal trait that would replace the alternatives, given sufficient evolutionary time.

than the alternatives, and "better" means that the net benefit associated with a true adaptation is greater than that associated with the nonadaptive alternatives that have been replaced, or are being replaced, by natural selection.

We can illustrate our first approach, which is based on **optimality theory**, by looking at the costs and benefits of four alternative hereditary behavioral phenotypes in a hypothetical species (Figure 6.30). Of these four phenotypes (W, X, Y, and Z), only one (phenotype Z) generates a net loss in fitness ($C > B$) and is therefore obviously inferior to the others. The other three phenotypes all are linked to a net fitness gain, but just one, phenotype X, is an adaptation, because it produces the greatest net benefit of the four phenotypes. Thus, the alleles for phenotype X will spread at the expense of the alternative alleles underlying the development of the other forms of this behavioral trait in this population. Phenotype X can be considered the optimal trait here because it is an adaptation, because the difference between $B$ and $C$ is greatest for this trait, and because this trait will spread while all others are declining in frequency (as long as the relationship between their costs and benefits remains the same).

If it were possible to measure $B$ and $C$ for a set of alternatives that might have existed within a population, one could predict that the phenotype with the greatest net benefit would be the one observed in nature. Unfortunately, it is often difficult to secure precise measures of $B$ and $C$ in the same fitness units. But if this can be done, then optimality theory makes it possible to produce a hypothesis founded on the premise that traits currently observed in a population will be optimal, and from this hypothesis can come quantitative predictions. Optimality theory is most strongly associated with the analysis of foraging behavior (as we will see in the next chapter), because in this context, it is sometimes possible to measure both benefits and costs in the same currency: calories gained from collected foods (the benefits) and calories expended in collecting those foods (the costs). But optimality theory has been applied to some antipredator behaviors as well.

For example, northern bobwhite quail spend the winter months in small groups, called coveys, which range in size from 2 to 22 individuals, but with a very strong peak around 11 individuals, in the midwestern United States. These coveys almost certainly form for antipredator benefits. For one thing, members of larger coveys are safer from attack, judging from the fact that overall group vigilance (percentage of time that at least one member of the covey has its head up and is scanning for danger) increases with increasing group size and then levels off around a group size of 10. Moreover, in aviary experiments, members of larger groups react more quickly than members of smaller ones when exposed to a silhouette of a predatory hawk.

The antipredator benefits of being in a large group are, however, almost certainly offset to some degree by the increased competition for food that occurs in larger groups.[1577] This assumption is supported by evidence that relatively large groups move more each day than coveys of 11. (Small groups also move more than groups of average size, probably because these birds are searching for other groups with which to amalgamate.) The mix of benefits and costs associated with coveys containing different numbers of individuals suggests that birds in intermediate-sized coveys have the best of all possible worlds, and in fact, daily survival rates are highest during the winter in coveys of this size (Figure 6.31). This work demonstrates not just that the quail form groups to detect their predators effectively, deriving a benefit from their social behavior, but that they attempt to form groups of the optimal size. If they succeed in joining such a group, they will derive a greater net benefit in terms of their survival than by joining groups of other sizes.

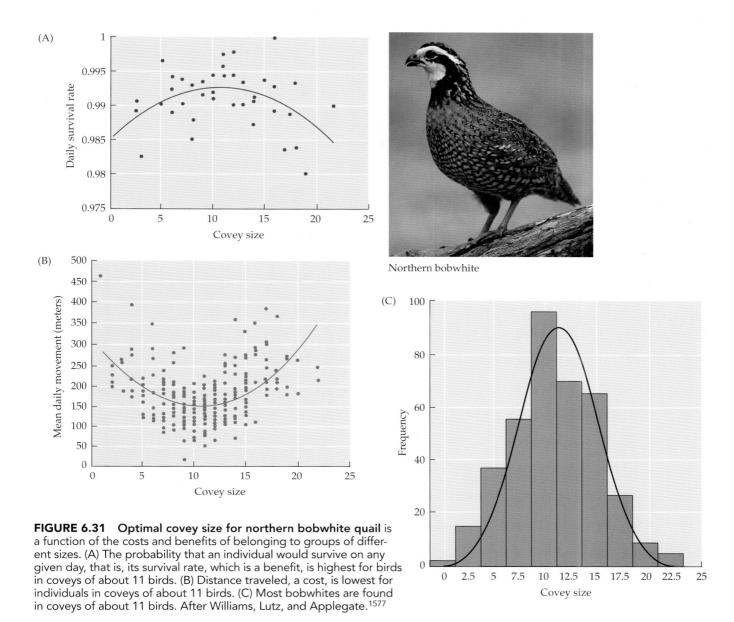

**FIGURE 6.31   Optimal covey size for northern bobwhite quail** is a function of the costs and benefits of belonging to groups of different sizes. (A) The probability that an individual would survive on any given day, that is, its survival rate, which is a benefit, is highest for birds in coveys of about 11 birds. (B) Distance traveled, a cost, is lowest for individuals in coveys of about 11 birds. (C) Most bobwhites are found in coveys of about 11 birds. After Williams, Lutz, and Applegate.[1577]

Northern bobwhite

## Game Theory Applied to Social Defenses

In addition to the optimality theory approach, we can look at behavioral evolution through the lens of **game theory**. Here too, both the costs and the benefits of behavioral decisions are considered, under the assumption that individuals are attempting unconsciously to maximize their reproductive success. But game theoreticians focus on cases in which individuals are competing with one another in such a way that the fitness consequences of a given behavioral option depend on the actions of the other competitors. Under this approach, decision making is treated as a game, just as it is by the economists who invented game theory in order to understand the choices made by people as they compete with one another for consumer goods or wealth. Game theoretical economists have shown that a strategy that works well in one situation may fail when matched against another way of making economic decisions. Figuring out what strategy will win often depends on what the other game players are doing.

The fact that all organisms, not just humans interested in how to spend their money or get ahead in business, are engaged in competitions of various sorts means that the game theory approach is a natural match for what goes on in the living world. The fundamental competition of life revolves around getting more of one's genes into the next generation than one's fellows. Winning this game almost always depends on what those other individuals are up to, which is why taking this factor into account makes a great deal of sense to evolutionary biologists.

One of the premier evolutionary biologists of all time, W. D. Hamilton, was a pioneer in thinking about evolution as a game between competing phenotypes. Hamilton argued that under some conditions, a behavioral strategy that caused individuals to be social could, over time, spread through a population in which the other individuals lived alone. The end result, according to Hamilton, could be a **selfish herd**[610] in which all the individuals were trying to hide behind others to reduce the probability of being selected by a predator. Imagine, for example, a population of antelope grazing on an African plain in which all individuals stay well apart, thereby reducing their conspicuousness to their predators. Now imagine that a mutant individual arises in this species, one that approaches another animal and positions itself so as to use its companion as a living shield for protection against attacking predators. The social mutant that employs this tactic would incur some costs; for example, two animals may be more conspicuous to predators than one and so attract more attacks than scattered individuals, as has been demonstrated in some cases.[1516] But if these costs were consistently outweighed by the survival benefit gained by social individuals, the social mutation could spread through the population. If it did, then eventually all the members of the species would be aggregated, with individuals jockeying for the safest position within their groups, actively attempting to improve their odds at someone else's expense. The result would be a selfish herd, whose members would actually be safer if they could all agree to spread out and not try to take advantage of one another. But since populations of individuals employing the solitary strategy would be vulnerable to invasion by an exploitative social mutant that uses a hide-behind-another strategy to take fitness from its companions, the exploitative tactic could spread through the species—a clear illustration of why we define adaptations in terms of their contribution to the fitness of individuals relative to that of other individuals with alternative traits.

Game theory, like its cousin optimality theory, can be used to generate mathematical models (hypotheses) from which can be drawn precise, quantitative predictions. But for the moment, let's apply the theory's general approach to a case involving Adélie penguins. These birds often wait on the ice near open water for some time until a group assembles, and only then do they all jump into the water more or less simultaneously to swim out to their foraging grounds. The potential value of this social behavior becomes clearer when one realizes that a leopard seal may be lurking in the water near the jumping-off point (Figure 6.32).[303] The seal can capture and kill only a certain small number of penguins in a short time. By swimming in a group through the danger zone, many penguins will escape while the seal is engaged in dispatching one or two of their unlucky fellows. If you had to run a leopard seal gauntlet, you would probably do your best to be neither the first nor the last into the water. If penguins behave as we would, then the groups that form at the water's edge qualify as selfish herds, whose members are engaged in a game whose winners are better than others at judging when to plunge into the water.

The selfish herd hypothesis generates testable predictions that can be applied to any prey species that forms groups. For example, redshanks, which are European sandpipers, feed in flocks. If these flocks are selfish herds and individuals are gaining a survival advantage by hiding behind others, we

**FIGURE 6.32  Selfish herds may evolve in prey species.** Adélie penguins have formidable predators, such as this leopard seal. While the seal is dispatching one penguin, others can more safely enter the water and escape to the open ocean. Photograph by Gerald Kooyman/Hedgehog House.

would expect that those individuals targeted by predatory sparrowhawks should be relatively far from the shelter provided by their companions. John Quinn and Will Cresswell collected the data needed to evaluate this prediction by recording 17 attacks on birds whose distance to a nearest neighbor was known. Typically, a redshank selected by a sparrowhawk was about five body lengths farther from its nearest companion than the companion was from *its* nearest neighbor (Figure 6.33).[1189] Thus, the birds that wandered off a bit from their fellow redshanks, presumably to forage with less competition for food, put themselves at risk by forgoing full participation in a selfish herd.

Game theory has many more applications, and we will deal with some of them in the chapters ahead. For now, it is enough to realize that although optimality theory and game theory are both derivatives of natural selection theory, they provide somewhat different tools to help us identify why animals behave the way they do.

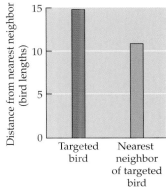

Redshank

Redshank flock

**FIGURE 6.33  Redshanks form selfish herds.** Redshanks that are targeted by hawks are usually standing farther from their nearest neighbors in the flock than *those* birds are standing from *their* nearest neighbors. After Quinn and Cresswell.[1189]

|  |  | Opponent | |
|---|---|---|---|
|  |  | Solitary | Social |
| Focal animal | Solitary | $P$ | $P-B$ |
|  | Social | $P+B-C$ | $P+\frac{B}{2}-\frac{B}{2}-C=P-C$ |

**FIGURE 6.34  A game theoretical model** in which the fitness gained by a solitary or social focal animal is dependent on the behavior of its opponent, which may be either a solitary or a social individual. Given the conditions laid out in the diagram, which trait is the adaptation: solitary or social behavior?

## Discussion Question

**6.13**  Consider Figure 6.34, a game theory diagram based on the concept of a selfish herd (with thanks to Jack Bradbury). In a population of prey animals, most individuals are solitary and stay well apart from others. But some mutant types arise that search out others and use them as living shields against predators. The mutants take fitness from the would-be solitary types by making them more conspicuous to their predators. We will set the fitness payoff for solitary living in a population composed only of solitary individuals at $P$. But when a solitary individual is found and used by a social type, the solitary animal loses some fitness ($B$) to the social type. There is a cost ($C$) to social individuals in terms of the time required to find another individual to hide behind, and there is a cost arising from the increased conspicuousness to predators of groups composed of two individuals rather than one. When two social types interact, we will say that they each have one chance in two of being the one that happens to hide behind the other when a predator attacks. If $B$ is greater than $C$, what behavioral type will come to predominate in the population over time? Now compare the average payoff for individuals in populations composed entirely of solitary versus social types. If the average fitness of individuals in a population of social types is less than that of individuals in a population composed of solitary types, can hiding behind others be an adaptation?

## Summary

1. An adaptation is a trait that has spread and been preserved by natural selection, which means that it is a hereditary trait that currently does a better job at promoting individual reproductive or genetic success than any available alternative form of that trait. Another way to say the same thing is that an adaptation has a better ratio of fitness benefits to fitness costs than any alternative characteristic that has happened to appear in the history of a species.

2. The adaptationist approach is the research procedure followed by persons who wish to test hypotheses on the possible adaptive values of attributes of interest to them. Traits that particularly intrigue adaptationists are those that have substantial fitness costs, which must therefore generate major fitness benefits if they are to spread and persist in populations under natural selection. Traits whose costs seem to exceed their benefits constitute Darwinian puzzles; the solutions to these puzzles are prized by adaptationists.

3. To test adaptationist hypotheses, a scientist must check the validity of predictions derived from these potential explanations. Relevant evidence on the fitness costs and benefits of traits can be gathered through field observations, controlled manipulative experiments, or natural experiments involving comparisons among living species. The comparative method of testing adaptationist hypotheses is based on two key assumptions: (1) that related species will exhibit differences in their attributes if they face different selection pressures, despite their having a common ancestor and thus a similar genetic heritage, and (2) that unrelated species that share similar selection pressures will converge on the same adaptive response, despite their having a different genetic heritage.

4. Two additional tools for the development and testing of adaptationist hypotheses are optimality theory and game theory. Optimality theory is most useful in cases in which it is possible to measure both benefits and costs of alternative traits in the same fitness currency. If this can be done, one can check whether an individual actually behaves in an optimal manner; that is to say, in such a way as to maximize its net benefit. Game theory comes into play when the benefit of a behavioral option to one individual depends on what the other members of its population are doing. This theory views evolution as a game in which the players are armed with different strategies that are in competition with one another, with the winners creating a population over time that cannot be invaded by a player with an alternative strategy.

## Suggested Reading

Two books, one by Wolfgang Wickler[1562] and the other by Rod and Ken Preston-Mafham,[1166] contain many amazing examples of animal coloration and behavioral defenses. John Endler has provided a modern review of the interrelationships between predator adaptations and prey counteradaptations.[444] The cost–benefit approach to antipredator behavior is described by Steven Lima and Lawrence Dill.[867] Tim Caro's papers on gazelle stotting[233, 234] illustrate the adaptationist approach very well, as does his review of how the pursuit deterrence hypothesis has been tested.[235] S. J. Gould and R. C. Lewontin's attack on adaptationism[557] is worth reading—critically. Bernie Crespi succinctly reviews the various definitions of adaptation and the several reasons for the occurrence of maladaptations in nature,[313] while Randolph Nesse revisits these issues largely in the context of understanding human disease.[1038] *The Selfish Gene* (second edition) by Richard Dawkins does a great job of explaining behavioral game theory.[369]

# 7

# The Evolution of Feeding Behavior

In our examination of antipredator behavior, we showed how a cost–benefit approach can be used to establish whether a behavioral trait has an adaptive survival-promoting function. Prey species have evolved any number of adaptations of this sort. Given that most animals are so good at postponing their demise, you may now possess a certain sympathy for foraging predators, which have to overcome a whole series of obstacles if they are to get something to eat. On the other hand, a predatory star-nosed mole (see Figure 4.26) requires as little as 120 milliseconds to process the information provided to it by its long and sensitive nose appendages. As a result, this predator, which feeds in complete darkness within underground tunnels, can almost instantly identify and consume the worms and other invertebrate prey that live abundantly in the swampy areas where the moles forage.[248] At night, in the skies above the swamps where star-nosed moles are hunting adeptly with their noses, bats are catching elusive moths in total darkness, thanks to their very different but equally amazing prey-detecting systems (see Figure 4.13). So perhaps we do not have to feel sorry for predators after all. In a way, the wonderful hunting skills that they exhibit are the product of their prey's defenses, which have favored the evolution of

◀ **Has this *bushtit* chosen this clump of red berries** *to maximize its caloric intake, or to acquire a key nutrient, or to avoid foraging areas where predators are more numerous? Photograph by Bruce Lyon.*

predatory counteradaptations in an arms race between the two antagonists (see Figure 6.9). As a result, predators often do get enough to eat.

This chapter examines hypotheses on the adaptive value of various elements of animal feeding behavior, with the goal of illustrating how behavioral ecologists use the theoretical tools at their disposal, including optimality theory and game theory. In addition to looking at how feeding behavior currently contributes to individual reproductive success, we shall consider some attempts to outline the historical sequence of events leading to modern feeding adaptations—the other component of an ultimate analysis of behavior.

## Optimal Foraging Behavior

A foraging crow has many decisions to make. Where should it search for food? At what time of day? For what prey? How long should it spend trying to process the prey that it has found? Users of optimality theory could analyze each of these foraging decisions in terms of their contribution to the crow's fitness by testing whether the bird is behaving in an optimal (fitness-maximizing) manner. Let's use this approach to examine the choices crows make about what kinds of shellfish to open.

When a beachcombing northwestern crow spots a clam, snail, mussel, or whelk, it sometimes, but not always, picks it up, flies into the air, and then drops its victim onto a hard surface. If the mollusk's shell shatters on the rocks below, the bird flies down to the prey and plucks out the exposed flesh. (A video of a crow opening a mussel is available at http://illuminations.nctm.org/java/Whelk/student/crows.html.) The adaptive significance of the bird's behavior seems straightforward: It cannot use its beak to open the extremely hard shells of certain mollusks. Therefore, it breaks the shells by dropping them on rocks. This seems adaptive. Case closed. But we can be much more ambitious in our analysis of the crow's foraging decisions. When a hungry crow is searching for food, it has to decide which mollusk to select, how high to fly before dropping the prey, and how many times to keep trying if the mollusk does not break on the first try.

Reto Zach made several observations while watching foraging crows:

1. The crows picked up only large whelks about 3.5 to 4.4 centimeters long.

2. The crows flew up about 5 meters to drop their chosen whelks.

3. The crows kept trying with each chosen whelk until it broke, even if many flights were required.

Zach sought to explain the crows' behavior by determining whether the birds' decisions were optimal in terms of maximizing whelk flesh available for consumption per unit of time spent foraging.[1634] The optimality hypothesis yielded the following predictions:

1. Large whelks should be more likely than small ones to shatter after a drop of 5 meters.

2. Drops of less than 5 meters should yield a reduced breakage rate, whereas drops of much more than 5 meters should not greatly improve the chances of opening a whelk.

3. The probability that a whelk will break should be independent of the number of times it has already been dropped.

Zach tested each of these predictions in the following manner: He erected a 15-meter pole on a rocky beach and outfitted it with a platform whose height could be adjusted and from which whelks of various sizes could be pushed off

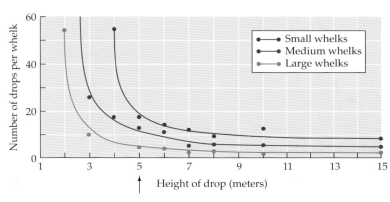

Northwestern crow

**FIGURE 7.1** **Optimal foraging decisions by northwestern crows when feeding on whelks.** The curves show the number of drops at different heights needed to break whelks of different sizes. Northwestern crows pick up only large whelks, increasing the calories available to them, and they drop their whelks from a height of about 5 meters, thereby minimizing the energy they expend in opening their prey. After Zach.[1634]

to drop onto the rocks below. He collected samples of small, medium, and large whelks and dropped them from different heights (Figure 7.1). He found, first, that large whelks required significantly fewer 5-meter drops before they broke than either medium-sized or small whelks. Second, the probability that a large whelk would break improved sharply as the height of the drop increased—up to about 5 meters—but going higher did little to improve the breakage rate. Third, the chance that a large whelk would break was not affected by the number of previous drops and was instead always about one chance in four on any given drop. Therefore, a crow that abandoned an unbroken whelk after a series of unsuccessful drops would not have a better chance of breaking a replacement whelk of the same size on its next attempt. Moreover, finding a new prey would take time and energy.

Zach went one step further by calculating the average number of calories required to open a large whelk (0.5 kilocalories), a figure he subtracted from the food energy present in a large whelk (2.0 kilocalories), for a net gain of 1.5 kilocalories. In contrast, medium-sized whelks, which require many more drops, would yield a net loss of 0.3 kilocalories; trying to open small whelks would have been even more disastrous. Thus, the crows' rejection of all but large whelks was adaptive, assuming that fitness is a function of energy gained per unit of time.[1634]

## Discussion Question

**7.1** In some places, American crows open walnuts by dropping them on hard surfaces. Unlike northwestern crows opening whelks, American crows reduce the height from which they drop walnuts from about 3 meters on the first drop to about 1.5 meters on the fifth drop. If this tendency is adaptive, what prediction follows about a difference between whelks and walnuts in the likelihood of breaking on successive drops? In addition, American crows tend to drop walnuts from lower heights when other crows are present. If this trait is an adaptation, what prediction must be true? Check your answers against data in Cristol and Switzer.[321]

**Discussion Question**

**7.2** A badger living in Oklahoma could hunt for either scorpions or ground squirrels (or both). Scorpions provide only 10 calories each, but they require only 2 minutes to find, on average, with an additional 3 minutes to remove the stinger; ground squirrels offer 1000 calories but take an average of 3 hours to find and an additional 90 minutes to capture, kill, and consume. If the badger's ultimate goal is to maximize its rate of caloric gain, should it forage for squirrels, scorpions, or both? Show your math. (Thanks to Doug Mock for this question.)

As just noted, a central assumption of the crow–whelk optimality study was that crows would achieve maximum reproductive success by maximizing the number of calories ingested per unit of time. This is an assumption that deserves to be checked. The relationship between whelk-opening efficiency and fitness has not been established for crows, but in an experiment with captive zebra finches, which were given the same kind of food but under different feeding regimes so that some birds had higher foraging costs than others, the individuals with the highest daily net caloric gains survived best and reproduced the most.[853] In a still more recent study, zebra finches that had decreased seed intake rates for 6 weeks (because they had to find seeds that had been mixed in with large quantities of chaff) took longer to lay their first eggs than other finches that had the same quantity of food presented with little or no chaff.[1569]

In an uncontrolled experiment of a somewhat similar sort, food intake has been shown to be related to the survival of the red knot (a sandpiper). In the spring, migrating red knots land on the shores of Delaware Bay in the eastern United States to feed on horseshoe crab eggs before completing their journey to northern Canada, where they breed. Horseshoe crab populations have crashed in recent years due to overharvesting by fishermen who use these unfortunate animals as bait.[60] As a result, the crabs no longer spawn in the huge numbers that they once did, and the knots find it much harder to gain the 60 to 100 grams of body weight that they need to complete their journey with enough in reserve to cope with bad weather, should it occur, on their Canadian breeding grounds.[204]

If there is a direct link between food and survival, as assumed in optimal foraging models, then we would expect a sharp drop in the population of red knots that stop in Delaware Bay on their migration north. This population did indeed decline from 51,000 in the year 2000 to 27,000 just 2 years later. In addition, researchers captured, weighed, and marked the sandpipers in Delaware Bay, then recaptured some of the same birds in subsequent years. The weights of knots that survived to return a second year were greater upon their initial capture than those of birds that were not recaptured or resighted a second time, a further indication that acquiring energy efficiently is directly associated with fitness.[60] Similar results have been secured for a different population of red knots, one that travels from Iceland to the high Canadian Arctic and back; in years in which breeding conditions in Canada were particularly bad, banded knots known to survive the summer had been unusually heavy when they left Iceland to start their migratory journey.[1011] These studies reveal that it may really only pay premigratory red knots to forage optimally in some years, not all, but the birds presumably cannot tell when conditions on the breeding grounds will make it essential to put on the maximum extra weight in the days before they head for Canada. They therefore always attempt to secure the fat reserves needed for the worst of times.

**7.3**  The Cape gannet, a seabird, normally feeds on oceanic fishes such as sardines, but during the nonbreeding season, the birds consume large quantities of fishery waste discarded by fishing vessels that process their catch at sea. Despite the fact that the birds do fine on a mixed diet of discards and sardines, when the breeding season comes round and there are young chicks to feed, gannets try to provide their youngsters with whole fish caught at sea rather than giving them the easily retrieved odds and ends thrown out of fishing boats. Even so, in recent years, the vast majority of chicks have died. Develop some hypotheses and predictions to account for the apparent failure of parental gannets to feed their chicks a diet that sustains fully grown birds and requires less energy to secure. Then read Grémillet et al.[577]

## How to Choose an Optimal Mussel

The Eurasian oystercatcher is a shorebird whose foraging decisions can also be matched against predictions taken from optimality models. Two Belgian researchers, P. M. Meire and A. Ervynck, developed a calorie maximization hypothesis to apply to oystercatchers feeding on mussels.[972] Like Reto Zach, they calculated the profitability of different-sized prey, based on the calories contained in the mussels (a fitness benefit) and the time required to open them (a fitness cost). Even though mussels over 50 millimeters long require more time to hammer or stab open, they provide more calories per minute of opening time than smaller mussels. Therefore, the model predicts that oystercatchers should focus primarily on the largest mussels. But in real life, the birds do not prefer the really large ones (Figure 7.2). Why not?

*Hypothesis 1*: The profitability of very large mussels is reduced because some cannot be opened at all, reducing the average return from handling these prey.

Oystercatcher

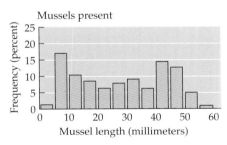

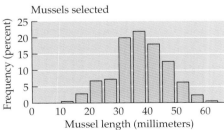

**FIGURE 7.2  Available prey versus prey selected.** Foraging oystercatchers choose mussels that are larger than the average available mussel, but they do not concentrate on the very largest mussels. After Meire and Ervynck.[972]

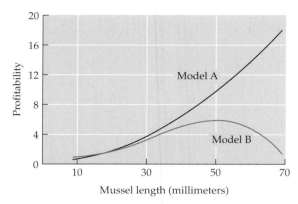

FIGURE 7.3   **Two optimal foraging models yield different predictions** because they calculate prey profitability differently. Model A calculates the profitability of a mussel based solely on the energy available in opened mussels of different sizes divided by the time required to open these prey. Model B calculates profitability with one added consideration, namely, that some very large mussels must be abandoned after being attacked because they are impossible to open. After Meire and Ervynck.[972]

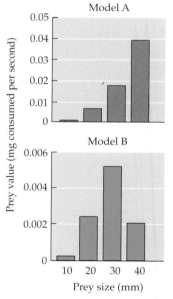

Pike cichlid

FIGURE 7.4   **Two optimal foraging models** of the value of hunting guppies of different sizes by a predatory fish, the pike cichlid. After Johansson, Turesson, and Persson.[728]

In their initial calculations of prey profitability, the researchers had considered only those prey that the oystercatchers actually opened (Figure 7.3, model A). As it turns out, oystercatchers occasionally select some large mussels that they find impossible to open, despite their best efforts. The handling time wasted on these large, impregnable mussels reduces the average payoff for dealing with this size class of prey. When this factor is taken into account, a new optimality model results, yielding the prediction that the oystercatchers should concentrate on mussels 50 millimeters in length, rather than the very largest size classes (Figure 7.3, model B). The oystercatchers, however, actually prefer mussels in the 30- to 45-millimeter range. Therefore, time wasted in handling large, invulnerable mussels fails to explain the oystercatchers' food selection behavior.

> *Hypothesis 2*: Many large mussels are not even worth attacking because they are covered with barnacles, which makes them impossible to open.

This additional explanation for the apparent reluctance of oystercatchers to feast on large, calorie-rich mussels is supported by the observation that oystercatchers never touch barnacle-encrusted mussels. The larger the mussel, the more likely it is to have acquired an impenetrable coat of barnacles, which eliminates these prey from consideration. According to a mathematical model that factors in (1) prey-opening time, (2) time wasted in trying but failing to open a mussel, and (3) the actual size range of realistically available prey, the birds should focus on 30- to 45-millimeter mussels—and they do. Note that these researchers used optimality theory to produce an initial hypothesis, which they rejected on the basis of evidence collected. They then modified their model and subjected it to a new test, with the result that they gained an improved understanding of oystercatcher feeding behavior.

Other researchers have also looked at oystercatcher foraging behavior from an optimality viewpoint. As a consequence we now know that oystercatchers prefer brown-shelled over black-shelled mussels, probably because of the lower water content in the mussels that produce brown shells. A lower water content means that more of this prey can be accommodated within the predator's esophagus, which enables the birds to have longer foraging bouts before having to stop to process the items that have been ingested,[1028] yet another way in which the oystercatcher can increase caloric gain per unit time.

## Discussion Question

**7.4**   The pike cichlid is a predatory fish that feeds on the guppy, a smaller fish. In the laboratory, at least, and probably in the rivers of Trinidad as well, this predator tends to attack and consume relatively large guppies. A Swedish research team measured the time it took for pike cichlids to detect, approach, stalk, and attack guppies of four size classes (10, 20, 30, and 40 millimeters long). They also recorded the capture success rate for each size class, as well as noting the time it took to handle and consume those prey that were actually captured. With these data, they constructed two models of prey value as measured by weight of prey consumed per unit of time.[728] Model A considered only the time to attack and the capture rate, whereas model B incorporated these two factors plus the post-capture handling time in calculating the weight of food consumed per unit of time (Figure 7.4). Which of the two models strikes you as the most realistic optimal foraging model, and why? In light of the two models, what evolutionary issues are raised by the actual preference of pike cichlids for 40 millimeter guppies?

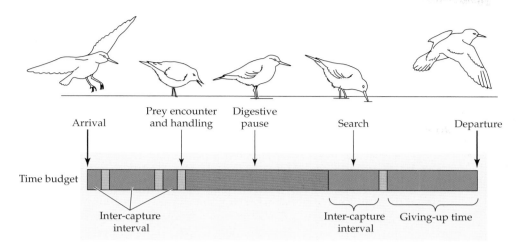

**FIGURE 7.5    A foraging bout by the red knot.** After a period of successful searching, the bird pauses to digest the prey it has found and eaten. When it resumes foraging, it must then decide how long to persist in a given patch before moving to a new one. After van Gils et al.[1484]

## Discussion Question

**7.5**   Red knots often feed on small shelled mollusks buried in scattered groups in tidal mudflats. The sandpipers enter a patch and probe for prey and usually swallow the shellfish they find whole, shell and all. If they acquire a number of thick-shelled prey, they eventually have to take a digestive break to process the shellfish. In order to feed optimally, should the birds consider short-term caloric intake as measured by the calories gained from mollusks of different sizes in relation to the time required to find and consume these items? Or should the birds make decisions based on the long-term caloric benefit, which requires consideration of calories gained in relation to digestive time? Under what conditions should the birds skip some large calorie-rich but thick-shelled prey in favor of smaller, less calorically valuable prey with thinner, more easily processed shells? At some point, the birds leave to go to another patch, where they engage in another round of hunting for mollusks (Figure 7.5).[1484] In order to determine whether the birds are foraging optimally, should you calculate their intake rate only when they are foraging, not when they are pausing to digest? When should a bird leave a patch?

## Criticisms of Optimal Foraging Theory

By developing and testing optimality models, researchers have concluded that northwestern crows and European oystercatchers, among other species, choose prey that provide the maximum caloric benefit in relation to time spent foraging. But some persons have criticized the use of optimality theory on the grounds that animals do not always hunt for food as efficiently as possible. As we have seen, however, optimality models are constructed not to make statements about perfection in evolution, but rather to make it possible to test whether one has correctly identified the variables that have shaped the evolution of an animal's behavior. As we have also seen, the factors included in an optimality model have a large effect on the predictions that follow. If an oystercatcher is assumed to treat every mussel in a tidal flat as a potential prey item, then it is predicted to make different foraging decisions than if the modeler assumes that oystercatchers simply ignore all barnacle-covered mussels. If the predictions of an optimality model fail to match reality, researchers can make progress by rejecting that model and going on to develop and test alternative hypotheses that take other factors into consideration.

   If ecological factors other than caloric intake affect oystercatcher foraging behavior, for example, then a caloric maximization model will fail its test, as it should. And for most foragers, foraging behavior does indeed have consequences above and beyond the acquisition of calories. If you suspected, for example, that predators had shaped the evolution of an animal's foraging behavior, then the kind of optimality model you might choose to construct and test would not focus solely on calories gained versus calories expended. If foraging exposes an animal to the risk of sudden death, then when that risk is high, we would expect foragers to sacrifice short-term caloric gain for long-term survival,[410, 733, 1510] but see Urban.[1478]

Such a sacrifice has been demonstrated for dugongs, which are large, relatively slow-moving marine mammals. In the aptly named Shark Bay of Western Australia, dugongs feed on sea grasses while attempting to avoid being consumed themselves by tiger sharks. Dugongs have two techniques for harvesting sea grasses: cropping, in which the herbivores quickly strip leaves from standing sea grasses, and excavation, in which the foragers stick their snouts into the sea bottom to pull out the sea grasses' underground components (rhizomes). Most of the time, the dugongs eat the sea grass rhizomes and all, which supplies them with more energy per time spent feeding. But the two techniques expose foraging dugongs to different levels of risk. When the animals have their heads partly buried in the sandy ocean floor, they cannot see well, whereas when they are cropping sea grass, they can more easily scan for approaching enemies. Dugongs in Shark Bay stop excavating sea grass rhizomes when tiger sharks are relatively common in the area, as measured by the shark catches made by local fisherman (Figure 7.6).[1606] An optimality model that failed to consider the trade-offs between foraging success and predation risk would fail to predict the dugongs' behavior accurately.

Dugongs are not the only animals to alter their behavior in the face of predation risk in ways that have costs as well as benefits for individuals. For example, elk living in Yellowstone National Park have changed their foraging behavior considerably following the reintroduction of wolves into the area. Instead of feeding comfortably in open meadows where their preferred foods are present in abundance, elk now leave the grasslands when wolves arrive and move into wooded areas where they are less easily spotted and chased by

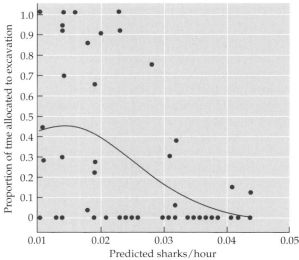

**FIGURE 7.6  Dugongs alter their foraging tactics** when dangerous sharks are likely to be present. The time these large, slow-moving sea mammals spend excavating sea grass from the ocean bottom declines in relation to the probability that tiger sharks will be in the area. After Wirsing, Heithaus, and Dill.[1606]

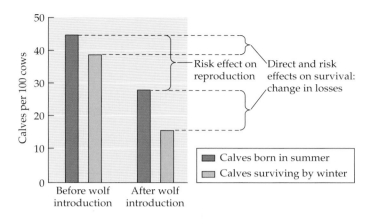

**FIGURE 7.7** **Even when predators do not directly kill their prey,** they can force individuals to change their behavior to avoid the risk of being killed in ways that reduce their reproductive output. After wolves were reintroduced into the Yellowstone ecosystem, elk modified their foraging behavior by spending more time hidden in woodlands rather than feeding in exposed meadows. This change reduced the production of calves by elk cows (blue bars) and decreased the survival of the calves they did have (pink bars). After Creel and Christianson.[312]

Elk mother and calf

their enemies.[311] Although this response promotes the survival of mature elk in a wolf-occupied environment, the prey pay a price for altering their foraging behavior. In the present era, when wolves hunt for elk, the probability that a given cow elk will give birth to a calf in the summer has fallen sharply, and the likelihood that her calf will still be with her by the time winter rolls around has also declined. Some of these changes in reproduction and calf survival can be attributed to what Scott Creel and his fellow students of elk behavior call a "risk effect" (Figure 7.7). In order to reduce the risk of being killed, the animals lower their energy intake, which reduces their chances of producing and tending a calf. If we were to fail to consider the fatal consequences of remaining in meadows visited by hunting wolves, we might conclude that elk were foraging in a suboptimal manner.

## Discussion Questions

**7.6** Various persons have proposed that when a horse switches from slower trotting to faster galloping, it changes gaits to minimize the energetic expense of locomotion, under the assumption that animals able to minimize the energetic costs of getting from point A to point B will enjoy greater reproductive success than individuals that squander their energy reserves in inefficient locomotion. The energy minimization hypothesis was tested by Claire Farley and Richard Taylor with the help of three cooperative horses willing to run on a treadmill while outfitted to provide data on their oxygen consumption, a factor directly related to energy use.[454] What scientific conclusion is justified on the basis of Figure 7.8? Does this study support those who claim that optimality theory is not useful because the assumptions underlying particular hypotheses are often oversimplified and incorrect?

**7.7** The golden-winged sunbird feeds exclusively on nectar from certain flowers during the winter in South Africa. Some birds are territorial at some patches of flowers but may abandon these defended areas at other times. Frank Gill and Larry Wolf devised a way to measure the rate of nectar production per bloom at a given site. They also secured previously published information on the caloric costs of the territorial chase flights of birds defending a feeding territory (2000 calories per hour), as well as the costs of foraging for nectar (1000 calories per hour) and resting (400 calories per hour). Table 7.1 shows the calories saved by birds by holding territories in relation to foraging in other, undefended flower patches, given various nectar production rates. Gill and Wolf assumed that the nonbreeding birds' goal was to collect sufficient nectar to meet their daily survival needs. (Why

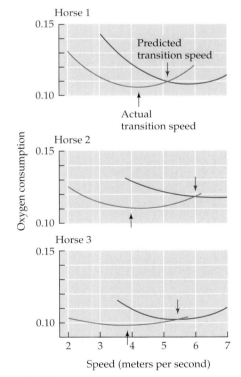

**FIGURE 7.8** **Records of energy consumption by three horses** in relation to two locomotory modes: trotting (red line) versus galloping (green line). Energy consumption was measured in terms of milliliters of oxygen consumed per kilogram of weight for every meter covered. After Farley and Taylor.[454]

**TABLE 7.1**  *Benefits of territoriality to golden-winged sunbirds under different conditions*

| Nectar production (microliters/blossom/day) | | 1 | 2 | 3 | 4 |
|---|---|---|---|---|---|
| Foraging time (hours) required to meet caloric need per day | | 8 | 4 | 2.7 | 2 |

| Nectar production in | | | |
|---|---|---|---|
| Territory | Undefended site | Hours of rest gained | Calories saved[a] |
| 2 | 1 | 8 − 4 = 4 | 2400 |
| 3 | 2 | 4 − 2.7 = 1.3 | 780 |
| 4 | 4 | 2 − 2 = 0 | 0 |

*Source*: Gill and Wolf[540]

[a]For each hour spent resting instead of foraging, a sunbird expends 400 calories instead of 1000, saving 600 calories. Total calories saved = 600 per hour of rest gained by not having to forage for nectar.

was this assumption reasonable?) They found that some sunbirds were territorial only when nectar production rates were higher in defended territories than in undefended areas. They also found that territory size contracted as the intrusion rate by nonterritorial birds increased; when the number of intruders was very high, the birds simply stopped defending their foraging areas altogether.[540, 541] Based on the data in Table 7.1, how many minutes of territorial defense would be worthwhile if a bird had access to a 2-microliters-per-blossom-per-day site while other flower patches had half that rate of nectar production? How many minutes of territorial defense would be worthwhile if a bird's choices were to defend a 3-microliters-per-blossom-per-day site versus foraging nonaggressively at another site where the blossoms each produced 2 microliters of nectar per day?

## Game Theory and Feeding Behavior

The preceding examples show how optimality theory can be used to develop mathematical models of foraging behavior that specify exactly what variables shape the behavior under investigation. Sometimes models of this sort permit the production of quantitative predictions that can be checked against reality with real precision, enabling researchers to reach firm conclusions about the validity of a hypothesis.

Another similar tool for exploring the possible adaptive value of feeding behavior is game theory, which as we outlined in the preceding chapter, is especially useful in cases in which individuals are in competition with one another for a valuable resource. This approach can be applied, for example, to cases in which two or more foraging techniques exist within the same species,[1357] including the garter snakes that eat slugs in one location and tadpoles in another, as described in Chapter 3. Whenever two or more distinctive foraging phenotypes are found within a species, the obvious question is, why hasn't the type associated with higher fitness replaced its rival over evolutionary time? The odds, for example, that rover and sitter fruit fly larvae (see page 85) will secure exactly the same amount of food per unit of time spent foraging seem vanishingly small. If one type of larva did even slightly better on average than the other, the genes specifically associated with that trait should spread and replace any alternative alleles linked to the less effective food-acquiring behavior. So why are *both* types still reasonably common in some places?

In the fruit fly case, the two kinds of behavioral phenotypes definitely differ genetically,[372] and so they can be said in game theory lingo to employ two different **strategies**.[589] In this context, strategies are not consciously adopted game plans of the sort humans often employ, but rather hereditary behavioral traits that differ among individuals. As noted, if one strategy confers higher fitness than the other in a population, eventually only the superior strategy will persist there. But under some special circumstances, two strategies can coexist indefinitely, thanks to the effects of what is called **frequency-dependent selection**. This kind of selection occurs when the fitness of one phenotype is a function of its frequency relative to the other phenotype. When the fitness of one type increases as that type becomes rarer, then that phenotype will become more frequent in the population—but only until such time as it has the same fitness as individuals playing the other strategy. Frequency-dependent selection will act against either type if it becomes even a little more common, pushing the proportion of that form back toward the equilibrium point at which both types have equal fitness. When this happens, the two phenotypes can coexist indefinitely.

In the case of rover and sitter fruit flies, experiments have shown that when food resources are scarce, the odds that an individual of the rarer phenotype will survive to pupation (a probable correlate of fitness) are greater than the corresponding odds for the more common type in the population (Figure 7.9).[470] The effect of this form of selection is to lead to an increase in the frequency of the rarer phenotype, maintaining its presence in the population.

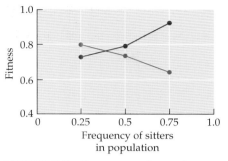

**FIGURE 7.9  Frequency-dependent selection.** When resources are scarce for fruit fly larvae, the fitness of a sedentary sitter (red line) versus a wandering rover (purple line) depends on which of the two types is the rarer. After Fitzpatrick et al.[470]

## Discussion Questions

**7.8**  Imagine a population of 1000 fruit fly larvae in which there are two hereditarily distinct foraging phenotypes, rover and sitter. Imagine that there are 195 rovers and 805 sitters. Let's say that both types survive to adulthood equally well and both types have 1.2 surviving offspring on average. What were the frequencies of the two phenotypes in the parental generation? What will the frequencies be in the generation composed of their offspring? What would happen if rovers had 1.1 surviving offspring on average, whereas sitters had 0.9? What's the point of this question?

**7.9**  When the frequency of sitters is 0.75 (see Figure 7.9), the fitness of rovers is much higher than that of the sitters. So why don't rovers quickly and completely eliminate the sitter phenotype in this population?

Another example of frequency-dependent selection in action involves an African cichlid fish, *Perissodus microlepis*. This cichlid comes in two forms, one with the jaw twisted to the right and the other with the jaw twisted to the left. The fish makes a living, believe it or not, by darting in to snatch scales from the bodies of other fish in Lake Tanganyika. Individuals with the jaw twisted to the right always attack the prey's left flank, while the other phenotype hits the right side (Figure 7.10). Right-jawed parents usually produce offspring that inherit their jaw shape and behavior; ditto for left-jawed parents, indicating that the difference between the two forms is hereditary.

So why do both phenotypes occur in this species? Michio Hori proposed that the fishes this predator attacks could learn to expect a raid on their scales from the left if most attacks were directed at the prey's left flank.[682] Thus, in a population of predators in which the right-jawed form predominated, a lefty would have an advantage because its victims would be less vigilant with respect to scale snatchers darting in on the right flank. This advantage would translate into higher reproductive success for the rarer phenotype and

**FIGURE 7.10  Two hereditary forms of an African cichlid fish.** Right-jawed and left-jawed cichlids have asymmetrical mouths that they employ to snatch scales from the left and right sides, respectively, of the fish they prey on.

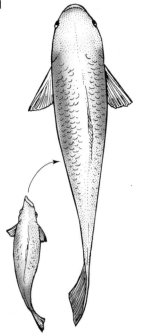

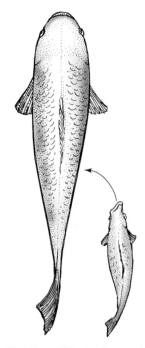

"Right-jawed" *Perissodus* attack prey from the left rear side

"Left-jawed" *Perissodus* attack prey from the right rear side

an increase in its frequency until left-jawed fish made up half the population. With a 50:50 split, the equilibrium point would have been reached.

If Hori's hypothesis was correct, he knew that he should find the frequency of any one phenotype oscillating around the equilibrium point. He confirmed that this prediction was true by measuring the relative frequencies of the two types over a decade (Figure 7.11).

But not every example of multiple foraging phenotypes resembles the *Perissodus* case. Take the ruddy turnstone. This small sandpiper has many different ways of finding prey items on beaches, ranging from pushing strands of seaweed to one side, to turning stones over, to probing in mud and sand for little mollusks. Some individuals specialize in one foraging method, while others prefer a different technique, but individual turnstones are rarely committed to just one way of finding food. This observation suggests that the differences between them are not caused by genetic differences, but instead they reflect an environmental difference of some sort.[1554] Philip Whitfield wondered if that environmental factor might be the dominance status of the foragers. Turnstones often hunt for food in small flocks, and the individuals in these groups form a pecking order. The birds at the top can displace subordinates merely by approaching them, thereby keeping them from exploiting the richer portions of their foraging areas. Dominants use their status to monopolize patches of seaweed on beaches, which they push about and flip over; subordinates keep their distance and are often forced to probe into the sand or mud instead of feasting on the invertebrates contained within the seaweed litter.

**FIGURE 7.11  The results of frequency-dependent selection in *Perissodus microlepis*.** The proportion of the left-jawed form in the population oscillates from slightly above to slightly below 0.5 because whenever it is more common than the alternative phenotype, it is selected against (and becomes less numerous); whenever it is rarer than the alternative phenotype, it is selected for (and becomes more numerous). After Hori.[682]

The turnstones exhibit flexibility in their foraging behavior, as individuals are apparently capable of adopting feeding methods in keeping with their ability to control different sources of food on the beach. The capacity to be flexible is provided by what game theoreticians have labeled a **conditional strategy**, which is an inherited mechanism that gives the individual the ability to alter its behavior adaptively in light of the conditions it confronts (such as having to deal with socially dominant competitors on a beach). Unlike the left- and right-jawed cichlids, which are locked into one particular phenotype, turnstones can almost certainly switch from one feeding **tactic** (or option) to another. But subordinate turnstones tend to stick with the probing technique because if they were to challenge stronger rivals for seaweed patches, the low-ranking birds would probably lose, in which case they would have wasted their time and energy for nothing while also running the risk of being badly hurt by an irate dominant. Instead, by knowing their place, these subordinates make the best of a bad situation and presumably secure more food than they would if they tried without success to explore seaweed patches being searched by their superiors. It can be adaptive to concede the better foraging spots to others if you are a turnstone with more powerful rivals.

We will have much more to say about conditional strategies in the chapters ahead, but note here that either genetic or environmental differences can provide a proximate explanation for why alternative behavioral phenotypes occur within a population (see Chapter 3). Behavioral variants can coexist either because of frequency-dependent selection or as the result of selection for individuals able to be flexible in their responses to variable environments.

## More Darwinian Puzzles in Feeding Behavior

Both optimality theory and game theory employ a cost–benefit perspective, which we will now use again to solve some other puzzling features of animal feeding behavior. For example, a great many animals form dense nesting colonies or nighttime roosts that bring large numbers of individuals into close contact. Being in groups could be more costly than beneficial, especially if the clustered animals rapidly deplete local food reserves. One possible benefit that might overcome the costs of being part of a flock or herd would be the ability of some individuals to take advantage of the food sources known to others. So, for example, the members of a colony of nesting seabirds might monitor returning foragers and follow the successful ones back to where they had been hunting. In this way, an animal that had been unsuccessful in its own attempts to detect food could find productive foraging areas.

### Discussion Question

**7.10**  Imagine that someone were to propose that nesting colonies of seabirds have formed to permit rapid information transfer about ephemeral food sources among their members, which in turn results in efficient collecting of food and maximization of the colonies' reproductive outputs. What theoretical objections would critical colleagues have about this person's hypothesis? What is the difference between this hypothesis and another, based on game theory,[77] that emphasizes the use of two tactics by the members of a colony: one a "producer" tactic, in which individuals seek out food patches, such as schools of fish, and the other a "scrounger" tactic, in which individuals exploit the searching successes of others by observing producers in action?

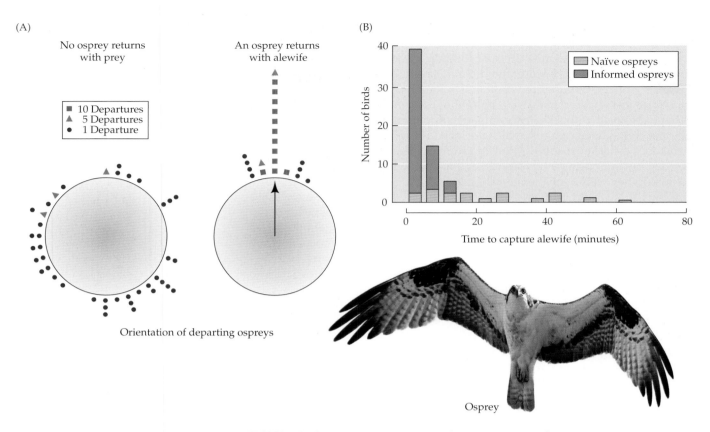

**FIGURE 7.12** **Do osprey nesting colonies serve as information centers?**
(A) When ospreys leave their nests at a time when no colonymate has returned with prey in the preceding 10 minutes, they fly off in all directions. But if another osprey has come back with an alewife (a schooling fish), the departing birds tend to fly off in the same direction (represented by an arrow) as the successful hunter. (B) Naïve ospreys (those that have not seen a prey-carrying colonymate prior to departure) take much longer to find alewife schools than informed ospreys (those that have just observed a returning neighbor carrying an alewife in its talons). After Greene.[572]

The idea that nesting colonies serve as information centers[1512] generates several predictions. One is that birds waiting in the colony should tend to follow successful hunters back out to productive foraging areas. But in an experimental study of a black-headed gull colony in which food platforms were set out at sea, the birds that found these sites were not followed by others on subsequent trips. Even when gulls landed by their nests with fish held conspicuously in their beaks, birds nesting nearby did not fly after the fish finder when it left to go on another foraging trip. Morever, when these neighbors did head out to sea, they did not often fly in the same direction as their successful colonymates.[32]

Although this test of the information center hypothesis yielded a negative result, Erick Greene found that fish-eating ospreys, which form loose nesting colonies in some coastal areas, do learn from others where to find fish species that appear here and there in large schools.[572] Ospreys not only are more likely to begin foraging shortly after a fellow osprey returns to the colony with a schooling alewife or smelt, but also tend to fly out in the direction taken by the successful forager. In addition, birds that see a successful hunter with prey are able to capture the same type of prey much more quickly than are ospreys that hunt without benefit of this kind of learning experience (Figure 7.12).

**FIGURE 7.13**  **Web ornament of an orb-weaving spider.** The large female spider has added four thick, conspicuous zigzagging lines of ultraviolet-reflecting silk to her web, which radiate out from her central resting point. Photograph by William Eberhard.

## Why Do Some Spiders Make Their Webs Conspicuous?

Another Darwinian puzzle is provided by orb-weaving spiders that incorporate striking zigzag lines of white ultraviolet-reflecting silk into their webs (Figure 7.13). These bright bands would seem to make the trap more obvious to prey, helping them evade the web, while at the same time making the spider on its web easier to find by its predators. However, a coterie of arachnologists have come up with several adaptationist hypotheses on how the benefits of web decorations might outweigh any negative effects on fitness that they might have.

One such explanation is that the silken ornaments are indeed seen by flying insects, but instead of repelling these prey, the decorations act as lures, drawing victims into the webs. Another possibility, not mutually exclusive, is that spiders sitting on web decorations gain an advantage in dealing with *their* predators, perhaps because the silken zigzags obscure the spider's body or make the spider look larger and more dangerous than it actually is.

Differences of opinion exist on the validity of the two hypotheses. On the one hand, the prey attraction hypothesis receives support from the discovery that in those spider webs in which only a single line of decorative silk had been laid down by the owner, insects were more likely to become trapped in the half of the web with the decorative silk (Figure 7.14).[309] However, an experimental study of another spider revealed that decorated webs caught fewer prey than undecorated ones. This conclusion was reached by letting orb-weaving spiders build webs in wooden frames prior to moving two such web frames to a spot side by side in a field. One of the two webs had had its decoration removed by excising the two web lines that supported the silk ornament (two web lines were also cut from the other web, just not those that held the decoration). Under these conditions, the reduced food gained from webs with decorations indicates that these devices come with a foraging cost, not a benefit, at least in some species under some conditions.[129]

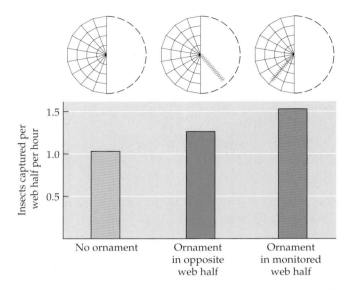

**FIGURE 7.14   Do web ornaments lure prey?** Garden spider webs without ultraviolet-reflecting ornaments capture fewer prey per hour than those containing ornaments. Furthermore, in webs with only one ornament, more flying insects are trapped in the half of the web containing the ornament than in the half lacking these structures. After Craig and Bernard.[309]

The same mixed results apply to the alternative hypothesis for web decorations, which is that they help protect the spider against predators, especially spider-hunting wasps. On the one hand, support for this idea comes from an outdoor study in which orb-weaving spiders that had built webs with or without extra silk decorations were exposed to wasps that hunt these animals to feed to their own offspring. In one experiment, only 32 percent of the spiders with web decorations were captured, as opposed to 68 percent of those without.[130] However, in another study, spiders with decorated webs were more likely to disappear than those with undecorated webs, suggesting that at least some predators may actually target the occupants of more conspicuous webs.[651] However, the disappearance of these spiders might have been caused by their more rapid growth (since spiders that eat more are more likely to decorate their webs), which enabled them to mature and reproduce more quickly, followed by their death or dispersal. The fact that spiders that have captured more prey are likely then to begin to decorate their webs is itself evidence against the prey attraction hypothesis.[128] If web decorations were to attract prey, you would predict that starved spiders would be especially likely to add these decorations to their webs, especially since they require relatively little extra silk and can be laid down very quickly.

The more recent papers on web decorations include one on *Argiope aemula*; the spiders were videotaped for hundreds of hours as they sat on decorated or undecorated webs. The mean rate at which prey were intercepted per unit area of spider web was substantially higher at the decorated webs, supporting the prey attraction hypothesis. During the same period of videotaping, several attacks by predator wasps on medium-sized spiders were recorded. Contrary to the predator defense hypothesis, more than twice as many attacks involved spiders on decorated webs as opposed to undecorated ones.[259]

On the other hand, William Eberhard has found two species that place silken decorations on one or two lines of silk where the spider appears to rest away from its prey-catching web. Since these "resting webs" are not sticky, they cannot function as prey-catching devices, and therefore the ornaments cannot function as prey-attracting lures. Instead, these silken constructions presumably help conceal the spider from visually hunting enemies.[428] The same seems likely for a third species studied by Eberhard, in which the spider often hides on a flattened cylinder of silken egg sacs and dried pieces of prey that the predator incorporates into its web. When egg sacs are removed experimentally, the spider sometimes replaces the missing items with strands of silk and, in so doing, enhances the concealing effect of its web ornament (Figure 7.15).[426]

Because of the conflicting evidence collected to date, no consensus exists on the adaptive value of web decorations,[201, 1381] although it is my impression that the anti-wasp hypothesis may be gaining the upper hand.

## Discussion Questions

**7.11**  Many spiders place the wrapped husks of consumed prey in their webs, which certainly make these webs more conspicuous to human beings (Figure 7.16). What hypotheses can you produce on the adaptive function of these web cemeteries?

**7.12**  Orb-weaving spiders can be divided into the kind that sits in the middle of the web waiting for prey and the kind that waits in a little hiding place next to the web, which it only leaves to go onto the web when an insect becomes entangled in its trap. Use the prey attraction and predator defense hypotheses to make a prediction about the proportions of species of the two types that add decorations to their webs. Check your prediction against the data in Herberstein et al.[651]

**FIGURE 7.15  Some spiders appear to hide on the large silken egg sacs in their webs.** Because these spiders will replace experimentally removed egg sacs with silky web ornaments on which they perch, some ornaments may help conceal the web-building spider. Photograph by William Eberhard.

## *Why Do Humans Consume Alcohol, Spices, and Dirt?*

One more selectionist mystery involves the apparently harmful or irrational food choices made by some people. For example, a great many persons in modern societies are addicted to alcohol, often at great fitness cost to themselves. The overconsumption of alcohol is clearly maladaptive. But is the trait an inexplicable artifact of human culture? Not according to Robert Dudley, who pointed out that since our closest relatives, the chimpanzees, derive most of their calories and nutrients from ripe fruits, it is likely that the ancestral species that gave rise eventually to modern chimpanzees and humans was also a frugivore.[416] Fruit-eating chimpanzees, and primates in general, prefer ripe fruits because these fruits have the highest concentrations of sugars. Ripe fruits also happen to contain a certain amount of ethanol, which is volatile and so provides a potential olfactory cue to the location of high-payoff foods, all the more so because ethanol itself is a calorie-rich food. Perhaps, therefore, an ancestral fondness for ethanol would have been adaptive even though currently such a proximate mechanism can be employed in a highly maladaptive manner in environments where beverages containing far more alcohol than a ripe fruit are easily available.

Dudley's historical hypothesis for the human proclivity to consume ethanol in large amounts predicts that those fruits with the highest ethanol concentrations will be most popular with fruit-eating mammals in

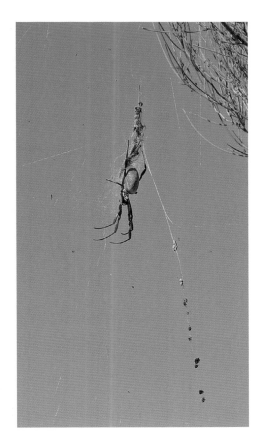

**FIGURE 7.16  Why does this very large Australian spider** place wrapped prey remains in a line above her resting place in the center of the web? Photograph by the author.

general. This prediction does not hold true, inasmuch as very ripe fruits are generally avoided by primates despite the fact that overripe fruits have the highest alcohol content;[988] likewise, fruit-eating bats appear to be deterred, not attracted by, high ethanol concentrations in very ripe fruit.[1274] Moreover, mice and rats, which are not frugivores, can easily become addicted to ethanol, which is not what would be expected if an attraction to ethanol is the evolutionary by-product of a fruit-dominated diet.[988]

Although we continue to be puzzled by why humans so often overindulge in alcohol, what about the use of spices in our food, another odd human indulgence.[120] Many spices are horribly expensive. The Countess of Leicester, who lived in England in the 13th century, was willing to pay as much for a pound of cloves as she would for a cow.[1177] Most of the major voyages of the early European explorers, Christopher Columbus included, were motivated by the desire for spices, as all North American schoolchildren learn. And yet the caloric and nutritive value of most spices is small, especially since they are often used sparingly. Perhaps, then, spice use is simply an arbitrary product of cultural invention—a plausible argument, since culinary traditions vary so greatly from culture to culture.

Jennifer Billing and Paul Sherman, however, proposed and tested the adaptationist hypothesis that spices served (and may continue to serve) a fitness-enhancing function by virtue of their antimicrobial properties. This hypothesis, needless to say, requires that spices kill the dangerous bacteria that sometimes contaminate our food, especially meats held under imperfect refrigeration, a prediction that has now been checked (Figure 7.17). The antimicrobial hypothesis also predicts that the extent to which spices are used should be a function not of their local cultivation, but of the risk of dangerous microbial contamination, which is related to local climate and the nature of

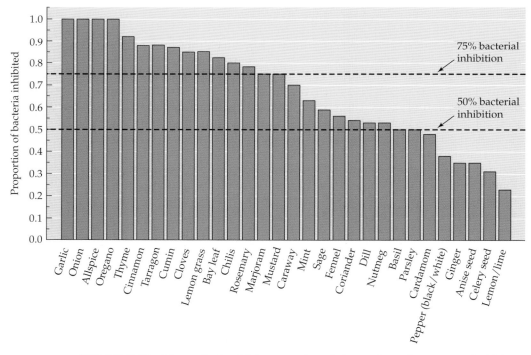

**FIGURE 7.17 The antimicrobial properties of the major spices,** most of which inhibit the growth of one-half or more of the bacterial types against which they have been tested. After Billing and Sherman.[120]

the food being prepared for consumption. As expected, traditional recipes in hot, tropical countries call for more bactericidal spices than do recipes in, say, Norway or Sweden.[120] Moreover, in a sample of cookbooks from 36 countries, recipes devised for meat-containing dishes, which are especially vulnerable to microbial contamination, call for more antimicrobial spices than vegetable dishes, which are less likely to support dangerous bacteria and other pathogens.[1318] Thus, we can conclude that spices have been used adaptively by humans even during the relatively short period of our evolution when these substances have been made widely available.

Finally, consider another strange behavior—the eating of dirt, a trait that is widespread among primates, including humans.[394] No one has ever claimed that the clayey soil consumed by chimpanzees or humans has any caloric value whatsoever. Among the hypotheses advanced to explain this habit in our species is the possibility that clay eating is a pathological, aberrant behavior of no functional significance. In contrast, an alternative adaptationist hypothesis proposes that clays are consumed to detoxify certain kinds of foods, thereby improving their nutritional value.[729]

The pathology hypothesis predicts that only a relatively few, possibly somewhat deranged, individuals will eat clay. This prediction does not withstand scrutiny, since clay eating is, or was, common practice in a diverse set of cultures, including the Aymara of Andean Bolivia, the Hopi of the American Southwest, and the natives of Sardinia, an island in the Mediterranean Sea. What the people in these cultures have in common is a dietary reliance on bitter alkaloid-laden varieties of potatoes or bitter tannin-rich acorns. When potatoes are dipped in a clay slurry and then baked, or when acorns are baked into a clay-containing bread, the tannins and alkaloids in these foods are either bound to the clays or otherwise altered chemically, rendering the foods more palatable and less toxic.[729]

As a comparative test of the detoxification hypothesis, we can predict that animals other than our closest relatives whose diets feature foods high in tannins or alkaloids will also seek out and eat certain clays. Rhesus monkeys and some other primates not especially closely related to humans do eat clay when feeding on tannin-rich vegetation.[806] In addition, two species of lemurs that live together in the same Malagasy rainforest but differ in their reliance on alkaloid-containing seeds also differ in their consumption of dirt in the predicted manner.[1163] Finally, even non-primates, such as African elephants, consume dirt when feeding on potentially toxic vegetation.[685]

And it is not just mammals that ingest dirt from time to time. Gorgeous red-and-green macaws and other parrots regularly visit certain South American riverbanks to gnaw off and eat chunks of clay. The diet of these birds includes some alkaloid-rich seeds, unripe fruits, and leaves (Figure 7.18).[539] But perhaps the parrots feed on soils to secure gravel for their gizzards, or alternatively, they could be eating dirt to acquire essential minerals absent in their vegetarian diets. Neither of these two hypotheses passes muster, given that the clays consumed by parrots at one famous source in the Peruvian jungle are composed of extremely fine particles, eliminating the gizzard grit hypothesis; moreover, the clays also contain very little in the way of minerals useful as dietary supplements.[394] Instead, the special clays selected by the parrots have negatively charged cation exchange sites that bind with positively charged alkaloids and other toxic chemicals found in unripe fruits and certain seeds. The clays line the birds' digestive tracts for hours, inactivating plant alkaloids and thereby protecting the cells of the gastrointestinal lining. When captive parrots were given a slug of an alkaloid with or without a dose of the clay they prefer, the subjects with clay-protected guts had 60 to 70 percent lower levels of the toxin in their blood when tested 3 hours later.[539] The fact that humans, lemurs, and macaws have converged on similar dietary solu-

**FIGURE 7.18  Clay eating has evolved in several species of parrots** that feed on foods rich in tannins or toxins, including these macaws, which regularly gather at this Amazonian riverbank to collect and consume a particular kind of clay.

tions to similar ecological problems shows again how the right comparisons among unrelated species can provide useful evidence for the evaluation of adaptationist hypotheses.

## Discussion Question

**7.13**  I suspect that most readers of this book consider eating dirt an unattractive proposition but that you would be even less thrilled at the idea of cannibalizing a fellow human being. Nevertheless, try to be dispassionate about the matter, and develop a cost–benefit analysis of human cannibalism as if you were an adaptationist. You should be able to make predictions about the circumstances under which an evolutionary biologist would expect the behavior to occur. Then read Jared Diamond's article on the subject[395] and reconstruct his argument: identify the question he seeks to answer, and produce the alternative hypotheses, the predictions, the tests, and the conclusion.

## The Adaptive Value and History of a Complex Behavior

The focus of this chapter up to this point has been primarily on the challenge of developing and testing adaptive explanations for behavioral attributes whose functions are not obvious. Here we will use the dances of honey bees to illustrate how one can explore questions about both the adaptive value of a complex behavioral trait and how the trait originated and was modified over evolutionary time.

The famous dances of honey bees are performed by workers when they return to their hive after having found a good source of pollen or nectar.[1500] As the dancers move in circuits on the vertical surface of the honeycomb in the complete darkness of the hive, they attract other bees, which follow them around as they move through their routines. Researchers watching dancing bees in special observation hives have learned that their dances contain a surprising amount of information about the location of a food source (such as a patch of flowers). If the bee executes a round dance (Figure 7.19), she has found food fairly close to the hive—say, within 50 meters of it. If, however, the worker performs a waggle dance (Figure 7.20), she has found a nectar or pollen source more than 50 meters away from the hive. By measuring the duration of the waggle-run portion of the circuit, a human observer can tell approximately how far away the food source is. The longer the waggle-run portion lasts, the more distant the food.

Moreover, by measuring the angle of the waggle run with respect to the vertical, an observer can also tell the direction to the food source. Apparently, a foraging bee on the way home from a distant but rewarding flower patch notes the angle between the flowers, hive, and sun. The bee transposes this angle onto the vertical surface of the comb when she performs the waggle-run portion of the waggle dance. If the bee walks directly up the comb while waggling, the flowers will be found by flying directly toward the sun. If the bee waggles straight down the comb, the flower patch is located directly away from the sun. A patch of flowers positioned 20° to the right of a line between the

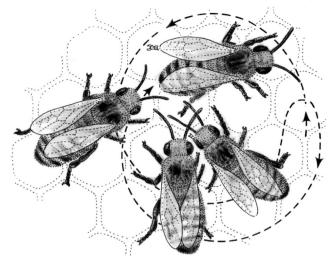

**FIGURE 7.19 Round dance of honey bees.** The dancer (the uppermost bee) is followed by three other workers, who may acquire information that a profitable food source is located within 50 meters of the hive. After von Frisch.[1500]

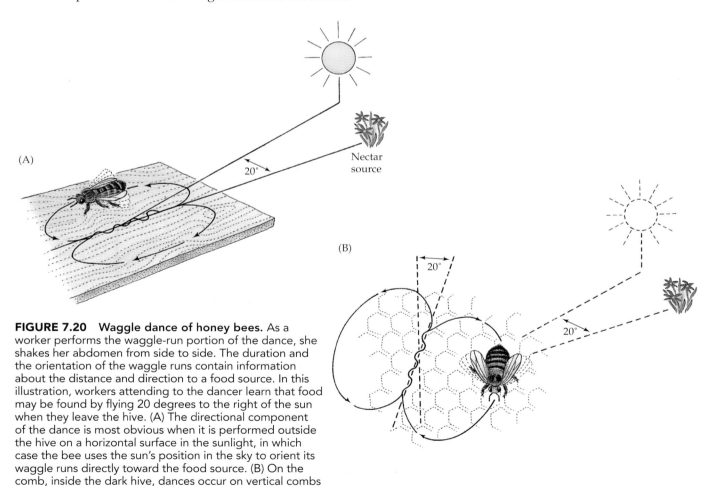

**FIGURE 7.20 Waggle dance of honey bees.** As a worker performs the waggle-run portion of the dance, she shakes her abdomen from side to side. The duration and the orientation of the waggle runs contain information about the distance and direction to a food source. In this illustration, workers attending to the dancer learn that food may be found by flying 20 degrees to the right of the sun when they leave the hive. (A) The directional component of the dance is most obvious when it is performed outside the hive on a horizontal surface in the sunlight, in which case the bee uses the sun's position in the sky to orient its waggle runs directly toward the food source. (B) On the comb, inside the dark hive, dances occur on vertical combs so that they are oriented with respect to gravity; the deviation of the waggle run from the vertical equals the deviation of the direction to the food source from a line between the hive and the sun.

(A)

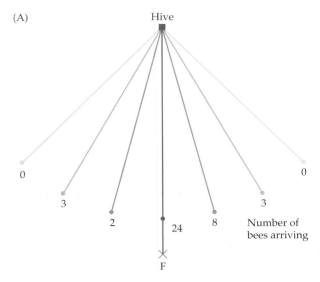

(B)

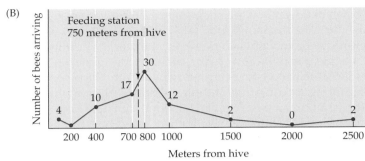

**FIGURE 7.21** **Testing directional and distance communication by honey bees.** (A) A "fan" test to determine whether foragers can convey information about the direction to a food source they have found. After training scout bees to come to a feeding station at F, von Frisch collected all newcomers that arrived at seven feeding stations with equally attractive sugar water. Most new bees arrived at the feeder in line with F. (B) A test for distance communication. After training scouts to come to a feeding station 750 meters from the hive, von Frisch collected all newcomers arriving at feeders placed at various distances from the hive. In this experiment, 47 newcomers were captured at the two feeders closest to 750 meters, far more than were caught at any two other feeders. After von Frisch.[1499]

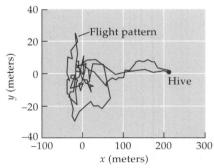

Honey bee

**FIGURE 7.22** **The response of a foraging honey bee** to the removal of a familiar food source is to return to the vicinity of the spot and then begin circling outward around the place where the bee had found food previously. The figure shows a path taken by one worker searching for food around the spot where a feeder had been placed in the past. After Reynolds et al.[1213]

hive and the sun is advertised with waggle runs pointing 20° to the right of the vertical on the comb. In other words, when outside the hive, the bees' directional reference is the sun, whereas inside the hive, their reference is gravity.

The conclusion that the dances of bees contain information about the distance and direction to good foraging sites was reached by Karl von Frisch after several years of careful experimental work.[1500] His basic research protocol involved training bees (which he daubed with dots of paint for identification) to visit feeding stations, which he stocked with concentrated sugar solutions. By watching the dances of these trained bees, he saw that their behavior changed in highly predictable ways depending on the distance and direction to a feeder. More importantly, his dancing bees were able to direct other bees to a feeder they had found (Figure 7.21), leading him to believe that bees use the information in the dances of their hivemates to find good foraging sites. Many years later, Jacobus Biesmeijer and Thomas Seeley were able to show that more than half the young worker bees that were just beginning their careers as pollen or nectar gatherers spent some time following dancing bees before launching their collecting flights.[119] Moreover, these bee biologists showed that experienced bees also followed dancers, particularly when resuming foraging after an interruption of some sort (such as a break caused by a rainstorm). These results suggest that worker followers gain useful information from their dancing hivemates.

## Discussion Question

**7.14** Bees that have learned the location of a good nectar source will return to it. But the source may become rapidly depleted, or the bee may miss its target by mistake. Figure 7.22 shows a typical flight path taken by a bee trained to come to a special feeder that was then removed.[1213] (The flight path was recorded through the use of radar, which honed in on a tiny radar transponder worn by the experimental subject.) How would you apply optimal foraging theory to determine whether this kind of response to a "missing" nectar source was actually adaptive?

Despite the appeal of the idea that the movements of dancing bees contained the information that guided foragers to food sources, some scientists argued that the workers ignored the dance movements per se and instead focused on the flower odors present on the dancing bee's body to guide their search.[1536] Contrary to this alternative hypothesis, however, when foragers from a hive were trained to two feeders that were given the same scent but were located in opposite directions from the hive and then only one feeder was advertised by waggle dances (because only that feeder contained a concentrated sugar solution), the bees recruited from the hive arrived primarily at the feeder advertised by the dancers. This finding indicates that workers learn something by attending to the movements of the dancers in their hive.[1300]

Additional evidence on this point comes from an experiment in which recruiter bees were trained to collect food at the end of an 8-meter-long tunnel set 3 meters away from the hive. When the recruiters advertised their find, they performed a waggle dance, not a round dance—which is surprising, given that the food source was only 11 meters distant, well within the range that would normally trigger a round dance. The reason for the recruiters' error has to do with the proximate mechanism by which bees determine the distance they have flown, which is based not on a direct calculation of the absolute distance flown, but instead on the total amount of image motion the bee's visual system has recorded while flying to a foraging site. Because the researchers forced foragers to travel through a narrow tunnel, the amount of image motion recorded by the bees' retinas was much greater than if the scouts had been traveling in the open air with objects far away from them. Thus, when they came back to announce their discovery, they danced as if they had found food about 70 meters away, judging from the mean duration of the waggle component of their dances, which was about 350 milliseconds. When the experimenters put out three empty feeders at 35, 70, and 140 meters from the hive, recruited workers showed up primarily at the 70-meter site, even though it contained neither food nor any special floral scent (Figure 7.23).[446] The fact that the recruits came to this site shows that they could "read" the dances of their hivemates that had been tricked into giving incorrect information about the location of a nectar source.

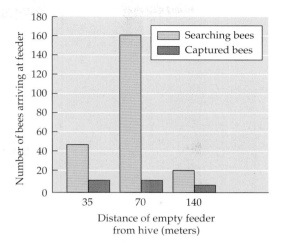

**FIGURE 7.23    Honey bee recruits** really do "read" the symbolic information in dances, as shown by their readiness to fly to empty feeders if misdirected by scouts that had been deceived into dancing as if the food source were farther away than it actually was. When workers attended to scouts whose dances announced (falsely) that food could be found 70 meters from the hive, more recruited bees (orange bars) showed up near an empty feeder 70 meters from the hive than at two feeders at other locations. Any recruit that actually landed on the feeder was collected (blue bars) to prevent these individuals from recruiting others to the site. After Esch et al.[446]

## Discussion Question

**7.15**  Wolfgang Kirchner and Andreas Grasser evaluated the performance of honey bee recruits from a special hive that could be turned on its side or held upright in the standard position.[775] They found that when the hive was on its side, bees continued to dance in the darkness, but on a horizontal surface, not a vertical one. Under these conditions, recruitment at distant feeders (more than 100 meters from the hive) that had been visited by dancing bees was very poor. When, however, the hive was returned to an upright position and the comb surface on which recruiters danced was vertical (as it would be in natural hives), most recruits appeared at the feeders that the scouts had visited. How do you interpret these results? What bearing do they have on the argument about whether recruits derive information from the dances of their colonymates? What prediction can you make about relative rates of recruitment to sites less than 50 meters from the hive when it is turned on its side as opposed to when it is upright?

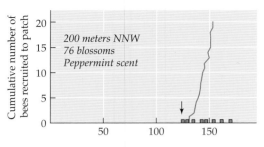
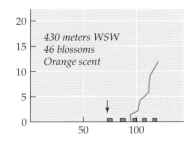
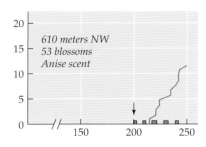

Time since flowers were set out (minutes)

**FIGURE 7.24** **Rapid buildup of recruited foragers** at experimental flower patches after their discovery by scout bees. Researchers placed pots of flowers, treated with different scents, at three locations on an island. At each of the three locations, considerable time passed before a scout found the patch (arrow), but shortly thereafter, many additional bees arrived at the flowers. The blue-green bars indicate the presence of the scout bee at the food patch. After Seeley and Visscher.[1302]

## The Adaptive Value of Honey Bee Dances

Tom Seeley and Kirk Visscher have examined how the time and energy costs of dancing might be offset by fitness benefits to the queen whose daughters perform this behavior.[1302] Since worker honey bees are sterile females, their activities cannot promote their own reproductive success, but their behavior could help their reproductively capable relatives, especially their mother (see pages 488–494). For example, if a hive's dancers could contribute to the rapid recruitment of a large forager workforce to a food patch, then the colony could harvest more resources there before other bee colonies and competitors arrived to deplete the food. If dancing has this effect, then the buildup of bees at a flower patch after a scout discovers it and begins advertising the fact should be faster than if each bee had to discover the patch without guidance from colonymates.

To test this prediction, Seeley and Visscher moved a colony of bees to a barren island off the coast of Maine, along with many pots of flowering plants. They then shifted these portable flower patches from place to place on the island and measured the time it took scout bees to locate each one. Scouts came by at the rate of about one bee every 1 to 3 hours. But once a scout returned to her hive and danced there, recruited bees quickly flooded the food source. Even though any one recruit took an average of 2 hours to find a newly advertised food source, some of the many bees following the scout's dances arrived at the advertised site quickly, resulting in a fairly rapid buildup of recruits at a given site (Figure 7.24).

Nevertheless, this work does not establish definitively that it was the symbolic dances per se, and not the flower scent information provided by the dancers, that led to the recruitment of workers to profitable food patches. A more direct way to examine the fitness consequences of dancing would be to weigh colonies during periods when they had access to dance recruitment information and when they did not. If dancing is adaptive, colonies should gain more weight (through improved pollen and nectar collection) when accurate dance information is available. To test this prediction, Gavin Sherman and Kirk Visscher devised an experiment in which they used four hives outfitted with horizontal dance platforms only. Half the hives had their dance platforms illuminated by diffuse light, while the other half were outfitted with a unidirectional light source. Under the diffuse light conditions, scouts continued to dance, but their movements were disoriented, since they had no one reference point against which to orient their dances. Under the unidirectional light con-

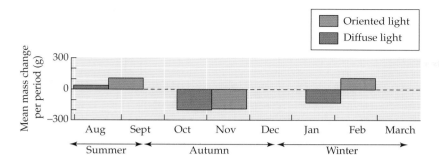

□ Oriented light
■ Diffuse light

**FIGURE 7.25  The adaptive value of the dance communication system.** A comparison of mean mass changes for four colonies over periods during the summer, fall, and winter when the bees did have access to accurate dance information (oriented-light conditions) and when they did not (diffuse-light conditions). The colonies made significant weight gains during the winter only when they could use their dance communication system. After Sherman and Visscher.[1313]

ditions, dancing recruiters used the lightbulb as a substitute for the sun such that their dances were oriented with respect to this cue, enabling recruits to derive useful information from their colonymates.

By alternating periods when the colonies were under diffuse-light versus unidirectional-light conditions, Sherman and Visscher could add up the gains or losses in mass during the two experimental treatments. The results (Figure 7.25) indicate that during the summer and fall, the occurrence of oriented dances did not have a statistically significant effect on colony weight. During the winter season, however, colonies gained mass during the oriented dancing periods and lost mass during the disoriented dancing periods.[1313] The ability of recruits to secure directional information about food location from others in the hive apparently has no effect during some times of year but a positive, fitness-enhancing effect at other times. As Sherman and Visscher point out, if honey bee colonies are locked into a boom-or-bust economy, then the ability to take full advantage (via dancing-directed foraging) of brief periods of resource abundance may be highly advantageous.

## The Origin and Modification of Honey Bee Dances

Having described honey bee dances and shown that they are almost certainly adaptive, we can now try to figure out how such complex behavior could have originated and changed over time. Martin Lindauer was the first to speculate on the history of the dances.[871] He began by looking at three other members of the genus *Apis*, which he found to perform dance displays identical to those of the familiar honey bee (*Apis mellifera*), except that in one species, *A. florea*, the bees dance on the horizontal surface of a comb built in the open over a tree branch (Figure 7.26). To indicate the direction to a food source, a worker of this species simply orients her waggle run directly toward the food's location. Because this is a less sophisticated maneuver than the transposed pointing done in the dark on vertical surfaces by *A. mellifera*, it may resemble a form of dance communication that preceded the dances of *A. mellifera*.

Lindauer then looked to stingless tropical bees not belonging to the genus *Apis* for recruitment behaviors that might provide hints about the steps preceding the first waggle dancing. Although debate exists on just how closely related stingless bees are to honey bees,[1047] different stingless bees employ different communication systems, which Lindauer organized into the following evolutionary hypothesis.

**FIGURE 7.26  The nest of an Asian honey bee,** *Apis florea*, is built out in the open around a branch. Dancing workers on the flat upper surface of the nest (two nests are shown here) can run directly toward the food source when performing the waggle dance. Photograph by Steve Buchmann.

**FIGURE 7.27 Communication by scent marking in a stingless bee.** In this species, workers that found food on the side of a pond opposite from their hive could not recruit new foragers to the site until Martin Lindauer strung a rope across the pond. Then the scouts placed scent marks on the vegetation hanging from the rope and quickly led others to their find. Photograph by Martin Lindauer.

*The possible first stage*: Workers of some species of stingless bees in the genus *Trigona* run about excitedly, producing a high-pitched buzzing sound with their wings, when they return to the nest from flowers rich in nectar or pollen. This behavior arouses their hivemates, which detect the odor of the flowers on the dancers' bodies. With this information, the recruits leave the nest and search for similar odors. The actions of the dancing bees do not provide specific signals indicative of the direction or distance to the desirable food. The same kind of behavior also occurs in the bumblebees, which form small colonies with "dancing" scouts that do not provide signals containing directional or distance information.[406]

*A possible intermediate stage*: Workers of other species of *Trigona* do convey information about the location of a food source. In these species, a worker that makes a substantial find marks the area with a pheromone produced by her mandibular glands. As the bee returns to the hive, she deposits pheromone on tufts of grass and rocks every few meters. Inside the hive entrance, other bees wait to be recruited. The successful forager crawls into the hive and produces buzzing sounds that stimulate her companions to leave the hive and follow the scent trail that she has made (Figure 7.27).

*A still more complex pattern*: A number of stingless bees in the genus *Melipona* convey distance and directional information separately. A dancing forager communicates information about the distance to a food source by producing pulses of sound; the longer the pulses are, the farther away the food is. In order to transmit directional information, she leaves the nest with a number of followers and performs a short zigzag flight that is oriented toward the food source. The scout returns and repeats the flight a number of times before flying straight off to the food source, with the recruited bees in close pursuit.

Please realize that when, for example, we speak of a *Trigona* bee exhibiting an "intermediate stage" in the evolution of communication about food location, we are not implying that this species has somehow failed to move along toward a more complex, more adaptive "final stage." For that *Trigona* bee in its current environment, trail marking may well be superior to any other option. The existence of trail marking in this modern species simply provides a clue

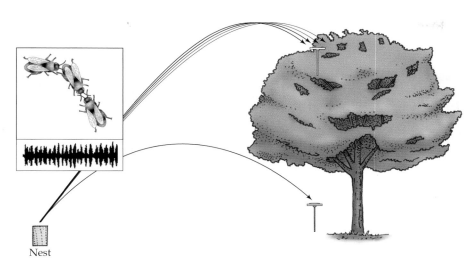

**FIGURE 7.28** **Acoustical communication of the height of a food source** by the bee *Melipona panamica*. When a scout that has been trained to collect food from a feeder high in a tree interacts with her nestmates (left panel), she produces a specific kind of acoustical signal (shown as a black sonogram in the panel beneath the bees) while unloading the food. After listening to these sounds, recruits are much more likely to locate the training feeder than a control feeder placed at the base of the tree. After Nieh.[1046]

about the possible behavior of a now extinct bee whose trail-marking behavior was modified in more recent species derived from this ancestral bee.

That all modern bee species exhibit communication systems well suited for their particular environment is a message supported by studies of tropical bees done since Lindauer's pioneering work. For some of these bees, the ability to communicate that a food source is high in the canopy of the forest, or that it is lower down, is ecologically highly relevant, and some species have evolved the requisite signals. Thus, the bee *Melipona panamica* produces one kind of buzzy acoustical signal when unloading food in a funnel-like entryway to the nest and then generates another kind of signal by dancing in the entryway. The sounds made while unloading food inform recruits where the resource is in the vertical plane (Figure 7.28).[1046] The dance conveys information about the distance to the food source advertised by the dancer. Finally, in order to convey information about the direction to the food source, the scout leaves the nest to guide recruits in the right direction. Only recruits that have access to all three sources of information about the location of a flower patch will get close enough to detect the scent marks deposited at the site by the scout before she returned to the hive to buzz, dance, and lead.[1045]

Although, judging from *M. panamica*, some stingless bees have communication systems every bit as complicated as that of the honey bee, not all do. The various behaviors of stingless bees suggest that communication about distance to a food source by an ancestor of the honey bee probably at first involved only agitated movements by a food-laden worker.[1048] Other workers that were stimulated by the activity of the returning forager would then have left the hive to find food, perhaps aided by memory of the odors associated with the food. In some species, selection may have subsequently favored standardization of the sounds and movements made by successful foragers, as in *Melipona*. These actions may have set the stage for further changes that have been incorporated into the round and waggle dances of *Apis* bees, which contain symbolic information about how far food is from the hive.[871, 1587]

In contrast, communication about the direction to a food source appears to have originated with personal leading, with a worker guiding a group of recruits directly to a nectar-rich area. Here the evolutionary sequence has involved less and less complete performance of the guiding movements as generations of queens have produced workers with a greater and greater tendency to perform incomplete leading. At first this may have taken the form

**FIGURE 7.29 Evolutionary history of the honey bee dance communication system.** Because bees in three of four closely related groups are social, the ancestor of all four groups may well have been social. If this ancestor evolved a simple form of "dancing," the trait may have been lost in the Euglossini along with the capacity for sociality, since euglossine bees do not form colonies with queens and workers. The retention of the dances in the Meliponini and Apini set the stage for evolutionarily more recent elaborations of this behavior, enabling dancers in some species to convey information about the distance to a food source and then both distance and directional information. The honey bee (*Apis mellifera*) possesses yet another modification to its dances, which encode symbolic, gravity-based information about the direction to a food source.

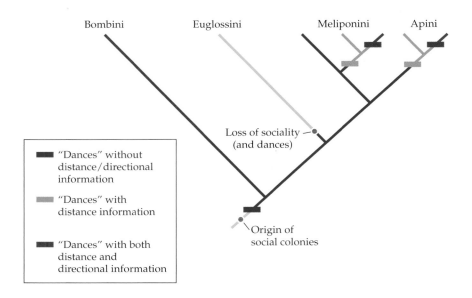

of partial leading (as in some *Melipona*) and later involved simply pointing in the proper direction with a waggle run on a horizontal surface (as in *A. florea*). From antecedents like this came the transposed pointing of *A. mellifera*, in which the direction of flight relative to the sun is converted into a signal (the waggle run) oriented relative to gravity. This evolutionary sequence of events can be plotted on a phylogeny of four closely related groups of bees, which include the honey bees (Figure 7.29). Thanks to the work of Lindauer and others, we now have ideas about both the adaptive value of bee dances and the historical sequence of events that resulted in their modern manifestation in the honey bee.

## Discussion Question

**7.16** Various persons have tried to think of reasons why some ancestral social bees made the transition from one form of communication to another. In light of this kind of hypothesis generating, what significance do you attach to the fact that currently a highly aggressive social bee, *Trigona spinipes*, can smell the odor marks placed near a food source by another social bee, *Melipona rufiventris*?[1049]

## Summary

1. Optimality theory and game theory have both played roles in the study of the possible adaptive value of feeding behavior. Optimality approaches often focus on the net caloric gains associated with a feeding decision. Game theoretical approaches consider the payoffs for individuals competing with others for valuable resources.

2. Optimality foraging theory has attracted some controversy, largely on the grounds that maximizing the rate of caloric intake does not always maximize fitness, counter to many optimal foraging hypotheses. The assumption that higher caloric intake rates boost fitness is false for those species in which there is a trade-off between energy maximization while foraging and reducing the risk of predatory attack. Tests of optimality hypotheses, however, are designed to help researchers identify the factors involved in the evolution of animal behavior. Hypotheses with incorrect assumptions will be rejected if properly tested.

3. Darwinian puzzles about feeding behavior include such things as the clumping of competitors for limited food resources in small areas, the addition of highly visible ornamental silk to the orb webs of spiders, and the consumption of dirt and other nonnutritive materials. All these actions would seem to have negative fitness consequences, leading adaptationists to look for special benefits that might outweigh the costs of the traits in question.

4. Ultimate studies of feeding behavior are not limited to research into the adaptive properties associated with getting enough to eat. In addition, evolutionary biologists have attempted to outline possible scenarios for the origin and subsequent modification of behaviors that have resulted in the complex attributes of living things, such as the astonishing dances of honey bees. Comparisons among living species can provide important clues in the development of these kinds of historical hypotheses.

## Suggested Reading

For a general review of predator decisions, see John Krebs and Alejandro Kacelnik's book chapter.[805] Their review also covers some mathematical models based on optimality theory, an approach that has often been applied to foraging behavior. The mathematics of optimality modeling are presented in detail by Dennis Lendrem[854] and by Marc Mangel and Colin Clark.[922] Reto Zach has written a paper on optimal foraging in crows that is a model of clarity.[1634] For a critique of optimality models, see a paper by C. J. Pierce and J. C. Ollason,[1135] which can be contrasted with Krebs and Kacelnik's view of the matter.[805] Bernd Heinrich's *Bumblebee Economics*[643] is a good companion to this chapter, as it deals clearly and simply with optimality theory as applied to bumblebees. To read more on the amazing lives of honey bees, see Tom Seeley's *The Wisdom of the Hive*.[1300] For reviews of the puzzles posed by web decorations, see Starks[1381] and Bruce.[201]

# 8

# Choosing Where to Live

When I was a teenager, my father and I regularly picked one weekend day in May to go on a bird-watching marathon with some fellow birders. As we searched through the fields and forests near our home in southeastern Pennsylvania, we usually found about 100 species, including yellow warblers singing high in the sycamore trees by White Clay Creek, blue-winged warblers flitting among the little saplings growing in abandoned farm fields, and common yellowthroats (another species of warbler) skulking in the marshy spots. The male warblers calling loudly from their perches not only had made the effort to find a very specific habitat, but also were prepared to defend their turf against others of their species, as they announced over and over again (see Chapter 2). In addition to the costly investments each bird had made in selecting a particular habitat and defending a territory, the warblers that my father and I saw had traveled hundreds or even thousands of miles from wintering grounds as far away as South America in order to set up a breeding territory in our neighborhood. In a few months, if they survived, they would head south again on another great trip, which their offspring would undertake as well.

Each of the decisions that these little song-birds make about where to live comes with obvious and substantial costs, which is why these decisions are interesting to evolutionary

◀ **Wildebeest do not let dangerous river crossings** *stop them from attempting their long distance migraions. Photograph by Suzi Eszterhaus.*

biologists. Why should a yellow warbler refuse to nest in a cattail marsh? Why should a common yellowthroat spend hours each day singing to keep rival males informed of his willingness to fight for space in the same marsh that yellow warblers ignore? And why should a blue-winged warbler born in Landenberg, Pennsylvania, fly all the way to Honduras, only to turn around in a few months and fly all the way back again? This chapter takes a look at Darwinian puzzles of this sort.

## Habitat Selection

The rule that certain species live in particular places applies to all groups of animals, not just warblers, perhaps because in so many cases, the opportunities for successful reproduction are so much better in habitat A than in habitat B for members of a given species. The importance of access to appropriate habitat has been dramatically illustrated by the link between habitat destruction and the declining populations of certain animals. Loss of isolated people- and dog-free beaches has, for example, endangered the western snowy plover and the least tern, both of which nest on open beaches and small sandy islands. In these places, their eggs are often inadvertently stepped upon and destroyed. To find out how frequently this occurred, researchers scattered quail eggs in unprotected parts of a public beach in California. Nearly 8 percent of the eggs disappeared or were squashed per day.[829]

If populations of some plovers and terns are falling because of human interference, then creation of artificial beaches inaccessible to people and their dogs ought to attract breeding plovers and terns. This experiment has been done in Batiquitos Lagoon in southern California and elsewhere. When material dredged off the sea bottom was used to produce artificial beaches and sandbars, both plovers and terns occupied the sites (Figure 8.1) and reproduced successfully.[921, 1161]

**FIGURE 8.1    Habitat selection and conservation.** Knowledge of the habitat preferences of nesting least terns has enabled conservation biologists to create suitable environments for this endangered species. (Left) A least tern on its nest in the open, sandy beach habitat the bird requires if it is to breed successfully. (Right) This large sandy island, built with dredged materials, has attracted many pairs of nesting seabirds, including the least tern. Left, photograph by D. Donohue; right, photograph by Troy Mallach.

Habitat restoration may help rebuild snowy plover and least tern populations. Habitat management is another conservation tool based on the premise that particular species are matched to particular environments. The Florida scrub jay, another endangered species, lives only in sandy, scrubby areas of Florida. Once this habitat burned irregularly but naturally as a result of lightning strikes, creating a mosaic of scrub oak woodland with open sandy patches surrounded by other plant associations. The scrub jays prefer to nest in substantial areas of open oak habitat, and their nesting success is highest there.[167] When natural wildfires are quickly put out, however, the scrub oak thickets eventually become tall and dense, egg-eating blue jays move in, and scrub jays lose out.[1618] Any management scheme devised to increase the Florida scrub jay population will surely have to employ fire as a tool to shape the habitat to match the preferences and needs of scrub jays.

Given the importance of being able to breed in specific habitat types, we would expect animals to have evolved strong preferences for some places over others, even if they are capable of reproducing in a variety of environments. The European great tit, for example, can nest either in mixed woodland or in hedgerows. But the birds prefer mixed woodland to hedgerows, as demonstrated by the shifts made by hedgerow birds into woodland sites upon the experimental removal of breeding pairs from the favored habitat.[803]

If habitat preferences are adaptive, then individuals that are able to fulfill their preferences ought to leave more descendants than those unable to acquire prime real estate. This proposition is true for great tits.[803] It is also consistent with the finding that in many species, some individuals occupy *source habitats* (where the population grows), while others are relegated to *sink habitats* (where the population declines). Poor-quality sink habitats are generally utilized by competitors that are unable to insert themselves into superior source habitats, often because they are excluded by older, more accomplished opponents[396] and so must make the best of a bad situation elsewhere.

The predicted relationship between habitat preference and reproductive success does not always hold, however. In the Czech Republic, for example, the blackcap (see Figure 3.14) has a choice between stream-edge deciduous forests and mixed coniferous woodlots away from water. The warbler prefers the stream-edge habitat, which attracts the initial colonists in spring. Yet the reproductive success of breeding pairs in the two environments is essentially the same.[1530] Why? One answer has been provided by Steve Fretwell and his colleagues, who used game theory to predict what animals should do when faced with a choice between alternative habitats of different quality and different levels of competition. They demonstrated mathematically that as the density of resource consumers in the superior habitat increased, there would come a point at which an individual could gain higher fitness by settling for a lower-ranked habitat that had fewer settlers of the same species and thus less competition for critical resources.[491] In fact, the Czech ornithologist Karel Weidinger found that the density of nesting blackcaps was four times higher in the preferred stream-edge forests than in the second-ranked habitat. Thus, these birds apparently make habitat selection decisions based not just on the nature of the vegetation and other markers of insect productivity, but also on the intensity of the competition from others of their species.

Pike also possess the ability to select habitats in relation to resource competition in ways that apparently maximize their fitness. This predatory fish lives in Lake Windermere in northern England, where it feeds primarily upon perch. Lake Windermere contains two basins, each with its own population of pike although individuals can move between the two bodies of water. For over 40 years, fishery biologists have been marking and recapturing pike in both basins, securing data that enabled them to measure survival and reproductive success of the fish in both populations. The recaptures of marked pike

(A)

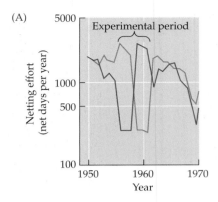

(B)

(C)

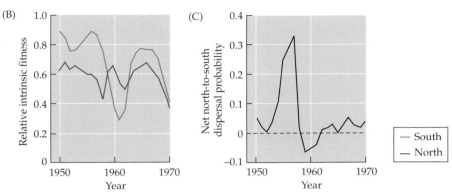

European pike

**FIGURE 8.2 Ideal free distribution of European pike** in a lake with two populations able to move between the two parts of the lake. (A) During the experimental period, many pike were removed, first, from the southern part of the lake and, then, from the northern part. (B) The fitness of individual survivors was at first high in the competition-reduced southern basin of the lake but then reversed when competition was reduced in the northern sector. (C) Fish tended to move into the superior southern habitat during the first 3 years of the experiment but then tended to move into the competition-reduced northern basin in the period from 1959 to 1962. After Haugen et al.[633]

also permitted the biologists to determine whether individuals moved from one basin to the next at times when a move would have been adaptive. After spawning, fish in the southern basin have generally experienced higher fitness than those in the north, and as predicted, the net movement of fish has been from the poorer-quality northern basin to the southern one until such time as the higher mortality rates in the increasingly dense southern population cancels the benefits of moving there.

In the period from 1956 to 1962, an experiment was done in which large numbers of pike were removed, first, from the southern basin for 3 years and, then, from the northern basin over the next 3 years. The reduction in competition for resources in first the southern pike population and then the northern population ought to have made it advantageous for pike to move, first into one, then into the other arm of the lake. In keeping with **ideal free distribution** theory (the theory that animals will, if given the chance, distribute themselves spatially in ways that maximize their reproductive success), the net movement of pike in the first 3 years was from north to south, whereas in the next 3 years, the pike tended to move from south to north (Figure 8.2).[633]

The capacity to evaluate habitat quality is also highly developed in the honey bee, a point that becomes evident when a colony splits in two. One half of the colony, which contains the old queen and half her worker force, fly off in a swarm, leaving the old hive and the remaining workers to a daughter queen. The departing swarm settles temporarily in a tree, where the workers hang from a limb in a mass around their queen (Figure 8.3). Over the next few days, scout workers search for chambers in the ground, in cliffs, or in hollow trees. Of the sites within range of the waiting swarm, only those with a volume of 30 to 60 liters cause returning scouts to perform a dance back at the swarm,[1299] which communicates information about the location of the potential new home (see pages 238–243). Other workers around a dancing scout may be sufficiently stimulated to fly out to the spot themselves. If it is attractive to them, they too will dance upon their return and send still more workers to the area.

**FIGURE 8.3 Finding a new home.** A swarm of honey bees waits while its scouts go off to search for a new hive site. Photograph by Kirk Visscher.

One highly significant feature of the dancing behavior of bees scouting for a nest site is that they typically make only a few trips out to a potential site, and each time they return, they dance for shorter periods before giving up altogether (Figure 8.4).[1495] As a result, if any one site is to build up a population of advertisers, it must be so attractive that the rate of recruitment of new scouts exceeds the drop-out rate of old scouts. Sites that are capable of inspiring energetic new recruiters will generate an exponentially increasing population of scouts, while those that are less appealing will produce an ever-decreasing, and eventually nonexistent, population of advertisers.[1027] As a result, at some point, many or all of the active recruiters will be advertising the same location, leading to the presence of many scouts investigating this potential new homesite (Figure 8.5).[1301] When a new scout encounters several dozen other workers at a site, she may return to the swarm and begin "piping," which is to say that she produces a vibrational signal that others in the swarm can sense. If enough scouts switch to the piping mode, all of their sisters in the swarm may get the message, which stimulates them to begin shivering their wing muscles. The muscle contractions raise the bees' internal temperature to the level necessary for flight. When the bees are ready, they take off, and the airborne swarm heads to the selected nest site, utilizing the information they have received from the swarm scouts.[1302]

What is especially interesting about this process from an optimality perspective is that when some bees were experimentally offered two identical high-quality nest sites at different distances from the home hive—say, 50 and 200 meters away—the bees chose the more distant of the two.[871] One might think that they would prefer the closer spot, since covering even a few hundred meters can exhaust their queen, a prodigious egg layer but a poor flier. However, because the value of a site is also affected by its proximity to other colonies, a dispersing swarm that moves a considerable distance probably reduces the likelihood that mother and daughter colonies will compete for the same flower patches.

We can test this hypothesis by predicting that a swarm's readiness to fly the extra distance should be correlated with the intensity of the potential competition between mother and daughter hives. As it turns out, northern European bee colonies are much larger, and thus more likely to compete for food, than those in southern Europe, a difference caused by the importance of having a large number of bees as a living insulation blanket for the queen and her core workers during cold winters.[716, 717] As expected, large honey bee colonies in cold Germany move farther when finding new hive sites than do the much smaller swarms of bees that live in warmer Italy.[556]

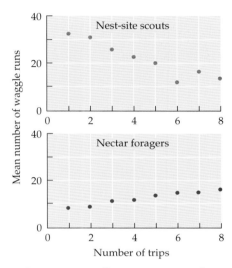

**FIGURE 8.4** **Different patterns of dancing behavior** exhibited by nest site scouts and foraging workers. Scouts perform dances with many waggle runs (see Figure 7.20) after their first few trips home from an attractive, potential new home for their colony, but then their dance activity falls off sharply. In contrast, the more trips a forager makes to and from a good nectar- or pollen-producing patch of flowers, the more waggle runs she is likely to incorporate into her recruitment dances. After Beering[96] and Visscher.[1495]

## Discussion Question

**8.1** The pattern of dancing by nest site scouts returning to a swarm differs markedly from the pattern exhibited by foragers returning to a hive (see Figure 8.4). The fact that foragers continue to recruit over many trips and gradually increase, rather than decrease, the number of waggle runs incorporated into their recruitment dances means that instead of having the entire population focusing on one foraging patch, several patches are likely to be exploited simultaneously. Does this fact suggest why the behavior of nest site scouts and foraging recruiters differs? Someone might say that we do not need evolutionary explanations for the collective decisions made by the foragers in a hive or the bees in a swarm, because these colony-wide decisions are the inevitable consequence of mindless behavioral rules used by individual members of the group. In other words, groups possess self-organizing properties that stem from the simple behaviors of their members, and knowledge of the "emergent properties" of groups negates the need for other explanations of their social activities. How would you respond to this claim?

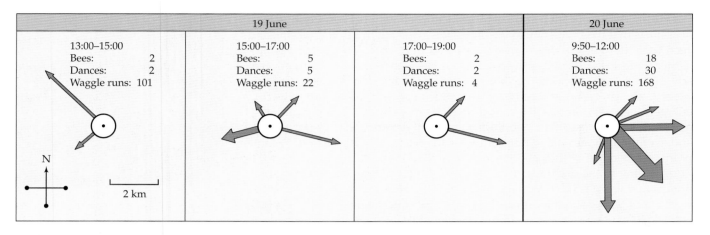

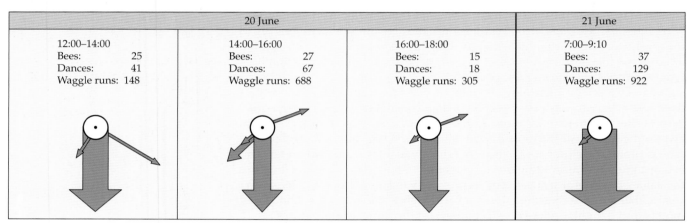

**FIGURE 8.5** **Reaching a decision about where to live.** Changes occurred over a 3-day period in the number of scouts advertising the locations of potential nest sites for their swarm (whose location is represented with a circle). The numbers in each panel represent the number of bees dancing for different sites during the time period, the total number of dances recorded during this time, and the total number of waggle runs made during these dances. The width of the arrows radiating from the circles is proportional to the number of different bees dancing for that potential nest site during the period shown at the top of each panel. The swarm left for its future home in the morning of 21 June. After Seeley and Buhrman.[1301]

## Habitat Preferences of a Territorial Aphid

Honey bee colonies do not attempt to defend a foraging area as their exclusive feeding preserve, and therefore dispersing swarms can settle wherever they choose without direct interference from other colonies. For other species, such as territorial blackcap warblers, a truly free choice among habitats is not possible because some individuals can exclude others from superior sites. Tom Whitham examined the effects of this kind of aggressive competition on habitat selection in a tiny insect, the poplar aphid.[1555–1557]

Each spring in Utah, vast numbers of aphid eggs hatch in crevices in the bark of cottonwood poplar trees. The emerging aphids walk to leaf buds on cottonwood branches. Each female—and there may be tens of thousands per tree—actively selects a leaf, settles by its midrib, almost always near the base, and in some way induces the formation of a hollow ball of tissue—a gall—in which she will live with the offspring she bears parthenogenetically (Figure 8.6). When her daughters are mature, the gall splits, and the aphids disperse to new plants.

(A)

(B)

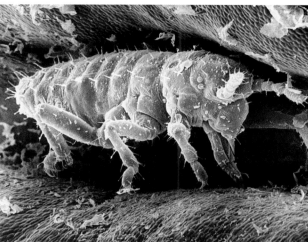

**FIGURE 8.6  Poplar galls are occupied by poplar aphids.** (A) A poplar aphid gall at the base of a leaf. (B) A scanning electron micrograph of a female aphid inside her gall. Courtesy of Tom Whitham.

Whitham found that newly emerged females quickly occupied all the very large leaves on poplars but that since there were about 20 aphids for every large leaf on the trees he studied, these females regularly encountered competitors, with whom they fought for as long as 2 days (Figure 8.7), sometimes dying in the attempt to secure a territory.[1558] Given the costly nature of these fights, we would expect that major benefits come from possessing large leaves—and they do (Figure 8.8).

Defeated females and small ones incapable of effective fighting are forced to accept inferior habitats. If the defeated individuals are able to make the best of a bad situation, they will settle on the best of the lower-value habitats available to them. The options for these individuals include finding a smaller, unoccupied leaf or settling with an established territorial foundress on a large leaf. If a leaf already has a resident female, the latecomer will have to form her gall farther out on the midrib, where it will provide fewer nutrients than the gall at the prime location near the leaf petiole. If the other aphid is already enclosed in a gall, the newcomer will not have to fight, but she will have to settle for a reduction in fitness. However, the second colonist on a medium-sized leaf can do just as well as a single aphid on a small leaf (Table 8.1). As predicted, when aphid females do double up, they choose medium and large leaves, despite their scarcity.[1557]

**FIGURE 8.7  Territorial dispute between two poplar aphids.** Females may spend hours kicking each other to determine who gets to occupy a preferred leaf or the superior location on a leaf. Courtesy of Tom Whitham.

**FIGURE 8.8   Territories and reproductive success.** The average number of progeny produced by aphids that succeed in monopolizing a poplar leaf versus those that are forced to share a leaf of the same size with a rival. After Whitham.[1557]

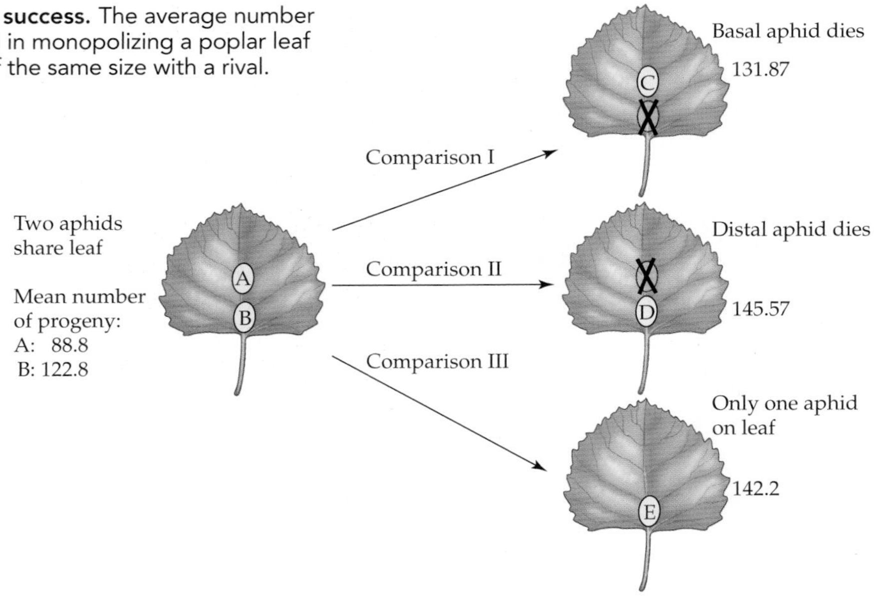

## Discussion Questions

**8.2**  In American robins, habitat selection is affected by experience. In particular, birds whose nesting attempts have failed in one year often do not return to the same spot to try again in the subsequent year. Several different hypotheses can account for this behavior. Deduce what some of these hypotheses might be, based on the following data from an experiment in which researchers destroyed the nests of a randomly selected subsample of nesting robins and then compared their return rate in the next year with that of other robins whose uninterrupted nesting attempts had resulted in the production of fledglings: of the subsample of robins with experimentally induced nest failures, only 18 percent came back, whereas 44 percent of the successful nesters returned.

**8.3**  In yellow-headed blackbirds studied in Illinois, males that left their territories had lower reproductive success in their new breeding territory the next year than they had in the old territory.[1511] Why would that result surprise an adaptationist? Does Figure 8.9 provide a possible resolution to this puzzle?

**TABLE 8.1**   *Effect of leaf size and position of the gall on the reproductive success of female poplar aphids*

| Number of galls per leaf | Mean leaf size (cm) | Mean number of progeny produced | | |
| --- | --- | --- | --- | --- |
| | | Basal female | Second female | Third female |
| 1 | 10.2 | 80 | | |
| 2 | 12.3 | 95 | 74 | |
| 3 | 14.6 | 138 | 75 | 29 |

*Source*: Whitham[1557]

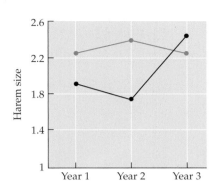

Yellow-headed blackbird

**FIGURE 8.9   Number of mates attracted by male yellow-headed blackbirds** that leave their old territory in year 1 (black line) versus those that stay put (tan line). Harem size actually declines in year 2 following the move. After Ward and Weatherhead.[1511]

## The Costs and Benefits of Dispersal

Once a poplar aphid finds a leaf to live on, she will spend the rest of her life there. But for other species, such as the honey bee, dispersal from one home base to another is a regular occurrence. In order to move from point A to point B, animals must expend calories not only while moving, but even before dispersal, when they invest in the development of locomotory muscles. Consider, for example, that if a cricket is to leave a deteriorating environment and move to a new and better place, it will need large flight muscles to fly away. The calories and materials that go into flight muscle development and maintenance presumably have to come out of the general energy budget of the animal. This means that other organ systems, such as a female's ovaries, cannot develop as rapidly as they could otherwise, which imposes a fitness cost on the flight-capable individual.

This trade-off hypothesis has been tested ingeniously in the following way. In some cricket species, two phenotypes occur, one that has the requisite muscular machinery for flying and another form with small wing muscles and low lipid (fuel) reserves that cannot get off the ground (see page 156). Females that can fly do not start producing eggs as quickly as flightless females, as one would predict from the trade-off argument.[1643] Researchers can also create flightless females in a cricket species that normally has only flight-capable individuals by dabbing young adults with a chemical similar to juvenile hormone (see Chapter 5). If it is true that the ability to fly comes at a reproductive price, then the experimentally produced flightless females should be able to invest more in their ovaries than control females not exposed to the hormone. Indeed, ovarian development occurs much more rapidly in the experimental females, which feed no more or less than the controls but whose flight muscles deteriorate while resources are directed to their ovaries (Figure 8.10).[1643]

Dispersing individuals not only have to pay energetic developmental and travel costs, but also might be easier prey for predators when in unfamiliar places, a possibility that James Yoder and his coworkers examined in a study of ruffed grouse. They followed birds that had been captured and outfitted with radio transmitters, which enabled them to map the movements of individuals precisely and to find any grouse whose transmit-

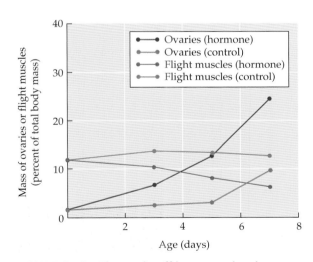

**FIGURE 8.10   The trade-off between development of locomotory muscles and reproductive equipment.** Female crickets treated with juvenile hormone, which results in flightless individuals with reduced wing muscle mass, followed different developmental curves than control (flight-capable) females. Note especially the difference in the growth rates of the ovaries for the two categories of females. (See also Figure 5.7.) After Zera, Potts, and Kobus.[1643]

Ruffed grouse

**FIGURE 8.11 Two patterns of movement by radio-tracked ruffed grouse.**
(A) This bird stayed pretty much within the same fairly small home range for many months. (B) Another individual alternated bouts of staying put with substantial dispersal movements through unfamiliar terrain, a risky business for a ruffed grouse. After Yoder, Marschall, and Swanson.[1629]

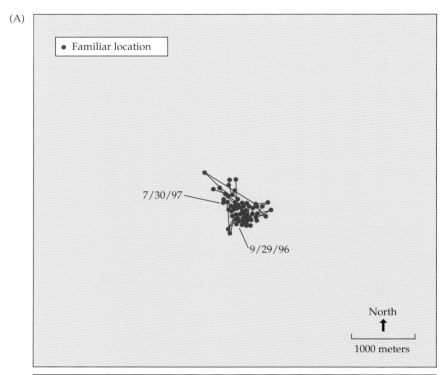

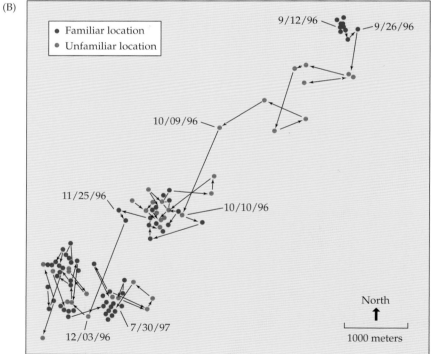

ter signaled that the bird had not moved for 8 hours, a very good indication that the bird was dead. Some birds stayed for months near the place where they had been captured, while others moved at considerable intervals from one location to another (Figure 8.11). Being in a new area boosted the risk of being killed by a hawk or a mammalian predator at least threefold compared with the risk facing birds in locations with which they had become familiar.[1629]

A finding like this raises the question, why are animals so often willing to leave home even when this means leaving a familiar, resource-rich location? This question is particularly pertinent for species in which some individuals disperse while others do not, or at least do not travel as far. In some cases these differences are linked to genetic differences between the two types. In the fire ant, for example, when a new generation of queens is produced, some fly away from their colony, mate with males at distant sites, and go on to seek out new locations where they will burrow into the ground and try to found nests on their own. Few females will succeed, but many try, especially those with a particular genotype that we will label *BB*. In contrast, queens with the genotype *Bb* leave the colony to mate but apparently often return home, attempting to rejoin their natal colonies to become egg-laying queens there.[379]

The simple genetic difference between the two kinds of queens provides a proximate explanation for the difference in their behavior. An ultimate reason that queens with the two genotypes behave differently is linked to what happens to *BB* queens if they try to join a colony where other queens are already present. Many of the workers in a colony will have the *Bb* genotype, and these individuals systematically dismember any new queen that lacks the *b* allele,[753] perhaps because they can recognize individuals with *BB* genotype by their odor.[555] Under these circumstances, it is not surprising that *BB* queens generally avoid established colonies and do the best they can to found colonies of their own.

Another species in which some individuals disperse farther than others is Belding's ground squirrel. A young male squirrel travels about 150 meters from the safety of his mother's burrow, whereas a young female usually settles down only 50 meters or so from the burrow in which she was born (Figure 8.12).[665] Why should young male Belding's ground squirrels go farther than their sisters?

According to one argument, dispersal by juvenile animals of many species may be an adaptation against **inbreeding depression**.[1181] When two closely related individuals mate, the offspring they produce are more likely to carry damaging recessive alleles in double dose than are offspring produced by unrelated pairs. The risk of associated genetic problems should in theory reduce the average fitness of inbred offspring, and high juvenile mortality does indeed occur in inbred populations of many animals.[1194] Thus, when inbred and non-inbred white-footed mice were experimentally released into a

Belding's ground squirrel

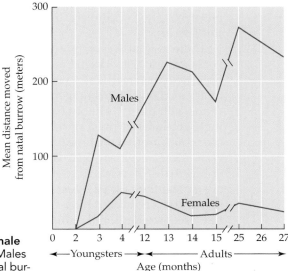

**FIGURE 8.12    Distances dispersed by male and female Belding's ground squirrels.** Males go much farther on average from their natal burrows than females. After Holekamp.[665]

Oldfield mouse

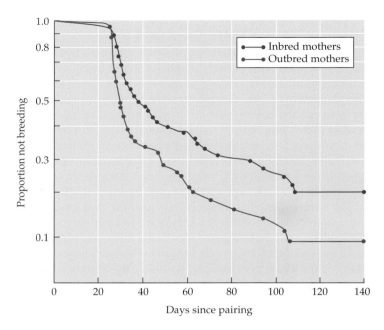

**FIGURE 8.13 Inbreeding depression in oldfield mice** may result in the failure of inbred females to reproduce as early as outbred females. Photograph of oldfield mouse by Mike Groutt, U.S. Fish and Wildlife Service, courtesy of Rob Tawes. After Margulis and Altmann.[933]

field from which their ancestors had been captured, the inbred mice survived only about half as well as their outbred companions.[725] And even if inbred mice manage to reach adulthood, they may be less likely to reproduce than outbred individuals (Figure 8.13).[933]

If avoidance of inbreeding is the point of dispersing, however, then one might expect as many female ground squirrels as males to travel 150 meters from their natal burrow. But they do not, perhaps because the costs and benefits of dispersal differ for the two sexes. As Paul Greenwood has suggested for mammals generally,[576] female Belding's ground squirrels may remain at or near their natal territories because their reproductive success depends on possession of a territory in which to rear their young. Female ground squirrels that remain near their birthplace enjoy assistance from their mother in the defense of their burrows against rival females. Thus, the benefits of remaining on familiar ground are greater for females than for males, and this difference has probably contributed to the evolution of sex differences in dispersal in this species.[665]

There may, however, be another reason why male mammals typically disperse greater distances than females. The usual rule is that males, not females, fight with one another for access to mates (see page 342), and therefore loser males may find it advantageous to move away from same-sex rivals that they cannot subdue.[1006] Although this hypothesis probably does not apply to Belding's ground squirrels, since young males have not been seen fighting with older ones around the time of dispersal, the idea deserves to be tested in other cases. Lions, as we saw in Chapter 1, live in large groups, or prides, from which young males disperse. In contrast, the daughters of the resident lionesses usually spend their entire lives where they were born (Figure 8.14).[1179] The sedentary females benefit from their familiarity with good hunting grounds and safe breeding dens in their natal territory, among other things.

The departure of many young male lions coincides with the arrival of new mature males that violently displace the previous pride masters and chase off the subadult males in the pride as well. These observations support the mate competition hypothesis for male dispersal. However, if young males are not

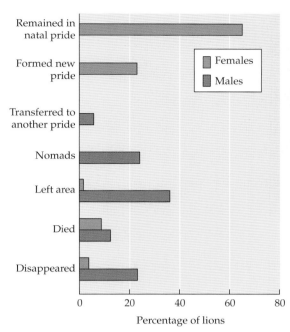

Lion pride

**FIGURE 8.14**   **Male and female lions differ in their dispersal patterns.** Lionesses on the Serengeti Plain of Africa tend to remain with their natal pride, whereas males move to other prides or become nomads. After Pusey and Packer.[1179]

evicted after a pride takeover, they often leave anyway, without any coercion from adult males and without ever having attempted to mate with their female relatives. Moreover, mature males that have claimed a pride sometimes disperse again, expanding their range to annex a second pride of females, at the time when their daughters in pride number one are becoming sexually mature. Proximate inhibitions against inbreeding apparently exist in lions and cause males to leave home. Ultimately, dispersing males may gain by mating with nonrelatives, even though the timing of their departure from their birthplace is not always under their control.[616]

## Discussion Question

**8.4**  In one study of brown bears (grizzlies) in Sweden, 15 of 16 males left their mothers and natal territories behind while only 13 of 32 females dispersed. Older and heavier females were less likely to be part of the dispersing cohort.[1636] So here, as in Belding's ground squirrels and many other mammals, males disperse while most females remain on or near their natal territories. Given the information above, do the explanations given for the pattern of ground squirrel dispersal also apply to brown bears? What other information would be useful in order to evaluate these hypotheses?

The standard mammalian pattern just described is exactly reversed in most birds. Why might this be so? In producing your hypotheses, consider the fitness value of a territory to a typical male mammal and a typical male bird.[576]

## Migration

A familiar, but nonetheless amazing, form of dispersal is **migration**, which often involves movement away from and subsequent return to the same location on an annual basis, although some persons also consider direct, long-distance dispersal from one place to another a form of migration. Many living birds, mammals, fishes, sea turtles (see Figure 4.46), and some insects engage

Arctic tern

**FIGURE 8.15** **The migratory route of arctic terns.** These birds fly from high in the Northern Hemisphere to Antarctica and back each year. Some young birds may spend 2 years circling Antarctica before returning to the northern breeding grounds (shown in dark blue).

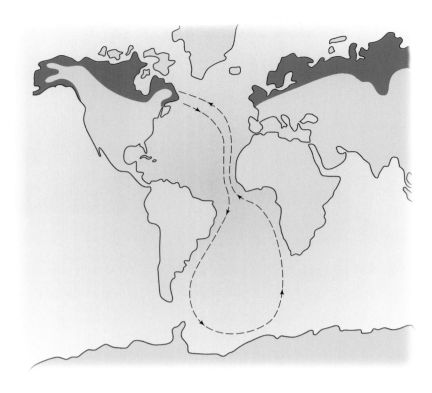

in this behavior, and so did some extinct creatures as well.[237, 295] In fact, nearly half of all the breeding birds of North America are migrants that take off in the fall for a trip to Mexico or Central America or South America, only to return in the spring.[307] Tiny ruby-throated hummingbirds, the weight of a penny, fly nonstop 850 kilometers across the Gulf of Mexico twice a year. Arctic terns breeding in Canada may complete a roughly 40,000-kilometer migration each year (the equivalent of seven trips across the continental United States) (Figure 8.15), much of it over the ocean,[1535] where the birds must stay airborne for many days and nights in a row. This is a journey that the average human would find exhausting even in a jumbo jet.

Migration poses a major historical problem: if sedentary species were ancestral to migratory ones, as they probably were, how could the ability to travel thousands of kilometers each year to specific destinations have evolved? Persons interested in this question have noted that many bird species in the tropics engage in short-range "migrations" of dozens to hundreds of miles, with individuals moving up and down mountainsides or from one region to another immediately adjacent one. The three-wattled bellbird, for example, has an annual migratory cycle that takes it from its breeding area in the mid-elevation mountainous forests of north-central Costa Rica to lowland forests on the Atlantic side of Nicaragua, then to coastal forests on the Pacific side of southwestern Costa Rica, from which the bird returns to its mountain breeding area (Figure 8.16).[1162] The distances traveled by migrating bellbirds are substantial (up to 200 kilometers), but not breathtaking.

Douglas Levey and Gary Stiles point out that short-range migrants occur in nine families of songbirds believed to have originated in the tropics. Of these nine families, seven also include long-distance migrants that move thousands of kilometers from tropical to temperate regions. The co-occurrence of short-range and long-distance migrants in these seven families suggests that short-range migration preceded long-distance migration, setting the stage for the further refinements needed for the impressive migratory trips of some

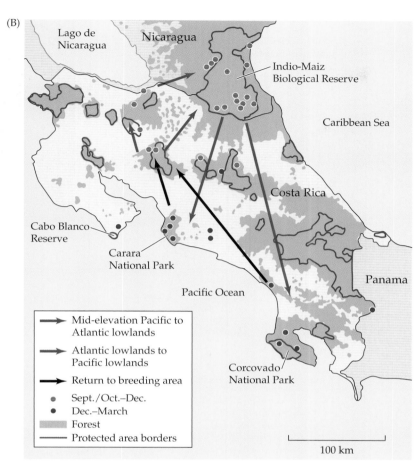

**FIGURE 8.16  Short-range migration in the three-wattled bellbird.** (A) A male three-wattled bellbird calling from a perch in a Costa Rican forest. (B) After breeding in the mountains of north-central Costa Rica, bellbirds first head to the north and east, then go south and west to reach forests on the Pacific coast before returning north to the mountains. A, photograph by Michael and Patricia Fogden; B, after Powell and Bjork.[1162]

species.[858] Thus, long-distance migrants are probably descended from species that moved far less on an annual basis.

One genus of birds, the *Catharus* thrushes, may shed some light on this theory of avian migration. The genus contains 12 species, 7 of which are resident in areas from Mexico to South America; the other 5 are migratory species that travel between breeding areas in northern North America and wintering zones to the south, especially in South America (Figure 8.17). These observations suggest that the ancestors of the migratory species lived in Mexico or Central America. Moreover, the most parsimonious interpretation of a phylogeny of this genus is that migratory behavior has evolved three times, with subtropical or tropical resident species giving rise to migratory lineages each time (Figure 8.18). Thus, the history of this genus supports the hypothesis that migratory species evolved from tropical nonmigratory ancestors.

The ornithologists Volker Salewski and Bruno Bruderer have also tried to identify the sequence of steps involved in the evolution of bird migration.[1273] Building on the ideas of Christopher Bell,[100] they propose that migratory tendencies began with the dispersal of birds away from current breeding areas to regions where nonbreeding individuals had a better chance of survival. Young birds of many species disperse in this way as they seek to find living space of their own. In some cases, a process of this sort could lead to a gradual extension of the range of the species over time as new dispersers from each generation left their natal areas and moved to other previously unoccupied places. Those dispersers that found areas with suitable resources would survive and later reproduce there, maintaining the hereditary basis for dispers-

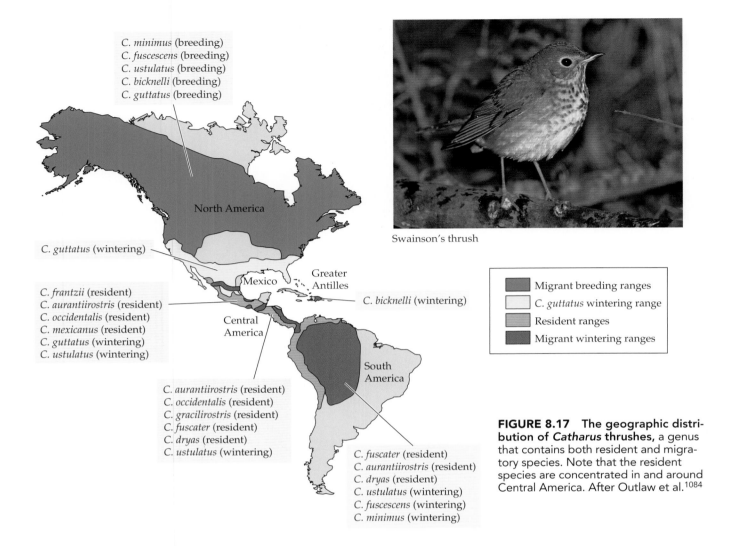

Swainson's thrush

| | |
|---|---|
| ■ | Migrant breeding ranges |
| □ | *C. guttatus* wintering range |
| ■ | Resident ranges |
| ■ | Migrant wintering ranges |

*C. minimus* (breeding)
*C. fuscescens* (breeding)
*C. ustulatus* (breeding)
*C. bicknelli* (breeding)
*C. guttatus* (breeding)

North America

*C. guttatus* (wintering)

Mexico

Greater Antilles

*C. frantzii* (resident)
*C. aurantiirostris* (resident)
*C. occidentalis* (resident)
*C. mexicanus* (resident)
*C. guttatus* (wintering)
*C. ustulatus* (wintering)

*C. bicknelli* (wintering)

Central America

*C. aurantiirostris* (resident)
*C. occidentalis* (resident)
*C. gracilirostris* (resident)
*C. fuscater* (resident)
*C. dryas* (resident)
*C. ustulatus* (wintering)

South America

*C. fuscater* (resident)
*C. aurantiirostris* (resident)
*C. dryas* (resident)
*C. ustulatus* (wintering)
*C. fuscescens* (wintering)
*C. minimus* (wintering)

**FIGURE 8.17  The geographic distribution of *Catharus* thrushes,** a genus that contains both resident and migratory species. Note that the resident species are concentrated in and around Central America. After Outlaw et al.[1084]

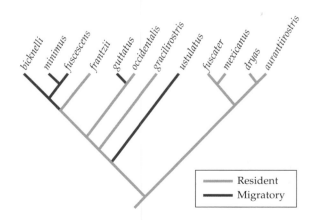

**FIGURE 8.18  The long-distance migratory trait overlain on the phylogeny of *Catharus* thrushes.** The phylogenetic tree was constructed on the basis of similarities among the species with respect to mitochondrial DNA. Another, more recent phylogeny based in part on nuclear DNA is not identical but yields the same conclusion about the three independent origins of migratory behavior in this cluster of species.[1601] After Outlaw et al.[1084]

ing. If, eventually, some dispersers moved into areas that were only suitable for survival and reproduction on a seasonal basis, then these individuals and the population they composed would die out—unless they happened to have the motivational and navigational mechanisms that caused them to abandon the site soon enough to return to a portion of their now expanded range where they could survive during the time when their breeding location was uninhabitable. The relationship between migration and seasonal changes associated with high latitudes has long been recognized.[1085]

One may reasonably wonder how some individuals or their offspring might come to have the sophisticated navigational equipment needed for travel away from their seasonally lethal habitats to safer regions within the range of their species. But if these mechanisms initially spread through a given species because they helped individuals travel efficiently over considerable distances to places where competition for food was less or resources more abundant, relatively modest modifications could produce the proximate basis for true migration. More work is clearly needed on this point. What does seem clear is that migratory behavior can change rapidly, as shown by the very recent shift in migratory paths followed by blackcaps discussed earlier (see page 76).[1273] Likewise, the close genetic similarities between southern and northern populations of the black-throated blue warbler (a North American species) indicates that these populations, which have different migratory destinations (see page 272), have evolved their distinctive routes within just the last 13,000 years, a short period indeed on the geological timescale.[363]

## The Costs of Migration

If, as some believe, the long-distance migratory abilities of many birds today evolved by degrees, as selection acted on the descendants of a sedentary resident ancestor, then each modification in migratory capacity that persisted must have had benefits that outweighed the costs associated with the new ability. The problem is that the costs of migration are not trivial. For birds, they include the extra weight that the migrant has to gain in order to build up energy reserves for the trip. Some migrant songbirds nearly double their body weights prior to their long flights. Even a moderate fuel load dramatically alters the takeoff angle of an individual startled by a predator, almost certainly increasing the chance that the predator will catch the fleeing songbird.[870] However, it is also true that fully loaded red knots, once under way, actually fly more efficiently, in terms of turning fuel into wing power, than when they are at their nonmigratory body weight.[823] But flight, no matter how efficient, still costs calories, and there is always the chance that a migrant will run out of gas before reaching its destination—not a fitness-enhancing event. (Lawrence Swan once saw a migrating hoopoe, a bird with the exquisite scientific name *Upupa epops*, forced by exhaustion to *hop* up a Himalayan pass at 20,000 feet.[1414])

An optimality approach to migration generates the prediction that migrants will evolve attributes that reduce the costs of the trip, which obviously include the energy expended in flight. Many persons have wondered, for example, whether the V-formation adopted by many large birds when migrating is an energy-saving adaptation. A team led by Henri Weimerskirch was able to test this proposition by enlisting the assistance of a group of great white pelicans that had been imprinted (see Chapter 3) on an ultralight aircraft.[1532] The pelicans' heart rates were monitored with electronic heart rate loggers attached to their backs, and their wingbeat rates were derived from film taken with a digital camera. The data on the pelicans' heart rates and wingbeat frequencies revealed that birds flying alone had to work harder than those flying in V-formation (Figure 8.19). The overall energy savings for pelicans that can take advantage of the updrafts created by the wingbeats of their companions in this way is about 11 to 14 percent, not a trivial sum for birds traveling long distances.

### Discussion Question

**8.5**  If an adaptationist were to consider the wingbeat frequencies of pelicans in formation flight (see Figure 8.19), that person would have some questions to ask. What are they?

If saving energy were the overriding goal of migrating birds, however, then we would not expect to find so many songbirds traveling east to west across all of Europe before crossing the Mediterranean at the narrow point between southern Spain and northern Africa (see Figure 3.14).[114] This route greatly lengthens the total journey for birds headed to central Africa but reduces the

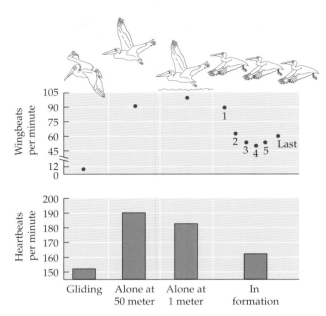

**FIGURE 8.19 Flying in V-formation is an energy saver.** Data on wingbeat frequency and heart rate are presented for various flight options available to the great white pelican. After Weimerskirch et al.[1532]

overwater component of the trip, perhaps decreasing the risk of drowning at sea, a survival advantage that might trump the extra calories expended.

If this hypothesis is true, then we would expect other songbirds to make migratory decisions that are sensitive in ultimate terms to the risks of mortality during the trip. Red-eyed vireos migrating in the fall from the eastern United States to the Amazon basin of South America must either cross a large body of water, the Gulf of Mexico, or stay close to land, moving in a southwesterly direction along the coast of Texas to Mexico and then south. The trans-Gulf flight is shorter, but vireos that cannot make it all the way to Venezuela are dead ducks, so to speak.

In light of this danger, Ronald Sandberg and Frank Moore predicted that red-eyed vireos that happened to have low fat reserves (for whatever reason) would be less likely than those with considerable body fat to risk the long journey due south across the Gulf of Mexico. They captured migrating vireos in the fall on the coast of Alabama, classified each individual as lean or fat, and placed the birds in orientation cages similar to those shown in Figure 3.15. Vireos with less than about 5 grams of body fat showed a mean orientation at sunset toward the west-northwest, whereas vireos that had been classified as having more fat tended to head due south, just as Sandberg and Moore had predicted (Figure 8.20).[1275]

Some songbirds even smaller than the red-eyed vireo attempt a still more impressive journey across water in the fall, one that requires a nonstop flight from Canada to South America over 3000 kilometers of ocean (Figure 8.21).[1582] At first glance, a blackpoll warbler that selects this migratory route would seem to have a death wish. Surely the birds should take the safer passage along the coast of the United States and down through Mexico and Central America. But migratory blackpolls commonly appear on islands in the Atlantic and the Caribbean, and many are in good condition when they arrive, demonstrating their capacity to make long nonstop ocean crossings.[844, 970]

Red-eyed vireo

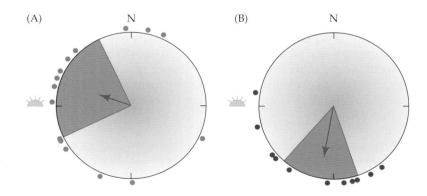

**FIGURE 8.20 Body condition affects the migratory route chosen by red-eyed vireos.** (A) Birds with low fat reserves do not head south toward the Gulf of Mexico but instead head west (sunset symbol) as if to begin an overland flight toward Mexico. (B) Birds with ample energy reserves orient due south. The central arrow shows the mean orientation for the birds tested in each group. After Sandberg and Moore.[1275]

Blackpoll warbler

**FIGURE 8.21  Transatlantic migratory path of blackpoll warblers** (arrows) from southeastern Canada and New England to their South American wintering grounds. Courtesy of Janet Williams.

The blindly courageous blackpoll warbler that manages this overwater trip has substantially reduced some of the costs of getting to South America. First, the sea route from Nova Scotia to Venezuela is about half as long as a land-based trek, although admittedly it requires an estimated 50 to 90 hours of continuous flight. Second, there are very few predators lying in wait in mid-ocean or on the islands of the Greater Antilles that the blackpolls may reach. Third, the birds leave the Canadian coast only when a west-to-east-traveling cold front can push them out over the Atlantic Ocean for the first leg of their journey, after which the birds use the westerly breezes typical of the southern Atlantic to help them reach an island landfall.

## Discussion Questions

**8.6**  When blackpolls return to Canada from South America in the spring, they do not retrace their fall migratory path but instead travel mostly over land. Why might this be?

**8.7**  Swainson's thrush (*C. ustulatus*), one of the species in the genus *Catharus* mentioned earlier, breeds in a large band right across North America. Those birds that live in the northwestern part of North America do not all follow the same migratory route. Some birds go right down the Pacific coast and winter in Central America. But others travel all the way to the eastern part of North America before flying south to winter in South America (Figure 8.22).[1251] One hypothesis to account for the behavior of the thrushes taking the long way south is that these birds are descendants of those that expanded the species' range from the East Coast far out to the west and north after the retreat of the glaciers about 10,000 years ago.[1252] What kind of evolutionary hypothesis is this? Is the behavior maladaptive? How can you account for the persistence of the trait?

**FIGURE 8.22   The two migratory routes of Swainson's thrush.** Although some birds travel more or less directly from the Pacific Northwest to Central America, others fly across North America before heading to South America to spend the winter. After Ruegg and Smith.[1251]

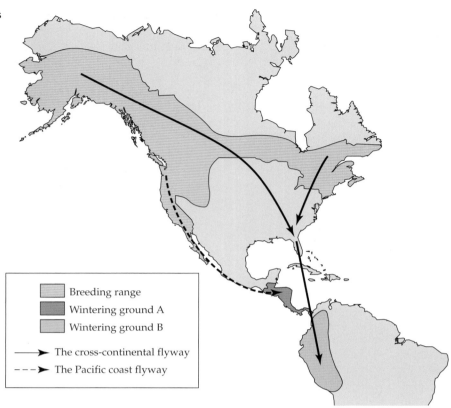

Breeding range
Wintering ground A
Wintering ground B
———▶  The cross-continental flyway
- - -▶  The Pacific coast flyway

## The Benefits of Migration

For all their navigational and meteorological skills, migrating blackpolls and other birds cannot eliminate the costs of their travels altogether. What ecological conditions might elevate the benefits of migration enough to outweigh these costs, leading to the spread of migratory abilities by natural selection? The answer for many migratory songbirds in the Americas may lie in the immense populations of protein-rich insects that appear in the northern United States and Canada in the summer, when long days fuel the growth of plants on which herbivorous insects feed.[191] Moreover, the many hours of summer daylight mean that breeding migrant songbirds can search for food longer each day than tropical bird species, which have only about 12 hours each day to harvest prey for their offspring. But a summertime food bonanza is not the only factor favoring migration, given that many migrants abandon areas where food is still plentiful in order to winter elsewhere.[100]

### Discussion Question

**8.8**  For some whale species that migrate from Arctic or Antarctic oceans to give birth in warmer water nearer the equator, food cannot provide an ultimate benefit, since the adults do not feed on the calving grounds. Therefore, other hypotheses for whale migration have been advanced, such as the idea that whale calves can gain weight more quickly in subtropical waters, where they need to invest less energy in keeping warm. Alternatively, some persons have suggested that infant whales in these waters are less likely to be attacked by predators, especially killer whales.[267, 298] How would you test these hypotheses, given the practical difficulties of directly measuring the metabolic costs of thermoregulation by whale calves or of actually observing killer whale attacks on other whales in any environment?

Resources other than food can also vary in availability seasonally, making migration adaptive. In the Serengeti National Park of Tanzania, over a million wildebeests, zebras, and gazelles move from south to north and back again each year. The move north appears to be triggered by the dry season, while the onset of the rains sends the herds south again. It might be that the herds are tracking grass production, which is dependent on rainfall. Eric Wolanski and his colleagues have established, however, that the most important factor sending the animals north is actually a decline in water supplies and an increase in the saltiness of the water in drying rivers and shrinking waterholes. If one knows the salinity of the water available to the great herds, one can predict when they will leave on their march north,[1609] although the precise route they follow north is influenced by the availability of vegetation, which in turn is influenced by rainfall patterns in the Serengeti.[1026]

The monarch butterfly is another species that does not migrate to find food. When monarchs fly in fall from the eastern half of North America, they head for central Mexico where they will spend the winter roosting (not feeding) in Oyamel fir forests high in the mountains (see Figure 4.42).[186, 1247, 1479] True, as they make their long journey south, monarchs must find enough flower nectar to fuel their flight, but unlike red knots (see page 265) that create and expend a large fat reserve on their migration, migrating monarchs carry relatively small amounts of lipids for much of their trip. Red knots use expensive flapping flight to get where they are going, while monarchs appear to use favorable winds to help them glide and soar toward their ultimate destination. Soaring uses far less energy than flapping flight. Only when monarchs get fairly close to their destination do they collect large amounts of nectar for conversion to the lipid energy reserves they will need for the long months of cold storage starvation in their winter roosts.[187]

But why go to the trouble of flying up to 3600 kilometers to reach a conifer tree in the cold high mountains of Mexico? Even if the butterflies keep the costs of the journey relatively low by migrating primarily on days when the winds facilitate inexpensive soaring flight, still one would think that they could spend the winter roosting in places much closer to the milkweed-producing areas that female monarchs will visit to produce their offspring the next spring and summer.

But perhaps not, since killing freezes occur regularly at night throughout eastern North America during winter. In contrast, freezes are very rare in the Mexican mountain refugia used by the monarchs. In these forests, at about 3000 meters elevation, temperatures rarely drop below 4°C, even during the coldest winter months. Occasionally, however, snowstorms do strike the mountains, and when this happens, as many as 2 million monarchs can die in a single night of subfreezing temperatures. The risk of freezing to death could be completely avoided in many lower-elevation locations in Mexico. But William Calvert and Lincoln Brower note that in warmer and drier areas, the monarchs would quickly use up their water and energy reserves. By remaining moist and cool—without freezing to death—the butterflies conserve vital resources, which will come in handy when they start back north after their 3 months in the mountains.[226]

The hypothesis that the stands of Oyamel fir used by the monarchs provide a uniquely favorable microclimate that promotes winter survival is being tested in an unfortunate manner. Even in supposedly protected reserves, an alarming amount of woodcutting and logging has occurred.[1196] Brower and his associates believe that timber removal causes butterfly mortality, even when some roosting trees are left in place. Opening up the forest canopy increases the chances that the butterflies will become wet and exposed, which increases the risk that they will freeze (Figure 8.23). Thus, even partial forest cutting may

(A)

**FIGURE 8.23** **Monarch butterfly habitat selection.** (A) Vast numbers of monarchs spend the winter resting in huge clusters on fir trees in a few Mexican mountain sites. (B) Habitat quality correlates with survival of overwintering monarchs. Protection from freezing in the high Mexican mountains depends on a dense tree canopy that reduces wetting of the butterflies by rain or snow and their exposure to open sky. A, photograph by Lincoln P. Brower; B, after Anderson and Brower.[29]

(B)

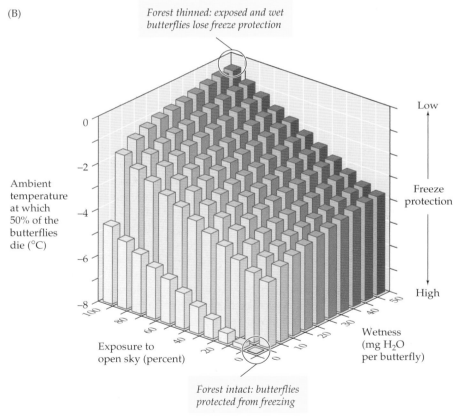

destroy the conditions needed for the survival of monarch aggregations. If the loss of a relatively small number of Oyamel firs causes the local extinction of overwintering monarch populations, it will be a powerful but sad demonstration of the value of a specific habitat for this migratory species.[29]

## Migration as a Conditional Tactic

Both migratory and nonmigratory individuals occur in some species, just as winged and wingless forms coexist in some cricket species. The European blackbird is an example, since in some portions of its breeding range, some individuals migrate in the fall, leaving the area to others that overwinter on the breeding grounds.[1291] As discussed earlier, if these two behaviors represent different hereditary strategies, then they would have to provide equal fitness in order to persist together within a population (see page 231). Therefore, a two-strategies hypothesis for blackbird behavior generates two predictions: (1) that the lifetime fitness of the two types should be the same, on average, and (2) that the differences between migratory and nonmigratory individuals are caused by differences in their genetic constitution.

Data on the fitness consequences of being migratory versus being a resident are not available, but individual blackbirds regularly change their behavior from one year to another (Figure 8.24). The fact that an individual bird can switch from the migrant strategy to the resident strategy indicates that the differences between the two behavior patterns are not hereditary. Because the two-strategies hypothesis seems unlikely, we shall consider a conditional strategy alternative (see page 231). All the blackbirds in a population could share a conditional strategy that enables them to choose between migrating or staying put, depending on the social conditions they encounter. If so, then we would expect individual birds to adopt whichever tactic yields higher fitness for them, given their social status. Socially dominant individuals should be in a position to select the better of the two options under the control of a conditional strategy, forcing subordinates to make the best of a bad situation by adopting the option with a lower reproductive payoff (but one that is greater than they could get from futile attempts to behave like dominants).[366] For example, perhaps an area can support only a few resident blackbirds during the winter. Under such circumstances, subordinates faced with a cadre of more powerful residents might do better to migrate away from the competition, returning in the spring to occupy territories made vacant by the deaths of some rivals over the winter.

Given the logic of this hypothesis, we can make several predictions: (1) blackbirds should have the ability to switch between tactics, rather than being

(A) Resident in preceding winter

(B) Migrant in preceding winter

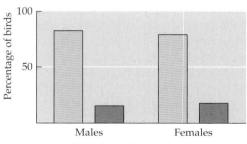

European blackbird

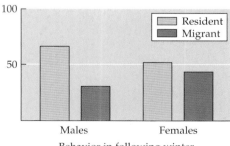

**FIGURE 8.24   Is migratory behavior the less preferred tactic for European blackbirds?** (A) Birds that were residents in the preceding winter tended to be nonmigratory the next winter as well. (B) In contrast, birds that were migrants in the preceding winter often switched to the resident option the following winter. After Schwabl.[1291]

locked into a single behavioral response, (2) socially dominant birds should adopt the superior tactic, and (3) when choosing freely between tactics, individuals should choose the option with the higher reproductive payoff (as poplar aphids do when choosing which leaf to colonize). In light of these predictions, it is significant that migratory European blackbirds head off in the fall at times when dominance contests are increasing in frequency.[896] Moreover, blackbirds do switch fairly often from the migratory option to staying put, typically when they are older and presumably more socially dominant.[897]

---

### Discussion Question

**8.9** Here's another way of thinking about the blackbird case, courtesy of Steve Shuster:[1328] Instead of most birds having one particular kind of genetic constitution that enables them to switch between migratory and resident behavior, perhaps any given population contains many different genotypes, each with its own developmental effect on decisions made about changing where one lives. The existence of multiple genotypes of this sort within a species could occur because, for example, a genotype favored in one place might not do as well as another in a different place. As a consequence of gene flow from one subpopulation to the next, genetic variation would be maintained within all the populations. The genetically different individuals in one area would do different things in response to various environmental circumstances. According to this view, conditional strategy theory is flawed by its assumption that there is genetic uniformity in a population with respect to the genes that influence the kinds of behavioral flexibility that we are talking about. What does such an approach predict about the degree of genetic variation underlying the "stay put" and "migrate" phenotypes in European blackbirds? Does Chapter 3 contain any helpful information on how much fitness-relevant genetic variation exists within animal populations? (You may want to reexamine the section on artificial selection.) How does the genetic diversity approach account for the existence of individuals whose behavior yields lower fitness than that of other individuals with different behavioral phenotypes? How might an advocate of conditional strategy theory respond to this counterview?

---

The migratory component of a species can also be composed of subgroups that vary in where they breed and overwinter, as in the case of the blackcap (see Chapter 3)[117] and Swainson's thrush.[1251] As more studies are done using stable isotope technology,[1247] more and more cases of subpopulations that utilize different portions of their species' geographic ranges are emerging. Stable isotope technology is based on the fact that when animals feed on different foods in different parts of the world, the ratios of different forms of key elements (i.e., isotopes) that they ingest may vary, which in turn affects the chemical composition of body parts into which these elements are incorporated. So, for example, the feathers of black-throated blue warblers that winter on islands in the western Caribbean are isotopically distinct from the feathers of birds of the same species that winter in the eastern Caribbean. These differences can be traced back to where the feathers were produced in the United States. Thus, it was discovered that black-throated blue warblers that breed in the mountains of the southern United States go to the eastern Caribbean for the winter, carrying their chemically distinctive feathers with them, whereas their fellow black-throated blues that reproduce in the northeastern United States migrate to the western Caribbean.[1246]

The occurrence of migratory variation of this sort poses the same kind of evolutionary question that we discussed in the context of the coexisting resi-

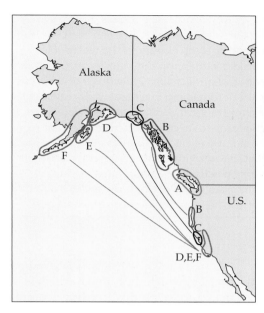

Fox sparrow

**FIGURE 8.25   The leapfrog migratory pattern of western North American populations of the fox sparrow.** The populations that breed farthest to the north migrate farthest to the south in the fall. (Population A is composed of birds that live year-round in the same area.) After Bell.[99]

dent and migrant strategies of blackbirds: Why do members of the same species vary in their choice of breeding and wintering sites? This problem can be illustrated by examining where six different populations of the fox sparrow breed and spend the winter in western North America (Figure 8.25).[99] The members of each population are sufficiently different in appearance to warrant placing them in six different subspecies (and in fact, some persons believe that the "fox sparrow" actually embraces a number of different species). One current subspecies (A) in the northwestern United States and southwestern Canada is essentially sedentary, but members of the other five populations migrate. They do not, however, select the same destination. Instead, the birds that breed in southern Alaska (subspecies D, E, and F) travel great distances (probably over the open ocean) to winter in southern California. Other populations (B and C), breeding closer to the United States, travel many fewer kilometers to wintering grounds in central to northern California. How can we account for this pattern?

One hypothesis is that birds in the different populations achieve equivalent fitness by adopting the optimal migration patterns for their particular breeding sites. Thus, the birds that form the southern Alaskan populations (D, E, and F) benefit by moving to southern California to exploit the surge of food available there in the mid to late spring, which enables these individuals to put on the large fat reserves needed for their long flight north. If sparrows from a breeding population in southern Canada (B or C) also went to southern California for the winter, they would have to compete for food with the Alaskan birds that were there, and they would be late in getting back to Canada, where the breeding season begins sooner than in Alaska. Late arrival in southern Canada would damage their chances in territorial competition with other fox sparrows that had migrated earlier over shorter distances. Thus, members of the southern Canadian breeding population winter closer to Canada than Alaskan fox sparrows, and they start their migration back earlier in the year as well. According to this view, members of each subspecies balance travel costs and arrival times in selecting the optimal sites in which to overwinter and breed.[99]

**Discussion Question**

**8.10** As an exercise, apply conditional strategy theory to the western fox sparrow case in the same way as we did with European blackbirds. Use your hypothesis to make predictions about what decisions individuals of different competitive abilities would make about remaining in an area year-round versus migrating various distances. For example, if your hypothesis were correct, what should happen if a bird improved its condition from one year to the next? What information presented above permits you to evaluate your predictions? Is it useful to know that in the white-crowned sparrow, resident and migrant birds have almost identical estimated annual survival rates, or that those sparrow species that migrate relatively long distances do not have lower survival rates than other species that travel shorter distances?[1276]

## Territoriality

Fox sparrows everywhere defend the areas in which they breed, which is to say that they are **territorial**. Migrant males arrive on the breeding grounds before females and stake out their claims, singing their special song lustily and charging after any other male fox sparrows that dare come close to their real estate. When females come north, they choose among the territorial males, each settling with a partner on his patch of brush and small trees, often next to a stream. After a bond has formed between male and female, she is prepared to help him chase away other fox sparrows, especially if they happen to be females. Although such aggressive behavior in defense of space is very widespread, occurring in creatures as different as poplar aphids and Belding's ground squirrels, many other animals, including honey bees and monarch butterflies, ignore or tolerate their fellows. So once again we have an evolutionary puzzle to resolve: Why spend the time and energy to be territorial?

A cost–benefit approach to territoriality requires that we consider the disadvantages of territorial defense, one of the most obvious of which is the time cost of the behavior. A territorial surgeonfish, for example, chases rivals away from its algae-rich turf on a Samoan reef an average of 1900 times each day.[310] The wear and tear of these territorial pursuits, to say nothing of the out-and-out fighting that occurs in some territorial species, can lead to a shortened life, as has been documented for a damselfly species with both territorial and nonterritorial individuals.[1472]

In addition to the risks of injury or exhaustion, other damaging effects can arise indirectly from the underlying mechanisms of territorial aggression. For example, in species in which testosterone promotes territorial defense, the effects of the hormone may exact a toll via a reduction in parental care or loss of immune function (see page 176).[1599] Moreover, the hormone may increase the activity level of males so much that they suffer as a result. Catherine Marler and Michael Moore demonstrated this point experimentally with Yarrow's spiny lizards.[934, 935] They inserted small capsules containing testosterone beneath the skin of some male lizards they captured in June and July, a time of year when the lizards are only weakly territorial. These experimental animals were then released back into a rock pile high on Mt. Graham in southern Arizona. The testosterone-implanted males patrolled more, performed more pushup threat displays, and expended almost a third more energy than controls (lizards that had been captured at the same time and given a chemically inert implant). As a result, the hyper-territorial males expended their energy reserves and died sooner than males with normal concentrations of testosterone (Figure 8.26).

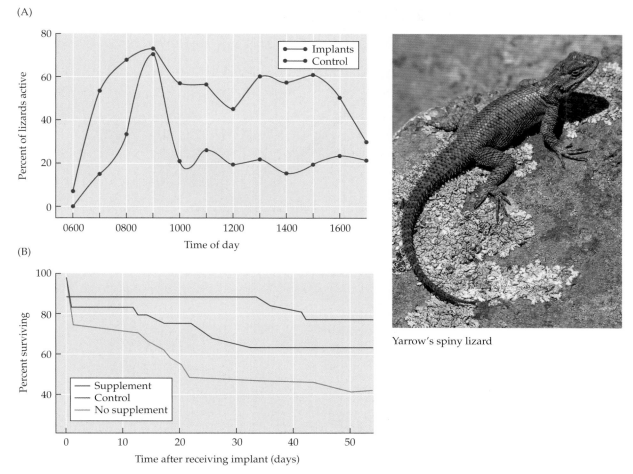

(A)

(B)

Yarrow's spiny lizard

**FIGURE 8.26    Energetic costs of territoriality.** (A) Males of Yarrow's spiny lizard that received an experimental implant of testosterone spent much more time moving about than did control males. (B) Testosterone-implanted males that did not receive a food supplement disappeared at a faster rate than did control males. Testosterone-implanted males that received a food supplement (mealworms) survived as well as or better than controls; thus the high mortality experienced by the unfed group probably stemmed from the high energetic costs of their induced territorial behavior. Photograph by the author; after Marler and Moore.[934, 935]

Because territoriality is exceedingly costly, we can predict that peaceful coexistence on an undefended living space or **home range** should evolve when the benefits of owning a valuable resource do not outweigh the costs associated with its monopolization. In Yarrow's spiny lizard, for example, males do live together in relative harmony in June and July, when females are not sexually receptive, but when fall comes and females are ready to mate, the males become highly aggressive, a pattern that applies to many other animals. This capacity to adjust the level of aggression up or down is also seen in the tropical pseudoscorpion *Cordylochernes scorpioides*. The life cycle of this arthropod involves periods of colonization of recently dead or dying trees, followed in a few generations by dispersal from the now older dead trees to fresh sites. Dispersing pseudoscorpions make the trip under the wing covers of a huge harlequin beetle (Figure 8.27), disembarking when the beetle touches down on an appropriate tree. The pseudoscorpions mate both on trees and on the backs of beetles, but male pseudoscorpions attempt to control real estate only

**FIGURE 8.27 A beetle and its pseudoscorpions.** (A) Two huge male harlequin beetles battle for position on a tree trunk. (B) Groups of the pseudoscorpion *Cordylochernes scorpioides* hitch a ride under the beetle's wing covers as it flies from one tree to another. Photographs by David and Jeanne Zeh.

(A)

(B)

when they occupy the small, and therefore economically defensible, patches on a beetle's back.[1640] Under these circumstances, the benefits of territoriality, in terms of number of mates monopolized, can exceed the costs of defense.

If individuals vary in their territory-holding abilities, then those with superior competitive ability should secure reproductive advantages over others excluded from favored habitat. A case in point is the American redstart. These warblers compete for territories during the nonbreeding season, when they are on their tropical wintering grounds in Central America and the Caribbean. In Jamaica, males tend to occupy black mangrove forests along the coast, while females are more often found in second-growth scrub inland. Field observations reveal that older, heavier males in mangrove habitat attack intruding females and younger males, apparently forcing them into second-rate habitats.[940] If these older, dominant males benefit from their investment in territorial aggression, then there should be some reproductive advantages for

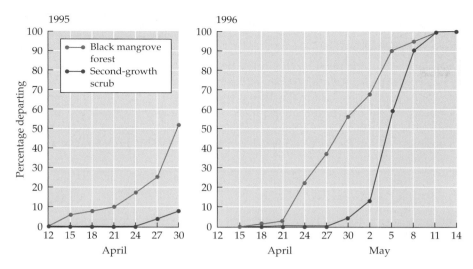

American redstart (female)

American redstart (male)

**FIGURE 8.28  Habitat quality and the date of departure** from Jamaican wintering grounds by American redstarts. Birds occupying the preferred black mangrove habitat are able to leave sooner for the northern breeding grounds than redstarts forced into second-rate, second-growth scrub. After Marra and Holmes.[940]

individuals that succeed in occupying the favored habitat. In fact, birds living in mangroves retained their weight over the winter, whereas redstarts in the apparently inferior scrub habitat generally lost weight.[940]

Probably because they have more energy reserves, territory holders in mangroves leave their wintering grounds sooner than birds living in second-growth scrub (Figure 8.28). In some migrant songbird species, early male arrivals secure a reproductive advantage by claiming the best territories and gaining quick access to females when they arrive (e.g., Hasselquist[630]). Female redstarts also have something to gain by reaching the Canadian breeding grounds early (and in good condition), because the short Canadian summer gives them only a few months to raise their chicks. Early birds get the fledglings in this species. Because the arrival time of a redstart in Canada is a function of where in Jamaica it spent the winter, overwintering habitat is a key predictor of reproductive success in this species (Figure 8.29).[1056]

The benefits of territoriality have been carefully measured in other animals as well. In the arctic ground squirrel, for example, males compete with one another for control of patches of meadows in the Canadian arctic. Female ground squirrels also live in these places, and at the start of the breeding season, they mate with one or (often) more than one male. Given that a territorial male does not necessarily monopolize sexual access to the females living within his plot of grassy meadow, it seems odd that he invests considerable time and energy keeping other males away from his home ground. But Eileen Lacey and John Wieczorek knew that even though female arctic ground squirrels regularly mate with several males, the male that is first to copulate with a given female is almost always the only male to fertilize her eggs. With this knowledge, they predicted that territorial males would be more likely to mate first with females in their territories, and this prediction was supported by the data (in 20 of 28 cases, a sexually receptive female was mated first by the owner of the patch of ground in which the female resided). These males gained a fertilization payoff for their investment in territorial behavior.[826]

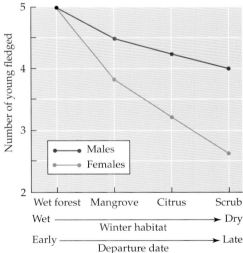

**FIGURE 8.29  Redstart habitat selection on the wintering grounds affects reproductive success.** The estimated number of fledglings produced by redstarts in relation to where they spent the winter. Birds that wintered in the most productive habitat, wet forest, are believed to enjoy the highest reproductive success. After Norris et al.[1056]

**Discussion Question**

**8.11** In some songbirds, a nonterritorial pair sets up housekeeping unobtrusively within the territory defended by another pair of the same species. However, nonterritorial pairs either do not breed at all[1356] or fledge fewer offspring on average than the owners of the territories they live in.[392] Develop a game theoretical hypothesis to analyze why there are any nonterritorial pairs of this sort. Come up with a two-strategies hypothesis and a conditional strategy hypothesis. What predictions follow from your two hypotheses?

## Territorial Contests

Studies of territoriality have found that winners in the competition for territories gain substantial indirect and direct reproductive benefits. Given this evidence, it seems paradoxical that when a territory holder is challenged by a rival, the owner almost always wins the contest—usually within a matter of seconds.

Why do the intruders give up so quickly? Game theory has supplied one possible answer via an algebraic demonstration that an arbitrary rule for resolving conflicts between residents and intruders could be an **evolutionarily stable strategy**—that is, one that cannot be replaced by an alternative strategy. One simple arbitrary rule might be "the resident always wins." If all competitors for territories were to adopt the "resident always wins" rule, so intruders always gave up and residents never did, a mutant with a different behavioral strategy could not spread its special allele. Therefore, the "resident always wins" strategy would persist indefinitely. (Similar algebraic maneuvers demonstrate that the contrary convention, "the intruder always wins," is also a possible evolutionarily stable strategy. Therefore, according to this hypothesis, any given territorial species should be as likely to have evolved an "intruder wins" rule as a "resident wins" rule.)

Attempts to locate species that employ an arbitrary "resident always wins" rule have failed, although at one time the speckled wood butterfly was believed to employ this strategy. Males of this insect defend small patches of sunlight on the forest floor, where they occasionally encounter receptive females. Males that successfully occupy sun-spot territories mate more frequently than those that do not, judging from an experiment in which males and females were released into a large enclosure. Under these conditions, sun-spot males secured nearly twice as many matings as their nonterritorial rivals.[110]

Based on these results, you would think that there would be stiff competition for sun spots, but Nick Davies found that territorial males *always* quickly defeated intruders, which invariably departed rather than engaging in lengthy territorial conflicts.[351] The butterflies were capable of such fights, as Davies showed by capturing and holding territorial males in insect nets until new males had arrived and claimed the seemingly empty sun-spot territories. When a prior resident was released, he returned to "his" territory, only to find it taken by a new male. This male, having claimed the site, reacted as if he were the resident, with the result that the two males engaged in what passes for a fight in the butterfly world. The combatants circled one another, flying upward, occasionally clashing wings, before diving back to the territory, sometimes repeating this maneuver. Eventually the previous resident gave up and flew away, leaving the site under the control of the replacement resident (Figure 8.30).

Although Davies's results were consistent with the hypothesis that male butterflies use an arbitrary rule to decide winners of territorial contests, Darrell Kemp and Christer Wiklund decided to repeat the experiment but without

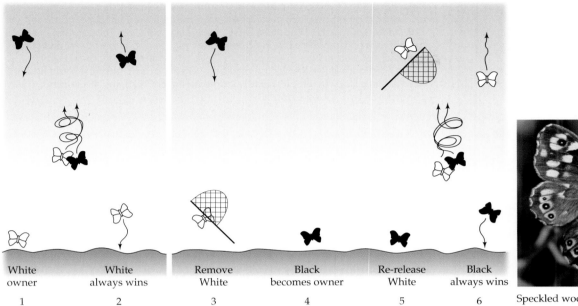

| White owner | White always wins | Remove White | Black becomes owner | Re-release White | Black always wins |
|---|---|---|---|---|---|
| 1 | 2 | 3 | 4 | 5 | 6 |

Speckled wood butterfly

**FIGURE 8.30  Does the resident always win?** An experimental test of the hypothesis that territorial residents always win conflicts with intruders was performed with males of the speckled wood butterfly. When one male ("White") is the resident, he always defeats intruders (1, 2). But when "White" is temporarily removed (3), permitting a new male ("Black") to settle on his sun-spot territory (4), then "Black" always defeats "White" upon his return after release from captivity (5, 6). After Davies.[351]

subjecting the captured initial resident to the potentially traumatic effects of being held in an insect net before release. Instead, they put their captives in a cooler, retrieving them after replacements had taken up residence in their sun-spot territories for about 15 minutes. After placing the original resident on the ground near his old territory, they tossed a wood chip over him. Butterflies pursue visual stimuli of this sort, so the new male could be steered toward his rival in this way. When the perched replacement male saw the incoming initial resident, he reacted in the manner of a territorial speckled wood butterfly by engaging his opponent in a spiral flight. But these flights lasted much longer than the ones that Davies had observed, and their outcomes were strikingly different, with the original sun-spot holders winning 50 of 52 contests with new residents.[756] So the current resident does not always win, a result that eliminates the arbitrary rule hypothesis. (Kemp and Wiklund believe that in Davies's experiment, the original resident males that were held in nets until their release just wanted to escape rather than fight for their old territories, and thus the relatively short interactions and the new resident "victories.")

If the arbitrary "resident always wins" rule does not explain why residents usually repel intruders, what hypothesis will do the job? Perhaps males that succeed in acquiring territories have some sort of nonarbitrary advantage over others that translates into superior **resource-holding power**. One such advantage might be large body size, which would provide a physical edge for the territory holder. In fact, in species ranging from rhinoceroses[1193] to fiddler crabs[719] to wasps,[1066] territorial individuals are relatively large individuals, which gives them the strength to drive smaller rivals away.

Nevertheless, being larger and so, presumably, stronger is not the key to territorial success in every species. In the red-shouldered widowbird, an African species that looks remarkably like the North American red-winged blackbird, males with larger and redder shoulder patches are more likely to hold territories than are rivals with smaller, duller epaulettes. The less gaudy males become nonterritorial "floaters," hanging around the territories of other males, conceding defeat whenever challenged, but ready to assume control of vacant territories should a resident disappear. Territorial males and floaters do not differ in size, but even so, when males in the two categories were captured and paired off in a cage unfamiliar to either individual, the ex-resident almost

Red-shouldered widowbird

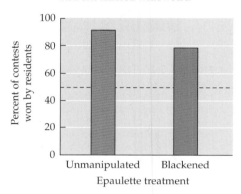

**FIGURE 8.31** **Territorial male red-shouldered widowbirds** have higher resource-holding power than nonterritorial floaters. When residents compete with floaters for food in captivity, residents usually win, even though they have been removed from their breeding territories for the experiment, and even if their red epaulettes have been painted black to eliminate this signal of dominance. The dotted line shows the result that would occur if residents and floaters were equally likely to win dominance contests. After Pryke and Andersson;[1173] photograph by P. Craig-Cooper.

always dominated the floater, even when the resident's red shoulders had been painted black (Figure 8.31).[1173] These results suggest that some intrinsic feature (other than body weight) is advertised by the size and color of the male epaulettes and is expressed even in the absence of this signal, which enables these males to win fights with others.

Just as is true for red-shouldered widowbirds, males of some damselflies can be divided into territory holders and floaters. Although territorial male damselflies are not necessarily larger or more muscular, those that win the lengthy back-and-forth aerial contests that occur over streamside mating territories almost always have a higher fat content than the males they defeat (Figure 8.32).[925, 1153] In these insects, contests do not involve physical contact, but instead consist of what has been called a war of attrition, with the winner outlasting the other, and the ability to continue displaying is related to the individual's energy reserves.

**FIGURE 8.32** **Fat reserves determine the winner of territorial conflicts in black-winged damselflies.** In this species, males may engage in prolonged aerial chases that determine ownership of mating territories on streams. (A) Larger males, as measured by dry thorax weight, do not enjoy a consistent advantage in territorial takeovers. (B) Males with greater fat content, however, almost always win. After Marden and Waage.[925]

(A) Thorax weight

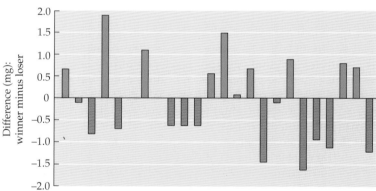

(B) Fat content

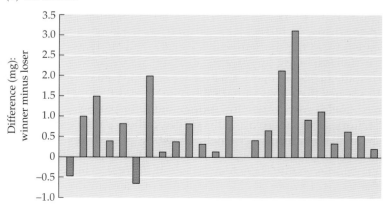

Black-winged damselfly

(A)

(B)

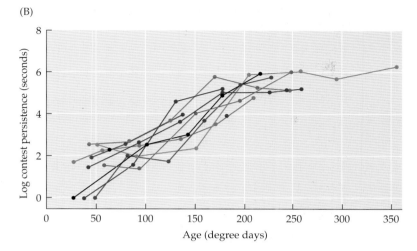

**FIGURE 8.33  Older males fight harder in the eggfly butterfly.** (A) An old male missing large chunks of his hindwings but still defending his perching territory in Queensland, Australia. (B) Results of a series of encounters in the laboratory between a hand-reared male of known age and a fresh, young rival placed in his cage. In the ensuing aerial clashes, the resident persisted longer as he grew older. Each color represents a different individual resident. A, photograph by Darrell Kemp; B, after Kemp.[755]

Differences in resource-holding power, however, whether based on size or on energy reserves, cannot account for every case in which territory owners seem to win at will. For example, older female Mediterranean fruit flies, who are almost certainly not as physically fit as younger flies, can keep younger females away from the fruits in which the insects lay their eggs.[1102] By the same token, older males of the song sparrow respond more intensely to simulated intrusion of their territories via song playback than younger ones,[704] and eggfly butterfly males are more willing to persist in their aerial war-of-attrition contests as they age than when they were younger (Figure 8.33).[755] Findings of this sort do not support the hypothesis that territorial winners are in better condition or more physically imposing than losers.

Therefore, let us consider a third hypothesis for resident success in territorial defense, namely, that when the fitness payoff for holding a territory increases over time, owners have more to fight for than intruders. This hypothesis has been invoked to explain why "experienced" pairs of convict cichlids have a strong advantage in territorial contests with "novice" pairs. When pairs of the fish had been together for 96 hours as opposed to 48 hours, they usually won their fights for nesting sites, maybe because they were closer to spawning and thus had more to gain by securing a suitable place for their eggs than the novice pairs, which still had time to find a breeding territory.[412]

More generally, as territory holders get older, their opportunities for future reproductive success tend to fall, and as a result, the costs of engaging in energetically expensive or risky contests also decrease. This fact of life history could in some cases boost the net gain from territorial defense by elderly residents. The net gain could also rise if the benefits of territorial possession grow over time because of the nature of interactions among territorial neighbors. An aggressive newcomer initially has to spend a lot of time dealing with boundary disputes with neighbors, but once territorial borders have been agreed upon, everyone calms down, producing what has been called the "dear enemy" effect.[467] For example, the mean distance at which territorial males of an African lizard will charge at a familiar neighbor is less than one-fifth that for strangers. Moreover, when chasing a neighbor, the resident pursues him for only a few centimeters, whereas he keeps after an unfamiliar intruder for a meter and a half on average.[1559]

Thus, once a territory owner and his neighbors have learned who is who, they no longer need to expend time and energy in lengthy chases. If a resident is ousted, however, the new resident will have to fight intensely for a time with his neighbors in order to get the borders of their territories right. The original resident therefore has more to gain by holding onto his present territory than the new intruder can secure by acquiring it, since any new boy on the block will still have to deal with his unfamiliar neighbors even after he has ousted the resident.

---

### Discussion Question

**8.12** The dear enemy effect has been explained in terms of familiarity (individuals learn who their neighbors are, and as they become familiar with these others, they become less aggressive toward them). An alternative explanation can be labeled the threat level hypothesis, which states that the dear enemy effect results from the reduced threat to the fitness of a territory holder offered by neighbors that no longer challenge the territory owner next door. The banded mongoose is a group-living, territorial mammal in which neighboring bands sometimes expand their territories into those held by adjacent bands, whereas transient strangers do not stay in established territories. In this species, individuals react *more* aggressively to members of neighboring bands than to strangers.[1017] If I were to claim that this result supports the threat level hypothesis for the dear enemy effect, and not the familiarity hypothesis, you might respond skeptically—in what way?

---

A logical extension of the dear enemy hypothesis is that neighbors may find it advantageous to combine forces to repel an intruder that might otherwise displace one of them, which would result in the remaining residents being left to cope with a rambunctious newcomer. This factor is believed to be behind the formation of occasional coalitions between two neighboring males in a territorial fiddler crab.[57] Every so often a male of *Uca mjoebergi* will leave his burrow in a mudflat and join a neighbor in fending off a wandering male that has challenged that neighbor. Typically the helper intervenes when the encroaching intruder is larger than his "dear enemy" neighbor but smaller than the helper. The two-on-one game plan usually works in that when a crab has a helper, the two are able to drive off the intruder 88 percent of the time, whereas defenders that have to cope alone with a takeover threat win only 71 percent of the time. The benefit to the helper appears to be the time and energy savings that come from having a familiar neighbor as opposed to an aggressive new neighbor intent on establishing himself in the neighborhood. Males that know one another almost never fight, but newcomers are far more combative, at least initially.[57]

The dear enemy effect could contribute mightily to an asymmetry between what an established territory holder has to lose in a territorial contest and what a challenger has to gain. Given this payoff asymmetry, we can predict that when a newcomer is permitted to claim a territory from which the original resident has been temporarily removed, the likelihood of the replacement resident's winning a fight against the original resident will be a function of how long the replacement has occupied the site. This experiment has been done with birds, such as the red-winged blackbird, as well as with some insects and fishes. In red-wings, when captive ex-territory holders are released from an aviary and allowed to return to their home marsh to compete again for their old territories, they are more likely to fail if the new males have been on the experimentally vacated territories for some time.[98]

The payoff asymmetry hypothesis also predicts that contests between an ex-resident and his replacement will become more intense as the tenure of the replacement increases, because longer tenure boosts the value of a site to the current holder and thus his motivation to defend it. This prediction has been supported in animals as different as tarantula hawk wasps[13] and songbirds.[804] If, for example, one removes a male tarantula hawk wasp (Figure 8.34) from the peak-top shrub or small tree that he is defending and pops him into a cooler, his vacant territory will often be claimed within a few minutes. If the ex-territory holder is quickly released, he usually returns promptly to his old site and chases the newcomer away in less than 3 minutes on average. But if the ex-territory holder is left in the cooler for an hour, then when he is released and hurries back to his territory, a battle royal ensues. The newcomer resists eviction, and the two males engage in a long series of ascending flights, in which they climb rapidly up into the sky side by side for many meters before diving back down to the territory, only to repeat the activity again and again until finally one male—usually the replacement wasp—gives up and flies away. The mean duration of these contests is about 25 minutes, and some go on for nearly an hour.[13]

**FIGURE 8.34**   **A male tarantula hawk wasp** (*Hemipepsis ustulata*) perched in a large creosote bush on the lookout for arriving females and intruder males. Photograph by the author.

Although cases of this sort support the payoff asymmetry hypothesis for why residents usually win, it is possible that lengthier contests occur after replacements have been on territories for some time because the removed residents have lost some resource-holding power while in captivity. This possibility was checked in a study of European robins by taking the first replacement away and permitting a second replacement to become established for a short time before releasing the resident. In this way, Joe Tobias was able to match 1-day replacements against residents that had been held captive for 10 days. Under these circumstances, the ex-territory holders always won, despite their prolonged captivity.[1451] In contrast, when ex-residents that had been caged for 10 days went up against replacements that had been on territories for 10 days, the original territory holders always lost. Therefore, contests between replacements and ex-residents were decided by how long the replacement had been on the territory, not by how long the ex-resident had been in captivity. The payoff asymmetry hypothesis appears to explain why the resident almost always wins in European robins. Perhaps it applies to many other species as well.

## Summary

1. In choosing where to live, many animals actively select certain places over others. If living in a certain kind of habitat enhances fitness, then individuals able to occupy preferred habitats should have higher fitness than the others living elsewhere, unless they are forced to share the more desirable places with more rivals of their species.

2. The selection of living space often occurs in the context of leaving one spot for another, as when juvenile animals leave the place where they were born to go in search of new homes. Cost–benefit analyses have helped to explain why young male mammals typically disperse farther than females of their species—probably because the costs of dispersal are greater for females than for males.

3. Migration is a form of dispersal in which the migrants move relatively long distances between two areas; sometimes migrants eventually return to the place they left. This phenomenon raises interesting questions about its evolutionary origin and subsequent modification as well as its adap-

tive value. The ability to migrate very long distances probably evolved in populations that had acquired short-range migratory capacity. Studies of the adaptive value of the behavior have focused on learning how migrants lower the costs of travel while maximizing the benefits gained.

4. The coexistence of resident versus migratory individuals (or territorial versus nonterritorial members of a species) has often been explained with conditional strategy theory. Possession of a conditional strategy confers behavioral flexibility on individuals, which are therefore able to choose adaptively among several options or tactics.

5. Territorial behavior is costly, which is why it is far from universal. Defense of living space evolves only when territorial individuals can gain substantial benefits, such as special access to mates or food. Territorial contests are usually won quickly by owners over intruders. The competitive edge held by territorial residents may stem from superior physical strength or energy reserves (which gives them high resource-holding power), or it may exist because residents have more to lose than intruders can gain (a payoff asymmetry) thanks to the dear enemy effect (in which familiar neighbors no longer fight intensely with one another over territorial boundaries).

## Suggested Reading

The concept of an evolutionary strategy has been made clear by Richard Dawkins[364, 368] and Mart Gross.[590] The related ability of organisms to develop or behave in different ways depending on their environment has been explored in depth by Mary Jane West-Eberhard.[1543] This concept is closely linked to conditional strategy theory, an approach that Steve Shuster and Michael Wade criticize in their book.[1328] For a mathematical critique of Shuster and Wade's criticism, see a short paper by Joseph Tomkins and Wade Hazel.[1455] Tom Whitham[1557] shows how to use evolutionary theory when studying habitat selection. Dispersal in Belding's ground squirrels is analyzed at the proximate and ultimate levels by Kay Holekamp and Paul Sherman.[666] The cost–benefit approach to territoriality was first used by Jerry Brown and Gordon Orians,[190] and the approach has been applied subsequently with special skill by Nick Davies and Alistair Houston.[354] The use of stable isotope technology, an important new tool in the study of migration, is clearly explained by D. Rubenstein and K. Hobson.[1247]

# 9

# The Evolution of Communication

ersons on an African safari sometimes have the good luck to see one spotted hyena present its large, engorged penis to a social companion, who then inspects the organ closely while perhaps also offering the presenter a chance to nuzzle, sniff, or lick its own erect penis (Figure 9.1). This rather indelicate operation catches the eye of most people who see it, but what really gets their attention is the news that the hyenas they have been watching are females, not males. In the spotted hyena, both sexes are endowed with a "penis" that can be erected, and both use the appendage in ways that appear to communicate something to their fellow hyenas.

We begin this chapter with the case of the female hyena's pseudopenis in order to illustrate two key points: first, the difference between questions about the evolutionary origin of a behavioral trait versus questions about the adaptive value of the behavior and, second, the ability of scientists to make progress by continually evaluating and reevaluating one another's hypotheses.

◀ **The frilled lizard can expand its neck frill.** *What is it communicating in this display? Photograph by Dave Watts.*

**FIGURE 9.1    The pseudopenis of the female spotted hyena** can be erected (A), in which condition it is often presented to a companion (B) in the greeting ceremony of this species. A, photograph courtesy of Steve Glickman; B, photograph by Heribert Hofer.

(A)

(B)

## The Origin and Modification of a Signal

At some point, a female ancestor of today's spotted hyenas must have been the first female ever to greet a companion by offering her a chance to inspect her "penis." How did this female ever come to behave in such an utterly bizarre way? To answer this question, we first turn to comparisons among the family Hyaenidae, which includes just four species, one of which is the spotted hyena. All four species engage in chemical communication, which includes the use of their anal scent glands to mark their home habitat.[1518] Male and female hyenas also inspect the anogenital regions of their companions, and therefore the ancestor of the spotted hyena probably engaged in similar behavior.[815] At some point, females of an immediate predecessor of spotted hyenas added an inspection of the erect penises of their fellow hyenas, male and female alike, to their anogenital analyses. But in order for this to happen, the female's "penis" had to evolve. Where did this organ come from?

In 1939, L. Harrison Matthews proposed that the pseudopenis might be the developmental result of high levels of male sex hormones in female hyenas.[953] In 1981, Stephen Jay Gould introduced this hypothesis to a large audience in one of his popular essays in the magazine *Natural History*.[558] As Gould told his readers, the male penis and the female clitoris of mammals develop from exactly the same embryonic tissues. The general mammalian rule seems to be that if these tissues are exposed early on to male sex hormones (testosterone and other androgens), as they almost always are in male embryos, a penis is the end result. If the same target cells do not interact with androgens, as is the case for the typical embryonic female mammal, then a clitoris develops. Indeed, when a female embryo of most mammals comes in contact with testosterone, either in an experiment or because of an accident of some sort, her clitoris becomes enlarged and looks like a penis.[450] This effect has been observed in our own species in the unlucky daughters of pregnant women who received medical treatment that inadvertently exposed their offspring to extra testosterone,[1003] as well as in females whose adrenal glands produce more testosterone than normal.[1152]

Given the general pattern of development of mammalian external genitalia, Gould thought it likely that female embryos in the spotted hyena must be exposed to unusual concentrations of male androgens. In support of this hypothesis, he pointed to a paper written in 1979 by P. A. Racey and J. D. Skinner,[1192] who reported that female spotted hyenas had circulating levels of testosterone equal to those of males, unlike other hyenas and unlike mammals generally. This discovery appeared to confirm the hypothesis that an unusual hormonal environment led to the origin of the spotted hyena's false penis, setting the stage for the origin of the dual-sex penis-sniffing greeting ceremony.

## Discussion Question

**9.1**   When Gould wrote his article on the spotted hyena,[558] he concluded that asking the question, what is the pseudopenis for? was unnecessary and indeed unwise in that it distracted attention from what he considered to be the more important question: How is it built? What do you think?

Work on the extra androgen hypothesis continued with some researchers checking on androgen levels in wild female spotted hyenas (not an easy task). These workers found to their surprise that testosterone levels in free-living females are actually lower than in adult males.[561] Nonetheless, pregnant female hyenas do have higher testosterone levels than lactating females (Figure 9.2).[403] This finding leaves the door open to some extent for the extra

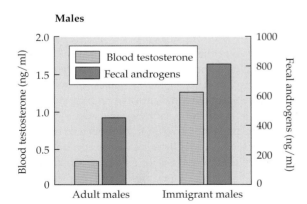
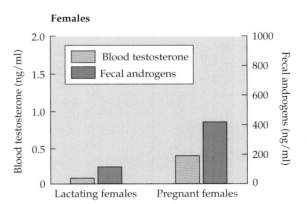

**FIGURE 9.2   Concentrations of testosterone in male and female spotted hyenas.** Testosterone was measured in the blood and in the feces of four categories of free-living hyenas: adult males born in the group where they live, adult males that had immigrated into a group, lactating females, and pregnant females. After Dloniak et al.[403]

androgen hypothesis,[414] as does the discovery that males and females have similar levels of a precursor of testosterone[561] called androstenedione (a substance made famous by baseball slugger Mark McGwire, who used this hormone regularly during 1998 when he hit 70 home runs). Because the placenta of a pregnant female converts androstenedione to testosterone, her female embryos could conceivably be exposed to masculinizing levels of testosterone.[545] However, it is also true that during the time that embryonic clitoral development is likely to be most sensitive to androgens, the mother hyena's placental cells are producing substantial amounts of an enzyme that inactivates androgens. In humans and other mammals, the same enzyme prevents masculinization of the genitalia of embryonic females.[291]

If the extra androgen hypothesis were correct, then the experimental administration of anti-androgenic chemicals to pregnant adult females should abolish the pseudopenis in their subsequent daughters while also feminizing the external genitalia of their newborn sons. In reality, however, when anti-androgens are administered to pregnant female hyenas, the daughters of the treated females retain their elaborate pseudopenis, albeit in altered form.[546] This result is also at odds with the view that early exposure to male sex hormones is sufficient in and of itself to cause female embryos of the spotted hyena to develop their unusual genital equipment.[413] Our understanding of the proximate basis for the pseudopenis is less certain than it once seemed.

If the hyena pseudopenis did originate as a result of a mutation that exposed the female fetus to high concentrations of androgens and if the development of the pseudopenis is now independent of fetal androgen exposure, then some major genetic changes must have occurred during the spotted hyena's evolution. These changes have evidently resulted in the partial or complete decoupling of the development of the pseudopenis from its original hormonal context, just as the activation of sexual behavior in red-sided garter snakes and white-crowned sparrows has been liberated from sex hormone control (see Chapter 5). In the spotted hyena, certain mutations may have spread, one after the other, in part because they lowered male-like androgen concentrations present in ancestral females, or modified their effects, thereby reducing the damaging developmental by-products of exposure to androgens for females (see below). Perhaps now an entire battery of altered genes contributes to the development of the secondary sexual characteristics of today's spotted hyena females. As a result, these creatures currently possess a less costly developmental mechanism for the production of an elaborate pseudopenis, which they apparently use to communicate with other hyenas.

### Discussion Question

**9.2** In studying the courtship behavior of the empid flies, E. L. Kessel was amazed to find a species, *Hilara sartor*, in which males gather together to hover in swarms, carrying empty silken balloons.[767] In this species, females fly to the swarm, approach a male, and receive a balloon, which they hold while mating occurs. In the overwhelming majority of fly species, including some other empid flies, courtship does not involve the transfer of any object from male to female. But in addition to (1) the empty balloon gifts of *H. sartor* and (2) the lack of courtship gifts in some empids, males of species in the group court females by offering (3) gifts of an edible food item (a freshly killed dead insect), (4) gifts of a dried insect fragment wrapped in a silken covering, or (5) gifts of an edible prey insect wrapped in silk. Construct a behavioral phylogeny that yields a hypothesis, consistent with the notion that evolutionary history involves a layering of change upon change, on the historical sequence leading to empty balloon gift-giving. How would you test your hypothesis? For papers on empid flies, see Cumming[326] and Sadowski, Moore, and Brodie.[1269]

In trying to understand how offering empty balloons to females ever began, what do you make of the fact that in one species some males offer large edible prey, while other males proffer small prey, and still others sometimes supply their mates an inedible dandelion-like tufted seed? When researchers took real prey from males of this species and substituted a small (inedible) cotton ball, females accepted this gift and copulated with males for about the same length of time as females given a small edible prey item.[824]

## The Adaptive Value of a Signal

Understanding the history of a trait with its origin and subsequent changes is one goal of evolutionary biologists, but another goal is figuring out why some changes have spread and persisted while others have disappeared over time. These goals come together in questions such as, why would the original mutant female hyena whose daughters had masculinized external genitalia have had more descendants than the other females of her era? Surely her masculinized daughters would have been harmed by the disruption of long-tested patterns of sexual development. In our species, females exposed to higher-than-normal amounts of androgens as embryos not only develop an enlarged clitoris, but also may be sterile as adults, illustrating just how reproductively damaging the developmental effects of male sex hormones can be for females. And even in modern populations of spotted hyenas, about 10 percent of all females die the first time they attempt to give birth, because their pups have to pass through the clitoris (Figure 9.3), which offers only a small birth canal, unlike the wider vaginal birth canal of most other mammals; moreover, about 60 percent of all first-born cubs of female hyenas die as well.[414] (But these mortality data come from a captive population; in their study of wild hyenas,[1519] Heather Watts and Kay Holekamp recorded no excess deaths among female hyenas in the first year they reproduced.) If there are high reproductive costs of masculinized genitalia for female spotted hyenas, a point that needs confirmation, the pseudopenis itself must provide substantial fitness benefits or else the developmental mechanism underlying the pseudopenis must do so.[490, 1018]

A focus on the proximate causes of the pseudopenis has led to a number of **by-product hypotheses** for this trait, in which selection for some other, truly adaptive attribute is thought to have as its by-product the development of an enlarged clitoris. For example, perhaps a mutant gene that boosted androgen levels in females spread because it helped to make adult females larger and more aggressive, not because it helped the females acquire a pseudopenis.[546] By all accounts, spotted hyena females are indeed unusually aggressive, at least by comparison with males, which are always subordinate and deferential to the opposite sex. Females not only keep males in their place, but also have a hierarchical system for their own sex, with high-ranking females (which have inherited their ranks from their mothers) gaining priority of access to the wildebeests and zebras that their clan kills or steals from lions (Figure 9.4).[489, 1518]

**FIGURE 9.3  A cost of the pseudopenis for female spotted hyenas.** The birth canal of this species extends through the pseudopenis, which greatly constricts the canal and may lead to fatal complications for the mother and her first-born pup during birth. Drawing courtesy of Christine Drea, from Frank, Weldele, and Glickman.[490]

**FIGURE 9.4** Competition for food is fierce among spotted hyenas, which may favor highly aggressive individuals. A hyena clan can consume an entire giraffe in minutes. Photograph by Andrew Parkinson.

With high rank comes dramatic gains in reproductive success. Top females are always able to hunt within the clan's territory, while subordinate females under some circumstances must travel outside their home ground to get enough to eat, a risky endeavor. The young of high-ranking females grow faster, are more likely to survive, and are more likely to become high-ranking individuals themselves than are the offspring of subordinate hyenas (Figure 9.5).[656] These outcomes appear to be related to the fact that dominant females produce more androgens during the second half of pregnancy than subordinate ones; the more androgens that reach the developing fetuses, the more likely they are to be aggressive as cubs, setting the stage for their ascension to dominance as adults.[404] All these benefits of exposure to androgens could possibly support some costly side effects, like having a pseudopenis if you are a female spotted hyena.

But there are problems here. For one thing, if having a pseudopenis was strictly a liability, selection surely would have favored mutant alleles that happened to reduce the harmful effects that androgens (or other developmentally potent chemicals) have on female embryos. So perhaps the pseudopenis has adaptive value in and of itself, despite the obvious costs of the thing. The first adaptationist hypothesis of this sort was offered by Hans Kruuk (the same biologist who studied mobbing in black-headed gulls; see page 185). In 1972 he suggested that the erect female genitalia were used to reduce tensions among the highly aggressive female members of a clan, promoting social bonds that kept the group together to work for the mutual benefit of all.[815] But Kruuk did not specify why an erect pseudopenis was required for bonding among females in this species when other hyenas have perfectly good greeting ceremonies based on inspection of their companions' anal glands alone. In these other species, females give birth via the vagina, with none of the complications that spotted hyenas regularly endure as a result of having to give birth through the elongated clitoris.

**FIGURE 9.5** Dominance greatly advances female reproductive success in the spotted hyena. A mother's social status is directly linked to cub survival to age 2 years and the dominance rank of her daughters as determined by observation of interactions between pairs of hyenas. After Hofer and East.[656]

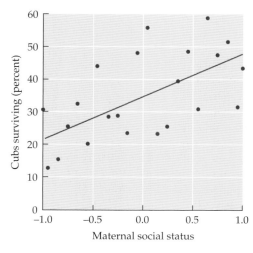

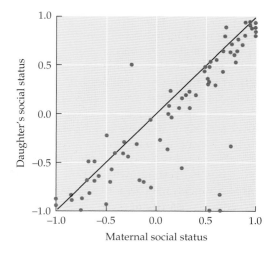

Because of the deficiencies in the social-bonding hypothesis, Martin Muller and Richard Wrangham offered another reason why the pseudopenis might be adaptive. They argued that the structural mimicry of males by females provides a "male-like" camouflage that diverts aggression away from them, presumably because females are tricked into treating an individual with a pseudopenis as if she were a male.[1018] As Muller and Wrangham themselves acknowledged, however, the evidence in favor of their idea is weak. For example, the hypothesis leads to the implausible prediction that aggressive females primarily use the visual cues associated with the external genitalia to identify the sex of a companion and therefore have difficulty telling the difference between males and females. But because hyenas, in particular, and mammals, in general, rely heavily on olfactory as well as acoustical cues to discriminate between males and females,[1018] adult females almost certainly can quickly tell whether a fellow adult is a male or a female, particularly when inspecting the other individual at very close range in a greeting ceremony.

All of this brings us back to this ceremony in which the pseudopenis currently plays a role of some sort. Could this role be so important as to outweigh the reproductive handicaps imposed by the female's enlarged clitoris?

## An Adaptationist Hypothesis

To explain why a female spotted hyena might benefit from having a mimetic penis to use in the greeting ceremony, perhaps we can unite elements of Kruuk's "social harmony" explanation with aspects of Muller and Wrangham's "sexual mimicry" hypothesis. This combination hypothesis requires us to consider the evolutionary context in which the pseudopenis may have originated (note the complementarity of the different levels of analysis offered by hypotheses on the origin of a trait and on its adaptive value). Imagine that the female-dominant, male-subordinate relationship evolved before the pseudopenis appeared. Furthermore, imagine that one of the cues that sexually motivated, subordinate male hyenas supplied to dominant females they were courting was an erect penis. Presentation of this organ to the female would clearly signal the suitor's male identity and thus his subordinate, nonthreatening status. Such a signal might encourage females on occasion to accept the male's presence rather than rejecting or attacking him.

Subsequent to the evolution of this kind of male–female communication system, mutant females with a pseudopenis would have been able to tap into an already established system for signaling subordination, a useful thing to be able to do in a highly competitive society with a strict **dominance hierarchy** among females.[1518] The benefit to a subordinate of being able to signal acceptance of her low status in such a hierarchy would come from being allowed to stay with the clan, rather than being tossed out to fend for herself in an environment in which lone hyenas do not live for long, let alone reproduce. Hyenas that live within a clan have a chance to reproduce eventually; hyenas that die young do not.

Evidence in support of this hypothesis comes from the observation that subordinate females and youngsters are far more likely than dominant animals to initiate interactions involving pseudopenile display, indicating that they gain by transferring information of some sort to dominant individuals.[424] Cooperation between subordinate and dominant hyenas might be advanced if the dominant female could accurately assess the physiological (e.g., hormonal) state of a subordinate by inspecting its erect, blood-engorged penis or clitoris. Perhaps truly submissive subordinates are signaling that they lack the hormones (or some other detectable chemicals) needed to initiate a serious challenge to the dominant individual, who can then afford to tolerate these others because they do not pose an immediate threat to her status. Note that this hypothesis resembles Kruuk's to the extent that "social harmony" follows from the subordinate's acceptance of its status relative to a more dominant individual. Moreover, male mimicry is also involved to the extent that subordinate females behave like males to demonstrate their rank relative to other females (but not to deceive their companions about their actual sex).

### Discussion Question

**9.3** How might you use comparative data to test the hypothesis that the use of a pseudopenis in the spotted hyena's greeting ceremony has adaptive value for the subordinate signaler as a means of demonstrating its willingness to submit to a dominant female? Take advantage of the fact that information exists on the behavior of the other, nonsocial species of hyenas.[987, 1086, 1518] In addition, another highly social mammal, the naked mole rat (see Figure 5.14), a member of the order Rodentia, is somewhat similar to spotted hyenas in that female reproduction in any one group is highly skewed. In the case of the mole rat, a single "queen" is dominant to all of the many other females in her group with whom she interacts, and the queen is the only reproducer in her group.[932] Although naked mole rat females do not have masculinized genitalia, an enlarged clitoris is characteristic of female lemurs in Madagascar,[1082, 1498] and the trait appears in a number of other mammals as well.[1152]

## The History of a Signal-Receiving Mechanism

The spotted hyena story illustrates, among other things, that behavioral biologists still have interesting puzzles to solve. Having gone as far as we can go for the moment with the possible origin and adaptive value of the pseudopenis signal, let us now focus on the evolution of the signal-receiving side of the equation. Our focal species will be the whistling moth, *Hecatesia exultans*, the males of which produce loud ultrasonic pulses of sound by beating their knobbed wings together (Figure 9.6A). Although we cannot hear these sounds,

(A)

(B)

**FIGURE 9.6 Ultrasonic communication.** (A) Male whistling moths produce ultrasounds by striking the knobby "castanets" (arrow) on their forewings together. (B) A male whistling moth has been attracted by the recorded calls of another male played from a small speaker in his territory, confirming that this species uses these sounds for communication. Photographs by the author.

other whistling moths can, as they demonstrate when they fly to a speaker playing a recording of whistling moth signals in the appropriate habitat (Figure 9.6B) in an attempt to find and interact with the signaling male.[12] Thus, the sounds made by males may play a role in defense of a calling territory as well as in attracting receptive females to the caller.

Both male and female whistling moths possess mechanisms for hearing their species' high-pitched calls.[1406] The sound-receiving system consists in part of an ear on either side of the thorax, near the point where the hindwing attaches to the thorax. The whistling moth is a noctuid moth, a member of the same family as the bat-detecting moths we met in Chapter 4. Like the ears of those moths, its outer ear is composed of a small ellipse of thin cuticle, called the tympanic membrane, which covers an air sac. When the membrane vibrates in response to ultrasound, the air sac also moves, supplying mechanical energy to associated sensory receptors. These mechanoreceptors then send signals to other parts of the nervous system. How might such a sophisticated device have evolved for detecting ultrasonic signals from other whistling moths?

Researchers interested in this question have operated with the same ground rules as those persons interested in the origin of the pseudopenis in spotted hyenas. They have assumed that in changing from the presumed ancestral state, which would have been no ears at all (even today, most moths lack ears), the whistling moth's ear probably did not arise via one or two mutations. The system is much too complex, and composed of too many interrelated parts, to have developed from a change in just one or two genes. Instead, the ear is much more likely to have been assembled gradually, with one small adaptive change layered on another, eventually producing the complex acoustical mechanism in its present form. But where do we go to test this hypothesis?

James Fullard and Jane Yack decided to look for clues about the past in the bodies of today's moths.[499] They examined some saturniid moths, members of a group that lacks ears and therefore cannot hear anything. They paid special attention to those parts of the saturniid thorax where one can find ears in noctuid moths (Figure 9.7). Fullard and Yack found mechanoreceptor cells attached to the saturniid thoracic cuticle that are remarkably similar in structure to the mechanoreceptor cells in the ears of noctuid moths, suggesting that the ears of noctuids were not built from scratch, but rather involved modifications of existing sensory cells that performed some other function for the earless moths.

What do these "non-acoustical mechanoreceptor cells" do for saturniids? They are attached to a part of the thorax that moves rhythmically as the wings move, particularly when the moth is vibrating its wings rapidly to generate metabolic heat prior to takeoff. These mechanoreceptors translate mechanical energy from the motion of the cuticle into messages that are relayed elsewhere in the moth's nervous system, where information about the alignment of various body parts is used to adjust the moth's

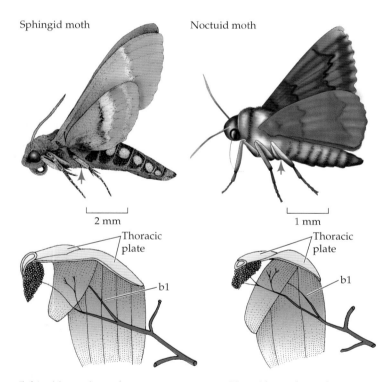

**FIGURE 9.7   Evolution of a sensory system.** Sphingid moths, like saturniid moths, cannot hear, but noctuid moths can. The red arrows in the upper panels point to the same location on the thorax of the two moths. At this site, sphingids do not have an ear with a tympanic cavity, but noctuids do. The lower panel shows the relationship between the thoracic plates (the two plain sheets at the top of each drawing), certain thoracic muscles (stippled bands), and a branched sensory nerve. Note the anatomical similarities between the two moths, especially with respect to the sensory nerve. The branch of the nerve labeled b1 carries information about the position of the hindwing in the sphingid; in the noctuid, this same branch relays acoustical information from the tympanic membrane to the moth's central nervous system. After Yack[1624] and Yack and Fullard;[1623] drawing of adult sphingid moth by Diane Scott.

position in space. A similar kind of mechanoreceptor in an ancestral moth in the noctuid lineage might have conferred a weak ability to hear certain loud sounds because loud sounds can make the insect exoskeleton vibrate.[499] Interestingly, an ancestral stretch receptor for monitoring body position has also evolved into an auditory receptor for detecting airborne sounds in other insects unrelated to moths,[1486] demonstrating that the function of receptors can indeed change over time.

Almost certainly, the first hearing noctuid moth had nowhere near the sophisticated sensory skills of the current descendants of that moth. But with the addition of other small changes, such as a slight thinning of the cuticle over the "acoustical" mechanoreceptors, an enlargement of the respiratory chamber behind this part of the thorax, and a mechanoreceptor design that was somewhat more sensitive to sounds of particular frequencies, these mutant moths could have heard bat cries more reliably. Currently the tympanum of at least some noctuid moths has clearly been modified in additional ways, because this region is structurally differentiated with a central "opaque zone" surrounded by a circle of thinner, clear cuticle. The receptors are attached to the opaque zone, which vibrates much more vigorously in response to ultrasonic stimulation than the surrounding cuticle.[1596] In other words, the tympanal region of modern noctuid moths is more complex (and presumably more effective in detecting ultrasound) than it appears at first glance.

If receivers with superior ultrasonic hearing survived better, and if as a result they reproduced more, then their genetic success would have set the stage for the next slight improvement to spread through the species via natural selection.

## The History of Insect Wings

Darwin recognized long ago that shifts in adaptive function occurred commonly over evolutionary time. In a wonderful book on the evolution of orchids, he wrote, "The regular course of events seems to be, that a part which originally served for one purpose, becomes adapted by slow changes for widely different purposes."[350] This rule appears to apply to whistling moth hearing mechanisms, which are derived from systems that once monitored wing vibrations rather than airborne sound vibrations. Moreover, whistling moth ears now provide individuals with information about the acoustical signals made by other whistling moths, whereas the ancestors of whistling moths almost certainly used their hearing ability for a different purpose: the detection of ultrasound produced by bats.

The point that structures take on different functions over evolutionary time can be reinforced by examining where insect wings came from and identifying what the antecedent structures did for their owners. In modern whistling moths, wings are used both for flight and for sound production. The designs of these wings and their control systems are extremely complex; they cannot have arisen in a single evolutionary step. Some persons believe that the precursors of insect wings were outgrowths of the respiratory apparatus—the gills—of an ancestral aquatic arthropod in which the gills were attached to the legs.[52] This ancestor endowed its modern descendants with a set of shared genes that now play a key role in the development of the body plan of crustaceans, insects, millipedes, horseshoe crabs, and spiders. Two of these genes, *pdm/nubbin* and *apterous*, are active in the development of the gills of modern crustaceans, which are also appendages attached to the legs of these animals. In horseshoe crabs, the same genes are turned on during the development of this animal's underwater breathing apparatus, the book gills. In two related groups of animals that left the water long ago, the spiders and the insects, these genes have been retained but now contribute to the development of book lungs in spiders and wings in insects (Figure 9.8).[346]

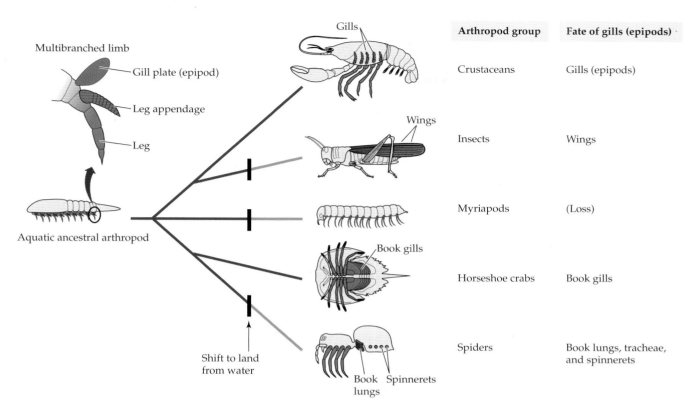

**FIGURE 9.8  Arthropod gills have evolved into many different structures with different functions.** In an aquatic ancestral arthropod, the gill plates served a respiratory function; they were appended to the legs of this animal. In the modern descendants of this ancient creature, the gills have retained their underwater respiratory role in crustaceans and have been modified somewhat as book gills in horseshoe crabs. In the three terrestrial lineages derived from the ancestral arthropod, the gill plates have been lost altogether from the myriapods, retained but modified as book lungs and spinnerets in spiders, and retained in highly modified form as wings in insects. After Damen, Saridaki, and Averof.[346]

Thus, one scenario for the evolution of insect wings sees their origin in insects that, in their immature aquatic life stages, had respiratory gill plates on the thorax that could also be moved as an aid to underwater locomotion (Figure 9.9).[818] If some of these creatures retained their movable plates when they metamorphosed into terrestrial adults, then the structures could have been used as sails to send the newly formed adults skimming across the water to land. A speedy trip may have been advantageous if there were fish predators eager to capture the new adults as they floated on the water. Even today, some flightless species of stoneflies—an ancient group of aquatic insects—stand on the water when they emerge as adults and use their stumpy wings as sails to catch the wind (Figure 9.10).[927] So does at least one mayfly species, a member of another ancient group of aquatic insects.[1253] (The mayfly is almost certainly descended from ancestors that flew, but it has lost this ability because it lives in waters where fish are extremely scarce; as the risk of predation has been reduced for this insect, the advantages of full flight have fallen correspondingly.) When Jim Marden and Melissa Kramer (an undergraduate at the time) clipped the wings of flying stonefly species so that the insects could not take off, the wing-clipped individuals could still sail along faster than they could swim as immatures. So here we have another demonstration that the winglets

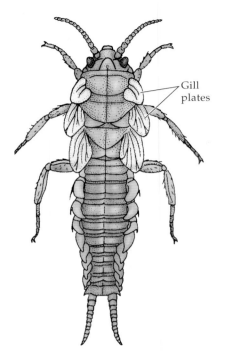

**FIGURE 9.9　Evolutionary precursors of insect wings?** The movable gill plates of this extinct stonefly-like insect may have been used to propel the insect underwater. Structures of this sort may then have been used as sails by adult aquatic insects as they skimmed the water surface. After Kukalová-Peck.[818]

Gill plates

of ancestral flightless insects might have had survival value and so could have spread through populations in the past, even though they did not confer the ability to fly.[926, 928]

Once sailing wings had become standard for some ancient insect species, mutant individuals able to move these appendages in a rowing motion might have been able to scull across the water, something that a living Chilean stonefly does.[930] Subsequently, other more powerful wing movements driven by larger muscles might have carried a rower or a skimmer faster across the water. These simple flappers would have set the stage for yet another modification, an increase in flight muscle capacity that could lift a powered skimmer off the water. Some modern stoneflies demonstrate how the transition between water skimming and full-fledged aerial flight might have come about. Adult stoneflies in the genus *Leuctra* beat their wings to achieve sufficient lift so that only the two hindlegs remain in contact with the water. This position stabilizes the insect while permitting it to move its wings over a much wider arc than is possible for those stoneflies that keep all six feet on the surface of the water. Hindleg skimmers travel about 40 percent faster than six-leg skimmers (Figure 9.11), showing how mutant individuals of some ancestral insect species that happened to use the hindleg technique might have enjoyed a selective advantage.[800] From hindleg skimming it is a short additional step to full flight, which would have carried the first true fliers away from insect-eating fish into the relative safety of the air. In fact, modern flying stoneflies employ almost exactly the same set of leg movements as they go from sitting on the water surface to jumping into the air as do the modern hindleg skimmers, which demonstrates the plausibility of the scenario in which a hindleg-skimming ancestral species gave rise to a fully flighted descendant.[929]

## Discussion Questions

**9.4**　The diversity of behavior among living stoneflies suggests a historical scenario that goes from floating on the water surface to jumping from the water surface and engaging in free flight (see Figure 9.11). This historical hypothesis can be tested by plotting the behavior of stonefly species on a phylogeny of this group (see Thomas et al.[1428] or http://www.bio.psu.edu/People/Faculty/Marden/skim.html). Given the available evidence, what is your assessment of the validity of this hypothesis?

(A)

(B)

**FIGURE 9.10　A surface-skimming stonefly** (A) standing on the water. (B) The stonefly has raised its short, stubby wings, which it uses strictly to catch the wind as it skitters over the water while keeping all six feet on the water's surface. Photographs courtesy of Jim Marden.

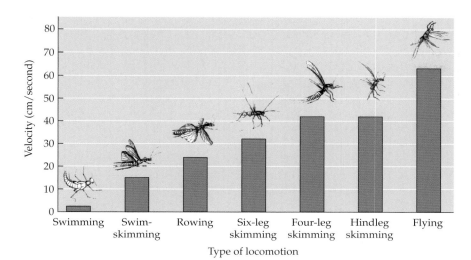

FIGURE 9.11   **A possible evolutionary pathway from swimming to full flight in the stoneflies.** Shown here are data on the mean speed of travel across or over water by living representatives of seven different kinds of stoneflies. The groups are organized by travel velocity from slowest to fastest. If evolution proceeded in the sequence shown here, each modified stonefly would have been able to move more rapidly across and then over water. After Marden and Thomas.[930]

**9.5**   What problem is posed for the hypothesis that all modern flying insects evolved from an aquatic water-skimming ancestor given that the closest relatives of the modern winged insects are believed to belong to a group called the Thysanura, which are exclusively terrestrial (the familiar silverfish that scuttle around in bookshelves belong to this group). Likewise, the insect group from which both the Thysanura and the modern winged insects (including stoneflies, of course) evolved is another essentially terrestrial group. Finally, the crustacean ancestor of all insects was apparently also terrestrial.[417]

At some point in the evolution of insect flight, the proto-wings of ancestral forms had become fully modified in ways that made powered flight possible. At this time, mechanoreceptors in the thorax that once provided sensory information about the position of sailing wings relative to the thorax could possibly have helped their owners control their wings by monitoring wing-flapping movements. When the ancestor of moths appeared on the scene, this species probably already possessed wing position sensors. Over time, these mechanoreceptors gradually evolved into ultrasound detectors with a new antipredator function (see page 114). However, ears that evolved because of their anti-bat advantages can also detect ultrasound coming from other sources, including members of one's own species. The whistling moth's ear was available to take on yet another function—the detection of ultrasonic signals produced by wing-beating male conspecifics.[1406]

## Sensory Exploitation of Signal Receivers

According to the scenario outlined above, male whistling moths use ultrasound to communicate with other whistling moths because an ancestor of *Hecatesia exultans* had bat predators, which made perception of ultrasound reproductively advantageous in the past. Once the ultrasonic detectors were in place, their existence affected the reproductive payoff to mutant individuals that happened to generate signals in the ultrasonic range, which receivers, because of their evolutionary heritage, could hear more easily than sounds of lower frequencies.

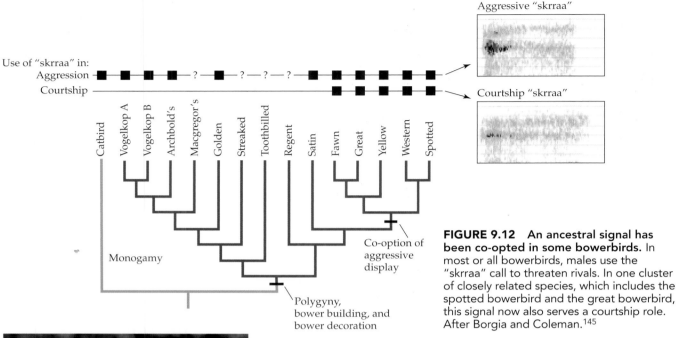

Use of "skrraa" in:
Aggression
Courtship

Aggressive "skrraa"

Courtship "skrraa"

Catbird · Vogelkop A · Vogelkop B · Archbold's · Macgregor's · Golden · Streaked · Toothbilled · Regent · Satin · Fawn · Great · Yellow · Western · Spotted

Monogamy

Co-option of aggressive display

Polygyny, bower building, and bower decoration

**FIGURE 9.12** **An ancestral signal has been co-opted in some bowerbirds.** In most or all bowerbirds, males use the "skrraa" call to threaten rivals. In one cluster of closely related species, which includes the spotted bowerbird and the great bowerbird, this signal now also serves a courtship role. After Borgia and Coleman.[145]

Great bowerbird

Likewise, I suggested earlier that we could better understand the female spotted hyena's use of her pseudopenis in greeting displays if this species already had evolved a male–female communication system that enabled a male to use his penis to signal his subordinate status to a female. By the same token, in one group of bowerbirds, males use a harsh and raspy "skrraa" call when courting females (a subject explored in more detail in the next chapter), almost certainly because this call has taken on a new role after originally having evolved as a male–female aggressive signal (Figure 9.12).[145] So here we have a preexisting male signal that is co-opted for another purpose because females can use it to make adaptive choices about mates.

The other side of the coin occurs when an animal happens to produce a novel signal that taps into preexisting perceptual mechanisms in a signal receiver, a phenomenon often called **sensory exploitation**.[79, 1262] In one of the better-studied examples, Heather Proctor hypothesized that modern courtship by male water mites began when males happened to "exploit" the predatory behavior of females waiting to ambush small aquatic invertebrates called copepods.[1169] While the predatory female is in her attack position, called the "net stance," the male vibrates a foreleg in front of her. She, in turn, may grab him, using the same response that she uses to capture passing copepods. However, she then releases the male unharmed. He turns around and places spermatophores (packets of sperm) near the female, which she picks up in her genital opening if she is receptive (Figure 9.13).

The apparent use of a predatory grab by the female in response to the male's trembling display suggested to Proctor that males were mimicking the stimuli produced by copepod prey. Perhaps the female's reaction identified her as a potential mate and showed the male where to position his spermatophores to best effect (water mites cannot see). If males trigger the prey detection response of females, then unfed, hungry female water mites held in captivity should be more responsive to male signals than well-fed females. They are, providing support for the contention that once the first ancestral male happened to use a trembling signal, the male's behavior and its heredi-

(A)                                    (B)

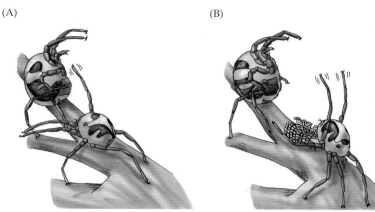

**FIGURE 9.13  Sensory exploitation and the evolution of a courtship signal in the water mite _Neumania papillator_.** (A) The female (on the left) is in her prey-catching position (the net stance). The male approaches and waves a trembling foreleg in front of her, setting up water vibrations similar to those a copepod might make. The female may respond by grabbing him, but she releases him unharmed. (B) The male then deposits spermatophores on the aquatic vegetation in front of the female before waving his legs over them. After Proctor.[1169]

tary basis spread because it effectively activated a preexisting prey detection mechanism in females.[1169] The case is not ironclad, however, for a number of reasons, including the possibility that hungry females may be more ready to solicit matings in order to acquire a nutrient-rich spermatophore to digest.[147]

Among the water mites, trembling courtship occurs in only a few species, namely, those in which females adopt the net stance. By measuring a large number of characteristics in several species of water mites, Proctor produced a phylogeny for these animals. The phylogeny indicates that the net stance of females originated in an ancestral water mite that eventually gave rise to eight descendant species. Within this lineage, male courtship trembling subsequently appeared twice, once in the ancestor of the genus _Unionicola_ and once in the ancestor that split into two species of _Neumania_ (Figure 9.14). If this phylogeny is correct, it would mean that male courtship trembling might have originated well after females had adopted a copepod-ambushing life-

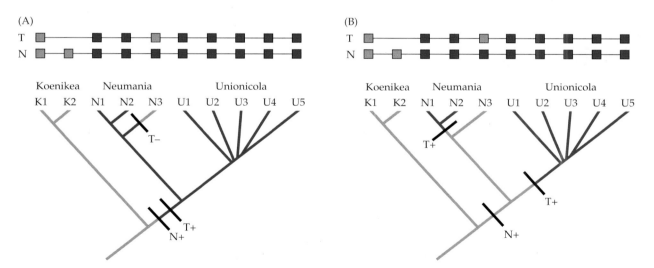

**FIGURE 9.14  Two evolutionary scenarios for the evolution of male courtship trembling in water mites** of the genera _Koenikea_ (two species), _Neumania_ (three species), and _Unionicola_ (five species). (A) Male courtship trembling (T) may have evolved at about the same time that predatory females adopted the net stance (N) in the ancestor of _Neumania_ and _Unionicola_ (N+, T+), with trembling being lost in the line leading to one species of _Neumania_ (T–). (B) Another equally parsimonious scenario calls for the evolution of the net stance in the ancestor of the two genera _Neumania_ and _Unionicola_, but instead of one origin for trembling, there might have been two origins, one in the ancestor of _Unionicola_ and another in the ancestor of _Neumania papillator_ (N1) and its close relative (N2). After Proctor.[1170]

**FIGURE 9.15**  A female cichlid fish (left) is attracted to the anal fin of a male by the orange spots on the fin.

style and had evolved sensitivity to underwater vibrations associated with those prey.[1170] Courting males that first happened to mimic those vibrations improved their chances of mating by taking advantage of a sensory mechanism that females had evolved for another reason.

### Discussion Questions

**9.6**  Females of an African cichlid fish lay their eggs on the lake bottom in depressions made by males.[1562] The female picks up her large orange eggs in her mouth almost as quickly as she lays them. (She will protect the fertilized eggs and the fry, when they hatch, by holding them in her mouth.) As the female lays her eggs, the male who made the "nest" may move in front of her and spread his anal fin (Figure 9.15), which is decorated with a line of large orange spots. The female moves toward him and attempts to pick up the objects on the fin. As she does, the male releases his sperm, which are taken up in the female's mouth, where they fertilize her eggs. Use sensory exploitation theory to explain the evolutionary origins of the male's behavior. Was the first male to use this signal taking advantage of his mate in the sense of reducing her fitness to benefit himself?

**9.7**  The body of the giant wood spider is extremely colorful (Figure 9.16). When the bright patches are painted black, the rate at which moths and other nocturnal prey fly into the spiders' webs declines sharply, especially at night.[262, 263] What relevance does this research have for persons interested in the sensory exploitation hypothesis for the evolution of courtship signals in animals?

The argument that has been developed for water mite courtship has also been applied to the signals and responses of male and female guppies. Females in some populations of this small fish prefer to mate with males that have bright orange spots on their skin (Figure 9.17).[578] Male guppies cannot synthesize the orange pigments that go into their body coloration but instead have to acquire these carotenoids from the plants they eat. Those that get enough carotenoids from their food become attractive to females, but why?

One hypothesis on the origin of the female preference for orange-spotted males suggests that when this mating preference first appeared, it was a byproduct of a sensory preference that had evolved in another context. This hypothesis has received support from observations of female guppies feeding avidly on rare but nutritionally valuable orange fruits that occasionally fall into the Trinidadian streams where guppies live.[1232] Thus, it is possible that females evolved visual sensitivity to orange stimuli because of the foraging

(A)

(B)

(C)

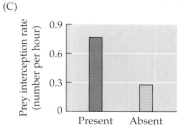

**FIGURE 9.16 The bright spots and stripes** in the color pattern of the spider *Nephila pilipes* appear to attract prey to the predator at night. (A) The upperside of the spider. (B) The underside of the spider. (C) The number of prey captured in the web of another brightly colored nocturnal spider when the spider was present (blue bar) and when the spider was absent (orange bar). A, photograph by Chih-Yuan Chang; B, photograph by Jin-Nan Huang; both photographs courtesy of I. Min-Tso; C, after Chuang, Yang, and Tso.[262]

benefits associated with this ability, not because of any fitness benefits from selective mate choice. If so, then female guppies are predicted to respond more strongly to orange stimuli than to other colors when feeding. In fact, female guppies do approach and try to bite orange discs far more often than discs of any other color. The research team that studied this phenomenon was able to take advantage of the fact that female guppies living in different streams differ in how strongly they prefer mates with orange spots. The strength of the mate choice preference was matched by the relative rates at which females pecked at orange discs presented to them underwater (Figure 9.18).

If sensory exploitation is a major factor in the origin of effective signals, then it should be possible to create novel experimental signals that generate responses from animals that have never encountered those stimuli before. To test this prediction, researchers have played sounds to frogs that contain

(A)

(B)

**FIGURE 9.17 Food, carotenoids, and female mate preferences in the guppy.** (A) Males have to acquire orange pigments from the foods they eat, like this *Clusia* fruit that has fallen into a stream where guppies live. Males that secure sufficient amounts of carotenoids incorporate the chemicals into ornamental color patches on their bodies. (B) Females (like the larger fish on the right) find males with large orange patches more sexually appealing than males without them. Photographs by Greg Grether.

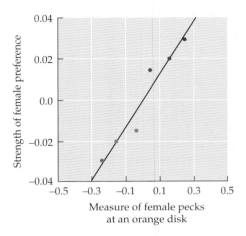

**FIGURE 9.18 Sexual preferences for orange spots** match foraging preferences for orange foods by female guppies. The strength of the female response to orange spots on males varies from population to population and is proportional to the strength of the foraging response to an orange disc. Each point represents a different guppy population living in a different stream. After Rodd et al.[1232]

acoustical elements not present in the species' natural calls,[1262, 1263] attached strips of yellow plastic to the tails of male platyfish,[79] added feather crests to the heads of least auklets,[737] and supplied male sticklebacks with bright red spangles to add to their nests.[1080] In all of these cases, researchers found that the artificial attributes elicited stronger reactions from females than the natural alternatives (Figure 9.19).

Although these data provide support for the sensory exploitation hypothesis, there is an alternative explanation for the response of the test species to these artificial courtship signals: perhaps the test species evolved from populations that used similar signals in the past. If so, then today's descendants of these ancestral populations might still retain the old sensory preferences even though they no longer exhibit the complex signals themselves.[1261] This conjecture is plausible, given the numerous cases in which elaborate male traits used in courtship and aggression have been lost after having once evolved, as indicated by the fact that in these cases all the close relatives of the ornament-less species possess the ornaments in question (which suggests that their mutual common ancestor did as well).[1568] An example is provided by the lizard *Sceloporus virgatus*, which lacks the large blue abdominal patches of many other members of its genus. These other species use their blue patches in the male threat posture, in which a lizard elevates and compresses its body laterally, making the abdomen visible. But although the signal has apparently been lost in *S. virgatus*, the behavioral response to the signal has not, as shown when some experimental subjects had blue patches painted on them. Lizards that observed a rival giving a threat display enhanced by the blue paint were far more likely to back off than were lizards that saw a displaying opponent that did not have blue patches added to its body (Figure 9.20).[1191]

(A)

(B)

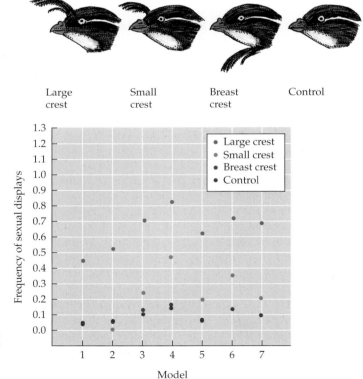

**FIGURE 9.19 The response of least auklets to three novel artificial signals.** (A) A stuffed male least auklet with an artificial crest of the sort used in an experiment on female sexual preferences in this species. (B) A diagram of some of the heads of models used in the experiment (from right to left): the control (which lacks a crest as do males of the least auklet), a breast crest model, a small crest model, and a large crest model. (C) The models with large crests elicited the highest frequency of sexual displays by the females during a standard presentation period. A, photograph by Ian Jones; B–C, after Jones and Hunter.[737]

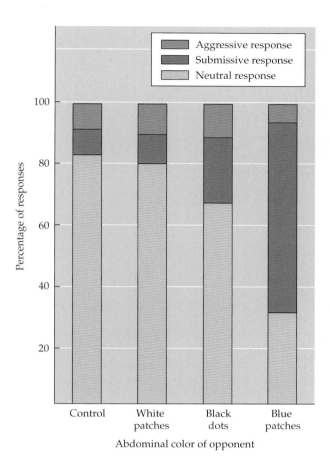

*Sceloporus virgatus*

**FIGURE 9.20    Receivers can respond to an ancestral signal not present in their species.** Lizards of a species whose relatives have blue patches on the abdomen are more likely to abandon a conflict when confronted by a conspecific rival with blue patches painted on its abdomen than when seeing the threat display of an unmanipulated control individual or one that has had white patches or black dots painted on the abdomen. After Quinn and Hews;[1191] photograph by Paul Hamilton.

But might some signal receivers have a preference for signals that neither they nor their ancestors ever possessed or responded to? One way to test this hypothesis is to provide subjects with a novel trait not likely to have been present in their ancestral species. This is why Nancy Burley and Richard Symanski outfitted zebra finches and long-tailed finches with rather ridiculous-looking white feathers glued to their heads. These Australian finches do not have crests, nor do any of their close relatives, so presumably, neither did their ancestors. Yet females of both species associated more with the novel white-crested males than with normal-looking males (Figure 9.21).[214]

The current preference for a colorful, elongate caudal fin (a "swordtail") in some species of *Xiphophorus* fishes could have originated simply because females happened by chance to be attracted to the stimuli offered by a somewhat longer, more colorful tail of a mutant male. In keeping with this hypothesis, the modern fish species *Priapella olmecae*, a close relative of *Xiphophorus*, does not have a swordtail, nor do other fish still more distantly related to *Xiphophorus*, suggesting that the ancestor of the genus almost certainly lacked a swordtail (Figure 9.22). And yet when Alexandra Basolo endowed males of *Priapella* and a swordless species of *Xiphophorus* with an artificial yellow sword, females found males with this novel trait highly attractive; furthermore, the longer the sword, the greater the female's desire to stay close to the amended male. Basolo concluded that a female preference for long tails preceded the eventual evolution of swordtails in some *Xiphophorus*.[80]

Not everyone agrees with the sensory bias hypothesis for female preferences in swordtails. For example, Alex Meyer and his colleagues argue on the basis of a detailed phylogeny of *Xiphophorus* that swords have evolved and been lost several times in this genus of fish.[978] Sexual display traits appear particularly

(A)

(B)

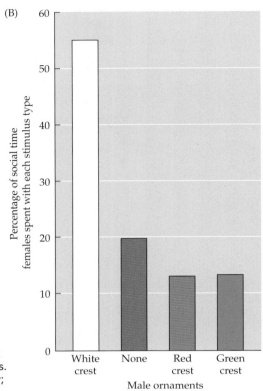

**FIGURE 9.21** **Mate preferences for a novel ornament.** (A) A male long-tailed finch (left) and a male zebra finch (right) have been outfitted with bizarre white plumes. (B) The addition of white plumes made male zebra finches more attractive to females than were control males without plumes or those given headdresses of red or green feathers. A, photograph by Kerry Clayman, courtesy of Nancy Burley; B, after Burley and Symanski.[214]

likely to change rapidly over evolutionary time, which could make it difficult to reconstruct evolutionary history via the standard comparative approach (if all surviving species in a group have lost the ancestral trait).[147] Moreover, some researchers have also asked whether what seems to be an arbitrary preference for a specific trait, such as an elaborate caudal extension, could actually be an adaptive preference for a more general trait, such as larger body size. A female preference for relatively large males could have been present in the ancestor of *Priapella* and *Xiphophorus*, and if so, a possibly adaptive and preexisting preference for healthy, active, large males could underlie what looks like a nonadaptive preference for a rather silly, purely ornamental tail.[506, 1207] (But see Basolo.[81])

Even if mutant males have appeared in some species with traits that activated hidden, preexisting aesthetic preferences in females, those males were probably "exploiting" the females only in the proximate sense of stimulating sensory equipment that had evolved for some other purpose. At the ultimate level, females that responded positively to males with the novel attribute could have gained fitness for any of several reasons. Perhaps their sons could inherit the sexually attractive trait and produce many offspring in turn, or perhaps males capable of producing exaggerated courtship signals were physiologically capable of helping to rear a choosy female's offspring. (We will have more to say about the adaptive value of female choice in the next two chapters.) If it had been reproductively harmful for females to respond positively to males with novel courtship signals, those females in the species that happened to have sensory equipment resistant to exploitation would presumably have had higher reproductive success, leading eventually to the replacement of females predisposed to be taken in by their sexual partners.

Göran Arnqvist has, however, pointed out that evolving resistance to damaging exploitative signals could be costly to females under some circumstances.[49] For example, male garter snakes "court" females by slithering on

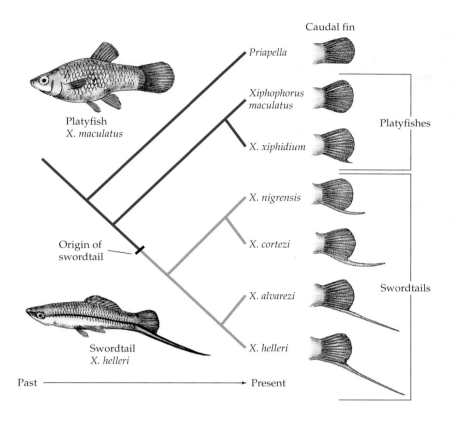

**FIGURE 9.22  Sensory exploitation and swordtail phylogeny.** The genus *Xiphophorus* includes the swordtails, which have elongated caudal fins, and the platyfishes, a group without tail ornaments. Because the closest relatives of the platyfishes and swordtails belong to a genus (*Priapella*) in which males lack long tails, the ancestor of *Xiphophorus* probably also lacked this trait. The long tail apparently originated in the evolutionary lineage that diverged from the platyfish line. Even so, females of the platyfish *X. maculatus* find males of their species with experimentally lengthened tails more attractive, suggesting that they possess a sensory bias in favor of these kinds of tails. After Basolo.[79]

top of them and then rhythmically moving their bodies in ways that make it difficult for females to breathe. In this state, females activate a response, cloacal gaping, that is almost certainly adaptive when they are grasped by predators that may be deterred by the repellent odorous chemicals and feces released from an open cloaca. Male garter snakes appear to take advantage of cloacal gaping, which makes it easier for them to insert their intromittent organ into the cloaca, a necessary prelude to copulation.[1321] And female garter snakes that happened not to respond to male precopulatory pressure might also fail to react to predators when captured, thereby reducing their chances for survival and future reproduction. If so, selection might favor retention of an otherwise valuable adaptive response even though males have evolved the ability to use female behavior strictly for their own reproductive ends.

The sensory exploitation hypothesis represents a special case of the historical argument that what has already evolved influences what kinds of additional changes are possible and which are not. If natural selection had been given the job of coming up with a jet plane, it would have had to start with what was already available, a propeller-driven plane, changing this "ancestral" form piece by piece, with each modification flying better than the preceding one, until a jet was built.[367] In nature, evolutionary transitions must occur this way because selection cannot start from scratch. As a result, the products of evolution often appear to be jury-rigged devices that might have been designed by Rube Goldberg, a cartoonist famous for his ridiculously overcomplicated inventions designed for everyday tasks. (Goldberg died in 1970 but his wacky cartoons can still be seen at http://www.rubegoldberg.com/). So, for example, the panda's thumb is not a "real" finger at all, but instead a highly modified wrist bone (the panda's hand has the same five digits that most vertebrates have plus a small thumblike projection from the radial sesamoid bone, which is present, although as a much smaller bone, in

**FIGURE 9.23** The panda principle is evident in the sexual behavior of a parthenogenetic whiptail lizard. In the left column, a male of a sexual species engages in courtship and copulatory behavior with a female. In the right column, two females of a closely related parthenogenetic species engage in very similar behavior. After Crews and Moore.[317]

the feet of bears, dogs, raccoons, and the like).[559] Why do pandas have their own special thumb? According to Stephen Jay Gould, pandas evolved from carnivorous ancestors whose first digit had become an integral part of a foot used in running. As a result, when pandas evolved into herbivorous bamboo eaters, the first digit was not available to be employed as a thumb in stripping leaves from bamboo shoots. Instead, selection acted on variation in the panda's radial sesamoid bone, which now can be used in a thumblike manner by the bamboo-eating panda.

The "panda principle," what Darwin spoke of as the principle of imperfection, can be seen in dozens of other cases. Consider the persistence of sexual behavior in parthenogenetic whiptail lizards (something that also occurs in certain fish species and salamanders). In some whiptails, the species is composed entirely of females, and yet if a female is courted and mounted by another female (and females do engage in such pseudomale sexual behavior, for reasons that are not fully understood), she is much more likely to produce a clutch of eggs than if she does not receive this sexual stimulation from a partner (Figure 9.23).[317] The relationship between courtship and female fecundity in unisexual lizards obviously exists because these reptiles had sexual ances-tors. The parthenogenetic females retain characteristics, such as a need for courtship, that their non-parthenogenetic ancestors possessed—characteristics that a biological engineer would surely eliminate if he or she could play God by designing an all-female species on paper and then creating it in one go. Natural selection, however, cannot play God, because it is a blind process with no goal in mind and no means to get to a predetermined endpoint.

## Discussion Question

**9.8** Although natural selection is blind, the products of this process are often amazingly complex. In order to explain how a blind process depen-dent on random events (mutations) can generate such complexity, Richard Dawkins has provided us with an analogy.[368] He invites us to imagine that a complex current trait is like an English sentence—for example, a line from Shakespeare's Hamlet: METHINKS IT IS LIKE A WEASEL. The odds that a monkey would produce this line by tapping at a typewriter are vanishingly small, one in 10,000 million million million million million million. These are not good odds. But instead of trying to get a monkey or a computer to get the "right" sentence in one go, let's change the rules so that we start with a randomly generated letter set, such as SWAJS MEIURNZMMVASJDNA YPQZK. Now we get a computer to copy this "sentence" over and over, but with a small error rate. From time to time, we ask the computer to scan the list and pick the sequence that is closest to METHINKS IT IS LIKE A WEASEL. Whatever "sentence" is closest is used for the next generation of copying, again with a few errors thrown in. The sentence in this group that is most similar to METHINKS ... is selected to be copied, and so on. Dawk-ins performed the experiment and found that it took 40 to 70 generations to reach the target sentence—a few seconds of computer time, not years. What was Dawkins's main point in illustrating what he called "cumulative selection"? In what sense might you say that the sentence analogy more closely resembles artificial selection than natural selection?

## Adaptationist Questions about Communication

Sensory exploitation theory has directed our attention to the importance of past changes in shaping the evolution of a new communication signal. Infor-mation on the origins of a trait does not, however, make it unnecessary to consider why a signal, once having appeared, was maintained in a species over time. For example, knowing that female guppies probably were sensi-tive to the color orange because of its association with a favored food does not eliminate the possibility that using this preference in a sexual context raised the reproductive success of the sensory-biased females. If female mate choice based on orange food-mimicking spots on male bodies increased female fit-ness, then this adaptive effect would have contributed to the selective mainte-nance of the color bias and its use in sexual encounters as well as when feed-ing. The fact that male guppies appear to boost their immune systems when they consume carotenoids suggests that by advertising their carotenoid intake via orange spots, they may be signaling their healthiness, making it advanta-geous for females that used this cue when choosing mates.[579]

More generally, an interest in the possible adaptive value of various ele-ments of communication systems has resulted in the solution of a class of puzzles that differ from those having to do with the origin and subsequent modification of signals. Darwinian puzzles in communication come to the surface when observers see signal givers or signal receivers apparently dis-

**FIGURE 9.24   A group of ravens feeding on a carcass** to which they were attracted by a yelling companion. Photograph by Bernd Heinrich.

advantaged by their actions. If giving a signal, or paying attention to a signal, lowered the fitness of the signalers or receivers relative to others, then their behaviors should eventually disappear.

Because he had this selectionist logic in mind, Bernd Heinrich knew he had something worth studying when he heard a mob of ravens yelling loudly while feeding together on a dead moose that a poacher had left in a Maine forest (Figure 9.24).[645] Ravens are uncommon birds in Maine, and yet somehow 15 of them had assembled at the hidden carcass, almost certainly because some of the first birds to find the moose had yelled loudly to the others. The behavior did not make sense to Heinrich. Why attract competitors to a food bonanza? Why not stay silent and eat moose meat all winter long by yourself instead of sharing the bounty with other ravenous ravens? In other words, why hadn't selection eliminated any hereditary tendency of ravens to yell at carcasses?

The puzzle might have been solved if the aggregated birds were one big family, with parents yelling to attract distant offspring to the feast. Heinrich felt that this was unlikely, since a pair of ravens produces a maximum of six offspring per breeding season. Subsequently, DNA fingerprinting studies showed unequivocally that these feeding flocks were indeed composed of unrelated individuals.[646]

All right, but perhaps ravens yell because the signals get the attention of a bear or coyote, which can open up a tough-skinned moose, eventually providing the caller with access to meat that it could not get in any other way. To test this hypothesis, Heinrich hauled a dead 150-pound goat through the Maine woods, taking it out to various sites during the day and storing the none-too-sweet-smelling carcass in his cabin at night to prevent its loss to a nocturnal scavenging bear. Ravens occasionally approached the decaying goat, though only after making Heinrich wait for hours in cramped and bitterly cold blinds. But contrary to his prediction, the birds never yelled when they found the bait. Moreover, Heinrich sometimes observed ravens yelling at carcasses that had already been ripped open. These findings forced him to abandon the "attract a carcass opener" hypothesis.

Instead, he switched his attention to an alternative explanation, an idea stimulated by seeing how cautiously some ravens approached carcasses when they first found them. Perhaps, he argued, a carcass discoverer yells to draw in other ravens so that if a predator lurks nearby, the other birds will provide possible targets, reducing the yeller's risk of being taken by a hiding coyote or fox. The incoming birds would be attracted because they would gain a wealth of food in return for taking a small risk of being the unlucky victim of a predator. However, this hypothesis generates the prediction that once a group has assembled and feeding has begun, the birds should shut up to avoid attracting more ravens, which would be unwanted and unnecessary for safety purposes. The observation that yelling continued at baits that had already acquired a retinue of actively feeding birds convinced Heinrich to discard the "diluted risk of predation" hypothesis.

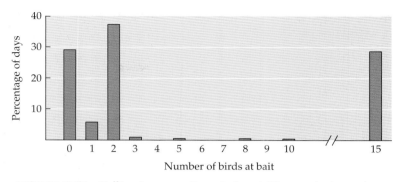

**FIGURE 9.25   Yelling is a recruitment signal.** The graph shows the percentage of days on which carcass baits were visited by various numbers of ravens. Carcasses were exploited either by nonyelling territorial singletons and pairs or by large groups of ravens, many of them yelling, most of them young, nonterritorial birds. After Heinrich.[645]

As Heinrich continued stoically lugging dead goats into the Maine woods, he came to realize that whenever he saw a single bird or a pair at a feeding site, those ravens were quiet. Yelling occurred only when three or more ravens were present, and it was then, and only then, that large numbers of other ravens came to the area. Heinrich knew that older adult ravens form pairs that defend a territory year-round. Unmated young birds usually travel solo over great distances in search of food. If a singleton attempts to feed in a resident pair's territory, the pair attacks it. Heinrich wondered if yelling could be a signal given by nonterritorial intruders, a signal that attracts other unmated wanderers to a food bonanza that they can exploit if they can overwhelm the defenses of the resident pair.

This "gang up on the territorial residents" hypothesis leads to a number of predictions: (1) resident territory owners should never yell, (2) nonresident ravens should yell, (3) yelling should facilitate a mass assault on a carcass by nonresident ravens, (4) resident pairs should be unable to repel a communal assault on their resources, and (5) a food bonanza should be eaten either by a resident pair alone or by a mob of ravens. Heinrich collected data that supported all of these predictions (Figure 9.25).[645] He concluded that when a young raven yells, the consequences of providing information ("Food bonanza here") to other nonterritorial birds can include benefits (personal access to food for the yeller) that outweigh the energetic costs of yelling, as well as the risk of attack by the resident ravens guarding the carcass.[646] Yelling therefore has the properties of an adaptation, because it can yield net fitness benefits to individuals who give the signal under the appropriate circumstances.

### Why Do Baby Birds Beg So Loudly?

Heinrich knew how to use the adaptationist approach to clear up a mystery, and other biologists have tried to do the same with other signals that at first glance seem to impose high costs on signalers. For example, you might think it would be suicidal for nestling songbirds to break into noisy begging calls whenever a parent returns to the nest with food. Their loud cheeps and peeps could give a listening hawk or raccoon all the information it needs in order to locate the nest and make off with the baby birds. In fact, when tapes of begging tree swallows were played at an artificial swallow nest containing a quail's egg, the egg in that "noisy" nest was taken or destroyed by predators before the egg in a nearby quiet control nest in 29 of 37 trials.[849]

Further evidence for the fitness costs of begging comes from a study of differences in the begging calls of warbler species that nest on the ground versus those that nest in the relative safety of trees.[629] The young of ground-nesting warblers produce begging cheeps of higher frequencies than their tree-nest-

(A)

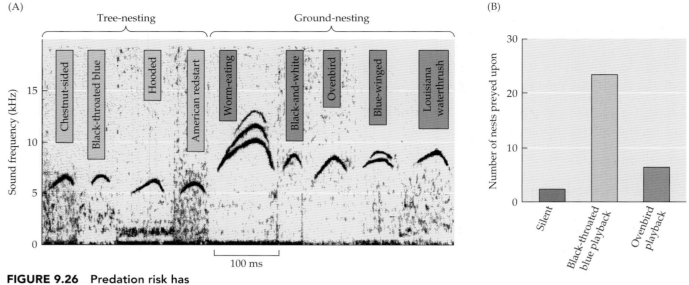

(B)

**FIGURE 9.26 Predation risk has affected the evolution of begging calls in warblers.** (A) The nestling begging calls of ground-nesting warbler species are higher in sound frequencies than the calls of young tree-nesting warblers. (B) Experimental playback of the begging calls of the black-throated blue warbler (a tree-nesting species) at artificial nests placed on the ground resulted in higher rates of discovery by predators than did playback of begging calls of the ground-nesting ovenbird. After Haskell.[629]

ing relatives (Figure 9.26A). These higher-frequency sounds do not travel as far and so may better conceal the individuals that produce them, which are especially vulnerable to predators in their low-lying nests. David Haskell created artificial nests with clay eggs and placed them on the ground by a tape recorder that played the begging calls of either tree-nesting or ground-nesting warblers. The eggs advertised by the tree nesters' begging calls were found and bitten significantly more often than the eggs associated with the ground nesters' calls (Figure 9.26B).

The warbler study suggests that begging calls have evolved properties that reduce their potential for attracting predators. If so, then perhaps baby birds of species that experience high rates of nest predation should also produce softer begging signals of higher frequencies than nestlings of other species less often victimized by nest predators. This prediction was supported by data collected in one survey of 24 species from an Arizona forest,[174] more evidence that predation pressure favors the evolution of begging calls that are hard to detect and pinpoint.

Finally, some parent birds give alarm calls when danger threatens the nest, and in at least one species, these calls can induce the baby birds to shut up, which provides another avenue for the reduction of the costs of the nestlings' vocalizations.[1155] But all the adjustments we have reviewed merely reduce, rather than eliminate, the risk that a predator will destroy a gaggle of baby birds whose calls seem to say "come and eat me." So why make noise at all if the consequence can be lethal?

One prominent hypothesis focuses on competition among sibling nestlings. Perhaps each young bird is using its vocalizations and other begging behaviors, like wing waving, to get its parents to deliver the amount of food needed to maximize the begging bird's fitness. Another explanation, the "honest signaling" hypothesis, looks at nestlings as producers of **honest signals** (messages that convey accurate information, in this case about the need for food) that enable their parents to deliver food as effectively as possible (that is, in ways that will maximize parental fitness). According to this idea, any attempt by nestlings to be deceptive or to otherwise disadvantage their parents should generally select for adults that ignore the manipulative "information" the offspring provide. Unfortunately, the two hypotheses yield some of the same predictions, including the expectation that parents should provide food in relation to the intensity of begging by their offspring.[1245] So, for example, the

(A)

(B)

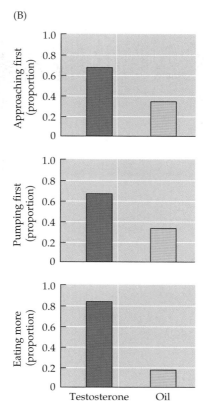

**FIGURE 9.27** **Testosterone affects begging rate and feeding rate in black-headed gull chicks.** (A) Adult black-headed gull with a chick. (B) Researchers took pairs of first-laid eggs produced on the same day and gave one egg a supplemental injection of testosterone while the other (the control) received an injection of oil. The two eggs were then given to a pair of nesting adults whose own clutch had been removed. When the chicks hatched, those that had received extra testosterone were more likely to approach the adults first, were more likely to "pump" first ("pumping" is an up-and-down motion of the head that is a part of the begging display), and wound up eating more than their oil-treated foster siblings. A, photograph by Corine Eising; B, after Eising and Groothuis.[432]

finding that larger nestling blue tits push their way closest to the male parent when he brings food to the nest[398] is compatible with both the competition hypothesis (with large nestlings depriving smaller ones of food delivered by their father) and the honest signal hypothesis if hungry large individuals are signaling to their father that they have a real need for food that, if satisfied, is likely to result in a healthy, well-fed offspring of high potential fitness.

### Discussion Question

**9.9** In the black-headed gull, the female lays three eggs but begins incubating before all of the eggs are laid. As a result, the first egg laid gets a head start and produces a senior chick that typically is larger than its siblings. It therefore begs more effectively and usually gets more food than its brothers and sisters. But mother gulls put more androgens into eggs that will hatch later, and the extra androgens enable junior gull chicks to beg more vigorously and claim more food than they would otherwise (Figure 9.27).[432] How does this example illustrate the difficulties in establishing who has the upper hand in the signaling interactions between young birds and their parents?

Little doubt exists that in some species nestling birds are capable of intense resource competition, so much so that siblings are prepared to kill one another in fights for food (see Chapter 12).[550] If nestling begging behavior is related to food competition among siblings, then it follows that youngsters should adjust their signals in relation to those produced by their nestmates. In keeping with this prediction, when experimentally food-deprived, loudly begging baby robins were placed in a nest with normally fed siblings, their better-fed brothers and sisters increased the intensity of their begging behavior too.[1350] Likewise, when young song sparrows wind up competing with a parasitic

(A)

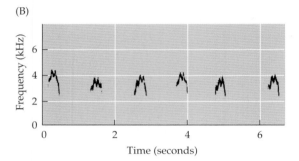

(B)

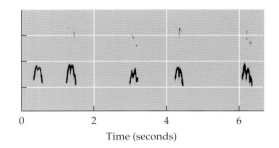

**FIGURE 9.28  An honest signal of hunger?** (A) A Wilson's storm-petrel chick. Although these chicks do not have nestmates with whom to compete, they do possess a special begging call, produced when a parent returns to the nest burrow with food. (B) Shown here are two records with six and five "long" calls, respectively. The more long calls given, the needier the chick. A, photograph courtesy of Petra Quillfeldt; B, after Quillfeldt.[1188]

cowbird nestling that has been slipped into their nest, they increase the intensity of their calls but reduce the frequency with which they give them when a parent shows up with food. This competition-induced change in baby sparrow signaling appears to work; they get fed as much as young sparrows of the same age that are fortunate enough to be reared in a nest without a hungry cowbird as a nestmate.[1097]

In some cases, we can eliminate competition as a factor influencing begging tactics by studying those bird species, like the Wilson's storm-petrel, in which parents leave their one and only chick in a nest burrow and provision it with regurgitated food at 2-day intervals. When Petra Quillfeldt examined begging by these chicks, she found that they produce a specific call that is used only when a parent comes to the burrow to deliver food (Figure 9.28). In order to evaluate the relationship between chick need and begging behavior, Quillfeldt kept track of the age and body weight of her subjects. These data enabled her to estimate the body condition of the young petrels. Some were in poor condition (well below the average weight for chicks of their age), some were in average condition (at about the mean weight for their age group), and some were in good condition (heavier than average). Chicks in poor condition were presumed to be hungrier than average, and these individuals produced begging calls at a faster rate, and gave more calls in all, than petrel chicks that were in better condition.[1188]

Thus, in at least some cases, begging appears to provide an honest signal of need (or perhaps chick viability) that parents could use to decide how much to invest in feeding that chick. The fact that parents can control who gets what no matter what their nestlings do is illustrated by the behavior of adult bluethroats (see Figure 4.34). These songbirds preferentially feed the largest (first-hatched) nestlings even though the smaller, later-hatched junior nestlings beg just as intensely and just as frequently when food deprived as their larger siblings. In this species at least, body size apparently trumps nestling begging behavior when it comes to parental decisions about who gets fed.[1347]

But the question arises, in those species in which parents do pay attention to nestling begging signals, why doesn't a chick beg especially vigorously even

(A)

**FIGURE 9.29**  **The European cuckoo chick's begging call matches that of four baby reed warblers.** (A) A cuckoo chick begging for food from its reed warbler foster parent. The calls shown below the photograph are those of (B) a single reed warbler chick, (C) a brood of four reed warblers, and (D) a single cuckoo chick. A, photograph by Roger Wilmshurst; B–D, after Davies, Kilner, and Noble.[360]

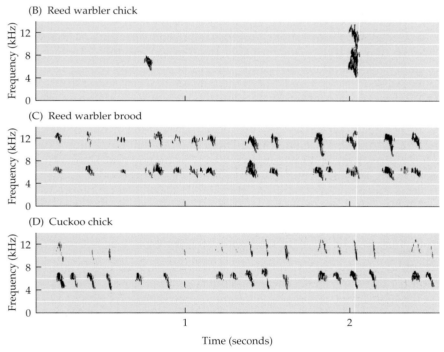

(B) Reed warbler chick

(C) Reed warbler brood

(D) Cuckoo chick

Time (seconds)

when it isn't all that hungry? In so doing, it might be able to secure food at a higher rate, which should result in more rapid growth or a larger size, either one of which could be advantageous. The young of brood parasites manage this trick and benefit greatly. A young European cuckoo, for example, is able to sound like an entire nest full of baby reed warblers (Figure 9.29), with the result that its host parents work as hard for it as they do for a complete brood of four warbler chicks.[770] If you place another nestling in a reed warbler's nest, choosing a species comparable in size to a cuckoo—say, a baby blackbird— the young blackbird begs less intensely than a cuckoo and, accordingly, is fed less rapidly. But play a recording of a cuckoo's begging calls when the warblers come to the nest, and the lucky blackbird receives a roughly 50-percent increase in the number of loads of food carried to it per hour by its super-stimulated host parents (Figure 9.30).[361]

(A)

(B)

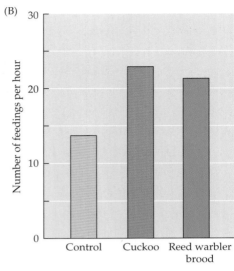

Playback

**FIGURE 9.30 The cuckoo's begging calls stimulate more frequent feeding by its host parents.** (A) A blackbird chick has been placed in a reed warbler nest. A speaker is visible in the background. (B) If the begging calls of a cuckoo chick are played to the reed warbler host parents when they visit the nest, the parents deliver more food to the blackbird, matching the amount they bring when a recording of an entire brood of reed warblers is played. A, photograph courtesy of Nick Davies; B, after Davies, Kilner, and Noble.[360]

Although vigorous begging must come with an energetic cost, measurements of the metabolic expenses incurred by begging nestlings demonstrate that this cost is probably small relative to the potential caloric gain.[56, 855] Therefore, the disadvantages of exaggerated begging may lie not in energy expended, but in the damage that any successful food hog might do to its siblings in species in which several offspring are cared for by the same two parents. An individual's success in propagating his or her genes can be affected by more than just his or her own personal reproductive success (see page 471). Those animals that harm their close relatives may, in effect, be destroying some of their own genes. Therefore, a begging nestling that happened to do very well at securing food from its parents but at the expense of its siblings might actually leave fewer copies of its genes overall than others that held back when begging for food.

## Discussion Questions

**9.10**  The parasitic young European cuckoo typically disposes of all competitors by pushing its host eggs and babies out of the nest. The cowbird, another parasitic bird, behaves differently, often coexisting with one or more of its hosts' offspring. Why is the cowbird's behavior puzzling? How might you account for this parasite's tolerance of nestmates in light of the analysis of begging behavior presented above? See Kilner, Madden, and Hauber[771] after completing your answer.

**9.11**  Birds are not the only animals in which offspring have special signals that appear to communicate need to a parent. Develop an evolutionary analysis of crying by human infants in which you consider the two theories outlined above for begging by baby birds. Then employ the following data in evaluating your hypotheses: (1) young infants expend considerable energy when crying, (2) the growth rate for typical infants is highest over the first 3 months of life, with smaller and smaller portions of the energy budget thereafter going to growth as opposed to maintenance, (3) consumption of breast milk peaks at 3 to 4 months of age and then declines, (4) crying peaks at about 6 weeks of age and occurs progressively less often after 3 months of age, except when the child is being weaned, (5) babies who are carried everywhere and nursed on demand (as in traditional societ-

(A)

(B)

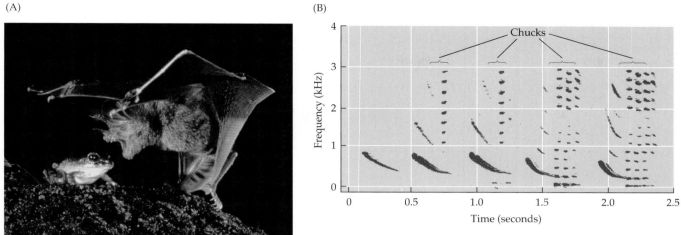

**FIGURE 9.31    Illegitimate receivers can detect the signals of their prey.** (A) A calling male túngara frog may inadvertently attract an illegitimate receiver—the fringe-lipped bat, a deadly predator. (B) The risk of attack is greater if the male's call includes one or more chucks (blue in the sonograms) as well as the introductory whine (purple in the sonograms). A, photograph by Merlin D. Tuttle, Bat Conservation International; B, sonograms courtesy of Mike Ryan.

ies) cry far less than babies in Western societies, and (6) the high-pitched cries of unhealthy babies are considered especially unpleasant by adult listeners.[508, 894, 1534]

## How to Deal with Illegitimate Receivers

The raccoon that listens in on the communication taking place between a brood of baby tree swallows and their parents is an **illegitimate receiver**, in the sense that it uses information from the signal to the fitness detriment of the legitimate signalers and receivers. Illegitimate receivers may have rapid and powerful effects on the evolution of communication systems. For example, Marlene Zuk and her colleagues have documented that almost no males of the field cricket *Teleogryllus oceanicus* now sing to attract mates on the Hawaiian island of Kauai. By remaining silent, thanks to their wing structure, these males avoid a parasitoid fly, *Ormia ochracea*, which has only recently been introduced to Kauai.[1648] Female flies locate male crickets by their calls and deposit lethal larvae on the unfortunate signalers (see page 126). The silent phenotype is therefore much less likely to be found by female flies. In less than 20 generations, silent males have come to dominate the cricket population, testimony to the power of illegitimate receivers to shape the evolution of a communication system.[1651]

An awareness of the possible evolutionary consequences of illegitimate receivers helped Mike Ryan explain why calling males of the túngara frog often give whining calls without chucks, despite the fact that mate-searching females prefer males who add the chuck element to their calls. As it turns out, calling male túngara frogs have another audience, which includes both the predatory fringe-lipped bat, which can find and kill the frogs (Figure 9.31), and certain blood-sucking flies.[111] Both the bat and the flies are attracted to the signals produced by their prey, especially those male frogs that give whine-chucks,[1259] which make localization of the callers easier.[1094] Thus, fringe-lipped bats were more than twice as likely to inspect, and even land on, a speaker broadcasting a whine-chuck than a whine alone.[1260] Therefore, we would expect the frogs to be more likely to give the two-component whine-chuck calls when the risk of predation is lower. The chance of becoming a meal for a bat declines for frogs calling in large groups because of the **dilution effect** (see page 196), and males in large assemblages are especially likely to produce whine-chuck calls.[111, 1259]

**FIGURE 9.32 Great tit alarm calls.** Sonograms of (A) the mobbing call and (B) the "seet" alarm call. Note the lower sound frequencies in the mobbing signal. A, courtesy of William Latimer; B, courtesy of Peter Marler.

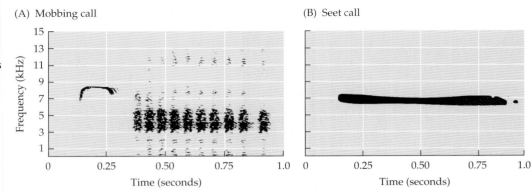

The risk of exploitation by an illegitimate receiver may also be responsible for the differences between the mobbing call and the "seet" alarm call of the great tit (Figure 9.32).[937] These small European songbirds sometimes approach a perched hawk or owl and give a loud mobbing call whose dominant frequency is about 4.5 kHz. This easily located acoustical signal helps other birds find the mobbers and join in the harassment of their mutual enemy (see page 184). If, however, a great tit spots a flying hawk, it gives a much quieter "seet" alarm call, which appears to warn mates and offspring of possible danger. This signal's dominant frequency lies within 7 to 8 kHz, so the sound attenuates (weakens) after traveling a much shorter distance than the mobbing signal. The rapid attenuation of the "seet" call compromises its effectiveness in reaching distant legitimate receivers, but it also lowers the chance that a dangerous predator on the hunt will be able to tell where the caller is. Moreover, the frequencies of the "seet" call lie outside the range that hawks can hear best, while falling within the range of peak sensitivity of the great tit (Figure 9.33). As a result, a great tit can "seet" to a family member 40 meters away but will not be heard by a sparrowhawk unless the predator is less than 10 meters distant.[781]

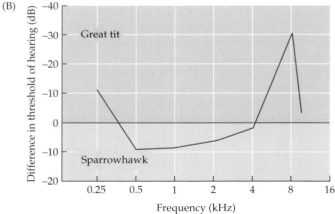

**FIGURE 9.33 Hearing abilities of a predator and its prey.** (A) A sparrowhawk made to appear as if it is attacking a great tit in this manipulated photograph. (B) The purple trace shows the difference between the softest sound of a given frequency that great tits and sparrowhawks can hear. A sparrowhawk can hear sounds in the 0.5 to 4 kHz range that are fainter (5 to 10 dB lower in intensity) than those that great tits can hear. But great tits can detect an 8 kHz sound (in the range of the "seet" call) that is fully 30 dB fainter than any 8 kHz sound that a sparrowhawk can detect. After Klump, Kretzschmar, and Curio.[781]

**FIGURE 9.34   Convergent evolution in a signal.** The great tit's high-pitched "seet" alarm call (see Figure 9.32B) is very similar to the alarm calls given by other, unrelated songbirds when they spot an approaching hawk. After Marler.[937]

Reed bunting

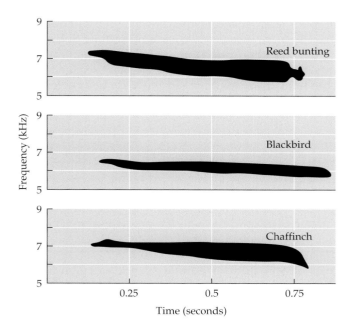

Blackbird

Chaffinch

If the "seet" call of the great tit has evolved properties that reduce the risk of detection by its enemies, then unrelated species should also have evolved alarm signals with similar properties. The remarkable convergence in the "seet" calls of many unrelated European songbirds suggests that predation pressure by bird-eating hawks has favored the evolution of alarm calls that are hard for hawks to hear (Figure 9.34).[937]

## Discussion Question

**9.12** Males of a fish called the northern swordtail attract mates with flashy displays that show off their elaborate tails. But this species has to share its habitat with a lethal predatory fish, the Mexican tetra. The bodies of male northern swordtails, especially their tails, reflect a considerable amount of ultraviolet radiation.[327] When researchers put male and female swordtails in tanks with and without ultraviolet filters, they found that female swordtails were more attracted to males when the ultraviolet channel was available to them. Given that these researchers were interested in how swordtails might reduce the risk of eavesdropping by Mexican tetras, what other experiment must they have done as well? Outline the science behind this research, starting with the causal question and ending with the possible conclusions the researchers might have reached.

## Why Settle Disputes with Harmless Threats?

We have been using an adaptationist approach to examine cases in which signalers appear, at first glance, to be lowering their genetic success. Let's now examine some instances in which receivers of signals seem to lose fitness by reacting to certain messages from others. First, why do so many animals resolve their disputes with highly ritualized threat displays? Rather than bludgeoning a rival, males of many bird species settle their conflicts over a territory or a mate with much singing and feather fluffing, but without ever touching one another (see Chapter 2). Even when genuine fighting does occur in the animal kingdom, the "fighters" often appear to be auditioning for a comic opera. After a body slam or two, a subordinate elephant seal generally lumbers off as fast as it can lumber, inchworming its blubbery body across the beach to the water, while the victor bellows in noisy, but generally harmless, pursuit.

### Discussion Question

**9.13** One sometimes hears that the reason why so many species resolve their contests via mostly harmless threat signals is to reduce the number of injuries and thereby protect the breeding adults who are needed to produce the next generation of offspring. What's the problem with this hypothesis?

Modern adaptationists have hypothesized that even losers gain fitness when contests are settled quickly without serious fighting.[352] Consider the European toad, *Bufo bufo*, whose males compete for receptive females. When a male finds another male mounted on a female, he may try to pull him from her back. The mounted male croaks as soon as he is touched, and often the other male immediately concedes defeat and goes away, leaving his noisy rival to fertilize the female's eggs. How can it be adaptive for the signal receiver in this case to give up a chance to leave descendants simply on the basis of hearing a croak?

Because European toad males come in different sizes, and because body size influences the pitch of the croak produced by a male, Nick Davies and Tim Halliday proposed that males can judge the size of a rival by his acoustical signals. If a small male can tell, just by listening, that he is up against a larger opponent, then the small male ought to give up without getting involved in a fight he cannot win. If this hypothesis is correct, then deep-pitched croaks (made by larger males) should deter attackers more effectively than higher-pitched ones (made by smaller males).[352] To test this prediction, the two researchers placed mating pairs of toads in tanks with a single male for 30 minutes. The mounted male, which might be large or small, had been silenced by looping a rubber band under his arms and through his mouth. Whenever the single male touched the pair, a tape recorder supplied a 5-second call of either low or high pitch. Small paired males were much less frequently attacked if the interfering male heard a deep-pitched croak (Figure 9.35). Thus, deep croaks do deter rivals to some extent, although tactile cues also play a role in determining the frequency and persistence of an attack.

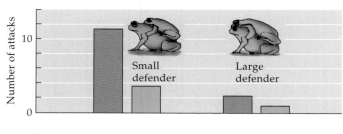
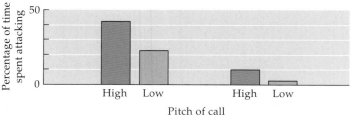

**FIGURE 9.35 Deep croaks deter rivals.** When a recording of a low-pitched croak was played to them, male European toads made fewer contacts with, and interacted less with, silenced mating rivals than when a higher-frequency call was played. Tactile cues also play a role, however, as one can see from the higher overall attack rate on smaller toads. After Davies and Halliday.[352]

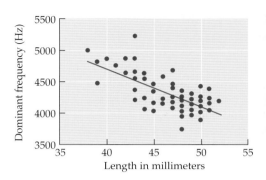

Barking gecko

**FIGURE 9.36 Vocal signals of the barking gecko** convey information about the size of the calling lizard to distant rivals. The longer the male, the lower the frequency of its calls. Photograph by Tony Hibbits; after Hibbitts, Whiting, and Stuart-Fox.[652]

So why don't small males pretend to be large by giving low-pitched calls? Perhaps they would, if they could, but they can't. A small male toad apparently cannot produce a deep croak, given that body mass and the unbendable rules of physics determine the pitch of the signal that a male toad can generate. Thus, toads have evolved a warning signal that accurately announces their body size. By attending to this honest signal, a male toad can determine something about the size of his rival, and thus his probability of winning an all-out fight with him. Barking geckos (which have been given the fine scientific name *Ptenopus garrulus*) employ the same kind of honest signal system as European toads to convey information to nearby would-be territorial usurpers; the lower the dominant frequency in their "barks," the bigger the gecko (Figure 9.36).[652] The same is true for scops owls, with heavier males producing hoots of lower frequency than lightweight rivals.[620]

You may recall that male collared lizards can tell just how hard a rival can bite by looking at the size of the ultraviolet-reflecting patches by the opponent's mouth,[842] an honest visual signal. Certain insects also use the visual channel to communicate honestly with opponents. For example, males of some flies with bizarre "antlers" and eyestalks (Figure 9.37A) compete for access to

(A)

Australian antlered fly

(B)

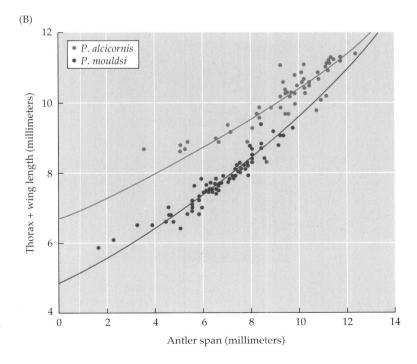

**FIGURE 9.37 Honest signals about body size?** (A) Males of this Australian antlered fly in the genus *Phytalmia* confront each other head to head, permitting each fly to assess his own size relative to the other's size. (B) Antler span in two New Guinean fly species provides accurate information about body size, enabling males to make sound decisions about an opponent's fighting ability. A, photograph by Gary Dodson; B, after Wilkinson and Dodson.[1576]

mates by standing directly in front of each other. In this position, rivals can compare the sizes of their antlers or stalked eyes, which are well correlated with body size (Figure 9.37B).[1576] Smaller males generally abandon the field of battle to larger opponents; the greater the difference in the size of their devices, the more rapidly smaller males depart, and the more energy they save.[1101]

When smaller toads or geckos or antlered flies withdraw upon hearing an honest signal from a larger individual, both the smaller and larger competitors gain: small males do not waste time and energy in a battle they are unlikely to win, and large males save time and energy that they would otherwise have to spend brushing aside annoying but persistent smaller opponents. To understand the value of a quick concession, imagine two kinds of aggressive individuals in a population, one that fought with each opponent until physically defeated or victorious, while the other checked out the rival's aggressive potential and then withdrew as quickly as possible from superior fighters. The "fight no matter what" types would eventually encounter a superior opponent who would administer a serious thrashing. The "fight only when the odds are good" types would be far less likely to suffer an injurious defeat at the hands of an overwhelmingly superior opponent.[956, 1542]

Further, imagine two kinds of superior fighters in a population, one that generated signals that other, lesser males could not produce, and another whose threat displays could be mimicked by smaller males. As mimics became more common in the population, natural selection would favor receivers that ignored the easily faked signals, reducing the value of producing them. This, in turn, would lead to the spread of the genetic basis for an honest signal that could not be devalued by deceitful signalers.

If honest signal theory is correct, we can predict that male threat displays should be relatively easy for large males to perform but more difficult for small, weak, or unhealthy males to imitate. For example, in the side-blotched lizard, the duration of a male's display and the number of pushups he performs fall markedly after he has been forced to run on a treadmill for a time (Figure 9.38). Conversely, males that first perform a series of displays before being placed on the treadmill run for a shorter time than when they have

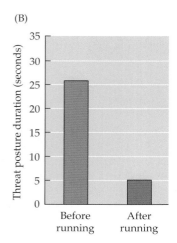
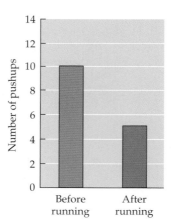

**FIGURE 9.38  Threat displays are energetically demanding** in (A) the side-blotched lizard. (B) Males that have been run on a treadmill to lower their endurance are not able to maintain their threat posture for as long as they can when they are not tired; they also generate fewer pushup displays. A, photograph by Paul Hamilton; B, after Brandt.[164]

**FIGURE 9.39   A dishonest signal of strength?** Males of the slender crayfish display their large front claws to opponents, and males with larger claws dominate those with smaller ones. However, claw size is not correlated with the strength of the claw's grip. Photograph courtesy of Robbie Wilson. From Wilson et al.[1593]

not been displaying. In other words, a male's current condition is accurately reflected by his pushup performance,[164] which enables him to advertise his fighting capacity accurately to rival males.

Treadmills have also been employed with fiddler crabs to test the prediction that in males, which have one greatly enlarged claw for use in male–male combat, a large claw comes with a price. As expected, males with intact major claws had higher metabolic rates and lower endurance than males whose major claws had been removed.[19] A male without a major claw, however, is in no position to deter rivals through visual displays alone.

### Discussion Question

**9.14**  When males of the Australian slender crayfish compete aggressively with one another (Figure 9.39), they begin by displaying their enlarged front claws. The larger the claw, the more likely the male is to dominate his rival, which may leave without grappling with the larger-clawed crayfish. However, the muscles in the claws of males generate only half the force of the claw muscles of females. In addition, the actual strength of the claw has no bearing on which male is dominant.[1593] Why do these results make us skeptical about the honest signal hypothesis for male displays in this species?

When red deer stags (known as elk in North America) compete for harems of does, the two males stand off at a considerable distance and roar loudly at each other (Figure 9.40).[278] These vocalizations are costly in terms of energy expended such that only males in top shape can roar often for many minutes[277] (see also Hack[601] and Hunt et al.[695]). Moreover, males of different sizes produce calls that differ consistently in certain acoustical properties;[255] females can and do discriminate between calls on the basis of their distinctive features. Therefore, just as in antlered flies and European toads, red deer stags could also gain honest and accurate information about the fighting ability of a rival by listening to him, and they could use this information to terminate encounters they are unlikely to win. In the same way, competing songbirds may be able to determine the condition of an opponent by evaluating the quality of the male's learned song. This would be true if a male's ability to learn his song is

**FIGURE 9.40 An honest signal.**
Only red deer (elk) stags in top condition can roar for long periods. Photograph courtesy of Tim Clutton-Brock.

impaired in individuals whose early life was marked by food deprivation or other hardships (see page 55 in Chapter 2), resulting in adults with reduced fighting ability.

## How Can Deception Evolve?

Although signal receivers benefit when they attend to the signals given by honest advertisers, deceptive signalers are not uncommon in nature. Recall the case of the orchid whose decoy lures male wasps into attempting to copulate with the flower (see Figure 3.36). Just as signal givers can lose fitness by providing information to illegitimate receivers, receivers can lose fitness by responding to signals generated by **illegitimate signalers**, which use deception to reduce the fitness of a receiver. A famous example of this phenomenon involves the firefly "femme fatales" studied by Jim Lloyd.[875] These predatory females belong to the genus *Photuris* but respond to the signals given by males of the genus *Photinus*. As a result, when traps were set out at night with light-emitting diodes programmed to mimic the mate-attracting flashes of *Photinus greeni*, they attracted and trapped vastly more *Photuris* females than identical traps that lacked the flashing diodes. Moreover, the shorter the interval between flashes, the more predators that came to the traps.[1616]

But *Photuris* females are not only illegitimate receivers; they are also illegitimate signalers. Once having approached a potential prey, the femme fatales answer the *Photinus* males' mating signals, which are designed to elicit flash signal replies from females of their own species. Some *Photuris* females can mimic the answering signals of up to three different *Photinus* species.[876] If a *Photuris* female succeeds in luring a male of a particular species of *Photinus* close enough, she will grab, kill, and eat him, at which point his reproductive potential drops to zero (Figure 9.41).

Deception of this sort poses a real puzzle for the adaptationist because the male firefly that pays attention to the wrong "come hither" signal can get himself killed. When an action is clearly disadvantageous to individuals, behavioral biologists have recourse to two main possibilities (see also Table 6.1):

1. *Novel environment theory*: The maladaptive response of the receiver is caused by a proximate mechanism that once was adaptive but is no longer. The current maladaptation occurs because modern conditions are different from those that shaped the mechanism in the past and because there has not been sufficient time for advantageous mutations to occur that would "fix the problem."

2. *Exploitation theory*: The maladaptive response of the receiver is caused by a proximate mechanism that results in fitness losses that on average reduce but do not eliminate the net fitness gain associated with reacting to a signal giver in a particular way.

Novel environments seem unlikely to account for the nature of interactions between *Photinus* males and their predatory relatives. This theory more often applies to cases in which very recent human modifications of the environment appear responsible for eliciting maladaptive behavior (see Figure 4.8). Instead, exploitation theory seems more likely to account for male *Photinus* behavior. The argument here is that, on average, the response of a male *Photinus* to certain light flashes increases his fitness, even though one of the costs of responding is the chance that he will be devoured by an exploitative *Photuris* signaler. Males that avoided these deceptive signals might live longer, but they would probably ignore females of their own species as well and would leave few or no descendants to carry on their cautious behavior.

This hypothesis highlights the definition of adaptation employed by most behavioral biologists. As noted earlier, an adaptation need not be perfect, but it must contribute more to fitness on average than other possible alternative traits. The male firefly that responds to the signals of a predatory female of another species possesses a mechanism of mate location that clearly is not perfect but is better than those alternatives that improve a male's survival chances at the cost of making him unlikely to reproduce.

If this adaptationist hypothesis is correct, then deception by an illegitimate signaler should exploit a response that has clear adaptive value under most circumstances.[261] We discussed this idea earlier when explaining why deception works for the orchids that take advantage of the generally adaptive sex drive of male wasps. Likewise, the anglerfish exploits smaller fish that possess a generally adaptive eagerness to attack visual stimuli associated with their prey. The predatory anglerfish provides the stimuli in question by waving a thin rod that projects out of the front of its head. On the end of the rod is a pale tip that looks to smaller fish like something to eat (Figure 9.42), luring them close enough so that the anglerfish can engulf them in its massive mouth.[1138] Although the deceived fish pay a heavy price for their interest in the anglerfish's lure, a complete failure to respond to these stimuli would probably doom them to starvation.

**FIGURE 9.41    A firefly femme fatale.** This female *Photuris* firefly is feeding on a male *Photinus* firefly that she lured to his death by imitating the flashes given by females of his species. Photograph by Jim Lloyd.

## Discussion Questions

**9.15**  Develop an honest signal hypothesis and a deception hypothesis to account for the fact that when copulating, males of some butterflies transfer a substance to their mates that makes these females sexually unappealing to other males.[31] What do you need to know to determine which explanation is correct?

**9.16**  The fork-tailed drongo, which perches in trees, sometimes gives alarm calls that warn of terrestrial predators when it is accompanying pied babbler flocks, which forage on the ground. When drongos give this kind of call, pied babblers often flee and sometimes leave insect prey behind that a drongo can snatch up.[1221] If we hypothesize that these alarm calls are often deceptive, what prediction can we make about the kind of alarm call produced by drongos vocalizing in the absence of babblers? Why might babblers find it adaptive to react to drongo warning calls if some, or even most, are false alarms?

**FIGURE 9.42   A deceptive signaler.** This orange angler-fish has a pale tip on the thin appendage on the front of its head; by moving the lure, the predatory fish attracts prey within striking distance. Photograph by Ann Storrie.

## Summary

1. A full understanding of the evolution of communication systems requires information about the origins of signals and the patterns of changes that took place in signalers and receivers over time, as well as information about the causal processes, especially those driven by natural selection, that made these changes occur. These two levels of evolutionary analysis complement each other.

2. The origins of signals and of responses to signals generally lie in small changes in behaviors or proximate sensory mechanisms that had different functions in the past. The gradual accumulation of many small adaptive changes over time can result in the formation of traits that differ greatly from the ancestral characteristics from which they are derived.

3. Each evolutionary change must be layered on that which has already evolved. In this sense, the traits already in place constrain or bias the pattern of evolution. One manifestation of this phenomenon is sensory exploitation, which occurs when mutant signalers happen to tap into pre-existing sensory biases of receivers that evolved because of their utility in some domain other than communication.

4. Natural selection can cause signal changes to spread through a species only if both signalers and receivers derive net fitness benefits from their participation in the system. Applying the adaptationist approach to communication systems has resulted in solutions to some puzzling cases of signalers and receivers whose behaviors seem to reduce, rather than increase, their fitness. In some instances, apparently self-sacrificing communication has been shown in reality to raise an animal's chances of reproducing. In other cases, costly behavioral traits have been shown to occur because certain individuals (illegitimate signalers or illegitimate receivers) are able to exploit an otherwise adaptive communication system for their own benefit.

## Suggested Reading

The huge field of animal communication has been reviewed by Jack Bradbury and Sandra Vehrencamp.[161] The way in which science works can be seen in the development of the spotted hyena story; start with Stephen Jay Gould's account,[558] and then read the reviews written by Stephen Glickman and colleagues.[546, 1152] Both Gerald Borgia[147] and Göran Arnqvist[49] have taken an often critical look at cases in which sensory exploitation has been invoked to explain male sexual ornaments or courtship behavior. Bernd Heinrich's *Ravens in Winter* illustrates how an adaptationist can use multiple hypotheses to solve an evolutionary puzzle in communication.[646] The same can be said for Jonathan Wells's review of why human babies cry.[1534] Marlene Zuk and Gita Kolluru survey the literature on how predators and parasitoids exploit the mate attraction signals of animals.[1650] For an adaptationist analysis of manipulation and deceit by communicating animals, read Dawkins and Krebs;[365] for the classic paper on honest signaling, see Zahavi.[1635]

# 10

# The Evolution of Reproductive Behavior

I was thrilled the first time I saw a male satin bowerbird fly down to his bower, which looked more like something a precocious child might have built than the handiwork of a bird not much larger than a robin (Figure 10.1). The arriving male held a bluish rubber band in his beak, which he dropped among the far more attractive blue parrot feathers that he had strewn about his bower. Although I did not stay to see a female satin bowerbird visit the bower and inspect the feathers and rubber bands there, Gerald Borgia tells me that when a female arrives, the male begins with a preamble of chortles and squeaks, followed by an elaborate courtship that has the male dancing across the entrance to his bower while opening and closing his wings in synchrony with a buzzing trill. This phase may be followed by another in which the male bobs up and down while imitating the songs of several species of other birds. (See for yourself at http://www.life.umd.edu/biology/borgialab.) Yet despite the apparent elaborateness of the male's displays, most courtships end with the abrupt departure of an evidently unreceptive female.[143]

**FIGURE 10.1  Bowerbird courtship revolves around the bower.** A male satin bowerbird, with a yellow flower in his beak, courts a female that has entered the bower that he has painstakingly constructed and ornamented with blue feathers. Photograph by Bert and Babs Wells.

In fact, female satin bowerbirds initially visit several bowers, scattered through the Australian forest, but not to mate with any bower builders on these inspections.[1481] After the first round, the female takes a break of about a week to construct a nest before returning to a number of bowers, during which time she usually observes the full courtship routine of several males. These inspections take several weeks before the female finally settles on one male. At this time, she enters the favored male's bower, where she is courted again before she crouches down to invite the male to copulate. After she flies off, she will have no further contact with her partner but will incubate her eggs and rear the young all by herself. Her mate stays at or near his bower for most of the 2-month breeding season, courting other females that come to inspect his creation and copulating with any that are willing.

Thus, not only do the two sexes of satin bowerbirds look different (see Figure 10.1), but their reproductive tactics are highly dissimilar. Although other animals obviously have their own special courtship routines and mating patterns, usually males do the courting and females do the choosing, whether we are talking about bowerbirds or belugas, aardvarks or zebras. This pattern is so widespread that biologists ever since Darwin have tried to provide an evolutionary explanation for it. This chapter reviews what we now know about the history and adaptive value of reproductive behavior.

## The Evolution of Differences in Sex Roles

Male satin bowerbirds build bowers; females do not. As always, we can employ two levels of evolutionary analysis in investigating this difference. First, what were the evolutionary origins of bower building by male satin bowerbirds? Second, why has bower building by males of this species been maintained by selection following its origin?

With respect to the first question, the satin bowerbird is one of 20 species in the bowerbird family, 17 of which build bowers.[494] No other bird species builds anything like these elaborate display structures, and so the trait appears to have evolved just once,[819, 820] although it is possible that avenue bower

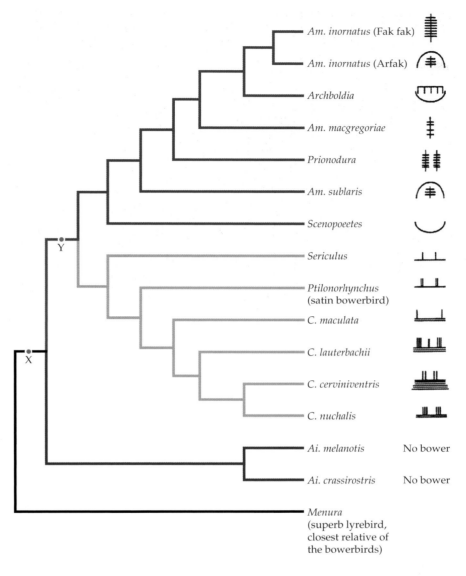

**FIGURE 10.2  Evolutionary relationships among 15 populations of bowerbirds,** based on similarities in their mitochondrial cytochrome *b* gene. The icons on the right represent the shape of each species' bower. Thus, *Scenopoeetes* builds a shallow bowl of twigs on the ground, while the satin bowerbird's avenue bower consists of a simple platform with two parallel rows of twigs, and *Chlamydera lauterbachii* constructs a more elaborate raised platform with two outer and one inner row of twigs. The really elaborate bowers are of the maypole variety, tall stacks of twigs built around a sapling. Note that two bowerbird species (*Ailuroedus melanotis* and *A. crassirostris*) do not build bowers at all, having retained the ancestral trait of extinct species X. Extinct species Y presumably built a simple avenue bower or a cleared display court; this species gave rise to all the current bower-building bowerbirds. After Kusmierski et al.[819]

building and maypole bower building evolved independently. Let's assume that the trait evolved once. If so, a single common ancestor of the modern bowerbirds gave rise to the cluster of bower-building species, endowing its descendants with the capacity for making bowers (Figure 10.2). Two early species derived from this ancestor subsequently gave rise to the two groups of modern bower-building species, the avenue bower builders (including the satin bowerbird) and the maypole bower builders (such as *Amblyornis inornatus*, whose bowers are shown in Figure 10.3). Among the two groups are descendant species whose bower-building behaviors vary greatly, with large differences even among some closely related species, especially the maypole-building bowerbirds. In fact, two geographically separated populations of the same species, *A. inornatus*, build very different maypole bowers (Figure 10.3), even though the birds in the two areas are genetically very similar.[1480] The changes that have occurred have apparently been so rapid and extensive that the bowers of many close relatives lack the kinds of shared features that would help us establish what the first bower might have looked like and what sequence of changes has taken place during the evolution of bower building.

**FIGURE 10.3** Different bowers in different populations of the same bowerbird species. (Above) The huge playhouse bower of one population of the bowerbird *Amblyornis inornatus*. (Left) The same species builds a quite different maypole bower in another location. Photographs by Will Betz and Adrian Forsyth.

Having provided the barest of sketches of the history of bowers, what can we say about the adaptive significance of bower building? In particular, what does a male satin bowerbird gain by spending so much time constructing his bower, gathering decorations (often taken from other males' bowers), and defending his display site from rival males? One possibility is that the original bower builder conveyed useful information about his quality as a mate to females and was rewarded when discriminating females copulated with him.

If this hypothesis is true, then we expect male mating success in modern satin bowerbirds to be correlated with some features of the bower that vary from male to male, such as the skillfulness with which the bower has been constructed and decorated, or perhaps the number of stolen blue feathers included among the bower decorations.[1608] In fact, even humans can detect differences among the bowers built by different males. Some contain neat rows of twigs, lined up to create a large, tidy, symmetrical bower, while others are obviously messier, less professionally assembled. Bowers also differ markedly in the number of decorative ornaments, such as feathers and rubber bands. Female bowerbirds evidently notice these differences too, because they are less likely to exhibit startle responses when visiting well-decorated bowers of high quality. The less a male satin bowerbird startles a female, the more likely she is to mate with the bower builder eventually,[1109, 1110] which may be why bower quality and the number of bower decorations correlate with male mating success in this[142] and other bowerbird species.[914, 915] In the satin bowerbird, male mating success translates directly into male genetic success, because females rarely mate with several males. Instead, a female typically uses the sperm of a single partner to fertilize her eggs.[1215]

If the mate quality advertisement hypothesis is correct, then attractive, well-decorated bowers should be built by males that are superior in some way to those birds that cannot erect a top-flight bower. Stéphanie Doucet and

Bob Montgomerie proposed that good bower builders might be healthier birds, unlikely to infect their mates with parasites or disease microbes and more likely to possess sperm with genes for disease resistance that could be passed on to their offspring. In keeping with this proposal, males that build better bowers do have fewer ectoparasitic feather mites than those that make less appealing display structures.[408] Likewise, adult males able to build and guard bowers are less infected as juveniles by an ectoparasitic louse than other males that lack bowers as adults.[146]

Another idea along these lines is that the bower's quality is in some sense an indicator of the developmental history of the male. For example, birds that had plenty of food as they matured should have well-constructed brains and should therefore be able to excel at the many demanding manipulative tasks needed to assemble a bower. (Note the parallel between this explanation for bower building and the hypothesis presented in Chapter 2 that males of other birds learn a complex song repertoire because in this way they can advertise their developmental past to choosy females.) Joah Madden recognized that if this hypothesis were correct, then the brains of bower-building bowerbirds should be proportionally larger than those of bird species that do not build bowers—a group that includes a few bowerbird species, such as the green catbird, that clear off a display court but build nothing on it. Madden went to the British Museum to X-ray the skulls of a series of stuffed bowerbirds and some of their relatives. Sure enough, bower builders are unusually brainy birds (Figure 10.4),[913] but see Healy and Rowe.[639]

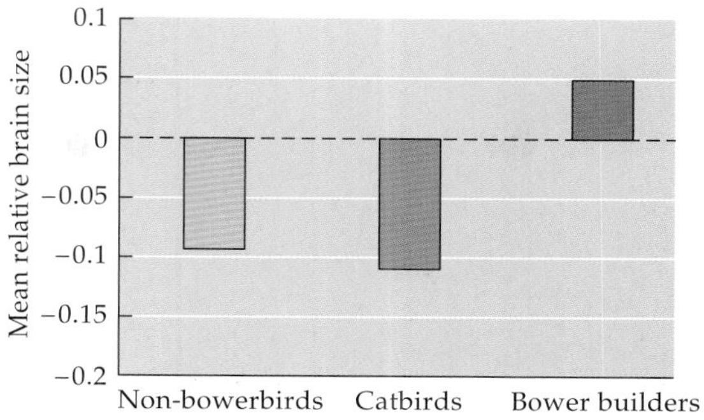

**FIGURE 10.4    Bower building may be an indicator of brain size.** Bowerbirds that build bowers have relatively large brains compared with other bowerbirds, the catbirds, that merely construct cleared display courts. The mean brain size (as determined by comparing brain cavity volumes against a measure of body size) of the bower-building bowerbirds also far exceeds that of a sample of other, unrelated bird species. After Madden.[913]

## Discussion Question

**10.1**  Female satin bowerbirds appear to favor males that give very intense courtship displays, which contain elements of aggression. Perhaps that is why females are often "jumpy" when in the presence of an intensely displaying male. Males differ in how they adjust their displays in response to female reactions. Females that often flinch as the male displays tend to leave the site without mating, whereas females that crouch down in the male's bower are more likely to stay and mate with the displaying male. Why might these observations have led Gerald Borgia and his coworkers to hypothesize that the bower originated because it enabled females to protect themselves against forced copulation attempts by displaying males? How could males gain by making it harder for themselves to force females to mate? What would you predict about a male's response to a model female bowerbird (a mechanical robot enclosed in the skin and plumage of a female) that could be controlled at a distance so as to remain standing throughout the male display, or crouch infrequently, or crouch often? See Patricelli et al.[1108]

Because most female bowerbirds mate with just one male—often the same individual that is popular with other females—male reproductive success in any given breeding season is very unequal (Figure 10.5).[142, 1481] This fact of reproductive life has great significance for a key question that we have not yet addressed: Why do male bowerbirds, but not females, build courtship display structures, and why do females evaluate male performance instead of the other way round? Indeed, it is common throughout the animal kingdom for males to try to mate with many females, while the objects of their desire are content with one or a few matings, albeit with a male or males they have

(A)

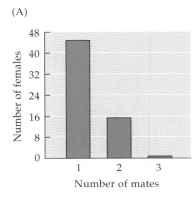

(B)

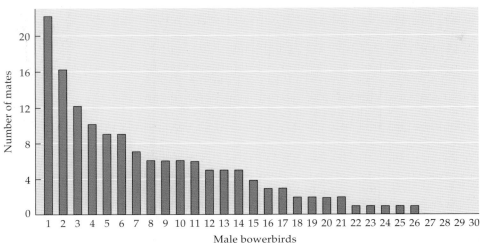

Male bowerbirds

**FIGURE 10.5** Variance in reproductive success is greater for males than for females in the satin bowerbird. (A) Very few female bowerbirds have more than two mates per breeding season, and few, if any, use the sperm of more than one male to fertilize their eggs. (B) Some male bowerbirds, however, mate with more than 20 females in a single season, while others do not mate at all. After Uy, Patricelli, and Borgia.[1481]

**FIGURE 10.6** Male and female gametes differ greatly in size. A hamster sperm fertilizing a hamster egg (magnified 4000 times) illustrates the trivial contribution of materials to the zygote by the male. Photograph by David M. Phillips.

carefully chosen. This widespread pattern is almost certainly related to a truly fundamental difference between the sexes, which is that males produce sperm and females produce eggs.

In sexual species, eggs are by definition larger than sperm, which are usually just big enough to contain the male's DNA and enough energy to fuel the journey to an egg. Even in species in which males produce oversized sperm—such as a fruit fly whose males make sperm that are (when uncoiled) nearly 6 centimeters long, or 20 times their body length[126]—the mass of an egg is still vastly greater than that of a sperm, and this is the typical story (Figure 10.6). A single bird egg may constitute 15 to 20 percent of the female's body mass, and some go as high as 30 percent.[827] By way of contrast, a male splendid fairy-wren, a very small bird indeed, may have as many as 8 billion sperm in his testes at any given moment.[1474] The same pattern applies to coho salmon, whose males shower about 100 billion sperm on a typical batch of 3500 eggs, according to Bob Montgomerie. Likewise, a woman has only a few hundred cells that can ever develop into mature eggs,[338] whereas a single man could theoretically fertilize all the eggs of all the women in the world, given that just one ejaculate contains on the order of 350 million minute sperm.

The critical point is that small sperm usually vastly outnumber the many fewer large eggs available for fertilization in any population. This sets the stage for competition among males to fertilize those eggs.[790] A male's contribution of genes to the next generation generally depends directly on how many sexual partners he has: the more mates, the more eggs fertilized, the more descendants produced, and the greater the male's fitness relative to less sexually successful individuals. The benefits to males of maximizing the number of inseminated partners have, for example, favored males that can discriminate between familiar mates and unfamiliar females, which may provide new opportunities for males to pass on their genes. Thus, in one *Anolis* lizard species, territorial males direct more than ten times as

**FIGURE 10.7   Parental investment takes many forms.** (Clockwise from top left) A male frog carries his tadpole offspring on his back. A male katydid has given his mate an edible spermatophore containing orange carotenoid pigments that will be incorporated in her eggs. A female eared grebe protects her young by letting them ride on her back. A female cicada killer wasp drags a cicada she has paralyzed with a sting to a burrow she has dug, where her offspring will feed on the unlucky cicada. Photographs by Roy McDiarmid; Klaus Gerhard-Keller; Bruce Lyon; and the author.

many courtship displays toward new females that have shown up on their territories for the first time, compared with their rather blasé response toward long-term female companions.[1077]

Whereas male animals usually try to have many sexual partners, females generally do not, because their reproductive success is typically limited by the number of eggs they can manufacture, not by any shortage of willing mates. Eggs are costly to produce because they are large, which means that females have to secure the resources to make them. Furthermore, after one batch of eggs has been fertilized, a female may spend still more time and energy caring for the resultant offspring. Thus, during the breeding season of the satin bowerbird, females are likely to spend more of their time foraging, building a nest, or caring for young than looking for mates, while adult male bowerbirds are on duty by their bowers every single day. Expenditures of time and energy and risks taken by a parent to help one offspring are considered **parental investment** in that offspring if they reduce the chance that the parent will reproduce successfully in the future.[1465]

Bob Trivers developed the concept of parental investment to emphasize the trade-offs for parents that make contributions to their progeny (Figure 10.7). On the plus side, parental investment may increase the probability that an existing offspring will survive to reproduce. But this fitness benefit comes at the cost of the parent's ability to generate additional offspring down the road. Females are more likely than males to derive a net benefit from taking care of existing offspring, for several reasons. First, the offspring they care for are extremely likely to carry their genes. In contrast, the male's paternity is often less certain, given that females of most species accept sperm from more than one individual. In addition, males have less incentive to be parental in those species in which paternal males would lose fertilization opportunities. If a male can mate with several females, it pays him to do so, particularly if he has attributes that give him an edge in the race to fertilize eggs.[1184] Unsuccessful males are trapped as a result of having inherited the nonparental attributes that work well only for more competitive rivals of the same sex. Because females that have already mated often have nothing to gain by copulating again, there are typically many fewer sexually active females than males at any given time, creating a male-biased **operational sex ratio** (the ratio of sexually receptive males to receptive females).[438]

Thus, key behavioral differences between the sexes have apparently evolved in response to the difference in the size and number of gametes they produce, a difference that is often amplified by differences in the degree to which females and males provide parental investment for their putative offspring. But why is it that the sexes differ in the resources they donate to a fertilized egg? Geoffrey Parker and his colleagues have argued that the evolution of the two kinds of gametes, and thus the two sexes, stemmed from divergent selection that favored either (1) individuals whose gametes were good at fertilizing other gametes because of their relatively small size and great mobility (a number-of-offspring strategy) or (2) individuals whose relatively large gametes were good at developing after being fertilized (a parental investment strategy).[1104] Sperm are superbly designed to race toward an egg when released by a male. Bluegill sperm, for example, dart along at up to five sperm lengths per second.[216] In contrast, eggs stay put because they are large, packed with nutrients useful for the development of the zygote following fertilization. No single kind of gamete could be equally good at both tasks, which may have led to the separate evolution of effective fertilization devices (the sperm of males) and effective development devices (the eggs of females).

## Testing the Evolutionary Theory of Sex Differences

We have reviewed a theory of sex differences that focuses on the role of gametic differences and inequalities between the sexes in parental investment (Figure 10.8). These differences favor males that compete vigorously with rivals for mates and pursue females avidly (Figure 10.9). In contrast, female fitness will rarely be advanced by receiving sperm from as many males as possible, so selection should favor instead those females that avoid the costs of additional matings after having chosen partners who have the most to offer in the way of resources or good genes (namely those genes that will help their offspring develop the traits associated with reproductive success). Satin bowerbirds exemplify this common pattern.

We can test this theory by finding unusual cases in which males make the larger parental investment or engage in other activities that reverse the typical operational sex ratio. So, for example, in some species, males make contributions other than sperm toward the welfare of their offspring (or their mates) because if they do not, they may not get a chance to fertilize any eggs at all. For species of this sort, we can predict female competition for mates and careful mate choice by males—in other words, a **sex role reversal** with respect to which sex competes for mates and which does the choosing. Such a reversal occurs in the mating swarms of certain empid flies, in which the operational sex ratio is heavily female biased because most males are off hunting for insect prey to bring back to the swarm as a mating inducement (see also Discussion Question 9.2).[1408] When a male enters the swarm bearing his **nuptial gift**, he may find females advertising themselves with (depending on the species) unusually large and patterned wings or decorated legs[600] or bizarre inflatable sacs on their abdomens (Figure 10.10).[507] The male gets to choose from among the many ornamented partners available to him.

Likewise, males of some fish species offer their mates something of real value, namely, a brood pouch in which the female can place her eggs. For example, in the pipefish *Syngnathus typhle*, "pregnant" males provide nutrients and oxygen to a clutch of fertilized eggs for several weeks, during which time the average female produces enough eggs to fill the pouches of two males. Females evidently compete for the opportunity to donate eggs to parental males, who pay a price when they are pregnant, because they feed and grow less while brooding eggs. As a result, at the start of the breeding season, males that are given a choice (in an aquarium experiment) between allocating time to feeding or responding to potential mates actually show more interest in feeding than mating.[109] In the lab and in the ocean, large males with free pouch space actively choose among mates, discriminating against small, plain females in favor of large, ornamented ones, which can provide selective males with larger clutches of eggs to fertilize.[108, 1241] In another pipefish, *Corythoichthys haematopterus*, a female-biased adult sex ratio generates a female-biased operational sex ratio. In this species, unpaired females compete aggressively and engage in courtship while males do not.[1363]

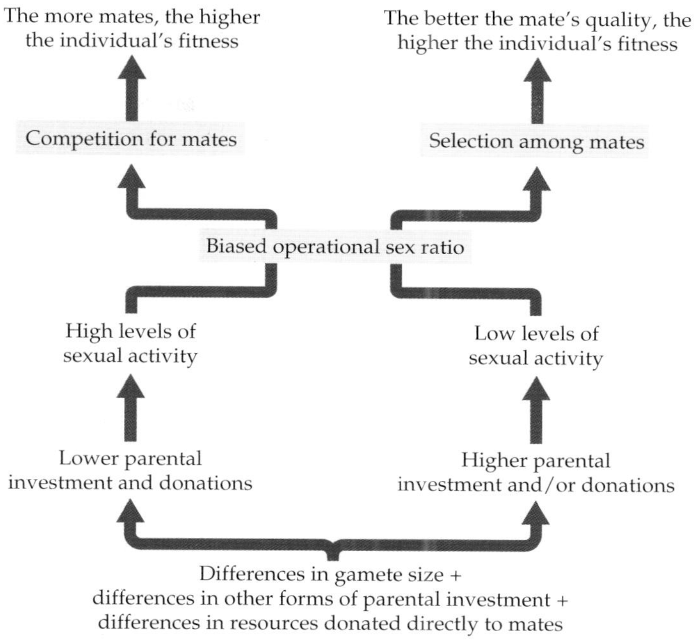

The more mates, the higher the individual's fitness

↑

Competition for mates

↑

High levels of sexual activity

↑

Lower parental investment and donations

The better the mate's quality, the higher the individual's fitness

↑

Selection among mates

↑

Low levels of sexual activity

↑

Higher parental investment and/or donations

Biased operational sex ratio

Differences in gamete size + differences in other forms of parental investment + differences in resources donated directly to mates

**FIGURE 10.8 Differences between the sexes in sexual behavior** may arise from fundamental differences in parental investment that affect the rate at which individuals can produce offspring. The sex that can potentially leave more descendants gains from high levels of sexual activity, whereas the other sex does not. An inequality in the number of receptive individuals of the two sexes leads to competition for mates within one sex, while the opposite sex can afford to be choosy.

**FIGURE 10.9 Male sex drive** is so intense that scientists were able to lure this male elephant seal to this area just by playing a tape recording of the call of a copulating female elephant seal.[386] Once there, the animal saw a moving dummy female seal composed of urethane foam covered with fiberglass. As the male pursued the moving model, he climbed onto a scale, enabling researchers to record his weight without having to sedate and immobilize the animal. Photograph by Chip Deutsch.

**FIGURE 10.10 A sex role reversal in which females, not males, advertise for mates.** In the long-tailed dance fly, *Rhamphomyia longicauda*, females fly to swarms where they wait for the arrival of a gift-bearing male. While waiting, the female inflates her abdomen and holds her dark, hairy legs around her body, making herself appear as large as possible to potential mates. Photograph by David Funk, from Funk and Tallamy.[507]

Another example of a species in which females sometimes compete for males is the flightless Mormon cricket, which despite its common name has no religious affiliation and is a katydid, not a cricket. When male Mormon crickets mate, they transfer to their partners another kind of nuptial gift, an enormous edible spermatophore (Figure 10.11).[598] Given that the spermatophore constitutes 25 percent of the male's body mass, most male Mormon crickets probably cannot mate more than once. In contrast, some females are able to produce several clutches of eggs, but they have to persuade several males to mate with them if they are to have their eggs fertilized.

Female competition for mates in Mormon crickets is evident when local populations have grown to such an extent that individuals cannot get enough protein and salt to eat. At this juncture, huge numbers leave home and march across the countryside, devouring farmers' crops (and one another).[1333] During these mass movements of cannibal katydids, some individuals take time out from feeding to copulate. When a male announces his readiness to mate, he begins to stridulate from a perch, at which time females come running. They often jostle for the opportunity to mount him, the prelude to insertion

**FIGURE 10.11 Mormon cricket males give their mates an edible nuptial gift.** Here a mated female carries a large white spermatophore just received from her partner. Photograph by the author.

of the male's genitalia and transfer of a spermatophore. Males, however, may refuse the chance to transfer a spermatophore to a lightweight female. A choosy male that rejects a 3.2-gram female in favor of one weighing in at 3.5 grams fertilizes about 50 percent more eggs as a result.[598] So here, too, when there are more receptive females than males, competition for mates takes place among females.

The theory of sex role differences also predicts that if the operational sex ratio were to change over the course of the breeding season, the sexual tactics of males and females should change as well. A test of this prediction was provided by a study of a skinny Australian katydid (Figure 10.12) whose food supply varies greatly over the course of the breeding season. When these katydids are limited to pollen-poor kangaroo paw flowers, the male's large spermatophore is both difficult to produce and valuable to females as a nuptial gift. Under these conditions, sexually receptive males are scarce and they are choosy about their mates, whereas females fight with one another for access to receptive, spermatophore-offering males. But when pollen-rich grass trees start to flower and males can produce spermatophores much more rapidly, the operational sex ratio can become male biased since the production of eggs by females is limited by the speed at which they can turn pollen into gametes. At this time, sex roles switch to the more typical pattern, with males competing for access to females and females rejecting some males.

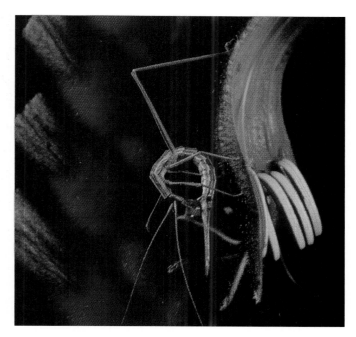

**FIGURE 10.12   A katydid that shifts sex roles in relation to the availability of spermatophores.** Female *Kawanaphila* katydids compete to mate with males that supply them with nutritious gelatinous spermatophores, but only when food is scarce. In this photograph, the female is eating a spermatophore that she received from her mate while she perches on a pollen-poor kangaroo paw flower. The female's two thin antennae and shorter, thicker ovipositor point downward. Photograph by Darryl Gwynne.

## Discussion Question

**10.2** In the two-spotted goby, a fish that lives on rocky shorelines in northern Europe, males provide parental care for one or more clutches of eggs. Males initially compete fiercely for territories (nest sites in seaweed or mussel shells) during the short breeding season (May to July). But over the course of the breeding season, males become scarce, and females begin to behave aggressively toward one another and to court males. Do the data in Table 10.1 permit you to test the theory outlined above for why the sexes differ in their reproductive tactics?

**TABLE 10.1**   *Seasonal changes in the reproductive biology of the two-spotted goby*

|  | May | June | Late June | Late July |
|---|---|---|---|---|
| Territorial males per square meter | 0.56 | 0.32 | 0.13 | 0.07 |
| Females ready to spawn | 0.15 | 0.39 | 0.29 | 0.33 |
| Nest space (number of additional clutches that each nest could accommodate) | 2.98 | 2.53 | 1.49 | 0.99 |

*Source*: Forsgren et al.[477]

## Sexual Selection and Competition for Mates

In most species, males, with their minute sperm, can potentially have a great many descendants, but if they are to achieve even a fraction of their potential, they must deal both with other males, who are attempting to mate with the same limited pool of receptive females, and with the females themselves, who often have a great deal to say about which males get to fertilize their eggs. The realization that the members of a species can determine who gets to reproduce and who fails to do so led Charles Darwin to propose that evolutionary change could be driven by **sexual selection**, which he defined as "the advantage which certain individuals have over others of the same sex and species, in exclusive relation to reproduction."[349] Darwin devised sexual selection theory specifically to account for the evolution of costly, survival-decreasing traits, such as the extravagant courtship displays of male bowerbirds and the ornamental plumes of peacocks. The costs of these traits include not only the time and energy required to perform demanding displays or to produce overblown feathers, but also heightened vulnerability to predators, and some subtle developmental trade-offs as well (Figure 10.13).[436] Darwin argued that although some mate-acquiring attributes surely shorten a male's life directly or indirectly, they may increase his lifetime reproductive success by enabling the well-ornamented male to secure mates in competition with others (Figure 10.14). Darwin contrasted these reproduction-boosting traits with those that enhance survival, which he explained in terms of conventional natural selection.

Today most evolutionary biologists emphasize the similarities between these two types of selection. Both processes require that individuals differ

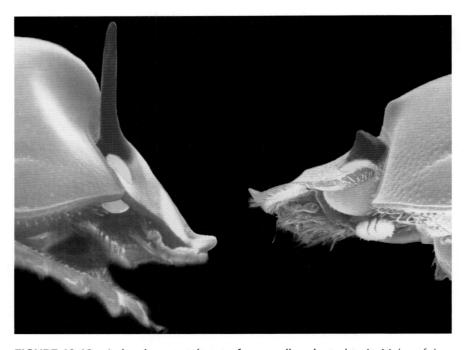

**FIGURE 10.13   A developmental cost of a sexually selected trait.** Males of the dung beetle *Onthophagus acuminatus* fight for mates. Beetles with long horns (shown in blue) have an advantage in these contests, but the tissues that go into horn construction are unavailable for the building of eyes (shown in yellow). As a result, males with long horns (left) have smaller eyes than rivals with short horns (right). From Emlen.[435]

(A)

(B)

(C)

**FIGURE 10.14   Sexually selected "ornaments" impose survival costs on males** but advance their success in the competition for mates. Darwin correctly believed that sexual selection via female choice was responsible for the evolution of elaborate plumage and remarkable displays in male birds such as the quetzal (A) and the sage grouse (B). Darwin argued mistakenly that the strange horns and snouts of certain beetles (C) also arose via female choice; males actually use these structures primarily as weapons when fighting with other males for mates. A, photograph by Bruce Lyon; B, photograph by Marc Dantzker; C, photograph by David McIntyre and Joan Gemme.

in their reproductive success; both will lead to evolutionary change only if these differences have a hereditary basis. However, although sexual selection is "just" a subcategory of natural selection, the distinction Darwin made between the two processes is useful because it focuses attention specifically on the selective consequences of sexual interactions within a species.

## Discussion Questions

**10.3**  Take one possibly survival-enhancing element of bowerbird behavior that could be the product of natural selection and another survival-decreasing trait that might be the evolutionary result of sexual selection. For each, list the essential conditions that had to occur in the past for natural selection and sexual selection to occur. Is the factor "differences among individuals in age at death" on your list? Why or why not?

**10.4** Male rats, sheep, cattle, rhesus monkeys, and humans that have copulated to satiation with one female are speedily rejuvenated if they gain access to a new female. This phenomenon is called the "Coolidge effect," supposedly because when Mrs. Calvin Coolidge learned that roosters copulate dozens of times each day, she said, "Please tell that to the President." When the President was told, he asked, "Same hen every time?" Upon learning that roosters select a new hen each time, he said, "Please tell that to Mrs. Coolidge." Provide a sexual selectionist hypothesis for the evolution of the Coolidge effect. Use your hypothesis to predict what kinds of male animals should lack the Coolidge effect.

One component of sexual selection (sometimes called intrasexual selection) arises when the members of one sex compete with one another for access to the other sex. Although competition among females for mates does occur,[276, 822] as we have already seen in pipefish, spotted hyenas, and Mormon crickets, males are much more likely than females to behave aggressively toward other members of the same sex. So, for example, male bowerbirds dismantle each other's bowers when they have a chance.[1171] Because males with destroyed bowers lose opportunities to copulate, males have been sexually selected to keep a close eye on their display territories and to be willing to fight with rivals. Outright fighting among males is one of the most common features of life on Earth (Figure 10.15) because winners of male–male competition generally mate more often. In previous chapters, we have made reference to aggressive elk, battling beetles, and fighting flies—and in all these cases, the combatants were males.

One of the most widespread effects of sexual selection of this sort is the evolution of large body size, thanks to the fact that larger males tend to be able to beat up smaller ones. In keeping with this hypothesis, when males regularly fight for access to mates, they tend to be larger than the females of their species.[131] We can also expect that sexual selection will produce males capable of performing at high levels during the conflicts that occur within their species. In this light, it is not surprising that males of a collared lizard, which chase intruders from their territory, are superb sprinters. The fastest males are the ones with the largest territories, and correspondingly, these lizards sire the most offspring (Figure 10.16).[701] In addition, males in animals ranging from spiders to dinosaurs, and from beetles to rhinoceroses, have evolved weapons in the form of horns, tusks, antlers, clubbed tails, and enlarged spiny legs, which they use when fighting with other males over females (Figure 10.17).[437]

Although males of many species battle for females, in other species, conflicts among males are not about an immediate mating opportunity but instead have to do with their standing in a dominance hierarchy. Once individuals have sorted themselves out from top dog to bottom mutt, the alpha male need only move toward a lower-ranking male to have that individual hurry out of the way or otherwise signal submissiveness. If the costly effort to achieve high status in a dominance hierarchy is adaptive, then high-ranking individuals should be rewarded reproductively. In keeping with this prediction, among mammalian species, dominant males generally mate more often than subordinates.[305]

The relationship between dominance and sexual access to mates has been especially well studied in groups of savanna baboons,

**FIGURE 10.15 Males of many species fight,** using whatever weapons they have at their disposal. Here two giraffes slam each other with their heavy necks and clubbed heads. Photograph by Gregory Dimijian.

(A)

(B)

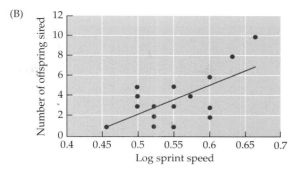

**FIGURE 10.16   Running speed in collared lizards is the product of sexual selection,** as it affects both the size of the male's territory and the number of offspring he sires. (A) A collared lizard basking within his domain. (B) The relationship between male running speed and offspring produced. A, photograph by the author; B, after Husak et al.[701]

where males compete intensely for social status. Opponents are willing to fight to move up the ladder, and as a result, individual males get bitten about once every 6 weeks, an injury rate nearly four times greater than that for females.[415] So, does it pay males to risk damaging bites and serious infections in order to secure a higher ranking? When Glen Hausfater first attempted to test this proposition, he counted matings in a troop of baboons and found that, contrary to his expectation, males of low and high status were equally likely to copulate.[636] Hausfater subsequently realized, however, that he had made a dubious assumption, which was that any time a male mated, he had an equal chance of fathering an offspring. This assumption would be wrong if matings really only "count" when the females have recently ovulated. When Hausfater reexamined the timing of the copulations, he found that dominant males had indeed monopolized females during the few days when they were fertile. The low-ranking males had their chances to mate, but typically only when females were in the infertile phase of their estrous cycles.

**FIGURE 10.17   Convergent evolution in male weaponry.** Rhinoceros-like horns have evolved repeatedly in males belonging to unrelated species thanks to competition among males within these species for access to females. The species illustrated here are 1, narwhal (*Monodon monoceros*); 2, chameleon (*Chamaeleo [Trioceros] montium*); 3, trilobite (*Morocconites malladoides**); 4, unicornfish (*Naso annulatus*); 5, ceratopsid dinosaur (*Styracosaurus albertensis**); 6, horned pig (*Kubanochoerus gigas**); 7, protoceratid ungulate (*Synthetoceras* sp.*); 8, dung beetle (*Onthophagus raffrayi*); 9, brontothere (*Brontops robustus**); 10, rhinoceros beetle (*Allomyrina [Trypoxylus] dichotomus*); 11, isopod (*Ceratocephalus grayanus*); 12, horned rodent (*Epigaulus* sp.*); 13, giant rhinoceros (*Elasmotherium sibiricum**);(*, extinct species). Courtesy of Doug Emlen.

(A)

(B)

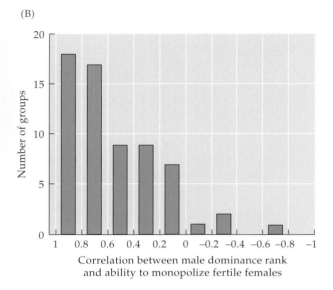

**FIGURE 10.18 Dominance and mating success in savanna baboons.** (A) Male baboons fight for social status. (B) In a Kenyan reserve in which many different troops were followed over many different breeding seasons, the relation between male dominance rank and ability to form consortships with fertile females generally yielded strongly positive correlation coefficients. A correlation coefficient of +1 would indicate a perfect match between the two variables. A, photograph by Joan Silk; B, after Alberts, Watts, and Altmann.[8]

Since Hausfater's pioneering research, others have followed up with more such studies. A summary of data gathered on troops of savanna baboons observed in a Kenyan reserve over many years reveals that male dominance is almost always positively correlated with male copulatory success with fertile females (Figure 10.18).[8] A male that copulates with an ovulating female while keeping all other males away from her would seem all but certain to be the father of any offspring that his partner subsequently produces. But we no longer need to guess about paternity in these cases, thanks to recent developments in molecular technology, including *microsatellite analysis*, which make it possible to determine with nearly complete certainty whether a given male has indeed fathered a given baby (see Box 11.1). Genetic analysis of 208 infant baboons in Kenya reveals that a male's dominance predicts not only his mating success but also his genetic success (Figure 10.19).[9] Dominant males sired more offspring than subordinate ones by virtue of their ability to identify receptive females that were highly likely to conceive,[529] and they also kept other males away from these fertile females,[9] as is true for other primate species.[1564]

## Discussion Questions

**10.5** In some species, like spotted hyenas and meerkats, females form dominance hierarchies in competition with others of their sex. In species of this sort, variance in reproductive success can be greater for females than for males because dominant females are able to suppress reproduction by subordinate females in their groups.[276, 283] Sexual selection theory is usually applied to males rather than females, but why is it legitimate to say that certain female traits in hyenas and meerkats, like status striving and aggressiveness, are the products of sexual selection? Why might sexual selection be stronger on female spotted hyenas and meerkats than on the males of these species?

**10.6** The manes of African lions vary in color from very dark to quite pale. When two models of male lions, one with a dark mane and the other with a light-colored mane, were put out in lion territories, the males that found the models first approached the light-colored one more often than the dark-maned model. In contrast, lionesses almost always walked close to the dark-maned model first. Peyton West and Craig Packer found that in general, mane color was darker in older males and in those with higher testosterone levels.[1545] Knowing what you know about lion social behavior (see Chapters 1 and 8) and sexual selection theory, how would you explain the differences in the reactions of males and females to the dark-maned models?

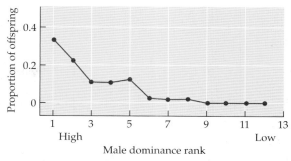

**FIGURE 10.19   Dominance affects male paternity in savanna baboons.** High-ranking males not only mate with receptive females more often, they tend to father more offspring than their subordinate rivals. After Alberts, Buchan, and Altmann.[9]

## Alternative Mating Tactics

Although research on baboons and many other animals has confirmed that high dominance status yields paternity benefits for males able to secure the alpha position, the benefits are often not as large as one would expect (Figure 10.20).[8] As it turns out, socially subordinate baboons can compensate to some extent for their inability to physically dominate others in their group. For one thing, lower-ranking males can and do develop friendships with particular females, relationships that depend not entirely on physical dominance, but rather on the willingness of a male to protect a given female's offspring (see Figure 12.12).[1099] Once a male, even a moderately subordinate one, has demonstrated that he is willing and able to provide protection for a female and her infant, that female may seek him out when she enters estrus again.[1398]

Male baboons also form friendships with other males. Through these alliances, they can sometimes collectively confront a stronger rival that has acquired a partner, forcing him to give her up, despite the fact that the high-ranking male could take out any one of his opponents mano a mano. Thus, for example, in one troop of savanna baboons that contained eight adult males, three low-ranking males (fifth through seventh in the hierarchy) regularly

(A)

(B)

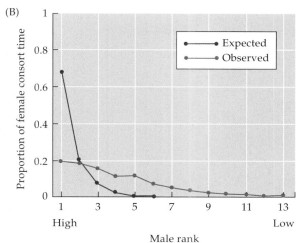

**FIGURE 10.20   Dominant male baboons do not control fertile females** as completely as one would expect if the top-ranked male always had priority of access to estrous females at times when only one was available in the troop. (A) A male guarding an estrous female. (B) Males of high rank did not remain close to an estrous female for as much time as expected based on male dominance status alone. A, photograph by Joan Silk; B, after Alberts, Watts, and Altmann.[8]

formed coalitions to oppose a single higher-ranking male when he was accompanying a fertile female. In 18 of 28 cases, the threatening gang of subordinates forced their lone opponent to relinquish the female.[1051]

Baboons are not the only species in which some males use special mating tactics in order to avoid getting thrashed by a stronger or more experienced opponent.[35, 590] While some gray seal males, for example, are fighting with one another to monopolize mating opportunities with females hauled out on beaches with their pups, other males evidently are looking for and finding mates swimming in the water near the colony. These noncombatants father many of the young produced in any one year, judging from tests of the paternity of the babies found in breeding colonies.[1584]

In another seal, the northern fur seal, almost all pups can be assigned by genetic tests to males that defend patches of beach used by their mothers as resting and birthing sites. But here, too, an alternative tactic is employed by beach masters whose territories have attracted significantly fewer females than their neighbors' real estate. A less successful individual of this sort sometimes attempts to abduct a female or two from his rivals, and he occasionally succeeds in forcing a female to stay on his side of the border, where she may eventually mate with her abductor.[779]

Yet another alternative tactic that may help less successful males do a little better than they would otherwise has evolved in the famous marine iguana, which lives in dense colonies on the Galápagos Islands. Male iguanas vary greatly in size, and when a small male runs over to mount a female, a larger male may arrive almost simultaneously to remove him unceremoniously from her back (Figure 10.21). Because it takes 3 minutes for a male iguana to ejaculate, one would think that small males whose matings were so quickly interrupted would not be able to inseminate their partners. However, the little iguanas have a solution to this problem, which involves having already ejaculated prior to any copulation attempt but retaining their sperm in their bodies. When, in the course of mounting a female, the small male everts his penis from an opening at the base of his tail and inserts it into the female's cloaca, these "old" sperm immediately begin to flow down a penile groove, so even if he does not have time to achieve another ejaculation, the small iguana can at least pass some viable sperm to his mate during their all-too-brief time together.[1571]

**FIGURE 10.21    Small males of the marine iguana must cope with sexual interference from larger rivals.** Two small males are mounted on the same female; a much larger male is trying to pull them off her back. Photograph by Martin Wikelski; from Wikelski and Baurle.[1571]

## Conditional Mating Strategies

Although the little iguanas' ability to inseminate without ejaculation does help them reproduce, odds are that the big boys, who are able to copulate and inseminate at their leisure, reproduce more successfully. So here we have another example of behavioral variation within a species in which some individuals employ a behavior with a high fitness payoff while others behave differently and appear to be making the best of a bad situation. This pattern is the sort associated with conditional strategies (see Page 231), which evolve when selection favors behaviorally flexible individuals that can opt for the tactic that provides them with the highest possible fitness return, given the constraints of their social standing.

Applying this approach to the competitively disadvantaged little iguanas, we can argue that these males do better by trying to inseminate females on the fly rather than fighting futilely for mates with larger, more powerful individuals. Likewise, males of the horned scarab beetle *Onthophagus nigriventris* make both developmental and behavioral decisions that reflect the reality that smaller individuals will invariably lose fights with males endowed with bigger horns. Therefore, at some point during its larval growth, when a male's developmental mechanisms sense that his body is likely to be relatively small, presumably because the beetle larva is poorly nourished, the male shifts his investment of resources away from growing horns and into what will become his sperm-producing testes. Such a "minor" male has small to nonexistent horn weaponry as an adult but larger testes than his bigger opponents. In keeping with his body plan, the minor male sneaks into burrows where a female is being guarded by a big-horned major male and attempts to inseminate the female on the sly, passing large amounts of sperm to her should he be successful in evading detection by her consort.[694] Large males cannot build both large horns *and* large testes (Figure 10.22),[1331] just as sand crickets and

(A)

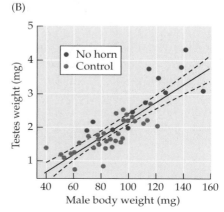

(B)

**FIGURE 10.22  Large horns means smaller testes.** When males of (A) the horned scarab beetle *Onthophagus nigriventris* are experimentally induced to develop as hornless males, (B) they tend to grow larger than horned controls and invest in relatively large testes for their body size. A, photograph by Doug Emlen; B, after Simmons and Emlen.[1331]

(A)

(B)

(C)

**FIGURE 10.23   Satellite male mating tactics.**
(A) A satellite male Great Plains toad crouches by a singing male (note the inflated throat pouch) whose signals he may be able to exploit by intercepting females attracted by the calls. (B) Six subordinate male bighorn sheep trail after a dominant male, who stands between them and the female with whom he will mate at intervals. (C) Two satellite male horseshoe crabs wait by a female on either side of a male that has attached himself to his mate. A, photograph by Brian Sullivan; B, photograph by Jack Hogg; C, photograph by Kim Abplanalp, courtesy of Jane Brockmann.

other insects that invest in wings and allied flight muscles have less to put into ovaries and other components of reproductive equipment (see Figure 5.7).[1236, 1642] A large-horned male *O. nigriventris* with his relatively small testes can lose egg fertilizations to a smaller rival, should that male copulate with the large male's mate and swamp the sperm she has received from her muscular consort (see Sperm Competition below).

In species with conditional strategies, the ability of disadvantaged individuals to switch to an alternative tactic yields a higher payoff than if these individuals were to try to use the tactics of their dominant opponents (Figure 10.23). In the horseshoe crab, for example, some males patrol the water off the beach, finding and grasping females heading toward shore to lay their eggs there. Other males are Johnny-come-latelies that swim onto the beach alone and crowd around paired couples there. As it turns out, an attached male fertilizes at least 10 percent more eggs than any competing **satellite male**.[179, 180] Although satellite males do not do as well as attached horseshoe crabs, they may do better than they would if they tried to attach themselves to a female at sea only to be pushed aside or displaced by males that are in better physical condition.

To test this proposition, Jane Brockmann removed samples of attached and unattached males from a beach and returned them to the sea with plastic bags over the claws that they use to attach themselves to females. Before releasing them, she marked each of her subjects and scored its body condition based on such things as whether the carapace was worn or smooth and whether the animal's eyes were covered by marine organisms or free from obstruction. She found that worn, fouled males were more likely to show up again on the beach as unattached individuals. In other words, males in poorer condition were prepared to accept the lower-payoff tactic of coming to shore unattached to search for satellite opportunities, while males in good condition evidently remained at sea, attempting to find a female to grab and defend, even though they were doomed to fail because of the bags over their claws.[181] The decision to be a satellite is an adaptive making-the-best-of-a-bad-job tactic if individuals that select this option gain more than they would by persisting in hopeless attempts to become attached males (Figure 10.24).

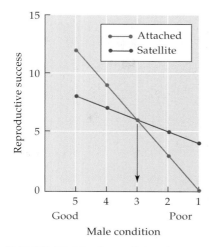

**FIGURE 10.24   A conditional strategy model** of the relationship between male condition and reproductive success in the horseshoe crab. This model predicts that when male condition has fallen to 3 on a 5-point scale, males will switch tactics because they will then gain more reproductive success from the satellite option than by trying to attach themselves to females. "Reproductive success" and "male condition" are given in arbitrary units. After Brockmann.[181]

### Discussion Questions

**10.7**  Satellite and attached male horseshoe crabs do not have the same reproductive success. Why hasn't sexual selection eliminated the low-payoff satellite option, if males exercising this option do not leave as many descendants as attached males do?

**10.8**  Roosters compete with one another for social dominance, and not surprisingly, dominant males have greater copulatory success than subordinate males. Use sexual selection theory to account for these differences among the two categories of males: dominant males produce more sperm than subordinates, and dominants transfer more and better (faster-moving) sperm to females with large red combs on their heads, whereas subordinates provide all their mates with sperm of the same quality (the same velocity).[299] In addition, use conditional strategy theory to predict what should happen if two dominant males were placed together until one became subordinate.

Thus, male horseshoe crabs appear to evaluate their body condition, and they use this information to adopt one or another mating tactic, evidence that they possess a conditional strategy. Let's apply the same theory to alternative mating tactics in a *Panorpa* scorpionfly (Figure 10.25) in which (1) some males aggressively defend dead insects, a food resource highly attractive to recep-

**FIGURE 10.25   A male *Panorpa* scorpionfly** with its strange scorpion-like abdomen tip, which it can use to grasp females in a prelude to forced copulation, one of three mating tactics available to males of this species. Photograph by Jim Lloyd.

tive females, (2) other males secrete saliva on leaves and wait for occasional females to come and consume this nutritional gift, and (3) still others offer females nothing at all but instead grab them and force them to copulate.[1434] In experiments with caged groups of ten male and ten female *Panorpa*, Randy Thornhill showed that the largest males monopolized the two dead crickets placed in the cage, which gave these males easy access to females and about six copulations on average per trial. Medium-sized males could not outmuscle the largest scorpionflies in the competition for the crickets, so they usually produced salivary gifts to attract females but gained only about two copulations each. Small males were unable to claim crickets, and they could not generate salivary presents either, so these scorpionflies forced females to mate but averaged only about one copulation per trial.

Thornhill proposed that in this case, all the males, large and small alike, possessed a conditional strategy that enabled each individual to select one of three options based on his social standing. This hypothesis predicts that the differences between the behavioral phenotypes are environmentally caused, not based on hereditary differences among individuals, and that males should switch to a tactic yielding higher reproductive success if the social conditions they experience made the switch possible.

To test these predictions, Thornhill removed the large males that had been defending the dead crickets. When this change occurred, some males promptly abandoned their salivary mounds and claimed the more valuable crickets. Other males that had been relying on forced copulations hurried over to stand by the abandoned secretions of the males that had left them to defend a dead cricket. Thus, a male *Panorpa* can adopt whichever of the three tactics gives him the highest possible chance of mating, given his current competitive status. These results clinch the case in favor of a conditional strategy as the explanation for the coexistence of three mating tactics in *Panorpa* scorpionflies.[1434]

### Distinct Mating Strategies

The conventional wisdom is that almost all cases of alternative mating tactics can be explained via conditional strategy theory (Gross,[590] but see Shuster and Wade[1328]). Exceptions to this general rule exist, however. The ruff, for example, is a sandpiper whose males are territorial "independents" or subordinate satellites or female mimics, traits that they pass on to their male offspring.[840] The territorial males claim display sites, where they attempt to attract receptive females. Satellites differ somewhat in their plumage from territorial males and are tolerated on the display courts, where they sometimes get to mate when the resident territory holder's attention is elsewhere. The rare female mimics look like females of their species, which may make territory holders less likely to attack them.[741] During a visit to a territorial site, a female mimic, which has very large testes, may sometimes be able to mount a visiting female and deliver large quantities of sperm to her, perhaps swamping the sperm she has received from a territory holder.

The marine isopod *Paracerceis sculpta* is another species with three coexisting reproductive strategies. This creature vaguely resembles the more familiar terrestrial sow bugs and pill bugs that live in moist debris in suburban backyards, but it resides instead in sponges found in the intertidal zone of the Gulf of California. If you were to open up a sufficient number of sponges, you would find females, which all look more or less alike, and males, which come in three dramatically different sizes: large (alpha), medium (beta), and small (gamma) (Figure 10.26), each with its own behavioral phenotype.

The big alpha males attempt to exclude other males from interior cavities of sponges that have one or more females living in them. If a resident alpha encounters another alpha male in a sponge, a battle ensues that may last hours before one male gives way. Should an alpha male find a tiny gamma male, however, the larger isopod simply grasps the smaller one and throws him out of the sponge. Not surprisingly, gammas avoid alpha males as much as possible while trying to sneak matings from the females living in their sponges.[1324] When an alpha and a medium-sized beta male meet inside a sponge cavity, the beta behaves like a female, and the male courts his rival ineffectually. Through female mimicry, the female-sized beta males coexist with their much larger and stronger rivals and thereby gain access to the real females that the alpha male would otherwise monopolize.

In this species, therefore, we have three different types of males, and one type has the potential to dominate others in male–male competition. If the three types represent three distinct strategies, then (1) the differences between them should be traceable to genetic differences, and (2) the mean reproductive success of the three types should be equal. If, however, alpha, beta, and gamma males use three different tactics resulting from the same conditional strategy, then (1) the behavioral differences between them should be induced by different environmental conditions, not different genes, and (2) the mean reproductive success of males using the alternative tactics need not be equal.

Steve Shuster and his coworkers collected the information needed to check the predictions derived from these two hypotheses.[1325, 1327] First, they showed that the size and behavioral differences between the three types of male isopods are the hereditary result of differences in a single gene represented by three alleles. Second, they measured the reproductive success of the three types in the laboratory by placing various combinations of males and females in artificial sponges. The males used in this experiment had special genetic markers—distinctive characteristics that could be passed on to their offspring—enabling the researchers to identify which male had fathered which of the baby isopods that each female eventually produced. Shuster found that the reproductive success of a male depended on how many females and rival males lived with him in a sponge. For example, when an alpha male and a beta male lived together with one female, the alpha isopod fathered most of the offspring. But when this male combo occupied a sponge with several females, the alpha male could not control them all, and the beta male outdid his rival, siring 60 percent of the resulting progeny. In still other combinations, gamma males outreproduced the others. For each combination, it was possible to calculate an average value for male reproductive success for alpha, beta, and gamma males.

Shuster and Michael Wade then returned to the Gulf of California to collect a large random sample of sponges, each one of which they opened to count the isopods within.[1325] Knowing how often alphas, betas, and gammas lived in various combinations with competitors and females enabled Shuster and Wade to estimate the average reproductive success of the three types of males, given the laboratory results gathered earlier. When the mathematical dust had settled, they estimated that alpha males in nature had mated with 1.51 females on average, while betas checked in at 1.35 and gammas at 1.37 mates. Since these means were not significantly different, statistically speaking, Shuster and Wade concluded that the three genetically different types of males had essentially equal fitnesses in nature. The requirements for a three-distinct-strategies explanation had been met (if the authors have their statistics right).

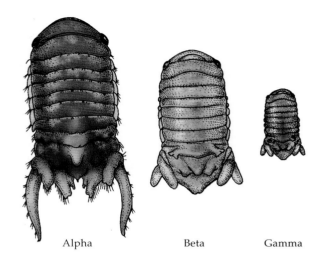

Alpha            Beta            Gamma

**FIGURE 10.26    Three different forms of the sponge isopod:** the large alpha male, the female-sized beta male, and the tiny gamma male. Each type not only has a different size and shape, but also uses a different hereditary strategy to acquire mates. After Shuster.[1326]

(A)

(B)

(C)

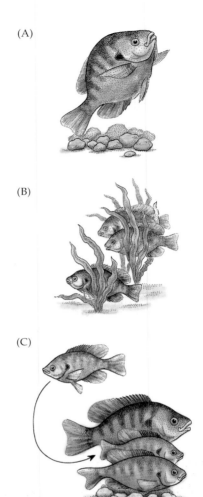

**FIGURE 10.27 Three different egg fertilization behaviors coexist in the bluegill sunfish.** (A) A territorial male guards a nest that may attract gravid females. (B) Little sneaker males wait for an opportunity to slip between a spawning pair, releasing their sperm when the territory holder does. (C) A slightly larger satellite male with the body coloration of a female hovers above a nest before slipping between the territorial male and his mate when the female spawns.

Note that in this example, the reason why three genetically distinct phenotypes can coexist is that at certain frequencies all three types have the same fitness. Therefore, any change in the proportion of one type in the population causes a change in the fitnesses of all three types. Imagine, for example, that for one reason or another, beta males became twice as frequent in a population as they were when Shuster and Wade did their collecting. If so, the "extra" betas would have to go somewhere, including sponges that already had one beta male co-residing with an alpha and some females. Then, instead of one beta, there would now be two beta males competing for the standard portion of the harem that alpha males cannot control, which would cut the mean beta fitness in this sponge by half, thereby also lowering the mean fitness of the beta phenotype overall. Relative to the beta phenotype, alphas and gammas would do better, and their relatively greater reproductive success would lead to an increase in the proportions of alphas and gammas in the next generation. Eventually, frequency-dependent selection (see page 231) could result in a population in which the three types had equal fitnesses on average.

### Discussion Questions

**10.9** Recall the case of the Hawaiian populations of a cricket in which stridulating males produce acoustical signals attractive to females—and to parasitic flies—as well as to silent satellite males that wait near callers to intercept receptive females.[1651] Produce a hypothesis that incorporates frequency-dependent selection to explain how it might be possible for both phenotypes to persist in a population even if the difference between the two kinds of males is hereditary.

**10.10** Figure 10.27 shows three different tactics used by male bluegill sunfish for fertilizing eggs: the territorial nest defense tactic used by large males, the sneaker option used by small males, and the female mimic option.[587] How would you test the competing three-different-strategies and one-conditional-strategy hypotheses to account for the existence of alternative mating tactics in this species? For test data, see DeWoody et al.[391]

### Sperm Competition

The alpha male and the beta male sponge isopod differ in how they go about securing copulatory partners. But the reproductive competition among them need not stop there. When females mate with more than one male in a short time, the two males may not divvy up a female's eggs evenly. In the case of bluegill sunfish, for example, older males nesting in the interior of the colony (see page 462) produce ejaculates with more sperm, and their sperm swim faster, suggesting that they would generally have a fertilization advantage over an average bluegill.[243] On the other hand, when a sneaker male bluegill and a guarding territorial male release their sperm more or less at the same time over a mass of eggs (see Figure 10.27), the sneaker male fertilizes a higher proportion of the eggs than the nest-guarding male.[497] What we have here is evidence of competition among males with respect to the fertilization success of their sperm. Such **sperm competition** is a very common phenomenon in the animal kingdom, no matter whether fertilization is external (as in bluegills and many other fishes) or internal (as in insects, birds, and mammals). If the sperm of some males have a consistent advantage in the race to fertilize eggs, then counting up a male's spawnings or copulatory partners will not measure his fitness accurately.[122]

## Discussion Question

**10.11** More evidence that matings do not necessarily translate into fitness for males comes from a study of a common European frog, *Rana temporaria*. In this species, some males find and grasp egg-laden females and then release their sperm as the female deposits a batch of eggs in a pond. Some other males locate floating egg masses soon after they have been laid. While grasping the clutch as if it were a female, these after-the-fact males release their sperm on the eggs, with the result in one pond that more than 80 percent of the clutches had split paternity.[1491] Some have interpreted this mating system as a means to ensure that the maximum amount of genetic diversity is passed on to the next generation, given that the sex ratio is heavily male biased. Devise another evolutionary explanation and evaluate the two alternatives.

Sperm wars occur in insects[1103, 1330] as well as in isopods and fishes (and most other animal groups) leading to the evolution of some remarkable attributes. Males of a scorpionfly, for example, have evolved the ability to estimate the number of sperm that a female has already stored in her body, the better to adjust the number of sperm that the estimator will transfer to his previously mated companion. The longer a female had mated with a previous partner (and thus, the more sperm she had received), the shorter the duration of copulation of her next mate (who had not been present during the initial copulation).[445]

Male scorpionflies can only add the right amount of their own sperm to those already received by a partner. Males of *Calopteryx maculata*, the black-winged damselfly of eastern North America, take sperm competition to a different level by physically removing rival gametes from their mate's body before transferring their own.[1502] Males of this species defend territories containing floating aquatic vegetation, in which females lay their eggs. When a female flies to a stream to lay her eggs, she may visit several males' territories and copulate with each site's owner, laying some eggs at each location. The female's behavior creates competition among her partners to fertilize her eggs,[1103] and the resulting sexual selection pressure on males has endowed them with an extraordinary penis.

To understand how the damselfly penis works, we need to describe the odd manner in which damselflies (and dragonflies) copulate. First, the male catches the female and grasps the front of her thorax with specialized claspers at the tip of his abdomen. A receptive female then swings her abdomen under the male's body and places her genitalia over the male's sperm transfer device, which occupies a place on the underside of his abdomen near the thorax (Figure 10.28). The male damselfly then rhythmically pumps his abdomen up and down, during which time his spiky penis acts as a scrub brush (Figure 10.29), catching and drawing out any sperm already stored in the female's sperm storage organ. Jon Waage found that a copulating male *C. maculata* removes between 90 and 100 percent of any competing sperm before he releases his own gametes,

**FIGURE 10.28  Copulation in the black-winged damselfly** enables the male to remove a rival's sperm before transferring his own. The male (on the right) has grasped the female with the tip of his abdomen; the female bends her abdomen forward to make contact with her partner's sperm-removing and sperm-transferring penis. Photograph by the author.

(A)

(B)

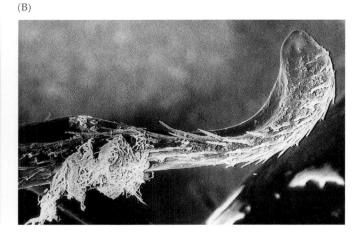

**FIGURE 10.29** **Sperm competition has shaped the evolution of the black-winged damselfly's penis.** (A) The male's penis has lateral horns and spines that enable him to scrub out a female's sperm storage organ before passing his own sperm to her. (B) A close-up of a lateral horn reveals rival sperm caught in its spiny hairs. Photomicrographs by Jon Waage, from Waage.[1502]

which he had earlier transferred from his testes on the tip of his abdomen to a temporary storage chamber very near the penis. After emptying the female's sperm storage organ, he lets his own sperm out of storage and into the female's reproductive tract, where they remain for use when she fertilizes her eggs—unless she mates with yet another male before ovipositing, in which case his sperm will be extracted in turn.[1502]

If the black-winged damselfly was to be our guide, we might conclude that sperm competition is basically something males do to one another. In reality, females often have an active role in deciding which of their partners' sperm will win the egg fertilization contest.[1151, 1330] Even a female of the black-winged damselfly could arrange for the removal of some sperm simply by mating with a second male after copulating with an individual that she did not favor. In other cases, females do not have to rely on males for sperm removal but can expel sperm themselves (Figure 10.30).[353]

**FIGURE 10.30** **Sperm competition in the dunnock requires female cooperation.** A male pecks at the cloaca of his partner after finding another male near her; in response, she ejects a droplet of sperm-containing ejaculate just received from the other male. After Davies.[353]

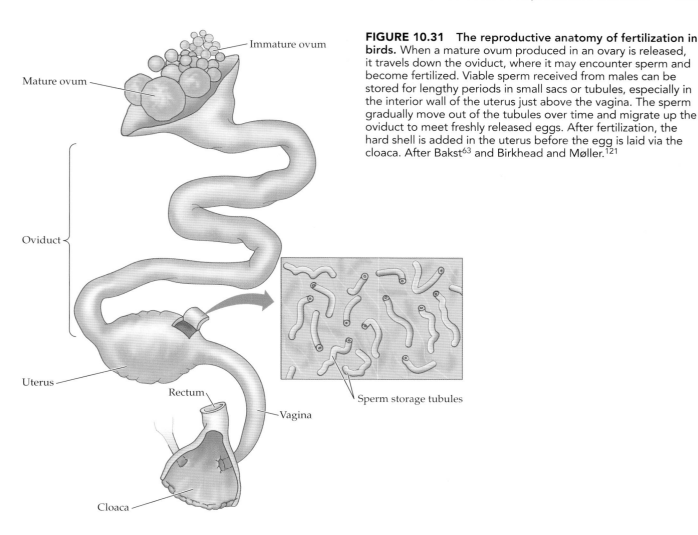

**FIGURE 10.31    The reproductive anatomy of fertilization in birds.** When a mature ovum produced in an ovary is released, it travels down the oviduct, where it may encounter sperm and become fertilized. Viable sperm received from males can be stored for lengthy periods in small sacs or tubules, especially in the interior wall of the uterus just above the vagina. The sperm gradually move out of the tubules over time and migrate up the oviduct to meet freshly released eggs. After fertilization, the hard shell is added in the uterus before the egg is laid via the cloaca. After Bakst[63] and Birkhead and Møller.[121]

## Discussion Question

**10.12**  The female black-legged kittiwake can eject sperm received from a partner.[1505] If you were told that this bird is monogamous, would you reject the hypothesis that sperm ejection in this species had evolved by sexual selection? Why or why not? In this species, monogamous pairs begin mating well before egg laying. If sperm ejection is related to an inability to keep sperm alive for long periods within the female's reproductive tract, when should females expel their partner's sperm?

Although we tend to think of birds as monogamous, that is not necessarily the case, as we'll see in the next chapter. Many birds form pair bonds with, and mate with, a social partner but may also engage in extra-pair matings with other individuals. Sperm is stored in female birds after copulation, and it becomes less viable over time (Figure 10.31). Sperm more recently received from an extra-pair male is therefore more likely to fertilize an egg than stored, older sperm that her social partner gave her previously. For this reason, females of the collared flycatcher, although they cannot physically remove unwanted male gametes as some other birds can, could bias male fertilization chances by copulating in ways that give one male's sperm a numerical advantage over another's. To keep female flycatchers from receiving sperm from their social partners, researchers glued anti-insemination rings around the males' cloacae,

**FIGURE 10.32 Female collared fly-catchers could bias egg fertilizations in favor of an extra-pair mate.** When females are experimentally prevented from receiving sperm from their social partners, some do not copulate with another male, and the number of sperm available for fertilizing eggs falls over time. But some other females do have an extra-pair mating in the middle or late part of the mating period, and these individuals go on to lay eggs that have access to "extra" sperm that must have come from an extra-pair male. After Michl et al.[980]

Collared flycatcher

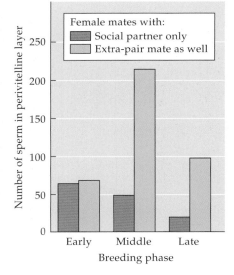

which did not keep them from mating with the females but did mean that their sperm could not enter the females' reproductive tracts. Some of these females did not mate with other males thereafter, and the amount of sperm they kept in storage gradually fell, reducing the sperm available to fertilize their eggs (Figure 10.32).[980] But some females that were paired with males incapable of insemination subsequently copulated once or twice with other males, boosting their sperm stores, which eventually were used to fertilize their eggs, as revealed by genetic analysis.

Under natural conditions, a female flycatcher could use her control of copulations to manipulate the number of sperm from different males within her reproductive tract. A female that stopped mating with her social partner for a few days and then mated with an attractive neighbor would have five times as many of her extra-pair mate's sperm available for egg fertilizations than she retained from earlier matings with her social partner. This imbalance would give the extra-pair male a big fertilization advantage. In fact, female collared flycatchers that are paired with a male with a small white forehead patch often do secure sperm around the time of egg laying by slipping away for a tryst with a nearby male sporting a larger white patch. Therefore, females of this species seem to play a major role in determining whose sperm will fertilize their eggs,[980] a conclusion that may apply to most animal species,[425] although not everyone agrees.[48]

## Mate Guarding

Males must compete for egg fertilization chances on a stage that has been designed to give females partial or complete control of the outcome. Nevertheless, males can sometimes give their sperm an advantage over those of their rivals. One way to do so would be to increase the number of sperm ejaculated into a female that had already mated with another male in the manner of certain hornless scarab beetles and sneaker bluegill fish. Male meadow voles can also boost the sperm count of an ejaculate by more than 50 percent when copulating with a female in a place where the odors of another male are present.[380]

Although adjustments in sperm donated to a mate can sometimes help males improve the odds that a female will use their gametes to fertilize her eggs, by standing guard over a mate, a male can sometimes prevent his partner from mating again, thereby keeping his sperm from having to compete with those of other males in a fertilization derby. Mate guarding after insemination, a common phenomenon, is achieved in many different ways. In some cases, males

(A)

(B)

**FIGURE 10.33 Mate guarding occurs in many animals.** (A) A red male damselfly grasps his mate in the tandem position so that she cannot mate with another male. (B) The male blueband goby, an Indonesian reef fish, closely accompanies his mate wherever she goes. A, photograph by the author.

keep their mates occupied after mating is finished;[20] in others, mated males deceptively lure new suitors away from the female,[463] while in still others, males may seal a partner's genitalia with various secretions.[399, 1158] A spectacularly final form of mate guarding is practiced by males of an orb-weaving spider, which die within minutes after having inserted both pedipalps (the sperm-transferring appendages) into a female's paired genital openings. The tips of the pedipalps inflate after insertion, promoting insemination while also blocking entry to the female's sperm-receiving apparatus. After his death, the male and his inserted pedipalps constitute a "whole-body mating plug" that other males cannot easily remove.[472]

Although many forms of mate guarding exist, most are less extreme than sacrificing one's life to serve as a chastity belt. In fact, the most common tactic by far is simply to stay with a mate in order to see off any other males, should they dare to approach (Figure 10.33).[35] So, for example, male meadow pipits that hear a tape of a male intruder in their territory early in the breeding season are especially likely to attack an artificial model of a male pipit if their mate is present (Figure 10.34) and thus potentially at risk of interacting sexually with the "new" male.[1126]

The playback technique has also been used to examine the adaptive value of mate guarding in savanna baboons. As mentioned earlier, dominant males often stay very close to an estrous female with whom they copulate at intervals. The tactic appears to have value because receptive females left unguarded quickly attract the attention of other males. Catherine Crockford and her colleagues demonstrated this point with a playback experiment. When "bachelor" males heard a taped copulation call of a female about 25 meters to one side

Meadow pipit

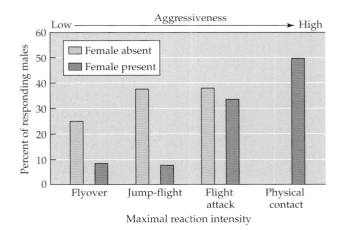

**FIGURE 10.34 Evidence of mate guarding.** Reactions to an intruder by territorial male meadow pipits become much more aggressive if the male's partner is present. After Petrusková et al.[1126]

and the grunt of her consort partner about 25 meters to their other side, they took notice and often approached the hidden speaker from which the female's call had come. They evidently recognized both individuals by their calls and deduced that the male was no longer guarding his recent partner, thereby enabling a fellow male to mount this female. If so, they too might have a chance to do the same in the absence of a dominant guarding male.[322] The ability of male baboons to monitor the sex lives of their companions means that dominant males must guard their fertile partners closely and constantly.

But is remaining with one partner in order to chase away other suitors really adaptive for the mate-guarding male? Any benefits of the behavior, such as fertilizing more of the eggs of one female, come with costs, including the loss of opportunities to seek out other mates. If true, we can predict that males of species in which the operational sex ratio varies will adjust their mate guarding accordingly. When the sex ratio of populations of the carrion beetle *Necrophila americana* were experimentally manipulated in the lab, the time males spent close to one mate increased under male-biased sex ratios and decreased when the sex ratio was female biased.[782]

The carrion beetle study shows that mate guarding can reduce opportunities to mate with other females. Janis Dickinson measured this opportunity cost in another beetle, the blue milkweed beetle, in which the male normally remains mounted on the female's back for some time after copulation. When Dickinson pulled pairs apart, about 25 percent of the separated males found new mates within 30 minutes. Thus, remaining mounted on a female after inseminating her carries a considerable cost for the guarding male, since he has a good chance of finding a new mate elsewhere if he just leaves his old one. On the other hand, nearly 50 percent of the females whose guarding partners were plucked from their backs acquired a new mate within 30 minutes. Since mounted males cannot easily be displaced from a female's back by rival males, mounted males reduce the probability that an inseminated partner will mate again, giving their sperm a better chance to fertilize her eggs. Dickinson calculated that if the last male to copulate with a female fertilized even 40 percent of her eggs, he would gain fitness by giving up the search for new mates in order to guard his current one.[400]

In general, the benefits of mate guarding increase with the probability that unguarded females will mate again and use the sperm of later partners to fertilize their eggs. But how do you figure out what unguarded females would do in a species in which all the females are guarded? Dickinson's technique involved the simple removal of a male from a female. Jan Komdeur and his colleagues achieved the same effect in their study of Seychelles warblers by tricking males into ending their guarding prematurely by placing a false warbler "egg" in a nest a few days before the male's partner was due to lay her one and only egg. As a result, male warblers used the cue of egg presence to stop mate guarding at a time when their partners were still fertile. In short order, many of these unguarded females copulated with neighboring males (Figure 10.35)[794] and used the sperm to fertilize their eggs. Indeed, the probability that a nestling would be sired by a male other than the female's social partner increased in relation to the number of days that her "mate" neglected to guard her during her fertile period.[796] Mate guarding provides clear fitness benefits for male Seychelles warblers.

Because mate guarding is costly, we can also predict that males will adjust their investment in mate guarding in relation to the risk of cuckoldry, which ought to be affected by how many males live in close proximity to a fertile female and her partner. In the Seychelles warbler, breeding pairs may be surrounded by up to six neighbors. As predicted, the more neighbors, the more time a male spends guarding his fertile partner, and the less time he invests in foraging (Figure 10.36).[795] In a similar vein, male red-winged blackbirds

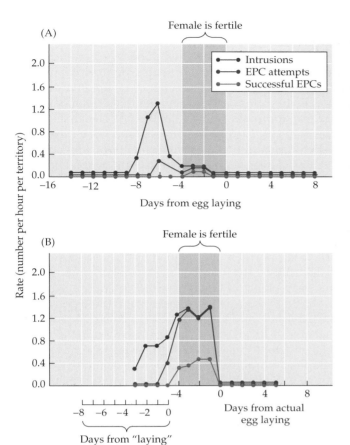

Seychelles warbler

**FIGURE 10.35   Adaptive mate guarding by the Seychelles warbler.** The graphs show the rate of intrusions and extra-pair copulations (EPCs) by males other than the female's social mate in relation to the female's fertile period (shaded area). (A) Control pairs, in which the female's mate was present throughout her fertile period. (B) Pairs in which the female's mate was experimentally induced to leave her unguarded by the placement of a false egg in the nest. Photograph by Cas Eikenaar; A and B, after Komdeur et al.[794]

are more vigilant and aggressive toward neighbors that are unusually sexually attractive, and therefore constitute a greater threat to the guarding male's paternity, than toward neighbors that are less appealing to their mates.[1068]

## Discussion Questions

**10.13** Mate guarding should be common in species in which females retain their receptivity after mating and are likely to use the sperm of their last mating partner when fertilizing their eggs. But there are species, including some crab spiders, in which males remain with immature, unreceptive females for long periods and fight with other males that approach these females.[405] How can "guarding" behavior of this sort be adaptive? Produce sexual selectionist hypotheses and allied predictions.

**10.14** When male bluethroats (see Figure 4.34) have their blue throats experimentally blackened, they become less attractive to females. Outline the costs and benefits to these males of guarding their mates, once they have secured a partner, and the costs and benefits of attempting to secure extra-pair matings with other females by visiting the territories of other pairs. Use conditional strategy theory (see page 231) to predict the tactics of experimental and control males. Discuss the significance of the finding that males with blackened throats fathered fewer of their social partners' offspring on average than did unaltered control males.

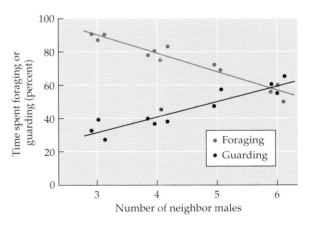

**FIGURE 10.36   Male Seychelles warblers adjust their mate guarding in relation to the risk of losing paternity to rivals.** The more male neighbors around a breeding pair, the more time male warblers spend guarding their partners. After Komdeur.[795]

(A)

(B)

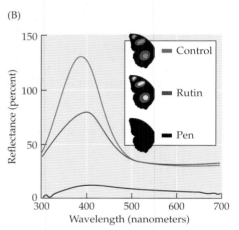

(C)

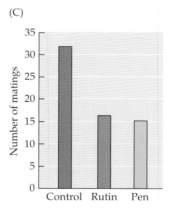

**FIGURE 10.37 Mate choice based on an iridescent color pattern.** (A) Males of the butterfly *Hypolimnas bolina* have large patches of iridescent blue on the upper surfaces of their wings. (B) If the iridescence is eliminated entirely by black pen ink or even partly by application of the compound rutin, (C) females are much less likely to accept the altered males (red and green bars) relative to controls that have retained their color ornament (blue bars). A, photograph by Darrell Kemp; B and C, after Kemp.[757]

## Sexual Selection and Mate Choice

Mate guarding is one of the many dramatic evolutionary consequences of sexual selection arising from competition for mates, but as we have also noted, mate choice can also create sexual selection pressure. In most species, mate choice is exercised primarily by females, a fact that has great evolutionary significance for males (Table 10.2).[1503, 1641] The attributes of males that females favor vary greatly from species to species, with females of some butterflies discriminating against males whose iridescent and ultraviolet wing patches are even slightly duller than those of other males (Figure 10.37),[757, 758] while female zebra finches detect and act upon the most subtle differences in the songs sung by potential mates.[1619]

It can be hard to imagine why females should choose males only slightly different from others of their species. In some cases, however, female preferences are based on male attributes of obvious practical utility, such as the ability of the male to supply them with a good meal. A classic example comes from Randy Thornhill's study of the black-tipped hangingfly, an insect in which female acceptance of a male depends on the nature of his nuptial gift (Figure 10.38). In this species, a male that tries to persuade a female that an unpalatable ladybird beetle is a good mating present is out of luck. Even males that transfer an edible prey item to their mates will be permitted to copulate only for as long as the meal lasts. If the nuptial gift is polished off in less than 5 minutes, the female will separate from her partner without having accepted a single sperm from him. When, however, the nuptial gift is large enough to keep the copulating female feeding for 20 minutes, she will depart with a full complement of the gift giver's sperm (Figure 10.39).[1433] Males of many other animals provide food presents before or during copulation,[1482] including male katydids and crickets that give their mates edible spermatophores (see Figure 10.11).

Some researchers have suggested that a special class of nuptial gift givers should be recognized, namely, those male mantids and spiders that wind up being eaten by their mates (see Figure 5.3). Indeed, it could be adaptive, under some special circumstances, for a male to conclude a copulation dramatically by becoming a meal for his recent sexual partner.[218, 1167]

**TABLE 10.2** *Ways in which females and males attempt to control reproductive decisions*

**A. Key reproductive decisions controlled primarily by females**

*Egg investment*: What materials, and how much of them, to place in an egg

*Mate choice*: Which male or males will be granted the right to be sperm donors

*Egg fertilization*: Which sperm to use to fertilize each egg

*Offspring investment*: How much maintenance and care goes to each embryo and offspring

**B. Ways in which males influence female reproductive decisions**

*Resources transferred to female*: May influence egg investment, mate choice, or egg fertilization decisions by female

*Elaborate courtship*: May influence mate choice or egg fertilization decisions by female

*Sexual coercion*: May overcome female preferences for other males

*Infanticide*: May overcome female decisions about offspring investment

*Source*: Modified from Waage[1503]

So, after a delicate and lengthy courtship,[1391] male redback spiders often make it easy for their mates to eat them. As he transfers sperm to a female, the male redback performs a somersault, throwing his body into his partner's jaws (Figure 10.40). About two-thirds of the time, the female accepts the invitation and devours her sexual companion.[37]

Since a male redback weighs no more than 2 percent of what a female weighs, he does not make much of a meal for a partner. Nevertheless, hunger may be part of the proximate basis for cannibalism, since females that are deprived of food are more likely to dine on males.[38] Once eaten, for whatever reason, the deceased male redback does in fact derive substantial benefits from what might seem like a genuinely fitness-reducing experience. Maydianne Andrade showed that eaten males fertilized more of their partners' eggs than uneaten males did, partly because a cannibalistic female spider is less likely to mate again promptly.[37] Moreover, the cost of being cannibalized is very low for male redbacks. Young adult males in search of mates are usually captured by predatory ants or other spider hunters long before they find webs with adult females. The intensity of predation on wandering male redbacks is such that fewer than 20 percent manage to locate even one mate, suggesting that the odds that a male could find a second partner if he should survive his initial mating are exceedingly low.[39] Moreover, when a male redback is finished transferring sperm, he may break off the tip of his sperm transferring appendage in the female's sperm-receiving opening. This kind of self-mutilation probably helps reduce the chance that another male can mate with the plugged female. The mutilated male, on the other hand, has lost one or both of his two copulatory pedipalps and so has little or no residual fertility. Under these circumstances, males need very little benefit from sexual suicide in order to make the trait an adaptive option.

This claim can be tested by predicting that males will make the ultimate sacrifice in other, unrelated spiders in which the male pedipalps become broken or altered during the male's first (and usually only) copulation. Jeremy Miller has found that males are complicit in their cannibalism or die of "natural causes" soon after mating in five or six lineages of spiders. In all but one of these, the male's genitalic pedipalps become broken or nonfunctional in the course of an initial mating,[985] support for the hypothesis that sexual suicide occurs when the costs to the male in terms of lost future mating opportunities are essentially nil.

**FIGURE 10.38  A potential nuptial gift.** A male hangingfly has captured a moth, a material benefit to offer to his copulatory partner. He advertises the availability of his gift by releasing a pheromone from abdominal glands. Photograph by the author.

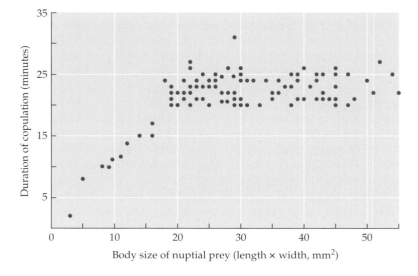

**FIGURE 10.39  Sperm transfer and the size of nuptial gifts.** In black-tipped hangingflies, the larger the nuptial gift, the longer the mating, and the more sperm the male is able to pass to the female. After Thornhill.[1433]

(A)

(B)

(C)

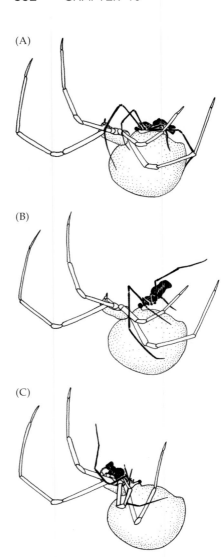

**FIGURE 10.40 Sexual suicide in the redback spider.** (A) The male first aligns himself facing forward on the underside of the female's abdomen while he inserts his sperm-transferring organ into her reproductive tract. (B) He then elevates his body and (C) flips over into the jaws of his partner. She may oblige by consuming him while sperm transfer takes place. After Forster.[478]

Although redbacks and some other spiders offer themselves up as nuptial gifts, males of other animals may care for the female's offspring, and in these species we would expect females to exert sexual selection in favor of males that provide more paternal care than average. In keeping with this expectation, females of the fifteen-spined stickleback, a small fish with nest-guarding males, associate more with courting males that shake their bodies relatively frequently. Males that can behave in this way also perform more nest fanning after courtship is over and eggs have been laid in their nests (see Figure 12.3). Nest fanning sends oxygenated water flowing past the eggs, which increases gas exchange and, ultimately, the rate of egg hatching.[1081]

The female fifteen-spined stickleback that evaluates the courtship display of a male sees him behaving in ways that are linked to his parental capacities. Males of a related species, the three-spined stickleback, also appear to offer their females cues of paternal helpfulness—in the form of a colorful belly. As a rule, males with redder bellies are more attractive to potential mates.[1243] The reddish pigment that colors a male's skin ornament comes from the carotenoids he consumes. Males with carotenoid-rich diets are able to fan their eggs for longer periods under low-oxygen conditions than males given low-carotenoid diets.[1141] In other words, a male stickleback's appearance can advertise his ability to supply oxygen for the eggs he will brood.

Could the carotenoid pigments in a male bird's plumage also be an indicator of his capacity for paternal behavior? Perhaps so, since females of many animal species are especially attentive to the reds and yellows of male color patterns,[578, 966] which might reveal something about the health of a male if, as has been argued, the quality of an individual's immune system is enhanced by a diet rich in carotenoids.

If so, then experiments in which some zebra finch males were provided with extra carotenoids should have produced two classes of males: males with improved immune systems and brighter beaks (which normally range from bright orange-red to dull orange) and males with inferior immune systems and duller beaks. When the experiment was done, the carotenoid-supplemented males had more carotenoids in their blood, brighter beaks, and stronger immune responses.[967] Moreover, female zebra finches find experimentally carotenoid-enhanced males more attractive than those that have been kept on normal diets.[133] In nature, female zebra finches that acquired a more brightly colored mate might benefit by having a healthy mate capable of providing superior care for his offspring.

Much the same thing may be happening in the blue tit, another small songbird with a carotenoid-based ornament, a bright yellow breast. Male blue tits collect and deliver food for their nestlings, usually in the form of carotenoid-containing caterpillars. If the amount or quality of food supplied by a male is related to how bright his yellow feathers are, then the offspring of brilliantly yellow males should be larger and healthier when they fledge than the youngsters of less brightly colored males. Indeed, the offspring of brighter parents are in better condition and have stronger immune systems than those of less yellow parents.[653]

But wait a minute. This same result might occur if bright yellow males were themselves large and healthy at fledging, thanks to their genetic makeup, in which case their offspring would simply inherit these traits from their parents. Because a team of Spanish behavioral ecologists recognized this problem, they wisely controlled for heredity through a cross-fostering experiment. They took complete clutches of eggs and transferred them between nests, moving the offspring of one set of parents to another pair's nest. The foster parents were willing to rear these genetic strangers, and when their adopted chicks had reached the age of independence, the size of the fledglings was a function of the brightness of their foster father's yellow plumage, not the color of their genetic father. Bright yellow foster males produced larger fledglings. If the parental effort of bright yellow males really is greater than that of duller individuals, females could benefit by choosing males on the basis of their plumage, although whether blue tit females actually use the quality of the yellow plumage of potential mates to make their choices has yet to be determined.[1307]

However, not every study of the effects of bright plumage on male parental care has produced identical results. Indeed, another, larger cross-fostering study found no relationship between male coloration and the degree of parental care in the blue tit.[602] In fact, in the common yellowthroat, a small songbird, males with more colorful plumage invest *less* in parental care than do duller males, leading to a rejection of the "good parent hypothesis" for the evolution of bright coloration in males of this species.[990]

One alternative explanation for the ability of some male yellowthroats to produce intensely yellow plumage is that these individuals gain by attracting and mating with females other than their primary mates. If true, we could account for why bright males are poor parents—because they invest their time and energy in securing extra-pair matings. Or it could be that the males with more yellow are signaling their excellent physiological condition to rival males,[990] which enables them to hold superior territories. According to this argument, even though males have to spend much of their time chasing off intruders, the superior quality of their territories might provide their social partners with extra food for their nestlings. In the case of the "good territory" hypothesis, we could rescue the argument that females prefer to pair off with brightly colored males because these males provide for their offspring, albeit indirectly through their control of real estate with superior food resources. The lesson to be drawn from the research reviewed here is that the good parent hypothesis needs to be tested and retested against alternative hypotheses in every case.

## Discussion Question

**10.15** Males of the barn swallow have thin outer tail feathers that are somewhat longer than those possessed by females. When Anders Møller analyzed the effect of tail length on male mating success in the barn swallow in Europe, he did an experiment in which he made some males' tail feathers shorter by cutting them and made other males' tail feathers longer by gluing feather sections onto their tails.[999] But he also created a group in which he cut off parts of the males' tail feathers and then simply glued the fragments back on to produce a tail of unchanged length. What was the point of this group? And why did he randomly assign his subjects to the shortened, lengthened, and unchanged tail groups? And why did a team of Canadian biologists repeat Møller's experiment on another continent?[1351] And why did yet another team of British ornithologists study the effect of the tail "streamers" on the maneuverability of male swallows, given their interest in female mate choice?[178]

## *Mate Choice without Material Benefits*

Some of the examples that we have just presented are consistent with **good parent theory**, which explains aspects of male color, ornamentation, and courtship behavior as sexually selected indicators of a male's capacity to provide parental care. Female choice based on these signals makes intuitive sense: the offspring of female sticklebacks or blue tits that pair off with a more-paternal-than-average male are going to be unusually well fed as a rule. Likewise, it is easy to see why females of some species might prefer males able to sup-

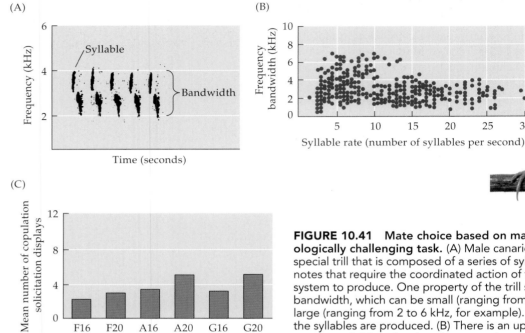

(A)

Frequency (kHz)

Syllable

Bandwidth

Time (seconds)

(B)

Frequency bandwidth (kHz)

Syllable rate (number of syllables per second)

(C)

Mean number of copulation solicitation displays

F16  F20  A16  A20  G16  G20

Trill type

**FIGURE 10.41  Mate choice based on male performance of a physiologically challenging task.** (A) Male canaries sing songs that contain a special trill that is composed of a series of syllables, each composed of two notes that require the coordinated action of the syrinx and the respiratory system to produce. One property of the trill syllables is their frequency bandwidth, which can be small (ranging from 2 to 4 kHz, for example) or large (ranging from 2 to 6 kHz, for example). Another is the rate at which the syllables are produced. (B) There is an upper limit on how fast males can sing syllables of a given frequency bandwidth. (C) Females prefer trills that are composed of broad-bandwidth syllables sung at a very fast rate. Tapes of three trill types were played to females: F had the narrowest bandwidth and G the greatest. Each trill type was played at two rates, 16 syllables per second and 20 syllables per second. The measure of female preference for the trills was the mean number of copulation solicitation displays given by listening female canaries. Females responded with significantly more displays to A20 and G20 trills. After Drăgăniou, Nagle, and Kreutzer.[411]

ply them personally with food or some other present before copulating. For example, a female black-tipped hangingfly that receives a large nuptial gift does not have to search for prey (the food she needs to live and produce eggs), a task that involves the very real possibility of flying into a spider's web.[1432] However, males of many species, such as the satin bowerbird, do not provide food or any other material benefit to their mates or their offspring. Even so, female bowerbirds prefer males with more ornaments (in their bowers) and the ability to court more intensely.[142, 1109] The same is true for many other animals.[785]

For example, despite the fact that male canaries do not help rear their young, a female canary's choice of a mate appears to be heavily influenced by his ability to sing a certain portion of the male song, the "A phrase," which is composed of many two-note syllables (Figure 10.41). Females that hear an A-phrase trill that packs many syllables into a second of song readily adopt the precopulatory position (if they have been primed with estradiol in advance of the experiment).[1483]

Passing a female canary's song test requires that males not only generate a rapid trill, but also make the individual syllables in the trill cover a relatively wide range of sound frequencies (the bandwidth of the trill). We know this because of the responses of females to tapes of artificial trills, including some that were impossibly exaggerated versions of the A phrase. The most extreme versions of the trill elicited the most copulation solicitation displays from listening female canaries. Because there is an upper limit for male canaries with respect to how rapidly they can sing syllables of a given bandwidth,

**FIGURE 10.42   A sexually selected ornament.** The extraordinary tail feathers of the long-tailed widowbird are displayed to choosy females while the male flies above his grassland territory.

the female preference in effect favors males able to sing at the outer edge of song capacity. Females may reward these individuals not just by mating with them, but also by adding testosterone to the eggs fertilized by their sperm, a maternal investment that other birds also make when mated by attractive males[892] and one that may enhance the chances of optimal development for these testosterone-boosted offspring.[537] (You may want to ask yourself why females do not add testosterone to all their eggs instead of just those fertilized by attractive males.)

Females of other species with nonparental males are also discriminating in their choice of mates. So it is that peahens apparently prefer peacocks with relatively large numbers of eyespots in their immense tails, which the males display in their famous spread-and-shake advertisements in front of potential mates.[1122] The importance of these decorations to females was demonstrated when Marion Petrie and Tim Halliday captured some peacocks and removed 20 of the outermost eyespots from certain males' tails. Birds so treated experienced a significant decline in mating success compared with their performance in the previous year. In contrast, control males, which were captured and handled but whose tails were left intact, were no less attractive to females afterwards,[1125] but see Takahashi et al.[1421]

If greater sensory stimulation from courting males is attractive to females, then experimentally augmenting a male's courtship ornaments should enhance his copulatory success. The relevant experiment was done first with the long-tailed widowbird, a species about the size of a red-winged blackbird that possesses an absurdly long tail (Figure 10.42). Males fly around their grassland territories in Kenya, displaying their magnificent tails to passing females. Malte Andersson took advantage of the wonders of superglue to perform an ingenious experiment. He captured male widowbirds, then shortened the tails of a group by removing from each member a segment of tail feathers, only to glue it onto another bird's tail, thereby lengthening that ornament.[33] The tail-lengthened males were much more attractive to females than those that had much-reduced ornaments. Moreover, the tail-lengthened males also did much better than controls, whose tails had been cut and then put back together.

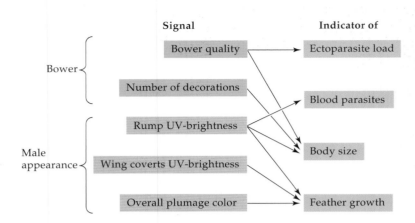

**FIGURE 10.43** Male satin bowerbirds provide females with multiple signals that are indicators of their health and physiological condition. After Doucet and Montgomerie.[408]

The elements of courtship that females may use to assess the quality of potential mates vary across the animal kingdom. Sexy songs and displays of extreme ornaments work for some species (Figure 10.43), but in others the male's performance during copulation itself might constitute a kind of test of male quality, an idea originally proposed by Bill Eberhard[425] (see also Hosken and Stockley[683]). If females evaluate partners on the basis of copulatory performance, then in species whose females mate with several males in a breeding cycle, males might be predicted to possess more elaborate penises, which would be capable of providing greater sensory stimulation during mating. In keeping with this prediction, the intromittent organs of many insects are remarkably complex, and genital variation among males can indeed affect their reproductive success. For example, male oriental beetles have a spiny hook on the aedeagus (penis); the larger the hook, the greater the proportion of eggs fertilized by the male following an initial copulation with a female that mates twice.[1537] The greater reproductive success of males endowed with large "spicules" could be due to some aspect of sperm competition among males, but it could also occur because of **cryptic female choice**, that is, choice generally hidden from view of researchers that is based on the internal workings of the female's reproductive machinery.[425, 1513]

### Discussion Question

**10.16** Peacocks and long-tailed widowbirds have truly extravagant ornamentation that could be fairly recently evolved (that is, derived) from an ancestral pattern in which male plumage was not nearly so extreme. In the peacock-pheasants, six species in a genus related to that of the peafowl, there is considerable variation in the degree of male plumage ornamentation. In four species, males have highly elaborate plumage featuring large eyespots, but in two species, they do not. Draw two phylogenies, one in which elaborate plumage is the derived trait and another in which a reduction in ornamentation is the derived condition. Then test your hypotheses with data available in Kimball et al.[772]

### Does Male Courtship Signal Mate Quality?

No matter what the basis for female preferences for one male's courtship or copulatory skill over another, a key question is, does female choosiness translate into fitness gains? This question carries real weight given that choosy females might pay a substantial price in terms of time and energy devoted to

**TABLE 10.3** *Four theories on why extreme male ornamentation and striking courtship displays have evolved in species in which males provide no material benefits to their mates*

| Theory | Females prefer trait that is: | Primary adaptive value to choosy females |
| --- | --- | --- |
| Healthy mates | Indicative of male health | Females (and offspring) avoid contagious diseases and parasites |
| Good genes | Indicative of male viability | Offspring may inherit the viability advantages of their father |
| Runaway selection | Sexually attractive | Sons inherit trait that makes them sexually attractive; daughters inherit the majority mate preference |
| Chase-away selection | Exploitative of preexisting sensory biases | No benefit received by female |

the mate selection process. For example, female marine iguanas are estimated to spend on average about 2 percent of their daily energy budget evaluating potential mates during the breeding season; females that inspect the most active displaying males spend even more.[1496] A few percentage points may not seem like a great deal, but choosy females do lose weight during the mate-choice period, and this could well reduce their survival chances if an El Niño year follows the breeding season, making it hard for the females to find the algae they need to eat. Likewise, female pronghorn antelope in estrus that visit a number of males before mating are estimated to expend the equivalent of a half-day's energy budget in the activity, another nontrivial amount.[225]

Now, it could be that females generally do not benefit by being choosy but instead are being manipulated by showy males, which gain if they are persuasive, while the females that select them actually lose fitness as a result. On the other hand, if male courtship and copulatory behavior are linked to some aspect of genuine male quality, then choosy females may leave more surviving descendants as a result of their preferences. We shall consider four major explanations for female preferences (Table 10.3).

One element of male quality that can affect female fitness is the health of a sexual partner. Unhealthy males could give their mates any of a variety of unpleasant sexually transmitted diseases, making it advantageous for females to avoid them. According to the **healthy mates theory**, female preferences are focused on a potential sexual partner's health or parasite load as indicated by his courtship displays and appearance.[1214] Females could use these traits to mate with males that are less likely to give them lice, mites, fleas, or bacterial pathogens, any one of which could harm them or their future offspring. As we noted earlier, male bowerbirds with high-quality bowers are less likely to carry transmissible ectoparasites in their feathers.[408]

Of course, females that mate with healthy males may not only steer clear of contagious parasites and diseases, but they may also acquire sperm that supply the genetic basis for good health in their offspring. The **good genes theory** proposes that preferences for certain male ornaments and courtship displays enable females to choose partners whose genes will help their offspring develop physiological mechanisms to combat infection and disease. In some species, for example, females might be able to evaluate (unconsciously) the strength of a male's immune system by his courtship displays. One possible example involves the cricket *Teleogryllus oceanicus*, whose females prefer to approach artificial male songs that have been manipulated to sound like those sung by males with strong immune systems as opposed to songs that sound like those sung by males with weak immune systems.[1463]

In other species, a male's appearance could be correlated with his hereditary resistance to parasites, a valuable attribute to pass on to offspring. Indeed W. D. Hamilton and Marlene Zuk predicted that selection for honest signals of noninfection would lead bird species with numerous potential parasites to evolve strikingly colored plumage. They argued that brightly colored feathers are difficult to produce and maintain when a bird is parasitized, because parasitic infection causes physiological stress. Hamilton and Zuk found the predicted correlation between plumage brightness and the incidence of blood parasites in a large sample of bird species, supporting the view that males at special risk of parasitic infection engage in a competition that signals their condition to choosy females.[611]

In addition, "good genes" derived from males could be involved in the development of other fitness-advancing traits besides resistance to parasites and diseases. For example, if females had a way to avoid genetically similar males or to identify males with high levels of heterozygosity, these selective females might well help their offspring avoid the developmental problems that can occur when individuals have two copies of certain recessive alleles, as may happen to offspring produced by inbreeding. Perhaps this is why female sedge warblers prefer males with larger song repertoires. Given that repertoire size is correlated with male heterozygosity in this species, female preferences for males with large repertoires should increase the heterozygosity of their offspring, particularly since females somehow ensure that their eggs are fertilized by those sperm that are genetically least like the eggs' genomes.[941]

A different view of what constitutes "good genes" is embodied in **runaway selection** theory.[46, 368] This approach argues that discriminating females acquire sperm with genes whose primary effect is to influence their daughters to prefer the male traits their mothers found attractive and to endow their sons with attributes that will be preferred by most females—even if those traits actually reduce the survival chances of individuals that possess them. For example, a preference for the elaborate song of male canaries could be adaptive for females if their sons inherit the capacity to sing attractive songs, even if they are costly to produce, because these songsters may be especially appealing to females in the next generation.

Because the runaway selection alternative is the least intuitively obvious explanation for extreme male courtship displays, let us sketch the argument underlying the mathematical models of Russell Lande[833] and Mark Kirkpatrick.[776] Imagine that a slight majority of the females in an ancestral population had a preference for a certain male characteristic, perhaps initially because the preferred trait was indicative of some survival advantage enjoyed by the male. Females that mated with preferred males would have produced offspring that inherited the genes for the mate preference from their mothers and the genes for the attractive male character from their fathers. Sons that expressed the preferred trait would have enjoyed higher fitness, in part simply because they possessed the key cues that females found attractive. In addition, daughters that responded positively to those male cues would have gained by producing sexy sons with the trait that many females liked.

Thus, female mate choice genes as well as genes for the preferred male attribute could be inherited together. This pattern could generate a runaway process in which ever more extreme female preferences and male attributes spread together as new mutations affecting these traits occurred. The runaway process would end only when natural selection against costly or risky male traits balanced sexual selection in favor of traits that appealed to females. Thus, if peahens originally preferred peacocks with larger-than-average tails because such males could forage efficiently, they might now favor males with extraordinary tails because this mating preference has taken on a life of its own, resulting in the production of sons that are exceptionally attractive to females and the production of daughters that will choose this kind of male for their own mates.

In fact, the Lande–Kirkpatrick models demonstrate that, right from the start of the process, female preferences need not be directed at male traits that are utilitarian in the sense of improving survival, feeding ability, and the like. Any preexisting preference of females for certain kinds of sensory stimulation (see page 300) could conceivably get the process under way. As a result, traits opposed by natural selection because they reduced viability could still spread through the population by runaway selection.[776, 833] Instead of mate choice based on genes that promote the development of useful characteristics in offspring, runaway selection could yield mate choice for arbitrary characters that are a burden to individuals in terms of survival, a disadvantage in every sense except that females mate preferentially with males that have them!

### Testing the Healthy Mate, Good Genes, and Runaway Selection Theories

Discriminating among these three alternative explanations for elaborate male courtship displays and female choice has proved very difficult, in part because these three hypotheses are not mutually exclusive. As just mentioned, female preferences and male traits that originated through a good genes process could then be caught up by runaway selection. Note, too, that males with hereditary resistance to certain parasites (good genes benefits) would also be less likely to infect their partners with those parasites (healthy mate benefits). Moreover, if at the end of a period of runaway selection, males had evolved extreme ornaments and elaborate displays, then only individuals in excellent physiological condition would be able to develop, maintain, and deploy their ornaments in effective displays. Males in such superb physiological condition would probably have to be highly effective foragers (good genes benefits) as well as parasite-free (healthy mate benefits), in which case females mating with such males would be unlikely to acquire parasites, and their daughters might well get some survival-benefiting genes while their sons received the pure attractiveness genes of their fathers.

Given the overlap among these three theories, let's just pick one and try to test it. As an example, Marion Petrie applied good genes theory to peacocks and derived the following predictions: (1) males should differ genetically in ways related to their survival chances, (2) male behavior and ornamentation should provide accurate information on the survival value of the males' genes, (3) females should use this information to select mates, and (4) the offspring of the chosen males should benefit from their mothers' mate choice. In other words, males should signal their genetic quality in an accurate manner, and females should pay attention to those signals because their offspring would derive hereditary benefits as a result.[784, 1635]

Petrie studied a captive but free-ranging population of peacocks in a large forested English park, where she found that males killed by foxes had significantly shorter tails than their surviving companions. Moreover, she observed that most of the males taken by predators had not mated in previous mating seasons, suggesting that females could discriminate between males with high and low survival potential, possibly on the basis of their ornamented tails.[1123] The peahens' preferences translated into offspring with enhanced survival chances, as Petrie showed in a controlled breeding experiment. She took a series of males with different degrees of ornamentation from the park and

Peacock

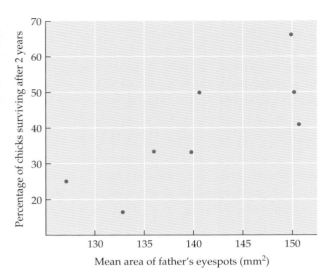

**FIGURE 10.44   Do male ornaments signal good genes?** Peacocks with larger eyespots on their tails produced offspring that survived better when released from captivity into an English woodland park. After Petrie.[1124]

paired each of them in a large cage with four females chosen at random from the population. The young of all the males were reared under identical conditions, weighed at intervals, and then eventually released back into the park. The sons and daughters of males with larger eyespots on their ornamented tails weighed more at day 84 of life and were more likely to be alive after 2 years in the park than the progeny of males with fewer eyespots (Figure 10.44).[1124]

This combination of results is consistent with the view that peahen mate preferences create sexual selection that is currently maintaining or spreading the genetic basis for those preferences because the offspring of the preferred males receive good genes from their fathers. In this regard, a team of French researchers studying peacocks have shown that the number of eyespots in the tail and the display rate are indeed linked to male health. Males with more eyespots had lower concentrations of a particular kind of white blood cell induced by infection, and these males, presumably healthy individuals, engaged in more courtship displays per hour than those with higher levels (Figure 10.45).[891] Because females prefer actively courting males with attractive tails, they seem likely to mate with males with genes for good health, assuming that a male's ability to resist infection is at least partly hereditary.

But, as noted above, perhaps healthy male benefits are involved as well if, for example, peahens choose to avoid parasitized males and so reduce their risk of acquiring nasty parasites that they would pass on to their offspring. Moreover, a demonstration of current selective advantages associated with female preferences and male traits does not rule out the possibility that these attributes originated as side effects of runaway selection in the manner described above.

Yet another complication comes from the discovery that females of some species invest more resources in offspring that are sired by preferred males than in those fathered by less attractive males. As noted earlier, in some species the female adjusts the amount of testosterone she contributes to an egg in relation to the attractiveness of her mate.[537, 892, 1292] Likewise, female mallards make larger eggs

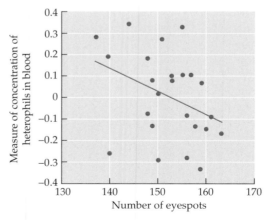

**FIGURE 10.45   Peacocks with many eyespots tend to be healthier than those with fewer eyespots,** judging from the lower concentrations of heterophils in their blood. Heterophils are white blood cells whose count goes up when a bird is fighting infection. After Loyau et al.[891]

after copulating with unusually attractive males,[328] while female black grouse produce and lay more eggs subsequent to mating with top-ranked males.[1225] All of these effects might easily be attributed to the "good genes" of the male, when in reality they arise from the female's manipulation of her own parental investment, not the genetic contribution of her partner. Untangling the contributions of the two parents to the welfare of their offspring is a challenging business.

## Sexual Conflict

Although cooperation between the sexes is often involved in sexually reproducing species, particularly in those in which both males and females rear the young together, nevertheless sexual conflict also appears to be common.[1105] Consider that females often turn down sexually motivated males, as seen in the typical response of female bowerbirds to displaying males. Much less frequently, males may also turn down potential sexual partners, as happens, for example, in the African topi. The mating restraint shown by this antelope's popular males, which occupy central positions in their leks (see spage 411), stems from the fact that these males can become sperm depleted as they copulate with a whole series of receptive females. They therefore may refuse to mate again with females they have already inseminated, in order to save sperm to donate to new partners. A jilted female of this antelope sometimes responds by attacking the male and disrupting his efforts to mount a newcomer (Figure 10.46).[177]

Sexual conflict can become even more unpleasant when, for example, males kill the infants of females in order to make them become sexually receptive sooner than otherwise (see page 23) or to force apparently unwilling females to mate with them. In these cases, it is difficult (but not impossible—see below) to believe that the male's behavior is advantageous to the female. When a male hangingfly grabs a female by a wing and mates with her without providing a food gift, she loses a meal. When orangutan females give in and mate

**FIGURE 10.46  Sexual conflict in a lekking species.** A female topi attacks a male on his display grounds after he refuses to copulate with her again. Males that can become sperm depleted may gain by refusing to mate with every possible copulatory partner. The male's refusal to mate may generate aggression on the part of the rejected female. Photograph by Jakob Bro-Jørgensen.[177]

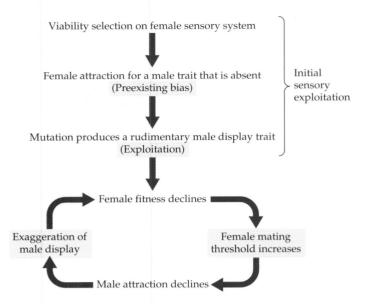

Viability selection on female sensory system

Female attraction for a male trait that is absent
(Preexisting bias)

Mutation produces a rudimentary male display trait
(Exploitation)

} Initial sensory exploitation

Female fitness declines

Female mating threshold increases

Male attraction declines

Exaggeration of male display

**FIGURE 10.47  Chase-away selection theory.** The evolution of extreme male ornaments and displays may originate with exploitation of females' preexisting sensory biases. If sensory exploitation by males reduces female fitness, the stage is set for a cycle in which increased female resistance to male displays leads to ever greater exaggeration of those displays. After Holland and Rice.[668]

with the sexually eager young males that have been harassing them for days on end, it looks as if they are mating against their will. Given a choice, female orangutans seek out huge, older adult males with whom they mate and remain for long periods while their partners deal with the apparently unwelcome advances of the small males.[485] Likewise, the female chimpanzees in a band do not synchronize their estrous cycles, perhaps so that there will usually be at least one dominant male available to guard them against lower-ranking males prone to harassment.[952] On the other hand, dominant and subordinate male chimpanzees alike often sexually assault females when they are fertile.[1020] Some of these violent interactions lead to the death of the female[1238] and the same is true for our own species as well.[199] Extreme conflict of this sort obviously does not benefit either the deceased female or the murderous male.

Sexual conflict plays a central role in a fourth general theory for why males have evolved extreme ornaments and elaborate courtship displays (see Table 10.3).[667] According to Brett Holland and Bill Rice, these traits could be the result of **chase-away selection**, a process that begins when a male happens to have a mutation for a novel display trait that manages to tap into a preexisting sensory bias that affects female mate preferences in his species. Such a male might induce females to mate with him even though he might not provide the material or genetic benefits offered by other males of his species.[1264] The resulting spread of such exploitative males over time would create selection on females favoring those that were psychologically resistant to the purely attractive display trait. As females with a higher threshold for sexual responsiveness to the exploitative trait spread, selection would then favor males able to overcome female resistance, which might be achieved by mutations that further exaggerated the original male signal. A cycle of increasing female resistance to, and increasing male exaggeration of, key characteristics could ensue, leading gradually to the evolution of costly ornaments of no real value to the female and useful to the male only because without them, he would have no chance of stimulating females to mate with him (Figure 10.47).

Chase-away selection theory illustrates how far some evolutionary biologists have come from the once-popular view of sexual reproduction as a gloriously cooperative enterprise designed to perpetuate the species. Instead, many behavioral biologists now see reproduction as an activity in which the two sexes battle for maximum genetic advantage, even if one member of a pair loses fitness as a result.

For example, because males of the fruit fly *Drosophila melanogaster* benefit by inducing females to use their sperm rather than the sperm of rivals, their seminal fluid contains chemicals that have damaging side effects on their partners. A prime culprit appears to be the protein Acp62F,[898] which boosts male fertilization success (perhaps by damaging rival sperm) at the expense of females, whose lives are shortened and whose fecundity is lowered.[1150] Despite the negative long-term effects that toxic protein donors have on their mates, males still gain because they are unlikely to mate with the same female twice. Under these circumstances, a male that fertilizes more of one female's current clutch of eggs can derive fitness benefits even though his chemical donations reduce the lifetime reproductive success of his partner.

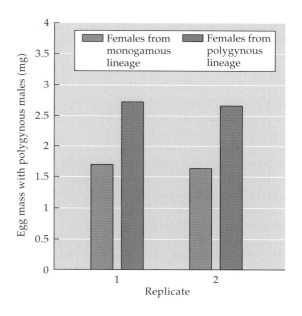

**FIGURE 10.48  Sexual selection and the evolution of male traits harmful to females.** Females from an experimental monogamous lineage of fruit flies have lost much of their biochemical resistance to the damaging chemicals present in the seminal fluids of polygynous male fruit flies. Therefore, monogamous females lay fewer eggs when mated with males from a control polygynous lineage than do control females that evolved with those males. The results of an experiment repeated twice are shown here as replicates 1 and 2. After Holland and Rice.[668]

If male fruit flies really do harm their mates as a consequence of their success in sperm competition, then placing generations of males in laboratory environments in which no male gets to mate with more than one female should result in selection against donors of damaging seminal fluid. Under these conditions, any male that happened not to poison his mate would reap a fitness benefit by maximizing his single partner's reproductive output. In turn, a reduction in the toxicity of male ejaculates should result in selection for females that lack the chemical counteradaptations to combat the negative effects of the spermicide protein. Indeed, after more than 30 generations of selection in a one male–one female environment, females from the monogamous population that were mated once with spermicide-donating "control" males taken from a typical multiple-mating population laid fewer eggs and died earlier than females who had evolved with polygynous, spermicide-donating males (Figure 10.48).[668]

## Discussion Question

**10.17**  Stuart Wigby and Tracey Chapman formed three populations of fruit flies with different sex ratios (female biased, even sex ratio, and male biased). Not surprisingly, the frequency with which females mated increased from female-biased to male-biased populations. After 18 and 22 generations of selection, fresh females from the three selected lines were taken from their environments and placed in cages with equal numbers of males. The mortality rate of females from the male-biased line was less than that of females from the even-sex-ratio line, and much less than that of females from the female-biased line.[1570] What do these results tell us about the evolutionary consequences of sexual conflict between the sexes in this species?

In the light of chase-away selection theory, it is revealing that female fruit flies actually reduce their fitness by preferring to mate with larger males, which they choose either because they find large body size an attractive feature in and of itself or because large body size is correlated with some other attractive feature, such as more persistent courtship. Whatever the reason for their preference, mate choice by females based on this characteristic lowers their

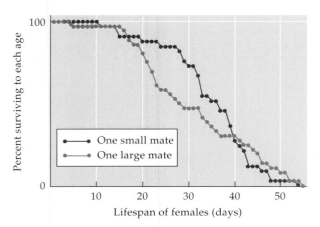

● One small mate

● One small mate
● One large mate

**FIGURE 10.49  Mating with large males reduces female fitness in fruit flies.** Even though females prefer to mate with large males, mortality rates are higher for females mated to larger males. After Friberg and Arnqvist.[492]

longevity (Figure 10.49) and also reduces the survival of their offspring.[492] These negative effects on female fitness may arise from the physiological costs of dealing with increased rates of courtship by "preferred" mates and possibly the increased quantities of toxins received from "attractive" males.[1150] These findings can be taken as evidence that, for the moment, large males are ahead in a chase-away arms race between the sexes in the fruit fly.

Another kind of arms race may be taking place in bedbugs. In this unpleasant insect, a male uses a knifelike intromittent organ to stab a female in her abdomen before injecting his sperm directly into her circulatory system (Figure 10.50).[1401] This highly unusual method of insemination presumably evolved originally as a result of males injecting their sperm in this way into sexually resistant females that had already acquired sperm from earlier partners via the traditional, less damaging mode of insemination. If so, then traumatic insemination ought to be costly to females, and it is, given that females mating at high frequencies live fewer days and lay fewer eggs.[1012, 1401]

But not everyone thinks that every case of apparent sexual conflict means that females are actually being harmed by males intent on reproducing at the expense of their mates. Let us accept that in some instances males do things that reduce female longevity. Let us also acknowledge that the genes of successful male fruit flies may be damaging when expressed in these males' daughters rather than in their sons.[1147] But these and other costs could be outweighed by the exceptional reproductive success of a female's adult sons.[1105] If so, the females that permitted the fathers of these sons to mate with them may be more than compensated by the increased reproduction of their male

(A)

(B)

**FIGURE 10.50  A genital product of conflict between the sexes?** (A) Male bedbugs have evolved a saberlike penis that they insert directly into the abdomen of a mate prior to injecting her with sperm. The trait may have originated when males benefited by employing traumatic insemination to overcome resistance to mating by unwilling females. (B) The white box shows the site on the female's abdomen that males penetrate with the penis. A, photograph by Andrew Syred; B, photograph by Mike Siva-Jothy.

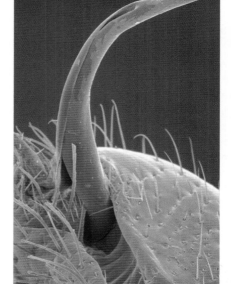

offspring. Females might even accept a reduction in their lifetime reproductive success if they produce sons that inherit effective manipulative, even coercive, tactics from their fathers—tactics that may make their sons unusually successful reproducers, thereby endowing their mothers with extra grandoffspring.[297] Apparent physical conflict among males and females could even be the way that females judge the capacity of males to supply them with sons that will be able to overcome the resistance to mating by females in the next generation.[427] Alternatively, aggressive courtship by males may be the way in which females choose males whose sons will do well in aggressive competition with other males.[147] In somewhat the same vein, although sneaker males (see page 352) are generally viewed as low-quality mates for females, females might benefit from mating with these males on occasion if their sperm increases the genetic diversity of the female's offspring.[1208] These kinds of hypotheses suggest that females could be "winning by losing" when they mate with males who appear to be forcing them to copulate or to be blocking their apparent preference for other males.

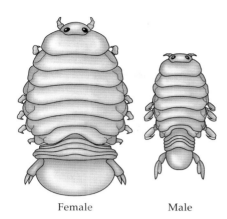

Female         Male

**FIGURE 10.51   A mutually cannibalistic species: the ultimate in sexual conflict.** Either the larger female or the smaller male of the freshwater isopod *Ichthyoxenus fushanensis* may kill and consume its partner. Drawing courtesy of C.-F. Dai.

## Discussion Question

**10.18**  In the parasitic isopod *Ichthyoxenus fushanensis* (Figure 10.51), a male lives with a female in a cavity they construct in their victim, a freshwater fish. Sexual cannibalism is not uncommon in this isopod.[1469] Females sometimes eat males early in their lengthy breeding season; males sometimes eat females later on. Replacement partners of both sexes are readily available at these times. In addition, this animal is one in which males can transform themselves into females, a phenomenon that occurs only when males have achieved a fairly large size. The larger the female, the more fecund she is. The smaller the difference in size between male and female, the fewer offspring produced by a pair. Why has conflict between the sexes reached such an extreme state in this species? What kinds of males are females expected to eat? What kinds of males are expected to eat their partners? How can you account for the differences in the timing of cannibalism by males and females?

## Summary

1. Sexual reproduction creates a social environment of competition among individuals as each strives to maximize its genetic contribution to subsequent generations. Males usually make huge numbers of very small gametes and often try to fertilize as many eggs as possible, while providing little or no care for their offspring. In contrast, females make fewer, larger gametes and often provide parental care as well. As a result, receptive females are scarce, and males typically compete for access to them, while females can choose among many potential partners.

2. Evolution by sexual selection occurs if genetically different individuals differ in their reproductive success because of differences in their ability (1) to compete with others of their own sex for mates, or (2) to attract members of the other sex.

3. The competition-for-mates component of sexual selection is behind the evolution of many elements of male reproductive behavior, including competition for social dominance, alternative mating tactics, and mate guarding after copulation. Although in a typical species, males exert selection pressure on one another in the competition for mates, females generally have the last word on reproduction, because they control the production and fertilization of eggs.

4. In a typical species, females choose among potential mates, creating the mate choice component of sexual selection. Males of some species seek to win favor with females by offering them material benefits, including nuptial gifts or parental care.

5. Mate choice by females occurs even in some species in which males provide no material benefits of any sort. Female preferences for elaborate ornaments may arise because males with these attributes are healthy and parasite-free, and so less likely to transmit disease or parasites to them (healthy mates theory). Or choosy females may gain by securing genes that enhance the viability of their offspring (good genes theory). On the other hand, extravagant male features could spread through a population in which even arbitrary elements of male appearance or behavior became the basis for female preferences. Exaggerated variants of these elements could be selected strictly because females preferred to mate with individuals that had them (runaway selection theory). A fourth possibility is that the extreme ornaments of males evolve as a result of an upward-spiraling cycle of conflict between the sexes, with males selected for ever-improved ability to exploit female perceptual systems and females selected to resist those males ever more resolutely (chase-away selection theory). The relative importance of these various mechanisms of sexual selection remains to be determined.

6. Interactions between the sexes can be viewed as a mix of cooperation and conflict as males seek to win fertilizations in a game whose rules are set by the reproductive mechanisms of females. Conflict between the sexes is widespread and includes sexual harassment and the transfer of damaging ejaculates, demonstrating that what is adaptive for one sex may be harmful to the other.

## Suggested Reading

To learn more about satin bowerbird sexual behavior, see the papers written by Gerry Borgia[143, 144] and an elegant book on bowerbirds by Clifford and Dawn Frith.[494] Malte Andersson's book on sexual selection is remarkably comprehensive,[35] while alternative mating tactics have been examined by Mart Gross[590] and, quite differently, by Steve Shuster and Michael Wade.[1328] You can learn about sperm competition from the classic paper by Geoff Parker,[1103] which focuses on insects, as does the book by Leigh Simmons.[1330] Sperm competition in other animal groups is covered by Tim Birkhead and Anders Møller,[122] while Bill Eberhard examines how females might control which of several males' sperm stored within them will fertilize their eggs.[425] A book on the house finch by Geoff Hill explores theories on the evolution of colorful ornaments as well as describing how behavioral ecologists actually use sexual selection theory in their research.[654] Finally, although this chapter did not explore the subject of how sexual reproduction evolved in the first place, you can learn about it from G. C. Williams[1579] and Bob Trivers,[1468] as well as Laurence Hurst and Joel Peck.[700]

# 11

# The Evolution of Mating Systems

M ale satin bowerbirds, as we saw in the previous chapter, are capable of copulating with dozens of females in a single breeding season, although they rarely have the good fortune to do so.[142] In other words, males of this species have the capacity to be polygynous. In contrast, female satin bowerbirds are almost always monogamous, typically mating with just one male per nesting attempt. But the satin bowerbird's mating system (males potentially polygynous, females monogamous) is only one of a variety of arrangements found in the animal kingdom. In fact, different mating systems can be found even among the bowerbirds. For example, both males and females of the monogamous green catbird, a close relative of the polygynous satin bowerbird, pair off one by one before rearing offspring together.[493] In some other birds, such as the spotted sandpiper, polyandry is the order of the day: females copulate with two or three males in a breeding season.[1073] A still more extreme version of this pattern is exhibited by the honey bee, whose young queens fly out from their hives into aerial swarms of drones that pursue, capture, and mate with them in midair. The average queen is highly polyandrous, coupling with many males and using the sperm of perhaps a dozen or so during her lifetime of egg laying.[1423] In contrast, drones never mate with more than one queen, because a drone violently propels his genitalia into his first and only mate, a suicidal act that ensures that he is both monogamous and, shortly thereafter, dead.[1620]

The diversity of mating systems offers a rich banquet of Darwinian puzzles for the evolutionist, including: (1) why are males ever voluntarily monogamous, (2) why do females of some species practice polyandry, and (3) why do males of different species exhibit so many different tactics to achieve polygyny? These puzzles are the central focus of this chapter.

## Is Male Monogamy Adaptive?

Although examples of monogamy by males are not common, they do occur, as we have just indicated. But why should a male honey bee (Figure 11.1) or a male prairie vole (see Figure 1.1) or the male of any other species restrict himself to a single mate? The standard rule, which was first discovered in a now classic study of *Drosophila* fruit flies,[83] is that the more females inseminated, the more eggs fertilized and the greater the reproductive success of a male (see page 340). As a result, males typically compete for mates, and the winners are polygynists that are likely to gain higher fitness than their monogamous rivals. For example, male house wrens that attract two females father about nine fledglings per year on average, whereas monogamous male wrens have fewer than six.[1374] Viewed in these terms, males that voluntarily restrict themselves to one sexual partner are a mystery.

However, we earlier presented an explanation for male monogamy in prairie voles (see page 5): if females remain receptive after mating, males that prevent a partner from accepting sperm from other males could leave more descendants by monogamously attending to one female rather than trying to mate with several partners.[1611] This same mate-guarding hypothesis was also used to explain monogamy in certain spiders whose males have little chance of finding a second female and therefore give their all, quite literally, to their first mate. Those males that commit sexual suicide by feeding themselves to a partner (see Figure 10.40) or by breaking off their genital appendages in a partner's reproductive tract appear to increase their fertilization success with that one female. This gain from what might be called the male's posthumous mate

(A)

(B)

**FIGURE 11.1**   **The monogamous honey bee drone dies after mating.** (A) An intact male. (B) A queen with the yellow genitalia of a deceased partner attached to the tip of her abdomen. Photographs by Christal Rau, courtesy of Nikolaus Koeniger.

**FIGURE 11.2   A monogamous mate-guarding shrimp.** When a male clown shrimp encounters a potential mate, he remains with her because receptive females are scarce and widely distributed. Here, a couple feeds on the severed arm of a starfish. Photograph by Stephen Childs.

guarding may exceed the cost of his actions if (1) his mate has the potential to remain receptive after one mating and (2) the male's probability of finding a second female is extremely low. These conditions appear to be met in the spiders discussed previously[985] and in some other species as well.[438]

For example, males of the beautiful clown shrimp, *Hymenocera picta* (Figure 11.2), spend weeks in the company of one female.[1563] As expected, the operational sex ratio (see page 336) is highly male biased because the scattered females are receptive for only a short period every 3 weeks or so. Because finding the right female at the right time is highly time-consuming, a male that encounters a potential mate guards her until she is willing to copulate. Here we have an example of a recurrent theme in this chapter, namely, the distribution of females in response to various ecological factors has a major effect on the evolution of male mating tactics.

A different explanation for male monogamy has been labeled the **mate-assistance** hypothesis, which proposes that males remain with a single female because paternal care and protection of offspring are especially advantageous.[438] In some environments, the additional offspring that survive because of paternal effort may more than compensate the monogamous male for giving up the chance to reproduce with other females. Readers will be most familiar with male parental care in birds, but the phenomenon occurs in other groups as well. Males of the seahorse *Hippocampus whitei*, for example, even take on the responsibility of "pregnancy," carrying a clutch of eggs in a sealed brood pouch for about 3 weeks (Figure 11.3). Each male has a durable relationship with one female, who provides him with a series of clutches. Pairs even greet one another each morning before moving apart to forage separately; they will ignore any others of the opposite sex they happen to meet during the day.[1492] Since a male's brood pouch can accommodate only one clutch of eggs, he gains nothing by courting more than one female at a time. And, in fact, in another species of *Hippocampus*, genetic data indicate that males do not accept eggs from more than one female (even though in this species, groups of females have been seen courting single males under some conditions[1585]).

A male seahorse may not benefit by switching mates if his long-term mate can supply him with a new complement of eggs as soon as one pregnancy is

**FIGURE 11.3   Mate-assistance monogamy in a seahorse.** A pregnant male is giving birth to his single partner's offspring. Several youngsters can be seen emerging, some tail-first, from their father's brood pouch.

over. Many females apparently can keep their partners pregnant throughout the lengthy breeding season but cannot produce batches of eggs so quickly that they have some to give to other males.[1493] By pairing off with one female, a male may be able to match his reproductive cycle with that of his partner. When the two individuals are in sync, the male will complete one round of brood care just as his partner has prepared a new clutch. Therefore, he need not spend time waiting or searching for alternative mates, but will be able to immediately secure a new clutch from his familiar mate. In keeping with this prediction, the experimental removal of a partner forced males of a paternal pipefish to change mates, which added over 8 days on average to their interspawning interval, relative to males permitted to retain their mates from one clutch to the next. The increase in the time between spawnings stemmed largely from the days the male spent waiting for his new mate to produce a clutch of mature eggs.[1362]

In other species, although males might gain by acquiring several mates, females may block their partners' polygynous interactions in order to monopolize their parental assistance, leading to **female-enforced monogamy**. This hypothesis is similar to the mate-guarding hypothesis, but with the female as the guardian of monogamy in this instance. So, for example, paired females of the burying beetle *Nicrophorus defodiens* are aggressive toward intruders of their own sex. In this species, a mated male and female work together to bury a dead mouse or shrew, which will feed their offspring once they hatch from the eggs the female lays on the carcass. But once the carcass is buried, the male may climb onto an elevated perch and release a sex pheromone to call a second female to the site. If another female added her clutch of eggs to the carcass, her larvae would compete for food with the first female's offspring, reducing their survival or growth rate. Thus, when the paired female smells her mate's pheromone, she hurries over to push him from his perch. These attacks reduce his ability to signal, as Anne-Katrin Eggert and Scott Sakaluk showed by tethering paired females so that they could not suppress their partners' sexual signaling.[430] Freed from a controlling spouse, the experimental males released scent for much longer periods than control males that had to cope with untethered mates (Figure 11.4). Thus, male burying beetles may often be monogamous not because it is in their genetic interest, but because females make it happen—another example of the kinds of conflicts that occur between the sexes (see Chapter 10).

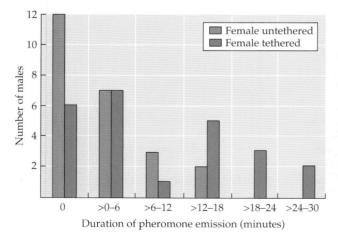

Burying beetle

**FIGURE 11.4  Female burying beetles combat polygyny.** When a paired female beetle is experimentally tethered so that she cannot interact with her partner, the amount of time he spends releasing sex pheromones rises dramatically. After Eggert and Sakaluk.[430] Photograph courtesy of C. F. Rick Williams.

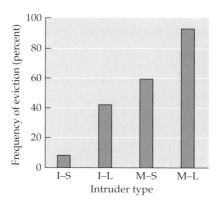

Emerald coral goby

FIGURE 11.5   Female-enforced monogamy may involve aggression by a breeding female toward other females. Breeding females of the emerald coral goby evict female intruders that are experimentally introduced into their coral territory, especially if the intruders are mature, large (M-L) individuals capable of reproducing. Immature, small (I-S) females are permitted to remain. Photograph by João Paulo Krajewski; after Wong et al.[1614]

Likewise, hostility toward other females by a dominant female emerald coral goby may force her partner to be monogamous even though in this species a number of females live with a single male in their coral retreat. The large dominant female in the group suppresses reproduction by the subordinate members, which accept their nonreproductive status because they may die if chased from their safe coral shelter. If they live long enough, they may eventually assume dominant status themselves. When they do, they will prevent other group members from breeding and will also attack intruder females, especially large mature females, which they usually force to leave the area (Figure 11.5).[1614] By keeping mature rivals from her home base, a female coral goby keeps her male partner all to herself.

A variation on this theme occurs when both males and females gain by guarding a partner of high quality, as, for example, when a male guards a large, highly fecund female and a female monopolizes a large, highly paternal male. Under these circumstances, **mate-guarding monogamy** provides benefits for both members of the pair, which makes its evolution easier to understand. Elizabeth Whiteman and Isabelle Côté argue that this form of monogamy is relatively common among marine fishes in which pairs defend territories against other duos (Figure 11.6).[1551]

FIGURE 11.6   A monogamous pair of cleaner wrasses. In this marine fish species, a male and a female cooperate in defense of a territory. Their territory attracts other fish species, from which the wrasses remove parasites and the like. Both individuals attempt to keep same-sex members of their own species away from their partners. Photograph by Liz Whiteman.

**Discussion Questions**

**11.1** We began this chapter with a mention of male monogamy in the honey bee. Try to explain the male's suicidal mating behavior in light of the alternative hypotheses on male monogamy outlined above. Also include in your list one hypothesis based on group selection theory. What predictions follow from the different explanations you have considered? What data are required to resolve the issue?

**11.2** In the starling, some males acquire several mates but do not assist them, whereas other, monogamous males work together with their sole partners to rear their broods. Sometimes when there are two females nesting on a male's territory, the first female to settle there attacks the clutch of the other female, piercing her companion's eggs with her beak.[1209] Why might she do so, and what kind of monogamy could result from her actions? Under what circumstances would the female's behavior actually qualify as a form of parental investment?

## Male Monogamy in Mammals

Although monogamy is not common in any group, this mating system is exceptionally rare in mammals, a lineage notable for the size of female parental investment in offspring. Sexual selection theory suggests that male mammals, which cannot become pregnant and do not offer milk to infants (with the exception of one species of bat[486]), should usually try to be polygynous—and they do. However, exceptions to the rule are useful for testing alternative hypotheses on male monogamy in this group.

For example, if the mate-assistance hypothesis for monogamy applies to mammals, then males of the rare mammalian species that exhibit paternal behavior[1615] should tend to be monogamous. One monogamous mammal with paternal males is the Djungarian hamster, whose males actually help deliver their partners' pups (Figure 11.7).[738] Male parental care contributes to offspring survival in this species and also in the monogamous California mouse, pairs of which were consistently able to rear a litter of four pups under

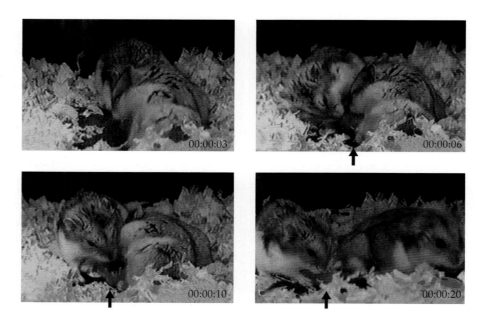

**FIGURE 11.7 An exceptionally paternal rodent.** A male Djungarian hamster may pull newborns from his mate's birth canal and then clear the infants' airways by cleaning their nostrils, as shown in these photographs (the male is the hamster on the left; arrows point to the pink newborn). From Jones and Wynne-Edwards.[738]

laboratory conditions, whereas single females did not do nearly as well.[227] The relationship also held under natural conditions, with the number of young reared by free-living California mice falling when a male was not present to help his mate keep the pups warm (Figure 11.8).[593]

In some mammals, one of the things a paternal male might do for his offspring is to protect them against infanticidal male intruders, which may destroy infants (see Figure 1.16) in order to mate with their mother. Earlier, we suggested that mate-guarding male prairie voles can and do defend their offspring against infanticidal rivals, so both mate guarding and mate assistance may contribute to the tendency toward monogamy in this species. If males and females form a social bond to reduce the risk of infanticide, then such monogamous associations should also evolve in primate species in which new mothers tend to travel with their offspring (rather than leaving them in a nest). Mothers with vulnerable young in tow need to have a protective male accompanying them if he is to defend the infant. In one group of primates, the prosimians, a nearly perfect correlation exists between infant carrying and year-round male–female pairings (Figure 11.9).[1485]

On the other hand, in a review of all primates, not just prosimians, Agustin Fuentes found little evidence in favor of the anti-infanticide hypothesis for monogamy. For example, in a number of primates in which a male forms a pair-bond with a female, females are larger than or dominant to males and are therefore in little need of help in dealing with aggressive males trying to cause trouble.[498] Moreover, a systematic comparative test of the proposition that mammalian monogamy goes hand in hand with male parental care produced completely negative results for primates, rodents, and all other groups.[797] So, for example, although male parental care occurs in half of the 16 primate taxa (species or groups of related species) known to be monogamous, it also characterizes 35 percent of the 20 polygynous taxa. The difference between the two categories is not statistically significant, contrary to the expectation that paternal primate species should be monogamous while nonpaternal species should be polygynous.

In fact, the only mammalian pattern that survives comparative analysis is that males tend to live with females in two-adult units more often when females live well apart from one another in small territories. The monogamous rock-haunting possum of northern Australia is a case in point. A female, her mate, and their young live along the edges of rock outcrops in territories of about 100 meters by 100 meters (Figure 11.10), an area about one-sixth of that occupied by other herbivorous mammals of equivalent weight.[1254] In their small territories, male possums can effectively monitor the activities of one female at relatively modest energetic expense. Any male that attempted to move between the home ranges of several females would run the risk that intruder males would visit those mates that he had temporarily left behind. Thus, ecological factors that enable females to live in small, defensible territories tilt the cost–benefit equation toward mate guarding, which then leads to male monogamy.

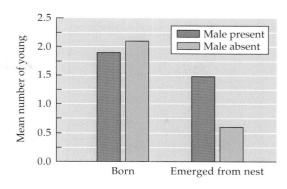

**FIGURE 11.8** **Male care of offspring affects fitness in the California mouse.** The mean number of offspring reared by female mice falls sharply in the absence of a helpful male partner. After Gubernick and Teferi.[593]

## Discussion Questions

**11.3** In a small African antelope called Kirk's dik-dik, most males and females live in monogamous pairs.[183] Evaluate alternative hypotheses for monogamy in this species in light of the following evidence: the presence of males does not affect the survival of their offspring; males conceal the female's estrous condition by scent-marking over all odors deposited by

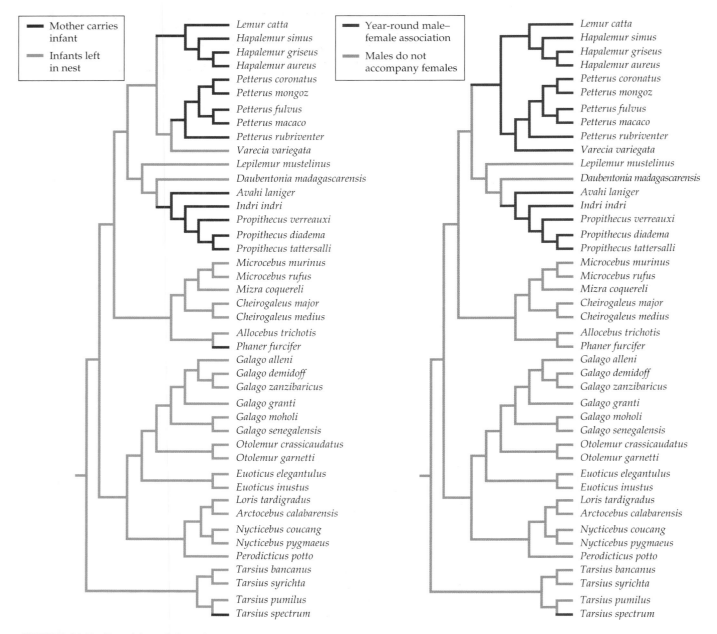

**FIGURE 11.9  Durable pair-bonds between males and females** have evolved independently four or five times in prosimian primates in which mothers carry their dependent infants with them. After van Schaik and Kappeler.[1485]

their mates in the pair's territory; males sire their social partner's offspring; females left unaccompanied wander from the pair's territory; some territories contain five times the food resources of others; the few polygynous associations observed do not occupy larger, or richer, territories than monogamous pairs of dik-diks.

**11.4**  In their classic paper on mating systems, Steve Emlen and Lew Oring suggested that two ecological factors could promote the evolution of monogamy: a high degree of synchrony in reproductive cycling within a population and a highly dispersed distribution of receptive females.[438] Try to reconstruct the logic of these predictions and then make counterarguments to the effect that synchronized breeding could facilitate acquisition of multiple mates while a relatively dense population of receptive females might actually promote monogamy.

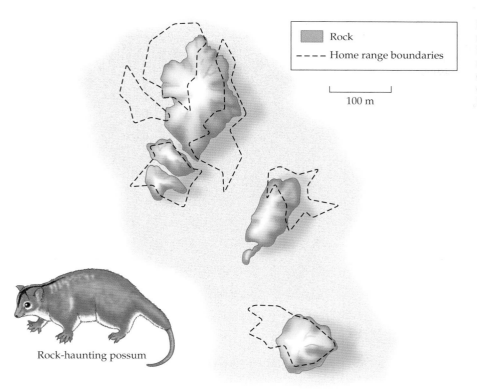

Rock

---- Home range boundaries

100 m

Rock-haunting possum

**FIGURE 11.10** **Mate-guarding monogamy in the rock-haunting possum** is facilitated by the small, discrete home ranges occupied by females of this species, which live along the edges of rock outcrops in northern Australia. After Runcie.[1254]

## Male Monogamy in Birds

In the vast majority of birds, males and females form long-term partnerships for one or more breeding seasons.[827] In keeping with the mate-assistance hypothesis, the males in these partnerships often contribute in a big way to the welfare of the offspring produced by their mates.[1072] In the yellow-eyed junco, for example, the male takes care of his mate's first brood of fledglings while the female incubates a second clutch of eggs. The paternal help provided is essential for the survival of these young "bumblebeaks," which are initially highly inept foragers.[1521] The value of male assistance has also been documented for starlings. In a population in which some males helped their mates incubate their eggs and others did not, the clutches with biparental attention stayed warmer (Figure 11.11) and so could develop more rapidly. Indeed, 97 percent of the eggs that had been incubated by both parents hatched, compared with 75 percent of those that had been cared for by their mother only.[1209]

Demonstrations of the importance of male parental care include some studies in which females have been experimentally "widowed" and left to rear their broods on their own. Widowed snow buntings, for example, usually produce three or fewer young, whereas control pairs often fledge four or more.[903] When male parental care is reduced rather than eliminated altogether, this too can have a negative effect on reproductive success. Give a male spotless starling some extra testosterone and he becomes less willing to feed nestlings, whereas males that receive an anti-androgenic chemical, which blocks the effects of naturally circulating testosterone, feed their offspring at increased rates. The mean number of fledged young per brood was lowest for starlings with extra testosterone and highest for those with the testosterone blocker (Figure 11.12).[1008]

Thus, we can safely conclude that paternal, pair-bonded males of at least some bird species really do increase the number of offspring their mates can produce in a breeding attempt. But have these paternal males actually fathered the offspring of their mates? Because the fitness gained by these males will

(A)

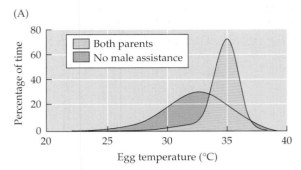

FIGURE 11.11 **Paternal male starlings keep their clutches warmer by helping their mates incubate their eggs.** (A) As the graph shows, eggs that were incubated by both parents were kept at about 35°C most of the time, whereas eggs incubated by the female alone were often several degrees cooler. (B) A starling with young—the eggs in this nest were successfully incubated. After Reid, Monaghan, and Ruxton.[1209]

(B)

Spotless starling

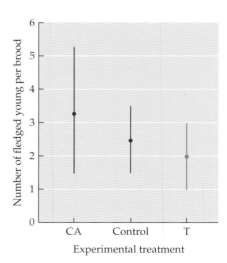

FIGURE 11.12 **Paternal care boosts reproductive success in the monogamous spotless starling.** Males whose testosterone levels were reduced by the anti-androgen cyproterone acetate (CA) provided more food for their broods and had the highest fledging rates per brood. Males given extra testosterone (T) provided less food and had the lowest fledging rates. Untreated controls were intermediate with respect to both feeding and fledging rates. After Moreno et al.[1008]

increase only to the extent that they care exclusively for their own genetic offspring, we expect monogamous males to sire all the offspring of their mates. This prediction has been confirmed for some species, such as the common loon: DNA fingerprinting of 58 young from 47 loon families revealed that all were the genetic offspring of the pair that raised them.[1146] A similar study of Florida scrub jays (see Figure 13.19) also showed that nestlings were always the offspring of the adults believed to be their parents.[1190]

But we now know, thanks to advances in molecular genetics (Box 11.1) that loons and scrub jays are exceptional cases. In most other birds, social monogamy (the pairing of male and female) does not equate with genetic monogamy (which happens when pairs produce and rear only their own genetic offspring). In nearly 90 percent of all bird species, some females engage in **extra-pair copulations** with males other than their social partners and use the sperm they acquire in this manner to fertilize some or all of their eggs.[582, 1547] In other words, socially monogamous males in most bird species run the risk of helping offspring other than their own, which clearly reduces the benefits of rearing the offspring of a social partner.

That social monogamy by males often coexists with genetic polyandry by their mates greatly surprised ornithologists who had always assumed that males helped their mates care for their mutual offspring, not for the progeny of other males. (Incidentally, the term "polyandry" often is used interchangeably with such terms as "promiscuity," "extra-pair copulation," and "multiple mating." However, promiscuity implies a lack of choosiness on the part of the female, which rarely applies in the real world. Extra-pair copulation is the means by which some females of some species achieve polyandry, namely by mating with more than one male, as in "multiple mating." **Polyandry** is the

## BOX 11.1   *Microsatellite analysis and behavioral ecology*

The discovery that social monogamy is not synonymous with genetic monogamy in birds came about primarily through the use of DNA fingerprinting technology,[210, 211, 718] which we have mentioned in earlier chapters. However, DNA fingerprinting has now been largely replaced by microsatellite analysis[1182, 1524] because mutations in the noncoding microsatellite regions of DNA (see below) tend to stay in populations because they are selectively neutral, that is, not subject to natural or sexual selection.[53] As a result, individuals differ greatly in the microsatellites they carry, which enables researchers to identify more readily whether a female has used the sperm from males other than her social partner to fertilize some of her eggs. Many studies of avian paternity now exist, and in about 70 percent of the cases, extra-pair copulations are known to occur, conferring extra-pair paternity on some philandering males.[582]

Microsatellite analysis takes advantage of the fact that scattered throughout the chromosomes are stretches of repeated sequences of DNA, such as

AAT AAT AAT AAT AAT

Here AAT is a sequence that is repeated five times, and therefore this microsatellite would be written as $(AAT)_5$. The number of times such a sequence is repeated at a given location on a chromosome often varies greatly, so one individual might carry five copies of AAT on one chromosome, whereas the other chromosome of the pair might have eight copies—$(AAT)_8$. Still other individuals might have other microsatellite alleles, such as $(AAT)_{10}$ or $(AAT)_{21}$. To perform microsatellite analyses, scientists employ procedures that enable them to identify a given nucleotide sequence, remove it from its chromosome, and copy this region over and over to make sufficient quantities to be detected by the appropriate technique. In the early days of microsatellite technology, a solution containing the amplified DNA was placed at one end of a gel (a thin sheet of acrylamide). When an electric field was run through the gel, microsatellite alleles of different sizes distributed themselves in relation to their mass as they moved up the gel. The lighter ones, with fewer repeats, moved farther

along than the heavier ones with more repeats. Today the procedure is usually performed in an automated DNA sequencer using gel in small capillary tubes with the fluorescently tagged alleles detected by lasers, rather than by the manual reading of large sheets of acrylamide gel.

Imagine that a female with the microsatellite genotype $(AAT)_5(AAT)_8$ mates with two males, one of which has the genotype $(AAT)_{10}(AAT)_{21}$ while the other has the genotype $(AAT)_9(AAT)_{33}$. Each offspring will carry a microsatellite allele from its mother, either $(AAT)_5$ or $(AAT)_8$, and a microsatellite allele present in the DNA of the sperm from its father. One need only establish an offspring's microsatellite genotype, usually for several loci, in order to determine who the real father is.

The table below shows the results of a microsatellite analysis of eight white-winged fairy-wrens with respect to four different microsatellite loci (Mcy3 to Mcy7). For each microsatellite locus, there are two columns. The numbers in each column are the alleles (labeled by size) of that DNA repeating sequence or locus possessed by the individuals being studied. The male and female at the top of the table were social partners who cared for three offspring in their nest. The nest was also attended by two helper males, who assisted the social pair in rearing the young. The territory of this group was adjacent to another territory, one of whose occupants was the "neighbor male." The microsatellite genotypes of baby 1 and baby 2 show that these were almost certainly the offspring of the social pair. The mother was homozygous for allele 274 of Mcy3, and so all of her offspring had to possess one maternally derived copy of this allele. The real father donated one copy of each microsatellite gene to each of his genetic offspring, so the female's social partner was the sire of babies 1 and 2. (Check out the alleles present in the male, female, baby 1, and baby 2 in the other microsatellites analyzed.) But baby 3 had an allele (266) at the Mcy3 locus that could not have been donated by either member of the pair at this nest. This allele was present in helper 2 and in the neighbor male, so either of these males could conceivably have been the donor of the haploid sperm

*(continued)*

| Bird | Microsatellite locus | | | | | | | |
|---|---|---|---|---|---|---|---|---|
| | Mcy3 | | Mcy4b | | Mcy5 | | Mcy7 | |
| Male | 274 | 258 | 168 | 168 | 130 | 98 | 106 | 104 |
| Female | 274 | 274 | 195 | 195 | 98 | 96 | 146 | 104 |
| Baby 1 | 274 | 274 | 195 | 168 | 96 | 96 | 104 | 104 |
| Baby 2 | 274 | 274 | 195 | 168 | 98 | 96 | 104 | 104 |
| Baby 3 | 274 | 266 | 195 | 140 | 102 | 98 | 112 | 104 |
| Helper 1 | 274 | 258 | 168 | 195 | 98 | 96 | 146 | 106 |
| Helper 2 | 274 | 266 | 184 | 184 | 98 | 96 | 106 | 104 |
| Neighbor male | 256 | 266 | 140 | 140 | 102 | 98 | 112 | 106 |

## BOX 11.1    *Microsatellite analysis and behavioral ecology (continued)*

that fertilized the egg that produced baby 3. Why then did the complete analysis of this data set rule out helper 2 as the father of baby 3 while leading to the conclusion

that the neighbor male was the genetic father of this youngster?

The photograph shows a male white-winged fairy-wren. Individuals can be captured and forced to donate a tiny drop of blood for microsatellite analysis. The idealized gas chromatogram (actual data are not quite so tidy) provides a visual record made by a gas chromatograph used in microsatellite analyses. The data shown here represent one individual found in the table. What is that individual's genotype? Which one has that genotype? Who was that individual's father? Printout and tabular data courtesy of Bob Montgomerie.

White-winged fairy-wren

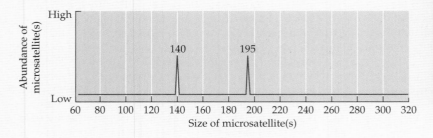

term we shall use, inasmuch as the word covers both multiple mating as well as the formation of pair-bonds with several males by a female.)

We shall deal with the puzzles associated with polyandry in the next section, but for the moment let's focus on the Darwinian puzzle for males that arises from polyandry, namely, why does a male attach himself to a single female only to run the risk that she will accept and use sperm from one or more extra-pair partners? One answer is that the male may be able to take advantage of the opportunity for extra-pair copulations with females other than his social partner. A polygynous male of this sort gets the best of both worlds by preventing his primary mate from mating with other males, while inseminating other females whose offspring will receive care, but not from him. These preferred males, which often exhibit ornaments and displays thought to be indicative of superior physical condition,[582] avoid some or most of the costs of monogamy. Therefore, it falls to the truly monogamous males, which are more likely to be cuckolded by their mates, to accept the disadvantages of monogamy. The question remains, why do these males fail to engage in fitness-boosting extra-pair copulations? One answer is that females have something to say about who gets to mate with them. If female mate choice revolves around parental contributions from a social partner and honest indicators of quality from extra-pair partners, then males that are unwilling to be paternal and are also unable to offer cues of good condition will likely find themselves between a rock and a hard place. They could drop out of the reproductive competition altogether but would obviously fail to leave descendants for as long as they remained on the sidelines. Their only real option is to help social partners who may fertilize at least some eggs with their sperm. When male life expectancy is short, handicapped individuals may leave more descendants by forming monogamous social partnerships than by adopting some other reproductive tactic.

**FIGURE 11.13    Males of the red phalarope may have to share a mate with other males.** In this polyandrous species, females secure first one male partner and then another, donating a clutch of eggs to each male in turn. The more brightly colored bird (on the left) is the female. Photograph by Bruce Lyon.

The existence of constraints imposed by competitors and choosy females may also help explain the evolution of mating systems in which most breeding females form social bonds with several monogamous males, rather than having one social partner and a variable number of extra-pair mates. The mating system of the Galápagos hawk, for example, ranges from monogamy to extreme polyandry.[136] Polyandry appears to be associated with a scarcity of suitable territories, which leads to a highly male-biased operational sex ratio since males outnumber the limited number of territorial, breeding females. The intense competition for these females and the territories on which they live has favored males capable of forming a cooperative defense team to hold an appropriate site. A breeding female may acquire as many as eight mates prepared to pair-bond with her for years, forming a male harem that will help her rear a single youngster per breeding episode.[451] In this and another odd bird, the purple swamphen, all the males that associate with one female appear to have the same chance of fertilizing her eggs.[711] If females do give all their mates equal fertilization opportunities, the mean fitness of each mate is almost certainly higher than that of males that fail to be part of a team.

In some other polyandrous species, the several mates of a female each receive a clutch of eggs from her, which forces them to share the reproductive output of the female with their fellow harem members. In the wattled jacana, for example, aggressive females fight for territories that can accommodate several males. These males mate with the territory owner, who then supplies each of them with one clutch of eggs, which will be cared for exclusively by that male. Males that pair monogamously with a polyandrous female are likely to care for offspring sired by other males, since 75 percent of the clutches laid by polyandrous female jacanas are of mixed paternity.[443] The same sort of thing happens in the red phalarope, a shorebird in which females are both larger and more brightly colored than males (Figure 11.13). Female phalaropes fight for males, which provide all the parental care for the clutch deposited in their nests, whether or not the eggs have been fertilized by other males.[334] Cases of this sort illustrate the disadvantages of being monogamously bonded with a polyandrous female, from the male fitness perspective, but male jacanas and phalaropes may have to accept the eggs their partners lay if they are to leave any descendants at all.

(A)

(B)

**FIGURE 11.14** **Female spotted sandpipers fight over males.** Two female spotted sandpipers (A) about to fight and (B) fighting for a territory that may attract several monogamous, paternal males to the winner. Photographs by Stephen Maxson.

However, a monogamous male under the control of a potentially polyandrous female may be able to do some things that at least reduce the probability that he will care for eggs fertilized by other males. For example, males of the red-necked phalarope, like those of its relative the red phalarope, care for broods on their own. Females of this species produce two clutches of eggs in sequence. Polyandrous females almost always draw their second mates from those males that have lost a first clutch to predators. Such a male, however, favors his original sexual partner over a novel female more than nine times out of ten.[1283] By copulating with this female again and then accepting her eggs, the male reduces the risk that the eggs he receives will have been fertilized by another male's sperm.

The same applies to the spotted sandpiper, whose females also behave like males in many ways.[1073] In addition to taking the lead in courtship, females are the larger and more combative sex, and they arrive on the breeding grounds first, whereas in most migrating birds, males precede females. Once on the breeding grounds, females fight with one another for territories (Figure 11.14). A female's holdings may attract first one and then later another male. The first male mates with the female and gets a clutch of eggs to incubate and rear on his own in her territory, which she continues to defend while producing a new clutch for a second mate, who may or may not fertilize all of the eggs he broods. As a result, a few females achieve higher reproductive success than the most successful males,[1075] an atypical result for animals generally (see page 337).

Lew Oring believes that an understanding of the mating system of spotted sandpipers is advanced by recognizing that in all sandpiper species, females never lay more than four eggs at a time, presumably because five-egg clutches cannot be incubated properly.[1074] Indeed, egg addition experiments have shown that sandpipers given extra eggs to incubate sometimes damage their clutch inadvertently and lose eggs more often to predators as well.[47] If female spotted sandpipers are "locked into" a four-egg maximum, then they can capitalize on rich food resources only by laying more than one clutch, not by increasing the number of eggs laid in any one clutch. To do so, however, they must acquire more than one mate to care for their sequential clutches, making this a rare case in which female fitness is limited more by access to mates than by production of gametes.

Male spotted sandpipers may be forced into monogamy by the confluence of several unusual ecological features.[839] First, the adult sex ratio is slightly biased toward males. Second, spotted sandpipers nest in areas with immense mayfly hatches, which provide superabundant food for females and for the young when they hatch. Third, a single parent can care for a clutch about as well as two parents, in part because the young are precocial—that is, they can move about, feed themselves, and thermoregulate shortly after hatching. The combination of excess males, abundant food, and precocial young means that female spotted sandpipers that desert their initial partners can find new ones without harming the survival chances of the first brood. Once a male has been deserted, he is stuck. Were he also to leave the nest, the eggs would fail to develop, and he would have to start all over again. If all females are deserters, then a male single parent presumably experiences greater reproductive success than he would otherwise, even if his partner acquires another mate to assist her with a second clutch. Furthermore, the first male to mate with a female spotted sandpiper may provide her with sperm that she stores and uses much later to fertilize some or all of her second clutch of eggs (see page 394). Thus, males that arrive on the breeding grounds relatively early and pair off quickly may not lose much when their mates acquire second copulatory partners.[1076] Once again, the males that get the short end of the stick are the less competitive individuals, the ones slower to arrive on the breeding grounds. They must deal with polyandrous females if they are to have any chance of reproducing, but the female's ability to control the reproductive process puts them at a major disadvantage.

## What Do Females Gain from Polyandry?

Having seen that a variety of factors permits females of some species to dictate the mating terms that more or less force some males into accepting monogamy, let's now turn our attention to the puzzling aspects of polyandry from the female perspective. Of course, it is not surprising that females are polyandrous when each has two or more social partners laboring on her behalf. A polyandrous female Galápagos hawk turns defense of her rare and vitally important territory over to a band of male protectors, which can surely do a better job than a single male. A polyandrous Kentish plover, like her counterparts among spotted sandpipers, takes advantage of male-biased sex ratios, when they occur, by abandoning her first partner, leaving him to take care of the chicks, while she lines up a second mate with whom she will rear a second brood.[1417]

Although this kind of polyandry makes obvious sense from the female perspective, what about those species in which a female has one social partner and a variable number of extra-pair mates that supply her with sperm but nothing else? Mating with these "extra" males exposes the female to a host of costs, including the time and energy spent searching for and mating with extra-pair males, the risk of losing parental care from a social partner that detects her extra-pair activities, and the chance of picking up a sexually transmitted disease.

If this last potential cost of polyandry is real, then we would expect the immune systems of highly polyandrous species to be stronger than those of species with a greater tendency toward monogamy. One research team led by Charles Nunn tested this hypothesis using comparative data from primates rather than birds. The team took advantage of the fact that there are extremely polyandrous primates, such as the Barbary macaque, whose females mate with as many as ten males in one day. At the other end of the spectrum, there are monogamous species, such as the gibbon. Nunn and his colleagues measured immune system strength across the spectrum of mating systems by looking at data on white blood cell counts from large numbers of adult females of 41 primate species, most of which were held in zoos, where their health is regu-

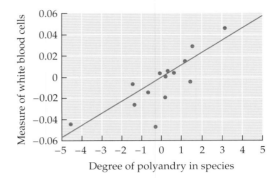

**FIGURE 11.15  Polyandry has fitness costs.** In primates, the number of mates accepted by females is correlated with the investment made in white blood cells, a component of an animal's immune system. (The higher the measure of white blood cells, the stronger the immune system.) This finding suggests that the more polyandrous the species, the greater the challenge to the female's immune system. Each data point represents a different primate species. After Nunn, Gittleman, and Antonovics.[1063]

larly monitored. (Naturally, only healthy specimens were included in the samples.) White blood cell counts were indeed higher (but within the normal range) in females of the more polyandrous species (Figure 11.15).[1063]

## Discussion Question

**11.5**  Why did Nunn and company also look at the relationship between the mean sizes of groups in the different primate species in their study, as well as the extent to which each species was terrestrial as opposed to arboreal?

Given the potential costs of extra-pair copulations for females, it is not too surprising that in at least some species, females attempt to avoid insemination by males other than their social partners. Female mallard ducks, for example, resist every copulatory attempt by males other than their long-term partners.[329] Their favored partners typically appear to be individuals of well-above-average condition, having molted into breeding plumage sooner than others. To overcome the severe energetic demands of the molt relatively early in the year indicates that these males are healthier and thus less likely to transmit venereal diseases to their mates (and waterfowl are at risk of sexually transmitted disease[1311]).

But although female mallards always try to say no to unsolicited copulatory attempts, females of other species accept, or even invite, matings from males other than their primary partners.[122] For example, 85 percent of female bluethroats (see Figure 4.34) whose social mates had been outfitted with a kind of avian condom went on to lay fertile eggs, demonstrating unequivocally that they were mating with other males.[479] Just what bluethroats and other socially monogamous songbirds might gain from extra-pair copulations initially puzzled behavioral biologists, but many hypotheses are now available (Table 11.1).

| TABLE 11.1 *Why do females mate voluntarily with more than one male?* | |
| --- | --- |
| **Genetic or indirect benefits polyandry** | |
| 1. Fertility insurance hypothesis | Mating with several males reduces the risk that some of the female's eggs will remain unfertilized because any one male may not have sufficient sperm to do the job. |
| 2. Good genes hypothesis | A female mates with more than one male because her social partner is of lower genetic quality than other potential sperm donors, whose genes will improve offspring viability or sexual attractiveness. |
| 3. Genetic compatibility hypothesis | Mating with several males increases the genetic variety of the sperm available to the female, increasing the chance that some sperm will have DNA that is an especially good match with the DNA of her eggs. |
| **Material or direct benefits polyandry** | |
| 1. More resources hypothesis | More mates mean more resources received from the sexual partners of a female. |
| 2. More care hypothesis | More mates mean more caregivers recruited for the female's offspring. |
| 3. Better protection hypothesis | More mates mean more time with protectors who will keep other males from sexually harassing a female. |
| 4. Infanticide reduction hypothesis | More mates mean greater confusion about the paternity of a female's offspring and thus less likelihood of losing offspring to infanticidal males. |

The possible benefits to polyandrous females have been categorized as genetic (or indirect) versus material (or direct). One genetic benefit, for example, is that extra-pair fertilizations could reduce the risk to a female of having an infertile partner as a social mate.[583] The **fertility insurance hypothesis** is supported by the observation that the eggs of polyandrous female red-winged blackbirds are somewhat more likely to hatch than the eggs of monogamous females.[569] Likewise, the rapid loss of stored sperm in the bearded tit may favor a female that copulates frequently with her own partner—or someone else's partner—in order to maintain her fertility.[1280] Moreover, in Gunnison's prairie dogs, polyandrous females become pregnant 100 percent of the time, whereas only 92 percent of monogamous females achieve this state.[678]

## Polyandry and Good Genes

Although fertility insurance may have contributed to the evolution of polyandry, a more frequently mentioned genetic benefit has to do with improved offspring quality for a female that mates with more than one partner. In one experiment, female live-bearing Trinidadian guppies that were given the chance to mate with four males produced offspring that stayed safely in schools, unlike the offspring of monogamous mothers, which tended to wander off on their own.[449] Likewise, females of a wild guinea pig (the yellow-toothed cavy) actively seek out copulations with more than one male when given the opportunity, and they favor relatively heavy males. The payoff for polyandry in this species appears to be a reduction in stillbirths and losses of babies before weaning (Figure 11.16).[662]

Polyandrous guppies and cavies that had several partners instead of just one apparently increased their odds of receiving some sperm of exceptional genetic quality. This version of the good genes hypothesis for extra-pair copulations suggests that a female seeks out several males in order to secure superior genes from at least one of these males—genes that make her offspring more fertile or likely to survive, for example. In keeping with this hypothesis, when a female guppy pairs with two males, the male that displays at a higher rate fertilizes more of the female's eggs than expected by chance. Thus, polyandrous females apparently permit or encourage sperm competition among males as a means of providing their offspring with genes from vigorous, active males.[449]

The superb fairy-wren is another animal for which polyandry and the acquisition of good genes seem to go together. In this gorgeous little Australian wren, a socially bonded pair lives on a territory with a number of

(A)

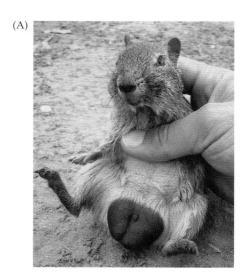

(B)

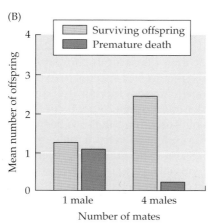

**FIGURE 11.16  Polyandry has fitness benefits.** (A) A male yellow-toothed cavy, showing the very large testes characteristic of this and other species in which intense sperm competition occurs because of a high frequency of multiple mating by females. (B) In this species, females that are experimentally restricted to a single mate have fewer surviving offspring and more offspring that die young than females allowed to mate freely with four males. A, photograph by Matthias Asher, courtesy of Norbert Sachser; B, after Hohoff, Franzen, and Sachser.[662]

**FIGURE 11.17** **Extra-pair paternity** contributes to the large annual variation in male reproductive success in the superb fairy-wren. Shown are the number of young produced per year by dominant males with social partners (blue bars) and auxiliary helper males (orange bars) that live in groups with a "breeding" pair. After Webster et al.[1525]

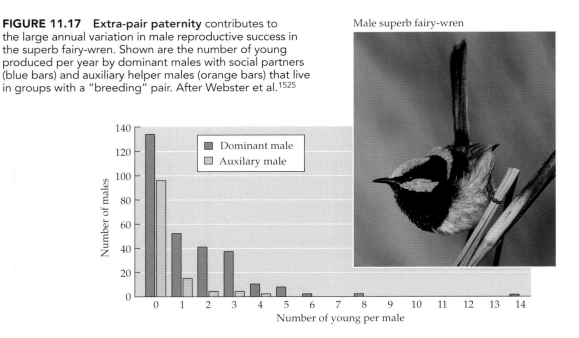

Male superb fairy-wren

subordinate helpers, usually males. When the breeding female is fertile, she regularly leaves her mate and auxiliary helpers before sunrise and travels to another territory, where she often mates with a dominant male before returning home.[407] Tiny radio transmitters attached to the females revealed what they were up to, and genetic analyses of offspring demonstrated that the females' social partners often lost paternity to distant rivals. Some "breeding males," that is, dominant males with a mate, sired no offspring at all in a given year, whereas others did very well. One male produced 14 fledglings in a single year, of which 10 were the result of extra-pair copulations; these youngsters were cared for by fairy-wrens in groups other than the one associated with their father. Overall, differences among males in the numbers of extra-pair offspring they sired contributed close to 50 percent of the variance (a statistical measure of variation) among males in their annual reproductive success (Figure 11.17).[1525]

Because female fairy-wrens are clearly careful in their choice of an extra-pair mate, we can ask what kind of male a female prefers and what benefit she gains from her preference. In this species, as in mallard ducks, females favor males that have managed to molt early, at least a month before the breeding season begins. Presumably females are selecting males that are in such good physical condition that they could meet the energetic demands of molting early, a capacity that these males might be able to pass on to their offspring.[422]

The idea that extra-pair copulations enable females to acquire good genes for their offspring is compatible with the discovery that older males appear to have greater success in extra-pair matings in some songbirds.[402, 730] Older males have demonstrated an ability to stay alive, which might be due in part to their genotype. If so, the offspring of such males might live longer as well, something that has been documented in the polyandrous blue tit.[759, 760] The conclusion that females go out of their way to mate with males likely to possess good genes has received support from studies of birds, fishes, and mammals—and invertebrates as well. The female sierra dome spider routinely accepts sperm from more than one male, especially from large and active partners capable of prolonged copulatory stimulation. When Paul Watson experimentally paired females with males of varying size and vigor, he found that the growth rates of the resulting offspring and their eventual sizes were correlated with their fathers' traits.[1515]

The females in the cases we have discussed thus far could have been mating with several males in order to secure valuable alleles that will make any female's offspring healthier, stronger, wiser—or sexier. Indeed, it could be that genes acquired from particularly attractive males may have only one beneficial effect, which is to enable polyandrous females to fertilize their eggs with these sperm and thereby produce "sexy sons." By inheriting the very traits that made their fathers sexually appealing, sexy sons increase the odds that their mothers will have many grand-offspring. As required by this hypothesis, male "attractiveness" is hereditary in some species.[1425] Indeed, the sons of sexually successful field crickets were more attractive to females than the sons of males that failed to acquire a mate when placed in a cage with a female and a competitor male (Figure 11.18).[1527] This result suggests that females can gain superior attributes for their sons by choosing some potential fathers over others.

In a study of the pied flycatcher, a bird in which some males are monogamous and others polygamous, a female mated to a monogamous male gets his undivided attention and parental care, whereas those females mated to bigamists receive somewhat less assistance on average. Fledgling success is correspondingly higher for females with monogamous mates. But in one study, there was no advantage for these females, in terms of grandoffspring, compared with the primary mates of bigamists (i.e., those females that acquired males who then went on to attract second partners) (Figure 11.19).[692] These data suggest that the first mate of an attractive bigamist was compensated for a reduction in assistance from her partner through the increased production of grandoffspring, perhaps because her sexy sons (see page 368) had inherited the traits that increased the odds that they would be able to acquire two mates, instead of just one. Remember, however, that in some cases, the improved performance of offspring may stem from the willingness of females to invest more resources in the offspring of favored males, not from genetic benefits the offspring have received from their fathers (see pages 360–363).

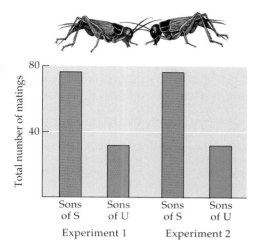

**FIGURE 11.18  A father's mating success can be transmitted to his sons.** In experiment 1, two male field crickets were given an opportunity to compete for a female; one male (S) mated successfully, while the other male (U) was unsuccessful. When the sons of male S were placed in competition for a female with the sons of male U (who had been given a female to mate with after failing to win the initial competition), the sons of S were about twice as likely to mate with the female as the sons of U. In experiment 2, a male that had won a mating competition was later allocated a female at random for breeding, as was a male that had lost the competition. The sons of the two males were then placed in an arena with a female, and as before, the sons of S were much more likely to mate with the female than the sons of U. After Wedell and Tregenza.[1527]

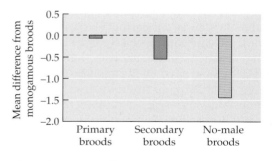

Pied flycatcher

**FIGURE 11.19  Differences in number of grandoffspring produced by three classes of female pied flycatchers** with polygynous partners compared against the mean number of grandoffspring generated by females mated to monogamous paternal males. Females with primary broods were those whose polygynous mates chose them first and only later acquired second females. The secondary broods of these second-choice females were sometimes assisted to some degree by their fathers, but they sometimes received no paternal care at all (as was true for all no-male broods). After Huk and Winkel.[692]

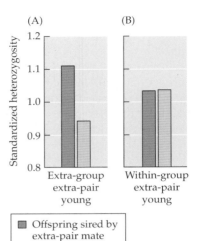

Superb starling

**FIGURE 11.20 Genetic heterozygosity of young superb starlings** produced by the mated pair (orange bars) versus those that came from extra-pair matings (red bars). (A) The comparison here shows that the heterozygosity of the mated pair's offspring is lower than that of the extra-pair offspring created when a bird in one group mated with a bird from another group altogether. (B) In contrast, the offspring of a mated pair were no less heterozygous than those produced by extra-pair copulations with another member of the same group of starlings. Photograph by Dustin Rubenstein; after Rubenstein.[1248]

Bluethroat

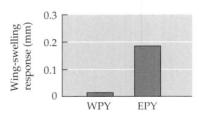

**FIGURE 11.21 Extra-pair matings can boost the immune responses of offspring in the bluethroat.** The mean wing-swelling response (an indicator of immune system strength) of youngsters sired by their mother's social partner (within-pair young, or WPY) was less than that of young birds sired by their mother's extra-pair partner (extra-pair young, or EPY) in broods of mixed paternity. Photograph by Bjørn-Aksel Bjerke; after Johnsen et al.[731]

It is also possible that females engage in polyandry not to hit the genetic jackpot with one especially sexy or long-lived male, but rather to increase the odds of receiving genetically compatible sperm. When eggs and sperm with especially complementary genotypes unite, they can result in especially viable progeny by, for example, having high levels of heterozygosity,[1637, 1639] something that is usually promoted by outbreeding and decreased by inbreeding. Individuals with two different forms of a given gene often enjoy an advantage over homozygotes (see Fossøy, Johnsen, and Lifjeld[480]). If genetically similar pairs are in danger of producing inbred offspring with genetic defects, then we can predict that females socially bonded with genetically similar individuals will be prime candidates to mate with other males who are either (1) genetically dissimilar or (2) highly heterozygous in their own right.

Dustin Rubenstein tested this possibility by comparing the degree of heterozygosity of male superb starlings whose mates sought out extra-pair partners from outside their social group versus those males whose females did not engage in this kind of extra-pair mating. Rubenstein found that, as predicted, a female whose partner had a genotype with less heterozygosity than she did was in fact more likely to copulate with a male from another group.[1248] The offspring fathered by the outsider were more heterozygous on average than those produced by the female's social partner (Figure 11.20).

Support for the **genetic compatibility** hypothesis also comes from studies of the bluethroat by Norwegian researchers who showed that extra-pair offspring were more heterozygous than those sired by the female's social partner.[480] Extra-pair progeny also enjoyed a stronger immune system, a point established by injecting a foreign substance into wings of nestling bluethroats. The strength of the immune response is measured by quantifying the degree of swelling observed at the injection site, which reflects the number of protective T cells available to disable the foreign chemical. Swelling at the injection site was greater in extra-pair nestlings than in their nestmates that had been produced by the female and her social partner (Figure 11.21). But females that engaged in extra-pair copulations were not acquiring universally good genes, because the extra-pair males and *their* mates produced youngsters with only average levels of immunocompetence.[731] Thus, it must have been the especially good match between the genes of the polyandrous female and her extra-pair mate that resulted in a more robust immune system for her offspring.

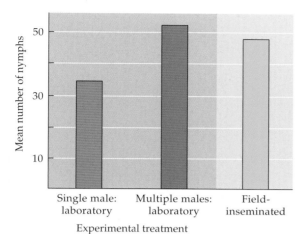

Pseudoscorpion

**FIGURE 11.22** Polyandry boosts female reproductive success in a pseudoscorpion. In laboratory experiments, female pseudoscorpions restricted to lone partners produced fewer nymphs than females that mated with several males. Paternity tests of the offspring of wild-caught females confirmed that females usually mate with several males under natural conditions. Photograph by Jeanne Zeh; after Zeh.[1638]

Still more evidence for genotype matching comes from a study of the fertilization success of pairs of roosters whose sperm were mixed together in equal amounts and then supplied via artificial insemination to a series of hens. The male whose sperm fertilized more eggs in female number 1 was in no way guaranteed to do likewise when his sperm and those of a rival were supplied to female number 2 or female number 10. This result suggests either that chickens have an internal mechanism for favoring some sperm genotypes over others or that some combinations of genes in embryos resulted in their early deaths.[123]

Reduction of embryo deaths has been identified as one reason why female pseudoscorpions mate with several males. Polyandrous pseudoscorpions had fewer failed embryos and more offspring surviving to the nymph stage than females that were experimentally paired with a single male (Figure 11.22).[1041, 1638] Yet when Jeanne Zeh compared the numbers of offspring different females produced from matings with the same male, she found no correlation. In other words, as is true for bluethroats and roosters, male pseudoscorpions could not be divided into studs and duds. Instead, the effect of a male's sperm on a female's reproductive success depended on the match between the two gametic genomes, as predicted by the genetic compatibility hypothesis. Because a female's chances of securing genetically compatible sperm increase with the number of males she mates with, we can predict that females of this pseudoscorpion should prefer to mate with new males rather than with previous partners. Indeed, when a female was given an opportunity to mate with the same male 90 minutes after an initial copulation, she refused to accept his spermatophore in 85 percent of the trials. But if the partner was new to her, she would usually accept his gametes.[1640]

## Discussion Questions

**11.6** Genetic incompatibility could explain why inbreeding is deleterious in so many species of animals, including field crickets (but see Thünken et al.[1439]). If so, what results would you expect from an experiment in which female crickets were each paired with two males, one a sibling and the other a nonrelative? (Imagine that you have the ability to determine the paternity of the offspring produced by a given female after her double mating.) In doing this experiment, why would you create two groups: one in which the first male to mate was a sibling and the other in which the first male was a nonrelative? Are the data in Figure 11.23[172] confirmatory or negative with respect to your expectation? What do these findings reveal about female mate choice even after mating has occurred?

**FIGURE 11.23** **Egg fertilizations in female crickets that have mated with a sibling and a nonrelative.** (A) The number of offspring sired by the nonrelative and the sibling when the first male to mate was the nonrelative. (B) The number of offspring sired by both males when the first male to mate was the sibling. Sample sizes were 19 and 18 females, respectively. After Bretman, Wedell, and Tregenza.[172]

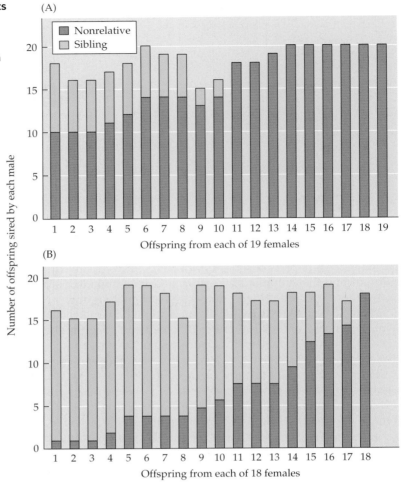

**11.7** The grey mouse lemur is a species in which females come into estrus and mate with several males on one night each year. Females accept any and all partners, but a genetic analysis of offspring produced by female grey mouse lemurs indicates that the actual fathers were more different from the mothers than males selected at random from the population with respect to the MHC genes, which code for proteins involved in the immune system.[1296] How do these data contribute to an evaluation of the various hypotheses on the indirect genetic benefits of polyandry? What do they suggest about the mechanism that enables a sexually promiscuous female mammal to exercise mate choice?

Some members of the genus *Apis*, a group to which the familiar honey bee belongs, are highly polyandrous, with females accepting and using sperm from between 10 and 63 males, depending on the species.[1303] As mentioned earlier, virgin queens of this group fly out from the hive on one or more nuptial flights, during which time they are "captured" and mated by a number of males in quick succession. What makes this business particularly puzzling is the finding that in most bees, including some of those from large social colonies, the queens mate just once.[1397] In the typical bee, a single mating suffices to supply the female with all the sperm she will need for a lifetime of egg fertilizations, thanks to her ability to store and maintain the abundant sperm she receives from her single partner. The widespread nature of bee monogamy indicates that this trait was the ancestral mating pattern, with polyandry a more recently derived characteristic. What selective pressures might have led to the replacement of monogamy by extreme polyandry in a relatively few bee species?

Benjamin Oldroyd and Jennifer Fewell were able to track down a full dozen potential answers to this question, ranging from rather mundane ideas, such as that polyandry guarantees that long-lived queens do not run out of sperm, to more inventive ones, such as that polyandry enhances the disease or parasite resistance of a colony. Alternatively, polyandry might increase worker genetic diversity, which could promote colony-level productivity by increasing the range of skills exhibited by the workforce.[1067]

Support for the anti-disease hypothesis includes the results of an experiment in which some honey bee colonies were set up with queens that had been artificially inseminated just once and some colonies were set up with queens that had been supplied with the sperm of ten drones. Both types of colonies were then infected with a bacterium that causes the disease American foulbrood, a killer of honey bee larvae. The colonies with polyandrous queens were not only less affected by foulbrood, but also more productive of offspring, and they had larger adult populations on average than the colonies headed by once-mated queens.[1303]

The experimental results described here are supportive of an anti-disease function associated with queen polyandry in the honey bee. However, the advantages shown by the colonies with polyandrous queens might have in part or in whole arisen from the ability of a more genetically diverse worker force to carry out the full range of worker tasks more efficiently than a less diverse worker force. To test this hypothesis, Heather Mattila and Thomas Seeley again used artificial insemination techniques to create two categories, single-insemination queens and queens with many male sperm donors. The colonies were kept free from diseases through antibacterial and antiparasitic treatments to eliminate this variable from the experiment. The genetically uniform and genetically diverse colonies were also monitored to measure such things as the total weight, the foraging rates of workers, and the mean area of comb in the hive (where offspring are reared and food is stored) (Figure 11.24).[954] None of the genetically uniform colonies made it through the winter, whereas about a quarter of the colonies led by polyandrous queens survived. These and other similar findings demonstrate that the colony's genetic diversity, which is boosted when the queen mates with several males, elevates the fitness of the queen. The puzzle of polyandry for the honey bee and its relatives is at least partly solved by the demonstration that a genetically diverse work force is better than one of limited genetic variety.

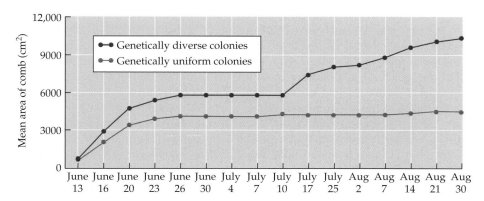

**FIGURE 11.24  Polyandry is adaptive in honey bees.** Genetically diverse colonies (the queens had mated with several males—shown in green) on average produced more comb than did workers in genetically uniform colonies (whose queens had mated with only one drone—shown in red). After Mattila and Seeley.[954]

**FIGURE 11.25 Polyandry can yield material benefits.** By mating with many males, females of this megachilid bee gain access to pollen and nectar in those males' territories. Photograph by the author.

## Polyandry and Material Benefits

Our focus thus far on genetic (indirect) gains for polyandrous females should not obscure the possibility that females sometimes mate with several males in order to secure certain direct benefits, often in the form of useful resources, rather than genes alone from these individuals. Thus, female red-winged blackbirds may be allowed to forage for food on the territories of males with whom they have engaged in extra-pair copulations, whereas truly monogamous females are chased away.[570] Similarly, females of some bees must copulate with territorial males each time they enter a territory if they are to collect pollen and nectar there (Figure 11.25).[11] It is even possible that males of some insects may pass sufficient fluid in their ejaculates to make it worthwhile for water-deprived females to copulate in order to combat dehydration.[429]

Female hangingflies and other insects also have an incentive to mate several times in order to receive **material benefits** in the form of food presents or nutritious spermatophores from their partners. The spermatophores of highly polyandrous butterfly species contain more protein than the spermatophores of generally monogamous species.[124] Males of polyandrous butterfly species could bribe females to mate with them by providing them with nutrients that can be used to make more eggs (in these species, the male's spermatophore can represent up to 15 percent of his body weight, making him a most generous mating partner). If this hypothesis is true, then the more polyandrous a female, the greater her reproductive output should be. Christer Wiklund and his coworkers used the comparative method to check this prediction. They took advantage of the fact that a number of closely related butterfly species in the same family (the Pieridae) differ substantially in the number of times they copulate over a lifetime. They used the mean number of lifetime copulations for females of each species as a measure of its degree of polyandry. To quantify reproductive output, they weighed the mass of eggs produced by a female in the laboratory and divided that figure by the weight of the female to control for differences among the species in body mass. As predicted, across these eight species, the more males a female mated with on average, the more spermatophores she received, and the greater her production of eggs (Figure 11.26).[1572]

In some butterflies, polyandry means more eggs produced. In some other animals, polyandry enables females to get more parental assistance from their several mates. Thus, in the dunnock, a small European songbird, a female that lives in a territory controlled by one male may actively encourage another, subordinate male to stay around by seeking him out and copulating with him when the alpha male is elsewhere. Female dunnocks are prepared to mate as many as 12 times per hour, and hundreds of times in all, before laying a complete clutch of eggs. The benefit to a polyandrous female of distributing her copulations between two males is that both sexual partners will help her rear her brood—provided that they have both copulated often enough. Female dunnocks ensure that this paternal threshold is reached by actively soliciting matings from whichever male, alpha or beta—usually the latter—has had less time in their company (Figure 11.27).[356] Likewise, a female superb starling that is pair-bonded with one male sometimes mates with another unpaired male in her flock. This individual then often joins the female and her primary partner in rearing the brood when

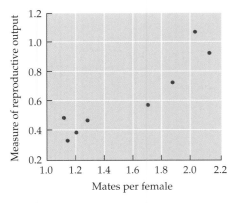

**FIGURE 11.26 Reproductive output is higher in polyandrous pierid butterfly species.** The mean number of mates during a female's lifetime and the mean reproductive output of females (measured as the cumulative mass of eggs produced divided by the mass of the female) is shown for eight pierid species. After Wiklund, Karlsson, and Leimar.[1572]

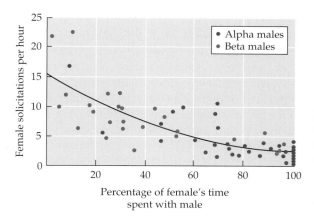

**FIGURE 11.27  Adjustment of copulation frequency by polyandrous female dunnocks.** Females living in a territory with two males solicit copulations more often from the male that has spent less time with them, whether that is the alpha or the beta male. After Davies.[356]

Dunnock

they appear.[1248] By copulating with the extra male and using his sperm to fertilize an egg or two, the polyandrous female has in effect made it advantageous for the "other male" to invest parentally in her offspring.

Alternatively, by mating with several males, a female may supply the proximate stimulation that encourages all her sexual partners to leave her next newborn alone. Males of the Hanuman langur generally do ignore a female's baby if they have mated with the mother prior to the birth of the infant.[149] On the other hand, if they have not copulated with the mother, they may kill her offspring (see Chapter 1). The fact that females will copulate with more than one male even when they are not ovulating—in fact, even when they are pregnant—suggests that polyandry promotes the female langur's interests by lowering the risk of infanticide.[649]

## Discussion Questions

**11.8**  The average proportion of extra-pair offspring in a brood varies among bird species from 0.0 to almost 0.8.[720] Consider how variation in the benefits and costs of extra-pair paternity might be responsible for variation in the willingness of females to engage in extra-pair copulations. For example, how might low variation in male genetic quality in a species affect the extra-pair paternity figure? What about differences among species in the risk of venereal disease, or in the likelihood that partners will detect and punish the sexual infidelity of a mate? Furthermore, if mating with extra-pair males provides females of a given species primarily with the opportunity to trade up to a genetically or materially superior partner,[251] what is the predicted relationship between extra-pair copulations and "divorce" among songbirds?

**11.9**  In many animals, females mate repeatedly with the same male, typically their social partner. Perform a cost–benefit analysis of this kind of multiple mating by females, and contrast it with that applied to polyandrous matings.

Having just now read the accounts that seem to indicate that females can gain any or all of an extremely wide range of indirect (genetic) benefits and/or direct (material) benefits by mating with more than one male, you may find it disconcerting to know that some biologists have concluded that none of these potential benefits actually amounts to much, at least when it comes to unfaithful female birds. Göran Arnqvist and Mark Kirkpatrick took the admittedly relatively few bird studies that have provided the kind of data that can be used to measure the strength of selection on female infidelity. They found that indirect (good genes) benefits were close to zero, while selection via the direct benefits side of the equation was actually negative.[48] In other words, their analysis shows that a female bird does not gain by mating with males other than her social partner. The fact that female birds of so many species do indeed copulate outside the pair-bond can be explained, according to Arnqvist and Kirkpatrick, as the result of positive selection on males (not females) to inseminate females other than their primary partners. Selection of this sort could lead males to do things that are not in the best interests of their partners, such as mate with them, because these actions are in the best interests of the sexually demanding males' genes. In fact, some behavioral biologists have concluded that there should be a category called "convenience polyandry" to account for cases in which females apparently do not benefit from mating with several males—except that it costs them less to mate with unwanted sexual harassers than to fight them off.[397, 1435]

Simon Griffith has disputed the conclusions of Arnqvist and Kirkpatrick,[583] but his critique has in turn been challenged by these researchers,[50] generating an exchange that is worth reading as an example of how scientists argue in print. I suspect that most behavioral biologists working in this area find it hard to believe that female birds typically gain nothing by participating in extra-pair copulations, but Arnqvist and Kirkpatrick's position is one that has to be considered by those trying to solve the persistent puzzle of female polyandry.

## The Diversity of Polygynous Mating Systems

We have now illustrated what interests evolutionary biologists about male monogamy and female polyandry. Some intriguing evolutionary issues also come up when we look at polygyny, despite the fact that male attempts to be polygynous are easily understood in terms of sexual selection theory. What is worth exploring from an evolutionary perspective is the great variation from species to species in the tactics employed by males to achieve polygyny. Consider the behavior of bighorn sheep, black-winged damselflies, and satin bowerbirds. Even though some males of all three species acquire many mates by defending territories of one sort or another, they do not all defend the same things. Bighorn rams go where potential mates are (see Figure 10.23), and they then fight with other males to monopolize females there (**female defense polygyny**). Male black-winged damselflies wait for females to come to them (see Figure 10.28), defending territories that contain the kind of aquatic vegetation in which females prefer to lay their eggs (**resource defense polygyny**). Male satin bowerbirds, on the other hand, defend territories containing only a display bower (see Figure 10.1), not food or other resources that females might use to promote their reproductive success (**lek polygyny**). Moreover, males of many other polygynous species skip combative territoriality altogether and instead try to outrace their rivals to receptive females (**scramble competition polygyny**).

How can we account for the variety of mating tactics among polygynous species? Steve Emlen and Lew Oring argued that the extent to which a male can monopolize mating opportunities depends on the distribution of females.[438] The degree to which receptive females are spatially clustered in their environment varies as a function of such things as predation pressure and food distribution. Females that live together or aggregate at resource patches can be monopolized economically by males, whereas widely scattered females cannot because as the size of a territory grows, the cost of defending it also increases.

### Female Defense Polygyny

The theory that differences in female distribution patterns underlie the diversity of mating systems generates many predictions, among them the expectation that when receptive females occur in defensible clusters, males will compete directly for those clusters and female defense polygyny will result. As predicted, social monogamy in mammals never occurs when females live in groups.[797] Instead, in animals as different as bighorn sheep and gorillas, groups of females, which have formed in part for protection against predators, attract males that compete to control sexual access to a group.[619, 658] Likewise, male lions compete to control prides of females, which have formed in part for defense of permanent hunting territories and protection against infanticidal males.[1091] Given the existence of these groups, female defense polygyny is the standard tactic (Figure 11.28).[960] Thus, males of the Montezuma oropendola, a tropical blackbird, try to control clusters of nesting females, which group their long, dangling nests in certain trees. The dominant male at a nest tree may secure up to 80 percent of all matings by driving subordinates away.[1522] These dominant males shift from nest tree to nest tree, following females rather than defending a nesting resource per se, demonstrating that they are employing the female defense tactic. This pattern applies especially to small colonies; in large colonies with many male competitors present, males that try to defend groups of potential mates are under constant assault by rival males (Figure 11.29). At large nesting colonies, therefore, males may leave to search for, defend, and mate with one female at a time elsewhere.[1523] Males that acquire more than one mate in this way can be said to be practicing sequential female defense polygyny.

FIGURE 11.28    Female defense polygyny in the greater spear-nosed bat. The large male (on the bottom) guards a roosting cluster of smaller females. A successful male may sire as many as 50 offspring with his harem females. Many of the bats have been banded so that they can be recognized as individuals. Photograph by Gary McCracken.

(A)

(B)

(C)

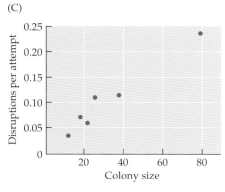

(D)

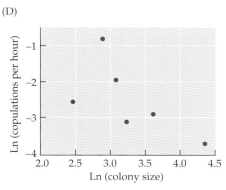

FIGURE 11.29    A bird that exhibits female defense polygyny. (A) Males of the Montezuma oropendola attempt to monopolize females in (B) small colonies of nesting females. But as colony size increases, (C) mating attempts are often disrupted by rivals and as a result, the (D) frequency of copulations per hour at the colony site decreases. After Webster and Robinson.[1523]

(A)

(B)

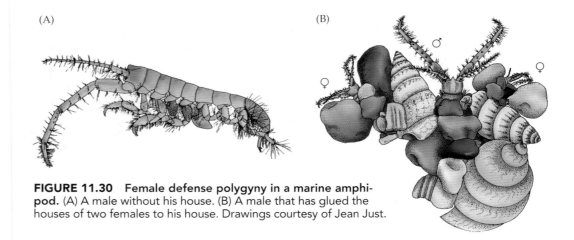

**FIGURE 11.30** Female defense polygyny in a marine amphipod. (A) A male without his house. (B) A male that has glued the houses of two females to his house. Drawings courtesy of Jean Just.

Groups of receptive females also occur in insects and other invertebrates, and female defense polygyny has evolved here as well.[1435] For example, nests of *Cardiocondyla* ants produce large numbers of virgin queens, leading to lethal competition among the males of the colony. The few individuals that survive the mayhem get to copulate with dozens of freshly emerged females.[648] But one does not have to be hyperaggressive in order to practice female defense polygyny. Thus, males of tiny siphonoecetine amphipods construct elaborate cases composed of pebbles and fragments of mollusk shells found in the shallow ocean bays where they live. They move about in these houses and capture females by gluing the females' houses to their own, eventually creating an apartment complex containing up to three potential mates (Figure 11.30).[742]

### Resource Defense Polygyny

In many species, females do not live permanently together, but a male may still become polygynous if he can control a rich patch of resources that females visit on occasion. Some male black-winged damselflies, for example, defend floating vegetation that attracts a series of sexually receptive females, each of which mates with the male and then lays her eggs in the vegetation he controls.[1501] Likewise, male antlered flies fight for small rotten spots on fallen logs or branches of certain tropical trees because these locations attract receptive, gravid females, which mate with successful territory defenders before laying their eggs (Figure 11.31).[405a]

A safe location for eggs constitutes a defensible resource for a host of animal species; the more of this resource a territorial male holds, the more likely he is to acquire several mates. In an African cichlid fish, *Lamprologus callipterus*, a female deposits a clutch of eggs in an empty snail shell, then pops inside to remain with her eggs and hatchlings until they are ready to leave the nest. Territorial males of this species, which are much larger than their tiny mates, not only defend suitable nest sites, but collect shells from the lake bottom and steal them from the nests of rival males, creating middens of up to 86 shells (Figure 11.32). Because as many as 14 females may nest simultaneously in different shells in one male's midden, the owner of a shell-rich territory can enjoy extraordinary reproductive success.[1279]

If the distribution of females is controlled by the distribution of key resources and if male mating tactics are in part dictated

**FIGURE 11.31 Resource defense polygyny in an Australian antlered fly.** Males compete for possession of egg-laying sites that can be found only on certain species of recently fallen trees. Photograph by Gary Dodson.

(A)

(B)

**FIGURE 11.32  Resource defense polygyny in an African cichlid fish.** (A) A territorial male bringing a shell to his midden. The more shells, the more nest sites are available for females to use. (B) The tail of one of the territorial male's very small mates can be seen in a close-up of the shell that serves as her nest. Photographs by Tetsu Sato.

by this fact of life, then it ought to be possible to alter the mating system of a species by moving resources around, thereby altering where females are located. This prediction was tested by Nick Davies and Arne Lundberg, who first measured male territories and female foraging ranges in the dunnock. In this drab little songbird, females hunt for widely dispersed food items over such large areas that the ranges of two males may overlap with those of several females, creating a polygynous–polyandrous mating system. However, when Davies and Lundberg gave some females supplemental oats and mealworms for months at a time, female home ranges contracted substantially.[355] Those females with the most reduced ranges had, as predicted, fewer social mates than other females whose ranges had not diminished in size (Figure 11.33). In other words, as a female's home range contracted, the capacity of one male to monitor her activities increased, with the result that female polyandry tended to be replaced by female monogamy. These results support the general theory that males attempt to monopolize females within the constraints imposed on them by the spatial distribution of potential mates, which in turn may be affected by the distribution of food or other resources.

## Discussion Question

**11.10**  The alpine accentor is a songbird with a mating system rather like that of its close relative, the dunnock, in that two males, an alpha and a beta, may live in the same area and copulate with the same female(s). However, in contrast to the dunnock, alpha male accentors reduce the amount of parental assistance they give to females as their mating share decreases, whereas the beta male does not make his paternal care donations dependent on his past mating frequency.[359] If female accentors have evolved a mating strategy designed to increase the parental assistance they receive from males, how should their polyandry differ from that of female dunnocks?

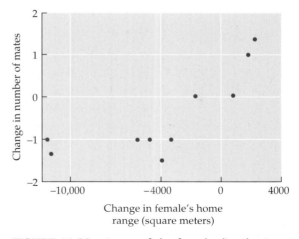

**FIGURE 11.33  A test of the female distribution theory of mating systems.** When food supplements reduce the size of a female dunnock's home range, a male can monopolize access to that female and reduce the number of males with whom she interacts. After Davies and Lundberg.[355]

What should a female do when the selection of an already mated male means that she must share the resources under his control with other females? According to the **polygyny threshold model**, there is a resource level at which she can gain more by mating with a polygynist on a good territory than by pairing off with a single male on a resource-poor, or predator-vulnerable, territory.[1072] If females really are sensitive to this choice threshold, then we can predict that females paired with polygynous and monogamous males should have about the same fitness. Michael Carey and Val Nolan checked this prediction by counting the number of young fledged in a population of indigo buntings, a small songbird whose parental males may attract either one or two mates to their territories in overgrown fields.[231] A monogamous female whose relationship with a male lasted the whole breeding season had only slightly greater reproductive success (1.6 young fledged) than a female that participated in a polygynous arrangement (1.3 young fledged). Therefore, females choosing an already occupied territory were not penalized heavily, if at all.

Stanislav Pribil and William Searcy put the polygyny threshold hypothesis to an experimental test. They took advantage of their knowledge that female red-winged blackbirds in an Ontario population almost always choose unmated males over mated ones. Their usual refusal to mate with a paired male is adaptive, judging from the fact that on those rare occasions when females do happen to select mated males, they produce fewer offspring than females that have their monogamous partners' territories all to themselves. Pribil and Searcy predicted that if they could experimentally boost the quality of territories held by already mated males while lowering the value of territories controlled by unmated males, then mate-searching females should reverse their usual preference. They tested this prediction by manipulating pairs of red-wing territories in such a way that one of the two sites contained one nesting female and some extra nesting habitat (added cattail reeds rising from underwater platforms) while the other territory had no nesting female but had supplemental cattail reed platforms placed on dry land. Female red-wings in nature prefer to nest over water, which offers greater protection against predators. For 14 pairs of territories, the first territory to be settled by an incoming female was the one where she could nest over water, even though this meant becoming part of a polygynous male's harem. Females that made this choice reared almost twice as many young on average as latecomers that had to make do with an onshore nesting platform in a monogamous male's territory.[1168]

Thus, when there is a free choice between a superior territory and an inferior one, it can pay a female to pick the better site even if she has to share it with another female. But why, then, do some female red-wings, and females of other birds as well, ever accept a second-rate territory? In the pied flycatcher, a second female generally does not do as well as the first female paired with a male,[6] especially if their partner devotes all his parental care to the brood of the first female, leaving the second female to fend for herself (see Figure 11.19).[692] The polygynous male of this species gets his second female by maintaining a second territory well away from the one that is used by his primary mate. Perhaps the second female never learns that her partner already has a mate until it is too late.

Svein Dale and Tore Slagsvold have shown, however, that mated male flycatchers spend less total time singing than unmated individuals, thus providing an important potential cue for perceptive females.[336] Because unmated females visit many males and do not rush into a paired relationship,[335] a male's mating status should usually become evident before a female makes a final decision. Thus, second females may not be deceived when they accept already mated males.[1426] Perhaps they choose them anyway because of the high costs of finding or evaluating other potential partners[1385] or because the remaining unmated males have extremely poor territories.[1426] In this case the real options for some females may be either to accept secondary status or not to breed at all, just as late-arriving female red-wings may have to make do with what is available in their neighborhood, even if it means making the best of a bad job.

## Scramble Competition Polygyny

Although female defense and resource defense tactics by competitive males make intuitive sense when females or resources are clumped in small, defensible areas, in many other species, receptive females and the resources they need are widely dispersed. Under these conditions, the cost–benefit ratio of mating territoriality usually falls, and males may instead simply try to find scarce receptive females before other males do. Females of a flightless *Photinus* firefly, for example, can appear almost anywhere over wide swaths of Florida woodland. Males of this species make no effort to be territorial; instead, they search, and search, and search some more. When Jim Lloyd tracked flashing males, he walked 10.9 miles in total, following 199 signaling males, and saw exactly two matings. Whenever Lloyd spotted a signaling female, a firefly male also found her in a few minutes.[877] Mating success in this species almost certainly goes to those searchers that are the most persistent, durable, and perceptive, not the most aggressive.

Male thirteen-lined ground squirrels behave like fireflies, searching widely for females, which become receptive for a mere 4 to 5 hours during the breeding season. The first male to find an estrous female and copulate with her will fertilize about 75 percent of her eggs, even if she mates again.[475] Given the widely scattered distribution of females and the first-male fertilization advantage, the ability to keep searching should greatly affect a male's reproductive success. In addition, male fitness may depend on a special kind of intelligence,[475] namely, the ability to remember where potential mates can be located. After visiting a number of females near their widely scattered burrows, searching males often return to those places on the following day. When researchers experimentally removed several females from their home sites, returning males spent more time searching for those missing females that had been on the verge of estrus. Moreover, the males did not simply inspect places that the females had used heavily, such as their burrows, but instead biased their searches in favor of the spots where they had actually interacted with about-to-become-receptive females.[1293, 1294] Individuals with superior spatial memory can probably relocate potential mates efficiently (Figure 11.34).

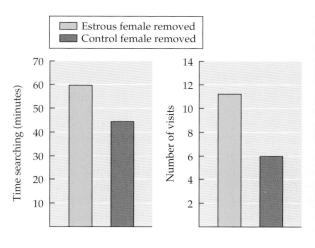

Thirteen-lined ground squirrel

**FIGURE 11.34 Scramble competition polygyny selects for spatial learning ability.** Male thirteen-lined ground squirrels remember the locations of females that are about to become sexually receptive. When males returned to an area where such a female had been the previous day but then had been experimentally removed, they spent more time searching for her than for a removed female not near estrus, and they returned to the near-estrus female's home range more often. After Schwagmeyer.[1294]

**FIGURE 11.35 An explosive breeding assemblage.** A male wood frog grasps a female (upper left) that he has found before rival males, two of which are near the mating pair. Numerous fertilized egg masses float in the water around the frogs. Photograph by Rick Howard.

Another form of scramble competition polygyny, the **explosive breeding assemblage**, occurs in species with a highly compressed breeding season. One such species is the horseshoe crab, whose females lay their eggs on just a few nights each spring and summer. Males are under the gun to be near the egg-laying beaches at the right times and to accompany a female to shore, where egg laying and fertilization occur[179] (see Figure 10.23). A race to find and pair off with a female also occurs in the wood frog, another species in which the opportunity to acquire a mate is restricted to one or a few nights each year. On that night or nights, most of the adult males in the population are present at the ponds that females visit to mate and lay their eggs. Just as in horseshoe crabs, the high density of rival males raises the cost of repelling them from a defended area. And because females are available only on this one night, a few highly aggressive territorial males cannot monopolize a disproportionate number of them. Therefore, male wood frogs eschew territorial behavior and instead hurry about trying to encounter one or more egg-laden females before the one-night orgy ends (Figure 11.35).[173]

### Lek Polygyny

In some species, males do not search for mates, nor do they defend groups of females or resources that several females come to exploit. Instead, they fight to control a very small area that is used only as a display arena; these mini territories may be clustered in a traditional display ground, or **lek**, or they may be somewhat scattered, forming a dispersed lek, as is true for satin bowerbirds.[158] Despite the fact that the males' territories do not contain food, nesting sites, or anything else of practical utility, females come to the leks anyway. When white-bearded manakin females arrive at a lek in a Trinidadian forest, they may find as many as 70 sparrow-sized males in an area only 150 meters square. Each male will have cleared the ground around a little sapling rising from the forest floor. The sapling and cleared court are the props for his display routine, which consists of rapid jumps between perches accompanied by loud sounds produced by snapping his clubbed primary wing feathers together. When a female is around, the male jumps to the ground with a snap and immediately back to the perch with a buzz, and then back and forth "so fast he seems to be bouncing and exploding like a firecracker."[1359] If the female is receptive and chooses a partner, she flies to his perch for a series of mutual displays followed by copulation. Afterward, she leaves to begin nesting, and the male remains at the lek to court newcomers. Alan Lill found that in a lek

**FIGURE 11.36  Mating success at leks.** Topi males in central positions at their leks mate with more females per capita than males forced to peripheral sites. The blue bars represent the averages for the three leks shown. After Bro-Jørgensen and Durant.[176]

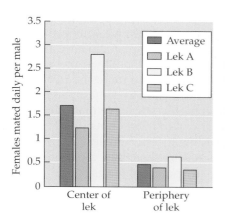

Topi male

with ten manakins, where he recorded 438 copulations, one male contributed nearly 75 percent of the total; a second male mated 56 times (13 percent), while six other males together accounted for a mere 10 matings.[864] Preferred males tend to occupy sites near the center of the lek, and they engage in more aggressive displays than less successful males.[1322]

Huge inequalities in male mating success are standard features of lekking species. Thus, in the topi antelope, the generally older males that occupy the center of leks located on African savannas copulate much more often than younger rivals forced to the periphery (Figure 11.36).[176] Even more extreme skew in mating success occurs in some other lekking mammals. For example, just 6 percent of the males in a lek of the bizarre West African hammer-headed bat (Figure 11.37) were responsible for 80 percent of the matings recorded by Jack Bradbury. In this species, males gather in groups along riverbanks, each bat defending a display territory high in a tree, where he produces loud cries that sound like "a glass being rapped hard on a porcelain sink."[157] Receptive females fly to the lek and visit several males, each of which responds with a paroxysm of wing-flapping displays and strange vocalizations (note the behavioral convergence with manakins and satin bowerbirds).

Why do male manakins, topi antelope, and hammer-headed bats behave the way they do? Bradbury has argued that lekking evolves only when other mating tactics do not pay off for males, thanks to a wide and even distribution of females.[158] Female manakins and hammer-headed bats do not live in permanent groups, but instead travel over great distances in search of widely scattered sources of food of unpredictable availability, especially figs and other tropical fruits. A male that tried to defend one tree might have a long wait before it began to bear attractive fruit, and when it did, the large amount of food would attract hordes of consumers, which could overwhelm the territorial capacity of a single defender. Thus, the feeding ecology of females of these species makes it hard for males to monopolize them, directly or indirectly. Instead, males display their merits to choosy females that come to leks to inspect them.

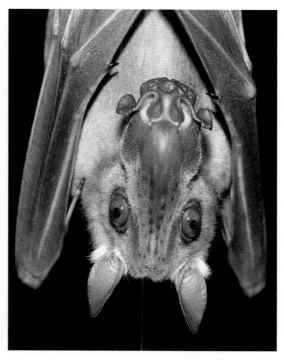

**FIGURE 11.37  A lek-polygynous mammal: the hammer-headed bat.**

But you may recall that the absence of defensible clusters of females was invoked as an explanation for nonterritorial scramble competition polygyny as well. Why some species subject to these conditions simply search for mates while others form elaborate leks is not at all clear. Nor is it known for certain why males of some lekking species congregate in small areas as opposed to displaying by themselves in a dispersed lek (a la the satin bowerbird), although one possibility is that predation favors adoption of a safety-in-numbers tactic.[536] Any reductions in the cost of displaying go some way toward explaining why lekking males congregate, but we still need to know what benefit males derive from defending a little display territory at a group lek site. Here we shall review three ideas: (1) the hotspot hypothesis, according to which males cluster in places (**hotspots**) where routes frequently traveled by receptive females intersect;[160] (2) the hotshot hypothesis, according to which subordinate males cluster around highly attractive males in order to have a chance to interact with females drawn to these "**hotshots**";[159] and (3) the **female preference hypothesis**, according to which males cluster because females prefer sites with large groups of males, where they can more quickly, or more safely, compare the quality of many potential mates.[158]

To test these hypotheses, Frédéric Jiguet and Vincent Bretagnolle managed to create artificial leks populated by plastic decoys painted to resemble males and females of the little bustard, a bird that exhibits a dispersed-lek mating system. The two researchers placed different numbers of decoys of the two sexes in different fields and then, over some time, counted the number of living little bustards attracted to their experimental leks. They found that female decoys failed to draw in males, leading to the rejection of the hotspot hypothesis. On the other hand, male decoys regularly attracted both females and males, particularly if those decoys had been painted to resemble individuals with highly symmetric plumage patterns. Therefore, the hotshot hypothesis may apply to little bustards. The fact that more females per decoy were attracted to clusters of four decoys than to smaller groups (Figure 11.38) is also consistent with the female preference hypothesis as well, although more than four decoys did not draw in additional females.[724]

Another way of testing the hotspot versus hotshot hypotheses is by temporarily removing males that have been successful in attracting females. If the

**FIGURE 11.38 An experimental test of alternative hypotheses for lek formation.** (A) A female little bustard visiting a decoy of a male of her species. (B) More females visit groups of four decoys than smaller (or larger) groups, as shown by the peak in the graph in per capita female number when the lek size was four. The sex ratio represents the number of male decoys divided by the total number of decoys. Photograph courtesy of Frédéric Jiguet; after Jiguet and Bretagnolle.[724]

(A)

(B)
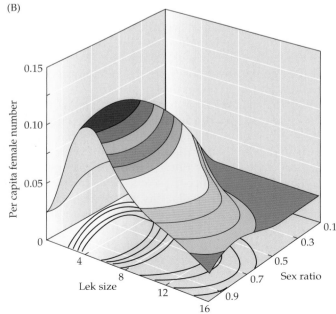

hotspot hypothesis is correct, then removal of these successful males from their territories will enable others to move into the favored sites. But if the hotshot hypothesis is correct, removal of attractive males will cause the cluster of subordinates to disperse to other popular males or to leave the site altogether.

In a study of the great snipe, a European sandpiper that displays at night, removal of central dominant males caused their neighboring subordinates to leave their territories. In contrast, removal of a subordinate while the alpha snipe was in place resulted in his quick replacement on the vacant territory by another subordinate. At least in this species, the presence of attractive hotshots, not the real estate per se, determines where clusters of males form.[659] Likewise, in the unrelated black grouse, although relatively large, centrally located display sites are associated with higher mating success,[1226] the exact location of the most successful territory can change somewhat from year to year within a durable lek, suggesting that a popular male, rather than a particular spot, most influences the behavior of other males (Figure 11.39A).[1224]

(A)

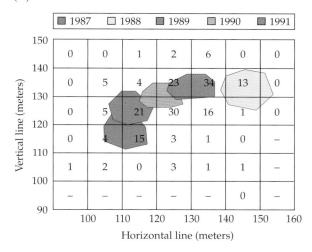

Black grouse

(B)

Marine iguana

**FIGURE 11.39   Hotspots or hotshots?** (A) Researchers divided a black grouse lek into 100-square-meter sectors and recorded the total number of copulations in each sector over a 5-year period from 1987 to 1991. The irregular polygons show the location of the top territory for each of the 5 years. The shifts in the preferred territory suggest that male attractiveness, rather than the territory itself, plays the key role in reproductive success in this species, as required by the hotshot hypothesis. (B) Similarly, in the marine iguana of the Galápagos Islands, the territory with the greatest number of recorded matings shifted over the years (in 1987, 1988, 1994, and 1995, four different territories among the 22 occupied sites were ranked highest by females). Sites are numbered; color indicates number of copulations. A, after Rintamäki et al.;[1224] B, after Partecke, von Haeseler, and Wikelski.[1106]

(A)

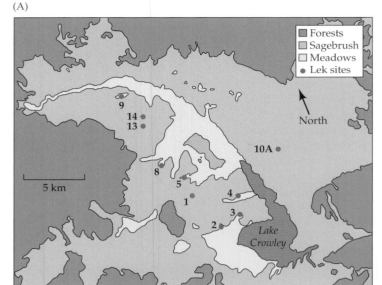

(B)

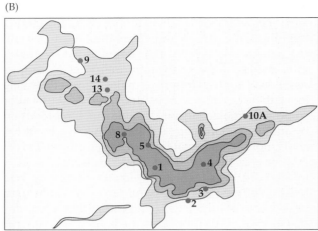

**FIGURE 11.40  A test of the hotspot hypothesis.** (A) The position of sage grouse leks (numbered red circles) in relation to sagebrush, meadows, forests, and a lake. (B) The distribution of nesting females in relation to the leks where males gather to display. The darker the shading, the more females present. After Gibson.[535]

Sage grouse

Much the same thing occurs in the lekking marine iguana of the Galápagos.[1106] In this species, too, the territory at which the most copulations occurred was not the same from one year to the next (Figure 11.39B). In the leks of both black grouse and marine iguanas, the most reproductively successful males are forced to fight most often by their neighbors. In the confusion following an attack when a female is present, a hotshot's neighbor may get a chance to mate (although molecular data for black grouse strongly suggest that the vast majority of females fertilize their eggs with the sperm of one male only[847]).

Although the hotshot hypothesis seems likely in some cases, the hotspot hypothesis has received support in others (Figure 11.40).[535] For example, a site at which fallow deer bucks gathered to display shifted when logging activity altered the paths regularly followed by fallow deer does.[43] Likewise, peacocks tend to gather near areas where potential mates are feeding (hotspots); the removal of some males has no effect on the number of females visiting leks of this species, as one would predict if females are inspecting those areas because of the food they provide, not because of the males there.[893]

The hotspot hypothesis cannot, however, apply to those ungulates in which estrous females leave their customary foraging ranges to visit groups of males some distance away, perhaps to compare the performance of many males

Kafue lechwe

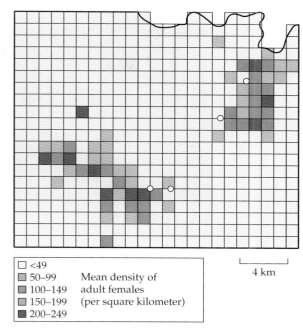

**FIGURE 11.41    Female density is not correlated with lek formation** in an African antelope, the Kafue lechwe. The four leks in this region (open circles) are not located in the areas of highest female density, providing evidence against the hotspot hypothesis in this species. After Balmford et al.[69]

| | Mean density of adult females (per square kilometer) |
|---|---|
| ☐ <49 | |
| 50–99 | |
| 100–149 | |
| 150–199 | |
| 200–249 | |

4 km

simultaneously (Figure 11.41).[69] A female preference for quick and easy comparisons might make it advantageous for males of the Uganda kob, an African antelope, to form large groups. If so, those leks with a relatively large number of males should attract proportionately more females than leks with fewer males. Contrary to this prediction, however, the operational sex ratio is the same for leks across a spectrum of sizes (Figure 11.42), so males are no better off in large groups than in small ones.[387] For this species at least, the female preference hypothesis can be rejected. The same is true for the barking treefrog (Figure 11.43). Here too, the more males chorusing in a pond, the more receptive females show up on a given night. But preventing males from coming to the pond to call does not reduce the number of females arriving, which is not what one would expect if a large number of calling males is needed to attract a large number of choosy females.[1025] On the other hand, in an Australian fruit

Uganda kob

(A)

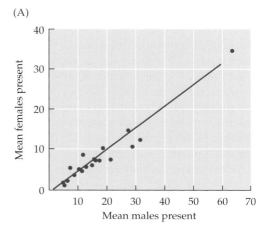

(B)

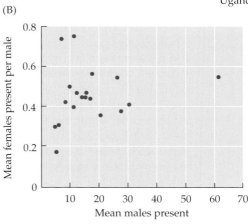

**FIGURE 11.42    Female Uganda kob do not aggregate disproportionately at leks with large numbers of males.** (A) Female attendance at leks is simply proportional to the number of males displaying there. (B) As a result, the female-to-male ratio does not increase as lek size increases. After Deutsch.[387]

Female barking treefrog

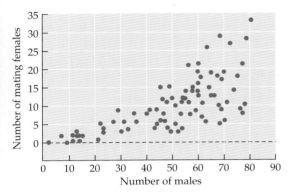

**FIGURE 11.43   The more female barking treefrogs** mating at a pond on a given night, the more males found chorusing there. But the correlation stems not from the ability of large numbers of males to attract more females with their calls, but from the fact that both sexes respond similarly to a set of environmental variables, including temperature and rainfall. After Murphy.[1025]

fly, the female encounter rate for males does increase as the number of males in a lek rises to 20, although it then falls off,[51] just as it does for the ruff, a lekking sandpiper (Figure 11.44).[1565]

Thus, no one hypothesis on why lekking males form groups holds for every species. Nevertheless, the interactions among males on a lek usually seem to enable individuals of high physiological competence to demonstrate their superior condition to their fellow males and to visiting females.[468] Whatever the basis for lek formation, lekking males are forced to compete in ways that separate the men from the boys, making it potentially advantageous for females to come to a lek to compare and choose, the better to select a partner of the highest quality. Indeed, one of the main themes of mating system theory is that in the vast majority of species, female reproductive tactics create the circumstances that determine which competitive and display maneuvers will provide payoffs for males. As a result, male mating systems are an evolved response to female mating systems—and to the ecological factors that determine the spatial distribution of females.

## Discussion Questions

**11.11**   Todd Shelly did an experiment with the Mediterranean fruit fly, a lekking species, in which he placed two cups with wire mesh tops and containing different numbers of males in a coffee bush.[1312] After the cups were in position, he released 400 females in another coffee bush about 10 meters away. He then counted, at 10-minute intervals, the number of males in each cup that were "calling," that is, releasing pheromones (signaling males evert an abdominal pheromone gland), and the number of females perched on or near each cup. Each trial lasted 80 minutes, and he did 30 trials in all. The data he collected are presented in Figure 11.45. Reconstruct the science underlying this research, working from the graphs in whatever direction makes sense until you have the causal question, hypothesis, prediction, test, and scientific conclusion. You should be able to relate the work to one of the hypotheses described in this section on why males aggregate at leks.

(A)

(B)

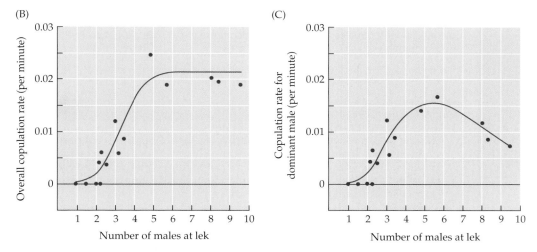

**FIGURE 11.44  Lek size and copulation rate in the ruff.** (A) Male ruffs fighting for dominance on a lek territory. (B) Up to a point, larger leks attract more receptive females than small leks. (C) At leks of six or more males, however, the average reproductive success of the dominant males falls because the number of arriving females levels off and some lower-ranking males copulate with some of the visiting females. After Widemo and Owens.[1565]

**11.12**  One of the possible benefits to females of participating in lek mating systems is the selection of mates of exceptional genetic quality. This form of sexual selection should tend to reduce genetic variation among males of lekking species over time. (Why?) If female choice did eliminate genetic variation among males over time, would this outcome support or undercut the argument that females go to leks to choose superior mates? Might the phenomenon of genetic compatibility help us out of this pickle? Is, however, the finding that a very few males monopolize matings at most leks consistent with the suggestion that females visit leks to secure sperm from genetically compatible partners? What would happen if female mate choice were to focus on different traits from year to year?[252]

(A)

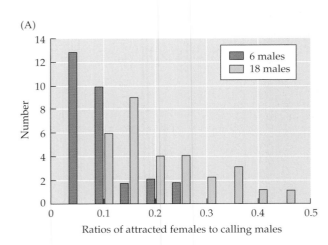

(B)

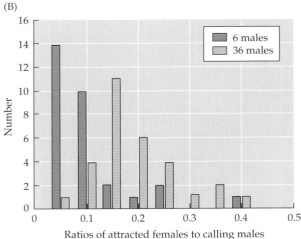

Mediterranean fruit fly

**FIGURE 11.45** **The ratios of attracted females to signaling males** of the Mediterranean fruit fly in two "leks" of (A) 6 and 18 males, and (B) 6 and 36 males. The color of the bar indicates the cup (lek); for example, in A the blue bars indicate results for the cup with 6 males. Each trial, of which there were 30 in all, generated a single ratio of females to males per "lek." After Shelly.[1312]

## Summary

1. Mating systems can be defined in terms of the number of sexual partners an individual acquires during a breeding season. Males can be monogamous (mating with a single female) or polygynous (mating with several females). Likewise, females can be monogamous (having a single partner) or polyandrous (mating with several).

2. Monogamy by males is a Darwinian puzzle because males that restrict themselves to a single female would seem likely to have fewer descendants than males that successfully mate with many females. Monogamous males can gain fitness, however, if there are large payoffs to parental males or to those that prevent their sole mates from accepting sperm from other males. Alternatively, conflict between females and males may thwart male attempts to be genetically polygynous, even if it would be advantageous from the male perspective.

3. Similarly, although females of most birds and other animals can secure all the sperm they need to fertilize their eggs from a single mate, polyandry is common, even in socially monogamous birds. Polyandrous females may secure various genetic or material benefits, ranging from superior genes for their offspring to greater amounts of parental assistance from their partners. It is also possible that females of some species mate with several males because their partners force them to do so. Both female choice and conflict between the sexes may play a part in the evolution of mating systems.

4. Ecological factors that affect males' ability to monopolize receptive females are also important influences on mating system diversity. In particular, the distribution of females may affect the profitability of different kinds of territorial mating tactics for males. When females or the resources they need are clumped in space, female defense or resource defense polygyny becomes more likely. If, however, females are widely dispersed or male density is high, males may engage in nonterritorial scramble competition for mates, or they may acquire mates by displaying at a lek. Many aspects of lek polygyny are still not completely understood, such as why it is that males of most lekking species gather in groups to display to females.

## Suggested Reading

Steve Emlen and Lew Oring's now classic paper changed the way we look at the ecology of mating systems.[438] For an especially thorough examination of a single species with multiple mating systems, read *Dunnock Behaviour and Social Evolution*.[358] Jacob Höglund and Rauno Alatalo have written a useful book on leks.[660] Malte Andersson's book on sexual selection explores mating system evolution in depth,[35] while Dave Ligon focuses exclusively on avian mating systems.[863] Extra-pair copulations by birds are the focus of two fine review articles.[582, 1547]

# 12

# The Evolution of Parental Care

A lthough most of us may think that both parents should contribute to child care, in a great many animal species neither the mother nor the father lifts a finger for their offspring, while in others just one parent—either the female or, much more rarely, the male—takes full responsibility for her or his progeny. This diversity in parental behavior has been hinted at in previous chapters, in which we mentioned the biparental care provided by the prairie vole, the strictly maternal care provided by most other mammals, and the fathers-only approach of the spotted sandpiper and red phalarope. Variety of this sort, as my readers know by now, always interests evolutionary biologists.

The key to explaining the diversity in parental behavior lies with the cost–benefit approach used by behavioral ecologists. The benefits of parental care are obvious, having to do with the improved survival of the assisted offspring. But the costs of being a devoted parent must also be considered if one is to deal with the major evolutionary questions that surround parental behavior. We have already seen how important it is to think about fitness costs if we want to explain why only some gulls mob predators hunting for their chicks (see Chapter 6). So let's begin this chapter with another look at both the costs and benefits of helping offspring survive.

◀ **The orange plumes on this baby** coot *may make all the difference in the way in which his parent treats him. Photograph by Bruce Lyon.*

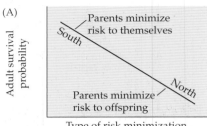

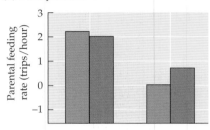

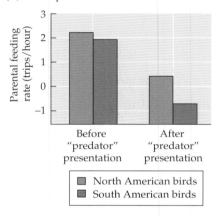

**FIGURE 12.1  Parental care is the evolutionary product of fitness costs and benefits.** (A) The reactions of parent birds to predatory threats to themselves and their offspring should vary in relation to the annual mortality rates of adults, which differ between South and North America. Shorter-lived North American birds are predicted to do more to reduce risks to their offspring; longer-lived South American birds are predicted to do more to promote their own survival. (B) The rate at which parents feed their nestlings falls more sharply in North American birds than in South American birds in response to an apparent risk of nest predation by jays, which can find nests by watching where parent birds go. (C) The rate at which parents feed their nestlings falls more sharply in South American birds than in North American birds after the adults see a hawk capable of killing them. After Ghalambor and Martin.[534]

# The Cost–Benefit Analysis of Parental Care

In many bird species, parents of both sexes work tirelessly to bring food to a brood of all but helpless offspring ensconced in the relative safety of a hidden nest. Without parentally supplied food, the baby birds would soon be dead. But there are risks attendant upon the parents' feeding trips. Predators can use their comings and goings to and from the nest to find it and feast on its occupants; alternatively, predators can lurk near the nest to intercept the parents as they return to their offspring with food. How does a parental bird balance the costs and benefits of its activities?

Cameron Ghalambor and Tom Martin predicted that parent birds should adjust their provisioning behavior adaptively in accordance with two key factors: the nature of the predator (whether it consumes nestlings or adults) and the annual mortality rate for breeding adults.[534] In birds with a generally low adult mortality rate, parents ought to minimize the risk of getting themselves killed by a predator, because they will probably have many more chances to reproduce in the future if a predator does not get them now. In breeding birds with a high annual mortality rate, however, parents should be less concerned with their own safety and more sensitive to the risks that their nestlings may confront from nest-raiding predators; these parents will have relatively few chances to reproduce in the future, so they can gain by doing more now for their current brood.

Ghalambor and Martin knew that birds that breed in North America tend to have shorter lives and produce larger clutches of eggs than their close relatives that breed in South America. So they matched up five pairs of these relatives, including, for example, two members of the same genus, the short-lived North American robin and the longer-lived Argentinian rufous-bellied thrush. When the ornithologists played tapes of Steller's jays to Arizonan robins and tapes of plush-capped jays to their Argentinian counterparts, the robins and thrushes both greatly reduced their visits to their nests for some time so as not to reveal to these nest predators where their nests were hidden. However, the robins curtailed their activity around the nest more strongly than the thrushes, presumably because they had more to gain by protecting their current crop of nestlings from the keen-eyed jays, given their relatively low probability of reproducing in subsequent years. When the ornithologists placed a stuffed sharp-shinned hawk, a killer of adult birds, near active nests and played recorded sharp-shin calls, the parent birds in the sampled species again reduced their visits to their nests for some time. In this round of tests, however, the potentially long-lived Argentinian birds delayed their return to the nest longer than the corresponding Arizonan species (Figure 12.1).[534] This case is merely one of many in which the parental strategies of North American birds appear to differ from those of South American species, probably because of differences in predation pressure on nesting adults in the two regions.[943] The costs of taking care of offspring, not just the benefits, have fine-tuned the evolution of parental behavior.

## Why More Care by Mothers than Fathers?

Although in many species of birds, including robins and thrushes, both fathers and mothers help their progeny survive, across the animal kingdom as a whole, females are much more likely to be maternal than males are to be paternal. Thus, for example, in some species of treehoppers, the females of these insects, but never the males, stand watch over their eggs day and night to protect them against predatory or parasitic insects that would destroy their brood. In some cases, females even stay on to protect their nymphs until they have become adults. Chung-Ping Lin and his coworkers explored the evolutionary history of egg-guarding parental care by mapping the trait on a phylogeny of the Membracinae that was derived from molecular comparisons

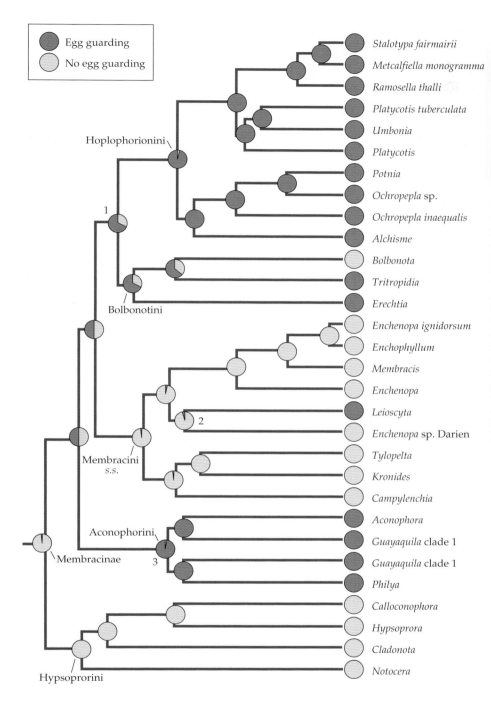

**Egg guarding**

**No egg guarding**

*Stalotypa fairmairii*
*Metcalfiella monogramma*
*Ramosella thalli*
*Platycotis tuberculata*
*Umbonia*
*Platycotis*
*Potnia*
*Ochropepla* sp.
*Ochropepla inaequalis*
*Alchisme*
*Bolbonota*
*Tritropidia*
*Erechtia*
*Enchenopa ignidorsum*
*Enchophyllum*
*Membracis*
*Enchenopa*
*Leioscyta*
*Enchenopa* sp. Darien
*Tylopelta*
*Kronides*
*Campylenchia*
*Aconophora*
*Guayaquila* clade 1
*Guayaquila* clade 1
*Philya*
*Calloconophora*
*Hypsoprora*
*Cladonota*
*Notocera*

Hoplophorionini
Bolbonotini
Membracini *s.s.*
Aconophorini
Membracinae
Hypsoprorini

**FIGURE 12.2 Parental care is provided by females in the Membracinae.** The genera that exhibit maternal care have been placed on a molecular phylogeny of this subfamily of treehoppers. The circles at the bases of the lineages represent ancestral species; the proportion of a given circle that is blue reflects the probability that females of this ancestral species cared for their eggs. The numbers mark the three probable originations of maternal care. The photograph shows a female treehopper in the genus *Guayaquila* standing guard over a cluster of eggs covered by a secretion applied by the female. After Lin, Danforth, and Wood;[868] photograph by Chung-Ping Lin.

among the genera in this group.[868] Their work indicates that while maternal care has probably originated three different times in this group (Figure 12.2), paternal care has never evolved in these insects.

## Discussion Question

**12.1** On the basis of Figure 12.2, how many times has maternal care been lost after evolving in the Membracinae? If you wanted to employ the comparative method to examine the evolved function of maternal care in this group, what genera would be of special interest to you, and why?

(A)

(B)

**FIGURE 12.3   Paternal care in fishes and opportunities for polygyny.** (A) A male Randall's jawfish holds his mate's eggs in his mouth. Mouth brooding limits a male to one clutch at a time. (B) In contrast, a male stickleback caring for a nest with a clutch of eggs can attract additional females, which add their eggs to the nest. This male is aerating the eggs in his nest, at the base of the aquatic plant, by drawing water through the nest.

Certain treehoppers therefore offer a clear example of the general rule that maternal care evolves more readily than paternal care. One intuitively appealing explanation for the female domination of parental caregiving takes the following form: because females (unlike males) have already invested so much energy in making eggs, they have a special incentive to make sure that their large initial gametic investment is not wasted. Females therefore continue to provide parental care after the eggs have been produced and fertilized. This hypothesis comes to grief, however, when we observe that females of a substantial number of species, including spotted sandpipers and many fishes, abruptly terminate their parental investment after laying their large and expensive-to-produce eggs, which they leave totally in the care of their male partners (Figure 12.3). These species show that a considerable initial investment in offspring does not automatically make it advantageous for females to invest still more in their broods. Instead, both the benefits and costs of each increment of maternal care will determine whether the investment is adaptive.

That maternal care costs the caregiver something can be illustrated by examining the effects of brood tending on female European earwigs, which often stay with a clutch of eggs laid in a burrow, waiting for them to hatch, in order to feed their larval offspring for some time (Figure 12.4) (for a video on this behavior, with a hideously cute narration but great photography, see http://www.nationalinsectweek.co.uk/smalltalk.php#sttitle). Females that provide this care help their little earwigs survive, an obvious benefit of their tending behavior, but they pay a price as well. For maternal females, the interval between laying a clutch of eggs is a week longer than for females that do not stay around to help one group of offspring get a good start in life.[791] If the costs of providing an additional unit of care were to exceed the fitness benefits gained from the act, females that did not make the added investment would leave more surviving descendants, on average, than females that supplied the extra care.

**FIGURE 12.4**  A maternal female earwig guards her clutch of eggs, which she protects against predators. In nature, as opposed to the laboratory, most females lay their eggs under leaf litter or some other shelter. Photograph by Mathias Kölliker.

Because there is no guarantee that females will always derive a net benefit from an extra dollop of parental care, we must find an explanation for the general pattern of female-only parental care that does not focus simply on the size of their gametes. David Queller has come to the rescue (see also page 336 in Chapter 10) by pointing out that if the costs of parental care were usually lower for females than for males, as they may well be, then this could tip the scales so that females would provide care more often than males.[1184] Let's assume for the sake of simplicity that one standard unit of parental care invested in a current offspring reduces the future reproductive output of a male and a female by the same amount. Let's also assume that we are looking at a species in which females sometimes mate with more than one male in a breeding season. In this case, the average benefit to a male from caring for a brood of offspring will be reduced to the extent that some of "his" offspring were actually fathered by other males. For example, if his paternity is, on average, 80 percent, then for every five offspring assisted, the male's investment can yield at best only four descendants, whereas all five youngsters could advance the female's reproductive success. In other words, when cuckolded paternal males waste some of their costly parental care on nonrelatives, they become less likely than their mates to experience a favorable benefit-to-cost ratio for parenting.

Not only are the benefits from paternal care likely to be less than the benefits from a comparable amount of maternal care, but the costs of parental care are likely to be greater for males than for females. As we noted when discussing sexual selection theory in Chapter 10, males that acquire many mates generally leave many descendants. These successful males would pay a steep price if they were to divert their efforts away from securing mates to caring for some of their offspring. Imagine a lek of black grouse in which the top male fertilizes most of the eggs of the 20 or so females that come to the lek to mate. Because regular attendance at the lek is one of the main correlates of male mating success, a grouse with a reasonable chance of becoming an alpha male would lose many potential offspring if he took time off from lekking to incubate a batch of eggs.[1184] The same rule probably applies to sexually attractive males of many other species.

## Discussion Questions

**12.2** Here's another way of looking at the effect of differences between males and females on the probability that their putative offspring are truly their genetic offspring. If a female lays a fertilized egg or gives birth to a baby, this offspring will definitely have 50 percent of her genes. In contrast, a male that mated with this female may or may not be the father of that offspring. Thus, the argument goes, males have less to gain by parental behavior, and this tilts the equation in favor of nonpaternal males that seek out multiple partners rather than limiting themselves to caring for the offspring of one or a few mates. But let's check the logic of this argument by imagining a hypothetical species in which males have a fairly low probability of siring the offspring of any given mate—say, a 40 percent chance. Further, imagine that there are two hereditarily different male phenotypes in the population, a paternal type and a nonpaternal type. The average paternal male mates with two females (each with an average of 10 eggs), whereas the average nonpaternal male mates with five females (which enables him to fertilize a proportion of 50 eggs). In addition, let's say that the paternal male boosts the survival chances of the eggs under his care to 50 percent versus 10 percent for the unprotected offspring of nonpaternal males. Which behavior is adaptive here? Show your math. What point does this example make about the evolution of male parental care?

**12.3** Among certain monkeys and apes with prolonged parental care, females live longer than males in species in which females provide most or all of the parenting, but males live longer than females in species in which males make the major contribution to offspring care.[24] In other words, adults of the parental sex tend to live longer than the nonparental sex. Does this finding indicate that parental care provides a fitness benefit for caretakers in the form of improved survival? Imagine that someone claims that the longer life span of the parental sex has been selected for because primate young are very slow to develop, and therefore parents must live long enough to get their offspring to the age of independence in order to maintain a stable population. Do you agree? Do you have an alternative explanation for the observed pattern?

## Exceptions to the Rule

The general rule that males are not parental has many exceptions. Male-only parental care is actually common among fishes (see Figure 12.3), despite the fact that male fish make vast quantities of sperm, like most other male animals. Because some male fish could potentially have many more offspring than the most fecund female of their species, they would seem, at first glance, to have much to lose by taking time and energy away from mating effort to be good parents. Upon mature reflection, however, we can see that there does not have to be a trade-off between parental care and mate attraction in mating systems in which female choice revolves around male care of eggs. Stickleback females are drawn to egg-guarding males, which are demonstrating their commitment to parenting. Moreover, the more eggs in a male's nest, the safer they are likely to be, thanks to the dilution effect (see Chapter 6).[1277]

In fact, stickleback males can care for up to ten clutches of eggs over the 2 weeks or so that it takes the eggs to hatch. In contrast, an average female stickleback can produce only seven clutches of eggs during this period, even without taking time out to guard her eggs.[281] Parenting would be far less advantageous for stickleback females than for males, given that a female would be caring for only one clutch at a time. Furthermore, while the female was so engaged, she would not be able to forage freely and so would not grow as rapidly as she might otherwise. This loss in growth could be especially damaging in those species in which female fecundity increases exponentially with increasing body size. In this case, for each unit of growth lost by being parental, a female may pay a particularly stiff price in loss of future egg production. Males that are parental also grow more slowly than they would otherwise, but since they must remain in a territory in any case if they are to attract mates, the decrease in their growth resulting exclusively from parental care is slight.

## Discussion Question

**12.4** Males of the golden egg bug are sometimes chosen by females to receive their eggs, which are glued to the male's back. Two possible hypotheses have been proposed for the male's willingness to accept this burden: (1) males carry eggs to attract gravid females, which may then copulate with them, or (2) males carry eggs (of their mates) to decrease the risk that their offspring will be afflicted by parasites. Given these alternatives, what significance do you attach to the following results: (a) males in an area where egg parasites are numerous are much more likely to carry eggs than those from a region where egg parasites are essentially absent, (b) eggs laid on plants, an alternative for egg-laden females, are up to ten times more likely to be destroyed by parasites than eggs laid on male bugs, and (c) when females were given a choice between mating with an egg-bearing male versus one unencumbered by eggs, they did not choose the egg bearers significantly more often than those without eggs.[549]

The costs of parental behavior for both males and females have been directly measured for a mouth-brooding cichlid, St. Peter's fish, in which either the male or the female may care for their young by orally incubating the fertilized eggs. Both sexes lose weight while mouth brooding, since they find it difficult to eat with a mouth full of eggs or baby fish. Furthermore, the interval between spawnings increases for parental fish of both sexes compared with individuals whose clutches are experimentally removed (Figure 12.5). However, parental females wait an extra 11 days between spawnings compared to nonparental females, whereas brooding males pay a smaller price—only

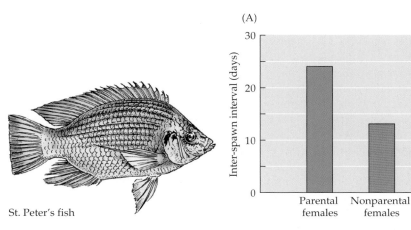
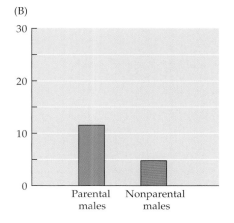

St. Peter's fish

**FIGURE 12.5 Parental care costs female St. Peter's fish more than it costs males.** (A) Females that have cared for a clutch of eggs are much slower to produce a new batch of mature eggs than are females that have not provided care to their previous clutch. (B) Parental males also spawn less often than nonparental males, but the difference between the two groups is less than for females. After Balshine-Earn.[70]

7 extra days between spawnings compared to nonparental males. Moreover, parental females produce fewer young in their next clutch than do nonparental females, whereas parental males are just as able to fertilize complete clutches of eggs in their next spawning as are nonparental males. Although both sexes pay a price for parental behavior, the costs of brood care to females are greater in terms of reduced fecundity.[70] Thus, in fishes, paternal behavior may evolve because males lose less from parenting than females do, which creates a more favorable benefit-to-cost ratio for paternal behavior than for maternal care of offspring.[588]

### Why Do Male Water Bugs Do All the Work?

Although exclusive paternal care of young is common among fishes, the trait is very rare among other animals, vertebrates and invertebrates alike.[280, 1422] Among the exceptional paternal insects are water bugs of the genus *Lethocerus*, which guard and moisten clutches of eggs that females glue onto the stems of aquatic vegetation above the waterline (see Figure 1.17).[1354] Males in some other genera of water bugs (e.g., *Abedus* and *Belostoma*) permit their mates to lay eggs directly on their backs (Figure 12.6), after which the male assumes responsibility for their welfare. A brooding male *Abedus* spends hours perched near the water surface, pumping his body up and down to keep well-aerated water moving over the eggs. Clutches that are experimentally separated from a male attendant do not develop, demonstrating that male parental care is essential for offspring survival in this case.

Bob Smith has explored both the history and the adaptive value of these unusual paternal behaviors.[1355] Since the closest relatives of the Belostomatidae, the family that contains these paternal water bugs, are the Nepidae, a family of insects without male parental care, we can be confident that the brooding species evolved from nonpaternal ancestors (Figure 12.7). Whether out-of-water brooding and back brooding evolved independently from such an ancestor, or whether one preceded the other, is not known, although some evidence suggests that out-of-water brooding came first. Notably, when *Lethocerus* females cannot find suitable exposed vegetation for their eggs, they

**FIGURE 12.6  Male water bugs provide uniparental care.** A male belostomatid broods eggs glued onto his back by his mate. Photograph by Bob Smith.

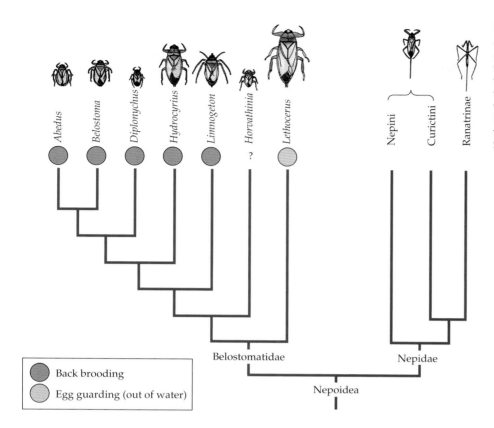

**FIGURE 12.7 Evolution of brood care by males in the Nepoidea,** the group that includes the belostomatid water bugs. The illustrations, drawn to scale, show the largest representatives of each group. Paternal care is widespread throughout the family Belostomatidae, but no species in the family Nepidae exhibits this trait. After Smith.[1355]

sometimes lay their eggs on the backs of other individuals, male or female. This unusual behavior indicates how the transition from out-of-water brooding to back brooding might have occurred. Females with the tendency to lay their eggs on the backs of their mates could have reproduced in temporary ponds as well as pools where emergent aquatic vegetation was scarce or absent.

But why do the eggs of water bugs require brooding? Huge numbers of aquatic insects lay eggs that do perfectly well without a caretaker of either sex. However, Smith notes that belostomatid eggs are much larger than the standard aquatic insect egg, with a correspondingly large requirement for oxygen needed to sustain the high metabolic rates underlying embryonic development. But the relatively low surface-to-volume ratio of a large aquatic egg leads to an oxygen deficit inside the egg. Since oxygen diffuses through air much more easily than through water, laying eggs out of water can solve that problem. But this solution creates another problem, which is the risk of desiccation that the eggs face when they are high and dry. The solution, brooding by males that moisten the eggs repeatedly, sets the stage for the evolutionary transition to back brooding at the air–water interface.

Wouldn't things be simpler if belostomatids simply laid small eggs with large surface-to-volume ratios? To explain why some water bugs produce eggs so large that they need to be brooded, Smith points out that water bugs are among the world's largest insects, an advantage when it comes to grabbing and stabbing large prey, like fish, frogs, and tadpoles. Water bugs, and all other insects, grow in size only during the immature stages. After the final molt to adulthood, no additional growth occurs. As an immature insect molts from one stage to the next, it acquires a new flexible cuticular skin that permits an expansion of size, but no immature insect grows more than 50 or 60 percent per molt. One way for an insect to grow large, therefore, would be to increase the number of molts before making the final transition to adulthood. However, no member of the belostomatid family molts more than six times. This observation suggests that these insects are locked into a five- or six-molt sequence, just as spotted sandpipers evidently cannot lay more than four eggs per clutch. If a water bug has just five or six molts in which to grow large enough to kill a frog, then the first instar (the nymph that hatches from the egg) must be large, because it will get to undergo only five or so 50-percent expansions. Thus, water bug development is an example of the "panda principle" (see page 308) in that evolutionary modifications of body size have to be layered on what has already evolved in this lineage. In order for the first-instar nymph to be large, the egg must be large. In order for large eggs to develop quickly, they must have access to oxygen, which is where male brooding comes into play. Thus, male brooding is an ancillary evolutionary development whose foundation lies in selection for a body size that enables a bug to take down relatively large prey.[1355]

Female water bugs, however, could conceivably provide care for their own eggs after laying them on exposed aquatic vegetation. Why is it that the males do the brooding, never the females? Here the situation parallels the fish story closely. First, male water bugs with one clutch of eggs sometimes attract a second female, perhaps because a male bug with a partial load of eggs is effectively advertising his capacity for parental care, just like a stickleback.[1422] Second, just as is also true for some fishes, the costs of parental care may be disproportionately great for females in terms of lost fecundity. In order to produce large clutches of large eggs, female belostomatids require far more prey than males do. Because brooding limits mobility and thus access to prey, parental care probably has greater fitness costs for females than for males, biasing selection in favor of male parental care.

## Discriminating Parental Care

No matter which parent provides care, an adaptationist would not expect a parent or parents to provide assistance freely to young animals that are not their own genetic offspring. But can parental animals always identify their own progeny? Consider the Mexican free-tailed bat, which migrates to certain caves in the American Southwest, where pregnant females form colonies in the millions. After giving birth to a single pup, a mother bat leaves her offspring clinging to the roof of the cave in a crèche that may contain 4000 pups per square meter (Figure 12.8). When the female returns to nurse her infant, she flies back to the spot where she last nursed her pup and is promptly besieged by a host of hungry baby bats.[961] Given the swarms of pups, early observers believed that mothers could not possibly identify their own offspring and instead had to provide milk on a first-come, first-served basis.

But do Mexican free-tailed bat mothers really nurse indiscriminately? To find out, Gary McCracken captured female bats and the pups nursing from them and took blood samples from both.[961] He then analyzed the samples using starch gel electrophoresis, a technique that can be used to determine whether two individuals have the same variant form of a given enzyme, and thus the same allele of the gene that codes for that enzyme. If female bats are indiscriminate care providers, then the enzyme variants of females and the pups they nurse should often be different. But if females tend to nurse their own offspring, then females and pups should tend to share the same alleles. McCracken focused on the gene for superoxide dismutase, an enzyme represented by six different forms in the population he sampled. Despite the chaotic conditions within the bat colony, the enzyme data indicated that females found their own pups at least 80 percent of the time. More recent direct observational studies indicate that females probably do better than that, almost always recognizing their own pups

**FIGURE 12.8** **Mexican free-tailed bat mothers recognize their pups,** despite the fact that they leave their infants in a dense mass of baby bats when they leave their caves to forage outside. When a female returns to nurse her pup, she can relocate her son or daughter among thousands of other individuals. Photograph by Gary McCracken.

through vocal and olfactory signals.[64, 962] Thus, mother free-tailed bats clearly deliver their parental care primarily to their own pups.

## Discussion Questions

**12.5** McCracken found that although female Mexican free-tailed bats usually feed their own pups, they do make occasional "mistakes," which they could have avoided by leaving the pup in a spot by itself instead of in a crèche with hundreds of other babies.[961] Does this mean that the parental behavior of this species is not adaptive? Use a cost–benefit approach to develop alternative hypotheses to account for these "mistakes."

**12.6** In some cases, males or females do care for young other than their own, as when certain male fish take over and protect egg masses being brooded by other males or when female ducks acquire ducklings that have just left someone else's nest (Figure 12.9). Devise alternative hypotheses to explain this phenomenon. Under what circumstances might adoptions actually raise the caregiver's reproductive success? Under what other circumstances might adopters be forced into helping nongenetic offspring as a cost of achieving some other goal?

If the ability to recognize one's genetic offspring is advantageous in proportion to the risk of misdirected parental care, then we would expect convergent evolution of parent–offspring recognition in colonial mammals other than Mexican free-tailed bats. Tests of this prediction have been positive in species as different as the degu, a roly-poly little rodent whose females rear young together in communal burrows,[723] and the subantarctic fur seal, whose females give birth on crowded island beaches. Female seals remain with their pups for about a week before leaving for an oceanic fishing trip that can last as long as 3 weeks. Playback experiments demonstrate that a baby seal takes no more than 5 days to learn its mother's call, while the females are also quick learners. When a mother seal returns to the beach, she calls out, and her infant calls back, leading to their reunion in less than 15 minutes as a rule.[256]

**FIGURE 12.9** **Adoption by a female goldeneye duck.** This mother escorts a mob of marked ducklings, some of whom were determined by their tags to have come from broods produced by other females. Photograph by Bruce Lyon.

Cliff swallow

Barn swallow

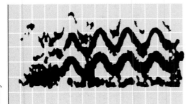

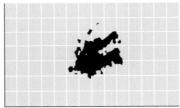

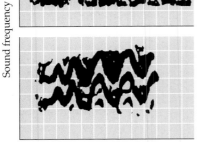

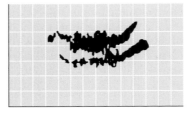

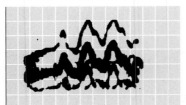

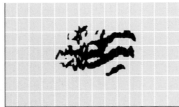

Sound frequency

**FIGURE 12.10 Call distinctiveness facilitates offspring recognition by parents.** The chicks of cliff swallows, a colonial species, produce highly structured and distinctive calls, helping their parents recognize them as individuals. The calls of barn swallow chicks, a less colonial species, are much less structured and more similar. The call frequencies of both species lie between 1 and 6 kHz; the durations of the calls range from 0.7 to 1.3 seconds for the cliff swallows and from 0.4 to 0.8 seconds for the barn swallows. After Medvin, Stoddard, and Beecher.[971]

Yet another comparative test of the hypothesis that offspring recognition functions to prevent misdirected parental care takes advantage of variation among swallows in the risk of making caregiving mistakes. Although both the bank swallow and the rough-winged swallow nest in clay banks, the bank swallow is colonial, whereas the rough-winged swallow nests by itself. Individual fledglings of the colonial bank swallow produce highly distinctive vocalizations, giving their parents a potential cue to use when making decisions about which individuals to feed, allowing them to distinguish between their own offspring and other fledglings that sometimes wind up in the wrong nests begging for food. Bank swallow parents rarely make mistakes, despite the high density of nests in their colonies.[87, 89] The solitary rough-winged swallow, on the other hand, never has a chance in nature to feed another's fledglings and so would not be expected to have evolved sophisticated offspring recognition mechanisms. Indeed, rough-winged swallow chicks produce calls that sound much more alike than those of bank swallows, a reflection of the fact that young rough-wings need not communicate their identity to their parents.[88]

Two other swallow species, the highly colonial cliff swallow and the less social barn swallow, should also differ in their chick recognition attributes. As expected, cliff swallow chicks produce calls containing about 16 times as much variation as the corresponding calls of barn swallow chicks (Figure 12.10).[971]

Therefore, it should be easier for cliff swallows to recognize their young than for barn swallows to discriminate among barn swallow chicks. In operant conditioning experiments that required adults of both species to discriminate between pairs of chick calls, cliff swallows reached 85-percent accuracy significantly faster than barn swallows. These results suggest that the acoustical perceptual systems of the cliff swallow, as well as its calls, have evolved to promote accurate offspring recognition,[881] just as is true of subantarctic fur seals.

Bluegill

## Discussion Questions

**12.7** Territorial male bluegill sunfish defend the eggs and fry in their nests against predatory fish such as largemouth bass (see Figure 13.7). Figure 12.11 shows how intensely males defended their nests in an experiment in which some territorial bluegills were exposed to potential cuckolds during the spawning season. Bryan Neff put sneaker males (see Figure 10.27) in plastic jars near the nests of his experimental subjects to provide the cues associated with a high risk of cuckoldry; he measured male brood defense by quantifying how intensely bluegill dads charged and threatened a predator of bluegill eggs and fry, a pumpkinseed sunfish, which Neff placed near bluegill nests in a clear plastic bag.[1032] How do you interpret the results shown in Figure 12.11? What is puzzling about them? Does it help to know that bluegill males can apparently evaluate the paternity of fry, but not eggs, by the olfactory cues they offer?

**12.8** When a male baboon intervenes in disputes between two juveniles, he tends to take the side of his genetic offspring (Figure 12.12). How might researchers have secured this information?

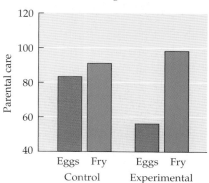

**FIGURE 12.11  Reactions of nest-defending bluegill males to potential egg and fry predators under two conditions.** Experimental males had been exposed to clear containers holding smaller male bluegills, mimicking the presence of rivals who might fertilize some of the eggs in the defender's nest; control males were not subjected to this treatment. "Parental care" was quantified using a formula based on the number of displays and bites directed at a plastic bag holding a predatory pumpkinseed sunfish. After Neff.[1032]

## Why Adopt Genetic Strangers?

The cases described thus far support the prediction that parents should recognize their own young and discriminate against others when the probability of being exploited by someone else's offspring is high. And yet some colonial, ground-nesting gulls occasionally adopt unrelated chicks. Although researchers initially reported that adults consistently rejected older, mobile

(A)

(B)

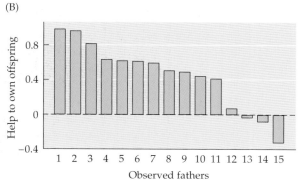

**FIGURE 12.12  Male baboons intervene on behalf of their own offspring** when young baboons start fighting with one another. (A) The adult male cradles his offspring, protecting it against an aggressive youngster. (B) Of 15 fathers whose behavior was monitored, 12 were more likely to help their own offspring than an unrelated juvenile. A, photograph by Joan Silk; B, after Buchan et al.[202]

Ring-billed gull and chick

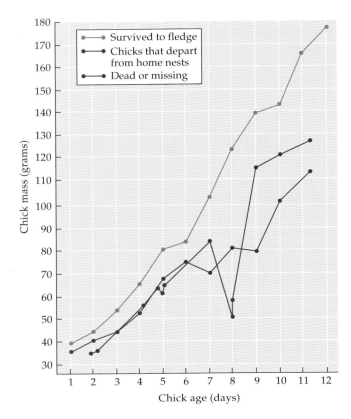

**FIGURE 12.13** **Why seek adoptive parents?** Gull chicks that abandoned their natal nests in search of foster parents weighed much less than average for chicks their age that were known to survive to day 12. But these potential adoptees were sometimes adopted by nonrelatives, and as a result they weighed more on average at day 11 than the subset of undefended gull nestlings that eventually died or disappeared. After Brown.[197]

chicks when they were experimentally transplanted between nests,[983] attacks by adults on these transferred youngsters apparently occurred because of the frightened behavior of the displaced chicks.[568] When juveniles voluntarily leave their natal nest—which they sometimes do if they have been poorly fed by their parents (Figure 12.13)—they do not flee from potential adopters, but instead beg for food and crouch submissively when threatened. These youngsters have a good chance of being adopted, even at the ripe old age of 35 days,[671] and if they are taken in, they are more likely to survive than if they had remained with the genetic parents that were failing to supply them with enough food.[197]

The adoption of strangers qualifies as a Darwinian puzzle, for here we have parents apparently failing to act in the best interests of their genes. As usual, resolving the paradox requires thinking about the costs, not just the benefits, of a trait of interest. And learned recognition of offspring carries costs as well as benefits, notably the risk of making a mistake by not feeding, or even attacking and killing, one's own offspring. Rather than erring on the side of harming their genetic offspring, gulls have evolved a readiness to feed any chicks in their nest that beg confidently when approached by an adult.[1136] Sometimes this rule of thumb permits a genetic outsider to steal food from a set of foster parents by slipping into a nest with other youngsters of its age and size.[783] When adoption occurs, the adoptive parents lose about 0.5 chicks of their own on average; however, adoption is rare, with fewer than 10 percent of adult ring-billed gulls taking in a stranger in any year.[197] The modest average annual fitness cost of a rule of thumb that results in occasional adoptions has to be weighed against the cost of rejecting one's own genetic offspring that would arise if parent gulls were more reluctant to feed chicks in their nests.

The argument here is that a less-than-perfectly-accurate rule of thumb can be selectively advantageous if the proximate mechanism responsible for that rule has a more favorable cost–benefit ratio than alternative psychological mechanisms that would result in different decision rules. This hypothesis has been used to explain why male western bluebirds adopt broods under some circumstances. When Janis Dickinson and Wes Weathers removed some nesting males, most of the experimentally widowed females soon attracted replacement partners, about half of whom fed the chicks of the females they had joined, even though at least some of those chicks had surely been fathered by the original male.[401] The observation that the paternal replacement males were those that found a female while she was still laying eggs (and thus both sexually receptive and potentially fertile) suggests that male western bluebirds have a proximate all-or-none decision-making mechanism that regulates their parental caregiving. If the male joins a female during her fertile phase, he exhibits all-out parental care; if he joins her after her fertile period is over, then the rule is "no parental care at all." Likewise, in the white-browed scrub wren, a polyandrous species with alpha and beta males, the rule for the subordinate is to be paternal only if he has copulated with the female.[1561] If male bluebirds or scrub wrens could run DNA fingerprinting tests on their putative offspring, then they could do a perfect job of discriminating between their own genetic progeny and those of other males. But since this technology is obviously out of their reach, selection can favor behavioral decision-making rules that are better than the possible alternatives, even though these rules do not always produce the "perfect" response.[861]

## Discussion Question

**12.9** Interspecific brood parasitism is very rare in birds, with only about 1 percent of all species practicing the trait.[827] Make a prediction about which group of birds, those with precocial young or those with altricial young, would be more likely to evolve into specialist brood parasites. (In altricial species, the eggs are small in relation to parental body weight, but the hatchlings are initially completely dependent on food supplied to them by a parent. In precocial species, the eggs are relatively large, but the youngsters can move about and feed themselves shortly after hatching.) Check your prediction against Lyon.[906]

## The History of Interspecific Brood Parasitism

But how did cowbirds, cuckoos, and the like come to specialize in parasitizing other bird species? In the case of cuckoos, one phylogenetic reconstruction based on molecular data indicates that specialized parasitism has arisen three times over the evolutionary history of this group (Figure 12.14).[44] Given this phylogeny and the overall rarity of specialist brood parasitic bird species, Oliver Krüger and Nick Davies hypothesized that the ancestor of the current parasitic cuckoos was a "standard" bird whose adults cared for their own offspring. By taking advantage of detailed information on the natural history of many of the 136 species of cuckoos worldwide, of which 83 are nonparasitic while 53 arrange to have other birds brood their young, Krüger and Davies were able to demonstrate that the ancestral state was represented not only by parental care, but also by the occupation of small home ranges and the absence of migra-

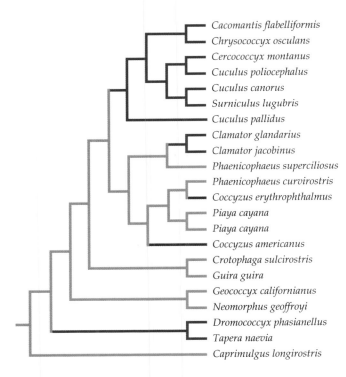

**FIGURE 12.14 Specialized brood parasitism by cuckoos has evolved three times,** based on a phylogeny of this family of birds, which contains both brood parasites and standard parental, nesting species. The blue branches indicate those lineages whose modern descendants are specialist parasites. The red branches are for two species that are occasional parasites. After Aragón et al.[44]

tion. The next stage of evolution involved species that provided parental care for their offspring but possessed relatively large ranges and a tendency to migrate. From lineages of this sort came the modern brood parasites, which also have large home ranges and are generally migratory.[813] Brood parasites presumably gain by roaming widely in search of appropriate hosts throughout an extensive home range; this readiness to move while hunting for ephemeral sources of hosts may have led to the evolution of migratory tendencies in this group.

But what about the transition between parental behavior and specialized brood parasitism? This shift could have taken place in stages, with an intermediate phase when the parasites of the day targeted nesting adults of their own species, with the shift to members of one or more other species occurring later. Alternatively, specialized interspecific parasitism may have arisen abruptly with the exploitation of adults of another species right from the start. The first, gradualist scenario leads to the prediction that females of some of today's birds should lay their eggs in nests of their own species, and indeed, parasitism of this sort has been documented in about 200 species and may occur in many others as well.[361]

A possible hint of a very early stage in the evolution of intraspecific brood parasitism comes from a study of wood ducks. In this species, suitable nest cavities in trees are scarce, with the result that two females sometimes lay eggs in the same nest before one duck evicts the other. The "winner" then cares for the eggs of the other female along with her own, having made the "loser" an involuntary parasite in the process.[1305] If the losers were to produce more offspring than they would have if they had retained the nest site, then selection could favor variants that voluntarily deposited and abandoned their eggs in the nests of others—a behavior that is fairly common in another hole-nesting duck, the Barrow's goldeneye.[423]

Another occasional brood parasite is the coot, a common waterbird, in which "floater" females that lack nests or territories of their own lay their eggs in the nests of other coots, apparently in an effort to make the best of a bad situation, since they cannot brood their own eggs themselves. But some fully territorial females with nests of their own also regularly pop surplus eggs into the nests of unwitting neighbors. Since there are limits to how many young one female coot can rear with her partner, even a territorial female can boost her fitness a little by surreptitiously enlisting the parental care of other pairs.[904] The exploitative nature of this behavior is revealed by the finding that older, larger females select younger, smaller ones to receive their eggs, presumably because this kind of host cannot easily prevent a larger female from gaining access to her nest.[906] Pressure of this sort has apparently shaped the evolution of coot behavior, judging from the fact that parasitized females tend either to bury the eggs of others or to keep their own eggs in the better brooding position in the center of the nest.[907]

The gradual shift hypothesis for the evolution of parasitism among species yields the prediction that when intraspecific brood parasites first began to exploit other species as hosts, they should have selected other related species with similar nestling food requirements. Currently, most specialized brood parasites take advantage of species that are not closely related to them, but perhaps most brood parasites have been evolving for many millions of years

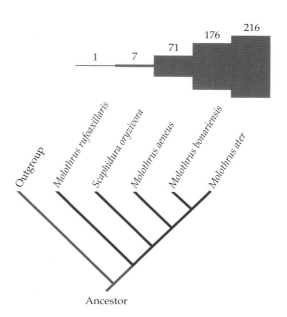

**FIGURE 12.15    Evolution of brood parasitism among cowbirds.** The phylogeny depicts the evolutionary relationships among cowbirds as determined by molecular genetic analyses. Above the phylogeny, the number of host species parasitized by each current living species of cowbird is shown. (The "outgroup" is a species of cowbird that does not engage in brood parasitism.) The pattern suggests that the first brood parasitic cowbird victimized only a single closely related host species, with increasingly generalized brood parasitism evolving subsequently. After Lanyon.[841]

since the onset of their interspecific parasitic behavior. Thus, to check this prediction, we need to find brood parasites that have a relatively recent origin. The familiar cowbirds of the Americas are one such group, its parasitic species having originated "only" 3 to 4 million years ago, whereas the parasitic cuckoos evolved about 60 million years earlier.[361] The living cowbird species believed to be closest to the ancestral brood parasite does indeed parasitize a single host species that belongs to the same genus that it does; the next closest species to the ancestral parasite parasitizes other birds belonging to its own family (Figure 12.15).[841] These data, if they have been interpreted properly, provide support for the gradual shift hypothesis.

Similarly, female widowbirds in the family Viduidae parasitize finch species that belong to the family Estrildidae. The two families are very closely related (Figure 12.16), which may be why both parasites and hosts share a number of important features in common, especially bright white eggs and an unusual form of begging behavior by nestlings, in which the baby birds turn the head nearly upside down and shake it from side to side, rather than stretching upward in the manner of most other nestling songbirds. Assuming that the ancestral parasitic widowbird also possessed these attributes, sensory exploitation (see Chapter 9) could account for the success the offspring of the original brood parasite had after hatching in the nest of an estrildid finch host.[1372] (But another contrarian hypothesis is that the young of the host species have rapidly evolved to match the parasite's behavior, whose novel and highly effective begging tactics put the young of the host species at a disadvantage until they evolved similar signals.[632])

On the other hand, the very large majority of living brood parasites take advantage of unrelated species much smaller than they are,[1343] a finding that could be explained if the ancestral parasites made an abrupt shift from normal parental care to exploiting one or more unrelated smaller host species. Such a shift might well have been more likely to succeed, given that, as already noted, brood parasite nestlings that become larger

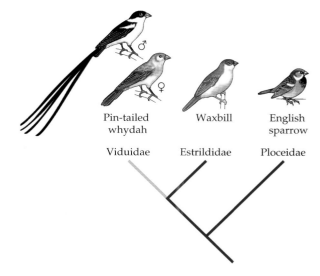

**FIGURE 12.16    Widowbirds parasitize closely related species.** Widowbirds (in the family Viduidae) parasitize nests of finches in the family Estrildidae (the family most closely related to the Viduidae). Adult male widowbirds, like the male pin-tailed whydah (*Vidua macroura*) shown here, look nothing like their hosts, but female pin-tailed whydahs and their nestlings do resemble adult and nestling waxbills (*Estrilda astrild*), an estrildid finch parasitized by this widowbird.

Great tit

Blue tit

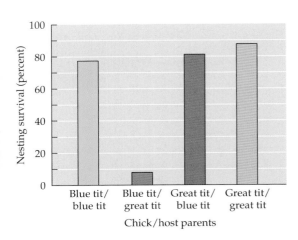

**FIGURE 12.17** **The size of an experimental "brood parasite" nestling** relative to its host species determines its survival chances. Larger great tit nestlings survived well when transferred to the nests of smaller blue tits, whereas blue tits did poorly in great tit nests. After Slagsvold.[1343]

than their hosts' offspring are more likely to be fed, another form of sensory exploitation that works to a parasite's advantage.

The importance of the size disparity between host and parasite has been demonstrated by creating some experimental brood parasites. When Tore Slagsvold shifted blue tit eggs into great tit nests, the experimentally produced brood-parasitic blue tit nestlings, which are smaller than those of great tits, did very poorly. In the reciprocal experiment, however, most of the great tit chicks cared for by blue tit parents survived to fledge (Figure 12.17).[1343] These findings suggest that unless the original mutant interspecific brood parasites happened to deposit their eggs in the nests of a smaller host species, the likelihood of success (from the parasite's perspective) was not great.

Yoram Yom-Tov and Eli Geffen have applied the comparative method to determine which historical scenario for the evolution of obligatory avian parasites is more likely—the indirect, or gradual, pathway with an intermediate intraspecific parasitism stage or the direct pathway in which standard parental behavior was quickly replaced by obligate parasitism (Figure 12.18). Their analysis indicates that, for a large group of altricial birds (those whose young are nearly helpless on hatching), the probability was much greater that the ancestral species of today's obligate parasites was a bird that did not engage in intraspecific parasitism and instead took advantage of members of an entirely different species (see, for example, Figure 4.7).[1630] Thus, proponents of both evolutionary scenarios for interspecific brood parasitism have at least some supportive evidence to which they can point, which leaves the rest of us in limbo.

## Discussion Question

**12.10** When great tits are experimentally reared in blue tit nests, they survive quite well, as just noted. But many of these experimental brood parasites did not associate with members of their own species when they became adults, and they often failed to mate with great tits (see Figure 3.6). The reproductive success of these individuals was consequently low.[1344] Explain these results in terms of the proximate mechanism of imprinting (see Chapter 3), and discuss both the negative and positive effects that this developmental process could have had on the evolution of interspecific brood parasitism.

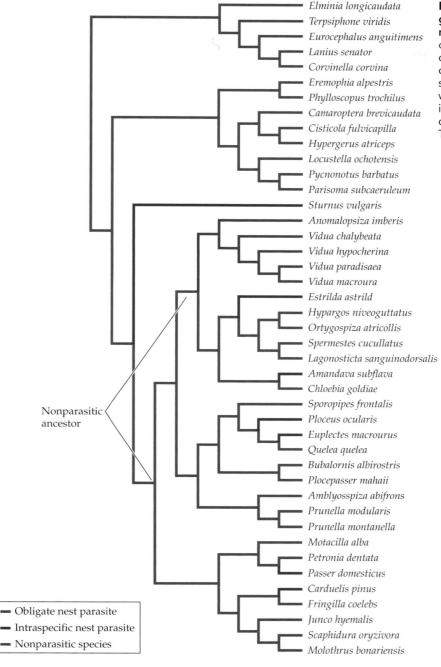

**FIGURE 12.18** **The transition to obligate parasitism was probably abrupt in most groups of birds.** Here a phylogeny of a large group of birds shows that the two clusters of obligate parasites (those that exclusively lay their eggs in the nests of other species) had ancestors that in all likelihood were completely nonparasitic (that is, some individuals did not lay eggs in the nests of other members of their species). After Yom-Tov and Geffen.[1630]

## Why Accept a Parasite's Egg?

Whatever the origin and subsequent history of interspecific brood parasitism, it is true that in order for a parasite nestling to take advantage of a host species' parental decision rules, the egg containing the parasite has to hatch. Why, then, don't host species take immediate action against the eggs of parasites? Some birds do, as noted already, by recognizing the foreign egg and burying it, or removing it from the nest, or abandoning the nest altogether along with the parasite's egg. However, each of these options has its disadvantages.[1597] Parent birds that made incubation of eggs dependent on their learned recognition of the eggs they had laid might sometimes abandon or destroy some of their

Prothonotary warbler

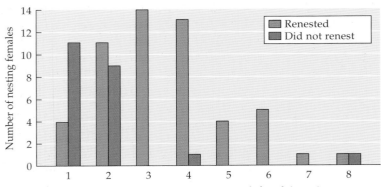

**FIGURE 12.19  The probability that a female prothonotary warbler will nest again** in her territory is a function of the number of potential nest sites in her territory. When only a few good nest holes are present, a female rarely makes a second nesting attempt on her territory; if relatively many sites are available, a female usually does renest there. After Petit.[1121]

own eggs by mistake, treating them as if they were foreign eggs. Indeed, reed warblers sometimes do throw out some of their own eggs while trying to get rid of cuckoo eggs.[357] If brood parasites victimize only a very small minority of the host population, then even a small risk of costly recognition errors by the host can make accepting parasite eggs the better option.[888]

Acceptance of the parasite egg is more likely to be adaptive when the host is a small species unable to grasp and remove large cowbird or cuckoo eggs.[1237] Such small-billed birds have two choices: abandon the clutch, either by leaving the site or by building a new nest on top of the old one, or stay put and continue brooding the clutch along with the parasite egg. The abandonment option imposes heavy penalties on the hosts, which must find a new nest site, build a new nest, and lay a new clutch of eggs. The costs of this option are especially high for hole-nesting species because suitable tree holes are generally rare. Perhaps the willingness of a female to start over after finding a cowbird egg in her nest is affected by how many other good nest holes are in her territory. Although prothonotary warblers are severely harmed by having baby cowbirds in their nests,[679] a female that does not have several nest holes in her territory often tolerates a cowbird in her nest for lack of a suitable alternative nest site (Figure 12.19).[1121] Likewise, yellow warblers tend to accept foreign eggs when their nests are parasitized near the end of the breeding season, when little time remains to rear a new brood from scratch.[1297]

Even if host birds could throw out or cover up a brood parasite's eggs without risk of error, the parasite might make this option unprofitable by returning to the nest to destroy or consume the host's eggs or young if it found that its egg had been harmed (Figure 12.20). This "mafia hypothesis" has been tested by examining the interactions between European magpies and parasitic great spotted cuckoos.[1366] Indeed, magpie nests from which cuckoo eggs had been ejected suffered a significantly higher rate of predation than nests with accepted cuckoo eggs (87 percent versus 12 percent in one sample). Furthermore, when researchers removed the cuckoo egg from a nest that was apparently being checked by a cuckoo and replaced the magpie's eggs as well with plasticine imitations, the cuckoo approached the nest after the researchers had finished and left its beak marks on the false magpie eggs. In nature, magpies that lose their clutches have to renest, which exposes them to all the

**FIGURE 12.20   Egg removal by a cuckoo.** Here a female European cuckoo destroys a reed warbler's egg prior to laying her own egg in this nest. Cuckoos could punish birds that harmed their eggs by destroying the entire brood in an uncooperative victim's nest. Photograph by Ian Wyllie.

negative effects that delays in breeding have in a seasonal environment. In this light, it is not surprising that acceptor magpies actually have somewhat higher reproductive success, as measured in terms of fledglings produced, than do ejectors of cuckoo eggs.[1366]

Brown-headed cowbirds are also avian mafiosi, as Jeffrey Hoover and Scott Robinson demonstrated in a study of one of their favorite victims, prothonotary warblers. The ornithologists worked with a large sample of warblers that had built their nests in boxes on top of greased poles, which made them immune to snakes and small mammals that often prey upon the eggs and young of prothonotary warblers. But these nests were still vulnerable, at least initially, to cowbirds, which could enter a nest and deposit an egg. Nests parasitized in this fashion were divided into three groups: (1) the researchers removed the cowbird egg and left the nest opening as it was—wide enough to permit the reentry of a cowbird, (2) the cowbird egg was left in place and the nest entrance was not modified, (3) the cowbird egg was removed and the nest entrance was made smaller so that only warblers, not cowbirds, could then enter the nest box. The results in the groups were as follows: (1) When the cowbird egg was removed, the nest was often subsequently visited by an avian predator, almost certainly the cowbird whose egg had been taken away, and the warbler eggs were destroyed. (2) When the parasite's egg was not ejected by the experimenters, the warbler's eggs were much less likely to disappear, even though adult cowbirds still had free access to the nests in this category. (3) When cowbirds were prevented from revisiting nests that they had parasitized, the loss of eggs to cowbirds and other predators did not occur (Figure 12.21). These results strongly suggest that cowbird females often come back to nests they have parasitized, in order to punish any host bold enough to get rid of the unwanted parasitic egg.[680]

The approach we have taken thus far is to examine the proposition that the costs of refusing to incubate a brood parasite's egg, or to feed a parasitic nestling after it hatched, could actually outweigh the benefits of refusal. Another complementary way to look at the interaction between host and parasite is to apply the perspective of evolutionary arms race theory. Whenever there are

**FIGURE 12.21 The mafia hypothesis as tested with parasitic cowbirds and prothonotary warblers.** (A) In treatment 1, a cowbird laid an egg in the nest, which was then removed by the experimenter. Subsequently, the warbler eggs in most nests in this treatment were destroyed, presumably by the thwarted cowbirds. In treatment 2, all nests were parasitized but the cowbird egg was left in the nests, which were largely untouched by predators thereafter. In treatment 3, the cowbird egg was removed from the parasitized nests, which were then made inaccessible to cowbirds; none of these nests was harmed after removal of the parasite's egg. (B) The warblers produced more offspring under treatments 2 and 3 than under treatment 1. After Hoover and Robinson.[680]

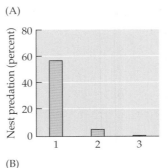

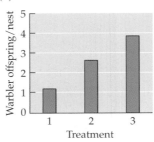

Brown-headed cowbird

two parties in conflict with each other, they exert reciprocal selection pressure on each other, with an adaptive advance made by one leading, in subsequent generations, to an adaptive counterresponse by the other. We have seen this phenomenon in the interplay between ultrasound-producing bats and ultrasound-detecting moths (see Chapter 4), in the sexual conflict that occurs between males and females of many species (see Chapter 10), and in the attempts of parasitic cuckoo nestlings to manipulate their adoptive parents by producing vocalizations that mimic those of an entire brood of reed warblers (see Figure 9.29).[360]

The arms race approach helps us make sense of the interaction between the parasitic Horsfield's bronze-cuckoo and its sole host, the superb fairy-wren.[838] If breeding fairy-wrens find an egg in their nest before they have begun to deposit their own eggs there, they almost always build over the intruder's egg, and they abandon nests altogether if a cuckoo drops an egg in after the wrens have begun to incubate their own complete clutch. So the wrens have defenses against cuckoos. Adult female cuckoos, however, have evolved a counterresponse; they are very good at slipping in to lay an egg when the host nest contains only a partial clutch of fairy-wren eggs. In these cases, the cuckoo egg is almost always accepted and incubated along with the hosts' eggs. But when the cuckoo chick hatches and pushes its wren nestmates out of the nest to their deaths, the fairy-wrens abandon the nest about 40 percent of time, leaving the baby cuckoo to die as well. In the remainder of cases, however, the fairy-wrens keep caring for the sole occupant of their nest and so waste their time and energy rearing the killer of their offspring.

If the semi-acceptance of Horsfield's bronze-cuckoo chicks stems from the parasite's ability to combat the evolved defenses of its hosts, then we would predict that these chicks must have some special way of stimulating parental care—which they do. Like European cuckoo chicks, they produce begging calls that sound very much like those made by their host's offspring (Figure 12.22).[837] Naomi Langmore and her colleagues believe that the vocal mimicry of Horsfield's bronze-cuckoo chicks is an evolved response to the host's discriminatory abilities. In support of this conclusion, they point to the fact that superb fairy-wrens always abandon nests that have been parasitized by another species, the shining bronze-cuckoo, whose nestlings produce vocal-

Horsfield's bronze-cuckoo

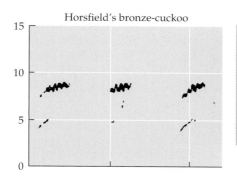

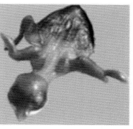

**FIGURE 12.22  A product of an evolutionary arms race?** Chicks of Horsfield's bronze-cuckoo, a specialist brood parasite on Australian fairy-wrens, mimic the calls of their host's chicks very closely, which may help them overcome the defenses of fairy-wren host parents. In contrast, the chicks of another bronze-cuckoo species, which rarely parasitizes fairy-wren nests, not only do not look like fairy-wren chicks, but lack a good facsimile of the begging calls of fairy-wren nestlings. After Langmore, Hunt, and Kilner.[837]

Splendid fairy-wren

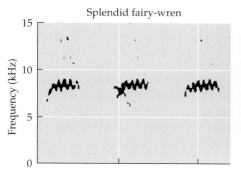

Shining bronze-cuckoo

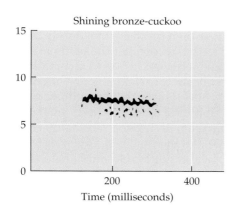

Time (milliseconds)

izations very different from those of the host chicks. (I suspect that you can guess how often shining bronze-cuckoos make the mistake of laying an egg in a superb fairy-wren nest.)

## Discussion Questions

**12.11**  Fairy-wrens are one of the very few bird species that abandon the chicks of brood parasites. Why don't other victimized species do the same? After all, many other birds exploited by cuckoos can identify and take action against the parasite's eggs, which they do by learning the distinctive visual features of their own eggs and then rejecting those that do not match. But the same species that are extremely good at learning to recognize egg features usually completely fail to recognize a cuckoo *chick*.[887] At first glance, this failure appears to be maladaptive, but consider the consequences of learned chick recognition for birds that were successfully parasitized in their first year of breeding by a single cuckoo chick that took over the nest and eliminated all the host's youngsters, as is the cuckoos' habit. How should the host adults respond to their own chicks in the next breeding attempt? (What do you predict about the response of fairy-wrens to their young after having once mistakenly reared a parasitic cuckoo?) How does this case illustrate the importance of cost–benefit analyses in behavioral ecology as well as the value of considering the proximate mechanisms by which individuals achieve behavioral goals?

**12.12**   About 15 to 20 percent of all nestling cuckoo parasites are abandoned and left to die by their reed warbler hosts after about 2 weeks of foster parent care. Tomas Grim suspected that reed warblers had evolved a proximate mechanism for avoiding extended care of a parasite, namely a time limit on parental care for a brood.[584] In order to test this idea, he performed experiments in which he manipulated broods of reed warbler chicks so as to extend the period of parental care needed if the young were to fledge. He transferred younger (and older) chicks between nests, creating various experimental broods of either one or four individuals. How did he expect the parent warblers would respond to each type of experimental brood if the time limit hypothesis was correct? What advantages are associated with a mechanism that does not require reed warblers to *learn* what kind of nestlings to care for and which ones to reject?

## The Evolution of Parental Favoritism

Even when parents invest only in their own progeny, they rarely distribute their care in a completely equitable fashion. Consider the parental tactics of the red mason bee, *Osmia rufa*, whose females nest in hollow stems and supply pollen and nectar for a series of brood cells, provisioning one after another. Initially, when the adult females are young and in good condition, they tend to give the first few offspring large amounts of food.[1304] These initial offspring are the products of fertilized eggs and so will develop into daughters of the red mason bee mothers (see page 483). But then as the season progresses and the females get older, their physiological condition declines, making it more and more difficult for them to forage efficiently. As this happens, females provide much less food per brood cell and they lay unfertilized eggs in these cells, which develop into sons that weigh much less than their sisters (Figure 12.23). Because females of this (and other) bee species are able to control both the sex of an egg and the amount of brood provisions the offspring will receive, they can invest more in a daughter than a son. In this way mothers give their daughters the resources needed to make their energy-demanding eggs and to do the hard work of foraging for their offspring. Sons can afford to be smaller, because they make tiny sperm and spend their time searching for receptive females, presumably less demanding endeavors than those tackled by females.

Other animals may lack the mechanisms needed to control the sex of their offspring as precisely as bees, ants, and wasps, but they can still invest more in some progeny than others. Adults of the burying beetle, *Nicrophorus vespilloides*, cooperate in parental behavior by burying a dead mouse or vole, removing the hair from the deceased animal, and fashioning a lump of flesh from the remains. The female then lays eggs near the brood ball. When the larvae hatch, they can feed themselves from the prepared carcass, or they can receive processed carrion regurgitated by a parent. The beetle grubs can differ markedly in size because some hatch out sooner than others, and under these circumstances, their parents give more food to the earlier-hatched (senior)

(A)

**FIGURE 12.23   Adjustment of investment in sons and daughters** by (A) the red mason bee *Osmia rufa*. (B) When females are young at the start of the breeding season, their provisioning efficiency is high, and (C) under these conditions, the sex ratio of their offspring is biased toward daughters. Small sons are produced when the females are older and their efficiency in filling their brood cells is low. Provisioning efficiency is measured in terms of the average increase in the mass of a larva per hour of bringing food to the nest. A, Photograph by Nicolas Vereecken; B and C, after Seidelmann.[1304]

(B)

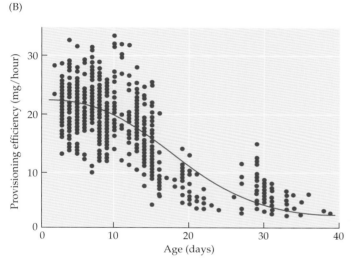

(C)

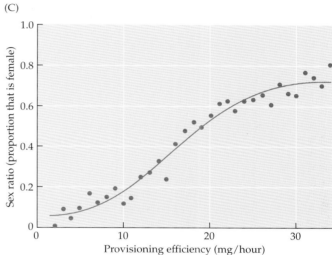

larvae than to later-hatched (junior) grubs (Figure 12.24). Given that only so many progeny can be supported by one mouse carcass, parents may gain by helping those offspring most likely to achieve adulthood, especially since the absolute amount of food needed to reach maturity is less for larvae that are farther along the road to adult metamorphosis.[1358]

Burying beetle adults distribute food unevenly to their offspring by directly feeding some larvae while refusing others. In the great egret, parents take a hands-off approach and instead let their sons and daughters fight for possession of the small fish their parents bring to the nest and drop in front of them (Figure 12.25).[991] Fighting among siblings can escalate to such an extent that a dominant nestling may bludgeon a brother or sister to death or push it out of the nest, thereby monopolizing the food brought to the nest. You may well wonder how behavior of this sort can possibly advance a parent's fitness when it leads to the outright loss of one or even two members from a brood of three or four.

Perhaps parents do not gain from **siblicide**, which may have evolved because of the fitness advantages enjoyed by offspring able to dispose of their siblings. Imagine a family with two offspring, each of which will eventually

**FIGURE 12.24 Discriminating parental care by the burying beetle *Nicrophorus vespilloides*.** (A) An adult beetle inspecting its brood, which live within a ball of carrion prepared by the parents for their offspring. (B) When the mother is present, the senior (older, larger) grubs are fed more and grow to a larger size than when the mother is absent. No such effect applies to the junior (younger, smaller) offspring. A, photograph by C. F. Rick Williams; B, after Smiseth, Lennox, and Moore.[1348]

(A)

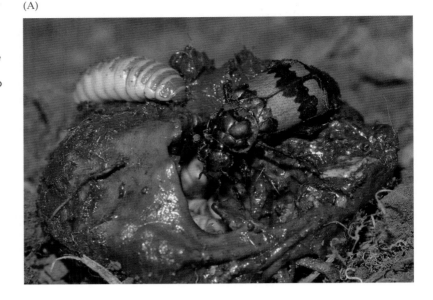

(B)

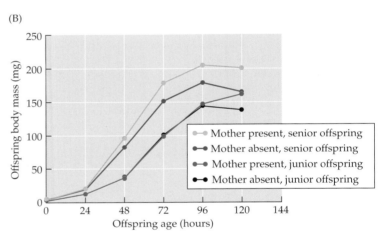

**FIGURE 12.25 Sibling aggression in the great egret.** Two chicks fight viciously in a battle that may eventually result in the death of one of them. The adult egret yawns while the fight goes on. Photograph by Douglas Mock, from Mock, Drummond, and Stinson.[993]

produce three surviving youngsters of its own on average. Now imagine that if one of the two eliminates its sibling, the siblicidal offspring boosts its output to five surviving offspring, thanks to its ability to hog the food its parents supply after the death of its brother or sister. Although the killer kid loses three nephews or nieces, which its sib would have produced were it not for its premature death, this genetic cost is more than compensated for, from the vantage point of the siblicidal individual, by the two extra offspring of its own. (Nephews and nieces share one-fourth of their genes with their uncles and aunts, whereas parents and offspring have one-half of their genes in common [see page 470]. You can do the math.) But the parent of a murderous offspring loses the three grandoffspring that its dead progeny would have had, a figure that is not matched by the two extra grandoffspring coming from the successful siblicidal survivor. This is the kind of situation predicted to lead to **parent–offspring conflict**, a concept developed by Bob Trivers after he realized that some actions can advance the fitness of an offspring while reducing the reproductive success of its parent, and vice versa.[1466]

In some animals in which siblicide occurs, parents can and apparently do resist their progeny's siblicidal behavior. Evidence for this claim comes from studies of seabirds called boobies, some of which exhibit "early siblicide,"

**FIGURE 12.26  Early siblicide in the brown booby.** A very young chick is dying in front of its parent, which continues to brood the larger, siblicidal chick that has forced its younger brother or sister out of the nest and into the sun. Photograph by the author.

in which an older "A" chick disposes of a younger "B" chick in the first few days of its sibling's short and unhappy life.[889] The A chick's ability to kill its brother or sister stems in part from the pattern of egg laying and incubation in boobies. In these birds, females lay one egg, begin incubating it at once, and then some days later lay a second egg. Because the first egg hatches sooner than the second, the A chick is relatively large by the time the B chick comes on the scene. In early siblicidal species, the A chick immediately begins to manhandle the B chick, soon forcing it out of the nest scrape, where it dies of exposure and starvation (Figure 12.26).

Early siblicide is standard practice for the masked booby but not for the blue-footed booby, whose A chicks engage in siblicide less often and generally later in the nesting period. If, however, you give a pair of blue-footed booby chicks the chance to be cared for by masked booby parents, which tolerate early sibling aggression, the A chick often quickly kills the B chick under the vacant gaze of one of its substitute parents. In contrast, blue-footed booby parents appear to keep their A chick under control during its initial days with its sibling. If so, then when masked booby chicks are given to blue-footed booby parents, the foster parents should sometimes be able to prevent them from immediately killing their siblings. As predicted, they do (Figure 12.27),[889] providing evidence that parents can interfere with lethal sibling rivalries, should it be in their interest to do so.

Observations of egrets reveal that parental intervention definitely does not occur when two juvenile egrets go at one another. Indeed, lethal sibling battles are actually promoted by earlier parental decisions about when to begin incubating eggs. Thus, as soon as a female egret lays her first egg, incubation begins, just as is true for boobies. Because 1 or 2 days separate the laying of each egg in a three-egg clutch, the young hatch out asynchronously, with the firstborn getting a head start in growth. As a result, this chick will be much larger than the third-born chick, which helps ensure that the senior chick will monopolize the small fish its parents bring to the nest. The senior chick is not only bigger, but also

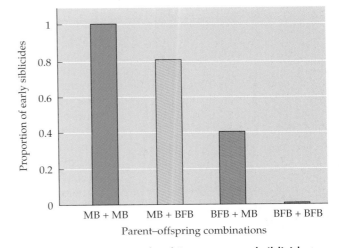

**FIGURE 12.27  Parent boobies can control siblicide to some extent.** The rate of early siblicide by masked booby (MB) chicks declines when they are placed in nests with intervention-prone blue-footed booby (BFB) foster parents. Conversely, the rate of early siblicide by blue-footed booby chicks rises when they are given laissez-faire masked boobies as foster parents. After Lougheed and Anderson.[889]

more aggressive because, judging from what happens in the cattle egret, the first-laid eggs receive relatively large quantities of androgens—a hormonal aggression facilitator. The unequal feeding rates that result further exaggerate the size differences among siblings, creating a runt of the litter that often dies from the combined effects of starvation and assault.[1292]

## Discussion Questions

**12.13** As noted above and elsewhere, some female birds make adjustments in the androgens they supply to their eggs in accordance with various factors. In the case of cattle egrets, the egg's position in the laying sequence is linked to its infusion with androgens. In the case of canaries, the ability of the father to sing an attractive song influences the mother's willingness to add male sex hormones to the eggs (see Chapter 10).[537] In other birds, females adjust the amount of feeding effort in relation to the quality of the offspring's father. Female blue tits, for example, provide less food for the offspring of partners whose crown feathers have been manipulated so that they reflect less ultraviolet light.[697] Why can these different kinds of allocation decisions all be considered examples of parental favoritism, and what do the three examples have in common, with respect to how maternal decisions advance the fitness of the mother?

**12.14** Birds are not the only animals in which intense, and sometimes fatal, sibling conflicts are known to occur. For example, spotted hyena females (see page 291, Chapter 9) often give birth to twins, and the pups compete aggressively for their mother's milk. The battles between pups sometimes lead to the death of one of the twins. Let's say that you want to test the proposition that these cases of occasional siblicide are adaptive. Develop one or more hypotheses and then make use of the following findings: (1) the total input from mothers to pairs of offspring in which siblicide eventually occurs is lower than from mothers to twins that do not commit siblicide, (2) females do not reduce the amount they provide after siblicide has occurred, (3) siblicide is more common when females have to travel great distances in search of prey, and (4) females sometimes separate fighting twins and may preferentially nurse the subordinate cub.[657, 1549]

Thus, parent egrets not only tolerate siblicide, they actually promote it. Why? Perhaps because parental interests are served by having the chicks themselves eliminate those members of the brood that are unlikely to survive to reproduce. Although in good years parents can supply a large brood with enough for all to eat, in most years food will be moderately scarce, making it impossible for the adults to rear all three offspring. When there is not enough food to go around, a reduction in the brood, accomplished by siblicide, saves the parents the time and energy that would otherwise be wasted on offspring with little or no chance of reaching adulthood even if their siblings had not killed them.

One way to test this hypothesis is to create unnaturally synchronous broods of cattle egrets.[992] In one experiment, synchronous broods were formed by putting chicks that had hatched on the same day in the same nest; normal asynchronous broods were assembled by bringing chicks together that differed in age by the typical 1.5-day interval. A category of exaggeratedly asynchronous broods was also created by putting chicks that had hatched 3 days apart into the same nests. If the normal hatching interval is optimal in promoting efficient brood reduction, then the number of offspring fledged per unit of parental effort should be highest for the normal asynchronous broods. This prediction was confirmed. Members of synchronous broods not only fought

**TABLE 12.1**   *The effect of hatching asynchrony on parental efficiency in cattle egrets*

|  | Mean survivors per nest | Food brought to nest per day (ml) | Parental efficiency[a] |
|---|---|---|---|
| Synchronous brood | 1.9 | 68.3 | 2.8 |
| Normal asynchronous brood | 2.3 | 53.1 | 4.4 |
| Exaggerated asynchronous brood | 2.3 | 65.1 | 3.5 |

*Source*: Mock and Ploger[992]

[a]The number of surviving chicks divided by the volume of food brought to the nest per day × 100.

more and survived less well, but required more food per day than normal broods, resulting in low parental efficiency (Table 12.1). The same result has been recorded for the blue-footed booby, in which experimental synchronous broods of two fought more and required much more food than control broods composed of an asynchronously hatched pair of chicks.[1079]

Cattle egret parents and others like them seem to know (unconsciously) what they are doing when they manipulate the hormone content of their eggs and incubate them in ways that lead to differences in size and fighting ability among their chicks. Sibling rivalry and siblicide actually help parents deliver their care only to offspring that have a good chance of eventually reproducing, while keeping their food delivery costs to a minimum. Although cases of this sort represent extreme examples of parental favoritism, even those nesting birds that bias their allocation of food resources toward vigorous offspring are really practicing infanticide, speeding the demise of those progeny unlikely to reproduce even if well fed.

## Discussion Question

**12.15** In species like boobies and egrets, parents' decisions about incubation and hormone allocation put their second or third offspring at great risk of being destroyed. If the second or third egg laid is destined to produce a chick that will die within a few days, why don't parents save the energy that goes into the superfluous egg, which would also save energy for their favored offspring? One possibility goes under the label of the insurance hypothesis: adults invest in a backup egg in order to have a replacement for a favored first-laid egg should something happen to that egg or to the nestling itself after the egg hatches.[273] How would you test this hypothesis experimentally?

## How to Evaluate the Reproductive Value of Offspring

Parents that have limited food resources at their disposal may use their offspring's behavior to decide how to allocate their parental investment toward individuals likely to reach adulthood. Consider an experiment in which some broods of house sparrows were given extra food by researchers while other broods did not receive bonus meals. You might think that parents blessed with food-supplemented broods would have been eager to cut back on their provisioning effort, but to the contrary, mothers kept up the same high rate of food deliveries and fathers actually increased their parental effort significantly.[995] Well-fed youngsters are likely to repay their parents for still more food by surviving to reproduce.

Parent sparrows surely judge the physiological state of their juvenile offspring by their appearance and by their begging behavior, factors that often influence parental behavior in birds (see page 311). One informative aspect

**FIGURE 12.28    An honest signal of condition?** The red mouth gape of nestling lark buntings is exposed when the birds beg for food from their parents. The brightness of a baby bird's gape could reveal something about the strength of its immune system. Photograph by Bruce Lyon.

of a nestling's appearance might be the bright red lining of its mouth, which is conspicuously displayed by many nestling songbirds as they stretch up to solicit food from a returning parent (Figure 12.28).[1272] Because the red color of the mouth lining is generated by carotenoid pigments in the blood, and because carotenoids are believed to contribute to immune function, a bright red gape could signal a healthy nestling that will make good use of whatever it is fed. In fact, in the barn swallow, chicks with redder mouths weighed more 6 days after hatching and had greater feather growth at 12 days of age than chicks with paler gapes.[371]

Parents that preferentially fed those members of their brood that had bright red mouths would be investing in nestlings of high **reproductive value**, namely healthy youngsters that are more likely to fledge and eventually reproduce than their sickly nestmates. If parents do indeed make adaptive parental decisions of this sort, then nestlings made ill by injection of a foreign material should have paler mouth linings than those that have not had their immune systems challenged. Furthermore, parents should feed offspring with artificially reddened gapes more than they feed offspring with unaltered mouth coloration. When Nicola Saino's research team tested both predictions in the barn swallow, the results were positive (Figure 12.29).[1272]

On the other hand, among the alternative explanations for bright gape coloration[274] is the straightforward possibility that young birds gain by having colored gapes simply because these make the begging bird's maw more visible to its parents, especially when the nest is placed in a dark tree hollow or other cavity.[642] Philipp Heeb and his coworkers found that when they painted the mouths of great tit nestlings, those birds with artificial yellow gapes were fed more often in relatively dark nest boxes than red-painted birds, whose mouths were less visible under low-light conditions. In nest boxes with clear Plexiglas windows on top, however, the red-mouthed nestlings suffered no begging handicap, as shown by their ability to achieve the

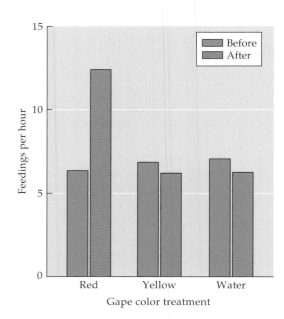

**FIGURE 12.29    The color of the mouth gape** affects the amount of food that nestling barn swallows are given by their parents. After experimenters colored the gapes of some nestlings with two drops of red food coloring, they received more food. In contrast, nestlings that received two drops of yellow food coloring or water were not fed more. After Saino et al.[1272]

same weight as their yellow-mouthed companions.[642] The fact that the mouths of some nestling birds strongly reflect ultraviolet radiation is also consistent with the hypothesis that mouth colors are designed simply to help parents find a receptacle into which food can be stuffed, because ultraviolet signals stand out strongly against the visual background of the typical nest.[698]

Perhaps, therefore, we should withhold judgment on whether nestling birds are communicating something about physical condition to their parents via the color of their mouth linings. Nevertheless, some aspect of the appearance of a young bird might provide information about the condition of the nestling, enabling its parents to make adaptive decisions about how to provide it with food. In the alpine swift, for example, a nestling's skin reflects ultraviolet light (UV) to varying degrees. The larger and heavier the nestling of a given age, the more UV-reflecting skin it possesses (Figure 12.30). Late-breeding parent swifts appear to use this information, because they bias their food deliveries to high-UV-reflective chicks while skimping on nestlings treated with UV-blocking Vaseline. This decision rule would help parents avoid complete nest failure when resources were declining by giving up on low-value offspring in order to concentrate on those few with the best chance of fledging.[125]

Alpine swift

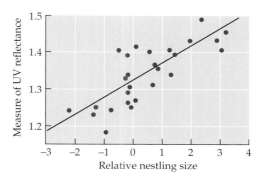

FIGURE 12.30 **Ultraviolet light reflected from the skin of alpine swift nestlings** could convey information to the parent about the physical condition of an offspring because UV-reflectance is greater from larger nestlings. After Bize et al.[125]

## Discussion Question

**12.16** Early in the breeding season, when nestling alpine swifts are just getting started in life outside the egg, many insects tend to be available for parent swifts to offer their offspring. At this time, how should the adults use information provided by the UV-reflecting skin of their nestlings to decide how to divvy up the abundant food available for their youngsters?

Likewise, Bruce Lyon and his colleagues suspected that the long, orange-tipped plumes on the backs and throats of baby coots might be the cues used by their parents to determine which individuals to feed and which to ignore. Coots produce large clutches of eggs, but soon after the young begin to hatch, the adult birds often begin what to humans seems an especially unpleasant process of brood reduction. As some babies swim up to beg for food from a parent, the parent may not only refuse to provide something to eat, but also aggressively peck the head of its youngster. Eventually, these chicks stop begging, and expire facedown in the water.

To test the link between chick ornaments and parental care decisions, Lyon and company trimmed the thin orange tips from these special feathers on half of the chicks in a brood, while leaving the other members of the brood untouched. The unaltered orange-plumed chicks were fed more frequently by their parents, and they grew more rapidly as well (Figure 12.31).[905] In control broods in which all of the chicks had had their orange feathers trimmed, the youngsters were fed as often, and survived as well, as control broods consisting only of untouched orange-feathered chicks. This result shows that the parents of the experimental mixed broods discriminated against the orange-deprived chicks because they were not as strongly ornamented as their feather-intact broodmates, not because the parents failed to recognize them as their offspring.[905]

There are other species in which juvenile offspring vary in the colorfulness of their feathers. In the well-studied great tit, parents do *not* base their feeding decisions on the intensity of the yellows in their offspring's plumage.[1471] One wonders why bird species differ in the extent to which their offspring's

**FIGURE 12.31 The effect of orange feather ornaments of baby coots on parental care.** (A) Baby coots have unusually colorful feathers near the head. (B) Control groups composed entirely of either unaltered (orange) chicks or chicks that had had the orange tips trimmed from their ornamental feathers (black) were fed at the same mean rate. (C) In experimental broods in which half the chicks were orange and half were black, the ornamented individuals received more frequent feedings from parent birds. (D) The relative growth rates of chicks in both control groups were the same, but (E) ornamented chicks grew faster in mixed broods compared with the experimentally altered chicks. A, photograph by Bruce Lyon; B–E, after Lyon, Eadie, and Hamilton.[905]

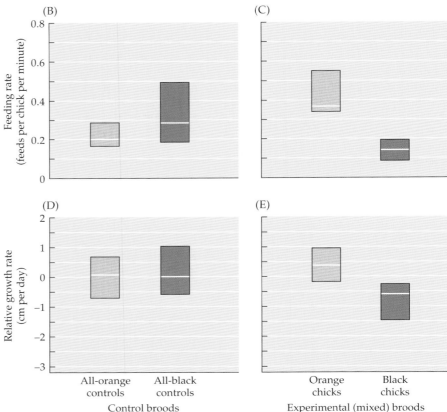

appearance affects parental decisions. Even so, the underlying message provided by coots, boobies, and egrets is that parents do not necessarily treat each offspring the same, but instead often help some survive at the expense of others. Cases of this sort remind us that selection acts not on variation in the number of offspring produced, but on the number that survive to reproduce and pass on the hereditary traits of their parents.

## Discussion Questions

**12.17** Magpie nestlings are increasingly likely to survive as they get older (Figure 12.32A).[1200] Given that fact, use a cost–benefit approach to explain why magpie parents change their defensive behavior in response to a biologist's approach to the nest over the nesting period (Figure 12.32B).

Eurasian magpie

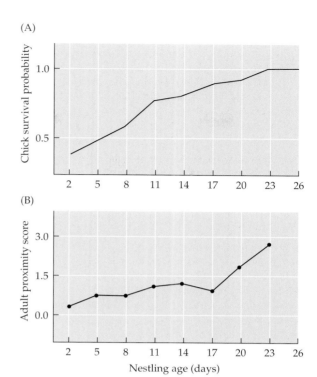

**FIGURE 12.32  Nestling survival probability and defense of the nest by parent magpies.** (A) The probability that nestlings of different ages will be alive at fledging. (B) The relation between nestling age and the intensity of nest defense by adult magpies, as measured by how close adult magpies came to a human observer (the higher the score, the closer the magpies came). After Redondo and Carranza.[1200]

**12.18**  Use the concept of reproductive value to make predictions about the escape decisions of incubating mallard ducks. These birds, which nest amid dense aquatic vegetation, are faced with a trade-off when a potential predator approaches. They can improve the odds of saving their own skin by quickly flushing from the nest, but their noisy departure will often give the nest location away, with the likely loss of all the eggs within. Or they can sit tight, remaining as inconspicuous as possible, improving the odds that the predator will pass by the nest and its contents but also increasing their personal risk of being killed. Make predictions about how mallards will respond in relation to the number of eggs being incubated, the mean size of the eggs in the clutch, and their stage of development. Check your predictions against data in Albrecht and Klvana.[10]

## Summary

1. The time, energy, and resources that parents devote to their offspring have costs, including reduced fecundity in the future and fewer opportunities to mate in the present, as well as the obvious benefit of improved survival of the assisted progeny. A cost–benefit approach helps explain why females are more likely than males to provide parental care, by focusing on the costs to males of helping broods of mixed paternity, costs that females rarely experience because they are usually genetically related to all the offspring in their broods.

2. Exceptions to the general rule occur. When males care for their young, these cases can also be analyzed by a cost–benefit approach. Thus, paternal care in fishes may often be favored because males caring for eggs laid in their territories can be more attractive to potential mates than are males without eggs to brood. In contrast, the costs of parental care to females may include large reductions in growth rate and consequent losses of fecundity.

3. An evolutionary approach to parental care yields the expectation that parents will be able to identify their own offspring when the risk of investing in genetic strangers is high. As expected, offspring recognition is widespread, particularly in colonial species in which opportunities for misdirected parental care are frequent. However, adults of many species do sometimes adopt nongenetic offspring, including specialist brood parasites, with consequent losses of fitness. Multiple hypotheses exist to account for these puzzling cases, including the possibility that highly discriminating host adults could lose fitness by sometimes erroneously rejecting their own offspring.

4. Another Darwinian puzzle is the indifference shown by some animals to lethal aggression among their young offspring. Cases like these may be explained as part of a parental strategy to let the offspring themselves identify which individuals are most likely to survive, and therefore which youngsters will provide a payoff for continuing parental investment. The more general principle is that selection rarely favors completely even treatment of offspring, because some youngsters are more likely than others to survive to reproduce.

## Suggested Reading

Good books on parental care include *The Evolution of Parental Care* by Timothy Clutton-Brock[280] and *Mother Nature* by Sarah Hrdy.[688] A vast literature exists on brood parasites and their interactions with their hosts; for a superb review, see *Cuckoos, Cowbirds and Other Cheats* by Nick Davies.[361] Avian siblicide is the subject of an excellent article by Doug Mock and his coworkers[993] as well as a fine book by Mock written for a general audience.[994]

# 13

# The Evolution of
# Social Behavior

**D**on't automatically reach for the Raid the next time you find a
*Polistes* paper wasp colony under an eave on your house
(see Figure 3.10). At least, this would have been the advice
of the great evolutionary biologist W. D. Hamilton, who
wrote:

> Social wasps are among the least loved of
> insects... Yet, where statistics will not alter a gen-
> eral impression, another approach might. Every
> schoolchild, perhaps as part of religious training,
> ought to sit watching a Polistes *wasp nest for just*
> an hour... I think few will be unaffected by what
> they see. It is a world human in its seeming mo-
> tivations and activities far beyond all that seems
> reasonable to expect from an insect: construc-
> tive activity, duty, rebellion, mother care, violence,
> cheating, cowardice, unity in the face of a threat—
> all these are there.[613]

Needless to say, if you choose to follow Hamilton's interesting suggestion, be very careful when approaching
a paper wasp nest, because these and other social wasps
have powerful stings. If you do not trigger a wasp attack,
you will be able to observe quasi-human melodramas at
the nest that involve females both competing and cooperating as they rear their offspring there. If I
were to tell you that only one of the several
females at the nest might be the mother of
all the eggs and grubs housed there, and

◀ **Emperor penguins are
intensely social animals**
*during the breeding season.
Tens of thousands of adults
have gathered at this Antarctic
rookery to rear their offspring.
Photograph by Nancy Pearson.*

that these youngsters were often fed with food collected by females other than their mother, I hope you would be at least mildly surprised. Although parental care can evolve when the benefits of the behavior exceed its costs (see Chapter 12), it is hard to imagine how an adult's fitness could be increased by behaving parentally toward someone else's youngsters. Yet helpers at nests are found not just in paper wasps, but in a host of other insects, as well as in some birds and mammals. These self-sacrificing individuals pose a wonderful Darwinian puzzle whose solution has been sought by some of the best evolutionary biologists in the world, including W. D. Hamilton and Charles Darwin himself.

This chapter focuses on how altruism and other helpful acts of social organisms can be analyzed from an adaptationist perspective. But first we must ask a more basic question: Why do any animals live in groups instead of living alone?

## The Costs and Benefits of Social Life

You may think that the reason so many animals join others of their species is that social creatures are higher up the evolutionary scale and so are fundamentally better adapted than animals that live solitary lives. You might hold this belief because you know that humans are highly social, and you would like to think that we, and perhaps a few other highly social species, represent the crowning achievements of the evolutionary process. But if you believed these things, you would be mistaken, because natural selection does not aim for preset endpoints (see Chapter 1). Instead, in each and every species, generation after generation, relatively social and relatively solitary types compete unconsciously with one another in ways that determine who has more surviving offspring on average. In some species, the more social individuals have won out, but in a large majority, it is the solitary types that have consistently left more descendants (and thus more copies of their genes) to the next generation.

Living alone is superior to living together when the cost–benefit ratio is better for solitary than social individuals (Table 13.1). The costs of living with others can be considerable. For example, in most social species, animals have to expend time and energy jockeying for social status. Those that do not occupy the top positions regularly have to signal their submissive state to their supe-

| TABLE 13.1   *Some potential costs and benefits of social living* | |
| --- | --- |
| **Costs** | **Benefits** |
| Greater conspicuousness of clumped individuals to predators | Defense against predators via the dilution effect or via mutual defense (see Chapter 6) |
| Greater transmission of disease and parasites among group members | Opportunities to receive assistance from others in dealing with pathogens |
| More competition for food among group members | Improved foraging via the information center effect (see Chapter 7) |
| Time and energy expended by subordinates in dealing with more dominant companions | Subordinates are granted permission to remain safely within the group |
| Greater male vulnerability to cuckoldry | Opportunity for some males to cuckold others |
| Greater female vulnerability to egg tossing, egg dumping, and other forms of reproductive interference by others | Opportunity to toss the eggs of others, to dump eggs in others' nests, and to interfere with rivals' reproduction |

(A)

Helpers

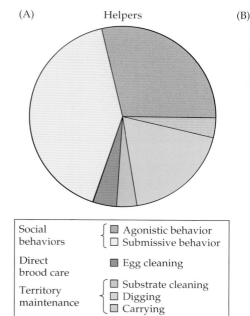

| Social behaviors | { ▨ Agonistic behavior<br>☐ Submissive behavior |
| Direct brood care | ▨ Egg cleaning |
| Territory maintenance | { ☐ Substrate cleaning<br>☐ Digging<br>☐ Carrying |

(B)

**FIGURE 13.1   The energy budget of subordinate, nonbreeding "helpers"** that associate with breeding pairs in the cichlid fish *Neolamprologus pulcher*. (A) The largest proportion of the subordinate fish's energy is expended on the performance of submissive behaviors—specifically, a tail-quivering display. Most of the remainder of the subordinate fish's energy budget is spent attacking intruders and removing sand and debris from the area defended by the breeding pair and their helpers. (B) A subordinate helper quivers his tail as the dominant fish approaches from behind. A, after Taborsky and Grantner;[1420] B, photograph by Michael Taborsky.

riors if they are to be permitted to remain in the group. All the kowtowing can take up a major share of a social subordinate's life (Figure 13.1).[1420]

## Discussion Question

**13.1**  In an African cichlid fish known for its cooperative breeding, helpers live with a breeding pair within the group's communal territory. In Figure 13.2, the locations occupied by five of these helpers are shown over a 3-day period; on day 3, the largest helper (1) was removed by researchers, but its territory outline remains in the figure.[1540] How would you interpret these data in light of the possibility that helpers are competing with one another while helping a breeding pair? What benefit might helpers derive from achieving dominance over other helpers?

Reproductive interference from others also boosts the price of sociality. Breeding males that live in close association with more attractive rivals may lose their mates to these individuals, while breeding females may wind up incubating eggs dumped in their nests by brood parasites of their own species.[684] Such costly reproductive penalties come with social living for the acorn woodpecker, a bird that forms breeding groups containing as many as three females and four males. The several females all lay their eggs in the same tree hole nest, perhaps because any female that tried to keep a nest to herself would have all her eggs destroyed by her vindictive companions.[788] Even when several females agree to use the same nest, the first eggs laid are almost always removed by another female member of the group (Figure 13.3).[1021] Eventually these "cooperatively breeding" females all lay eggs on the same day, at which time they finally stop tossing one another's eggs out and incubate the clutch. By this time, however, more than a third of the eggs laid by the woodpeckers may have been destroyed. Having your eggs tossed is an unadulterated cost of social living for females of this species.

*Neolamprologus brichardi*

**FIGURE 13.2** **Effect of removal of the top-ranked subordinate helper** in a cooperatively breeding group of cichlid fish (*Neolamprologus brichardi*). The removal occurred on day 3 in an aquarium that housed a breeding pair and five helpers. (The different-colored dots represent the different fish and show where these individuals were seen on a given day.) After Werner et al.; photograph by Michael Taborsky.[1540]

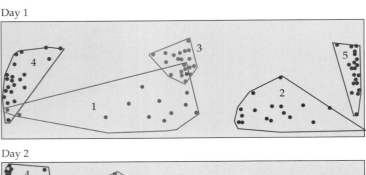

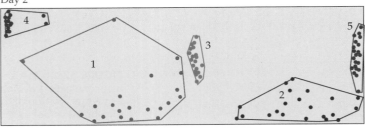

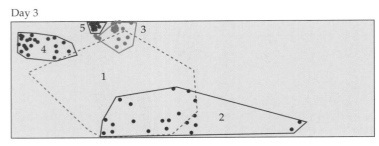

**FIGURE 13.3** **Reproductive interference in a social animal.** One member of a breeding group of acorn woodpeckers removes an egg from a nest that she shares with several companions. Photograph courtesy of Walt Koenig.

In addition to these direct reproductive costs, sociality has two other potential disadvantages. The first is heightened competition for food, which occurs in animals as different as colonial fieldfares (Figure 13.4)[1573] and prides of lions, whose females are often pushed from their kills by hungry male partners.[1284] The second is increased vulnerability to parasites and pathogens, which plague social species of all sorts[16] (but see Rosengaus et al.[1240]). Evidence on the importance of this second point comes from tests of the prediction that the larger the group, the greater the risk of infection by damaging microbes. This prediction is supported by the finding that the degree of sociality among bees is linked to the ability of species to combat staph bacteria. To demonstrate this point, researchers washed the cuticles of bees, ranging from solitary species that nest in isolation to highly social species whose members crowd together by the thousands, even millions, in their colonies. The resulting solution contained protective chemicals from the bees' bodies. The body wash taken from highly social bees was over 300 times more effective in destroying staph bacteria than the comparable antibacterial fluid derived from solitary species.[1393] If we assume that the defensive compounds produced by the bees are costly, then we have evidence that individuals in large groups pay a special price to combat the greater risk of bacterial infection associated with their social nature.

The fact that some social animals have evolved counteracting responses to pathogens and parasites may enable those animals to reduce the damage they cause but cannot totally eliminate the burden they impose. Thus, honey bees warm their hives in response to an infestation by a fungal pathogen, which apparently helps kill the heat-sensitive fungus but at the price of time and energy expended by the heat-producing workers.[1379] Similarly, termites can reduce the lethal effect of a different fungal invader in their nest mounds because unexposed colony members can acquire some protection simply by associating with others of their group that are already immune to the pathogen.[1462] Even so, the existence of special responses to fungal infection speaks to the high probability that a colony will become infected, perhaps because it offers such a large target. Moreover, the antifungal mechanisms themselves do not come for free, but require physiological expenditures on the part of the termites.

A heightened probability of contagious infection clearly applies to cliff swallows, which pack their nests side by side in colonies composed of anywhere from a handful of birds to several thousand pairs. The more swallows nesting together, the greater the chance that at least one bird will be infested with blood-sucking swallow bugs, which can then readily spread from one nest to another.[188] Charles and Mary Brown demonstrated that the bugs were guilty of harming swallow nestlings; they fumigated a sample of nests in an infested colony while leaving other, control nests untreated. The nestlings doused with insecticide weighed much more, and were more likely to survive, than those plagued by growth-stunting parasites (Figure 13.5).

The parasites, bacteria, and fungi that make life miserable for swallows and other social creatures demonstrate that if sociality is to evolve, the assorted costs of living together must be outweighed by compensatory benefits. Cliff swallows may join others to take advantage of the improved foraging that comes from following companions to good feeding sites (see Chapter 7),[189, 572] while other animals, such as egg-brooding male emperor penguins, save thermal energy by huddling shoulder to shoulder during the brutal Antarctic winter.[28] Still others, such as lionesses, join forces to fend off enemies of their own species, including infanticidal males.[1386]

Fieldfare

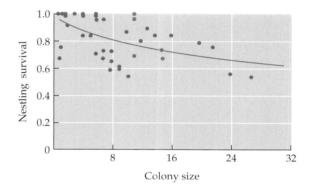

**FIGURE 13.4  Competition for food is a cost of sociality in the fieldfare,** a songbird that nests in loose colonies in woodlands. The larger the colony, the lower the survival rate of nestlings, due to increased juvenile mortality caused largely by starvation. After Wiklund and Andersson.[1573]

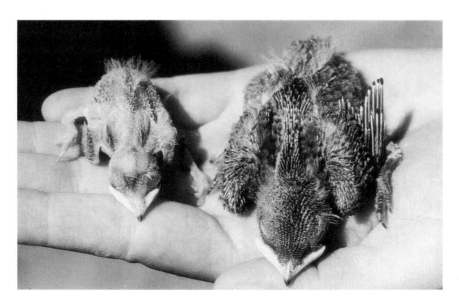

**FIGURE 13.5  Effect of parasites on cliff swallow nestlings.** The much larger nestling on the right came from an insecticide-treated nest; the stunted baby of the same age on the left occupied a nest infested with swallow bugs. From Brown and Brown.[188]

**FIGURE 13.6  Social living with defensive benefits?** The members of this dense school of small (5-centimeter-long) striped catfish living on a coral reef near Sulawesi have almost certainly joined forces to improve their chances of survival. Schooling in this and other species can enhance the survival of individual fish either by intimidating some predators through the collective size of the massed school or by amplifying their defenses, if the fish are protected by spines or chemical repellents.

The most widespread fitness benefit for social animals, however, seems to be improved protection against predators (Figure 13.6).[16] Many studies have shown that animals in groups gain by diluting the risk of being captured, or by spotting danger sooner, or by ganging up on their enemies (see Chapter 6). Thus, in the maomao, a reef fish with a memorable scientific name (*Abudefduf abdominalis*), individual males in large nesting groups chase other egg-eating predatory fish only about one-fourth as often as males in small nesting aggregations. And when a nest-defending male is removed from a small group, his eggs are attacked by a predator sooner than the eggs of a male removed from a large group, indicating that maomao males definitely derive mutual antipredator benefits by nesting together.[1476]

Males in nesting colonies of the bluegill sunfish also cooperate in driving egg-eating bullhead catfish away from their nests at the bottoms of freshwater lakes (Figure 13.7).[586] If the social behavior of the bluegill has indeed evolved

**FIGURE 13.7  Mutual defense in a society of bluegills.** Each colonial male defends a territory bordered by the nest sites of other males, while bass (above), bullhead catfish (left), snails, and pumpkinseed sunfish (right foreground) roam the colony in search of eggs. Drawing courtesy of Mart Gross.

in response to predation, then closely related species that nest alone should suffer less from predation. As predicted, the solitary pumpkinseed sunfish, a member of the same genus as the bluegill, has powerful biting jaws and so can repel egg-eating enemies on its own, whereas bluegills cannot bite hard with their small, delicate mouths.[586] Pumpkinseed sunfish are in no way inferior to or less well adapted than bluegills because they are solitary; they simply gain less through social living, which makes solitary nesting the adaptive tactic for them.

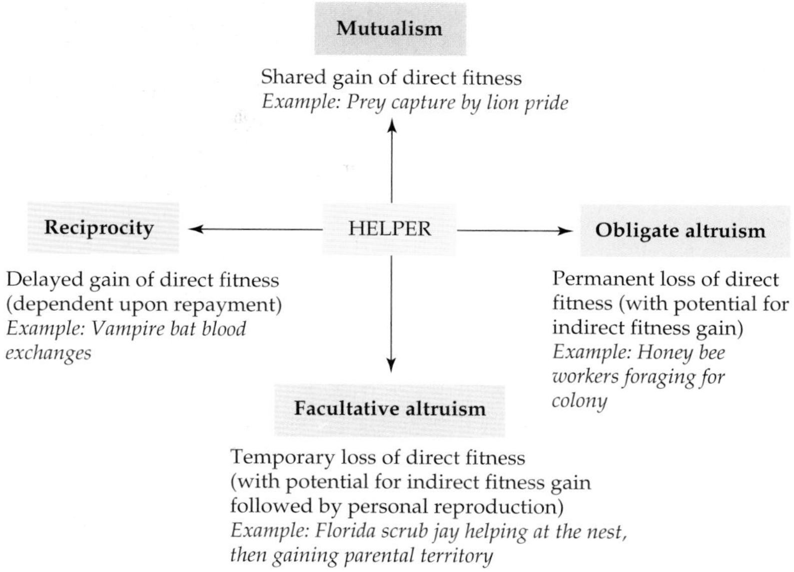

## The Evolution of Helpful Behavior

Animals that live together have the potential to assist one another, and they often do, as maomao and bluegill males demonstrate. Until the mid-1960s, biologists took helpful behavior of this sort more or less for granted because they assumed that animals should help one another for the benefit of the species as a whole. But when George C. Williams pointed out the defects of this assumption (see page 21), helpful actions, especially self-sacrificing ones, suddenly became much more interesting to evolutionary biologists.

Social interactions can vary with different payoffs for the two interactors (Figure 13.8, Table 13.2). Sometimes two individuals help one another, in which case they are said to be engaged in a **mutualism**. When one lioness drives a wildebeest into a lethal ambush set by her fellow pride members,[1377] the cooperative driver will usually get some meat, even if she did not personally pull the antelope down and strangle it herself. Likewise, if several male bluegills succeed in fending off a bullhead catfish that has entered their part of the nesting colony, the eggs in all the males' nests are more likely to survive to hatch. When both parties enjoy large reproductive gains from their interaction, their mutualism, or **cooperation**, generally requires no special evolutionary explanation.

This is not to say that mutualism is uninteresting. Consider the coalitions of male lions that form to oust rival males living with a pride of females. When

**Mutualism**
Shared gain of direct fitness
*Example: Prey capture by lion pride*

**Reciprocity**
Delayed gain of direct fitness (dependent upon repayment)
*Example: Vampire bat blood exchanges*

HELPER

**Obligate altruism**
Permanent loss of direct fitness (with potential for indirect fitness gain)
*Example: Honey bee workers foraging for colony*

**Facultative altruism**
Temporary loss of direct fitness (with potential for indirect fitness gain followed by personal reproduction)
*Example: Florida scrub jay helping at the nest, then gaining parental territory*

**FIGURE 13.8  The different categories of helping behavior.** Cooperative helpers can be placed into four groups based on the fitness consequences of their actions.

| | **TABLE 13.2** *The direct reproductive success of individuals that engage in different kinds of social interactions* | |
|---|---|---|
| | **Effect on direct reproductive success of** | |
| **Type of interaction** | **Social donor** | **Social recipient** |
| Mutualism (cooperation) | + | + |
| Reciprocity | + (delayed) | + |
| Altruism | − | + |
| Selfish behavior | + | − |
| Spiteful behavior[a] | − | − |

[a]You should not be surprised that spiteful behavior is almost never observed in nature; you should be surprised that altruism is not uncommon despite the loss of reproductive success experienced by altruists.

**FIGURE 13.9 Cooperation among competitors.** Yearling male lazuli buntings range in color from dull brownish to bright blue and orange. (Their plumage scores range from less than 16 to more than 32.) Bright yearling males permit dull males, but not males of intermediate brightness, to settle on territories neighboring their own. As a result, brownish males often pair off with females in their first year, whereas yearling males of intermediate plumage typically remain unpaired. After Greene et al.;[574] photographs courtesy of Erick Greene.

cooperating males are successful, they may gain sexual access to a large group of receptive females. When Craig Packer and his associates analyzed lion coalitions, they found that partnerships of two or three males shared access to the females fairly evenly.[1092] Even so, some males in these groups do not do as well as others. Why do the disadvantaged males tolerate their situation? Probably because if they went solo, their chance of acquiring and defending a pride would be next to zero because one male has little chance against two or three rivals. Thus, some males may be more or less forced to cooperate with domineering companions if they are to have any chance of mating.[786]

Likewise, subordinate yearling male lazuli buntings, which have dull brown plumage, engage in an interesting mutualism with brightly colored, dominant yearling males (Figure 13.9). The bright males aggressively drive other males with bright or intermediate plumage away from top-quality territories with good shrub cover, but they tolerate dull-plumaged neighbors, which are allowed to settle in superior habitat right next to their brightly colored companions. One hypothesis for the bright males' surprising behavior is that they gain by having low-ranking birds next door because they can mate with these males' females. In nests sampled by Erick Greene and his coworkers, dull-plumaged males often did indeed care for between one and two extra-pair young, which were probably the genetic offspring of their more brightly colored neighbors.[574]

Given the costs of trying to raise a family next to dominant male lazuli buntings, why do dull males accept the offer to live near them? Perhaps because

**FIGURE 13.10    Cooperative courtship of the long-tailed manakin.** The two males are in the cartwheeling portion of their dual display to a female, who is perched on the vine to the right.

the subordinate buntings at least get to hold high-quality territories, which enables them to acquire a social mate more often than males of intermediate plumage brightness. Those in-between males are often pushed by dominant rivals into habitat so poor that no female will join them. Although drab yearling males may often rear the chicks of other males, they also produce some of their own on occasion, achieving a modest amount of reproductive success, unlike most yearlings of intermediate plumage and in-between social status, which have to wait another full year to breed.[574]

### Discussion Question

**13.2**  Given the differences in reproductive success for the three categories of male lazuli buntings, how can we account for the evolutionary persistence of males with dull and, especially, intermediate plumage?

The fact that both dull and bright yearling neighbors gain some fitness from their interactions means that their social arrangement constitutes a mutualism. But what about the male coalitions of the long-tailed manakin studied by David McDonald, in which only one of two cooperative males appears to reproduce? In this bird, males form pairs and sing loud duets over and over to attract females to a display court.[482, 963] Visiting females land on the pair's display perch, often a horizontal section of liana that lies a foot or so above the ground. In response, the two males dart in and land close to the prospective mate before performing an astonishing cartwheel display (Figure 13.10). After a series of these moves, the males may then flutter slowly back and forth in front of the female, showing off their beautiful plumage in the "butterfly flight." Should a female visitor start jumping excitedly on the perch in response to these displays, one member of the duo discreetly leaves, while the

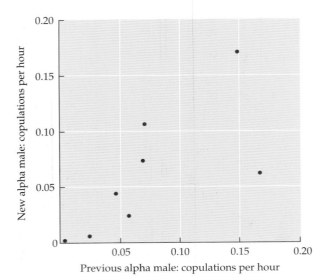

**FIGURE 13.11    Cooperation with an eventual payoff.** After the death of his alpha male partner, the beta male long-tailed manakin (now an alpha) copulates about as frequently as his predecessor did, presumably because the females attracted to the duo in the past continue to visit the display arena when receptive. After McDonald and Potts.[963]

remaining male stays to copulate with her. The female then flies off, after which the mated male calls for his display partner, who hurries back to resume his duties.

By marking the males at display perches, McDonald and his manakin watchers found that each site has only one mating male. This alpha manakin may have several display companions, but not one of them gets to reproduce, not even the alpha's favorite colleague, a beta male, who in turn is dominant to any other part-time cooperators.[963] How can it be adaptive for the subordinates to work so hard on behalf of the sexually monopolistic alpha male? By patiently following males year after year, McDonald found that subordinate males spent as long as 10 years trying to be accepted as the main dancing companion of an alpha male. Young socially active males that managed to do some displaying with many other males, especially the ones most attractive to females, had a better chance of working their way up through the ranks to become a beta male eventually.[964] Only when a subordinate achieved this level did he have a chance to make the final step up to alpha-hood upon the death or disappearance of his more dominant partner. By conceding all females to the alpha male, a beta manakin is permitted to establish his claim to be next in line, keeping other (mostly younger) birds at bay. When a beta male becomes an alpha, he usually gets to mate with many of the same females that copulated with the previous alpha (Figure 13.11).[963] Thus, beta males form a mutualism with their exclusionary partners because this is the only way to join a queue to become a reproducing alpha male—eventually.

### Discussion Question

**13.3** In several ant species, two or more unrelated females may join forces to found a colony after they have mated. The females may cooperate in digging the nest and producing the first generation of workers, but then they start fighting until only one is left alive.[112] How can it pay to join such an association? What prediction can you make about the survival rates and average productivity of colonies founded by lone females? Under what conditions would it be accurate to call this social system a mutualism? Develop at least one cost–benefit hypothesis to account for the timing of the switch from cooperation to aggressive behavior. If the behavior of the two queens is the product of natural selection, not group selection, what prediction can you make about the interactions between them during the colony establishment phase prior to the fight-to-the-death phase?

### The Reciprocity Hypothesis

The study of long-tailed manakins shows us that some superficially self-sacrificing actions actually advance the reproductive chances of helpful individuals. Another possible case of this sort involves the meerkat, a small African mammal that forages in groups. From time to time, one meerkat will stop digging for insects in the soil and climb a tree or a termite mound to look around for approaching predators (Figure 13.12).[282] Should a goshawk come swooping in, the elevated sentinel is usually the first to give an alarm, which sends all the still-foraging meerkats dashing for cover. One explanation for this behavior is that sentinels help others at personal cost now because they will be repaid later by their teammates when they take their turns at being lookouts. Bob Trivers called this kind of social relationship "reciprocal altruism" (it is also known as **reciprocity**) because helped individuals eventually return the favors they receive.[1464] If the initial cost of helping is modest but the benefit from receiving the returned favor is great, then selection can favor making the initial gesture. Imagine, for example, that a meerkat lookout has a 2-percent chance of being killed for every 100 hours spent scanning for enemies. But imagine that for every hour he spends on his perch, another companion will pay him back. If having others keep an eye open for danger for 100 hours improves the helpful lookout's chance of survival by any amount more than 2 percent, then the benefit is greater than the cost, which makes it more likely for reciprocity to spread through a population (but see the discussion of "the prisoner's dilemma" below).

However, consider an alternative explanation for sentinel behavior. Perhaps the lookouts are sated and do not need to forage for food, so they climb trees in order to better spot danger to themselves. Rather than offering costly assistance to others in their band, the "sentinels" could be securing personal fitness benefits, especially if an approaching goshawk is likely to chase one of the lookout's fleeing companions rather than the alert sentinel. (Note that this argument requires that the sated meerkats be safer on a lookout perch than in a burrow. Moreover, the sentinel must be less likely to be attacked if its companions are running for cover than if they are not. Finally, this hypothesis also requires that the signaler's companions gain more by dashing for a burrow than by remaining frozen in place in an effort to avoid detection by the onrushing predator.)

How can we test the reciprocity hypothesis against the personal safety alternative? The reciprocity hypothesis predicts that meerkats should follow a regular rotation of sentinel duty and that sentinels should run some risk of predation. However, in reality, sentinel duty is established haphazardly, and lookouts are usually closer to an escape burrow than are their fellows, suggesting that lookouts do not put themselves in special danger. The personal safety hypothesis also receives support from the finding that solitary meerkats spend about the same proportion of each day in sentinel behavior as do the members of a band. Moreover, when meerkats are given supplemental food, which reduces the cost of taking time out to look around for predators, they increase the amount of time spent on a lookout perch. Thus, what initially appears to be a rotation of lookouts may actually be the product of individuals spending as much time as possible during the day in a relatively safe position.[282]

This is not to say that reciprocity is absent from nature.[1089, 1575] When pied flycatchers observed two sets of neighbors, both mobbing stuffed tawny owls placed simultaneously near their nests, in 30 of 32 trials the flycatchers chose to help the neighbors that had assisted them an hour earlier.[802] (The other pair that the test flycatchers ignored had not been able to help them previously because they had been captured and held for the period when the tawny owl had been placed near the subject pair's nest.) In other words, pied flycatchers appear to remember who has helped them and who has not, and they use this information to pay back those who have been helpful while ignoring those that have not been cooperative. Note that pied flycatchers can choose particular individuals to help (or to withhold help from), unlike meerkats, whose behavioral decisions affect the entire group, not just a couple of group members.

The capacity for reciprocity also appears to exist in another primate, the cotton-top tamarin, as Marc Hauser and his fellow researchers demonstrated

**FIGURE 13.12** **A meerkat sentinel on the alert for approaching predators.** Photograph by Nigel J. Dennis.

(A)

(B)

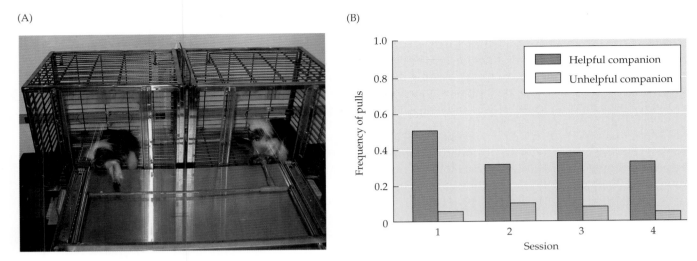

**FIGURE 13.13 Experimental demonstration of reciprocity in cotton-top tamarins.** (A) A double-compartmented cage with a pull tool that one cotton-top subject (the actor, on the right) could use to drag food down the tray toward its companion (on the left, reaching for a food reward). (B) The proportion of trials during which a cotton-top reciprocated by pulling food to a helpful companion (which had been trained to always pull a food item within reach of the other monkey) and an unhelpful companion (which had been trained never to pull a food item within reach of another individual). A, photograph courtesy of Marc Hauser; B, after Hauser et al.[634]

experimentally in the following way.[634] They constructed a special cage with a separate compartment for each of two monkeys (Figure 13.13), one of whom had access to a pull bar that could be used to drag food to within reach of either the puller or his companion (depending on the placement of the food item by a researcher). The question was, would a cotton-top repay a puller that used the tool to deliver food to it? Hauser and company conditioned one monkey to always pull the food within reach of a companion. This invariant puller was then paired with a genetically unrelated individual, which we shall call the actor. The actor and the trained altruist were given opportunities to take turns pulling food for each other over 24 test trials. The actor repaid the trained-to-pull-every-time companion somewhere between a third and a half of the time, much more so than when the actor was paired with a "defector" monkey that had been trained never to use the pull tool to deliver food to a cagemate. In other words, when paired with an apparently helpful companion, cotton-tops reciprocated, but when given an opportunity to assist an unhelpful companion, the tamarins refused.

Although at least some animals have the capacity for reciprocity, the behavior is not particularly common, perhaps because a population composed of reciprocal altruists would be vulnerable to invasion by individuals happy to accept help but eager to forget about the payback. "Defectors" reduce the fitness of "helpers" in such a system, which ought to make reciprocity less likely to evolve. The problem can be illustrated with a game theoretical model called the **prisoner's dilemma** (Figure 13.14), which is based on a human situation (see also Figure 6.34). Imagine that a crime has been committed by two persons who agreed not to squeal on each other if caught. The police have brought them in for interrogation and have put them in separate rooms. The cops have enough evidence to convict them both on lesser charges but need to have the criminals implicate each other in order to jail them for a more serious

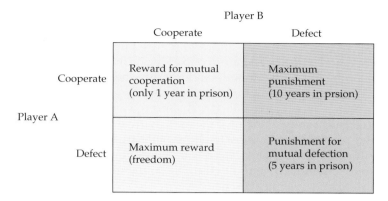

**FIGURE 13.14**  **The prisoner's dilemma.** The diagram lays out the payoffs for player A that are associated with cooperating or not cooperating with player B. Defection is the adaptive choice for player A given the conditions specified here (if the two individuals will interact only once).

crime. The police therefore offer each suspect freedom if he will squeal on his pal. If suspect A accepts the offer ("defect") while B maintains their agreed-upon story ("cooperate"), A gets his freedom (the maximum reward) while B gets hit with the maximum punishment—say, 10 years in prison (the "sucker's payoff"). If together they maintain their agreement (cooperate + cooperate), then the police will have to settle for conviction of both on the lesser charges, leading to, say, a 1-year prison term for each suspect. And if each one fingers the other, the police will use this evidence against both and renege on their offer of freedom for the snitch, so both A and B will be punished quite severely with, say, a 5-year prison sentence each.

In a setting in which the payoffs for the various responses are ranked "defect while other player cooperates" > "both cooperate" > "both defect" > "cooperate while other player defects," the optimal response for suspect A is always to defect, never to cooperate. Under these circumstances, if suspect B maintains their joint innocence, A gets a payoff that exceeds the reward he could achieve by cooperating with a cooperative B; if suspect B squeals on A, defection is still the superior tactic for A, because he suffers less punishment when both players defect than if he cooperates while his companion squeals on him. By the same token, suspect B will always come out ahead, on average, if he defects and points the finger at his buddy.

This model predicts, therefore, that reciprocal cooperation should never evolve. How, then, can we account for the cases of reciprocity that have been observed in nature? One answer comes from examining scenarios in which two players interact repeatedly, not just once. Robert Axelrod and W. D. Hamilton have shown that when this condition applies, individuals that use the simple decision rule "do unto individual X as he did unto you the last time you met" can reap greater overall gains than cheaters who accept assistance but do not return the favor.[54] When multiple interactions are possible, the rewards for back-and-forth cooperation add up, exceeding the short-term payoff from a single defection. In fact, the potential accumulation of rewards can even favor individuals who "forgive" a fellow player for an occasional defection because that tactic can encourage maintenance of a long-term relationship with its additional payoffs.[1526]

Vampire bats appear to meet the required conditions for adaptive multiple-play reciprocity. These animals must find scarce vertebrate victims from which to draw the blood meals that are their only food. After an evening of foraging, the bats return to a roost where individuals known to one another regularly assemble. A bat that has had success on a given evening can collect a large amount of blood, so much that it can afford to regurgitate a life-sustaining amount to a companion who has had a run of bad luck. Under these circumstances, the cost of the gift to the donor is modest, but the potential benefit to the recipient is high, since vampire bats die if they fail to get food three nights in a row. Thus, a cooperative, blood-transferring vampire bat is really buying insurance against starvation down the road. Individuals that establish durable "give and take" relationships with one another are better off over the long haul than those cheaters that accept one blood gift but then renege on repayment, thereby ending a potentially durable cooperative arrangement that could involve many more meal exchanges.[1575]

## Altruism and Indirect Selection

Reciprocity is really a special kind of mutualism in which the helpful individual endures a short-term loss until its help is reciprocated, at which time it earns a net increase in fitness. In contrast, there are some cases in which a donor really does permanently lose opportunities to produce offspring of its own as a result of helping another individual. In evolutionary biology, this kind of helpful self-sacrificing behavior is called **altruism** (see Table 13.2). Altruistic actions, if they exist, are an especially exciting Darwinian puzzle for adaptationists because they violate the "rule" that traits cannot spread over evolutionary time if they lower an individual's reproductive success relative to that of other individuals (see Chapter 1).

In order to explain how altruism could evolve, W. D. Hamilton developed a special explanation that did not rest on for-the-good-of-the-group arguments.[609] Instead, Hamilton's theory was based on the premise that individuals reproduce with the unconscious goal of propagating their alleles more successfully than other individuals. Personal reproduction contributes to this ultimate goal in a direct fashion. But helping genetically similar individuals—that is, one's relatives—survive to reproduce can provide an indirect route to the very same end.

To understand why, the concept of the **coefficient of relatedness** comes in handy. This term refers to the probability that an allele in one individual is present in another because both individuals have inherited it from a recent common ancestor. Imagine, for example, that a parent has the genotype $Aa$, and that $a$ is a rare form of the $A$ gene. Any offspring of this parent will have a 50-percent chance of inheriting the $a$ allele because any egg or sperm that the parent donates to the production of an offspring has one chance in two of bearing the $a$ allele. The coefficient of relatedness ($r$) between parent and offspring is therefore 1/2, or 0.5.

The coefficient of relatedness varies for different categories of relatives. For example, an uncle and his sister's son have one chance in four of sharing an allele by descent because the man and his sister have one chance in two of having this allele in common, and the sister has one chance in two of passing that allele on to any given offspring. Therefore, the coefficient of relatedness for an uncle and his nephew is $1/2 \times 1/2 = 1/4$, or 0.25. For two cousins, the $r$ value falls to 1/8, or 0.125. In contrast, the coefficient of relatedness between an individual and another, unrelated individual is 0.

With knowledge of the coefficient of relatedness between altruists and the individuals they help, we can determine the fate of a rare "altruistic" allele that is in competition with a common "selfish" allele. The key question is whether the altruistic allele becomes more abundant if its carriers forgo reproduction and instead help relatives reproduce. Imagine that an animal could potentially have one offspring of its own or, alternatively, invest its efforts in the offspring of its siblings, thereby helping three nephews or nieces survive that would have otherwise died. A parent shares half its genes with an offspring; the same individual shares one-fourth of its genes with each nephew or niece. Therefore, in this example, personal reproduction yields $r \times 1 = 0.5 \times 1 = 0.5$ genetic units contributed directly to the next generation, whereas altruism directed at relatives yields $r \times 3 = 0.25 \times 3 = 0.75$ genetic units passed on indirectly in the bodies of relatives. In this example, the altruistic tactic is adaptive because it results in more shared alleles being transmitted to the next generation.

## Discussion Question

**13.4** If an altruistic act increases the genetic success of the altruist, then in what sense is this kind of altruism actually selfish? In everyday English, words like "altruism" and "selfishness" carry with them an implication about the motivation and intentions of the helpful individual. Why might everyday usage of these words get us into trouble when we hear them in an evolutionary context? Consider the proximate–ultimate distinction here. If an individual inadvertently helped another at reproductive cost to itself, could the behavior be called altruistic under the evolutionary definition?

Another way of looking at this matter is to compare the genetic consequences for individuals who aid others at random versus those who direct their aid to close relatives. If aid is delivered randomly, then no one form of a gene is likely to benefit more than any other, and the altruism allele pays a price for the help that raises the fitness of carriers of other forms of the gene. But if close relatives aid one another selectively, then any rare family alleles they possess may survive better, causing those alleles to increase in frequency compared with other forms of the gene in the population at large. When one thinks in these terms, it becomes clear that a kind of natural selection can occur when genetically different individuals differ in their effects on the reproductive success of close relatives. Jerry Brown calls this form of selection **indirect selection**, which he contrasts with **direct selection** for traits that promote success in personal reproduction (Figure 13.15A).[192]

A brief digression is necessary to deal with yet another term, **kin selection**, which was originally defined by John Maynard Smith to embrace the evolutionary effects of both parental aid given to descendant kin (offspring) and altruism directed to **nondescendant kin** (relatives other than offspring). Currently, however, kin selection is more widely used to explain altruism supplied to relatives other than offspring. In other words, *kin selection* usually is a synonym for *indirect selection*, a term that keeps the focus clearly on the

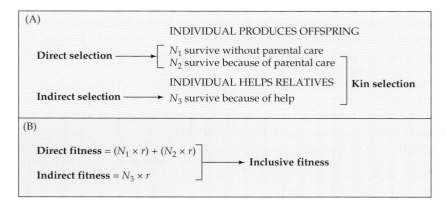

**FIGURE 13.15  The components of selection and fitness.** (A) Direct selection acts on variation in individual reproductive success. Indirect selection acts on variation in the effects individuals have on their relatives' reproductive success. (B) Direct fitness is measured in terms of personal reproductive output; indirect fitness is measured in terms of genetic gains derived by helping relatives reproduce. Inclusive fitness is the sum of the two measures and represents the total genetic contribution of an individual to the next generation. $N$ represents the number of offspring in each category; $r$ is the coefficient of relatedness. After Brown.[192]

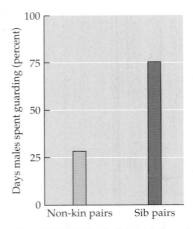

*Pelivicachromis taeniatus*

**FIGURE 13.16** **Sibling pairs of the cichlid fish *Pelivicachromis taeniatus* cooperate more than unrelated males and females when it comes to guarding a nest containing their eggs.** After Thünken et al.[1439]

distinction between parental effects on offspring and an aid giver's effects on nondescendant kin,[192] which is why we will speak of indirect selection in place of kin selection here.

### The Importance of Relatedness

An awareness of the evolutionary consequences that follow when social interactors are genetically related has shaped the way that biologists view social behavior. So, for example, the discovery that huge numbers of Argentine ant colonies covering large parts of California interact peacefully generated the prediction that these colonies should be genetically very similar—and they are, probably because the enormous numbers of ants belonging to the same supercolony are all descendants of a population that underwent a severe loss of genetic diversity in the past.[1382] But there are places where members of two different supercolonies come into contact and, as predicted, these genetically different ants attack each other ferociously.[1429]

The behavior of the Argentine ants of California is similar, albeit on a far grander scale, to that of groups of anemones, which form competing clones of genetically identical individuals. The anemone clones contain specialized castes, including scouts and warriors as well as reproductives.[487] Here we have another case of genetically identical individuals capable of developing into different phenotypes with extreme altruists prepared to sacrifice for their genetically identical, but phenotypically different, relatives. Scouts move across the border between two groups, locating the "enemy" that warriors can engage, armed with white-tipped fighting tentacles that can do real damage to the opposition.[55]

Yet another illustration of the importance of relatedness for the evolution of cooperative behavior comes from a study of a cichlid fish in which mating pairs sometimes consist of a brother and sister, while other couples are two unrelated individuals. Sibling pairs assist one another significantly more than the outbred couples when it comes to defense of a nest where the eggs are monitored by the mother fish (Figure 13.16).[1439]

### Indirect Selection and the Alarm Call of Belding's Ground Squirrel

We now have many, many examples of animals, from anemones to zebras,[1157] that factor in their relatedness to others when making unconscious decisions about how to respond to these other individuals. Not so long ago, however, the possibility that indirect selection had shaped behavior was considered novel. One of the persons who was first to consider the possible effects of indirect selection on social interactions was Paul Sherman in his study of the Belding's ground squirrel.[1314] This rat-sized North American rodent produces a staccato whistle (Figure 13.17) when a coyote or badger approaches. The sound of one Belding's ground squirrel whistling sends other nearby ground squirrels rushing for safety, just as the alarm call of a meerkat alerts companions to danger. But are the ground squirrels like meerkats in behaving purely in their own self-interest?

Sherman answered this question in the negative by discovering that alarm-calling squirrels are tracked down and killed by weasels, badgers, and coyotes at a higher rate than non-callers. Moreover, the possibility that alarm calling evolved as a form of reciprocity is also unlikely, because the probability that an individual will give an alarm call is not correlated with familiarity or length of association between the caller and the animals that benefit from its signal.[1315] Remember that reciprocity is more likely to evolve when reciprocators belong to long-term associations.

Sherman's observation that adult female squirrels with relatives nearby are more than twice as likely as males to give costly alarm calls is consistent with both a parental care hypothesis (based on direct selection) and an altruism hypothesis (based on indirect selection). You may recall that female Belding's ground squirrels tend to settle near their mothers, whereas males disperse some distance away from their natal burrow (see Figure 8.12). If the parental care hypothesis is correct, we would expect females to give more alarm calls than males because only female squirrels live near their offspring. If the altruism hypothesis is correct, we can also predict a female bias in alarm calling because females not only live near their genetic offspring, but also are surrounded by other female relatives, such as sisters, aunts, and female cousins. When self-sacrificing females warn their nondescendant kin, they could be compensated for the personal risks they take by an increased probability that kin other than offspring will survive to pass on shared genes, resulting in indirect fitness gains for the altruists. Females with offspring living nearby as well as females with only nondescendant kin as neighbors are in fact more likely to call when they detect a predator than are females who lack relatives in their neighborhood. These findings suggest that both direct and indirect selection contribute to the maintenance of alarm calling behavior in this species.[1314]

**FIGURE 13.17** A Belding's ground squirrel gives an alarm call after spotting a terrestrial predator. Photograph by George Lepp, courtesy of Paul Sherman.

## The Concept of Inclusive Fitness

Because fitness gained through personal reproduction (**direct fitness**) and fitness achieved by helping nondescendant kin survive (**indirect fitness**) can both be expressed in identical genetic units, we can sum up an individual's total contribution of genes to the next generation, creating a quantitative measure that can be called **inclusive fitness** (see Figure 13.15B). Note that an individual's inclusive fitness is not calculated by adding up that animal's genetic representation in its offspring plus that in all of its other relatives. Instead, what counts is an individual's own effects on gene propagation (1) directly in the bodies of its surviving offspring that owe their existence to the parent's actions, not to the efforts of others, and (2) indirectly via nondescendant kin that would not have existed except for the individual's assistance. For example, if the animal we mentioned earlier successfully reared one of its own offspring and also adopted three of its sibling's progeny, then its direct fitness would be $1 \times 0.5 = 0.5$, and its indirect fitness would be $3 \times 0.25 = 0.75$; the union of these two figures provides a measure of the animal's inclusive fitness, or genetic success ($0.5 + 0.75 = 1.25$).

The concept of inclusive fitness, however, is typically used not to secure absolute measures of the lifetime genetic contributions of individuals, but rather to help us compare the evolutionary (genetic) consequences of two alternative hereditary traits.[1183] In other words, inclusive fitness becomes important as a means to determine the relative genetic success of two or more competing behavioral strategies. If, for example, we wish to know whether an altruistic strategy is superior to one that promotes personal reproduction, we can compare the inclusive fitness consequences of the two traits. In order for an altruistic trait to be adaptive, the inclusive fitness of altruistic individuals has to be greater than it would have been if those individuals had tried to reproduce personally. In other words, a rare allele "for" altruism will become more common only if the indirect fitness gained by the altruist is greater than the direct fitness it loses as a result of its self-sacrificing behavior. This statement is often presented as **Hamilton's rule**: a gene for altruism will spread only if $r_b B > r_c C$. Spelling this out, we calculate the indirect fitness gained by multiplying the extra number of relatives that exist thanks to the altruist's actions ($B$) by the mean coefficient of relatedness between the altruist and those extra individuals ($r_b$); we calculate the direct fitness lost by multiplying the number of offspring not produced by the altruist ($C$) by the coefficient of relatedness between parent and offspring ($r_c$). For example, if the genetic cost of an altruistic act were the loss of one offspring ($1 \times r_c = 1 \times 0.5 = 0.5$ genetic units) but that altruistic act led to the survival of three nephews that would have otherwise perished ($3 \times r_b = 3 \times 0.25 = 0.75$ genetic units), the altruist would experience a net gain in inclusive fitness, thereby increasing the frequency of any distinctive allele associated with its altruistic behavior.

### Discussion Question

**13.5** Let's say that in calculating the inclusive fitness of a male in a coalition of lions, you measured his direct fitness by multiplying by 0.5 the number of offspring produced by the male, and then you added as his indirect fitness the total number of all offspring produced by the other members of his coalition times the mean value of $r$ between those offspring and the male in question. Your calculation of his inclusive fitness would be challenged on what grounds?

### Inclusive Fitness and the Pied Kingfisher

The value of Hamilton's rule can be illustrated by Uli Reyer's study of the pied kingfisher.[1212] These attractive African birds nest colonially in tunnels in banks by large lakes and rivers. Some year-old males are unable to find mates and instead become *primary helpers* that bring fish to their mothers and their nestlings while attacking predatory snakes and mongooses that threaten the nest. Are these males propagating their genes as effectively as possible by helping to raise their siblings? They do have other options: they could help unrelated nesting pairs in the manner of *secondary helpers*, or they could simply sit out the breeding season, waiting for next year in the manner of *delayers*.

To learn why primary helpers help, we need to know the costs and benefits of their actions. Primary helpers work harder than delayers and the more laid-back secondary helpers (Figure 13.18). The greater sacrifices of primary helpers translate into a lower probability of their surviving to return to the breeding grounds the next year (just 54 percent return) compared with secondary helpers (74 percent return) or delayers (70 percent return). Furthermore, only two in three surviving primary helpers find a mate in their second year and reproduce personally, whereas 91 percent of returning secondary helpers succeed in breeding. Many one-time secondary helpers breed with the females they helped the preceding year (10 of 27 in Reyer's sample), suggesting that improved access to a potential mate is the ultimate payoff for their initial altruism.

These data enable us to calculate the direct fitness cost to the altruistic primary helpers in terms of reduced personal reproduction in their second year of life. For simplicity's sake, we shall restrict our comparison to solo primary helpers that help their parents rear siblings in the first year and then breed on their own in the second year, if they survive and find a mate, versus secondary helpers that help nonrelatives with no other helpers present in the first year and then reproduce on their own in the second year, if they survive and find a mate.

Primary helpers throw themselves into helping their parents produce offspring at the cost of having less chance of reproducing personally in the next year. Although primary helpers do better than delayers in the second year

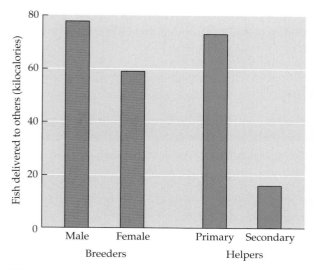

Pied kingfisher

**FIGURE 13.18  Altruism and relatedness in pied kingfishers.** Primary helpers deliver more calories per day in fish to a nesting female and her offspring than do secondary helpers, which are not related to the breeders they assist. After Reyer.[1212]

(0.41 versus 0.29 units of direct fitness), secondary helpers do better still (0.84 units of direct fitness) because they have a higher survival rate and a greater probability of securing a partner (Table 13.3).

But is the cost to primary helpers of 0.43 lost units of direct fitness (0.84 − 0.41 = 0.43) in the second year offset by a gain in indirect fitness during the first year? To the extent that these males increase their parents' reproductive success, they create siblings that would not otherwise exist, indirectly propagating their genes in this fashion. In Reyer's study, the parents of a primary helper gained an extra 1.8 offspring, on average, when their son was present. Some primary helpers assisted their genetic mother and father, in which case the extra 1.8 siblings were full brothers and sisters, with a coefficient of relatedness of 0.5. But in other cases, one parent had died and the other had remated, so the offspring produced were only half siblings ($r = 0.25$). The average coefficient of relatedness for sons helping a breeding pair was thus between one-fourth and one-half ($r = 0.32$). Therefore, the average gain for primary helper sons was 1.8 extra sibs × 0.32 = 0.58 units of indirect fitness, a figure higher than the mean direct fitness loss experienced in their second year of life.

**TABLE 13.3**  *Calculations of inclusive fitness for male pied kingfishers*

| Behavioral tactic | First year | | | Second year | | | | |
|---|---|---|---|---|---|---|---|---|
| | $y$ | $r$ | $f_1$ | $o$ | $r$ | $s$ | $m$ | $f_2$ |
| Primary helper | 1.8 × 0.32 = 0.58 | | | 2.5 × 0.50 × 0.54 × 0.60 = 0.41 | | | | |
| Secondary helper | 1.3 × 0.00 = 0.00 | | | 2.5 × 0.50 × 0.74 × 0.91 = 0.84 | | | | |
| Delayer | 0.0 × 0.00 = 0.00 | | | 2.5 × 0.50 × 0.70 × 0.33 = 0.29 | | | | |

*Source*: Reyer[1212]

*Symbols*: $y$ = extra young produced by helped parents; $o$ = offspring produced by breeding ex-helpers and delayers; $r$ = coefficient of relatedness between the male and $y$, and between the male and $o$; $f_1$ = fitness in first year (indirect fitness for the primary helper); $f_2$ = direct fitness in the second year; $s$ = probability of surviving into the second year; $m$ = probability of finding a mate in the second year.

Reyer used Hamilton's rule to establish that primary helpers sacrifice future personal reproduction in year 2 in exchange for increased numbers of nondescendant kin in year 1.[1212] Because these added siblings carry some of the helpers' alleles, they provide indirect fitness gains that more than offset the loss in direct fitness that primary helpers experience in their second year relative to secondary helpers.

---

### Discussion Questions

**13.6** Given the results of our calculations of inclusive fitness for male pied kingfishers, why are there ever any delayers? Is it maladaptive to be a delayer? Can you use conditional strategy theory (see page 231) to analyze this case? Can you use the theory to explain why there are no helper male pied kingfishers that are completely sterile?

**13.7** Another lion problem: Let's say that a lion pride typically consists of 10 reproductively mature females. Imagine that a male working by himself has a 30-percent chance of acquiring and defending a pride for 1 year. However, a pair of males has an 80-percent chance of holding a pride for this same period. (1) Assuming that all the females mate with the male or males that control their pride and produce one youngster apiece during the year and that both males in a two-male pride sire an equal number of young, should two unrelated males get together to secure this harem of females? (2) What if the males are cousins? (3) Now imagine that the males are half brothers, but the dominant male manages to get 80 percent of all matings, so 80 percent of the offspring are his. Should the subordinate join a coalition with his dominant half brother? In all your calculations of inclusive fitness, identify the direct and indirect fitness components. (This question is courtesy of Mike Beecher, to whom any complaints should be directed.)

---

### Inclusive Fitness and Helpers at the Nest

In the pied kingfisher, primary helpers raise their fitness indirectly through their increased production of nondescendant kin, whereas secondary helpers raise their fitness directly by increasing their future chances of reproducing personally. Primary helpers demonstrate that altruism can be adaptive; secondary helpers show that helping need not be altruistic, but instead may generate direct fitness benefits to helpers. Thus, this one species offers support for two very different adaptationist hypotheses on the evolution of helping behavior. These hypotheses can be tested for other cases of helpers at the nest, which are found in a variety of other birds, as well as fishes, mammals, and insects.[192, 1186, 1419]

Each case of helpers at the nest presents a separate puzzle that deserves to be analyzed in light of a full range of hypotheses, including the possibility that caring for another's offspring is a nonadaptive side effect of other adaptive traits. As Ian Jamieson has pointed out, helping may have originated in some bird species as an incidental by-product of genetic or ecological changes that made it adaptive for young adults to delay their dispersal from their natal territory.[709, 710] If these stay-at-home birds were exposed to the nestlings being cared for by their parents, the begging behavior of the baby birds might have activated parental behavior in the young nonbreeding adults. This behavior could then be maintained over evolutionary time as a by-product of two adaptive traits, delayed dispersal and the tendency to care for one's own offspring, even if feeding someone else's young reduced the fitness of nonbreeding helpers.

**FIGURE 13.19   Cooperation among scrub jay relatives.** In the Florida scrub jay, helpers at the nest provide food for the young, defense of the territory, and protection against predators. Based on a drawing by Sarah Landry, from Wilson.[1588]

   This nonadaptive by-product hypothesis assumes that selection could not eliminate the helper's tendency to feed its parents' offspring without also destroying the capacity of the stay-at-home bird to invest in its own nestlings at a later date. This proposition is testable. One of its key predictions is that the underlying mechanisms of parental care should be no different in species with helpers than in species without helpers. The group of birds known as jays provides the necessary comparative test. In the Mexican jay (*Aphelocoma ultramarina*) and Florida scrub jay (*Aphelocoma coerulescens*), some nonbreeding birds help their parents rear additional siblings (Figure 13.19). The western scrub jay (*Aphelocoma californica*) is a member of the same genus but lacks helpers at the nest. In this species, only breeding individuals have high prolactin levels, while nonbreeders have low levels of this hormone, which appears to regulate parental care in many birds. In contrast, nonbreeding helpers in the Mexican jay and Florida scrub jay have prolactin levels that match those of their breeding parents.[1290, 1497] Moreover, the prolactin in nonbreeding Mexican jays rises to peak levels before there are young to feed in the nest (Figure 13.20), suggesting that selection has favored nonbreeding individuals of this species that become hormonally primed to rear their siblings.[1497] These results are at odds with the nonadaptive by-product explanation for helping.

   But if helping is adaptive, do helpers derive inclusive fitness gains via the direct or indirect route, or both? In the Mexican jay and Florida scrub jay, some stay-at-home helpers inherit their natal territories from their parents—a direct fitness benefit. But indirect fitness benefits are also possible for helpers if parents with helpers rear more offspring than parents without non-reproducing assistants (Table 13.4).[192, 1618] For example, in the cooperative-breeding pied babbler, the higher the ratio of adults to fledglings (a figure affected by the number of helpers in the group), the more likely the offspring of the breeding adults were to receive prolonged assistance, which increased the likelihood that they would become heavier, forage more efficiently, and disperse relatively successfully from their group. Successful dispersal was associated with earlier personal reproduction, so helpers may have had their shared genes added to the next generation sooner than otherwise.[1222]

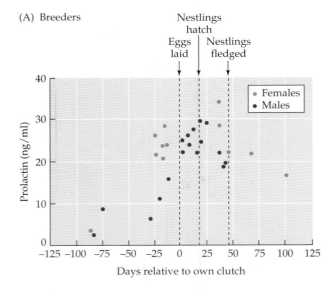

(A) Breeders

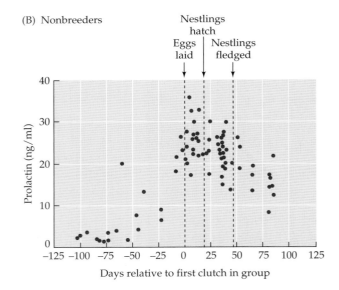

(B) Nonbreeders

**FIGURE 13.20 Seasonal changes in prolactin concentrations** in (A) breeders and (B) nonbreeding helpers at the nest in the Mexican jay. Nonbreeding birds in a group exhibit the same pattern of increased prolactin production prior to the hatching of eggs as do breeding adults. After Brown and Vleck.[195]

Another more roundabout way in which helpers could add to their indirect fitness account is by extending the lives of their parents. So, by providing extra food to nestlings, helpers at the nest in the superb fairy-wren enabled their breeding mother to make lighter eggs that contained significantly less fat and protein than those laid by a breeding female without a helpful retinue.[1256] By cutting down their gametic investment per breeding attempt, females with helpers may live longer and, thus, have more lifetime opportunities to reproduce. If so, helpers may be boosting the lifetime output of a relative, adding to their own indirect fitness.[284]

Nonetheless, we have to consider the possibility that any apparent benefits supplied by helpers may actually be the effect of occupying superior territories, which provide more food or superior nesting sites. Thus, differences in the numbers of offspring fledged by parents with and without helpers might be due to differences in the quality of the territories held by pairs in the two categories, not because one set had helpers at the nest and the other did not. Ronald Mumme tested this hypothesis by capturing and removing the non-

**TABLE 13.4** *Effect of Florida scrub jay helpers at the nest on the reproductive success of their parents and on their own inclusive fitness*

| | Parents without breeding experience[a] | Parents with breeding experience |
|---|---|---|
| Average number of fledglings produced with no helpers | 1.03 | 1.62 |
| Average number of fledglings produced with helpers | 2.06 | 2.20 |
| Increase in reproductive success due to help | 1.03 | 0.58 |
| Average number of helpers | 1.70 | 1.90 |
| Indirect fitness gained per helper | 0.60 | 0.30 |

*Source*: Emlen[439]

[a]Includes pairs in which one parent has reproduced, which is why some pairs in this category acquire a helper at the nest.

breeding helpers from some randomly selected breeding pairs of Florida scrub jays, while leaving other helpers untouched. The experimental removal of helpers reduced the reproductive success of the experimental pairs by about 50 percent, as measured by the number of offspring known to be alive 60 days after hatching (Figure 13.21). Helpers apparently really do help in this species.[1023]

In fact, helper scrub jays also improve the chances that their parents will live to breed again another year, as suggested for the superb fairy-wren as well. Improved parental survival means that the scrub jay helpers are responsible for still more siblings in the future; these extra siblings yield an average of about 0.30 additional indirect fitness units for helpers.[1022] The total indirect fitness gains from helping at the nest can potentially exceed its costs in terms of lost direct fitness, especially if the young birds have almost no chance of reproducing personally. When very few openings are available for dispersing young adults, helping is more likely to be the adaptive option for individuals that have the potential to be either altruistic or to reproduce on their own. Individuals of this sort can be said to be *facultative altruists* (see Figure 13.8) because they are not locked into the helper role.

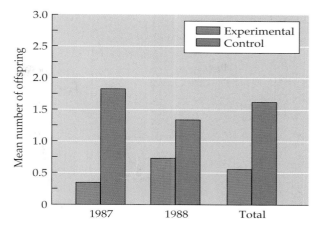

**FIGURE 13.21** **Helpers at the nest help parents raise more siblings in the Florida scrub jay.** The graph shows numbers of offspring alive after 60 days in experimental nests that lost their helpers and in unmanipulated control nests during a 2-year experiment. After Mumme.[1023]

Whether saturated nesting habitats contribute to the maintenance of helping at the nest is also testable. If young birds remain on their natal territories because they cannot find suitable nesting habitat, then yearlings given an opportunity to claim good open territories should promptly become breeders. Jan Komdeur did the necessary experiment with the Seychelles warbler, a drab little brown bird that has played a big role in testing evolutionary hypotheses about helping at the nest. When Komdeur transplanted 58 birds from one island (Cousin) to two other nearby islands with no warblers, he created vacant territories on Cousin, and helpers at the nest there immediately stopped helping in order to move into open spots and begin breeding. Since the islands that received the transplants initially had many more suitable territorial sites than warblers, Komdeur expected that the offspring of the transplanted adults would also leave home promptly in order to breed elsewhere on their own. They did, providing further evidence that young birds help only when they have little chance of making direct fitness gains by dispersing.[793]

Moreover, the sophisticated conditional strategy that controls the dispersal decisions made by young Seychelles warblers is sensitive to the quality of their natal territory. Breeding birds occupy sites that vary in size, vegetational cover, and insect supplies. By using these variables to divide warbler territories into categories of low, medium, and high quality, Komdeur showed that young helpers on good territories were likely to survive there while also increasing the odds that their parents would reproduce successfully. Young birds whose parents had prime sites often stayed put, securing both direct and indirect fitness gains in the process. In contrast, young birds on poor natal territories had little chance of making it to the next year, nor could they have a positive effect on the reproductive success of their parents. They left home and tried to find a breeding opportunity of their own.[792]

The probability that a warbler would disperse was also influenced by whether both its genetic parents were alive and in control of the family territory or whether one or both had been replaced by new stepparents (Figure 13.22).[431] Thus, the dispersal of helpers becomes more likely if opportunities to help close kin have been reduced or lost by changes in the breeding pair controlling a territory.

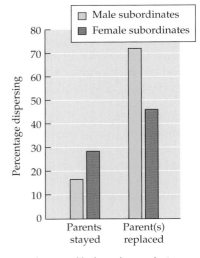

Seychelles warbler

**FIGURE 13.22  Helper Seychelles warblers** are much more likely to leave their home territory if they lose one or both of the parents they have been helping. Note the colored leg bands that enable researchers to track individuals over time. Photograph by Cas Eikenaar; data from Eikenaar et al.[431]

## Discussion Question

**13.8** Helpers at the nest have been found in only about 3 percent of all bird species.[631] One attribute of this small minority of birds that has often been linked to the evolution of helpers is the delayed dispersal of juveniles, as we have just illustrated for Florida scrub jays and Seychelles warblers. But another factor that might have promoted the evolution of helping is a very low adult mortality rate. These two ideas have sometimes been presented as competing hypotheses, but how might they both reflect the same ecological pressure that makes helping at the nest an adaptive making-the-best-of-a-bad-job option for young birds?

The flexibility of behavior exhibited by Seychelles warblers is not unique to that species. Consider how young female white-fronted bee-eaters make adaptive conditional decisions about reproducing (Figure 13.23). This African bird nests in loose colonies in clay banks. Like male pied kingfishers, young female white-fronted bee-eaters can choose to breed, or to help a breeding pair at their nest burrow, or to sit out the breeding season altogether. If an unpaired, dominant, older male courts her, a young female almost always leaves her family and natal territory to nest in a different part of the colony, particularly if her mate has a group of helpers to assist in feeding the offspring they will produce. Her choice usually results in high direct fitness payoffs. But if young, subordinate males are the only potential mates available to her, the young female will usually refuse to set up housekeeping. Young males come with few or no helpers, and when they try to breed, they are often harassed by their fathers, who may force their sons to abandon their mates and return home to help rear their siblings.

A female that opts not to pair off under unfavorable conditions may choose to slip an egg into someone else's nest or to become a helper at the nest in her natal territory—provided that the breeding pair there are her parents, to whom she is closely related. If one or both of her parents have died or moved away, she is unlikely to help rear the chicks there, which are at best half siblings, and instead will simply wait, conserving her energy for a better time in

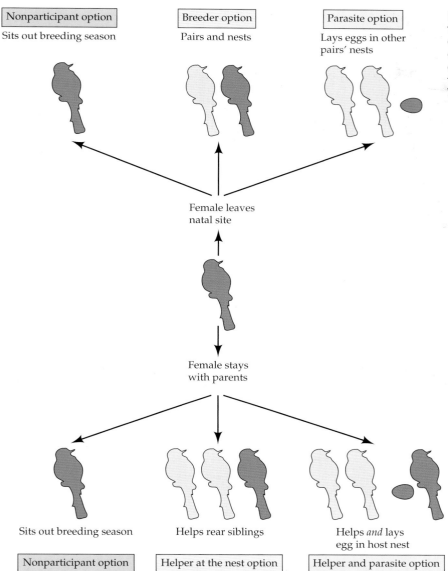

| Nonparticipant option | Breeder option | Parasite option |
|---|---|---|
| Sits out breeding season | Pairs and nests | Lays eggs in other pairs' nests |

Female leaves
natal site

Female stays
with parents

| Sits out breeding season | Helps rear siblings | Helps *and* lays egg in host nest |
|---|---|---|
| Nonparticipant option | Helper at the nest option | Helper and parasite option |

**FIGURE 13.23 Conditional reproductive tactics of female white-fronted bee-eaters.** Females of this species have many options, of which helping at the nest is only one. Females select a given tactic suited for their special circumstances. After Emlen, Wrege, and Demong.[442]

which to reproduce.[442] Thus, although daughter bee-eaters have the potential to become helpers at the nest, they choose this option only when the indirect fitness benefits of helping are likely to be substantial.

## The Evolutionary History of Helping at the Nest

Helpers at the nest are not all that common among birds, which leads us to ask about the special ecological circumstances responsible for the evolution of helping. Dustin Rubenstein and Irby Lovette explored the issue through a comparative analysis of African species of starlings. Among the dozens of species within this group, some have helpers at the nest while others do not. Rubenstein and Lovette considered the social systems of the species with respect to their habitats, and they found that helping was strongly associated with starlings that live in African grasslands rather than deserts or forests (Figure 13.24). Grassland, or savanna, habitats in Africa are places where rainfall is both seasonal and highly variable, according to climatological records going back over a century. As species have occupied grasslands, they have indepen-

**FIGURE 13.24 Cooperative breeding in African starlings** is associated with species that live in savanna grasslands, where rainfall is seasonal and erratic. After Rubenstein and Lovette.[1249]

dently evolved cooperative breeding systems time and again, suggesting a causal connection between an environmental factor (grassland habitat) and a social system (helping at the nest).[1249]

This scenario, which focuses on ecological unpredictability as the driver for prolongation of family life, differs from another explanation, offered by Rita Covas and Michael Griesser.[304] For Covas and Griesser, the fact that long-lived species are well represented in the species with slow-to-leave young indicates a causal connection between the two features. These biologists argue that delayed dispersal and non-reproduction by young birds requires *environmental predictability* in the form of guaranteed access to the resources controlled by their territorial parents. Their conclusion, however, has been challenged by Daniel Blumstein and Anders Møller, whose test of the longevity hypothesis, involving over 250 species of North American birds, yielded no causal relationship between longevity and social living in this group, once they had controlled for the independent effects of body size, mortality rate, and age at first reproduction on social living.[134] In other words, perhaps it is that social living arises because in some environments delayed dispersal and reproduction are advantageous, which makes it possible for individuals to live longer, rather than that living longer causes cooperative breeding to evolve.

(A)

## Insect Helpers at the Nest

Although some birds provide impressive examples of adaptive helping at the nest, the phenomenon also occurs in highly sophisticated forms in certain insects, including the paper wasps mentioned at the beginning of this chapter. Paper wasp colonies usually consist of one or more reproductively active females and a number of helpers at the nest, or workers, which are always females, never males. In order to generate a set of female helpers early in the nesting cycle, egg-laying female wasps fertilize some eggs by releasing sperm from a sperm storage organ as the eggs pass down the oviduct. These **diploid** eggs will develop into females, whereas to produce a son, a queen need only lay an unfertilized (**haploid**) egg (Figure 13.25).

Paper wasp queens make daughters early in the nesting cycle because many will remain at the nest to help rear more offspring, which later in the season will include both brothers and sisters that are destined to reproduce, not help at the nest. Nonetheless, the helper daughters of paper wasps have functional ovaries and so are capable of reproducing. Even if they have not mated, they could lay unfertilized eggs and so have sons, contributing to their direct fitness if these offspring were to survive. But despite their apparent capacity for personal reproduction, workers usu-

(B)

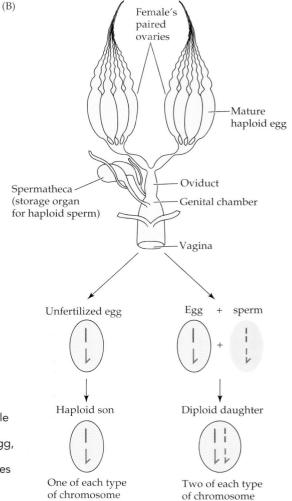

**FIGURE 13.25  Haplodiploid sex determination in Hymenoptera.** (A) When a haploid male bee copulates with a diploid adult female, he provides her with a quantity of haploid sperm, which are stored in the female bee's spermatheca. (B) When the female "decides" to produce a son, she releases a mature haploid egg from an ovary and lays it as an unfertilized egg, which will develop into a male. In order to produce a daughter, the female fertilizes the mature egg with sperm from her spermatheca as the egg passes down the oviduct. For simplicity's sake, only two chromosomes are shown here. Photograph by the author.

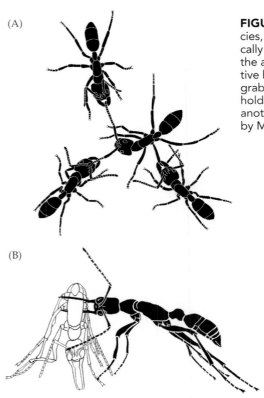

(A)

(B)

**FIGURE 13.26 Conflict within ant colonies over reproduction.** In two ant species, individuals that are about to reproduce are detected by colonymates and physically prevented from doing anything for hours or days. (A) The queen has smeared the ant in the center with her sting. Three workers grasp the would-be reproductive by assorted appendages, preventing her from moving. (B) The ant in black has grabbed and immobilized a nestmate whose ovaries had begun to develop. After holding her nestmate captive for 3 or 4 days, a worker may turn her prisoner over to another worker to continue the imprisonment. A, from Monnin et al.;[1004] B, drawing by Malu Obermayer, from Liebig, Peeters, and Hölldobler.[862]

ally do not exercise this option and instead work, often tirelessly, on behalf of their queen or the queen's reproducing sons and daughters. The standard explanation for the workers' refusal to reproduce personally has been that by voluntarily giving up personal reproduction in favor of helping increase the reproductive success of relatives, the workers experience a net gain in inclusive fitness.

A somewhat different but related hypothesis for worker altruism, however, states that this behavior is forced upon them by the policing actions of the queen or other workers in their colony. Typically this punishment takes the form of destroying any eggs laid by workers. But the behavior can be more elaborate, as in the ant *Dinoponera quadriceps*, where the dominant reproducing female (effectively the queen) smears a potential competitor with a chemical from her stinger, after which lower-ranking workers immobilize the unlucky pretender queen for days at a time (Figure 13.26A).[1004] Likewise, workers of *Harpegnathos saltator* punish nestmates that are developing their ovaries by holding the offenders in a firm grip (Figure 13.26B), preventing them from doing anything. These tactics inhibit further development of an immobilized ant's ovaries.[862]

Tom Wenseleers and Francis Ratnieks undertook to test whether sanctions imposed on workers that try to reproduce can make it more profitable for this class of individuals to behave altruistically. By securing quantitative data from 20 species of social insects on (1) the effectiveness of policing efforts within colonies and (2) the proportion of workers that laid unfertilized eggs in their colonies, Wenseleers and Ratnieks were able to show that the more likely it was that the eggs of workers would be destroyed, the lower the proportion of worker reproductives in the colony (Figure 13.27A). This result supports the enforced altruism hypothesis.[1539]

Another way of looking at this issue is to consider what percentage of the males in a colony of social insects are the sons of workers, a figure that varies from 0 to 100 percent. If worker policing is responsible for the cases in which few males are the offspring of workers, then these cases should involve species in which the workers are more closely related to the queen's sons than to the sons of other workers. This prediction is also correct (Figure 13.27B).

### Discussion Question

**13.9** Given the conclusion from the data in Figure 13.27B, why might it be that the more drones that mate with a queen honey bee, the more attractive the queen is to the workers in her colony and the longer she retains a workforce that acts in her interests, not their own personal reproductive interests?[1219]

In contrast, a key prediction from the voluntary altruism hypothesis is not met. If a high degree of relatedness promotes altruistic behavior, then the more closely related the worker force is to one another (and more importantly, to the

(A)

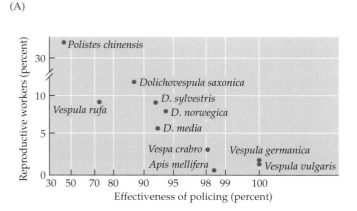

(B)

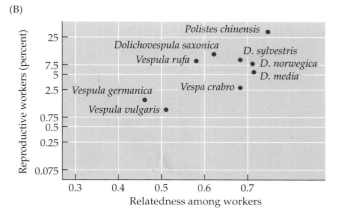

**FIGURE 13.27  Are workers in social insect colonies forced to be altruistic?**
(A) The percentage of reproducing workers in the nests of social wasps decreases as
the effectiveness with which their fellow workers destroy their eggs increases. (B) The
percentage of reproducing workers in social wasp colonies increases as the relatedness among the workers increases. The data in both graphs support the involuntary
altruism hypothesis. After Wenseleers and Ratnieks.[1538]

future queens and kings being produced within their colony), the less likely
a worker would be to produce sons of her own. In reality, the exact opposite
is true (Figure 13.28).[1187, 1538]

We can accommodate both explanations for altruism by noting that one is
proximate and the other ultimate. At the proximate level, workers are forced
to give up reproduction by the colony police. At the ultimate level, as the
likelihood of reproducing declines for potential workers, the probability that
involuntary worker altruism will become adaptive for them increases because
if $rC$ is very low, then $rB$ need not be great to favor individuals that help their
relatives. But there still need to be some benefits for indirect selection to result
in the spread of altruism of any sort. Worker altruism will yield indirect fitness
gains to the extent that altruists help their mother produce "extra" reproducing sons and daughters of the queen. Given the higher rate of loss of nests

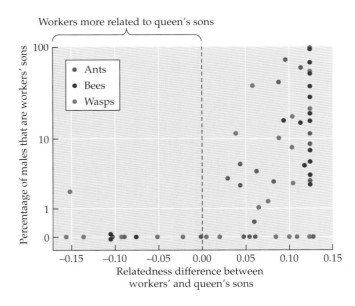

**FIGURE 13.28  The proportion of males produced by work-
ers varies among ants, social bees, and social wasps.** The
greater the difference in the relatedness of workers' and the
queen's sons (values greater than zero), the greater the proportion of males that were produced by the colony's workers. After
Wenseleers and Ratnieks.[1539]

built and defended by a single paper wasp to raiding female wasps and nest predators,[511, 1054] we can be quite confident that helpers usually do boost the reproductive success of the queen they help. Our confidence in this conclusion is further enhanced by the results of some experimental studies involving the removal of subordinate female wasps by researchers. In one such study, nests that lost helpers did not survive as well as those that retained them; in another, the role of subordinates in producing extra brood was documented experimentally.[1323, 1441]

---

### Discussion Question

**13.10** Colonies of social Hymenoptera sometimes lose their queens and may remain queenless for a time. What prediction can you make about the proportion of workers that rear their own sons in queenless colonies? What prediction can you make about the proportion of workers that will continue to behave altruistically in these colonies, which vary in the degree to which the workers are related to one another?

---

In some social insects, subordinate females undertake to help queens to whom they are unrelated. Under these circumstances, no indirect fitness gains are possible for the worker females. Understanding their puzzling behavior has been tackled with the help of a **transactional theory** of social behavior. This theory views animal societies as arenas in which dominants and subordinates "negotiate" their reproductive rights within the group.[734, 1204] One derivative of transactional theory is a concessions model, in which a powerful dominant member of the group concedes a certain amount of reproduction to lower-ranking individuals to make it advantageous for them to stay in the group, thereby boosting the fitness of the concession-making animal. Although we are going to apply the theory primarily to social insects, it can be employed with other social organisms in which one individual dominates a number of subordinates. So, for example, male chimpanzees compete intensely for dominance in their groups, but the dominant male may allow some subordinates to mate on occasion, provided that they help him remain on top (Figure 13.29).[418]

In paper wasps when unrelated females help a dominant queen, they cannot derive indirect fitness from their assistance, as the queen's relatives do, so these females should require some direct fitness payoffs for staying and helping. One would predict, therefore, that queens allied with nonrelatives should make larger concessions to keep helpers on hand than queens fortunate enough to have a bevy of relatives available. Indeed, when dominant

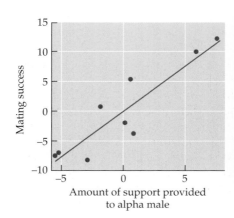

**FIGURE 13.29 Alpha male chimpanzees concede matings to supportive members of the group.** When the effect of dominance ranking was controlled statistically, the mating success of a given male was highly correlated with the number of times the male had supported the dominant individual in competitive interactions with other males in the band. After Duffy, Wrangham, and Silk.[418]

Male chimpanzees

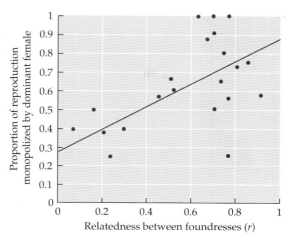

*Polistes fuscatus* queen

**FIGURE 13.30   A test of a hypothesis based on transactional theory.** When a queen of *Polistes fuscatus* is not closely related to the other foundress females that join her to set up a colony, she apparently concedes some reproductive chances to her companions, which reduces the degree to which a single female monopolizes reproduction at the nest. After Reeve et al.[1206]

females of the wasp *Polistes fuscatus* associate with unrelated helpers, the reproductive differences among dominants and subordinates are less than when all the females in a colony of this species are related to one another (Figure 13.30).[1206]

Alternative transactional hypotheses are, however, available for cases in which subordinates assist unrelated queens. Perhaps instead of accepting concessions from a queen, subordinate helpers reproduce to the extent that they can by circumventing the controlling dominant female, something they might be quite good at. In other words, there might be a tug-of-war going on within a colony, whose members attempt to take as many opportunities as possible to reproduce rather than merely accepting whatever a dominant queen is willing to offer.

The fact is that there are numerous examples of insect societies that do not neatly fit the concessions model. In one Australian social bee, for example, nesting groups that are composed of two unrelated individuals exhibit a greater degree of reproductive skew (with the dominant taking the lion's share) than is found in duos made up of close relatives.[835] This result is not consistent with expectations from a concessions hypothesis but does match predictions from the alternative tug-of-war scenario, in which dominants are confronted by difficult-to-control subordinates. When the members of a social unit spend time fighting with one another, they should have less time to contribute to colony productivity. Thus, under a tug-of-war scenario, we would expect groups of unrelated individuals to produce fewer offspring overall than groups of relatives, which have an indirect fitness incentive to moderate their conflicts with their sisters.

Subordinates unrelated to a dominant group member might in some instances tolerate a reduction in personal reproduction because (1) they have little chance of founding a successful nest elsewhere on their own and (2) they have some chance of inheriting the nest where they are helping upon the demise of the current queen, at which time their reproductive payoff could be substantial. (Note the similarity between this hypothesis and those proposed to account for secondary helpers in pied kingfishers and beta display partners in long-tailed manakins.) This kind of facultative altruism (see Figure 13.8) can be considered yet another kind of social transaction, with some individuals helping others in exchange for permission to remain at the colony where

White-browed scrub wren

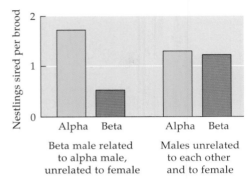

**FIGURE 13.31 The effect of relatedness on equality of reproductive opportunities** in the cooperatively breeding white-browed scrub wren. After Whittingham, Dunn, and Macgrath.[1560]

they have some chance of eventually becoming dominant breeders. In support of this hypothesis, observations of 28 nests of *Polistes dominulus* yielded 13 records of a change in ownership within the nesting season, 10 of which were accomplished by resident helpers.[1186] Much the same thing occurs in a tropical social wasp in a different subfamily in which one dominant female typically takes total charge of reproduction while an assortment of subordinates help whether they are related to the queen or not. In effect, the non-reproducers, which have their own dominance hierarchy, are lined up in a queue awaiting the demise of the queen, at which point the top subordinate will have a chance to become an independent reproducer. Moreover, because this now dominant individual inherits a coterie of workers, her offspring are likely to survive even if she dies before completion of a brood cycle.[1403] The point is that an abundance of alternative hypotheses exists to account for helping at the nest in the Hymenoptera, and the evidence suggests that many of them are valid for one or another ant, bee, or wasp.

## Discussion Questions

**13.11** When a new queen of the paper wasp *P. dominulus* does take control of a nest, other females in the colony often disperse.[1443] Explain why (in ultimate terms) dispersers might opt to leave a nest where they had been living.

**13.12** Figure 13.31 contains data on the relative reproductive success of alpha and beta males of the white-browed scrub wren in two different kinds of cooperative breeding associations.[1560] Is the observed pattern consistent with any hypothesis derived from transactional theory?

**13.13** Coalitions of male lions appear to be of two types: large groups of closely related individuals and smaller groups of nonrelatives. Use two hypotheses based on transactional theory to make predictions about the degree of reproductive monopolization that should be observed in the two kinds of coalitions. In one hypothesis, dominants (which are in total control of the group) concede reproduction to subordinates to retain their services as group members; in another hypothesis, dominants (which lack complete control) and subordinates are engaged in a tug-of-war over who gets to reproduce. Check the actual data presented by Craig Packer and colleagues.[1092]

## The Evolution of Eusocial Behavior

In attempting to trace the evolutionary history of **eusociality**, W. D. Hamilton[609] and Richard Alexander[16] proposed that the first altruistic, non-reproducing workers must have made use of traits that had evolved for another function, probably the care and nurturing of brood produced by the female herself. This hypothesis produces a testable prediction: the genes for maternal behavior that are expressed in reproducing foundress queens of living eusocial insects will also be the ones that affect the behavior of non-reproducing worker females.[27] Foundress queens initiate colonies (Figure 13.32), building the nests that will accommodate the eggs and then feeding the resultant larvae with food collected away from the nest. Most of the larvae are likely to become workers, which expand the nest and attend to the colony's offspring. In contrast, queens of established

**FIGURE 13.32   A foundress female paper wasp.** This female *Polistes* has begun a nest on her own. If she is fortunate, the daughters she has begun to rear in the brood cells of her nest will help her produce still more offspring in the weeks ahead. Photograph by the author.

colonies and future queens-in-waiting do not devote themselves to maternal activities, but instead rest on the nest while workers take care of the colony's offspring.

In order to compare genetic activity in workers relative to foundresses, established queens, and future queens, a team of researchers examined 32 genes expressed in the brains of the paper wasp *Polistes metricus*.[1457] These genes were similar to a set found in honey bees, where they are known to influence the behavior of adult females. As expected from the historical hypothesis on the evolution of worker behavior, workers and foundresses exhibit very similar patterns of gene expression in brain cells, a pattern that is substantially different from that of the other two categories of female paper wasps (Figure 13.33). This result suggests that selection acted on genes already available in female wasp brains, such that non-reproducing individuals activated genes that in the past would have only been expressed in females with offspring of their own. Note the similarity between this explanation for the behavior of worker insects and that given earlier for the historical underpinnings of helping at the nest among birds (see page 476). In both cases, individuals that have not reproduced behave like parents caring for their own offspring.

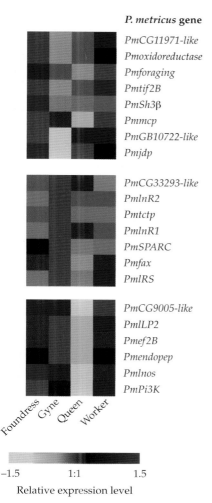

*P. metricus* gene

*PmCG11971-like*
*Pmoxidoreductase*
*Pmforaging*
*Pmtif2B*
*PmSh3β*
*Pmmcp*
*PmGB10722-like*
*Pmjdp*

*PmCG33293-like*
*PmlnR2*
*Pmtctp*
*PmlnR1*
*PmSPARC*
*Pmfax*
*PmlRS*

*PmCG9005-like*
*PmlLP2*
*Pmef2B*
*Pmendopep*
*Pmlnos*
*PmPi3K*

Foundress  Gyne  Queen  Worker

−1.5          1:1          1.5
Relative expression level

**FIGURE 13.33   Foundress females and workers of the paper wasp *Polistes metricus*** have a similar pattern of gene activity, whereas future reproductives (gynes) and queens exhibit very different patterns. The genes chosen for analysis are known to be expressed in honey bee worker brains; the levels of gene expression range from low (−1.5) to high (+1.5). After Toth et al.[1457]

## Discussion Question

**13.14**   Figure 13.34 shows a phylogeny of a portion of the bee species in the genus *Lasioglossum*, with the social system of the species superimposed on the evolutionary tree. What is the minimum number of times that eusociality has evolved in this group? How many times has eusociality been lost, either completely or partially? (Partial loss of eusociality has occurred in those species that are labeled "polymorphic," which means that in these cases some populations are eusocial and others are not.) How is this phylogeny relevant to the common view that complex social systems are generally superior, and more recently evolved, than simpler ones? For a paper on *when* sociality evolved in *Lasioglossum*, see Brady et al.[162]

**FIGURE 13.34  Eusociality has an evolutionary history.** Among the many species of bees belonging to the genus *Lasioglossum*, some are eusocial, others are solitary, and still others are polymorphic (exhibit both patterns in different populations). The social parasitic species exploits established colonies of other bees. The photograph shows two females of *L. (Sphecodogastra) oenotherae*, a solitary species whose closest relatives are mostly eusocial bees. After Danforth, Conway, and Ji;[347] photograph by Bryan Danforth.

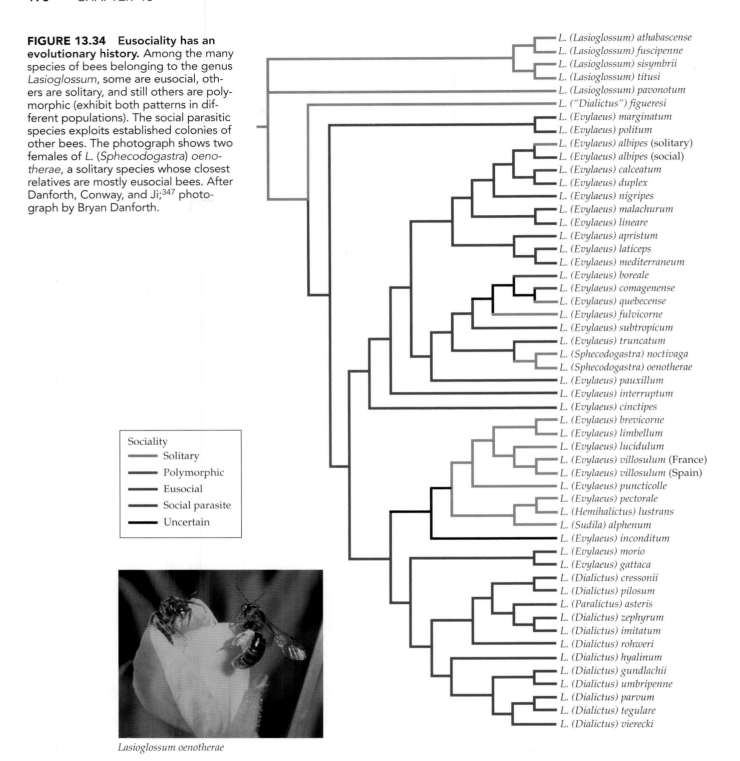

*Lasioglossum oenotherae*

## Haplodiploid Sex Determination and the Evolution of Extreme Altruism

In many social insects, helpers are, as we noted already, capable of "selfish" reproduction, while in others helpers have so little chance for personal reproduction that they can be considered *obligate altruists* (see Figure 13.8). The anatomy and behavior of these sterile castes with their undeveloped ovaries often reveal extreme specialization for self-sacrifice. Honey bee workers,

(A)

(B)

**FIGURE 13.35** **Sacrifices by social insect workers.** (A) In a nasute termite colony, soldiers incapable of reproducing attack colony intruders and spray these enemies with sticky repellents stored in glands in their heads. (B) When a honey bee stings a vertebrate, she dies after leaving her stinger and the associated poison sac attached to the body of the victim. A, photograph by the author; B, photograph of Bernd Heinrich's knee by Bernd Heinrich.

for example, have a barbed sting designed to catch in the skin of vertebrate enemies threatening their hive, the better to deter them, even though it means the death of the defender (Figure 13.35). Soldier ants of some species have immense jaws capable of piercing the bodies of invading insect predators. Other ant species have grenade soldiers, whose suicidal mission is to rush at enemies entering their colony while simultaneously constricting their abdominal muscles so violently that they burst a large abdominal gland, spilling a disabling glue over their enemies.[944]

Charles Darwin was well aware of the eusocial insects—those with an essentially sterile worker caste—and his solution to the problem of their evolution was closely related to the indirect fitness hypothesis already discussed at length in this chapter. Darwin noted that social insect colonies are extended families, so when sterile members of the group help others survive to reproduce, the helpers (even if they die in the process) are helping to maintain family traits—including the ability of the reproducing members of the family to generate sterile helpers.[348]

The next major advance on Darwin's explanation came 120 years later, when W. D. Hamilton developed his now famous genetic cost–benefit analysis of worker altruism. Remember that according to Hamilton's rule, altruism can evolve when the altruist's loss in personal reproduction (C) times the degree of relatedness of a parent to its offspring ($r_c$) is less than the added number of reproducing relatives that owe their existence to the altruist (B) times the degree of relatedness between the altruist and the helped individuals ($r_b$). Hamilton realized that if $r_b$, the relatedness of the altruist to the relatives it helped, was particularly high, then the indirect fitness side of the equation would be increased. He was also the first to point out that $r_b$ for sisters could indeed be unusually high in the Hymenoptera because of the haplodiploid system of sex determination in this group.[609]

We are going to take a long look at the haplodiploid hypothesis for the evolution of eusociality because it is an ingenious idea that has played an important role in the study of social behavior. However, even if the method of sex determination in the Hymenoptera had little to do with the evolution of extreme altruism in the group, this would not mean that indirect (or kin) selection was irrelevant for the evolution of eusociality. The haplodiploid hypothesis only identifies one factor that could make indirect selection particularly effective in one order of insects. Other factors could also promote indirect selection for altruism in the Hymenoptera and other groups of animals.

Hamilton focused on haplodiploidy in Hymenoptera because he realized the significance of male haploidy with respect to the coefficient of relatedness among a male's daughters. Since male ants, bees, and wasps have only one set of chromosomes, not two, all the haploid sperm a male makes are chromosomally (and thus genetically) identical. So if a female ant, bee, or wasp mates with just one male, all the sperm she receives will have the same set of genes. When the female uses those sperm to fertilize her eggs, all her diploid daughters will carry the same set of paternal chromosomes and genes, which make up 50 percent of their total genotype. The other set of chromosomes carried by daughter hymenopterans comes from their mother. The mother's haploid eggs are not genetically uniform, because she is diploid; gamete formation by a parent with two sets of chromosomes involves the production of a cell with just one set drawn at random from those in that parent. An egg made by a female bee, ant, or wasp will share, on average, 50 percent of the alleles carried within her other eggs. Thus, when a queen bee's eggs unite with genetically identical sperm, the resulting offspring will share, on average, 75 percent of their alleles: 50 percent from their father and 0 to 50 percent from their mother (Figure 13.36).

Under the haplodiploid system of sex determination, hymenopteran sisters may therefore have a coefficient of relatedness of 0.75, higher than the 0.5 figure for a mother and her daughters and sons. As a consequence of this genetic fact, $r_c \times C$ should be less than $r_b \times B$ more often in the Hymenoptera than in other groups, all other things being equal, which would facilitate the evolution of eusociality in these insects. If sisters really are especially closely related, indirect selection could more easily favor hymenopterans that, so to speak, put all their eggs (alleles) in a sister's basket rather than reproducing personally. Perhaps not coincidentally, the Hymenoptera have the greatest number of eusocial species with female-only castes of any insect order.

### Testing the Haplodiploid Hypothesis

Note, however, that there is an alternative explanation for certain special features of hymenopteran eusociality that do not have anything to do with the haplodiploid system of sex determination. Yes, there are many eusocial Hymenoptera with female, not male, workers, but female-only parental care is unusually widespread in this group, providing many evolutionary opportunities for female hymenopterans to utilize their parental capacities on behalf of relatives other than their offspring. The point is that the haplodiploid explanation, brilliant though it is, needs to be tested, which has been done. For example, if a female worker in a eusocial bee, ant, or wasp colony is to cash in on her potentially high degree of relatedness to her fellow females, she should bias her help toward reproductively competent sisters rather than toward male siblings. Although sister hymenopterans share up to 75 percent of their genes, a sister shares only 25 percent of her genes with her haploid brother (see Figure 13.25). Males do not receive any of the paternal genes their sisters possess. The remaining half of the genome that sisters and brothers both receive from their mother ranges from 0- to 100-percent identical, averaging

(A) Mother–daughter genetic relatedness

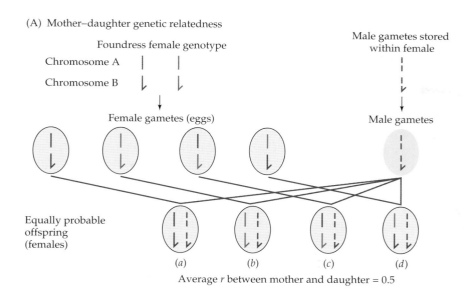

Average $r$ between mother and daughter = 0.5

**FIGURE 13.36  Haplodiploidy and the evolution of eusociality in the Hymenoptera.** The degree of genetic relatedness of a female wasp (A) to her daughters and (B) to her sisters. For the sake of simplicity, only two chromosomes are shown. A queen's sons develop from unfertilized eggs with only one copy of each chromosome, so mothers share 50 percent of their genes with their sons.

(B)  Sister–sister genetic relatedness

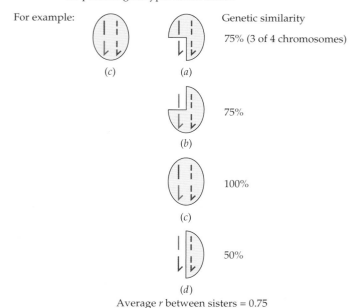

Average $r$ between sisters = 0.75

50 percent; 50 percent of a half means that a sister shares, on average, only one-fourth of her genes with her brothers ($r = 0.25$). In part because sisters are three times more closely related to one another than to their brothers, Bob Trivers and Hope Hare realized that worker hymenopterans are expected to favor the production of three times as many sisters as brothers.[1467]

A worker bias in favor of sisters would bring workers into conflict with their mothers because queens generally have nothing to gain by having their colonies produce more of one sex than the other. Why? Because, as standard sex ratio theory tells us,[1578] a queen donates 50 percent of her genes to each offspring, whether male or female, she therefore gains no genetic advantage by having more daughters than sons, or vice versa. Imagine a hypothetical population of, say, an ant species in which queens did tend to make more of one sex than the other. In this situation, any mutant queens that did the opposite and had more offspring belonging to the rarer sex would be handsomely repaid in grandoffspring. If males were scarce, for example, then a queen that used her parental capital to generate sons would create offspring with an abundance of potential mates, and thus many more opportunities to reproduce than a comparable number of daughters. The greater fitness of son-producing queens would effectively add more males to the next generation, moving the sex ratio back toward equality. If, over time, the sex ratio overshot and became male biased, then daughter-producing queens would gain the upper hand, shifting the sex ratio back the other way. When the investment ratio for sons and daughters is 1:1, there is no advantage for a son-producing specialist or a daughter-producing specialist. Therefore, an equal investment strategy is favored by frequency-dependent selection (see page 229) acting on queens.

There seems little doubt that honey bee queens, to pick one social insect, could lay eggs in a ratio that benefited their genes, not those of their worker daughters. This conclusion comes from an experiment in which queens were confined to parts of their hive where the brood comb only had smaller cells of the sort that accommodate worker grubs (her daughters). Under these circumstances, the queens laid only fertilized eggs, whose sex is female. (See also Ratnieks and Keller.[1199]) But when the confined queens were subsequently released and had access to empty cells both small and large, the queens then compensated for their recent overproduction of daughters by seeking out the larger cells, which they filled with unfertilized eggs destined to become their sons.[1548] Therefore queen honey bees clearly have the capacity to determine the sex of the offspring produced in their hives by laying unfertilized haploid eggs (sons) in large cells and fertilized diploid eggs (worker daughters) in smaller cells, although they depend upon the workers to build brood cells of the two types for both sexes of offspring.

Queens of other social insects also have control over the developmental fate of their offspring,[1470] with older queens of a harvester ant,[1295] for example, producing daughter queens and sons only after having experienced a period of winter cold. When workers were experimentally subjected to cold while the queen was not, only workers, not new queens, were produced.

But even though queens can control the sex of their eggs, perhaps workers withhold food from larval brothers in order to nourish sister larvae instead. If workers indeed attempt to maximize their own inclusive fitness, then the combined weight of all the adult female reproductives (a measure of the total resources devoted to the production of females) raised by the colony's workers should be three times as much as the combined weight of all the adult male reproductives. When Trivers and Hare surveyed the literature on the ratio of total weights of the two sexes produced in colonies of different species of ants, they found a 3:1 investment ratio, favorable to the workers, not the 1:1 ratio expected if queens were in complete control of offspring production.[1467]

## Discussion Question

**13.15** Precisely why would a worker ant gain if the colony invested three times as much in the production of reproductively capable females as males? Illustrate your answer with a case in which male and female offspring cost exactly the same amount to produce, so 100 units of investment (such as food for larvae) will yield the same number of adult males as females.

Consider the population-wide sex ratio, and explain why it might not pay workers to force their colony to produce only female reproductives, even though the workers would share many more genes with those individuals than with their brothers.

The haplodiploid hypothesis generates other predictions as well. Hymenopteran workers should bias their production of reproductives toward females only if their mother mates just once. Queens that copulate with two or more haploid males have two or more sperm genotypes with which to fertilize their eggs. Daughters with different fathers (i.e., half sisters) will not be particularly closely related. Only when females have the same father will they share 75 percent of their genes (see Figure 13.25). In reality, in some hymenopterans, queens do mate with several males, and for good reason in those species in which the more polyandrous queens produce more reproductively capable offspring.[552]

Another factor that reduces the relatedness between workers and the individuals they are helping produce is the coexistence of a number of unrelated queens in the same nest, a not uncommon phenomenon in the social insects. Both polyandrous queens and multi-queen nests can help us understand the evolution of worker behavior.[138, 1053] For example, consider a species of *Formica* ant whose queens may be either polyandrous or monogamous. Liselotte Sündstrom realized that this species provided a wonderful opportunity to find out whether workers did indeed take $r_b$ into account when allocating food to future reproductive sisters and brothers. They did. The daughters of single-mating mothers heavily biased their investment toward producing sister queens. But workers in colonies with queens that mated multiple times behaved quite differently. For them, brothers were as genetically valuable as sisters, and they did not bias the colony's production toward females.[1404]

Ulrich Mueller has also shown that worker hymenopterans alter their investment in colonymates according to their coefficient of relatedness.[1016] He experimentally manipulated colonies of a eusocial bee, removing the foundress queen from some nests but leaving her in others. When a colony has its foundress queen, the usual asymmetry in relatedness persists between workers and their sisters ($r = 0.75$) and their brothers ($r = 0.25$). Under these conditions, a bias toward investment in female progeny is expected under the haplodiploid hypothesis. But in a colony from which the foundress has been removed, a daughter assumes reproductive leadership. Under these conditions, her sister–workers are now helping her produce nieces ($r = 0.375$) and nephews ($r = 0.375$), rather than additional siblings. Thus, the relatedness asymmetry disappears, and workers ought to treat male progeny more favorably in these colonies. In fact, workers in the experimental colonies did invest more in males (the combined weight of which equaled 63 percent of the total weight of all reproductives) than did workers in colonies that retained their foundress queen (in which males constituted 43 percent of the total weight).

## Discussion Questions

**13.16** If a female of a monogamous social wasp species could help to produce more sisters with an *r* of 0.75, why would she ever reproduce personally, given that reproducers are related to their offspring by just 0.5? In addition, use Hamilton's rule to explain why some "worker" paper wasps produced early in the year leave their natal nests to wait for opportunities to "adopt" nests containing unrelated individuals, rather than becoming helpers at the nests of their mothers.[1380]

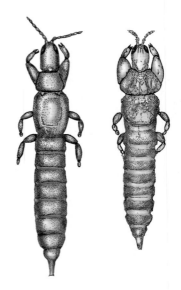

**FIGURE 13.37** A sterile thrips soldier (right) next to a reproductive foundress female (left). Note the large forelegs of the soldier, which it uses to defend its gall-occupying relatives. Drawing based on a photograph, courtesy of B. Kranz and Bernie Crespi.

**13.17** As we have seen, helpers and queens of social insects can have disputes about any number of things, despite being members of one family (e.g., Heinze, Hölldobler, and Peeters[648]). Why, for example, might a worker with a once-mated mother let her mother produce daughters but attempt to produce sons herself (assuming that in this social species, workers have functional ovaries)? On the other side of the coin, why would an unmated future queen ant be very aggressive toward egg-laying workers in the period before she mated but then back off after copulating? (Queens of some ants lay eggs in their natal colony both before and after mating.[332])

The haplodiploid hypothesis is based on the premise that exceptionally close relatedness between helper and recipient promotes the evolution of eusociality. If this is true, then other mechanisms that result in extremely close genetic relatedness among members of social groups should also be associated with sterile caste formation and extraordinary self-sacrificing behavior. Both inbreeding and clonal, or asexual, reproduction can result in very high coefficients of relatedness among family members. As mentioned earlier, anemones form clones in which some individuals sacrifice for others by developing into risk-taking scouts or aggressive warriors.[55] The same sort of thing has evolved in certain species of thrips with brother–sister mating systems; as predicted, a sterile self-sacrificing caste has evolved in some of these highly inbred species.[253] The altruistic soldiers possess enlarged, spiny forelegs (Figure 13.37), the better to grasp and stab enemies that would enter the plant galls in which the altruists live with their relatively helpless siblings.

Extreme self-sacrifice also occurs in some aphids in which mothers reproduce asexually; all their daughters get carbon copies of the same genotype, which means that the *r* value for sisters is 1.0. A number of species of asexual aphids, like the gall-forming thrips, have the ability to form clones composed of reproductively capable females and non-reproducing soldier sisters.[707, 1387] In at least one species, the soldiers aggregate around an opening into the gall, the better to repel enemies that seek to enter the gall chamber in search of the soldiers' sisters.[1139] The aphid amazons use their powerful spiny forelegs (or thickened swordlike mouthparts) to dispatch predators, such as syrphid fly larvae, when these insects approach the gall opening (Figure 13.38). Certain aphid soldiers do more than simply stab their enemies; they inject a poisonous insecticidal protein through their piercing tubular mouthparts into the body of a predator, thereby hastening its demise.[821]

In some aphid species, many soldiers die in defense of their sisters. In experiments conducted by William Foster, an average of about 20 soldiers of *Pemphigus spyrothecae* fell in battle while dispatching one syrphid fly larva. In the absence of soldiers, however, one predatory larva could eat all of the 100 non-soldier aphids that Foster brought together for his tests.[483] In additional experiments, all the aphids occupying a sample of galls were removed before being returned to their homes in reassembled groups that either included some soldiers or were without defenders. The soldierless galls were ten times as likely to be attacked by syrphids or other insect predators as those with soldiers present.[484] Thus, in nature, soldier aphids do not die in vain, because they derive an indirect fitness payoff when the beneficiaries of their actions survive to reproduce.

Because of the nature of asexual reproduction in aphids, it had always been simply assumed that apparent family groups in a gall were indeed a clone. But when Patrick Abbot and his colleagues used DNA fingerprinting technology to examine that assumption in *Pemphigus obesinymphae*, they made the surprising discovery that, on average, about 40 percent of the inhabitants of a given gall were intruders from nearby galls. These aphids had left the cottonwood leaf gall in which they were born and had moved in with the neighbors to take advantage of someone else's food resources and the protection offered by someone else's soldiers. When these colonies of mixed parentage were experimentally challenged by a pseudopredator, a fruit fly larva, aphids living in their natal colony attacked just as if the grub were a syrphid larva. But although about 40 percent of the aphids in mixed colonies were outsiders, they contributed a mere 2 percent to the soldier workforce that responded to the potential predator (Figure 13.39). Instead of helping with defense, the intruders were free to develop relatively rapidly, reaching the reproductive stage long before the more slowly maturing soldiers, which in this species move from the defensive form to the reproductive form over time. The selfish behavior of the genetically unrelated gall freeloaders reduced the benefits gained by altruistic soldiers defending their natal colony.[1]

Soldiers occur in species other than anemones, thrips, social insects, and aphids, including the very small parasitic wasps in the genus *Copidosoma*. When a female wasp of this sort parasitizes a cabbage looper, she inserts two eggs into the unlucky caterpillar. As the eggs divide over and over, they ultimately create thousands upon thousands of descendant eggs. At some point, the mass of eggs derived from one of the original eggs goes on to become a clone of genetically identical daughters of the mother wasp while the other set of eggs becomes a clone of sons. Among the daughters, all of which are genetically identical, two different phenotypes develop. One form eventually resembles the mother wasp. Individuals of this type usually mate with brothers when they metamorphose into adults and emerge from the caterpillar corpse. The other type, however, will turn into killer larvae that never reach maturity and instead seek out and destroy the offspring of other female *Copidosoma*—as well as their own baby brothers.[544]

Because brothers and sisters of *Copidosoma* developed from genetically different eggs, an indirect selectionist explanation immediately suggests itself for why female soldiers target their brothers rather than their sisters. Nonetheless, the brothers and sisters still share a substantial number of genes in common, and so there is an inclusive fitness cost associated with siblicide in this species. The counterbalancing benefit derives from the mating system of the wasp, which, as noted, involves brother–sister copulations. Because one brother can inseminate many sisters, the destruction of many brothers by the specialist soldiers does not reduce the likelihood that their non-soldier sisters will be mated effectively upon emergence. Therefore, by removing some brothers, soldiers benefit their identical sisters by reducing competition for the food contained within the caterpillar victim. If for every brother dispatched ($r = 0.5$), an extra sister ($r = 1$) can grow to maturity, a killer female experiences a net gain in genes transmitted to the next generation.

## Eusociality in the Absence of Very Close Relatedness

Although the studies described above are generally consistent with the contention that a high coefficient of relatedness can facilitate the evolution of altruism, many unresolved issues remain. Note, for example, that altruistic soldier aphids are extremely rare, appearing in only 1 percent of the thousands

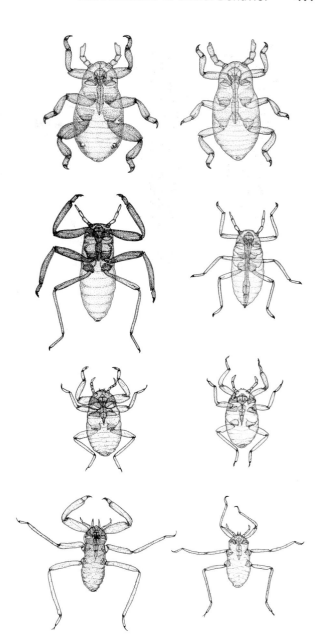

**FIGURE 13.38  Altruism in aphids.** Four species of aphids in which obligately sterile soldiers (left) with enlarged grasping legs and short stabbing beaks protect their more delicate colonymates (right), which have the potential to reproduce when mature. The species were drawn at different scales by Christina Thalia Grant. After Stern and Foster.[1388]

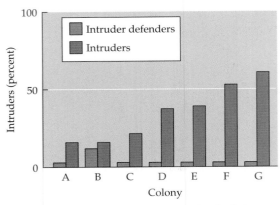

**FIGURE 13.39 Selfish behavior of aphid clone invaders.** Gall-defending aphid colonies may contain intruders that have come from elsewhere. Although these outsiders may make up a substantial proportion of the group, almost no newcomers take part in the defense of the gall. Instead, defenders are supplied almost exclusively by the colony's native population. After Abbot, Withgott, and Moran.[1]

of asexually reproducing aphid species worldwide. Even among the many gall-forming aphids, soldiers are still uncommon, having evolved in only about 50 species,[1387] generally ones whose colonies develop slowly and occupy galls that have openings through which predators can enter.[1140] As we have just seen, soldier altruism in these few aphids can sometimes persist in colonies containing unrelated intruders, a condition that greatly lowers the average relatedness of colonymates.

Similarly, among the eusocial Hymenoptera, sisters are often not especially closely related,[34] for reasons already discussed. Direct measurements of $r$ for female offspring of two species of polyandrous eusocial wasps produced mean $r$ values no greater than 0.40.[1242] Likewise, in *Polistes* wasp colonies, the average $r$ of nestmates almost never reaches the 0.75 maximum value and often falls under 0.5.[1396] These results could arise if the current queen were a fairly recent replacement for the preceding one, or if several females simultaneously contributed to brood production, or, as noted earlier, if the sole queen had mated with more than one male, all phenomena known to occur in some social insects.

The point is that the haplodiploid system of sex determination does not guarantee that workers in the colonies of current eusocial hymenopterans will be very closely related. Of course, it is possible that high coefficients of relatedness were a necessary precondition for the origins of eusocial systems long ago. William Hughes and his colleagues checked the validity of this proposition by constructing a phylogeny in which they showed the distribution of species that were highly polyandrous, slightly polyandrous, and monogamous.[691] This tree of life diagram shows that polyandry, strong or weak, evolved independently many times within the Hymenoptera but that it always did so from an apparently monogamous ancestor (Figure 13.40). Subsequent to its origins in monogamous ancestors, eusociality may have been retained even as polyandry began to be practiced in some lineages and high $r$ values were consequently lost. In the honey bee, for example, polyandry may have spread secondarily because of the benefits of having high genetic diversity within a colony to promote disease resistance[1424] or worker specialization and task efficiency,[1095] as discussed in Chapter 11 (see page 401). Thus, the highly eusocial lifestyle of the now polyandrous honey bee may currently be maintained by selection pressures that differ from those that were responsible for the origin of eusociality in this species.

Even so, we can say with certainty that the haplodiploid method of sex determination is not absolutely essential for either the origin or the maintenance of a complex social system. The termites, for example, have diploid males and diploid females, but they are as eusocial as honey bees and paper wasps. Another diplodiploid, but definitely eusocial, organism is the bizarre-looking naked mole rat.[208, 1317] This little hairless, sausage-shaped mammal (Figure 13.41) lives in a complex maze of burrows under the African plains that can house more than 200 individuals. The impressive size of their subterranean home stems from extraordinary cooperation among chain gangs of colony members, which work together to move tons of earth to the surface each year while burrowing about in

**FIGURE 13.40 Indirect selection and the origin of eusociality in the** ▶ **Hymenoptera.** In this group, there are many different eusocial species, and among these, polyandry has often evolved independently. But based on this phylogeny, the ancestral species were in every case monogamous, the condition required for there to be a high degree of relatedness among sister hymenopterans. After Hughes et al.[691]

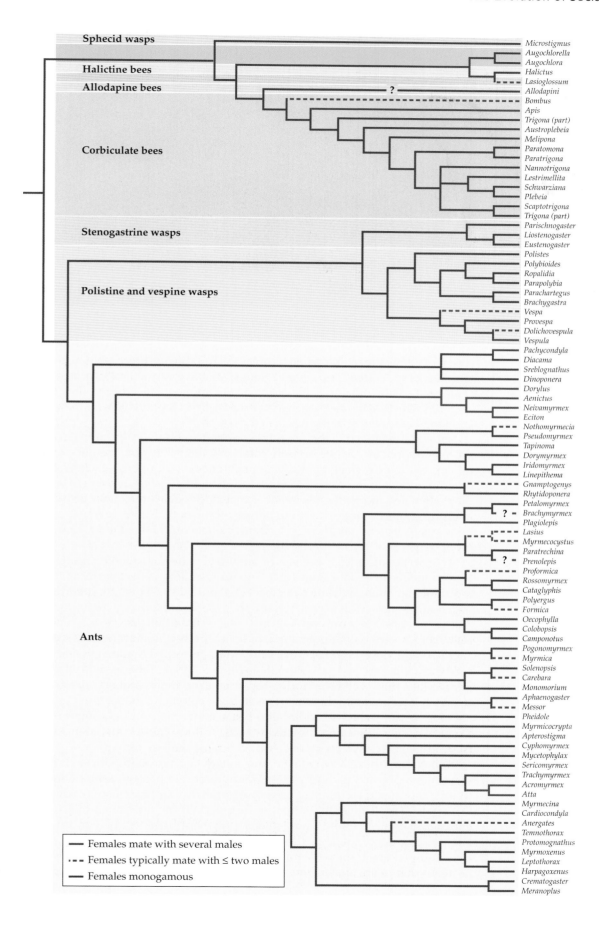

Sphecid wasps
- Microstigmus
- Augochlorella
Halictine bees
- Augochlora
- Halictus
- Lasioglossum
Allodapine bees
- Allodapini
- Bombus
- Apis
- Trigona (part)
- Austroplebeia
Corbiculate bees
- Melipona
- Paratomona
- Paratrigona
- Nannotrigona
- Lestrimellita
- Schwarziana
- Plebeia
- Scaptotrigona
- Trigona (part)
Stenogastrine wasps
- Parischnogaster
- Liostenogaster
- Eustenogaster
- Polistes
- Polybioides
Polistine and vespine wasps
- Ropalidia
- Parapolybia
- Parachartegus
- Brachygastra
- Vespa
- Provespa
- Dolichovespula
- Vespula
- Pachycondyla
- Diacama
- Sreblognathus
- Dinoponera
- Dorylus
- Aenictus
- Neivamyrmex
- Eciton
- Nothomyrmecia
- Pseudomyrmex
- Tapinoma
- Dorymyrmex
- Iridomyrmex
- Linepithema
- Gnamptogenys
- Rhytidoponera
- Petalomyrmex
- Brachymyrmex
- Plagiolepis
- Lasius
- Myrmecocystus
- Paratrechina
- Prenolepis
- Proformica
- Rossomyrmex
- Cataglyphis
- Polyergus
- Formica
- Oecophylla
- Colobopsis
- Camponotus
- Pogonomyrmex
- Myrmica
- Solenopsis
- Carebara
- Monomorium
- Aphaenogaster
- Messor
- Pheidole
- Myrmicocrypta
- Apterostigma
- Cyphomyrmex
- Mycetophylax
- Sericomyrmex
- Trachymyrmex
- Acromyrmex
- Atta
- Myrmecina
- Cardiocondyla
- Anergates
- Temnothorax
- Protomognathus
- Myrmoxenus
- Leptothorax
- Harpagoxenus
- Crematogaster
- Meranoplus

Ants

— Females mate with several males
--- Females typically mate with ≤ two males
— Females monogamous

**FIGURE 13.41** **A mammal with an effectively sterile caste.** Naked mole rats live in large colonies made up of many workers who serve a queen and one or a few breeding males. Photograph by Raymond Mendez.

search of edible tubers. Yet when it comes to reproducing, breeding is restricted to a single big "queen" and several "kings" that live in a centrally located nest chamber. Females other than the queen do not even ovulate. Instead, they serve as sterile helpers at the nest, consigned to specialized support roles for the queen and kings, as are most of the males in the colony.[825]

Is the altruism seen in this eusocial vertebrate voluntary, or is it instead the product of policing by one or more colony members? In fact, the queen mole rat appears to be the chief policewoman as she takes it upon herself to shove other members of the colony around, inducing high levels of stress in subordinate females and males. (Remember that aggressive interactions also occur in groups of birds where there are helpers at the nest, and these involve such things as destruction of the eggs of some females and forced eviction from a group territory.[637])

At a proximate level, the effects of bullying by the queen mole rat suppress the production of sex hormones in her underlings, so they become incapable of reproducing. Thus, at the proximate level, the altruism shown by subordinate naked mole rats occurs in part because they are forced to forego reproduction. At this juncture, their only options are to leave the colony to try to reproduce elsewhere (a very risky proposition) or to accept their nonreproductive status and assist the queen mother sufficiently to be permitted to remain within the safety of their group. At an ultimate level, the decision to stay with the bullying queen may well be adaptive because naked mole rat workers are very closely related to their fellow workers in those colonies in which their parents are siblings.[1202] When a brother and sister mate, their inbred offspring are likely to have coefficients of relatedness above 0.5 because both parents share rare family alleles as a result of having the same mother and father.

Nonetheless, not every mole rat king and queen are siblings, or even cousins; some reproducing mole rats evidently prefer to pair off with a nonrelative.[266] Moreover, colonies of the naked mole rat and its close relative the Damaraland mole rat both produce some especially fat individuals that apparently leave home to found a new colony elsewhere, presumably with an unrelated individual of the opposite sex from another group.[165, 1281] The fact that mole rat workers in some colonies are not extraordinarily closely related tells us again that the evolution of eusociality is not utterly dependent on excep-

(A)

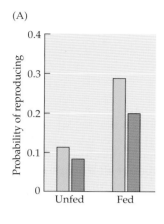

Meerkats

(B)

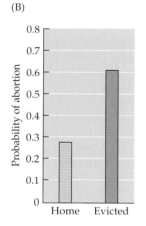

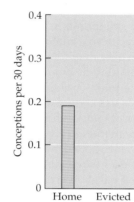

**FIGURE 13.42    Research on sociality in meerkats.** (A) The effect of supplemental food on the probability that young females (light green) and young males (dark green) would achieve some reproductive success during their lifetime. (B) Subordinate females that were temporarily driven from their band were more likely to abort their fetuses and less likely to conceive during the period of eviction than those females that avoided being evicted. A, after Russell et al.;[1257] B, after Young et al.[1631]

tionally high $r$ values. Indeed, breeding pairs of the eusocial Damaraland mole rat have a mean coefficient of relatedness of 0.02—no inbreeding here. As a result, the mean relatedness among members of the same colony is very close to 0.5, the standard coefficient of relatedness for siblings of outbred unions.[213] In the naked mole rat, the Damaraland mole rat, and many other species, helpers usually care for siblings, and that is apparently enough to accommodate the evolution of complex sociality.

## Discussion Question

**13.18**  In meerkats, another cooperatively breeding African mammal, helpers supply insects to youngsters, which helps the juveniles become heavier more quickly. In light of this result, why did a research team supply one group of young meerkats with supplemental food while refraining from doing so with another equivalent group? See Figure 13.42A.[1257] And why did researchers collect the data shown in Figure 13.42B?[1631]

## The Ecology of Eusociality

Hamilton's rule that altruism can evolve when $r_c \times C$ is less than $r_b \times B$ contains more elements than $r_b$. In fact, altruism can spread even when $r_b$ is close to zero, provided that $C$ (the number of offspring the helper gave up to be an altruist) is also very low. In other words, if the ecology of the species is such that young

**FIGURE 13.43 Fortress living space may provide an incentive against dispersal in many eusocial insects.** The probability that a young spinifex termite will found a colony that eventually builds a mound of the size shown here is exceedingly low. Millions of workers laboring over many years in Western Australia produced this safe castle made of red clay as hard as concrete. Photograph by the author.

dispersing adults have little chance of reproducing successfully, then non-dispersers that remain at their natal site to help their relatives are likely to secure sufficient indirect fitness to make helping at the nest the adaptively superior option—an argument that we presented earlier when discussing avian helpers at the nest.

For many social animals, especially social insects, the chance that a dispersing individual will ever succeed in building something like the natal colony is vanishingly small (Figure 13.43).[16] Consider that a foundress female of a eusocial ambrosia beetle takes about half a year to gnaw her way just 5 centimeters into a eucalyptus tree.[765] Most foundresses die long before they get to this point, which is only the first phase of building a tunnel complex in the tree. Once a large tunnel network is established, however, a colony can persist for decades, with helper daughters assured of a safe home in which to assist their mother in rearing reproductive males and new foundresses, a very few of which may disperse successfully, raising the indirect fitness of their stay-at-home sisters.

One of the main functions of helpers is the defense of their mother's valuable nest against predators and potential nest usurpers of their own species. Defense of a fortress nest where colony members can safely feed occurs in the ambrosia beetle just mentioned, as well as in a few thrips and many termite species. Likewise, soldier aphids tend to belong to species whose galls are unusually hard and durable.[1216] But defense of a large or long-lasting nest is not the only way in which sterile workers can increase the reproduction of their relatives. David Queller and Joan Strassmann argue that in many of the eusocial ants, bees, and wasps, the sterile workers' most important service is gathering food for their larval relatives.[1185] Whereas fortress defenders such as termites and aphids typically live amid a wealth of digestible plant material, the standard ant, bee, or wasp must roam far from the nest site in search of scarce food. In so doing, it runs a gauntlet of predators. Because mortality rates are high under these conditions, a female that lived and foraged alone would often die before her brood achieved independence. If, however, a nesting female can enlist the aid of others, care will continue to be provided for her young even if she should die prematurely. Under these circumstances, helpers related to the primary reproductive female gain considerable indirect fitness by bringing the brood along the rest of the way to adulthood.[462]

Thus, although certain genetic factors may give the evolution of altruism a boost, ecological factors that increase the positive effect of helping on the survival of relatives are equally important. Nonetheless, our understanding of complex sociality is still incomplete. For example, there are seven species of African mole rats, all of which are burrowing, parental animals. For all these species, the costs of leaving a safe underground burrow with ample supplies of food would seem to be high, and the benefits of helping great, given the value of the burrow, the care required by helpless youngsters, and the advantages of communal tunneling. Therefore, we might expect to observe eusocial life in all seven species. However, direct evidence of eusociality is available for only two species, the naked mole rat and the Damaraland mole rat (although some persons believe that the other mole rats may also exhibit some elements of eusociality).[208] More generally still, worker castes are extremely rare in burrowing rodents, of which there are many species.[713] In fact, joint occupation of a burrow by several adults has been reported for only a handful of rodents.[825] These facts pose uncomfortable questions for the argument that fortress nests promote the evolution of eusociality.[713] The use of subterranean tunnels and the associated high costs of dispersal must be only part of the ecological story behind the evolution of the eusocial mammals.

In general, it is easier to offer a tentative explanation for why a species has evolved a particular trait than for why a species has not evolved a particular trait. For example, researchers have accounted convincingly for the social life of the Florida scrub jay. But what about its close relative, the western scrub jay, a nonsocial species? Why haven't members of this species become social? There must be years when young western scrub jays have little chance of finding a suitable vacant territory. Why haven't they evolved the ability to remain as helpers at the nest under these conditions? Much more remains to be learned about the genetic and ecological bases of altruism and social living before we can close the book on this great evolutionary puzzle.

## Summary

1. In animal societies, individuals often tolerate the close presence of other members of their species despite the reproductive interference, increased competition for limited resources, and heightened risk of disease that social living entails. Under some ecological circumstances, the advantages of sociality (often improved defense against predators) are great enough to outweigh the many and diverse costs of social living. The common view that social life is always evolutionarily superior to solitary life is incorrect.

2. Animals that live together may help one another in various ways. Some cooperative acts may immediately elevate the personal reproductive success of both cooperators (mutualism). Still others may be performed at some cost that will be more than repaid when the recipient pays the helper back at a later date (reciprocity). Finally, some helpful actions are considered altruistic because they reduce the reproductive success of the helper while at the same time raising the reproductive output of another individual.

3. Mutualism and reciprocity can spread through a population via the action of direct natural selection. If, however, a helper really does permanently reduce its direct fitness while helping to raise the fitness of another, its altruism poses an evolutionary puzzle. One solution may be that the direct fitness costs of certain kinds of altruism are outweighed by the indirect fitness gains generated when an individual increases the number of its surviving nondescendant kin.

4. An allele for altruism can spread in competition with an alternative form of a gene that promotes personal reproduction, provided that altruistic individuals increase their relatives' reproductive success enough that the indirect fitness they gain more than compensates for any reduction in their direct fitness. As expected, the overwhelming majority of cases of altruism found in nature fit this description. Self-sacrificing helpers, which may even be sterile, almost always assist close relatives, thereby boosting their inclusive fitness (the sum of their indirect and direct fitness).

5. Although the indirect fitness gained by helping is increased if the coefficient of relatedness between helper and beneficiary is high, sterile castes can evolve even when the degree of relatedness is not great among group members, provided that the ecology of the species is such that helpers can improve the direct fitness of relatives, especially if the odds are against successful personal reproduction by individuals dispersing from a safe natal nest. But many questions about extreme altruism remain unanswered.

## Suggested Reading

W. D. Hamilton's work[609] initiated a revolution in the understanding of social behavior; see also reviews by Richard Alexander,[16] Mary Jane West-Eberhard,[1541] and Steve Emlen.[441] I have relied heavily on Jerry Brown's *Helping and Communal Breeding in Birds* as a guide for understanding the kinds of selection that affect the evolution of social behavior.[192] David Sloan Wilson and Edward O. Wilson argue that group selection theory must replace kin selection theory if we are to understand the evolution of social behavior. Although some disagree (e.g., Foster et al.[481] and West, Griffiths, and Gardner[1546]), you can read about the Wilsons' position in "Rethinking the theoretical foundation of sociobiology."[1586]

For most of us, "the social insects" is synonymous with "the social bees, wasps, and ants" (see Bourke and Franks,[154] Hölldobler and Wilson,[669] Michener,[979] and Wilson[1587]), but there are other social invertebrates as well. Their behavior and its evolutionary basis are explored in *The Other Insect Societies* by James Costa.[302] The social behavior of birds has been written about extensively; for examples, see Uli Reyer's study of the social pied kingfisher,[1212] Jan Komdeur and colleagues' work on the Seychelles warbler,[792, 796] and Walt Koenig and Ron Mumme's review of acorn woodpecker sociality.[787] Mammalian social behavior has also attracted much attention: see Packer[1090] on lions and Wolff and Sherman[1613] on rodents, as well as Sherman, Jarvis, and Alexander[1317] on naked mole rats in particular.

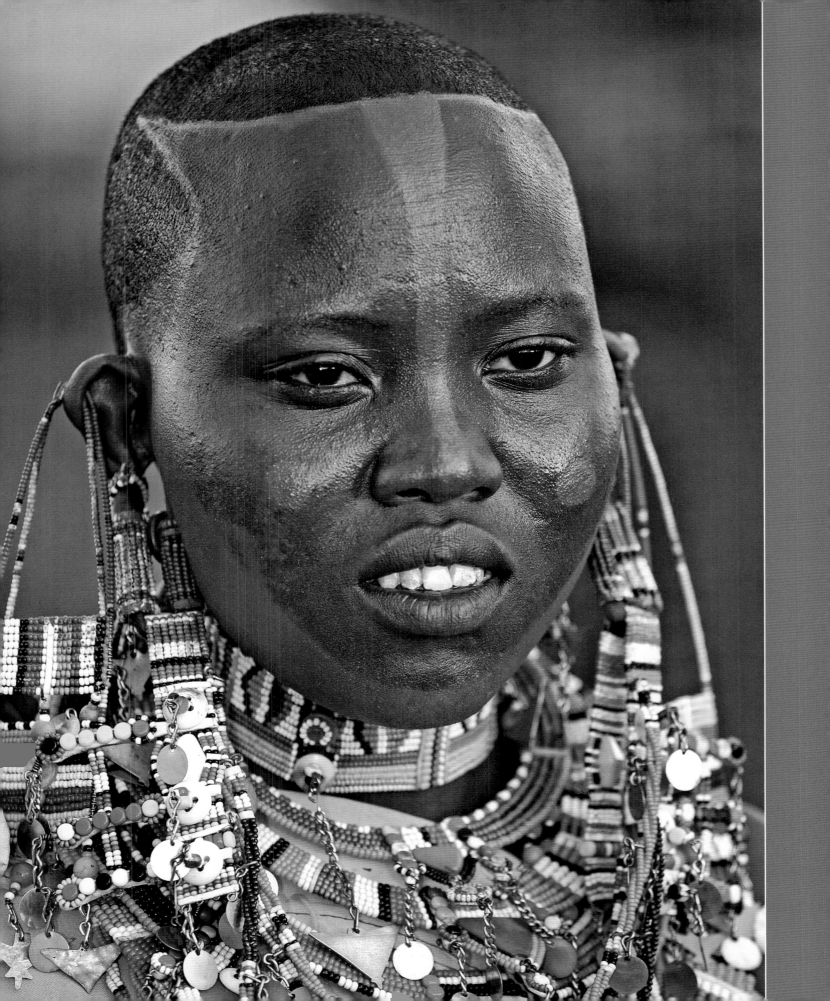

# 14

# The Evolution of Human Behavior

Humans are an animal species with an evolutionary history. Yes, we are unusual and quite wonderful in our own way, but so are kittiwakes and hangingflies. Having applied evolutionary thinking to kittiwakes and hangingflies, let's see if we can do the same for humans. If our evolution has been shaped by natural selection, then there should be some connection between our behavior and the capacity to pass on our genes to the next generation. This proposition has, however, been rejected by those who believe that our behavior is essentially "cultural" as opposed to "biological." And, of course, our cultural traditions do influence our behavior. If, for example, I had been born and reared by tribesmen in Papua New Guinea, I would consider it natural to be seen in public completely naked except for a long, thin, hollow gourd covering my penis (Figure 14.1). Having not grown up in Papua New Guinea, I have no desire whatsoever to wander around in a similar outfit here in Tempe, Arizona, or anywhere else, for that matter. The obvious differences in human behavior in cultures around the world might lead us to think that we cannot use evolutionary theory to help us understand our species' behavior. But because human cultures are the product of an evolved brain, an understanding of our evolution could cast light on why we do what we do, whether in New Guinea

◄ **This Masai woman** *follows the unique traditions of her culture in her choice of ornaments. Photograph by Martin Harvey.*

or New York. Therefore, we shall use natural selection theory to tackle some of the Darwinian puzzles associated with our behavior. Our first step is to analyze why I have on occasion given blood to a blood bank. Could my donation really have anything to do with human evolution?

## The Adaptationist Approach to Human Behavior

When I gave blood, I was not alone. Those of us hooked up to our little plastic blood bags were behaving altruistically because we were taking a tiny risk of a medical complication in order to provide a substantial fitness benefit to an unknown recipient in need of a transfusion. You may recall from the preceding chapter that one ultimate hypothesis to account for helpful, self-sacrificing acts is that they generate indirect fitness for the helper. Blood donation, however, cannot produce such gains when the helper and the recipient are not related, as is usually the case. Moreover, the action can hardly be a standard case of reciprocity with eventual repayment to the donor, given that most people who receive a blood transfusion do not even know who the donor was.

Some persons have claimed, therefore, that blood donation is immune to evolutionary analysis and is instead a kind of "pure" altruism practiced for no possible fitness advantage. According to Peter Singer, "Common sense tells us that people who give blood do so to help others, not for a disguised benefit to themselves."[1335] Maybe so, but Richard Alexander disagreed with Singer. Perhaps, he said, the donor is repaid, not by the recipient of his blood, but by the donor's everyday companions, who may be impressed by, and wish to cooperate with, someone "so altruistic that he is willing to give up a most dear possession for a perfect stranger."[18]

In Alexander's view, blood donation occurs at the proximate level because our psychological mechanisms make us feel better when we do any of a wide range of conspicuous good deeds, perhaps especially for people unrelated to us. According to Alexander, our brains must have a special system that rewards us for certain actions as a means of encouraging us to behave adaptively. Such a mechanism would be similar to the human brain's face recognition center, which, as you may recall (see pages 135–136), gives us the adaptive ability to identify individuals visually very quickly and accurately. The putative "cheap altruism" mechanism in our brains motivates some of us in Western societies to give blood once in a while or to exhibit other low-cost forms of culturally approved charity. In the past, when our psychological systems were evolving, our hominid ancestors could not donate blood or give used clothing to the Salvation Army, but they could potentially help their companions in various small ways. In a social environment in which an individual's reproductive success surely depended on building and maintaining good relationships with other people, those persons who performed visible acts of charity, which everyone knew were unlikely to be repaid, may have acquired a reputation for generosity, stimulating others to join them in beneficial alliances and thereby generating a fitness payoff for their behavior.

This **indirect reciprocity** hypothesis for blood donation, according to which the donor gains fitness from others who see his or her good deed, yields many

**FIGURE 14.1 Cultural traditions are powerful influences on human behavior.** Men in many parts of New Guinea, including this Dani man from Irian Jaya, have traditionally worn large penis sheaths, but little else.

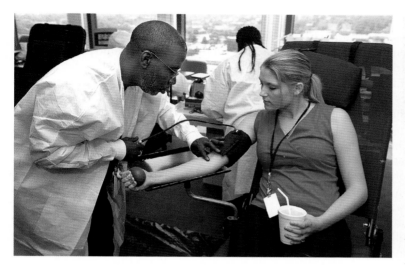

**FIGURE 14.2    Blood donors often advertise their altruism.** They may sometimes expect (and receive) a reputational reward for their behavior.

testable predictions. For one thing, unpaid blood donors should almost always let their friends know that they have given blood. Bobbi Low and Joel Heinen report that students at the University of Michigan are significantly more likely to donate to fund-raising drives if they receive a pin or tag that advertises their participation. Such devices are routinely distributed by the American Red Cross during their blood drives (Figure 14.2).[890]

A second prediction is that people will be reluctant to broadcast their refusal to give blood. In one study in Germany, 34 percent of those interviewed said that although they had never given blood, they were nevertheless willing to do so, whereas only 17 percent said that they simply would not be donors.[1223] In a similar project in the United States, 30 percent of the students given a survey form checked a box indicating that they probably would not give blood in a drive to be held in the following week; the other 70 percent left the box blank, which meant, according to the instructions they received, that they probably would give blood the next week. In reality, a mere 17 percent of the students actually followed through, showing that many of those who failed to check the "not a donor" box did not wish to acknowledge, even on an anonymous survey form, that they were not going to participate in the blood drive.[265]

A third prediction is that when people have a choice of other people to help, they will prefer to assist those who are known to be generous.[1039] To test this prediction, Klaus Wedekind and Manfred Milinski devised an experimental game that involved many rounds of giving and (potentially) receiving. Players were supplied with a certain amount of cash at the outset of the game and were offered the opportunity to give some of it to other players. A donor could specify that a recipient, unknown to the donor, would receive one Swiss franc (or two), in which case the recipient actually received four francs, with the extra money coming from the experimenters. Thus, the donor paid a small price in order to give the recipient a larger benefit. As each player made his decision whether to give up a franc or two, he knew the past record of giving by each potential recipient, but again, he did not know the identity of this individual, nor did he know whether this person had given him money in a previous round. Even so, individuals with a record of giving were more likely to receive cash donations from the other players, as predicted by the indirect reciprocity hypothesis for altruism.[1526]

Milinski and some others have used a similar kind of game to look at the effects of giving to charity. They found that individuals known (under a pseudonym) to donate both to members of their group of game players and to UNICEF (the highly regarded children's fund of the United Nations) not only received more money in return from their fellow players, but also received more votes in a student council election (admittedly staged by the experimenters). In other words, public donations appeared to boost the social reputation of the generous person, making it more likely that others would believe that he or she had the best interests of the group in mind and so was deserving of financial and political support.[981]

Yet another variant on this theme involved groups of undergraduates who attended a class in which a professional charity organizer discussed opportunities to help the charity that she represented. In some groups, students were permitted to indicate their willingness or refusal to contribute in front of their fellow students, while in others, students filled out a questionnaire where they could indicate their response to the organizer without letting the other students know their intentions. A significantly higher fraction of the students in the "public groups" made charity offers, compared with those who responded in groups where they did not have to reveal their decisions publicly. Moreover, when the students were asked at the end of the class period to score their fellow class members in terms of such things as their trustworthiness, something they had done at the start of the class as well, the scores of the "generous" students went up substantially, unlike those who had made no offer—but only in the groups where the responses of the participants were public knowledge.[107]

One potential reward for being visibly generous may not just be in developing a good reputation, but also in directly advertising one's value to members of the opposite sex. Because women are typically interested in the resources controlled by prospective mates (see below), unmarried or potentially polygynous men are predicted to say that they would give money to charity organizations, provided that female observers are made aware of their generosity. A team of social psychologists found that undergraduate men were indeed free with the imaginary funds made available to them, assigning much larger amounts to charitable entities than the young women who were also participants.[585] Similar games have revealed that charity can be much more self-serving than is often suspected, provided that donors can be identified and their altruism compared by observers.[76, 621]

---

**Discussion Question**

**14.1** Almost all of us have, at one time or another, been approached by a person in search of a donation who gave us something, a pamphlet or pen or ribbon, prior to or during his or her request for money.[264] Why, in evolutionary terms, is this such a common tactic by those in search of charity?

---

### The Sociobiology Controversy

The studies I outlined above have not established a direct link between acquiring a reputation for generosity and an increase in reproductive success. Indeed, it would be difficult to do so in the modern world, given the access we have to effective birth control. In order to test the prediction that indirect reciprocity confers fitness benefits on generous individuals, we would presumably have to examine the relationship between these variables in a traditional society, one in which people live in relatively small groups where everyone's behavior is under constant scrutiny and where no one has birth control pills or condoms. But I have no doubt that even if studies of this sort showed that persons whose reputations had been enhanced by their generosity to nonrelatives did indeed have higher fitness, a great many people would still object to the conclusion that our capacity to be nice is the product of natural selection. Even W. D. Hamilton, whose research on inclusive fitness was founded on the principle that individuals were expected to behave in the interests of their genes, wrote that he had a "dislike for the idea that my own behaviour or the behaviour of my friends illustrates my own theory of sociality or any other. I like always to imagine that I and we are above all that, subject to far more mysterious laws."[612]

While Hamilton exhibited some discomfort with evolutionary analyses of human behavior, others have been downright hostile, as illustrated by the response to the 1975 publication of E. O. Wilson's *Sociobiology: The New Synthesis*.[1588] This book's final chapter on human behavioral evolution generated a furious controversy led by some of Wilson's own colleagues at Harvard University, who accused him of promoting a discredited but still dangerous view of human behavior.[21] Wilson and others effectively answered these charges,[1589] and now, many years later, human **sociobiology** is flourishing under the labels of human behavioral ecology, **evolutionary psychology**, and evolutionary anthropology.[14] Although this controversy has faded, one sometimes still hears the very same criticisms that Wilson's opponents published in 1975. Here are three of these complaints, with a rebuttal for each one.

***"We humans don't do things just because we want to raise our inclusive fitness."*** Some opponents of evolutionary studies of human behavior have pointed out that, although humans desire a great many things, the wish to maximize our inclusive fitness rarely, if ever, is at the top of our list.[1238] If you had asked Picasso why he wanted to paint, or Bill why he wanted to marry Jane, Picasso and Bill would not have explained that they wanted to increase their genetic success. But if a baby cuckoo could talk, it would not tell you that it rolled its host's eggs out of the nest "because I want to propagate as many copies of my genes as possible." Neither cuckoos nor humans need be aware of the ultimate reasons for their activities in order to behave adaptively. The human brain's costly decision-making mechanisms were shaped by natural selection to enhance our fitness, not to provide us with the capacity to monitor the reproductive consequences of each and every action. It is enough that proximate mechanisms, like a well-developed sex drive, motivate individuals to do things, like copulate, that are correlated with fitness—the production of offspring. On the proximate level, we have sex, we enjoy sweet foods, and we derive satisfaction from our charitable actions because we possess physiological mechanisms that facilitate these behaviors. Because honey tastes good, we want to eat it, and when we do, we acquire useful calories that may contribute to our survival and reproductive success without our ever being aware of the evolved function of our fondness for sweets.

***"But not all human behavior is biologically adaptive!"*** Over the years, critics have claimed that an array of cultural practices, such as circumcision, prohibitions against eating perfectly edible foods, and the use of birth control devices to limit fertility, seem unlikely to advance individual fitness. If some humans do things that reduce their fitness, these persons argue, then sociobiology cannot tell us anything about human behavior. Note, however, that the claims of these critics are based on the assumption that natural selection theory requires that every aspect of every organism be currently adaptive.[557] This assumption is incorrect, as noted before (see Table 6.1). The adaptationist approach is used by persons who want to identify interesting puzzles, produce plausible hypotheses, and test alternative explanations. If, for example, an adaptationist were to examine religiously motivated celibacy, he or she would be fully aware of the possibility that the trait was a maladaptive by-product of certain brain modules that control other generally adaptive abilities. Nonetheless, a sociobiologist might still try to produce a testable hypothesis on how acceptance of celibacy could paradoxically enable priests to leave more copies of their genes than if they were heterosexually active. Needless to say, this would be a challenge, but perhaps not an insuperable one for someone aware of indirect selection.

Even if 20 adaptationist hypotheses on celibacy were developed, there is no guarantee that any would withstand testing. This is as it should be. T. H. Huxley, the great defender of Darwinian theory, wrote, "There is a wonderful truth in [the] saying [that] next to being right in this world, the best of all things is to be clearly and definitely wrong, because you will come out somewhere."[703] If our sociobiological hypotheses about the celibate priesthood were incorrect, effective tests would tell us so, which would enable us to remove some ideas from the table.

*"Evolutionary approaches to human behavior are based on a politically reactionary doctrine that supports social injustice and inequality."* The original critics of sociobiology denounced it as politically dangerous because they feared it would provide scientific cover for immoral social policies of the sort advanced by racist and fascist demagogues in the past.[21, 22] According to this view, the claim that such and such a trait is adaptive implies that it is both genetically determined and good, and therefore cannot and should not be changed. After all, if one claims that male dominance is adaptive, isn't this saying that the status quo is desirable and that feminist claims to the contrary fly in the face of what is genetically fixed and morally necessary? The implication of the critics was that sociobiologists were right-wing types who let their conservative politics affect their "science." The persons making this claim never made any effort to test this charge, which has only been done recently by a team of researchers at the University of New Mexico. These persons surveyed the political attitudes of a large group of doctoral candidates in psychology, 31 of whom said they were enthusiastic about sociobiology versus 137 others who identified themselves with one or another nonadaptationist theory of behavior. The two groups did not differ overall in their political positions, which across the board were far more liberal than that of the average American.[1475] Admittedly this is just one small study, but it offers no support for the contention that academic sociobiologists are distinctively reactionary in their political positions.

But even if the original sociobiologists did not intend to block progressive social policies, could their research have been used by malevolent nonscientists? Perhaps. Darwin's theory of evolution has been misunderstood and misused by some persons to defend the principle that the rich are evolutionarily superior beings, as well as to promote unabashedly racist plans for the "improvement of the human species" by selective breeding. We can hope that political perversions of evolutionary theory have been so discredited that they will not happen again. The critical point here, however, is that sociobiologists attempt to *explain* why social behavior exists, not to justify any particular trait. This distinction is easily understood in cases involving other organisms. Biologists who study infanticide in Hanuman langurs or how a small marine copepod feeds on the eye of the Greenland shark are never accused of approving of infanticide or the blinding of sharks. To say that something is biologically or evolutionarily adaptive means only that it tends to elevate the inclusive fitness of individuals with the trait—nothing more.

Moreover, a hypothesis that a behavioral ability is adaptive does not mean that the characteristic is developmentally inflexible. All sociobiologists understand that development is an interactive process involving both genes and environment. Change the environment, and you will change the gene–environment interactions underlying a behavioral phenotype, with the result that the phenotype may change. A classic example in human biology involves language acquisition, a clearly adaptive ability that rests on vast numbers of extraordinarily complex gene–environment interactions. Change the cultural environment that a baby is exposed to so that, say, a child of English-speaking parents is raised by adults who speak Urdu, and we all know what will happen. But if infants are to adopt the local language, they need very special genes to do so, especially those genes that code for proteins that promote the development of the brain modules that underlie language learning. These neural units have evolved specifically in the context of language acquisition, rather than as a side effect of some sort of generalized intelligence, as we can see from the existence of two rare human phenotypes. On the one hand, certain profoundly retarded individuals chatter away, producing completely grammatical sentences with little inherent meaning. On the other hand, some English-speaking individuals with normal to above average intelligence have great trouble with the rules of grammar, often failing, for example, to add *ed* to verbs when they wish to speak of past events.[1143] This kind of evidence tells us that selection is behind the evolution of the special mechanisms that make language acquisition easy and effective for the vast majority of us.

## Discussion Questions

**14.2** In 1981, the first vocational school for deaf children was opened in Nicaragua. Although the children at the school had never been taught a sign language, they invented one of their own that became increasingly complex.[1308] This unique language has many of the fundamental properties of any language, including the breakdown of information into discrete units and a grammatical presentation of words (gestures in this case). What is the significance of this case study for discussions about the evolutionary basis of language-learning ability in human beings?

**14.3** Philip Kitcher states that "socially relevant science," such as sociobiology, demands "higher standards of evidence" because if a mistake is made (a hypothesis that is presented as confirmed when it is false), the societal consequences may be especially severe. For example, a hypothesis that men are more disposed to seek political power and high status in business and science than women is dangerous because it "threaten(s) to stifle the aspirations of millions."[778] How do you suppose a sociobiologist would respond to Kitcher's claim?

## Arbitrary Culture Theory

One popular alternative to an evolutionary approach to human behavior is the theory that human cultural traditions arise from accidents of history and the almost unlimited inventiveness of the human mind. According to this theory, most of our activities have little or nothing to do with fitness maximization but, instead, reflect the more or less arbitrary processes by which traditions have originated and been passed down from generation to generation.[1144] The contrast between the adaptationist and the arbitrary cultural approaches can be illustrated by the differences between sociobiologists and cultural anthropologists in their analysis of adoption, which is common in some societies but much less so in others. At one time, for example, an amazing 30 percent of all children became adoptees in Oceania (the islands of the central Pacific Ocean), leading to the claim that this peculiar feature of life in this part of the world is a cultural anomaly immune to evolutionary analysis.[1271] However, when

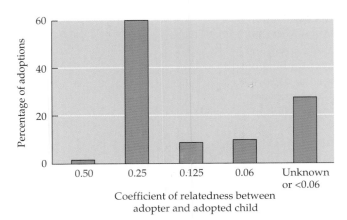

FIGURE 14.3 **The indirect fitness hypothesis for adoption** can be tested by examining the coefficient of relatedness between adoptive parents and their adopted children. In 11 island societies in Oceania, adopter and adoptee were usually close relatives, yielding indirect fitness gains for the adopters. After Silk.[1329]

Joan Silk analyzed data on the relationships between moderately large samples of adopters and adoptees in 11 different cultures in Oceania,[1329] she found that most adopters were actually caring for their close relatives (Figure 14.3). The highly kin-biased nature of adoption in these societies casts doubt on the **arbitrary culture** hypothesis while supporting an indirect fitness hypothesis for this form of human altruism.

Interestingly, 96 percent of 400-plus adults recently surveyed in the Marshall Islands, a small country in Oceania, felt that adopted children should be raised by their relatives. When asked if they would be willing to adopt, 69 percent indicated a readiness to adopt related children, while far fewer (41 percent) said that they were prepared to take in a nonrelative.[1231]

The sensitivity of people to their relatedness to those they might help goes well beyond a simple division of others into "relatives" versus "nonrelatives." Joonghwan Jeon and David Buss made this point by asking about 200 undergraduates to evaluate their willingness to help four different kinds of cousins: father's brother's children, father's sister's children, mother's brother's children, and mother's sister's children. (Before reading more, you might want to evaluate your own willingness to help with respect to these four groups.) To simplify things, Jeon and Buss told their undergraduate subjects to assume that mothers and their siblings, and fathers and their siblings, are indeed full siblings. Jeon and Buss realized that a potential altruist had the greatest possibility of not being an actual cousin of the child of an uncle on the father's side because (1) the altruist's father might not really be his or her father (due to an extra-pair mating by the mother) and (2) the altruist's cousin might not really be the uncle's offspring (because the uncle's wife had an extra-pair partner). In contrast, a potential altruist dealing with his or her mother's sister's children is certain to be related to these cousins (given the assumption made above). The certainty of maternity is always higher than the certainty of paternity. The other two categories of cousins are intermediate in probable relatedness because in each case there is one link in which a putative father might not be the actual genetic father, if his mate had engaged in copulation outside the pair bond.[722] As predicted, the willingness to help the four categories of cousins was proportional to the probability that they were full cousins (Figure 14.4).

Although people tend to favor genuine relatives when it comes to behaving altruistically, sometimes we behave generously toward nonrelatives without apparent regard to personal benefit. Thus, a minority of adopters in Oceania in the past and in the present have taken in the children of strangers, their genetic competitors, and some have offered these children the same love and affection that parents typically supply to their own offspring. Are these exceptions to the typical adoption pattern in Oceania impossible to explain from a sociobiological perspective? Not necessarily. Silk suggests that small families in some agricultural cultures might benefit from gaining adoptees, even if they were nonrelatives, because adopted persons could contribute to the family workforce, raising the economic productivity of the family unit and improving the survival chances of the adopters' genetic offspring. This direct fitness hypothesis produces the prediction that small families in Oceania should be more likely to adopt than large ones, a prediction that Silk showed was correct.

An alternative evolutionary hypothesis for adoption among nonrelatives, which occurs in many other societies besides those in Oceania, recognizes

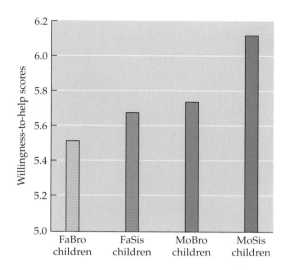

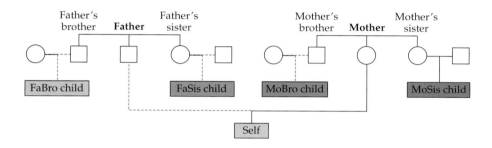

**FIGURE 14.4  The probability of being related to others affects the willingness of people to help their cousins.** Potential altruists are most likely to be fully related to their aunt's children on their mother's side of the family, and respondents say that these individuals merit more assistance than the other three categories of cousins. In these other types, the chance that the mother's brother, or the altruist's own father, or both the altruist's father and his brother have been cuckolded reduces the average degree of relatedness between putative cousins. The dashed lines in red show cases in which the relatedness between "self" and a cousin could be lowered if the indicated parent's spouse had engaged in extramarital reproduction. After Jeon and Buss.[722]

that some decisions may be the maladaptive by-product of otherwise adaptive proximate mechanisms. Adoption of a nonrelative might, for example, be a consequence of motivational systems that cause adult humans to want to have children and raise a family. According to this hypothesis, although adults who adopt infant strangers may reduce their fitness, the urge to have and care for children is often adaptive because it is usually applied to one's genetic offspring. Because such psychological mechanisms tend to elevate fitness, they are maintained in human populations even though they sometimes induce people to behave maladaptively.

The point is that in the past, selection has favored individuals whose behavior has been guided by desires that steered them toward reproductive success. As noted earlier, these proximate desires, such as a strong sex drive and eagerness to have a family, are what really control our behavior, not dispassionate mathematical calculations of the fitness consequences of various options open to us. Therefore, persons unable to satisfy their proximate wishes will sometimes do things that do not boost their inclusive fitness, as when a couple unable to have a family adopts a child who is not genetically related to them.

This side-effect hypothesis for adoption generates testable predictions, one of which is that husbands and wives who have lost an only child, or who have been unable to produce children themselves, should be especially likely to adopt nonrelatives. In fact, infertility is by far the most common reason given by Californian couples who want to adopt; moreover, women who have been treated for infertility are significantly more likely to attempt to adopt than are women who have not been treated for this condition.[466]

**FIGURE 14.5** **Adoption occurs in nonhuman animals,** often when adults have just lost an offspring but encounter a substitute. Here several emperor penguins compete for "possession" of a youngster. Photograph by Kim Westerskov.

Another prediction from this side-effect hypothesis is that adoption should also sometimes occur in nonhuman animals when adults have lost their offspring and fortuitously encounter a substitute. Over 80 percent of the king penguins observed feeding others' youngsters were failed reproducers, having lost their eggs or chicks during the breeding season;[848] members of other penguin species that have lost chicks are also eager adopters (Figure 14.5).[740]

The likelihood of a maladaptive application of the parental drive may be especially high in modern human societies because these environments are so very different from the ones in which our proximate psychological mechanisms evolved. In Western cultures, babies are routinely made available to nonrelatives who do not know the parents of the adoptee, something that almost never occurred in the distant past. Under these novel conditions, our motivational and emotional systems, which evolved long ago, are especially likely to cause us to behave maladaptively and, in so doing, reveal something about the naturally selected features of our psyches.

The abundance of testable sociobiological hypotheses on adoption (and we have not discussed the possibility that adoption can be a form of charitable behavior that raises the social reputation of the adopters) speaks to the productive nature of evolutionary theory. No one evolutionary hypothesis explains every case of adoption, just as no one hypothesis resolves every aspect of a complex anatomical or physiological phenomenon. However, the adoption example demonstrates that an evolutionary approach can more than hold its own against the competing arbitrary culture theory.

## Discussion Questions

**14.4** Marshall Sahlins has argued that sociobiology is contradicted because people in most cultures do not even have words to express fractions. Without fractions, a person cannot possibly calculate coefficients of relatedness, and without this information (Sahlins claims), people cannot determine how to behave in order to maximize their indirect fitness.[1271] Has Sahlins delivered a knockout blow to sociobiological theory?

**14.5** A common view in some circles is that the behavioral differences between men and women stem largely from the effects of cultural pressures to conform to stereotypical male and female roles. Thus, young boys receive certain toys to play with, such as guns and miniature trucks, and young girls get different toys, such as dolls and baby carriages. The sex stereotyping involved in the selection of toys for boys and girls is said to push boys into "masculine roles" while guiding girls into culturally approved "feminine roles." Two researchers gave young male and female monkeys both types of toys and measured the amount of time they spent with each kind of toy. Why did they do this? What prediction based on an arbitrary culture hypothesis were they able to test? Check their results in Alexander and Hines.[15]

## Cultural Evolution Theory

In addition to the adaptationist approach and its polar opposite, arbitrary culture theory, another kind of analysis of human behavior is available based on cultural evolution theory.[1220] Adherents of this approach accept that human behavior rests on many adaptive proximate psychological mechanisms, but they also are struck by the very large differences among the behaviors exhibited in different cultures, an indicator of the power of cultural influences on our behavior. They also argue that cooperation within large groups of people requires an explanation that the standard adaptationist approach cannot provide, particularly with respect to actions that seem both irrational and maladaptive. So, for example, when individuals are asked to engage in experimental games where they have an opportunity to accept or refuse a sum offered to them by a co-player who has received, say, ten dollars to play with, people often reject rewards that they consider unfairly low, say a dollar instead of three or four. In so doing, the rejecters deny the other player the remaining portion of the ten dollar amount and, in effect, cut off their noses to punish the other person for his stingy offer. A rational player would take whatever was offered and in this way come out ahead of those who were willing to forego cash in order to punish greedy players (see review in Gaulin and McBurney[526]).

Indeed, most people possess a very strong sense of what is just and what is not, and these individuals are often keen to impose sanctions on those who behave unfairly.[458] Such actions exact a price from the persons doing the policing for the benefit of others (just as is true for those social insects in which some females take it upon themselves to force others to conform to a particular role[1538]). Adherents of cultural evolution theory argue that actions of this sort need a special explanation, as can be supplied, for example, in terms of an evolutionary competition among cultures with different traditions.[650] Those traditions that foster effective group action by, for example, controlling selfish, self-centered, or antisocial individuals have a better chance of surviving the process of cultural selection. These traditional rules can shape the behavior of people so strongly that they can even induce maladaptive responses in strictly Darwinian terms.

Cultural evolution theory has strong advocates who wish to analyze human behavior, especially as it applies to economics and morality.[543, 977] This chapter, however, will remain focused on the adaptationist approach, in part because I believe that irrational human behavior is not in and of itself reason to abandon adaptationism. For example, costly policing by persons eager to punish social transgressors could yield hidden benefits for the moralists in question by demonstrating to others that these persons are willing to pay a price to uphold *all* of their views and beliefs, some of which may well be personally advantageous to the policers and their relatives. Such an adaptationist argument parallels that made to account for costly religiously motivated behaviors, like self-mutilation during religious celebrations.[206, 1373] The fanatically religious individual identifies himself as someone willing to suffer for his beliefs; such a person is advertising that he can be counted on by others of a similar orientation, the better to join his coreligionists in mutually beneficial endeavors. In other words, some costly actions may have subtle but large fitness benefits for individuals, which, if true, would mean that the adaptationist approach suffices for an ultimate explanation for these actions, no matter what their proximate (i.e., cultural) basis.

---

### Discussion Question

**14.6** Use the maladaptive by-product approach to explain why persons will give substantial tips to waiters and waitresses whom they do not know and whom they will never meet again. Use the same argument to explain why persons enlisted to play in games by cultural evolutionists will often behave in a manner counter to their economic interest even when they do not know who their co-players are or how these individuals have behaved with respect to, say, monetary offers made to others in these games.

---

## Adaptive Mate Preferences

Having examined some adaptationist hypotheses on donating blood and adopting children, let's examine the application of the approach to human reproductive behavior. Tackling this component of our behavior is a challenge because the cultural rules and regulations surrounding human sexual behavior are extraordinarily diverse. There are monogamous, polygynous, and polyandrous societies, some in which you cannot marry an unrelated person who belongs to your clan, others in which adult men can marry prepubescent girls, some in which males and their relatives provide payment for a bride, and others in which women must bring a costly dowry with them to their wedding.

Despite all this cultural variety, certain biological facts of life still apply to our species as a whole. For one thing, women are typical female mammals in that they retain control of reproduction by virtue of their physiological investments in producing eggs, nurturing embryos, and providing infants with breast milk after they are born. Although men are also able and often willing to make large parental investments in their offspring, their reproductive decisions nevertheless take place in a setting defined by female physiology and psychology.[338, 527] For example, consider the reproductive significance for men of variation in female fertility, which is a function of a woman's age, health, and body weight, among other things. Preadolescent and postmenopausal women obviously cannot become pregnant. Women in their twenties are more likely to become pregnant than women in their forties. Healthy women are more fertile than sick ones. Women that are substantially overweight or underweight are less likely to become pregnant than women of average weight.[1458]

Given the reality that women differ in their likelihood of conception, many evolutionary biologists have predicted that men should possess psychological mechanisms that enable them to accurately evaluate female fertility. One way to check this prediction is to look for a positive correlation between female fertility and what men consider to be "good looks." Note that the evolutionary expectation here is very different from a frequently cited alternative explanation for male preferences, which is that men and women have been culturally indoctrinated into acceptance of a battery of social norms. These cultural pressures are often said to set nearly impossible standards of feminine beauty as a means of keeping most women on the defensive about their looks and age.[1610]

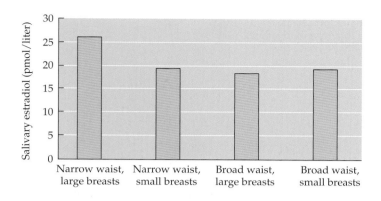

**FIGURE 14.6   Body shape is correlated with fertility in women.** Women with features that men often claim they prefer—namely, narrow waists and large breasts—are more fertile than women with other body shapes, based on a sample of healthy Polish women between 24 and 37 years of age who were not taking birth control pills. After Jasiénska et al.[714]

So, are standards of beauty in Western society essentially arbitrary, perhaps even imposed on women by malicious men, or do they reflect a largely unconscious, evolved male interest in the fertility of potential mates? In Western cultures, males generally prefer the following feminine features: full lips, small noses, large breasts, a waist circumference that is substantially smaller than hip circumference (the hourglass figure),[1336] and an intermediate weight rather than extreme thinness or obesity.[1458] Men also prefer certain female scents[300] and vocal stimuli;[290, 693] women that produce the preferred odors and vocalizations are also visually attractive to men. The favored attributes are associated with developmental homeostasis, a strong immune system, good health, high estrogen levels, and, especially, youthfulness,[527, 1437] all of which together constitute a recipe for high fertility. Thus, for example, the level of circulating estrogen in a sample of healthy Polish women was related to their body shapes (Figure 14.6). The added estrogen present in the large-breasted, narrow-waisted women in this study meant that they were perhaps three times as likely to conceive compared with the other participants.[714]

## Discussion Question

**14.7**  We have just offered the fertility hypothesis for the male preference for women with an hourglass figure. Produce a different hypothesis for this sexual preference, based on the following facts: (1) the body fat stored in a pregnant woman's lower body is of a type that promotes the growth of the fetal brain whereas (2) upper body (abdominal) fat differs in its composition and is not used for the development of the embryo's brain. Once you have come up with your hypothesis, use it to produce at least one testable prediction. Compare your explanation and prediction with that of William Lassek and Steve Gaulin.[843]

At least some of the physical features that men in Western societies tend to find sexually appealing may be linked to a woman's potential to become pregnant. But even highly fertile women can conceive only during the few days each month when they are ovulating. If men evaluate women in terms of their immediate reproductive value, we would expect males to prefer the body odor of a female during her fertile phase over the odor of the same woman during her nonovulatory phase. Devendra Singh and Matthew Bronstad showed that, as predicted, men find the smell of a T-shirt worn by an ovulatory woman to be "more pleasant and sexy" than the scent of a T-shirt worn by the same woman when she is not fertile.[1337] Subsequently, another research group checked the prediction that women's faces should be more attractive during the fertile phase than during the nonfertile phase of the menstrual cycle. They were. Men shown photographs of the face of a woman when she was and when she was not ovulating usually voted for the "fertile face."[1228]

**FIGURE 14.7** **Dominant chimpanzee males prefer to copulate with old parous females** (those that have had infants previously). Nulliparous females (those that have yet to give birth) are the least preferred. After Muller, Thompson, and Wrangham.[1019]

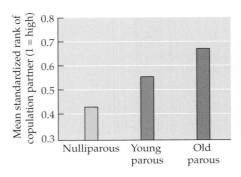

Old chimp mother with youngster

Geoffrey Miller and his colleagues used an unusual method to test the hypothesis that male behavior has been shaped by an unconscious ability to assess female fertility. They recruited a number of lap dancers working in Albuquerque, some of whom were on the pill and others who were not. These women were willing to cooperate with the researchers by telling them the amount they received as tips after their erotic performances. Before reading on, you may want to identify what predictions the adaptationist Miller and his coworkers must have had in mind.

Because the research team considered it likely that men are more capable than often suspected at detecting the cues of female fertility and because men ought to find these cues attractive, the team predicted that lap dancers who were not on the pill and were ovulating would receive more money from their customers than would those women who were not ovulating, either because they were taking birth control pills or because they were in the nonovulatory phase of the menstrual cycle. The actual results of their research matched the predicted ones, with women taking in about twice as much in tips when they were most fertile compared with when they were menstruating or if they were on the pill.[984]

- ●—— Most attractive (minus own age)
- ●—— Maximum difference preferred
- ●—— Minimum difference preferred

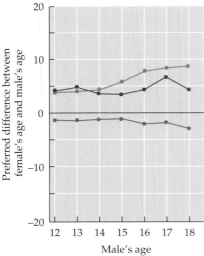

**FIGURE 14.8** **Preferred dating partners by age for male teenagers.** The age of preferred dates is shown in relation to the age of the male subjects. After Kenrick et al.[764]

## Discussion Questions

**14.8** In our species, males typically prefer to mate with youthful women for the reasons discussed above. But in our closest relative, the chimpanzee, older females in estrus are approached more often by sexually motivated males, and they copulate more frequently with dominant males (Figure 14.7).[1019] What prediction can you make therefore about the fertility of older female chimpanzees as well as the occurrence of menopause in this species? Check your prediction against the data in Thompson et al.[1430]

**14.9** Some persons believe that men in Western societies generally prefer sexual partners who are younger than they are because they have been taught what is an arbitrary cultural convention from an early age. What prediction follows from this hypothesis with respect to the dating preferences of teenage males for females of different ages? What conclusion follows from the data in Figure 14.8?[764]

## Adaptive Mate Preferences of Women

We have presented the argument that because women vary in fertility and because female fertility has a major effect on male reproductive success, men should find cues associated with high fertility sexually appealing. Although men also vary in fertility to some extent, the degree of variation is much less than in women, if only because there is no male equivalent of the menstrual cycle or of menopause. On the contrary, most men of any age can supply plenty of functional sperm on any day of the month to any woman willing to receive their gametes. Not surprisingly, therefore, evolutionary biologists have never bothered to explore whether female mate choice revolves around an assessment of male fertility. Instead, adaptationists have focused on two other factors: the capacity of men to supply good genes to their offspring and their ability (and willingness) to provide resources for their partner's offspring. Both of these traits appear to vary markedly among men, and both could greatly affect a woman's reproductive success.

From an evolutionary perspective, then, we would expect women to find attractive those male physical attributes that indicate high genetic quality or parental ability. Some researchers have found that women do indeed prefer men with "masculine" facial features, namely, a prominent chin and strong cheekbones (Figure 14.9).[735] In addition, facial symmetry has been identified as a plus (and not just in Western societies[874]), and so are an athletic, muscular upper body[527] and a deep voice.[459] This combination has been linked to high testosterone levels, good current health, and perhaps most importantly, good health during juvenile development. Despite developmental homeostasis (see page 89), early deficits in nutrition can have some long-lasting negative effects on survival and reproductive success in our species.[895] Male development is at special risk because of the potentially damaging side effects of the male sex hormone, testosterone. Therefore, the ability of a man to develop normally, despite high levels of circulating testosterone, is a possible indicator of a strong immune system capable of overcoming the handicap imposed by

(A)    (B)

**FIGURE 14.9   Computer-modified facial images** provide a way to test the preferences of subjects for (A) masculinized faces versus (B) feminized ones. Photographs by Ben Jones and Lisa De Bruine of the Face Research Laboratory, University of Aberdeen; from Jones et al.[735]

the hormone.[474] If strong, healthy men can pass on defense against disease to their offspring, their mates will benefit.

In addition, these men probably can compete effectively with rivals for dominance within their group. In this context, it is relevant that the men rated as having the most dominant, masculine, and attractive faces are those with the greatest handgrip strength, which correlates well with overall physical strength.[464] Moreover, men with high handgrip strength do develop sexual relationships sooner and have more (self-reported) sexual partners than individuals with less powerful grips.[510] Over the course of human evolution, dominant, powerful men probably have been able to protect their partners, as well as supply them and their offspring with the resources that usually come to males of high social and political status.[220, 562]

---

### Discussion Question

**14.10** How might the following findings be understood in terms of the adaptive value of female mate preferences? Deep-voiced men have more children in a traditional hunter-gatherer culture, the Hazda of Tanzania.[42] Taller men are more likely to be chosen in speed-dating competitions than their shorter rivals.[102] In yet another study, about two-thirds of the women interviewed said that they had ended at least one potentially romantic relationship after an unsatisfying first kiss with their date.[1509] Finally, in what was billed as a political study, researchers found that viewers were able to pick winners of gubernatorial races about 70 percent of the time after a 100-millisecond glimpse at the faces of the two main candidates, who were both unknown to the subjects.[68]

---

The importance for women of having a good provider as a mate has been established in studies showing that females in cultures without birth control who secure relatively rich husbands do tend to have higher fitness than females whose partners cannot offer many material benefits. Among the Ache of Paraguay, the children of men who were good hunters were in fact more likely to survive to reproductive age than the children of less skillful hunters.[749] Likewise, several studies of traditional societies in Africa and Iran have revealed a positive correlation between a woman's reproductive success and her husband's wealth, as measured by land owned or number of domestic animals in the husband's herds.[141, 706, 911] Even in modern societies, household income is correlated with children's health, with the effect growing larger as children get older. Chronic illnesses in childhood can reduce the earning power of children who reach adulthood, thus perpetuating poverty across generations,[242] with all of the reproductive consequences this has for human beings.

Evidence of this sort has convinced some evolutionary psychologists to predict that females will usually put wealth, social status, and political influence ahead of good looks when it comes to selecting mates. This evolutionary prediction has been supported by many questionnaire or interview studies, as in Buss.[219] However, even when researchers have found clear differences between the sexes in the value they attach to "good financial prospects" versus "good looks," the absolute measures of importance given to these attributes have not necessarily been especially high for either sex. But the men and women in these studies typically have not had to specify which items among a list of attributes are absolutely essential to their choice of mates versus which would be nice to have in a partner but are not crucial. Therefore, a team of social psychologists led by Norm Li attempted to put constraints on the choices made by the people they interviewed by giving them a limited budget to expend on designing a hypothetical ideal mate.[860] A subject was given a list of traits and told to decide how many of his or her limited supply of "mate dollars" to use when purchasing any one item, such as physical attractiveness, creativity, yearly income, and so on. To secure a partner at the second level of attractiveness, or creativity, or yearly income (the highest level was 10) would require 2 mate dollars; to get someone at the eighth level would require 8 mate dollars. When the persons interviewed had only 20 mate dollars to work with, their investments differed greatly according to their sex. Men devoted 21 percent of their total budget toward the acquisition of a physically attractive partner; women spent 10 percent of the same total amount to the same end. On the other hand, women on this tight budget devoted 17 percent of their money to boost the yearly income of an ideal mate, whereas men invested just 3 percent of their mate dollars on this attribute.

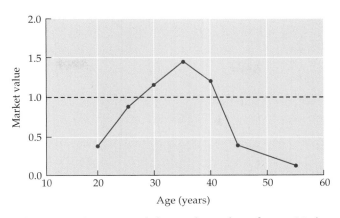

**FIGURE 14.10   Age and the market value of men.** Market value is measured by the number of advertisements in the personals sections of newspapers by women requesting men of certain age classes divided by the number of men of those age classes announcing their availability. After Pawłowski and Dunbar.[1111]

After having specified how they would spend their first 20 mate dollars, the participants were given two additional 20-dollar increments. By the time they reached the third 20 mate dollars, the sexes did not differ markedly with respect to the attributes they were buying. Having already purchased what they really valued, they could and did spend on other attributes. This experiment tells us that people view some mate characteristics as essential items and others as mere luxuries, add-ons if you will. The essential elements are not the same for men and women, as predicted by an evolutionary approach.[860]

Personal ads, whose cost limits the number of words used by the advertisers, also provide relevant evidence on what people consider fundamentally important in a mate. So, for example, women seeking partners through newspapers are far more likely than men to specify that they are looking for someone who is relatively rich.[1520] In keeping with this goal, women advertisers in both Arizonan and Indian newspapers also often specify an interest in someone older than they are;[762] older men usually have larger incomes than younger men.[219]

If women really are highly interested in a partner's wealth and capacity to provide for offspring, then men in their thirties should be most desirable because men of this age have relatively high incomes and are likely to live long enough to invest large amounts in their children over many years. One can calculate the "market value" of men of different ages by using samples of personal ads and dividing the number of women requesting a particular age class of partner in their advertisements by the number of men in that age class who are advertising their availability; this measure thus combines both demand and supply. Men in their late thirties have the highest market value (Figure 14.10).[1111]

## Discussion Question

**14.11**   If an evolutionary approach to human reproductive behavior is useful, what should the market value curve for women look like? Add your predicted data points to the graph shown in Figure 14.10. Check your prediction against data in Pawłowski and Dunbar.[1111]

It could be, however, that women's interest in the earning power of potential mates is a purely rational response to the fact that males in almost every culture control their society's economy, making it difficult for a woman to achieve material well-being on her own. If this nonevolutionary hypothesis explains why females favor wealthy males, then women who are themselves well off and not dependent on a partner's resources should place much less importance on male earning power. Contrary to this prediction, several sur-

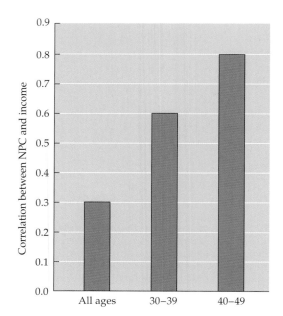

**FIGURE 14.11    Higher income increases male copulatory success.** Income is positively correlated with the number of potential conceptions (NPC) in the preceding year for unmarried Canadian men of various age groups, but especially for older men. After Perusse.[1120]

veys have shown that women with relatively high expected incomes actually put more emphasis, not less, on the financial status of prospective mates.[1459, 1566] For example, relatively wealthy female undergraduates value wealth and status in potential long-term mates more than do less wealthy females.[224]

Mating preferences are one thing, but mating behavior is another. Are women actually more likely to copulate with men who possess the attributes they prefer than with men who lack them? Among the Ache of eastern Paraguay, good hunters with high social status are more likely to have extramarital affairs and produce illegitimate children than poor hunters, suggesting that females in this society find skillful providers sexually attractive.[749] Likewise, in Renaissance Portugal, noblemen were more likely to marry more than once, and more likely to produce illegitimate children, than men of lower social rank. These results are consistent with the prediction that females use possession of resources as a cue when selecting a father for their children.[139]

If we bear the imprint of past evolution on our psyches, then women in modern Western societies should also use resource control and its correlates, such as high social status, when deciding which men to accept as sexual partners. In order to study the relationship between male income and copulatory success in modern Quebec, Daniel Perusse secured data from a large sample of respondents on how often they copulated with each of their sexual partners in the preceding year. With this information, Perusse was able to estimate the number of potential conceptions (NPC) a male would have been responsible for, had he and his partner(s) abstained from birth control. Male mating success, as measured by NPC, was highly correlated with male income, especially for unmarried men (Figure 14.11). Perusse concluded that single Canadian men attempt to mate often with more than one woman, but their ability to do so is affected by their wealth and social standing. These findings have been replicated in great detail with a much larger random sample of men living in the United States.[745] Thus, the striving for high income and status exhibited by males in North America may be the product of past selection by choosy females, which occurred in environments in which potential conceptions had an excellent chance of being actual ones.[1120]

## Discussion Question

**14.12**  Although women seem to prefer men who bring wealth to the relationship, in most modern cultures, high family income is not positively correlated with the number of children produced (Figure 14.12). Indeed, poor couples often have more surviving children than rich ones. Does this finding invalidate an evolutionary analysis of human behavior, as has been

**FIGURE 14.12    Fertility declines as family income increases in the United States** (and many other industrial societies). Shown here is the average number of children per family in relation to the amount of income per family member. The families are grouped by income decile, so the bottom 10 percent are shown at D1, and the top 10 percent are shown at D10. Data are for households in the United States in 1994 as collected by the Institute for Social Research at the University of Michigan.

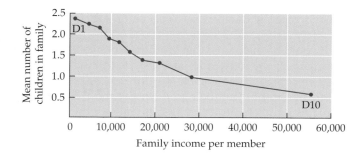

claimed?[1494] You might want to contrast aspects of the current human environment with our ancestors' environment. Can you make use of the finding that in preindustrial Finland, the number of *surviving* offspring was less for women of high fecundity in resource-poor landless families? In contrast, in wealthy landowning families women who had many babies also had higher fitness.[542] In addition, fit the following finding into your analysis: in a survey of modern data from 145 countries, human fertility was negatively linked to population density.[900]

## Conditional Mate Preferences of Men and Women

The kind of research we have reviewed thus far tells us what people want in an ideal mate, but most people know that ideal mates are in short supply. Thus, you will not be surprised to learn that not every man is paired off with an exceedingly fertile gorgeous model, nor is every woman married to an extremely wealthy, highly parental, loving individual with outstanding genes.[223] In reality, although many men evaluate the sexual attractiveness of women in much the same way, their actual mate choices often diverge considerably from what would seem to be evolutionarily desirable. The same holds true for women. Why? One line of argument holds that there are trade-offs involved in any pairing. Men that marry extremely attractive women may lose paternity to other men who are attracted to their partners. Women that marry extremely strong, powerful men may lose resources to other women attracted to their partners. Perhaps this sort of consideration underlies findings of the sort shown in Figure 14.13.[1411]

What does seem certain is that people's real-world mate choices are not set in stone but can vary in many ways depending on a host of factors.[513] One of these factors that may influence the mating decisions of premenopausal women is the menstrual cycle, which provides only a small window of fertility each month. Even though women are generally said not to know

(A)

(B)

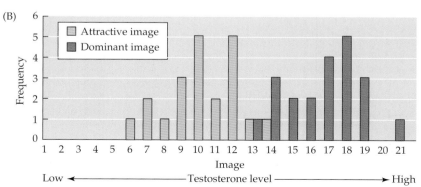

**FIGURE 14.13  Females differ with respect to facial features they associate with dominant men versus attractive men.** (A) In this study, digital images of the same male's face were altered to reflect the developmental effects of testosterone from low to high levels (left to right). (B) When young women were asked to judge these photographs for physical attractiveness, they tended to pick images like the one in the middle, whereas when they were asked to rate the images in terms of social dominance, they tended to pick those similar to the right-hand image. After Swaddle and Reierson.[1411]

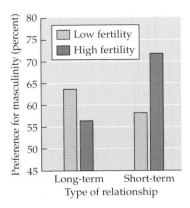

**FIGURE 14.14 Menstrual cycles affect female ratings for masculinized male bodies** (those with broad shoulders and narrow hips). Women who rate two male body images, one relatively feminized and the other masculinized, evaluate them differently depending on the phase of their menstrual cycle. Fertile (ovulating) females (blue bars) who are asked to judge the two images for a prospective long-term relationship favor the feminized image, whereas they pick the masculinized image when asked about a short-term relationship preference. After Little, Jones, and Burriss.[873]

when they are about to ovulate, some evolutionary psychologists have used the adaptationist approach to predict that women should change their mate preferences over the course of the menstrual cycle. The argument here is that women might gain fitness by being particularly sexually selective during the few days of peak fertility each month.

Considerable evidence now exists that the sexual preferences of women do change over the course of the menstrual cycle,[515, 735] particularly with respect to preferences for a short-term sexual relationship. For example, when asked to evaluate a pair of potential partners for a brief sexual encounter, fertile women tend to favor the smell and appearance of the more masculine male[873] (Figure 14.14) while also becoming more favorably inclined toward men with symmetrical faces.[874] This change in mate preference supports the view that women with social partners of average or low quality are unconsciously attempting to become pregnant with sexual partners endowed with universally "good genes," which might make their male children especially attractive or dominant. Alternatively, selective women might be more likely to secure superior "complementary genes" from an extra-pair mate, which will generate better offspring genotypes, perhaps particularly with respect to immune system development (see also Chapter 11, pages 398–399). This second hypothesis is supported by the finding that women whose partners are similar to them with respect to MHC genes (which play a critical role in defense against pathogens [see Brown and Eklund[193]]) report lower satisfaction with their mates and a greater number of extra-pair partners than do those women in relationships with men who are genetically dissimilar with respect to these genes.[521] If these women were to become pregnant after making a choice based on MHC dissimilarity, their offspring would tend to be more heterozygous with respect to these genes, a factor that may promote improved immunity against disease.[194]

In addition to cyclical changes in mate evaluation, women apparently also adjust their reproductive strategy in accordance with a variety of factors, including a realistic assessment of their own market value as a mate. A research team headed by Anthony Little has established that women who rate themselves as "highly attractive" show a stronger preference for both relatively masculine and relatively symmetrical faces than women who believe that they are of average or low attractiveness (Figure 14.15).[872] David Buss and

(A)

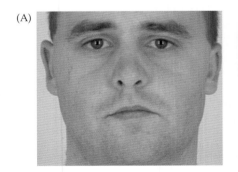 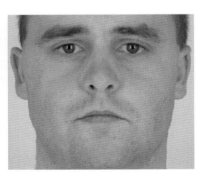

(B)

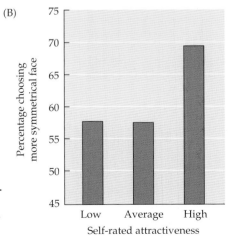

**FIGURE 14.15 Women who think they are highly attractive prefer more attractive men.** (A) When women were given a choice between a pair of digitally altered photographs of men, one of which was more symmetrical than the other (right versus left), they differed in the degree to which they said they preferred the more symmetrical face. (B) Women who rated themselves highly attractive chose the more symmetrical face nearly 70 percent of the time, whereas women of lower self-rated attractiveness did so less than 60 percent of the time. After Little et al.[872]

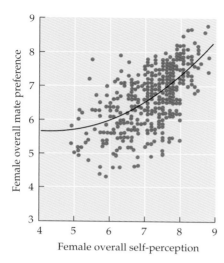
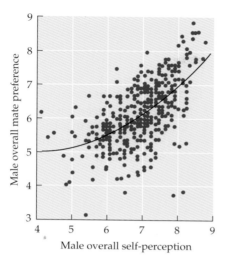

**FIGURE 14.16    Self-perception of attractiveness affects mate preferences in both sexes.** The degree to which women and men consider themselves attractive is correlated with their mate preferences. Less attractive individuals are willing to settle for less in a partner. The values for mate preferences are mean scores of the subjects' answers to questions about the importance of ten attributes in their dating decisions, such as: On a scale of 1 to 9, how important is physical attractiveness to you? The values for self-perception are the mean scores of the respondents' own evaluations of how they would score on these ten attributes. After Buston and Emlen.[224]

Todd Shackelford report that not only are physically attractive women desirous of a highly attractive mate, they also have higher standards when it comes to partner wealth, commitment, and parental abilities.[223] Finally, Peter Buston and Steve Emlen found that *both* men and women who consider themselves high-ranking prospects for a long-term relationship expressed a preference for an equally high-ranking partner, whereas those individuals with a lower self-perception of market value were less demanding (Figure 14.16).[224]

### Discussion Question

**14.13** What fitness benefits might accrue to a person of modest market value in selecting a partner of more or less equal value instead of aiming for a person of much higher fertility or wealth?

The conditional reproductive strategy of females evidently also factors in the potential duration of a relationship. Thus, only women in a long-term partnership with a committed companion feel a change in sexual desire over the menstrual cycle, with a peak occurring around the time of ovulation. An increase in sexual desire at that time increases the probability that the woman will conceive a child with a male who is likely to be there to assist later in rearing that child.[1142] The quite different sexual psychology of unattached women reduces the risk of conception at a time when a long-term partner is not available.

The conditional reproductive strategy of men also appears to enable individuals to select different tactics for different kinds of sexual relationships. Short-term flings require relatively little commitment on the part of a male because any offspring produced will be cared for by someone else. In contrast, a long-term relationship, as in marriage, requires costly resource transfers to a woman and parental investment for her offspring. A cost–benefit analysis yields the prediction that men should have far lower standards for a one-night stand than for a marriage partner. Indeed, when social psychologist Doug Kenrick surveyed a group of undergraduates about the minimum acceptable level of intelligence they would require in a partner for interactions ranging from a first date to marriage, both men and women adjusted the minimum level upward similarly in relation to degree of long-term commitment involved. But men and women differed strikingly when considering their standards for partners in a casual sexual encounter (Figure 14.17).[761] The conditional strategies of the two sexes are not identical.

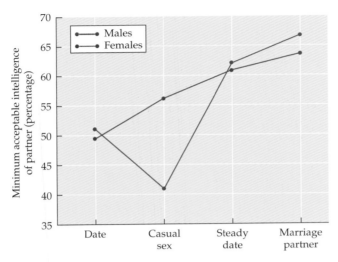

**FIGURE 14.17 Sex differences in mate selectivity.** College men differ from college women in the minimum intelligence that they say they would require in a casual sexual partner. However, men and women have similar standards with respect to the minimum intelligence they say is essential for a marriage partner. "Intelligence" was scored on an IQ scale so that a score of 50 meant that the acceptable individual had an IQ higher than half the population. After Kenrick et al.[761]

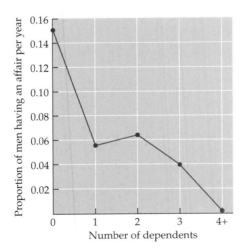

**FIGURE 14.18 Men in a traditional society are less likely to engage in extramarital affairs** in a given year if they have a relatively large number of dependents to care for with their wives. After Winking et al.[1602]

Thus, game theory, and specifically conditional strategy theory, offers an explanation for why it is that there is so much variation among people in their reproductive tactics. We, like Seychelles warblers and white-fronted bee-eaters, are able to pick and choose among a restricted range of responses based on an analysis of many variables, such as our social status, our looks, the attributes of potential partners and competitors of the same sex, and the degree of investment required for a sexual relationship.

## Sexual Conflict

Because the mate preferences and genetic interests of men and women are not the same, we would expect to observe a good deal of sexual conflict in our species, as we do in most other animals (see Chapter 10). One significant source of sexual conflict would arise if men are, on average, more interested in acquiring multiple sexual partners than women are, as just suggested. The basis for this view is that although monogamous men may advance their reproductive success by helping one mate, polygynous men with several wives or an extramarital partner or two can potentially produce still more offspring. Polygyny has almost certainly been an option for men with substantial resources throughout our history as a species, judging from the fact that the acquisition of several wives was culturally sanctioned in 83 percent of all preindustrial societies.[1024] Extramarital affairs, which could also have been part of the ancestral reproductive pattern, have the potential to increase the fitness of an adulterous male substantially, especially if his illegitimate children are cared for by his extramarital partners and their husbands.

However, extramarital activity has potential costs as well as benefits for males if their unfaithfulness results in a diversion of resources from existing children in order to pursue an additional partner (or two). Jeffrey Winking and his colleagues attempted to test whether this cost would moderate the polygynous tendencies of men in a traditional society, the Tsimane of Bolivia.[1602] If extramarital activity reduces the chances that existing offspring will achieve their maximum reproductive potential, due to loss of paternal investment, then the frequency with which men have sexual affairs will decline as a man and his primary mate have more children. The predicted pattern does occur (Figure 14.18).

Despite the costs of extramarital activity for men, many are still motivated, at least under some conditions, to seek out sexual variety, which expresses itself in the willingness of some men to patronize female prostitutes; women, on the other hand, almost never pay men to copulate with them. Moreover, males, not females, also support a huge pornography industry in Western societies because men, not women, are willing to pay just to look at nude women. Note that in modern societies these particular aspects of male sexual behavior are surely maladaptive; prostitutes almost universally employ effective birth control or undergo abortions when pregnant, and payment for pornography is unlikely to boost a man's reproductive success. The prostitution and pornography industries take advantage of the male psyche, which evolved prior to modern birth control and the publication of *Playboy*.[1416]

David Buss and David Schmitt illuminated these differences between the sexes simply by asking a sample of college undergraduates how many sexual partners they would like to have over different periods of time.

(A)

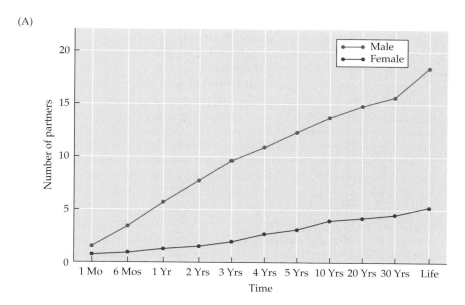

(B)

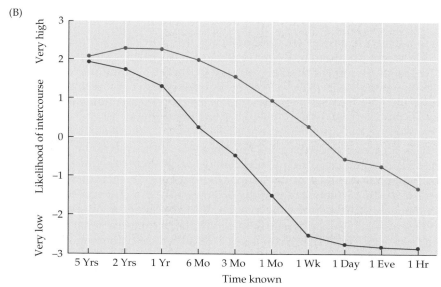

**FIGURE 14.19   Sex differences in the desire for sexual variety.** (A) Men and women differ in the number of sexual partners that they say they would ideally like to have over different periods of time. (B) Men and women also differ in their estimates of the likelihood that they would agree to have sexual intercourse with an attractive member of the opposite sex after having known that individual for varying lengths of time. After Buss and Schmidt.[221]

The men in their study wanted many more mates than the women did (Figure 14.19A). Moreover, when Buss and Schmitt asked their subjects to evaluate the likelihood that they would be willing to have sex with a desirable potential mate after having known this person for periods ranging from 1 hour to 5 years, the differences between men and women were also dramatic (Figure 14.19B): "After knowing a potential mate for just 1 hour, men are slightly disinclined to consider having sex, but the disinclination is not strong. For most women, sex after just 1 hour is a virtual impossibility."[221]

The typically greater enthusiasm of males than females for sexual activity is reflected in the results of another study conducted by Martie Haselton.[628] She asked about 100 undergraduate men and 100 undergraduate women whether they had had encounters with the opposite sex in which the other person evidently thought they were more (or less) interested sexually in this person than they actually were. Men reported about an equal number of encounters during the preceding year in which women had "overperceived" and "underperceived" the males' romantic intentions. Women, on the other hand, claimed that men were far more likely to think that they were sexually interested in

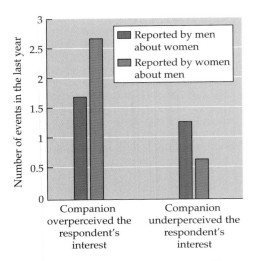

**FIGURE 14.20 Men and women differ in their tendency to overperceive sexual interest on the part of a companion.** Men are more likely to think that women have romantic inclinations toward them when they do not. After Haselton.[628]

them, when in fact the women were not, than to make the opposite mistake of underperception of sexual intent (Figure 14.20).

This kind of bias was documented in another way by two social psychologists who sent confederates, an attractive young man and an attractive young woman, on the following mission. They were to approach strangers of the opposite sex on a college campus, asking some of them, "Would you go to bed with me tonight?" Not one woman agreed to the proposition, but 75 percent of the men said yes. Remember that the male subjects had known the woman in question for about a minute.[269]

Now, it is possible that all the women in this study who said no did so because they sensibly feared becoming pregnant or did not wish to risk injury or disease from a sexual encounter with a male stranger. If so, then homosexual women should have fewer inhibitions about casual sex, since sexual interactions between two women cannot result in pregnancy and are unlikely to lead to physical assault. But homosexual women are no more interested in having multiple partners than are heterosexual females.[59]

What are the fitness consequences for women whose partners are able to satisfy their desire for sexual variety? Polygyny carries the risk for a female that her partner will divert resources from her and her children to another woman and her offspring. Over evolutionary history, females of our species have probably gained fitness when they had exclusive access to a male's resources and parental care. In 19th-century Utah, for example, women married monogamously to relatively poor Mormon men had more surviving children on average (6.9) than women married to rich polygynous Mormons (5.5).[640] Even though the polygynists' wives had lower fitness, the polygynists themselves did much better than monogamous men, because they had children by several wives, not just one.

The potential benefits of polygyny to males increase the likelihood of conflict between husbands and wives. On the other side of the coin, some women are receptive to extramarital affairs, which may enable them to acquire additional material goods, or more protection, or better genes for their offspring from their extra-pair partners. If a wealthy or powerful extra-pair partner becomes a woman's primary partner, she may be able to exchange a low-ranking husband for a socially superior one, with all the positive effects on her fitness that trading up affords.[1287] The main point is that because both women and men can potentially elevate their fitness by mating with more than one partner, it is possible for a husband to reduce his wife's fitness, and vice versa.

This fact of life may be responsible for any number of features of human behavior. For example, men are well-known to be concerned about the paternity of the children of their wives, and as a result, they pay special attention to the resemblance between themselves and their putative children. Wives are well aware of this interest on the part of their husbands, and they are often quick to suggest that their newborns look very much like their husbands (even though impartial judges detect a greater similarity between a baby's appearance and that of the mother).[26] Moreover, a father's evaluation of the similarity between himself and his child is a major factor in his investment in the child, as self-reported in surveys.[41]

## Discussion Question

**14.14** If two blue-eyed persons marry and have children, their offspring will all have the blue-eyed phenotype, whereas brown-eyed individuals that reproduce may have children with various eye colors. Blue-eyed men find blue-eyed women more attractive than brown-eyed women.[828] How might an evolutionary biologist interpret this finding?

Another result of the potential fitness conflicts between man and woman may be a capacity for sexual jealousy, which is an emotional state that helps individuals detect and interfere with the fitness-damaging sexual behavior of a mate. However, the nature of sexual jealousy should differ between the sexes, according to evolutionary psychologists, because the reproductive harm done by a multiply mating partner differs for men and women. A wife whose husband secures an additional mate (in a polygynous society) or goes on to divorce her in favor of a new partner usually loses some or all of her access to her husband's wealth, and thus the means to support herself and her children. Therefore, a woman's sexual jealousy, so the argument goes, should be focused on the possible loss of an attentive provider and helpmate, which is more likely to occur when a man becomes emotionally involved with another woman. On the other hand, a husband whose wife mates with another man may eventually care unknowingly for offspring fathered by that man. A man's sexual jealousy should therefore revolve around the potential loss of paternity and parental investment arising from a wife's extramarital sexual activity rather than the potential loss of resources and emotional commitment.[337]

If this view is correct, then if men and women were asked to imagine their responses to two scenarios, one in which a partner develops a deep friendship with another individual and one in which a partner engages in sexual intercourse with another individual, women should find the first scenario more disturbing than the second, whereas men should be more upset at the thought of a mate copulating with another man. Data from several cultures confirm these predictions.[220, 337] For example, in a study involving Swedish university students, who live in a fairly sexually permissive culture, 63 percent of the women found the prospect of emotional infidelity more troubling, whereas almost exactly the same percentage of the men deemed sexual infidelity more upsetting.[1567]

One effect of male sexual jealousy and possessiveness is to reduce the possibility that another male will supply a man's partner with sperm. In this light, marriage, including the honeymoon, may be a cultural institution that serves the function of mate guarding (see page 531). Everywhere, men aspire to monopolize or restrict sexual access to their mates, although they do not necessarily succeed. Marriage institutionalizes these ambitions. Although one sometimes hears of societies in which complete sexual freedom is the norm, the notion that such cultures actually exist appears to have been a (wistful?) misinterpretation on the part of outside observers. In all cultures studied to date, adultery committed by a woman, or even suspicion of it, is considered an offense against her husband and often precipitates violence.[341] Presumably, suspicion of lost paternity explains why pregnant women in the United States are at double the risk of violent domestic assault compared with nonpregnant women.[207] In some other societies, a woman known to have cuckolded her husband may be legally killed by her aggrieved partner.[337]

## Discussion Questions

**14.15** Fortunately, men in the United States are not permitted under any circumstances to kill their wives. Unfortunately, husbands do commit that crime on occasion, with nearly 14,000 homicides of this sort found in an FBI database assembled for the years 1976 to 1994. In some cases, a lovers' triangle was known to be involved.[1309] In this subset of homicides, young women were far more likely to be victims than older ones. Analyze this result as dispassionately as you can in terms of the potential fitness costs and benefits to the killer husband. With these fitness effects in mind, consider the possibility that wife killing in the context of potential infidelity is an evolved adaptation. Contrast this possibility with an alternative explanation, namely, that wife killing occurs as a maladaptive by-product of the psychological mechanisms that inspire violent sexual jealousy in males who think that their partners are unfaithful.

**14.16**  Mate guarding is an evolved response to sperm competition (see Chapter 10). If sperm competition has been a factor in human evolution, we can make some predictions about the relative investment of men in their testes, the sperm-producing organs, compared with that made by our two closest relatives, chimpanzees and gorillas. Chimpanzee females regularly mate with several males in the same estrous cycle, whereas gorilla females almost never do, since they typically live in bands, each controlled by a single, powerful male. How large (as a proportion of body size) should the testes of chimpanzee males be relative to gorilla testes? If the testes of men are more similar to those of chimpanzees, what might this tell us about the intensity of sperm competition during our evolutionary past? If, on the other hand, humans resemble gorillas, what conclusion is justified? Check your predictions against Harcourt et al.[619]

## Coercive Sex

The fact that the murder of women suspected of adultery is still condoned in some parts of the world is one of the least attractive manifestations of sexual conflict in our species. Another is the regular occurrence of forced copulation, a phenomenon that is not limited to human beings, by the way (Figure 14.21). Despite the fact that human rapists are often severely punished, rape occurs in every culture studied to date.[1438] Although most persons find the topic highly unpleasant and therefore find it difficult to discuss the behavior calmly, if we were to understand the phenomenon more fully, we might be in a better position to reduce the frequency of rape in our society.

Efforts to analyze the causes of rape include the work of Susan Brownmiller in her highly influential book *Against Our Will*.[199] In her view, rapists act on behalf of all men to instill fear in all women, the better to intimidate and

(A)

(B)

**FIGURE 14.21  Rape occurs in animals other than humans.** (A) In the beetle *Tegrodera aloga*, a male (right) can court a female (left) decorously by repeatedly drawing her antennae into grooves on his head; copulation ensues only if the female responds to this courtship. (B) Alternatively, a male (below) can force a female (above) to mate by running to her, grasping her, throwing her on her side, and inserting his everted genitalia as the female struggles to break free. Photographs by the author.

control them, thus keeping them "in their place." Brownmiller's intimidation hypothesis implies that some males are willing to take the substantial risks associated with rape in order to provide a benefit for the rest of male society. This argument suffers from all the logical problems inherent in "for the good of the group" hypotheses (with the added difficulty that groups composed of only one sex cannot be the focus of any realistic sort of group selection), but we can test it anyway. If the evolved function of the trait is to subjugate all women, then the rapist element in male society can be predicted to target older, dominant women (or young women who aspire to positions of power) to demonstrate the penalty that comes from stepping outside the traditional subordinate role. This prediction is not supported: most rape victims are young, poor women.[1436]

An alternative evolutionary hypothesis proposed by Randy and Nancy Thornhill is that rape is an adaptive tactic in a conditional sexual strategy.[1436] According to the Thornhills, sexual selection has favored males with the capacity to commit rape under some conditions as a means of fertilizing eggs and leaving descendants. In this view, rape of strangers by men is analogous to forced copulation in *Panorpa* scorpionflies (see pages 349–350), in which males unable to offer nuptial gifts use the low-gain, last-chance tactic of trying to make females copulate with them. According to this hypothesis, human males unable to attract willing sexual partners might also use rape as a reproductive option of last resort. (Incidentally, note that there are other kinds of rapists besides losers who use violence to force strangers to mate with them.[969])

The proposition that rape might serve an adaptive sexual function has angered many people, including Brownmiller, who wrote, "It is reductive and reactionary to isolate rape from other kinds of violent antisocial behavior and to dignify it with adaptive significance."[200] This response, however, confuses efforts to explain rape with attempts to excuse or justify the behavior. As noted earlier, when evolutionary biologists examine the adaptive value of a trait, whether it be rape in the human species, forced copulation in scorpionflies, or brood parasitism by cowbirds, their goal is to explain the evolutionary causes of the behavior, not to condone rape, or brood parasitism, or anything else.

The explanatory hypothesis that human rape is an evolved reproductive tactic controlled by a conditional strategy generates the prediction that some raped women will become pregnant, which they do, even in modern societies in which many women take birth control pills.[1438] In fact, copulatory rape apparently is more likely to result in pregnancy than consensual sex.[554] In the past, in the absence of reliable birth control technology and abortion procedures, rapists would have had a still higher probability of fathering children through forced copulation. Furthermore, if rape really is a product of an evolved reproductive mechanism, then rapists should more often target women of high fertility, just as bank swallows and other birds identify fertile (egg-laying) females and try to force those individuals to copulate with them.[86, 1371] In contrast, Brownmiller's view is that rape has nothing to do with reproduction but is merely another form of violent antisocial behavior driven by a desire of men to dominate women. Note that this hypothesis focuses on the proximate cause of rape, not its ultimate reproductive consequences; it is possible for rapists to be motivated purely by a desire to hurt women and yet to have children as a result of their aggression. The notion, however, that rape has no proximate sexual component at all leads to the prediction that the age distribution of rape victims should be the same as that of women murdered by male assailants. Crime data are at odds with this prediction (Figure 14.22).

Although these findings suggest that rape may increase the fitness of some men, it is also entirely possible that rape is not adaptive per se but is instead a maladaptive by-product of the male sexual psyche, which causes quick sexual arousal, a desire for variety in sexual partners, and an interest in impersonal

**FIGURE 14.22 Testing alternative hypotheses for rape.** If rape were motivated purely by the intent to attack women violently (a proximate hypothesis), we would expect that the distribution of rape victims would match that of female murder victims. Instead, rape victims are especially likely to be young (fertile) women, a result consistent with ultimate hypotheses proposing that rape is linked to male reproductive tactics. Data on rape victims come from 1974–1975 police reports for 26 U.S. cities. After Thornhill and Thornhill.[1436]

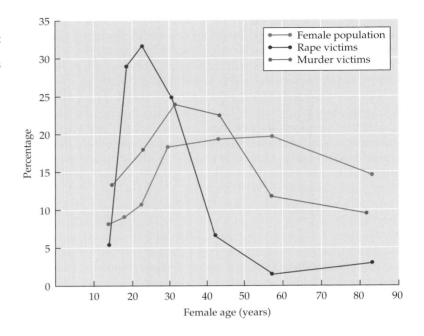

sex, all attributes that generate many adaptive (i.e., fitness-enhancing) consequences while also incidentally leading some men to rape some women.[1438] After all, men have been known to engage in many decidedly non-reproductive sexual activities, including masturbation, homosexual rape, and rape of postmenopausal women and prepubertal girls, just as males of many nonhuman species also exhibit sexual activity that cannot possibly result in offspring, such as the copulatory mounting of weaned pups by male elephant seals.[1239] Moreover, one attempt to estimate the reproductive consequences of rape for men in a traditional society yielded the conclusion that the fitness cost to a rapist exceeded the benefit by a factor of about ten,[1349] a result that strongly suggests rape is not an adaptation, at least in this society.

Even if coercive copulation usually reduces the fitness of its practitioners, the rape as by-product hypothesis would be tenable if the systems motivating male sexual behavior had a net positive effect on fitness. One prediction unique to the by-product hypothesis is that rapists will have unusually high levels of sexual activity with consenting as well as nonconsenting partners. Some evidence supports this hypothesis,[832, 1098] but as is true of many issues in human sociobiology, more data are required. Nevertheless, in this case as in so many others, the adaptationist approach has generated novel hypotheses that are entirely testable in principle and practice. The evolutionary angle on rape is now available for skeptical scrutiny, and as a result, we may eventually gain a better understanding of the ultimate causes of the behavior. When we do, we will not in any way be obliged to be more understanding of the illegal and immoral activity of rapists.[1438]

## Discussion Questions

**14.17** Discussions of rape are invariably emotionally charged. From an evolutionary perspective, why might women have an especially strong visceral response to the topic and an intense desire to punish rapists? Why might most men also wish to deter rape? Would an understanding of evolutionary theory have led to a revision in a legal ruling by the U.S. Supreme Court that contained the following claim: "Rape is without doubt deserving of serious punishment; but in terms of moral depravity and of the injury to the person and to the public, it does not compare with murder … [Rape] does not include … even the serious injury to another person" (quoted in Jones[739]).

**14.18** Natalie Angier states that married men have the same probability of fertilizing an egg per copulation with their wives as rapists do when forcing copulation on a victim.[40] In the past, the probability that an offspring of a married man would survive to reproduce was almost certainly much higher than the probability that a rapist's child would reach reproductive age, because married men assist their children whereas rapists do not. Is Angier correct, therefore, in claiming that rape cannot be an adaptive tactic? (Remember that *adaptive* means "reproductively useful.") What do you make of the fact that low-status men are more likely to rape women unknown to them whereas high-status men dominate the category of acquaintance or partner rape?[1487]

## Adaptive Parental Care

You may have concluded by now that evolutionary biologists are interested only in human sexual activities. Although it is true that adaptationists have intensively studied the reproductive component of our behavior, this does not mean that other aspects of human behavior have been ignored. The entire spectrum of human actions is fair game for researchers using an evolutionary approach.[526] As we saw at the outset of this chapter, issues of reciprocity, reputation, and social judgment can be analyzed in terms of their ultimate significance. Now we shall turn to an evolutionary analysis of parental behavior.

As noted before, in our species both males and females can provide parental care for their offspring, which has benefits (improved survival of offspring) and costs (including increased mortality of the parents). These costs are not trivial, especially for women in preindustrial societies. In a study of over 20,000 couples that were married between 1860 and 1895 and whose reproductive histories are recorded in the Utah Population Database, women were much more likely than men to die within a year after delivering their last child (Figure 14.23).[1116] For both sexes, however, death rates increased as family size increased.

Because childbearing and child rearing are costly exercises, parents can be expected to be discriminating with respect to how much care they will offer a given offspring.[1605] For example, the parents of an infant that produces abnormally high-pitched cries do not respond as rapidly as they would to a baby that cries in a more typical manner.[508] But infants with normal crying patterns are also ignored, or even killed, on occasion by a parent. Here would seem to be the kind of behavior that must defeat evolutionary analysis.

But maybe not. Remember that an individual's impact on the next generation's gene pool is determined not by the number of babies conceived or born, but by the number of offspring that reach reproductive age. If babies with high-pitched cries have serious birth defects that make it unlikely that they will survive to reproduce, then parents that reduce or end their investment in these children may have more resources to give to other, more viable offspring now or in the future. (And babies unable to cry normally often do have serious illnesses or congenital defects.[508]) Likewise, if caring for a newborn or carrying a fetus to term threatens to reduce the lifetime reproductive success of a woman, then ending her investment in that offspring can potentially increase, not reduce, her fitness.[340] Throughout most of human history, when single women have

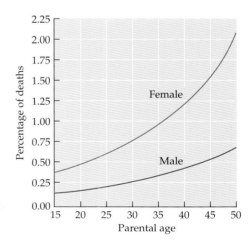

**FIGURE 14.23** Mortality rates are higher for women than men in the year after the birth of their last child, indicating that the costs of parental effort are especially severe for females. After Penn and Smith.[1116]

(A)

(B)

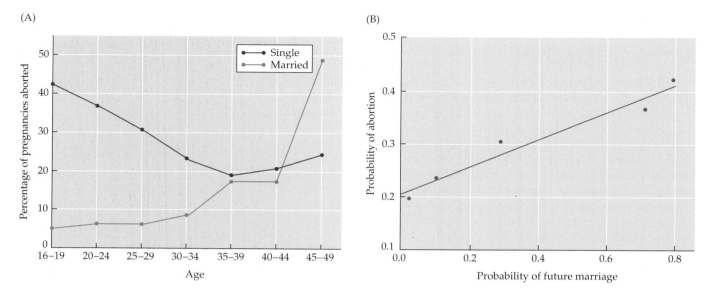

**FIGURE 14.24  Do women employ abortion adaptively?** (A) Single women in England and Wales are less likely to abort pregnancies as they grow older, whereas married women are more likely to abort pregnancies as they age. These differences reflect the different costs and benefits of investing in an offspring when a woman has or does not have a husband's support. (B) The probability that a single woman will undergo an abortion is a function of the age-related probability that she will attract a husband in the future. After Lycett and Dunbar.[901]

attempted to rear a child unassisted, they have probably failed. An evolutionary prediction, therefore, is that single pregnant women should be more likely to end their pregnancies than married women, who have partners to help them raise the offspring.

However, the evolutionary prediction that single pregnant women will seek abortions more often than married ones should apply more strongly to younger than to older women because young unmarried women have a greater chance of securing a husband and his fitness-enhancing support in the future than older women do. Older *married* women are also expected to terminate pregnancies relatively often, given the increased probability of medical complications for older pregnant women, which could threaten their ability to care for their other still-dependent children or grandchildren. When the willingness of British women to carry a fetus to term was examined in relation to age and marital status, the results were entirely in accord with these predictions (Figure 14.24).[901]

Men, as well as women, can potentially provide or withhold parental investment in their offspring. Because men run the risk of caring for a partner's child that was fathered by someone else, the sociobiological prediction is that married men should have evolved psychological mechanisms that protect them against that risk. As we have noted, men are alert to signs of infidelity, as in the mismatch between their appearance and that of a child said to be their offspring. Men are also sexually jealous and if they learn of a wife's adultery, they are likely to react extremely negatively.[1416] Male concern about paternity is so obsessive that husbands of rape victims in many cultures may legally divorce their unfortunate wives.[199]

Stepfathers are another category of males who are placed in the position of caring for offspring who are not their own. To test the prediction that stepfathers should favor their genetic offspring over stepchildren, Mark Flinn monitored some Trinidadian families in which a stepfather lived with children of his own as well as those his wife had by another man. In these families, the percentage of interactions that involved conflicts of one sort or another was about twice as high for interactions between a stepfather and a stepchild as for those between a father and a genetic offspring.[471] Likewise, in modern American society, a man's genetic offspring by a current mate are far more likely to receive money for college than are his stepchildren with either a current or a previous mate (Figure 14.25).[30]

The same pattern applies to stepmothers, who are less likely to provide for their stepchildren than for their own genetic offspring. Households in which a mother cares for stepchildren, adopted children, or foster children spend less on food than households in which mothers reside with their genetic offspring.[240] Moreover, in blended families, the mother's own genetic offspring receive, on average, one more year of schooling than the stepchildren who reside with her. Note that by studying blended families, researchers have eliminated the possibility that stepchildren receive reduced educational opportunities because women who remarry are on average less capable of providing for the education of any children in their household.[241]

Stepchildren not only tend to receive fewer resources from their stepparents, but also are at greater risk of being physically assaulted, a discovery made by Martin Daly and Margo Wilson when they tested the evolutionary hypothesis that our evolved psychological mechanisms encourage us to bias our parental care toward our genetic offspring.[342] A by-product of these mechanisms might be a greater tendency of stepparents to maltreat children that are not their own. In keeping with this prediction, Daly and Wilson found that in Hamilton, Ontario, children 4 years old or younger were 40 times more likely to suffer abuse in families with a stepparent than in families with both genetic parents present! Note that for both categories of Canadian parents, the absolute likelihood that a child would be abused was small (Figure 14.26), but the relative risk was far greater for children in households with a stepparent.[339] Daly and Wilson argued that the psychological systems that promote selective parental care in Canada and elsewhere[343] generally lead to increased fitness by encouraging adults to invest in their offspring, but these same aspects of parental psychology may occasionally have maladaptive effects for a very few parents with blended families.

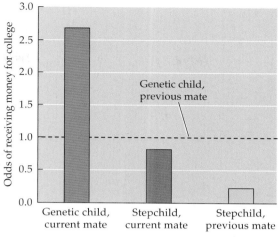

**FIGURE 14.25    Parental favoritism.** The odds that a man will give money to a child for college are much higher if the man is the genetic father of the potential recipient than if he is a stepparent to the child. The four categories of offspring examined in this study were genetic offspring living with their father, genetic offspring living with their father's previous mate, and the man's stepchildren living either with him or with a previous mate. The amount given to a genetic offspring living with a previous mate was used as a standard against which the other donations were measured. After Anderson, Kaplan, and Lancaster.[30]

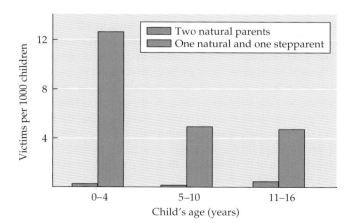

**FIGURE 14.26    Child abuse and the relatedness of parents and offspring.** Child abuse is far more likely to occur in households with a stepparent than in households with two genetic parents. After Daly and Wilson.[339]

**Discussion Question**

**14.19**   Daly and Wilson found that stepchildren incurred an approximately 100-fold increase in the risk of fatal child abuse compared with children reared by two genetic parents.[342] Apply their explanation for the abuse of children by their stepfathers to these cases of infanticide. How does Daly and Wilson's hypothesis differ from the hypothesis for infanticide by male langurs and lions (see Chapter 1)? What would you need to know to test the two alternative explanations for infanticide by stepfathers? Now imagine the following hypothesis: Child abuse by stepparents occurs because the new caretakers typically come into the family some time after the child's birth and so lack the physiological mechanisms, perhaps hormonal ones, that promote tolerance of an infant's difficult behaviors, such as colicky crying or whining. In the absence of these psychological buffering devices, things can go terribly wrong, and they do, albeit very rarely. How does this hypothesis stand up to the other two? What significance would there be in comparing the parental behavior of a genetic father—say, one serving in the military—who has not been present for the first year of his child's life with the paternal responses of a stepfather who moves in with a partner who has a one-year-old toddler?

## Helping Children Marry

We have seen that adults care for children in a selective fashion, usually favoring genetic offspring over nonrelatives. But even in families without stepchildren, parents do not always treat their progeny identically when it comes to dishing out parental benefits. An example involves the material sacrifices parents make to help their offspring acquire spouses. In some societies, the groom and his family are called upon to donate resources such as cattle or money or labor—the bridewealth—to the bride's family (Figure 14.27); in other societies, the family of the bride sends their daughter off to marry with a special donation—the dowry—to her new husband or his family.

If bridewealth or dowry payments were purely arbitrary cultural traditions, then we would predict that the two forms of payment should be equally represented among cultures worldwide. They are not.[525] (Before reading further, use sexual selection theory to predict whether bridewealth or dowry should be the more common cultural practice worldwide.)

Because males typically compete for access to females, we can predict that bridewealth payments should be far more common than dowries. This is indeed the case. Bridewealth payments have been recorded in 66 percent of the 1267 societies described in the *Ethnographic Atlas*,[1024] whereas dowry is standard practice in just 3 percent of these societies. Bridewealth payments are found particularly frequently in polygynous cultures, occurring in more than 90 percent of those societies classified under the heading "general polygyny," in which more than 20 percent of married men have more than one wife (Table 14.1). When some males monopolize several females, marriageable females become an especially scarce and valuable commodity. One way to secure multiple wives, and so achieve exceptional reproductive success, is to provide payments to their parents or other relatives. To do so often requires much wealth.

**FIGURE 14.27   Marriage requires bridewealth payments in many traditional African cultures,** such as the Masai of Kenya and Tanzania. The father who stands here with his about-to-be-married daughter will receive cattle from her husband-to-be and his relatives. Photograph by Jason Lauré.

**TABLE 14.1**  *The relationship between the mating systems of human cultures, bridewealth payments, and inheritance systems that favor sons*

| Mating system | Bridewealth payment | | Sons favored | |
|---|---|---|---|---|
| | No | Yes | No | Yes |
| Monogamy | 62% | 38% | 42% | 58% |
| Limited polygyny | 46% | 54% | 20% | 80% |
| General polygyny | 9% | 91% | 3% | 97% |

*Source*: Hartung[625]

*Note*: The data are from Murdock's *Ethnographic Atlas*[1024] for 112 monogamous cultures, 290 cultures that practice limited polygyny (less than 20 percent of men are polygynous), and 448 that practice general polygyny (more than 20 percent of men are polygynous).

Rich men in polygynous societies can produce a great many children, which enables wealthy parents to secure more descendants by putting their resources primarily in the hands of sons rather than daughters, whose direct fitness is limited by the number of embryos they can personally produce. Biased parental investment can even occur posthumously, when wealthy parents pass most of their wealth to a son or sons, enabling these individuals to become successful polygynists.[140] Inheritance rules that favor sons are in fact associated with the practice of polygyny (see Table 14.1).[625]

Guy Cowlishaw and Ruth Mace confirmed that this pattern was real after controlling for the nonindependence of cultures—that is, after dealing with the statistical problem of how to treat information from several cultures that may have inherited similar practices as a result of sharing a recent common cultural ancestor. First, they constructed a cultural phylogeny based on linguistic information; this diagram identified which cultures were closely related and which were not. They then superimposed the mating system of each culture on the phylogeny to identify those cultures whose mating systems had changed to polygyny from a monogamous predecessor and vice versa. Independent cultural changes to polygyny were much more likely to be associated with inheritance rules favoring sons than were changes to monogamy.[306]

Even in supposedly monogamous Western societies, rich men may have opportunities for unusual copulatory success because, as we noted earlier, their wealth makes them attractive to women. If parents in modern societies retain an ancestrally selected bias that causes them to favor those offspring with the highest reproductive potential, we can predict that even today, very wealthy parents should be inclined to give their sons more inheritance than their daughters—a prediction that has been shown to be correct (Figure 14.28).[1353]

In contrast to the prevalence of biased parental investment in male offspring, societies in which parents give significantly more to their daughters are rare—as one might expect, given that females typically are in demand as marriage partners. However, Lee Cronk has found one tribal society, the Mukogodo of Kenya, in which parents often provide more food and more medical care (from a local Catholic mission clinic) to their daughters than to their sons.[323] The Mukogodo have only recently abandoned their traditional hunting–gathering lifestyle in favor of an economy built on sheep and goat herding. Their herds are small, and their

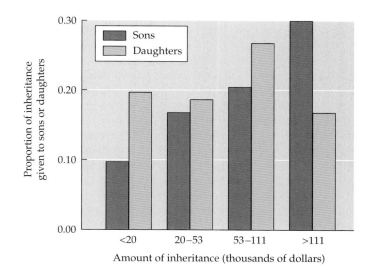

**FIGURE 14.28    Inheritance decisions.** Wealthy Canadian parents bias their legacies toward their sons, who are more likely than daughters to convert exceptional wealth into exceptional mating success. After Smith, Kish, and Crawford.[1353]

standing with other pastoral tribes in the area is very low. A son of an impoverished Mukogodo family is unlikely to acquire a large enough flock to pay the bride price for a wife from any of the surrounding tribes. In contrast, a Mukogodo daughter has a good chance of marrying a member of a higher-status tribe, since polygyny is standard among these groups and women are in short supply. As a result of the greater ease of marriage for daughters than for sons, the average number of offspring of a Mukogodo daughter is nearly four, whereas a son's direct fitness is only about three. This inequality favors families that have more daughters than sons, an outcome that is achieved through greater investment in young girls than in young boys.

This case nicely illustrates the point that human behavior is adaptively flexible, not arbitrarily or infinitely variable. Whatever psychological mechanisms control parental solicitude toward offspring, these evolved systems permit parents to favor sons under some circumstances and daughters under others, while encouraging equal treatment under yet another set of conditions. The option chosen tends to enhance the fitness of parents in their particular environment. The differences between the daughter-favoring Mukogodo and those nearby herding cultures in which sons receive preferential treatment are surely not genetic. Instead, these parental differences reflect our ability to use evolved conditional strategies to select among a limited set of options, choosing the one with the highest fitness payoff in a given setting.

Among the Mukogodo and most other tribal groups, men pay a price to acquire a bride. Why do a few contrary cultures sanction payments that help a woman secure a husband? One answer is that in monogamous societies in which males typically invest materially in their children, parents that help their daughters "buy" the right kind of man gain fitness as a result. Males in socially stratified monogamous cultures vary greatly in their status and wealth. In such cultures, women married to elite males should generally enjoy a reproductive advantage because a monogamous husband's wealth will not be divided among a bevy of wives, but instead will go to his only wife and her children. To the extent that wealth and high status translate into reproductive success, a woman's parents may gain by competing with the parents of other families for an "alpha" husband, even if this requires that they offer a material inducement to the right male or to his family. Steven Gaulin and James Boster found that dowry payments occur in substantially less than 0.5 percent of nonstratified societies, whereas 9 percent of stratified societies and 60 percent of monogamous stratified cultures permit or encourage dowries.[525] This highly nonrandom set of associations supports the hypothesis that dowry practices are not a random cultural artifact, but part of an adaptive parental strategy. Thus, we have yet another example of how sociobiologists, far from being flummoxed by cultural diversity, can make use of it to test evolutionary hypotheses about human behavior.

## Discussion Question

**14.20** In the camel-herding Gabbra, an African tribe, the number of camels owned by a household boosts a man's reproductive success more than that of a woman. In the horticultural Chewa, the more land under cultivation, the higher the fertility of an individual, but the effect is the same for men and women (Figure 14.29).[664] In the polygynous Gabbra, inheritance is male biased, and in the Chewa, in which individuals belong to their mother's lineage, inheritance is effectively female biased. Explain why, keeping in mind that certainty of maternity (the probability that a mother's child is genetically related to her) is always greater than the certainty of paternity (the probability that a putative father's child is actually his genetic offspring). In

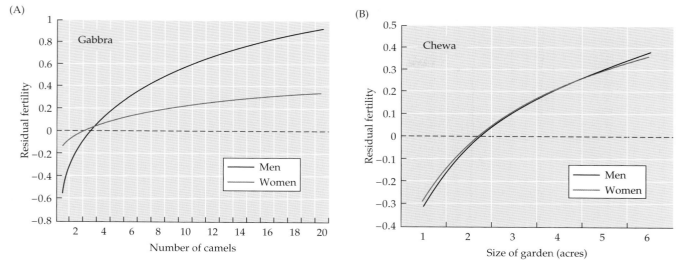

**FIGURE 14.29   Wealth affects male and female fertility differently in two African cultures.** In the Gabbra, residual fertility, a measure of reproductive success, increases more rapidly with wealth (number of camels owned) for men than it does for women. In the Chewa, the sexes do not differ in the effect that wealth (size of garden) has on their fertility. After Holden, Sear, and Mace.[664]

other words, under what conditions would grandparents gain more genetic grandchildren by investing in their daughters than in their sons? How do the data from these two cultures help answer this question?

## Applications of Evolutionary Psychology

Although the ability of evolutionary theory to contribute to a purely academic understanding of our behavior is enough to delight many biologists and psychologists, one can also argue that there is a hardheaded, applied side to this kind of approach. For example, criminal investigators working on child abuse cases in North America have surely benefited from an awareness of the statistical relationship between this kind of crime and the presence of a stepparent in the family, a connection first uncovered not by criminologists, but by two evolutionary psychologists. Likewise, I believe that young women have much to gain from an understanding of the evolutionary bases of coercive sex, especially the reality of the reproductive component of the crime, which derives from the evolved sexual psychology of males. If women recognized that coercive sex, including rape, is not practiced solely by classic criminal sociopaths, they might be more on guard against more ordinary men, whose intense sex drive enables them to feel sexual desire even when their "partners" do not.

Moreover, everyone should know that our perfectly natural, highly adaptive drive to have children is directly linked to an exploding human population that has passed 6 billion and is still climbing, much to the detriment of the Earth. Our efforts at getting things under control might be advanced if we all knew why we think fat, healthy babies are wonderful and why we believe nothing is too good for our children and grandchildren (Figure 14.30), even though the effect of having lots of plump babies and affording them a middle-class upbringing has pushed the life support system of our globe ever closer to collapse.[1115]

**FIGURE 14.30** **The author with his one and only granddaughter,** Abigail Alcock, a recent and much admired addition to an already crowded world. Photograph by Nick Alcock.

Equally in need of correction is our evolved tendency not just to reproduce enthusiastically, but to consume the Earth's resources as if there were no tomorrow. This human attribute is driven in part by the production of large families but perhaps even more so by the desire of human beings to acquire wealth and use it ostentatiously, a key component of the reproductive strategy of men. In addition, people often discount the future when it comes to resource use, which is to say that most of us tend to spend what we have now at the expense of setting something aside for the long term.[1591] Some persons have linked this aspect of human psychology to the high mortality rates that have characterized most of human evolution until very recently. Given the reality that the odds were against a long life, it would have been generally adaptive for people during our history as a species to consume resources at once and to produce children as soon as they could.

This argument can be tested by comparing the behavior of people in modern communities where life expectancy varies. Margo Wilson and Martin Daly have done the test, taking advantage of the fact that during a period around 1990, male life expectancy in different Chicago neighborhoods varied by about 25 years. If men possess a conditional strategy whose operating rules are affected by the likelihood of early death, then we would predict that individuals in high-mortality neighborhoods should be prepared to take greater risks than those living in places where death from natural causes is likely to be postponed. One dramatic manifestation of risk taking by men is their willingness to engage in extreme violence, which in turn affects homicide rates. As predicted, the rate at which homicides occurred in a neighborhood was tightly linked to male life expectancy (with the values for this figure adjusted by removal of the effects of homicide itself) (Figure 14.31).[1592] In other words, when men perceive that their lives may well be short, they are more likely to take up a life of crime and violence—a high-risk tactic, but one with a potentially high immediate payoff for the successful risk taker.

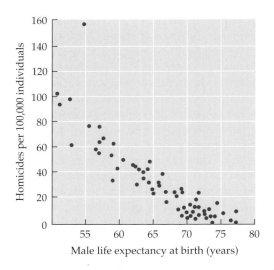

**FIGURE 14.31** **Homicide rates are highly correlated with male life expectancy.** These data came from a number of neighborhoods in Chicago in which males could expect to live anywhere from under 55 to more than 75 years (after the effects of homicides themselves on life expectancy are removed). After Wilson and Daly.[1592]

## Discussion Question

**14.21** Often teenage pregnancies are treated as a pathology of some sort. But can you make a case that women should possess a conditional strategy that takes life expectancy into account when making decisions about the onset of reproduction? If so, then what is the predicted pattern for the mean age of first pregnancy in the different Chicago neighborhoods just mentioned? See Wilson and Daly[1592] to check your answer.

The point is that not one product of natural selection, whether it be our wish to have children or our desire to have a good reputation, is guaranteed to produce socially or morally desirable results, either for us or for our species as a whole. Take our capacity for cooperation. On the one hand, humans do often band together for goals that seem admirable: making a computer program work better, trying to eliminate polio, striving to reduce white-collar crime, putting on a play or opera—the list is all but endless. But the ability to identify with a group and to form strong bonds with its members can also be employed in profoundly aggressive ways against other groups formed of people with similarly powerful cooperative abilities.

Indeed, Richard Alexander has argued that violent competition among human groups for resources led to the evolution of the extraordinary cooperation and the intensely group-centered sense of morality that people exhibit today.[17] The adaptively unpleasant nature of human morality can be illustrated by examining what the injunction "thou shalt not kill" has meant to people who have accepted it. This commandment has almost always been interpreted to mean "thou shalt not kill members of one's own tribe or community or nation—but the destruction of others living outside the band, city, or country will be accepted, even encouraged, if these others pose any threat to the communal welfare or if the resources they control can be taken from them with only modest risk of personal injury."[626] Highly selective morality of this sort can be seen in the historically frequent and widespread practice of genocide.[393]

## Discussion Question

**14.22** In writing about genocide, Stephen Jay Gould reviewed the adaptationist hypothesis that the capacity for large-scale murder evolved as a result of intense competition for resources or mates between small bands during our evolutionary history. Gould dismissed this hypothesis, saying the following: "An evolutionary speculation can only help if it teaches us something we don't know already—if, for example, we learned that genocide was biologically enjoined by certain genes, or even that a positive propensity, rather than a mere capacity, regulated our murderous potentiality. But the observational facts of human history speak against determination and only for potentiality. Each case of genocide can be matched with numerous incidents of social benevolence; each murderous clan can be paired with a pacific clan."[560] Evaluate Gould's argument critically in the light of what you know about (1) the proximate–ultimate distinction, (2) conditional strategies, and (3) the differences between adaptationist and arbitrary culture hypotheses.

Richard Wrangham has pointed out that this special kind of moral behavior practiced by humans may well be traced to an ancestral primate in the lineage that gave rise to both the common chimpanzee and *Homo sapiens*. These two species are the only living ones in which males regularly band together with the intent of assaulting, or even killing, their neighbors before

**FIGURE 14.32 Male chimpanzees cooperate in attacks on other bands.** This species and our own are the only mammals in which male teams leave their home range to assault members of other groups and then return home after the attack.

returning to their home base (Figure 14.32).[1621] Male chimpanzees patrol the borders of their territory in groups apparently searching for members from adjacent groups.[989] Should they find these other chimpanzees, they will attack them viciously, especially other adult males, but even adult females are not safe,[1517] suggesting that the adaptive function of male hostility is to keep all other chimpanzees away from the group's territory. By doing so, the attackers protect or expand their access to food within their territorial borders, which the band's females need if they are to reproduce successfully.[1431] One study found that the size of the defended area was correlated with the frequency with which males mated with sexually receptive females belonging to their group, which led to increases in male fitness.[1581]

The shared components of group aggression in common chimpanzees and humans, two very closely related species, suggest that solicitous support for one's own group and often profound and violent antipathy for all others may be traced to an extinct forest-dwelling great ape, which endowed the two modern species with the hereditary rudiments of our tribalism. These traits, modified over time, have presumably been maintained until the present because of the selective advantages they have conferred upon individuals capable of working together to defeat competitors for the limited resources upon which individual reproductive success depends.

The attempts by evolutionary biologists to understand our moral systems and other features of our behavior are far from complete, although progress has been made.[18, 635] One thing is certain—evolutionary analyses have demonstrated beyond doubt that just because something is "adaptive" or "natural" or "evolved" does not mean that it is "good" or "moral" or "desirable." Once you know that our genes have the capacity to make us work on their behalf without regard for our welfare or that of most other people, then it may be easier to fight against our evolved impulses.[1580] Knowledge of evolutionary outcomes could, to name just one thing, make us less susceptible to exploitation by unscrupulous individuals.[264, 288] Informed persons could also better avoid the moral certainty and self-righteousness that enables us to demonize and dehumanize our opponents prior to attacking and killing them. When Pogo declared, "We have met the enemy and he is us," he was correct, and those of us who understand the role of natural selection in shaping our evolutionary history know why he was right. Perhaps some of you can use that understanding to help make the human species less an enemy to itself.

## Summary

1. Human beings are an evolved animal species. Human sociobiology, a field of study that includes evolutionary psychology and evolutionary anthropology, employs natural selection theory to generate testable hypotheses about the possible adaptive value of our species' behavior.

2. The study of sociobiology has been marked by intense controversy, in part because some persons have misunderstood the goals and foundations of the discipline. Contrary to the views of some critics, sociobiologists are not motivated by a political agenda, nor are sociobiological hypotheses based on the premise that elements of human behavior are genetically determined or morally desirable. Sociobiological hypotheses are designed to explain, not justify, our behavior. Testing these hypotheses can help us understand why we have psychological mechanisms that motivate us to behave in certain ways.

3. Unlike arbitrary culture theory, which proposes that human behavior is the arbitrary product of cultural traditions constrained only by the limits of our imagination, sociobiological theory views human behavior as the product of natural selection. It assumes that human beings have conditional strategies that can be used to increase, rather than decrease, the fitness of individuals.

4. Many sociobiological hypotheses have been advanced for elements of human reproductive behavior, as illustrated by evolutionary analyses of the different kinds of cues important to men and women in evaluating potential mates. Males prefer those characteristics associated with high fertility; females typically place more emphasis on the resources held by potential partners. The decisions that people make based on these psychological mechanisms would have generally increased the fitness of individuals living in pre-contraceptive environments. The same mechanisms can, however, also lead to behavior that most people would condemn, such as coercive sex and marital infidelity.

5. Elements of parental behavior, like elements of mate choice, can be explained in terms of their contribution to individual reproductive success. Even apparently maladaptive actions, such as child abuse by stepfathers, can be better understood as by-products of fitness-maximizing mechanisms that evolved because of their other consequences, such as the generally adaptive desire to limit one's parental investment to one's genetic offspring.

6. When a claim is made that something is adaptive or natural, this does not mean that it is desirable, moral, or necessary, only that it tends to propagate our genes. Widespread acceptance of this point might enable us to understand that trusting our impulses is more likely to help us pass on our genes than to produce wise decisions that maximize either our personal happiness or the general good.

## Suggested Reading

The debate on sociobiology began with E. O. Wilson; see the final chapter of *Sociobiology*[1588] and an attack on the book in Allen et al.[21] and Wilson's reply.[1589] I summarize the debate and its outcome in *The Triumph of Sociobiology*.[14] The misconceptions that many persons have about human sociobiology have been outlined by two of my colleagues, Doug Kenrick[763] and Owen Jones.[739] Martin Daly and Margo Wilson critically examine some critiques of the evolutionary explanation for child abuse by stepparents, one of the best examples of sociobiological research.[345] Martin Daly and Margo Wilson have also written a compact chapter on evolution and human behavior.[344] Most of the classic papers on the evolutionary analysis of human behavior can be found, along with updates and critiques, in *Human Nature*, edited by Laura Betzig.[118] A textbook treatment of an evolutionary approach to human behavior comes from Robert Boyd and Joan Silk.[155] Textbooks that focus on evolutionary psychology include those by David Buss[222] and Steve Gaulin and Donald McBurney.[526] Readers can also find many interesting articles on human behavior in the journal *Evolution and Human Behavior*.

# Glossary

**Action potential**   The neural signal; a self-regenerating change in membrane electrical charge that travels the length of a nerve cell.

**Adaptation**   A characteristic that confers higher inclusive fitness to individuals than any other existing alternative exhibited by other individuals within the population; a trait that has spread or is spreading or is being maintained in a population as a result of natural selection or indirect selection.

**Adaptationist**   A behavioral biologist who develops and tests hypotheses on the possible adaptive value of a particular trait. Persons using an adaptationist approach test whether a given trait enables individuals to propagate their special genes more effectively than if they were to behave in some other manner.

**Adaptive value**   The contribution that a trait or gene makes to inclusive fitness.

**Allele**   A form of a gene; different alleles typically code for distinctive variants of the same enzyme.

**Altruism**   Helpful behavior that raises the recipient's direct fitness while lowering the donor's direct fitness.

**Arbitrary culture theory**   The view that human behavior is the arbitrary product of whatever cultural traditions people are exposed to within their society; thus, our actions are not expected to be explicable in evolutionary terms.

**Artificial selection**   See *Selection*.

**Associated reproductive pattern**   A seasonal change in reproductive behavior that is tightly correlated with changes in the gonads and hormones, in contrast to a dissociated reproductive pattern, in which the onset of reproductive behavior is apparently not triggered by a sharp change in circulating hormones.

**Biological clock**   An internal physiological mechanism that enables organisms to time any of a wide assortment of biological processes and activities.

**Brood parasite**   An animal that exploits the parental care of individuals other than its parents.

**By-product hypothesis** An explanation for a maladaptive or nonadaptive attribute that is said to occur as a by-product of a proximate mechanism that has some other adaptive consequence for individuals.

**Central pattern generator** A group of cells in the central nervous system that produces a particular pattern of signals necessary for a functional behavioral response.

**Chase-away selection** See *Selection*.

**Circadian rhythm** A roughly 24-hour cycle of behavior that expresses itself independent of environmental changes.

**Circannual rhythm** A annual cycle of behavior that expresses itself independent of environmental changes.

**Coefficient of relatedness** The probability that an allele present in one individual will be present in a close relative; the proportion of the total genotype of one individual present in the other, as a result of shared ancestry.

**Command center** A neural cluster or an integrated set of clusters that has primary responsibility for the control of a particular behavioral activity.

**Comparative method** A procedure for testing evolutionary hypotheses based on disciplined comparisons among species of known evolutionary relationships.

**Conditional strategy** See *Strategy*.

**Convergent evolution** The independent acquisition over time through natural selection of similar characteristics in two or more unrelated species.

**Cooperation** A mutually helpful action.

**Cost–benefit approach** A method for studying the adaptive value of alternative traits based on the recognition that phenotypes come with fitness costs and fitness benefits; an adaptation has a better cost–benefit ratio than alternative versions of that trait.

**Cryptic female choice** The ability of a female in receipt of sperm from more than one male to choose whose sperm get to fertilize her eggs.

**Darwinian puzzle** A trait that appears to reduce the fitness of individuals that possess it; traits of this sort attract the attention of evolutionary biologists.

**Developmental homeostasis** The capacity of developmental mechanisms within individuals to produce adaptive traits, despite potentially disruptive effects of mutant genes and suboptimal environmental conditions.

**Dilution effect** Safety in numbers that comes from swamping the ability of local predators to consume prey.

**Diploid** Having two copies of each gene in one's genotype.

**Direct fitness** See *Fitness*.

**Direct selection** See *Selection*.

**Display** A stereotyped action used as a communication signal by individuals.

**Dissociated reproductive pattern** See *Associated reproductive pattern*.

**Divergent evolution** The evolution by natural selection of differences among closely related species that live in different environments and are therefore subject to different selection pressures.

**Dominance hierarchy** A social ranking within a group, in which some individuals give way to others, often conceding useful resources to others without a fight.

**Entrainment** Involves the resetting of a biological clock so that an organism's activities are scheduled in keeping with local conditions.

**Ethology** The study of the proximate mechanisms and adaptive value of animal behavior.

**Eusociality** A social system in which specialized nonreproducing castes work for the reproductive members of the group.

**Evolutionarily stable strategy** See *Strategy*.

**Evolutionary psychology** The study of the adaptive value of psychological mechanisms, especially of human beings; a key component of sociobiology.

**Explosive breeding assemblage** The temporary formation of large groups of mating individuals.

**Extra-pair copulation** A mating by a male or female with someone other than his or her primary partner in a seemingly monogamous species.

**Female defense polygyny** See *Polygyny*.

**Female-enforced monogamy** See *Monogamy*.

**Female preference hypothesis** An explanation for the formation of leks in which females prefer to choose mates from a group rather than by inspecting potential mates one by one.

**Fertility insurance hypothesis** An explanation for why females might mate with more than one male per breeding cycle, with the benefit being an increase in egg fertilization rate.

**Fitness** A measure of the genes contributed to the next generation by an individual, often stated in terms of the number of surviving offspring produced by the individual.

    **Direct fitness** The genes contributed by an individual via personal reproduction in the bodies of surviving offspring.

    **Indirect fitness** The genes contributed by an individual indirectly by helping nondescendant kin, in effect creating relatives that would not have existed without the help of the individual.

**Inclusive fitness**   The sum of an individual's direct and indirect fitness.

**Fitness benefit**   That aspect of a trait that tends to raise the inclusive fitness of individuals.

**Fitness cost**   That aspect of a trait that tends to reduce the inclusive fitness of individuals.

**Fixed action pattern**   An innate, highly stereotyped response that is triggered by a well-defined, simple stimulus; once the pattern is activated, the response is performed in its entirety.

**Free-running cycle**   The cycle of activity of an individual that is expressed in a constant environment.

**Frequency-dependent selection**   See *Selection.*

**Game theory**   An evolutionary approach to study of adaptive value in which the payoffs to individuals associated with one behavioral tactic are dependent on what the other members in the group are doing.

**Gene**   A segment of DNA, typically one that encodes information about the sequence of amino acids that makes up a protein.

**Genetic compatibility**   The ability of genes present in some sperm to complement the genes present in some eggs, resulting in an increased likelihood of the development of superior offspring.

**Genotype**   The genetic constitution of an individual; may refer to the alleles of one gene possessed by the individual or to its complete set of genes.

**Good genes theory**   The argument that mate choice advances individual fitness because it provides the offspring of choosy individuals with genes that promote reproductive success by advancing the offspring's chances of survival or reproductive success.

**Good parent theory**   An explanation for female preferences for males whose appearance or behavior signals that these potential mates are likely to provide above-average parental care for their offspring.

**Group selection**   See *Selection.*

**Hamilton's rule**   The argument made by W. D. Hamilton that altruism can spread through a population where $rB > C$ (with $r$ being the coefficient of relatedness between the altruist and the individual helped, $B$ being the fitness benefit received by the helped individual, and $C$ being the cost of altruism in terms of the direct fitness lost by the altruist due to his or her actions).

**Haploid**   Having only one copy of each gene in the genotype, as for example in the sperm and eggs of diploid organisms.

**Healthy mates theory**   An explanation for preferences by females for males whose appearance or behavior signals that these potential mates are unlikely to transmit communicable diseases or parasites to them.

**Home range**   An area that an animal occupies but does not defend, in contrast to a territory, which is defended.

**Honest signals**   Signals that convey accurate information about the signaler's real fighting ability or true value as a potential mate.

**Hotshot**   A male whose attributes are especially appealing to sexually receptive females.

**Hot spot**   A location whose properties attract sexually receptive females to the male able to hold the site against rival males.

**Hypothesis**   A tentative explanation that will require testing before acceptance.

**Ideal free distribution**   The spatial distribution of individuals that are free to choose where to go in ways that could maximize individual fitness.

**Illegitimate receiver**   An individual that listens to the signals of others, thereby gaining information that it uses to reduce the fitness of the signaler.

**Illegitimate signaler**   An individual that produces signals that may deceive others into responding in ways that reduce the fitness of the signal receiver.

**Imprinting**   A form of learning in which individuals exposed to certain key stimuli, usually early in life, form an association with an object (or individual) and may later show sexual behavior toward similar objects.

**Inbreeding depression**   The tendency of inbred organisms to have lower fitness than non-inbred members of their species.

**Inclusive fitness**   See *Fitness.*

**Indirect fitness**   See *Fitness.*

**Indirect reciprocity**   See *Reciprocity.*

**Indirect selection**   See *Selection.*

**Innate releasing mechanism**   A hypothetical neural mechanism thought to control an innate response to a sign stimulus.

**Instinct**   A behavior pattern that reliably develops in most individuals, promoting a functional response to a releaser stimulus the first time the behavior is performed.

**Interneuron**   A nerve cell that relays messages either from receptor neurons to the central nervous system (a sensory interneuron) or from the central nervous system to neurons commanding muscle cells (a motor interneuron).

**Kin selection**   See *Selection.*

**Learning**   A durable and usually adaptive change in an animal's behavior traceable to a specific experience in the individual's life.

**Lek**  A traditional display site that females visit to select a mate from among the males displaying at their small resource-free territories.

**Lek polygyny**  See *Polygyny*.

**Mate-assistance monogamy**  See *Monogamy*.

**Mate guarding**  Actions taken by males (usually) to prevent a sexual partner from acquiring sperm from other males. See also *Mate-guarding monogamy*.

**Mate-guarding monogamy**  See *Monogamy*.

**Material benefits hypothesis**  An explanation for why females of some species might mate with several males per breeding cycle, with benefits coming to polyandrous females because they gain access to the material benefits controlled by the several males.

**Migration**  The regular movement back and forth between two relatively distant locations by animals that use resources concentrated in these different sites.

**Mobbing behavior**  When prey closely approach and attempt to harass a predator.

**Monogamy**  A mating system in which one male mates with just one female, and one female mates with just one male, in a breeding season.

> **Female-enforced monogamy**  A mating system in which females prevent their mates from copulating with more than one individual, resulting in partnerships of one male and one female.

> **Mate-assistance monogamy**  Monogamy that arises because a male gains more fitness by offering parental care for the offspring of his mate than by seeking out additional sexual partners.

> **Mate-guarding monogamy**  The mating system that occurs when one or the other member of a pair guards his or her partner in ways that prevent that partner from acquiring an additional mate.

**Mutualism**  A mutually beneficial relationship or cooperative action.

**Natural selection**  See *Selection*.

**Neuron**  A nerve cell.

**Nondescendant kin**  Relatives other than offspring.

**Nuclei**  Dense clusters of cell bodies of neurons within nervous systems.

**Nuptial gift**  A food item transferred by a male to a female just prior to or during copulation.

**Operant conditioning**  A kind of learning based on trial and error, in which an action, or operant, becomes more frequently performed if it is rewarded.

**Operational sex ratio (OSR)**  The ratio of receptive males to receptive females over a given period.

**Optimality theory**  An evolutionary theory based on the assumption that the attributes of organisms are optimal, i.e., better than others in terms of the ratio of fitness benefits to costs; the theory is used to generate hypotheses about the possible adaptive value of traits in terms of the net fitness gained by individuals that exhibit these attributes.

**Parent–offspring conflict**  The clash of interests that occurs when parents can gain fitness by withholding parental care or resources from some offspring in order to invest in others now or later, even though the deprived offspring would gain more fitness by receipt of parental care or resources.

**Parental investment**  Costly parental activities that increase the likelihood of survival for some existing offspring but that reduce the parent's chances of producing offspring in the future.

**Phenotype**  Any measurable aspect of an individual that arises from an interaction of the individual's genes with its environment.

**Pheromone**  A volatile chemical released by an individual as a scent signal for another.

**Photoperiod**  The number of hours of light in 24 hours.

**Phylogeny**  An evolutionary genealogy of the relationships among a number of species or clusters of species that can be used to develop hypotheses on the evolutionary history of a given trait.

**Pleiotropy**  The capacity of a gene to have multiple developmental effects on individuals.

**Polyandry**  A mating system in which a female has several partners in a breeding season.

**Polygyny**  A mating system in which a male fertilizes the eggs of several partners in a breeding season.

> **Female defense polygyny**  Polygynous males directly defend several mates.

> **Lek polygyny**  Polygynous males attract several mates to a display territory.

> **Resource defense polygyny**  Polygynous males acquire several mates attracted to resources under the males' control.

> **Scramble competition polygyny**  Polygynous males acquire several widely scattered mates by finding them first.

**Polygyny threshold model**  An explanation for polygyny based on the premise that females will gain fitness by mating with an already paired male if the resources controlled by that male greatly exceed those under the control of unmated males.

**Polyphenism**  The occurrence within a species of two or more alternative phenotypes whose differences are induced by key differences in the environments experienced by individual members of this species.

**Prisoner's dilemma**   A game theory construct in which the fitness payoffs to individuals are set such that mutual cooperation between the players generates a lower return than defection, which occurs when one individual accepts assistance from the other but does not return the favor.

**Proximate cause**   An immediate, underlying cause based on the operation of internal mechanisms possessed by an individual.

**Reciprocity**   Also known as **reciprocal altruism** in which a helpful action is repaid at a later date by the recipient of assistance.

   **Indirect reciprocity**   A form of reciprocity in which a helpful action is repaid at a later date by individuals other than the recipient of assistance.

**Releaser**   A sign stimulus given by an individual as a social signal to another.

**Reproductive value**   A measure of the probability that a given offspring will reach the age of reproduction, or the potential of an individual to leave surviving descendants.

**Reproductive success**   The number of surviving offspring produced by an individual; direct fitness.

**Resource defense polygyny**   See *Polygyny*.

**Resource-holding power**   The inherent capacity of an individual to defeat others when competing for useful resources.

**Runaway selection**   See *Selection*.

**Satellite male**   A male that waits near another male to intercept females drawn to the signals produced by the other male or attracted by the resources defended by the other male.

**Scientific conclusion**   In the scientific method, a hypothesis that has been tested and then accepted or rejected on the basis of the test results.

**Scramble competition polygyny**   See *Polygyny*.

**Selection**   The effects of differences among individuals in their ability to transmit copies of their genes to the next generation.

   **Artificial selection**   A process that is identical with natural selection, except that humans control the reproductive success of alternative types within the selected population.

   **Chase-away selection**   The reciprocal, spiraling effects of males attempting to exploit female mate choice mechanisms while females are evolving resistance to these attempts.

   **Direct selection**   A synonym for natural selection acting on hereditary differences among individuals in the production of surviving offspring.

**Frequency-dependent selection**   A form of natural selection in which those individuals that happen to belong to the less common of two types in the population are the ones that are more fit because of their lower frequency in the population.

**Group selection**   The process that occurs when groups differ in their collective attributes and the differences affect the survival chances of the groups.

**Indirect selection**   The process that occurs when individuals differ in their effects on the survival of nondescendant kin, creating differences in the indirect fitness of the individuals interacting with this category of kin.

**Kin selection**   The process that occurs when individuals differ in ways that affect their parental care or helping behavior, and thus the survival of their own offspring or the survival of nondescendant kin.

**Natural selection (direct selection)**   The process that occurs when individuals differ in their traits and the differences are correlated with differences in reproductive success. Natural selection can produce evolutionary change when these differences are inherited.

**Runaway selection**   A form of sexual selection that occurs when female mating preferences for certain male attributes create a positive feedback loop favoring both males with these attributes and females that prefer them.

**Sexual selection**   A form of natural selection that occurs when individuals vary in their ability to compete with others for mates or to attract members of the opposite sex. As with natural selection, when the variation among individuals is correlated with genetic differences, sexual selection leads to genetic changes in the population.

**Selfish herd**   A group of individuals whose members use others as living shields against predators.

**Sensory exploitation**   The evolution of signals that happen to activate established sensory systems of signal receivers in ways that elicit responses favorable to the signal sender.

**Sex role reversal**   A change in the typical behavior patterns of males and females as when, for example, females compete for access to males, and when males choose selectively among potential mates.

**Sexual selection**   See *Selection*.

**Siblicide**   The killing of a sibling by a brother or sister.

**Sign stimulus**   The effective component of an action or object that triggers a fixed action pattern in an animal.

**Sociobiology**   A discipline that uses evolutionary theory as the foundation for the study of social behavior; often used to refer to studies of this sort involving human beings.

**Sperm competition**    The competition between males that determines whose sperm will fertilize a female's eggs when both males' sperm have been accepted by that female.

**Stimulus filtering**    The capacity of nerve cells and neural networks to ignore stimuli that could potentially elicit a response from them.

**Strategy**    A genetically distinctive set of rules for behavior exhibited by individuals.

**Conditional strategy**    A set of rules that enables individuals to use different tactics under different environmental conditions; the inherited behavioral capacity to be flexible in response to certain cues or situations.

**Evolutionarily stable strategy**    That set of rules of behavior that when adopted by a certain proportion of the population cannot be replaced by any alternative strategy.

**Synapse**    The point of near contact between one nerve cell and another.

**Tactic**    A behavior pattern that is enabled by an underlying hereditary mechanism of some sort; tactics are often referred to when someone wishes to distinguish between an option available to an individual as opposed to a strategy, which specifies a fixed response to a particular situation.

**Territorial**    Exhibiting a readiness to defend an area against intruders.

**Transactional theory**    The view that social units form as a result of the ability of individuals to negotiate for reproductive opportunities with one another; concession theory is derived from the transactional approach based on the premise that dominant members of the group concede some reproductive rights to others in return for their cooperation while staying with the group.

**Ultimate cause**    The evolutionary, historical reason why something is the way it is.

# Bibliography

1. Abbot P, Withgott JH, Moran NA (2001) Genetic conflict and conditional altruism in social aphid colonies. *Proceedings of the National Academy of Sciences USA* 98: 12068–12071.

2. Acharya L, McNeil JN (1998) Predation risk and mating behavior: The responses of moths to bat-like ultrasound. *Behavioral Ecology* 9: 552–558.

3. Adkins-Regan E (1981) Hormone specificity, androgen metabolism, and social behavior. *American Zoologist* 21: 257–271.

4. Airey DC, Castillo-Juarez H, Casella G, Pollak EJ, DeVoogd TJ (2000) Variation in the volume of zebra finch song control nuclei is heritable: Developmental and evolutionary implications. *Proceedings of the Royal Society of London, Series B* 267: 2099–2104.

5. Airey DC, DeVoogd TJ (2000) Greater song complexity is associated with augmented song system anatomy in zebra finches. *NeuroReport* 11: 2339–2344.

6. Alatalo RV, Lundberg A (1984) Polyterritorial polygyny in the pied flycatcher *Ficedula hypoleuca*—evidence for the deception hypothesis. *Annales Zoologici Fennici* 21: 217–228.

7. Alaux C, Robinson GE (2007) Alarm pheromone induces immediate-early gene expression and slow behavioral response in honey bees. *Journal of Chemical Ecology* 33: 1346–1350.

8. Alberts SC, Watts HE, Altmann J (2003) Queuing and queue-jumping: long-term patterns of reproductive skew in male savannah baboons, *Papio cynocephalus*. *Animal Behaviour* 65: 821–840.

9. Alberts SC, Buchan JC, Altmann J (2006) Sexual selection in wild baboons: from mating opportunities to paternity success. *Animal Behaviour* 72: 1177–1196.

10. Albrecht T, Klvana P (2004) Nest crypsis, reproductive value of a clutch and escape decisions in incubating female mallards *Anas platyrhynchos*. *Ethology* 110: 603–614.

11. Alcock J, Eickwort GC, Eickwort KR (1977) The reproductive behavior of *Anthidium maculosum* and the evolutionary significance of multiple copulations by females. *Behavioral Ecology and Sociobiology* 2: 385–396.

12. Alcock J, Bailey WJ (1995) Acoustical communication and the mating system of the Australian whistling moth *Hecatesia exultans* (Noctuidae: Agaristidae). *Journal of Zoology* 237: 337–352.

13. Alcock J, Bailey WJ (1997) Success in territorial defence by male tarantula hawk wasps *Hemipepsis ustulata*: The role of residency. *Ecological Entomology* 22: 377–383.

14. Alcock J (2001) *The Triumph of Sociobiology*. Oxford University Press, New York.

15. Alexander GM, Hines M (2002) Sex differences in response to children's toys in nonhuman primates (*Cercopithecus aethiops sabaeus*). *Evolution and Human Behavior* 23: 467–479.

16. Alexander RD (1974) The evolution of social behavior. *Annual Review of Ecology and Systematics* 5: 325–383.

17. Alexander RD (1979) *Darwinism and Human Affairs*. University of Washington Press, Seattle, WA.

18. Alexander RD (1987) *The Biology of Moral Systems*. Aldine de Gruyter, Hawthorne, NY.

19. Allen BJ, Levinton JS (2007) Cost of bearing a sexually selected ornamental weapon in a fiddler crab. *Functional Ecology* 21: 154–161.

20. Allen GR, Kazmer DJ, Luck RF (1994) Post-copulatory male behaviour, sperm precedence and multiple mating in a solitary parasitoid wasp. *Animal Behaviour* 48: 635–644.

21. Allen L, Beckwith B, Beckwith J, Chorover S, Culver D, Daniels N, Dorfman D (1975) Against "sociobiology." *New York Review of Books* 22 (Nov. 13): 43–44.

22. Allen L, Beckwith B, Beckwith J, Chorover S, Culver D, Duncan M, Gould SJ (1976) Sociobiology—another biological determinism. *BioScience* 26: 182–186.

23. Allison T, Puce A, McCarthy G (2000) Social perception from visual cues: The role of the STS region. *Trends in Cognitive Sciences* 4: 267–278.

24. Allman J, Rosin A, Kumar R, Hasenstaub A (1998) Parenting and survival in anthropoid primates: Caretakers live longer. *Proceedings of the National Academy of Sciences USA* 95: 6866–6869.

25. Allman J (1999) *Evolving Brains*. Scientific American Library, New York.

26. Alvergne A, Faurie C, Raymond M (2007) Differential facial resemblance of young children to their parents: who do children look like more? *Evolution and Human Behavior* 28: 135–144.

27. Amdan GV, Csondes A, Fondrk MK, Page RE (2006) Complex social behavior derived from maternal reproductive traits. *Nature* 439: 76–78.

28. Ancel A, Visser H, Handrich Y, Masman D, Maho YL (1997) Energy saving in huddling penguins. *Nature* 385: 304–305.

29. Anderson JB, Brower LP (1996) Freeze-protection of overwintering monarch butterflies in Mexico: Critical role of the forest as a blanket and an umbrella. *Ecological Entomology* 21: 107–116.

30. Anderson KG, Kaplan H, Lancaster J (1999) Paternal care by genetic fathers and stepfathers I: Reports from Albuquerque men. *Evolution and Human Behavior* 20: 405–432.

31. Andersson J, Borg-Karlson A-K, Wiklund C (2000) Sexual cooperation and conflict in butterflies: A male-transferred antiaphrodisiac reduces harassment of recently mated females. *Proceedings of the Royal Society of London, Series B* 267: 1271–1275.

32. Andersson M, Gotmark F, Wiklund CG (1981) Food information in the black-headed gull, *Larus ridibundus*. *Behavioral Ecology and Sociobiology* 9: 199–202.

33. Andersson M (1982) Female choice selects for extreme tail length in a widowbird. *Nature* 299: 818–820.

34. Andersson M (1984) The evolution of eusociality. *Annual Review of Ecology and Systematics* 15: 165–189.

35. Andersson M (1994) *Sexual Selection*. Princeton University Press, Princeton, NJ.

36. Andersson S, Amundsen T (1997) Ultraviolet colour vision and ornamentation in bluethroats. *Proceedings of the Royal Society of London, Series B* 264: 1587–1591.

37. Andrade MCB (1996) Sexual selection for male sacrifice in the Australian redback spider. *Science* 271: 70–72.

38. Andrade MCB (1998) Female hunger can explain variation in cannibalistic behavior despite male sacrifice in redback spiders. *Behavioral Ecology* 9: 33–42.

39. Andrade MCB (2003) Risky mate search and male self-sacrifice in redback spiders. *Behavioral Ecology* 14: 531–538.

40. Angier N (1999) *Woman, An Intimate Geography*. Houghton Mifflin Company, New York.

41. Apicella CL, Marlowe FW (2004) Perceived mate fidelity and paternal resemblance predict men's investment in children. *Evolution and Human Behavior* 25: 371–378.

42. Apicella CL, Feinberg DR, Marlowe FW (2007) Voice pitch predicts reproductive success in male hunter-gatherers. *Biology Letters* 3: 682–684.

43. Apollonio M, Festa-Bianchet M, Mari F, Bruno E, Locati M (1998) Habitat manipulation modifies lek use in fallow deer. *Ethology* 104: 603–612.

44. Aragón S, Møller AP, Soler JJ, Soler M (1999) Molecular phylogeny of cuckoos supports a polyphyletic origin of brood parasitism. *Journal of Evolutionary Biology* 12: 495–506.

45. Arnold SJ (1980) The microevolution of feeding behavior. In: Kamil A, Sargent T (eds) *Foraging Behavior: Ecology, Ethological and Psychological Approaches*. Garland STPM Press, New York.

46. Arnold SJ (1983) Sexual selection: The interface of theory and empiricism. In: Bateson PPG (ed) *Mate Choice*. Cambridge University Press, Cambridge.

47. Arnold TW (1999) What limits clutch size in waders? *Journal of Avian Biology* 30: 216–220.

48. Arnqvist G, Kirkpatrick M (2005) The evolution of infidelity in socially monogamous passerines: The strength of direct and indirect selection on extrapair copulation behavior in females. *American Naturalist* 165: S26–S37.

49. Arnqvist G (2006) Sensory exploitation and sexual conflict. *Philosophical Transactions of the Royal Society of London, Series B* 361: 375–386.

50. Arnqvist G, Kirkpatrick M (2007) The evolution of infidelity in socially monogamous passerines revisited: A reply to Griffith. *American Naturalist* 169: 282–283.

51. Aspi J, Hoffman AA (1998) Female encounter rates and fighting costs of males are associated with lek size in *Drosophila mycetophaga*. *Behavioral Ecology and Sociobiology* 42: 163–170.

52. Averof M, Cohen SM (1997) Evolutionary origins of insect wings from ancestral gills. *Nature* 385: 627–630.

53. Avise JC (2004) *Molecular Markers, Natural History and Evolution*. Sinauer Associates, Sunderland, MA.

54. Axelrod R, Hamilton WD (1981) The evolution of cooperation. *Science* 211: 1390–1396.

55. Ayre DJ, Grosberg RK (2005) Behind anemone lines: factors affecting division of labour in the social cnidarian *Anthopleura elegantissima*. *Animal Behaviour* 70: 97–110.

56. Bachmann GC, Chappell MA (1998) The energetic cost of begging behaviour in nestling house wrens. *Animal Behaviour* 55: 1607–1618.

57. Backwell PRY, Jennions MD (2004) Coalition among male fiddler crabs. *Nature* 439: 414–417.

58. Badyaev AV, Foresman KR, Fernandes MV (2000) Stress and developmental stability: Vegetation removal causes increased fluctuating asymmetry in shrews. *Ecology* 81: 336–345.

59. Bailey JM, Gaulin S, Agyei Y, Gladue BA (1994) Effects of gender and sexual orientation on evolutionarily relevant aspects of human mating psychology. *Journal of Personality and Social Psychology* 66: 1081–1093.

60. Baker AJ, Gonzalez PM, Piersma T, Niles LJ, do Nascimento IDS, Atkinson PW, Clark NA, Minton CDT, Peck MK, Aarts G (2004) Rapid population decline in red knots: fitness consequences of decreased refuelling rates and late arrival in Delaware Bay. *Proceedings of the Royal Society of London, Series B* 275: 875–882.

61. Baker MC, Bottjer SW, Arnold AP (1984) Sexual dimorphism and lack of seasonal-changes in vocal control regions of the white-crowned sparrow brain. *Brain Research* 295: 85–89.

62. Baker MC, Cunningham MA (1985) The biology of bird-song dialects. *Behavioral and Brain Sciences* 8: 85–133.

63. Bakst MR (1998) Structure of the avian oviduct with emphasis on sperm storage in poultry. *Journal of Experimental Zoology* 282: 618–626.

64. Balcombe JP (1990) Vocal recognition of pups by mother Mexican free-tailed bats, *Tadarida brasiliensis mexicana*. *Animal Behaviour* 39: 960–966.

65. Balda RP (1980) Recovery of cached seeds by a captive *Nucifraga caryocatactes*. *Zeitschrift für Tierpsychologie* 52: 331–346.

66. Balda RP, Kamil AC (1992) Long-term spatial memory in Clark's nutcracker, *Nucifraga columbiana*. *Animal Behaviour* 44: 761–769.

67. Ballentine B, Hyman J, Nowicki S (2004) Vocal performance influences female response to male bird song: an experimental test. *Behavioral Ecology* 15: 163–168.

68. Ballew CC, Todorov A (2007) Predicting political elections from rapid and unreflective face judgments. *Proceedings of the National Academy of Sciences USA* 104: 17948–17953.

69. Balmford A, Deutsch JC, Nefdt RJC, Clutton-Brock T (1993) Testing hotspot models of lek evolution: Data from three species of ungulates. *Behavioral Ecology and Sociobiology* 33: 57–65.

70. Balshine-Earn S (1995) The costs of parental care in Galilee St Peter's fish, *Sarotherodon galilaeus*. *Animal Behaviour* 50: 1–7.

71. Balthazart J, Baillien M, Charlier TD, Cornil CA, Ball GF (2003) The neuroendocrinology of reproductive behavior in Japanese quail. *Domestic Animal Endocrinology* 25: 69–82.

72. Baptista LF, Petrinovich L (1984) Social interaction, sensitive phases and the song template hypothesis in the white-crowned sparrow. *Animal Behaviour* 32: 172–181.

73. Baptista LF, Petrinovich L (1986) Song development in the white-crowned sparrow: Social factors and sex differences. *Animal Behaviour* 34: 1359–1371.

74. Baptista LF, Morton ML (1988) Song learning in montane white-crowned sparrows: From whom and when. *Animal Behaviour* 36: 1753–1764.

75. Barber JR, Conner WE (2007) Acoustic mimicry in a predator-prey interaction. *Proceedings of the National Academy of Sciences USA* 104: 9331–9334.

76. Barclay P, Willer R (2007) Partner choice creates competitive altruism in humans. *Proceedings of the Royal Society of London, Series B* 274: 749–753.

77. Barta Z, Giraldeau LA (2001) Breeding colonies as information centers: a reappraisal of information-based hypotheses using the producer-scrounger game. *Behavioral Ecology* 12: 121–127.

78. Bartlett TQ, Sussman RW, Cheverud JM (1993) Infant killing in primates: A review of observed cases with special reference to the sexual selection hypothesis. *American Anthropologist* 95: 958–990.

79. Basolo AL (1990) Female preference predates the evolution of the sword in swordtail fish. *Science* 250: 808–810.

80. Basolo AL (1995) Phylogenetic evidence for the role of a pre-existing bias in sexual selection. *Proceedings of the Royal Society of London, Series B* 259: 307–311.

81. Basolo AL (1998) Evolutionary change in a receiver bias: A comparison of female preference functions. *Proceedings of the Royal Society of London, Series B* 265: 2223–2228.

82. Bass AH (1996) Shaping brain sexuality. *American Scientist* 84: 352–363.

83. Bateman AJ (1948) Intra-sexual selection in *Drosophila*. *Heredity* 2: 349–368.

84. Baum DA, Larson A (1991) Adaptation reviewed: A phylogenetic methodology for studying character macroevolution. *Systematic Zoology* 40: 1–18.

85. Baylies MK, Bargiello TA, Jackson FR, Young MW (1987) Changes in abundance or structure of the *Per* gene-product can alter periodicity of the *Drosophila* clock. *Nature* 326: 390–392.

86. Beecher MD, Beecher IM (1979) Sociobiology of bank swallows: Reproductive strategy of the male. *Science* 205: 1282–1285.

87. Beecher MD, Beecher IM, Hahn S (1981) Parent-offspring recognition in bank swallows, *Riparia riparia*: II. Development and acoustic basis. *Animal Behaviour* 29: 95–101.

88. Beecher MD (1982) Signature systems and kin recognition. *American Zoologist* 22: 477–490.

89. Beecher MD, Medvin MB, Stoddard PK, Loesche P (1986) Acoustic adaptations for parent-offspring recognition in swallows. *Experimental Biology* 45: 179–193.

90. Beecher MD, Stoddard PK, Campbell SE, Horning CL (1996) Repertoire matching between neighbouring song sparrows. *Animal Behaviour* 51: 917–923.

91. Beecher MD, Campbell E, Nordby JC (2000) Territory tenure in song sparrows is related to song sharing with neighbours, but not to repertoire size. *Animal Behaviour* 59: 29–37.

92. Beecher MD, Campbell SE, Burt JM, Hill CE, Nordby JC (2000) Song-type matching between neighbouring song sparrows. *Animal Behaviour* 59: 21–27.

93. Beecher MD, Burt JM (2004) The role of social interaction in bird song learning. *Current Directions in Psychological Science* 13: 224–228.

94. Beecher MD, Campbell E (2005) The role of unshared song in the singing interactions between neighbouring song sparrows. *Animal Behaviour* 70: 1297–1304.

95. Beecher MD, Burt JM, O'Loghlen AL, Templeton CN, Campbell SE (2007) Bird song learning in an eavesdropping context. *Animal Behaviour* 73: 929–935.

96. Beering M (2001) *A Comparison of the Patterns of Dance Language Behavior in House-hunting and Nectar-foraging Honey Bees*. University of California, PhD. thesis. Riverside, CA.

97. Behrmann M, Winocur G, Moscovitch M (1992) Dissociation between mental imagery and object recognition in a brain-damaged patient. *Nature* 359: 636–637.

98. Beletsky LD, Orians GH (1989) Territoriality among male red-winged blackbirds. III. Testing hypotheses of territorial dominance. *Behavioral Ecology and Sociobiology* 24: 333–339.

99. Bell CP (1997) Leap-frog migration in the fox sparrow: Minimizing the cost of spring migration. *Condor* 99: 470–477.

100. Bell CP (2000) Process in the evolution of bird migration and pattern in avian ecogeography. *Journal of Avian Biology* 31: 258–265.

101. Bell DA, Trail PW, Baptista LF (1998) Song learning and vocal tradition in Nuttall's white-crowned sparrows. *Animal Behaviour* 55: 939–956.

102. Belot M, Fancesconi M (2006) Can anyone be "The One:" evidence on mate selection from speed dating. *CEPR Discussion Papers* 5926.

103. Ben-Shahar Y, Robichon A, Sokolowski MB, Robinson GE (2002) Influence of gene action across different time scales on behavior. *Science* 296: 741–744.

104. Benkman CW (1990) Foraging rates and the timing of crossbill reproduction. *Auk* 107: 376–386.

105. Bennett ATD, Cuthill IC, Norris KJ (1994) Sexual selection and the mismeasure of color. *American Naturalist* 144: 848–860.

106. Bentley D, Hoy RR (1974) The neurobiology of cricket song. *Scientific American* 231 (Aug.): 34–44.

107. Bereczkei T, Birkas B, Kerekes Z (2007) Public charity offer as a proximate factor of evolved reputation-building strategy: an experimental analysis of a real-life situation. *Evolution and Human Behavior* 28: 277–284.

108. Berglund A, Rosenqvist G, Svensson I (1986) Mate choice, fecundity and sexual dimorphism in two pipefish species (Syngnathidae). *Behavioral Ecology and Sociobiology* 19: 301–307.

109. Berglund A, Rosenqvist G, Robinson-Wolrath S (2006) Food or sex—males and females in a sex role reversed pipefish have different interests *Behavioral Ecology and Sociobiology* 60: 281–287.

110. Bergman M, Gotthard K, Berger D, Olofsson M, Kemp DJ, Wiklund C (2007) Mating success of resident versus non-resident males in a territorial butterfly. *Proceedings of the Royal Society of London, Series B* 274: 1659–1665.

111. Bernal XE, Page RA, Rand AS, Ryan MJ (2007) Cues for eavesdroppers: Do frog calls indicate prey density and quality? *American Naturalist* 169: 409–415.

112. Bernasconi G, Strassmann JE (1999) Cooperation among unrelated individuals: The ant foundress case. *Trends in Ecology and Evolution* 14: 477–482.

113. Berthold P (1991) Genetic control of migratory behaviour in birds. *Trends in Ecology and Evolution* 6: 254–257.

114. Berthold P, Helbig AJ, Mohr G, Querner U (1992) Rapid microevolution of migratory behaviour in a wild bird species. *Nature* 360: 668–670.

115. Berthold P, Pulido F (1994) Heritability of migratory activity in a natural bird population. *Proceedings of the Royal Society of London, Series B* 257: 311–315.

116. Berthold P, Querner U (1995) Microevolutionary aspects of bird migration based on experimental results. *Israel Journal of Zoology* 41: 377–385.

117. Berthold P (2003) Genetic basis and evolutionary aspects of bird migration. *Advances in the Study of Behavior* 33: 175–229.

118. Betzig L (ed) (1997) *Human Nature, A Critical Reader*. Oxford University Press, New York.

119. Biesmeijer JC, Seeley TD (2005) The use of waggle dance information by honey bees throughout their foraging careers. *Behavioral Ecology and Sociobiology* 59: 133–142.

120. Billing J, Sherman PW (1998) Antimicrobial functions of spices: Why some like it hot. *Quarterly Review of Biology* 73: 3–49.

121. Birkhead TR, Møller AP (1992) *Sperm Competition in Birds: Evolutionary Causes and Consequences*. Academic Press, London.

122. Birkhead TR, Møller AP (eds) (1998) *Sperm Competition and Sexual Selection*. Academic Press, San Diego, CA.

123. Birkhead TR, Chaline N, Biggins JD, Burke T, Pizzari T (2004) Nontransitivity of paternity in a bird. *Evolution* 58: 416–420.

124. Bissoondath CJ, Wiklund C (1995) Protein content of spermatophores in relation to monandry/polyandry in butterflies. *Behavioral Ecology and Sociobiology* 37: 365–372.

125. Bize P, Piault R, Moureau B, Heeb P (2006) A UV signal of offspring condition mediates context-dependent parental favouritism. *Proceedings of the Royal Society of London, Series B* 273: 2063–2068.

126. Bjork A, Dallai I, Pitnick S (2007) Adaptive modulation of sperm production rate in *Drosophila bifurca*, a species with giant sperm. *Biology Letters* 3: 517–519.

127. Bjorksten TA, Fowler K, Pomiankowski A (2000) What does sexual trait FA tell us about stress? *Trends in Ecology and Evolution* 15: 163–166.

128. Blackledge TA (1998) Stabilimentum variation and foraging success in *Argiope aurantia* and *Argiope trifasciata* (Araneae: Araneidae). *Journal of Zoology* 246: 21–27.

129. Blackledge TA, Wenzel JW (1999) Do stabilimenta in orb webs attract prey or defend spiders? *Behavioral Ecology* 10: 372–376.

130. Blackledge TA, Wenzel JW (2001) Silk mediated defense by an orb web spider against predatory mud-dauber wasps. *Behaviour* 138: 155–171.

131. Blanckenhorn WU (2005) Behavioral causes and consequences of sexual size dimorphism. *Ethology* 11: 977–1016.

132. Bloom G, Sherman PW (2005) Dairying barriers affect the distribution of lactose malabsorption. *Evolution and Human Behavior* 26: 301–312.

133. Blount JD, Metcalfe NB, Birkhead TR, Surai PF (2003) Carotenoid modulation of immune function and sexual attractiveness in zebra finches. *Science* 300: 125–127.

134. Blumstein DT, Møller AP (2008) Is sociality associated with high longevity in North American birds? *Biology Letters* 4: 146–148.

135. Boggess J (1984) Infant killing and male reproductive strategies in langurs (*Presbytis entellus*). In: Hausfater G, Hrdy S (eds) *Infanticide: Comparative and Evolutionary Perspectives*. Aldine, Chicago.

136. Bollmer JL, Sanchez T, Cannon MD, Sanchez D, Cannon B, Bednarz JC, De Vries T, Struve MS, Parker PG (2003) Variation in morphology and mating system among island populations of Galapagos Hawks. *Condor* 105: 428–438.

137. Boncoraglio G, Saino N (2007) Habitat structure and the evolution of bird song: a meta-analysis of the evidence for the acoustic adaptation hypothesis. *Functional Ecology* 21: 132–142.

138. Boomsma JJ, Grafen A (1990) Intraspecific variation in ant sex ratios and the Trivers-Hare hypothesis. *Evolution* 44: 1026–1034.

139. Boone JL, III (1986) Parental investment and elite family structure in preindustrial states: A case study of late medieval-early modern Portuguese genealogies. *American Anthropologist* 88: 859–878.

140. Borgerhoff Mulder M (1988) Reproductive consequences of sex-biased inheritance. In: Standen V, Foley R (eds) *Comparative Socioecology of Mammals and Man*. Blackwell, London.

141. Borgerhoff Mulder M (1990) Kipsigis women's preferences for wealthy men: Evidence for female choice in mammals. *Behavioral Ecology and Sociobiology* 27: 255–264.

142. Borgia G (1985) Bower quality, number of decorations and mating success of male satin bowerbirds (*Ptilonorhynchus violaceus*). *Animal Behaviour* 33: 266–271.

143. Borgia G (1986) Sexual selection in bowerbirds. *Scientific American* 254 (June): 92–100.

144. Borgia G (1995) Why do bowerbirds build bowers? *American Scientist* 83: 542–547.

145. Borgia G, Coleman SW (2000) Co-option of male courtship signals from aggressive display in bowerbirds. *Proceedings of the Royal Society of London, Series B* 267: 1735–1740.

146. Borgia G, Egeth M, Uy JAC, Patricelli GL (2004) Juvenile infection and male display: testing the bright male hypothesis across individual life histories. *Behavioral Ecology* 15: 722–728.

147. Borgia G (2006) Preexisting traits are important in the evolution of elaborated male sexual display. *Advances in the Study of Behavior* 36: 249–303.

148. Borries C (1997) Infanticide in seasonally breeding multimale groups of Hanuman langurs (*Presbytis entellus*) in Ramnagar (South Nepal). *Behavioral Ecology and Sociobiology* 41: 139–150.

149. Borries C, Launhardt K, Epplen C, Epplen JT, Winkler P (1999) Males as infant protectors in Hanuman langurs (*Presbytis entellus*) living in multimale groups—defence pattern, paternity and sexual behaviour. *Behavioral Ecology and Sociobiology* 46: 350–356.

150. Borries C, Launhardt K, Epplen C, Epplen JT, Winkler P (1999) DNA analyses support the hypothesis that infanticide is adaptive in langur monkeys. *Proceedings of the Royal Society of London, Series B* 266: 901–904.

151. Bouchard TJ, Jr. (1997) IQ similarity in twins reared apart: Findings and responses to critics. In: Sternberg R, Grigorenko E (eds) *Intelligence: Heredity and Environment*. Cambridge University Press, New York.

152. Bouchard TJ, Jr., McGue M (1981) Familial studies of intelligence: A review. *Science* 212: 1055–1059.

153. Boulcott PD, Walton K, Braithwaite VA (2005) The role of ultraviolet wavelengths in the mate-choice decisions of female three-spined sticklebacks. *Journal of Experimental Biology* 208: 1453–1458.

154. Bourke AFG, Franks NR (1995) *Social Evolution in Ants*. Princeton University Press, Princeton, N J.

155. Boyd RS, Silk JB (2006) *How Humans Evolved*, Fourth ed. University of California Press, Los Angeles, CA.

156. Braaten RF, Reynolds K (1999) Auditory preference for conspecific song in isolation-reared zebra finches. *Animal Behaviour* 58: 105–111.

157. Bradbury JW (1977) Lek mating behavior in the hammer-headed bat. *Zeitschrift für Tierpsychologie* 45: 225–255.

158. Bradbury JW (1981) The evolution of leks. In: Alexander RD, Tinkle DW (eds) *Natural Selection and Social Behavior*. Chiron Press, New York.

159. Bradbury JW, Gibson RM (1983) Leks and mate choice. In: Bateson P (ed) *Mate Choice*. Cambridge University Press, Cambridge.

160. Bradbury JW, Vehrencamp SL, Gibson RM (1989) Dispersion of displaying male sage grouse. I. Patterns of temporal variation. *Behavioral Ecology and Sociobiology* 24: 1–14.

161. Bradbury JW, Vehrencamp SL (1998) *Principles of Animal Communication*. Sinauer Associates, Sunderland, MA.

162. Brady SG, Sipes S, Pearson A, Danforth BN (2006) Recent and simultaneous origins of eusociality in halictid bees. *Proceedings of the Royal Society of London, Series B* 273: 1643–1649.

163. Brakefield PM, Liebert TG (2000) Evolutionary dynamics of declining melanism in the peppered moth in The Netherlands. *Proceedings of the Royal Society of London, Series B* 267: 1953–1957.

164. Brandt Y (2003) Lizard threat display handicaps endurance. *Proceedings of the Royal Society of London, Series B* 270: 1061–1068.

165. Braude S (2000) Dispersal and new colony formation in wild naked mole-rats: Evidence against inbreeding as the system of mating. *Behavioral Ecology* 11: 7–12.

166. Breed MD, Guzmán-Novoa E, Hunt GJ (2004) Defensive behavior of honey bees: Organization, genetics, and comparisons with other bees. *Annual Reviews of Entomology* 49: 271–298.

167. Breininger DR, Larson VL, Duncan BW, Smith RB, Oddy DM, Goodchild MF (1995) Landscape patterns of Florida scrub jay habitat use and demographic success. *Conservation Biology* 9: 1442–1453.

168. Brenowitz EA (1991) Evolution of the vocal control system in the avian brain. *Seminars in the Neurosciences* 3: 399–407.

169. Brenowitz EA, Lent K, Kroodsma DE (1995) Brain space for learned song in birds develops independently of song learning. *Journal of Neuroscience* 15: 6281–6286.

170. Brenowitz EA, Margoliash D, Nordeen KW (1997) An introduction to birdsong and the avian song system. *Journal of Neurobiology* 33: 495–500.

171. Brenowitz EA, Beecher MD (2005) Song learning in birds: diversity and plasticity, opportunities and challenges. *Trends in Ecology and Evolution* 28: 127–132.

172. Bretman A, Wedell N, Tregenza T (2004) Molecular evidence of post-copulatory inbreeding avoidance in the field cricket *Gryllus bimaculatus*. *Proceedings of the Royal Society of London, Series B* 271: 159–164.

173. Breven KA (1981) Mate choice in the wood frog, *Rana sylvatica*. *Evolution* 35: 707–722.

174. Briskie JV, Martin PR, Martin TE (1999) Nest predation and the evolution of nestling begging calls. *Proceedings of the Royal Society of London, Series B* 266: 2153–2159.

175. Britt EJ, Hicks JW, Bennett AF (2006) The energetic consequences of dietary specialization in populations of the garter snake, *Thamnophis elegans*. *Journal of Experimental Biology* 209: 3164–3169.

176. Bro-Jørgensen J, Durant SM (2003) Mating strategies of topi bulls: getting in the centre of attention. *Animal Behaviour* 65: 585–594.

177. Bro-Jørgensen J (2007) Reversed sexual conflict in a promiscuous antelope. *Current Biology* 17: 2157–2161.

178. Bro-Jørgensen J, Johnstone RA, Evans MR (2007) Uninformative exaggeration of male sexual ornaments in barn swallows. *Current Biology* 17: 850–855.

179. Brockmann HJ, Penn D (1992) Male mating tactics in the horseshoe crab, *Limulus polyphemus*. *Animal Behaviour* 44: 653–665.

180. Brockmann HJ, Colson T, Potts W (1994) Sperm competition in horseshoe crabs (*Limulus polyphemus*). *Behavioral Ecology and Sociobiology* 35: 153–160.

181. Brockmann HJ (2002) An experimental approach to alternative mating tactics in male horseshoe crabs (*Limulus polyphemus*). *Behavioral Ecology* 13: 232–238.

182. Brooks DR, McLennan DA (1991) *Phylogeny, Ecology, and Behavior*. University of Chicago Press, Chicago, IL.

183. Brotherton PNM, Manser MB (1997) Female dispersion and the evolution of monogamy in the dik-dik. *Animal Behaviour* 54: 1413–1424.

184. Brower JVZ (1958) Experimental studies of mimicry in some North American butterflies. 1. The monarch, *Danaus plexippus*, and viceroy, *Limenitis archippus*. *Evolution* 12: 3–47.

185. Brower LP, Calvert WH (1984) Chemical defence in butterflies. In: Vane-Wright RI, Ackery PR (eds) *The Biology of Butterflies*. Academic Press, London.

186. Brower LP (1996) Monarch butterfly orientation: Missing pieces of a magnificent puzzle. *Journal of Experimental Biology* 199: 93–103.

187. Brower LP, Fink LS, Walford P (2006) Fueling the fall migration of the monarch butterfly. *Integrative and Comparative Biology* 46: 1123–1142.

188. Brown CR, Brown MB (1986) Ecto-parasitism as a cost of coloniality in cliff swallows (*Hirundo pyrrhonota*). *Ecology* 67: 1206–1218.

189. Brown CR, Brown MB (1996) *Coloniality in the Cliff Swallow: The Effect of Group Size on Social Behavior*. University of Chicago Press, Chicago, IL.

190. Brown JL, Orians GH (1970) Spacing patterns in mobile animals. *Annual Review of Ecology and Systematics* 1: 239–262.

191. Brown JL (1975) *The Evolution of Behavior*. W. W. Norton, New York.

192. Brown JL (1987) *Helping and Communal Breeding in Birds: Ecology and Evolution*. Princeton University Press, Princeton, NJ.

193. Brown JL, Eklund A (1994) Kin recognition and the major histocompatibility complex: An integrative review. *American Naturalist* 143: 435–461.

194. Brown JL (1997) A theory of mate choice based on heterozygosity. *Behavioral Ecology* 8: 60–65.

195. Brown JL, Vleck CM (1998) Prolactin and helping in birds: Has natural selection strengthened helping behavior? *Behavioral Ecology* 9: 541–545.

196. Brown JR, Ye H, Bronson RT, Dikkes P, Greenberg ME (1996) A defect in nurturing in mice lacking the immediate early gene *fosB*. *Cell* 86: 297–309.

197. Brown KM (1998) Proximate and ultimate causes of adoption in ring-billed gulls. *Animal Behaviour* 56: 1529–1543.

198. Brown WM, Cronk L, Grochow K, Jacobson A, Liu CK, Popovi Z, Trivers RL (2005) Dance reveals symmetry especially in young men. *Nature* 438: 1148–1150.

199. Brownmiller S (1975) *Against Our Will*. Simon and Schuster, New York.

200. Brownmiller S, Merhof B (1992) A feminist response to rape as an adaptation in men. *Brain and Behavioral Sciences* 15: 381–382.

201. Bruce MJ (2006) Silk decorations: controversy and consensus. *Journal of Zoology* 269: 89–97.

202. Buchan JC, Alberts SC, Silk JB, Altmann J (2003) True paternal care in a multi-male primate society. *Nature* 425: 179–181.

203. Buchanan KL, Catchpole CK (2000) Song as an indicator of male parental effort in the sedge warbler. *Proceedings of the Royal Society of London, Series B* 267: 321–326.

204. Buehler DM, Piersma T (2008) Travelling on a budget: predictions and ecological evidence for bottlenecks in the annual cycle of long-distance migrants. *Philosophical Transactions of the Royal Society of London, Series B* 363: 247–266.

205. Bugoni L, Krause L, Petry MV (2001) Marine debris and human impacts on sea turtles in southern Brazil. *Marine Pollution Bulletin* 42: 1330–1334.

206. Bulbulia J (2004) The cognitive and evolutionary psychology of religion. *Biology & Philosophy* 19: 655–686.

207. Burch RL, Gallup GG (2004) Pregnancy as a stimulus for domestic violence. *Journal of Family Violence* 19: 243–247.

208. Burda H, Honeycutt RL, Begall S, Locker-Grutjen O, Scharff A (2000) Are naked and common mole-rats eusocial and if so, why? *Behavioral Ecology and Sociobiology* 47: 293–303.

209. Burger J, Gochfeld M (2001) Smooth-billed ani (*Crotophaga ani*) predation on butterflies in Mato Grosso, Brazil: risk decreases with increased group size. *Behavioral Ecology and Sociobiology* 49: 482–492.

210. Burke T, Bruford MW (1987) DNA fingerprinting in birds. *Nature* 327: 149–152.

211. Burke T (1989) DNA fingerprinting and other methods for the study of mating success. *Trends in Ecology and Evolution* 4: 139–144.

212. Burkhardt RW (2004) *Patterns of Behavior: Konrad Lorenz, Niko Tinbergen, and the Founding of Ethology.* University of Chicago Press, Chicago, IL.

213. Burland TM, Bennett NC, Jarvis JUM, Faulkes CG (2002) Eusociality in African mole-rats: new insights from patterns of genetic relatedness in the Damaraland mole-rat (*Cryptomys damarensis*). *Proceedings of the Royal Society of London, Series B* 269: 1025–1030.

214. Burley NT, Symanski R (1998) "A taste for the beautiful": Latent aesthetic mate preferences for white crests in two species of Australian grassfinches. *American Naturalist* 152: 792–802.

215. Burmeister SS, Jarvis ED, Fernald RD (2005) Rapid behavioral and genomic responses to social opportunity. *PLoS Biology* 3: 1–9.

216. Burness G, Casselman SJ, Schulte-Hostedde AI, Moyes CD, Montgomerie R (2004) Sperm swimming speed and energetics vary with sperm competition risk in bluegill (*Lepomis macrochirus*). *Behavioral Ecology and Sociobiology* 56: 65–70.

217. Burt JM, Campbell SE, Beecher MD (2001) Song type matching as threat: a test using interactive playback. *Animal Behaviour* 62: 1163–1170.

218. Buskirk RE, Frolich C, Ross KG (1984) The natural selection of sexual cannibalism. *American Naturalist* 123: 612–625.

219. Buss DM (1989) Sex differences in human mate preferences: Evolutionary hypothesis tested in 37 cultures. *Behavioral and Brain Sciences* 12: 1–149.

220. Buss DM, Larsen RJ, Westen D, Semmelroth J (1992) Sex differences in jealousy: Evolution, physiology, and psychology. *Psychological Science* 3: 251–255.

221. Buss DM, Schmitt DP (1993) Sexual strategies theory: An evolutionary perspective on human mating. *Psychological Review* 100: 204–232.

222. Buss DM (2007) *Evolutionary Psychology, The New Science of the Mind*, 3rd edn. Allyn and Bacon, Boston, MA.

223. Buss DM, Shackelford TK (2008) Attractive women want it all: Good genes, economic investment, parenting proclivities, and emotional commitment. *Evolutionary Psychology* 6: 134–146.

224. Buston PM, Emlen ST (2003) Cognitive processes underlying human mate choice: The relationship between self-perception and mate preference in Western society. *Proceedings of the National Academy of Sciences USA* 100: 8805–8810.

225. Byers JA, Wiseman PA, Jones L, Roffe TJ (2005) A large cost of female mate sampling in pronghorn. *American Naturalist* 166: 661–668.

226. Calvert WH, Brower LP (1986) The location of monarch butterfly (*Danaus plexippus* L.) overwintering colonies in Mexico in relation to topography and climate. *Journal of the Lepidopterists' Society* 40: 164–187.

227. Cantoni D, Brown R (1997) Paternal investment and reproductive success in the California mouse, *Peromyscus californicus. Animal Behaviour* 54: 377–386.

228. Carde RT, Staten RT, Mafra-Neto A (1998) Behaviour of pink bollworm males near high-dose, point sources of pheromone in field wind tunnels: insights into mechanisms of mating disruption. *Entomologia Experimentalis et Applicata* 89: 35–46.

229. Cardoso GC, Mota PG, Depraz V (2007) Female and male serins (*Serinus serinus*) respond differently to derived song traits. *Behavioral Ecology and Sociobiology* 61: 1425–1436.

230. Carew TJ (2000) *Behavioral Neurobiology: The Cellular Organization of Behavior.* Sinauer Associates, Sunderland, MA.

231. Carey M, Nolan V, Jr. (1975) Polygyny in indigo buntings: A hypothesis tested. *Science* 190: 1296–1297.

232. Carey S (1992) Becoming a face expert. *Philosophical Transactions of the Royal Society of London, Series B* 335: 95–103.

233. Caro TM (1986) The functions of stotting in Thomson's gazelles: Some tests of the predictions. *Animal Behaviour* 34: 663–684.

234. Caro TM (1986) The functions of stotting: A review of the hypotheses. *Animal Behaviour* 34: 649–662.

235. Caro TM (1995) Pursuit-deterrence revisited. *Trends in Ecology and Evolution* 10: 500–503.

236. Caro TM, Graham CM, Stoner CJ, Vargas JK (2004) Adaptive significance of antipredator behaviour in artiodactyls. *Animal Behaviour* 67: 205–228.

237. Carpenter SJ, Erickson JM, Holland FD (2003) Migration of a Late Cretaceous fish. *Nature* 423: 70–74.

238. Carroll SB, Grenier JK, Weatherbee SD (2005) *From DNA to Diversity: Molecular Genetics and the Evolution of Animal Design.* Blackwell Publishing, Malden, MA.

239. Carter R (1998) *Mapping the Brain.* Weidenfeld and Nicolson, London.

240. Case A, Lin I-F, McLanahan S (2000) How hungry is the selfish gene? *Economic Journal* 110: 781–804.

241. Case A, Lin I-F, McLanahan S (2001) Educational attainment of siblings in stepfamilies. *Evolution and Human Behavior* 22: 269–289.

242. Case A, Lubotsky D, Paxson C (2002) Economic status and health in childhood: The origins of the gradient. *American Economic Review* 92: 1308–1334.

243. Casselman SJ, Montgomerie R (2004) Sperm traits in relation to male quality in colonial spawning bluegill. *Journal of Fish Biology* 64: 1700–1711.

244. Catania KC, Kaas JH (1996) The unusual nose and brain of the star-nosed mole. *BioScience* 46: 578–586.

245. Catania KC, Kaas JH (1997) Somatosensory fovea in the star-nosed mole: Behavioral use of the star in relation to innervation patterns and cortical representation. *Journal of Comparative Neurology* 387: 215–233.

246. Catania KC (2000) Cortical organization in insectivora: The parallel evolution of the sensory periphery and the brain. *Brain Behavior and Evolution* 55: 311–321.

247. Catania KC, Remple MS (2002) Somatosensory cortex dominated by the representation of teeth in the naked mole-rat brain. *Proceedings of the National Academy of Sciences USA* 99: 5692–5697.

248. Catania KC, Remple FE (2005) Asymptotic prey profitability drives star-nosed moles to the foraging speed limit. *Nature* 433: 519–522.

249. Catania KC, Henry EC (2006) Touching on somatosensory specializations in mammals. *Current Opinion in Neurobiology* 16: 467–473.

250. Catchpole CK, Slater PJB (1995) *Bird Song, Biological Themes and Variations*. Cambridge University Press, Cambridge.

251. Cézilly F, Nager RG (1995) Comparative evidence for a positive association between divorce and extra-pair paternity in birds. *Proceedings of the Royal Society of London, Series B* 262: 7–12.

252. Chaine AS, Lyon BE (2008) Adaptive plasticity in female mate choice dampens sexual selection on male ornaments in the lark bunting. *Science* 319: 459–462.

253. Chapman T, Crespi B (1998) High relatedness and inbreeding in two species of haplodiploid eusocial thrips (Insecta: Thysanoptera) revealed by microsatellite analysis. *Behavioral Ecology and Sociobiology* 43: 301–306.

254. Chapman T, Bangham J, Vinti G, Lung O, Wolfner MF, Smith HK, Partridge L (2003) The sex peptide of *Drosophila melanogaster*: Female post-mating responses analyzed by using RNA interference. *Proceedings of the National Academy of Sciences USA* 100: 9923–9928.

255. Charlton B, Reby D, McComb K (2007) Female red deer prefer the roars of larger males. *Biology Letters* 3: 382–385.

256. Charrier I, Mathevon N, Jouventin P (2003) Vocal signature recognition of mothers by fur seal pups. *Animal Behaviour* 65: 543–550.

257. Cheng K (2000) How honeybees find a place: Lessons from a simple mind. *Animal Learning and Behavior* 28: 1–15.

258. Cheng MY, Bullock CM, Li CY, Lee AG, Bermak JC, Belluzzi J, Weaver DR, Leslie FM, Zhou QY (2002) Prokineticin 2 transmits the behavioural circadian rhythm of the suprachiasmatic nucleus. *Nature* 417: 405–410.

259. Cheng R-C, Tso I-M (2007) Signaling by decorating webs: luring prey or deterring predators? *Behavioral Ecology* 18: 1085–1091.

260. Chilton G, Lein MR, Baptista LF (1990) Mate choice by female white-crowned sparrows in a mixed-dialect population. *Behavioral Ecology and Sociobiology* 27: 223–227.

261. Christy JH (1995) Mimicry, mate choice, and the sensory trap hypothesis. *American Naturalist* 146: 171–181.

262. Chuang C-Y, Yang E-C, Tso I-M (2008) Deceptive color signaling in the night: a nocturnal predator attracts prey with visual lures. *Behavioral Ecology* 19: 237–244.

263. Chuang CY, Yang EC, Tso I-M (2007) Diurnal and nocturnal prey luring of a colorful predator. *Journal of Experimental Biology* 210: 3830–3837.

264. Cialdini RB (2001) The science of persuasion. *Scientific American* 284 (Feb.): 76–81.

265. Cioffi D, Garner R (1998) The effect of response options on decisions and subsequent behavior: Sometimes inaction is better. *Personality and Social Psychology Bulletin* 24: 463–472.

266. Ciszek D (2000) New colony formation in the "highly inbred" eusocial naked mole-rat: Outbreeding is preferred. *Behavioral Ecology* 11: 1–6.

267. Clapham J (2001) Why do baleen whales migrate? A response to Corkeron and Connor. *Marine Mammal Science* 17: 432–436.

268. Clarac F, Pearlstein E (2007) Invertebrate preparations and their contribution to neurobiology in the second half of the 20th century. *Brain Research Reviews* 54: 113–161.

269. Clark RD, Hatfield E (1989) Gender differences in receptivity to sexual offers. *Journal of Psychology and Human Sexuality* 2: 39–55.

270. Clayton DF (1997) Role of gene regulation in song circuit development and song learning. *Journal of Neurobiology* 33: 549–571.

271. Clayton NC, Krebs JR (1994) Hippocampal growth and attrition in birds affected by experience. *Proceedings of the National Academy of Sciences USA* 91: 7410–7414.

272. Clayton NS (1998) Memory and the hippocampus in food-storing birds: A comparative approach. *Neuropharmacology* 37: 441–452.

273. Clifford LD, Anderson DJ (2001) Experimental demonstration of the insurance value of extra eggs in an obligately siblicidal seabird. *Behavioral Ecology* 12: 340–347.

274. Clotfelter ED, Schubert KA, Nolan V, Ketterson ED (2003) Mouth color signals thermal state of nestling dark-eyed juncos (*Junco hyemalis*). *Ethology* 109: 171–182.

275. Clucas B, Owings DS, Rowe MP (2008) Donning your enemy's cloak: ground squirrels exploit rattlesnake scent to reduce predation risk. *Proceedings of the Royal Society of London, Series B* 275: 847–852.

276. Clutton-Brock T (2007) Sexual selection in males and females. *Science* 318: 1882–1885.

277. Clutton-Brock TH, Albon SD (1979) The roaring of red deer and the evolution of honest advertisement. *Behaviour* 69: 145–170.

278. Clutton-Brock TH, Albon SD, Gibson RM, Guinness FE (1979) The logical stag: Adaptive aspects of fighting in red deer. *Animal Behaviour* 27: 211–225.

279. Clutton-Brock TH, Harvey PH (1984) Comparative approaches to investigating adaptation. In: Krebs JR, Davies NB (eds) *Behavioural Ecology: An Evolutionary Approach*. Blackwell, Oxford.

280. Clutton-Brock TH (1991) *The Evolution of Parental Care*. Princeton University Press, Princeton, NJ.

281. Clutton-Brock TH, Parker GA (1992) Potential reproductive rates and the operation of sexual selection. *Quarterly Review of Biology* 67: 437–456.

282. Clutton-Brock TH, O'Riain MJ, Brotherton PNM, Gaynor D, Kansky R, Griffin AS, Manser M (1999) Selfish sentinels in cooperative mammals. *Science* 284: 1640–1644.

283. Clutton-Brock TH, Hodge SJ, Spong G, Russell AF, Jordan NR, Bennett NC, Sharpe LL, Manser MB (2006) Intrasexual competition and sexual selection in cooperative mammals. *Nature* 444: 1065–1068.

284. Cockburn A, Sims RA, Osmond HL, Green DJ, Double MC, Mulder RA (2008) Can we measure the benefits of help in cooperatively breeding birds: the case of superb fairy-wrens *Malurus cyaneus*? *Journal of Animal Ecology* 77: 430–438.

285. Cohen L, Lehéricy S, Chochon F, Lemer C, Rivaud S, Dehaene S (2002) Language-specific tuning of visual cortex? Functional properties of the Visual Word Form Area. *Brain* 125: 1054–1059.

286. Cohen ML (1992) Epidemiology of drug resistance: implications for a post-antimicrobial era. *Science* 257: 1050–1055.

287. Colantuoni C, Purcell AE, Bouton CM, Pevsner J (2000) High throughput analysis of gene expression in the human brain. *Journal of Neuroscience Research* 59: 1–10.

288. Colarelli SM, Dettmann JR (2003) Intuitive evolutionary perspectives in marketing practices. *Psychology & Marketing* 20: 837–865.

289. Collins JP, Cheek JE (1983) Effect of food and density on development of typical and cannibalistic salamander larvae in *Ambystoma tigrinum nebulosum*. *American Zoologist* 23: 77–84.

290. Collins SA, Missing C (2003) Vocal and visual attractiveness are related in women. *Animal Behaviour* 65: 997–1004.

291. Conley AJ, Corbin CJ, Browne P, Mapes SM, Place NJ, Hughes AL, Glickman SE (2006) Placental expression and molecular characterization of aromatase cytochrome P450 in the spotted hyena (*Crocuta crocuta*). *Placenta* 28: 668–675.

292. Conover MR (1994) Stimuli eliciting distress calls in adult passerines and response of predators and birds to their broadcast. *Behaviour* 131: 19–37.

293. Conroy CJ, Cook JA (2000) Molecular systematics of a holarctic rodent (Microtus: Muridae). *Journal of Mammalogy* 81: 344–359.

294. Cook LM (2003) The rise and fall of the *Carbonaria* form of the peppered moth. *Quarterly Review of Biology* 78: 399–417.

295. Coombs WP, Jr (1990) Behavior patterns of dinosaurs. In: Weishampel DB, Dodson P, Ósmólska H (eds) *The Dinosauria*. University of California Press, Berkeley.

296. Cooper WE, Pérez-Mellado V, Baird TA, Caldwell JP, Vitt LJ (2004) Pursuit deterrent signalling by the bonaire whiptail lizard *Cnemidophorus murinus*. *Behaviour* 141: 297–311.

297. Cordero C, Eberhard WG (2003) Female choice of sexually antagonistic male adaptations: a critical review of some current research. *Journal of Evolutionary Biology* 16: 1–6.

298. Corkeron PJ, Connor RC (1999) Why do baleen whales migrate? *Marine Mammal Science* 15: 1228–1245.

299. Cornwallis CK, Birkhead TR (2007) Changes in sperm quality and numbers in response to experimental manipulation of male social status and female attractiveness. *American Naturalist* 170: 758–770.

300. Cornwell RE, Boothroyd L, Burt DM, Feinberg DR, Jones BC, Little AC, Pitman R, Whiten S, Perrett DI (2004) Concordant preferences for opposite-sex signals? Human pheromones and facial characteristics. *Proceedings of the Royal Society of London, Series B* 271: 635–640.

301. Coss RG, Goldthwaite RO (1995) The persistence of old designs for perception. *Perspectives in Ethology* 11: 83–148.

302. Costa JT (2006) *The Other Insect Societies*. Harvard University Press, Cambridge, MA.

303. Court GS (1996) The seal's own skin game. *Natural History* 105(8): 36–41.

304. Covas R, Griesser M (2007) Life history and the evolution of family living in birds. *Proceedings of the Royal Society of London, Series B* 274: 1349–1357.

305. Cowlishaw G, Dunbar RIM (1991) Dominance rank and mating success in male primates. *Animal Behaviour* 41: 1045–1056.

306. Cowlishaw G, Mace R (1996) Cross-cultural patterns of marriage and inheritance: A phylogenetic approach. *Ethology and Sociobiology* 17: 97–98.

307. Cox GW (1985) The evolution of avian migration systems between temperate and tropical regions of the New World. *American Naturalist* 126: 452–474.

308. Coyne J (1998) Not black and white. *Nature* 396: 35–36.

309. Craig CL, Bernard GD (1990) Insect attraction to ultraviolet-reflecting spider webs and web decorations. *Ecology* 71: 616–623.

310. Craig P (1996) Intertidal territoriality and time-budget of the surgeonfish, *Acanthurus lineatus*, in American Samoa. *Environmental Biology of Fishes* 46: 27–36.

311. Creel S, Winnie J, Jr, Maxwell B, Hamlin K, Creel M (2005) Elk alter habitat selection as an antipredator response to wolves. *Ecology* 86: 3387–3397.

312. Creel S, Christianson D (2008) Relationships between direct predation and risk effects. *Trends in Ecology and Evolution* 23: 194–201.

313. Crespi BJ (2000) The evolution of maladaptation. *Heredity* 84: 623–629.

314. Crews D (1975) Psychobiology of reptilian reproduction. *Science* 189: 1059–1065.

315. Crews D, Greenberg N (1981) Function and causation of social signals in lizards. *American Zoologist* 21: 273–294.

316. Crews D (1984) Gamete production, sex hormone secretion, and mating behavior uncoupled. *Hormones and Behavior* 18: 22–28.

317. Crews D, Moore MC (1986) Evolution of mechanisms controlling mating behavior. *Science* 231: 121–125.

318. Crews D, Hingorani V, Nelson RJ (1988) Role of the pineal gland in the control of annual reproductive behavioral and physiological cycles in the red-sided garter snake (*Thamnophis sirtalis parietalis*). *Journal of Biological Rhythms* 3: 293–302.

319. Crews D (1991) Trans-seasonal action of androgen in the control of spring courtship behavior in male red-sided garter snakes. *Proceedings of the National Academy of Sciences USA* 88: 3545–3548.

320. Crews D (1992) Behavioral endocrinology and reproduction: An evolutionary perspective. *Oxford Reviews of Reproductive Biology* 14: 303–370.

321. Cristol DA, Switzer PV (1999) Avian prey-dropping behavior. II. American crows and walnuts. *Behavioral Ecology* 10: 220–226.

322. Crockford C, Wittig RM, Seyfarth RM, Cheney DL (2007) Baboons eavesdrop to deduce mating opportunities. *Animal Behaviour* 73: 885–890.

323. Cronk L (1993) Parental favoritism toward daughters. *American Scientist* 81: 272–279.

324. Cullen E (1957) Adaptations in the kittiwake to cliff nesting. *Ibis* 99: 275–302.

325. Cumming GS (1996) Mantis movements by night and the interactions of sympatric bats and mantises. *Canadian Journal of Zoology* 74: 1771–1774.

326. Cumming JM (1994) Sexual selection and the evolution of dance fly mating systems (Diptera: Empididae; Empidinae). *Canadian Entomologist* 126: 907–920.

327. Cummings ME, Rosenthal GG, Ryan MJ (2003) A private ultraviolet channel in visual communication. *Proceedings of the Royal Society of London, Series B* 270: 897–904.

328. Cunningham EJA, Russell AF (2000) Egg investment is influenced by male attractiveness in the mallard. *Nature* 404: 74–77.

329. Cunningham EJA (2003) Female mate preferences and subsequent resistance to copulation in the mallard. *Behavioral Ecology* 14: 326–333.

330. Curtin R, Dolhinow P (1978) Primate social behavior in a changing world. *American Scientist* 66: 468–475.

331. Cutler DM, Miller G, Norton DM (2007) Evidence on early-life income and late-life health from America's Dust Bowl era. *Proceedings of the National Academy of Sciences USA* 104: 13244–13249.

332. Cuvillier-Hot V, Gadagkar R, Peeters C, Cobb M (2002) Regulation of reproduction in a queenless ant: aggression, pheromones and reduction in conflict. *Proceedings of the Royal Society of London, Series B* 269: 1295–1300.

333. Dagg AI (1998) Infanticide by male lions hypothesis: A fallacy influencing research into human behavior. *American Anthropologist* 100: 940–950.

334. Dale J, Montgomerie R, Michaud D, Boag P (1999) Frequency and timing of extrapair fertilisation in the polyandrous red phalarope (*Phalaropus fulicarius*). *Behavioral Ecology and Sociobiology* 46: 50–56.

335. Dale S, Rinden H, Slagsvold T (1992) Competition for a mate restricts mate search of female flycatchers. *Behavioral Ecology and Sociobiology* 30: 165–176.

336. Dale S, Slagsvold T (1994) Polygyny and deception in the pied flycatcher: Can females determine male mating status? *Animal Behaviour* 48: 1207–1217.

337. Daly M, Wilson M, Weghorst ST (1982) Male sexual jealousy. *Ethology and Sociobiology* 3: 11–27.

338. Daly M, Wilson M (1983) *Sex, Evolution and Behavior*, 2nd edn. Willard Grant Press, Boston.

339. Daly M, Wilson M (1985) Child abuse and other risks of not living with both parents. *Ethology and Sociobiology* 6: 197–210.

340. Daly M, Wilson M (1988) *Homicide*. Aldine de Gruyter, Hawthorne, NY.

341. Daly M, Wilson M (1992) The man who mistook his wife for a chattel. In: Barkow J, Cosmides L, Tooby J (eds) *The Adapted Mind*. Oxford University Press, New York.

342. Daly M, Wilson M (1998) *The Truth about Cinderella*. Yale University Press, New Haven, CT.

343. Daly M, Wilson M (2001) An assessment of some proposed exceptions to the phenomenon of nepotistic discrimination against stepchildren. *Annales Zoologici Fennici* 38: 287–296.

344. Daly M, Wilson M (2005) Human behavior as animal behavior. In: Bolhuis JJ, Giraldeau LA (eds) *Behavior of Animals: Mechanisms, Function, and Evolution*. Blackwell Publishing, Oxford.

345. Daly M, Wilson M (2008) Is the "Cinderella Effect" controversial?: A case study of evolution-minded research and critiques thereof. In: Crawford C, Dennis K (eds) *Foundations of Evolutionary Psychology*. Lawrence Erlbaum Associates, New York.

346. Damen WGM, Saridaki T, Averof M (2002) Diverse adaptations of an ancestral gill: A common evolutionary origin for wings, breathing organs, and spinnerets. *Current Biology* 12: 1711–1716.

347. Danforth BN, Conway L, Ji SQ (2003) Phylogeny of eusocial *Lasioglossum* reveals multiple losses of eusociality within a primitively eusocial clade of bees (Hymenoptera: Halictidae). *Systematic Biology* 52: 23–36.

348. Darwin C (1859) *On the Origin of Species*. Murray, London.

349. Darwin C (1871) *The Descent of Man and Selection in Relation to Sex*. Murray, London.

350. Darwin C (1892) *The Various Contrivances by which Orchids are Fertilised by Insects*. D. Appleton, New York.

351. Davies NB (1978) Territorial defence in the speckled wood butterfly (*Pararge aegeria*): The resident always wins. *Animal Behaviour* 26: 138–147.

352. Davies NB, Halliday TR (1978) Deep croaks and fighting assessment in toads *Bufo bufo*. *Nature* 275: 683–685.

353. Davies NB (1983) Polyandry, cloaca-pecking and sperm competition in dunnocks. *Nature* 302: 334–336.

354. Davies NB, Houston AI (1984) Territory economics. In: Krebs JR, Davies NB (eds) *Behavioural Ecology: An Evolutionary Approach*. Blackwell Scientific Publications, Oxford.

355. Davies NB, Lundberg A (1984) Food distribution and a variable mating system in the dunnock, *Prunella modularis*. *Journal of Animal Ecology* 53: 895–912.

356. Davies NB (1985) Cooperation and conflict among dunnocks, *Prunella modularis*, in a variable mating system. *Animal Behaviour* 33: 628–648.

357. Davies NB, de L. Brooke M (1988) Cuckoos versus reed warblers: Adaptations and counteradaptations. *Animal Behaviour* 36: 262–284.

358. Davies NB (1992) *Dunnock Behaviour and Social Evolution*. Oxford University Press, Oxford.

359. Davies NB, Hartley IR, Hatchwell BJ, Langmore NE (1996) Female control of copulations to maximize male help: A comparison of polygynandrous alpine accentors, *Prunella collaris*, and dunnocks, *P. modularis*. *Animal Behaviour* 51: 27–47.

360. Davies NB, Kilner RM, Noble DG (1998) Nestling cuckoos *Cuculus canorus* exploit hosts with begging calls that mimic a brood. *Proceedings of the Royal Society of London, Series B* 265: 673–678.

361. Davies NB (2000) *Cuckoos, Cowbirds and Other Cheats*. T & A D Poyser, London.

362. Davis-Walton J, Sherman PW (1994) Sleep arrhythmia in the eusocial naked mole-rat. *Naturwissenschaften* 81: 272–275.

363. Davis LA, Roalson EH, Cornell KL, McClanahan KD, Webster MS (2006) Genetic divergence and migration patterns in a North American passerine bird: implications for evolution and conservation. *Molecular Ecology* 15: 2141–2152.

364. Dawkins R (1977) *The Selfish Gene*. Oxford University Press, New York.

365. Dawkins R, Krebs J (1978) Animal signals: Information or manipulation? In: Krebs JR, Davies NB (eds) *Behavioural Ecology: An Evolutionary Approach*. Blackwell, Oxford.

366. Dawkins R (1980) Good strategy or evolutionarily stable strategy? In: Barlow GW, Silverberg J (eds) *Sociobiology: Beyond Nature/Nurture?* Westview Press, Boulder, CO.

367. Dawkins R (1982) *The Extended Phenotype*. W.H. Freeman, San Francisco.

368. Dawkins R (1986) *The Blind Watchmaker*. W.W. Norton, New York.

369. Dawkins R (1989) *The Selfish Gene*. Oxford University Press, Oxford.

370. Dawson JW, Dawson-Scully K, Robert D, Robertson RM (1997) Forewing asymmetries during auditory avoidance in flying locusts. *Journal of Experimental Biology* 200: 2323–2335.

371. de Ayala RM, Saino N, Møller AP, Anselmi C (2007) Mouth coloration of nestlings covaries with offspring quality and influences parental feeding behavior. *Behavioral Ecology* 18: 526–534.

372. de Belle JS, Sokolowski MB (1987) Heredity of *rover/sitter*: Alternative foraging strategies of *Drosophila melanogaster* larvae. *Heredity* 59: 73–83.

373. de Belle JS, Hilliker AJ, Sokolowski MB (1989) Genetic localization of *foraging* (*for*): A major gene for larval behavior in *Drosophila melanogaster*. *Genetics* 123: 157–163.

374. De Block M, Stoks R (2007) Flight-related body morphology shapes mating success in a damselfly. *Animal Behaviour* 74: 1093–1098.

375. de Kort SR, Clayton NS (2006) An evolutionary perspective on caching by corvids. *Proceedings of the Royal Society of London, Series B* 273: 417–423.

376. de Renzi E, di Pellegrino G (1998) Prosopagnosia and alexia without object agnosia. *Cortex* 34: 403–415.

377. DeCoursey PJ, Buggy J (1989) Circadian rhythmicity after neural transplant to hamster third ventricle: Specificity of suprachiasmatic nuclei. *Brain Research* 500: 263–275.

378. Dediu D, Land DR (2007) Linguistic tone is related to the population frequency of the adaptive haplogroups of two brain size genes, *ASPM* and *Microcephalin*. *Proceedings of the National Academy of Sciences USA* 104 (26): 10944–10949.

379. DeHeer CJ, Goodisman MAD, Ross KG (1999) Queen dispersal strategies in the multiple-queen form of the fire ant *Solenopsis invicta*. *American Naturalist* 153: 660–675.

380. delBarco-Trillo J, Ferkin MH (2004) Male mammals respond to a risk of sperm competition conveyed by odours of conspecific males. *Nature* 431: 446–449.

381. Delhey K, Peters A, Johnsen A, Kempenaers B (2007) Fertilization success and UV ornamentation in blue tests *Cyanistes caeruleus*: correlational and experimental evidence. *Behavioral Ecology* 18: 399–409.

382. Dennett DC (1995) *Darwin's Dangerous Idea*. Simon & Schuster, New York.

383. Dennis TE, Rayner MJ, Walker MM (2007) Evidence that pigeons orient to geomagnetic intensity during homing. *Proceedings of the Royal Society of London, Series B* 274: 1153–1158.

384. Dethier VG (1962) *To Know a Fly*. Holden-Day, San Francisco.

385. Dethier VG (1976) *The Hungry Fly: A Physiological Study of the Behavior Associated with Feeding*. Harvard University Press, Cambridge, MA.

386. Deutsch CJ, Haley MP, Le Boeuf BJ (1990) Reproductive effort of male northern elephant seals: Estimates from mass loss. *Canadian Journal of Zoology* 68: 2580–2593.

387. Deutsch JC (1994) Uganda kob mating success does not increase on larger leks. *Behavioral Ecology and Sociobiology* 34: 451–459.

388. Deviche P, Sharp PJ (2001) Reproductive endocrinology of a free-living, opportunistically breeding passerine (White-

winged Crossbill, *Loxia leucoptera*). *General and Comparative Endocrinology* 123: 268–279.

389. DeVoogd TJ (1991) Endocrine modulation of the development and adult function of the avian song system. *Psychoneuroendocrinology* 16: 41–66.

390. DeWolfe BB, Baptista LF, Petrinovich L (1989) Song development and territory establishment in Nuttall's white-crowned sparrow. *Condor* 91: 297–407.

391. DeWoody JA, Fletcher DE, Mackiewicz M, Wilkins SD, Avise JC (2000) The genetic mating system of spotted sunfish (*Lepomis punctatus*): Mate numbers and the influence of male reproductive parasites. *Molecular Ecology* 9: 2119–2128.

392. Dhondt AA, Schillemans J (1983) Reproductive success of the great tit in relation to its territorial status. *Animal Behaviour* 31: 902–912.

393. Diamond JM (1992) *The Third Chimpanzee.* HarperCollins Publishers, New York.

394. Diamond JM (1999) Dirty eating for healthy living. *Nature* 400: 120–121.

395. Diamond JM (2000) Talk of cannibalism. *Nature* 407: 25–26.

396. Dias PC (1996) Sources and sinks in population biology. *Trends in Ecology and Evolution* 11: 326–330.

397. Dibattista JD, Feldheim KA, Gruber SH, Hendry AP (2008) Are indirect genetic benefits associated with polyandry? Testing predictions in a natural population of lemon sharks. *Molecular Ecology* 17: 783–795.

398. Dickens M, Berridge D, Hartley IR (2008) Biparental care and offspring begging strategies: hungry nestling blue tits move towards the father. *Animal Behaviour* 75: 167–174.

399. Dickinson JL, Rutowski RL (1989) The function of the mating plug in the chalcedon checkerspot butterfly. *Animal Behaviour* 38: 154–162.

400. Dickinson JL (1995) Trade-offs between postcopulatory riding and mate location in the blue milkweed beetle. *Behavioral Ecology* 6: 280–286.

401. Dickinson JL, Weathers WW (1999) Replacement males in the western bluebird: Opportunity for paternity, chick-feeding rules, and fitness consequences of male parental care. *Behavioral Ecology and Sociobiology* 45: 201–209.

402. Dickinson JL (2001) Extrapair copulations in western bluebirds (*Sialia mexicana*): female receptivity favors older males. *Behavioral Ecology and Sociobiology* 50: 423–429.

403. Dloniak SM, French JA, Place NJ, Weldele ML, Glickman SE, Holekamp KE (2004) Non-invasive monitoring of fecal androgens in spotted hyenas (*Crocuta crocuta*). *General and Comparative Endocrinology* 135: 51–61.

404. Dloniak SM, French JA, Holekamp KE (2006) Rank-related maternal effects of androgens on behaviour in wild spotted hyaenas. *Nature* 440: 1190–1193.

405. Dodson GN, Beck MW (1993) Pre-copulatory guarding of penultimate females by male crab spiders, *Misumenoides formosipes*. *Animal Behaviour* 46: 951–959.

405a. Dodson GN (1997) Resource defense mating system in antlered flies, *Phytalmia* spp. (Diptera: Tephritidae). *Annals of the Entomological Society of America* 90: 496–504.

406. Dornhaus A, Chittka L (2001) Food alert in bumblebees (*Bombus terrestris*): possible mechanisms and evolutionary implications. *Behavioral Ecology and Sociobiology* 50: 570–576.

407. Double MC, Cockburn A (2003) Subordinate superb fairy-wrens (*Malurus cyaneus*) parasitize the reproductive success of attractive dominant males. *Proceedings of the Royal Society of London, Series B* 270: 379–384.

408. Doucet SM, Montgomerie R (2003) Multiple sexual ornaments in satin bowerbirds: ultraviolet plumage and bowers signal different aspects of male quality. *Behavioral Ecology* 14: 503–509.

409. Doupe AJ, Solis MM (1997) Song- and order-selective neurons develop in the songbird anterior forebrain during vocal learning. *Journal of Neurobiology* 33: 694–709.

410. Downes S (2001) Trading heat and food for safety: Costs of predator avoidance in a lizard. *Ecology* 82: 2870–2881.

411. Drăgăniou TI, Nagle L, Kreutzer M (2002) Directional female preference for an exaggerated male trait in canary (*Serinus canaria*) song. *Proceedings of the Royal Society of London, Series B* 269: 2525–2531.

412. Draud M, Lynch PAE (2002) Asymmetric contests for breeding sites between monogamous pairs of convict cichlids (*Archocentrus nigrofasciatum*, Cichlidae): pair experience pays. *Behaviour* 139: 861–873.

413. Drea CM, Weldele ML, Forger NG, Coscia EM, Frank LG, Licht P, Glickman SE (1998) Androgens and masculinization of genitalia in the spotted hyaena (*Crocuta crocuta*). 2. Effects of prenatal anti-androgens. *Journal of Reproduction and Fertility* 113: 117–127.

414. Drea CM, Place NJ, Weldele ML, Coscia EM, Licht P, Glickman SE (2002) Exposure to naturally circulating androgens during foetal life incurs direct reproductive costs in female spotted hyenas, but is prerequisite for male mating. *Proceedings of the Royal Society of London, Series B* 269: 1981–1987.

415. Drews C (1996) Contests and patterns of injuries in free-ranging male baboons (*Papio cynocephalus*). *Behaviour* 133: 443–474.

416. Dudley R (2000) Evolutionary origins of human alcoholism in primate frugivory. *Quarterly Review of Biology* 75: 3–15.

417. Dudley R, Byrnes G, Yanoviak SP, Borrell B, Brown RM, McGuire JA (2007) Gliding and the functional origins of flight: biomechanical novelty or necessity? *Annual Review of Ecology, Evolution and Systematics* 38: 179–201.

418. Duffy KG, Wrangham RW, Silk JB (2007) Male chimpanzees exchange political support for mating opportunities. *Current Biology* 17: R585–R587.

419. Dufour KW, Weatherhead PJ (1998) Bilateral symmetry as an indicator of male quality in red-winged blackbirds: Associations with measures of health, viability, and parental effort. *Behavioral Ecology* 9: 220–231.

420. Duncan J, Seitz RJ, Kolodny J, Bor D, Herzog H, Ahmed A, Newell FN, Emslie H (2000) A neural basis for general intelligence. *Science* 289: 457–460.

421. Dunlap AS, Chen BB, Bednekoff PA, Greene TM, Balda RP (2006) A state-dependent sex difference in spatial memory in pinyon jays, *Gymnorhinus cyanocephalus*: mated females forget as predicted by natural history. *Animal Behaviour* 72: 401–411.

422. Dunn PO, Cockburn A (1999) Extrapair mate choice and honest signaling in cooperatively breeding superb fairy-wrens. *Evolution* 53: 938–946.

423. Eadie JM, Fryxell JM (1992) Density dependence, frequency-dependence, and alternative nesting strategies in goldeneyes. *American Naturalist* 140: 621–641.

424. East ML, Hofer H, Wickler W (1993) The erect "penis" is a flag of submission in a female-dominated society: Greetings in Serengeti spotted hyenas. *Behavioral Ecology and Sociobiology* 33: 355–370.

425. Eberhard WG (1996) *Female Control: Sexual Selection by Cryptic Female Choice.* Princeton University Press, Princeton, NJ.

426. Eberhard WG (2003) Substitution of silk stabilimenta for egg sacs by *Allocyclosa bifurca* (Araneae: Araneidae) suggests that silk stabilimenta function as camouflage devices. *Behaviour* 140: 847–868.

427. Eberhard WG (2005) Evolutionary conflicts of interest: are female conflicts of interest different? *American Naturalist* 165: S19–S25.

428. Eberhard WG (2006) Stabilimenta of *Philoponella vicina* (Araneae: Uloboridae) and *Gasteracantha cancriformis* (Araneae: Araneidae): evidence against a prey attraction function. *Biotropica* 39: 216–220.

429. Edvardsson M (2007) Female *Callosobruchus maculatus* mate when they are thirsty: resource-rich ejaculates as mating effort in a beetle. *Animal Behaviour* 74: 183–188.

430. Eggert A-K, Sakaluk SK (1995) Female-coerced monogamy in burying beetles. *Behavioral Ecology and Sociobiology* 37: 147–154.

431. Eikenaar C, Richardson DS, Brouwer L, Komdeur J (2007) Parent presence, delayed dispersal, and territory acquisition in the Seychelles warbler. *Behavioral Ecology* 18: 874–879.

432. Eising CM, Groothuis TGG (2003) Yolk androgens and begging behaviour in black-headed gull chicks: an experimental field study. *Animal Behaviour* 66: 1027–1034.

433. Ellegren H (2001) Hens, cocks, and avian sex determination. *EMBO Reports* 2: 192–196.

434. Ellis AW, Young AW (1996) *Human Cognitive Neuropsychology*. Psychology Press, East Sussex, UK.

435. Emlen DJ (2000) Integrating development with evolution: A case study with beetle horns. *BioScience* 50: 403–418.

436. Emlen DJ (2001) Costs and the diversification of exaggerated animal structures. *Science* 291: 1534–1536.

437. Emlen DJ (2008) The evolution of animal weapons. *Annual Review of Ecology, Evolution and Systematics* 39: 387–413.

438. Emlen ST, Oring LW (1977) Ecology, sexual selection and the evolution of mating systems. *Science* 197: 215–223.

439. Emlen ST (1978) Cooperative breeding. In: Krebs JR, Davies NB (eds) *Behavioural Ecology: An Evolutionary Approach*. Blackwell, Oxford.

440. Emlen ST, Demong NJ, Emlen DJ (1989) Experimental induction of infanticide in female wattled jacanas. *Auk* 106: 1–7.

441. Emlen ST (1991) Evolution of cooperative breeding in birds and mammals. In: Krebs JR, Davies NB (eds) *Behavioural Ecology: An Evolutionary Approach*. Blackwell Scientific, Oxford.

442. Emlen ST, Wrege PH, Demong NJ (1995) Making decisions in the family: An evolutionary perspective. *American Scientist* 83: 148–157.

443. Emlen ST, Wrege PH, Webster MS (1998) Cuckoldry as a cost of polyandry in the sex-role-reversed wattled jacana, *Jacana jacana*. *Proceedings of the Royal Society of London, Series B* 265: 2359–2364.

444. Endler JA (1991) Interactions between predators and prey. In: Krebs JR, Davies NB (eds) *Behavioural Ecology: An Evolutionary Approach*. Blackwell Scientific, Oxford.

445. Engqvist L (2007) Male scorpionflies assess the amount of rival sperm transferred by females' previous mates. *Evolution* 61: 1489–1494.

446. Esch HE, Zhang SW, Srinivasan MV, Tautz J (2001) Honeybee dances communicate distances measured by optic flow. *Nature* 411: 581–583.

447. Evans HE (1966) *Life on a Little Known Planet*. Dell, New York.

448. Evans HE (1973) *Wasp Farm*. Anchor Press, Garden City, NY.

449. Evans JP, Magurran AE (2000) Multiple benefits of multiple mating in guppies. *Proceedings of the National Academy of Sciences USA* 97: 10074–10076.

450. Ewer RF (1973) *The Carnivores*. Cornell University Press, Ithaca, NY.

451. Faaborg J, Parker PG, DeLay L, de Vries TJ, Bednarz JC, Paz SM, Naranjo J, Waite TA (1995) Confirmation of cooperative polyandry in the Galapagos hawk (*Buteo galapagoensis*). *Behavioral Ecology and Sociobiology* 36: 83–90.

452. Fadool DA, Tucker K, Perkins R, Fasciani G, Thompson RN, Parsons AD, Overton JM, Koni PA, Flavell RA, Kaczmarek LK (2004) Kv1.3 channel gene-targeted deletion produces "super-smeller mice" with altered glomeruli, interacting scaffolding proteins, and biophysics. *Neuron* 41: 389–404.

453. Falls JB (1988) Does song deter territorial intrusion in white-throated sparrows (*Zonotrichia albicollis*)? *Canadian Journal of Zoology* 66: 206–211.

454. Farley CT, Taylor CR (1991) A mechanical trigger for the trot-gallop transition in horses. *Science* 253: 306–308.

455. Farner DS (1964) Time measurement in vertebrate photoperiodism. *American Naturalist* 95: 375–386.

456. Farner DS, Lewis RA (1971) Photoperiodism and reproductive cycles in birds. *Photophysiology* 6: 325–370.

457. Farries MA (2001) The oscine song system considered in the context of the avian brain: Lessons learned from comparative neurobiology. *Brain Behavior and Evolution* 58: 80–100.

458. Fehr E, Schmidt KM (1999) A theory of fairness, competition, and cooperation. *Quarterly Journal of Economics* 114: 817–868.

459. Feinberg DR (2008) Are human faces and voices ornaments signaling common underlying cues to mate value? *Evolutionary Anthropology* 17: 112–118.

460. Ferguson JN, Young LJ, Hearn EF, Matzuk MM, Insel TR, Winslow JT (2000) Social amnesia in mice lacking the oxytocin gene. *Nature Genetics* 25: 284–288.

461. Fernald RD (1993) Cichlids in love. *The Sciences* 33: 27–31.

462. Field J, Shreeves G, Sumner S, Casiraghi M (2000) Insurance-based advantage to helpers in a tropical hover wasp. *Nature* 404: 869–871.

463. Field SA, Keller MA (1993) Alternative mating tactics and female mimicry as post-copulatory mate-guarding behaviour in the parasitic wasp *Cotesia rubecula*. *Animal Behaviour* 46: 1183–1189.

464. Fink B, Neave N, Seydel H (2007) Male facial appearance signals physical strength to women. *American Journal of Human Biology* 19: 82–87.

465. Fink S, Excoffier L, Heckel G (2006) Mammalian monogamy is not controlled by a single gene. *Proceedings of the National Academy of Sciences USA* 103: 10956–10960.

466. Fisher AP (2003) Still "not quite as good as having your own"? Toward a sociology of adoption. *Annual Review of Sociology* 29: 335–361.

467. Fisher J (1954) Evolution and bird sociality. In: Huxley J, Hardy AC, Ford EB (eds) *Evolution as a Process*. Allen & Unwin, London.

468. Fiske P, Rintamäki PT, Karvonen E (1998) Mating success in lekking males: A meta-analysis. *Behavioral Ecology* 9: 328–338.

469. Fitzpatrick MJ, Ben-Shahar Y, Smid HM, Vet LEM, Robinson-Wolrath S, Sokolowski MB (2005) Candidate genes for behavioral ecology. *Trends in Ecology and Evolution* 20: 96–104.

470. Fitzpatrick MJ, Feder E, Rowe L, Sokolowski MB (2007) Maintaining a behaviour polymorphism by frequency-dependent selection on a single gene. *Nature* 447: 210–212.

471. Flinn MV (1988) Step-parent/step-offspring interactions in a Caribbean village. *Ethology and Sociobiology* 9: 335–369.

472. Foellmer MW, Fairbairn DJ (2003) Spontaneous male death during copulation in an orb-weaving spider. *Proceedings of the Royal Society of London, Series B* 270: S183–S185.

473. Follett BK, Mattocks PW, Jr, Farner DS (1974) Circadian function in the photoperiodic induction of gonadotropin secretion in the white-crowned sparrow, *Zonotrichia leucophrys gambelli*. *Proceedings of the National Academy of Sciences USA* 71: 1666–1669.

474. Folstad I, Karter AJ (1992) Parasites, bright males, and the immunocompetence handicap. *American Naturalist* 139: 603–622.

475. Foltz DW, Schwagmeyer PL (1989) Sperm competition in the thirteen-lined ground squirrel: Differential fertilization success under field conditions. *American Naturalist* 133: 257–265.

476. Ford EB (1955) *Moths*. Collins, London.

477. Forsgren E, Amundsen T, Borg AA, Bjelvenmark J (2004) Unusually dynamic sex roles in a fish. *Nature* 429: 551–554.

478.   Forster LM (1992) The stereotyped behaviour of sexual cannibalism in *Latrodectus hasselti* Thorell (Araneae: Theridiidae), the Australian redback spider. *Australian Journal of Zoology* 40: 1–11.

479.   Fossøy F, Johnsen A, Lifjeld JT (2006) Evidence of obligate female promiscuity in a socially monogamous passerine. *Behavioral Ecology and Sociobiology* 60: 255–259.

480.   Fossøy F, Johnsen A, Lifjeld JT (2008) Multiple genetic benefits of female promiscuity in a socially monogamous passerine. *Evolution* 62: 145–156.

481.   Foster KR, Wenseleers T, Ratnieks FLW, Queller DC (2006) There is nothing wrong with inclusive fitness. *Trends in Ecology and Evolution* 21: 599–600.

482.   Foster MS (1977) Odd couples in manakins: A study of social organization and cooperative breeding in *Chiroxiphia linearis*. *American Naturalist* 111: 845–853.

483.   Foster WA (1990) Experimental evidence for effective and altruistic colony defence against natural predators by soldiers of the gall-forming aphid *Pemphigus spyrothecae* (Hemiptera: Pemphigidae). *Behavioral Ecology and Sociobiology* 27: 421–439.

484.   Foster WA, Rhoden PK (1998) Soldiers effectively defend aphid colonies against predators in the field. *Animal Behaviour* 55: 761–765.

485.   Fox EA (2002) Female tactics to reduce sexual harassment in the Sumatran orangutan (*Pongo pygmaeus abelii*). *Behavioral Ecology and Sociobiology* 52: 93–101.

486.   Francis CM, Elp A, Brunton JA, Kunz TH (1994) Lactation in male fruit bats. *Nature* 367: 691–692.

487.   Francis L (1976) Social organization within clones of the sea anemone *Anthopleura elegantissima*. *Biological Bulletin* 150: 361–375.

488.   Francis RC, Soma KK, Fernald RD (1993) Social regulation of the brain-pituitary-gonadal axis. *Proceedings of the National Academy of Sciences USA* 90: 7794–7798.

489.   Frank LG, Holekamp HE, Smale L (1995) Dominance, demographics and reproductive success in female spotted hyenas: A long-term study. In: Sinclair ARE, Arcese P (eds) *Serengeti II: Research, Management, and Conservation of an Ecosystem*. University of Chicago Press, Chicago.

490.   Frank LG, Weldele ML, Glickman SE (1995) Masculinization costs in hyaenas. *Nature* 377: 584–585.

491.   Fretwell SD, Lucas HK, Jr. (1969) On territorial behavior and other factors influencing habitat distribution in birds. I. Theoretical development. *Acta Biotheoretica* 19: 16–36.

492.   Friberg U, Arnqvist G (2003) Fitness effects of female mate choice: preferred males are detrimental for *Drosophila melanogaster* females. *Journal of Evolutionary Biology* 16: 797–811.

493.   Frith CB, Frith DW (2001) Nesting biology of the spotted catbird, *Ailuroedus melanotis*, a monogamous bowerbird (Ptilonorhynchidae), in Australian Wet Tropics upland rainforests. *Australian Journal of Zoology* 49: 279–310.

494.   Frith CB, Frith DW (2004) *Bowerbirds*. Oxford University Press, London.

495.   Frost WN, Hoppe TA, Wang J, Tian LM (2001) Swim initiation neurons in *Tritonia diomedea*. *American Zoologist* 41: 952–961.

496.   Froy O, Gotter AL, Casselman AL, Reppert SM (2003) Illuminating the circadian clock in monarch butterfly migration. *Science* 300: 1303–1305.

497.   Fu P, Neff BD, Gross MR (2001) Tactic-specific success in sperm competition. *Proceedings of the Royal Society of London, Series B* 268: 1105–1112.

498.   Fuentes A (2002) Patterns and trends in primate pair bonds. *International Journal of Primatology* 23: 953–978.

499.   Fullard JH, Yack JE (1993) The evolutionary biology of insect hearing. *Trends in Ecology and Evolution* 8: 248–252.

500.   Fullard JH (1997) The sensory coevolution of moths and bats. In: Hoy RR, Popper AN, Fay RR (eds) *Comparative Hearing: Insects*. Springer, New York.

501.   Fullard JH, Dawson JW, Otero LD, Surlykke A (1997) Bat-deafness in day-flying moths (Lepidoptera, Notodontidae, Dioptinae). *Journal of Comparative Physiology A* 181: 477–483.

502.   Fullard JH (2000) Day-flying butterflies remain day-flying in a Polynesian, bat-free habitat. *Proceedings of the Royal Society of London, Series B* 267: 2295–2300.

503.   Fullard JH, Otero LD, Orellana A, Surlykke A (2000) Auditory sensitivity and diel flight activity in Neotropical Lepidoptera. *Annals of the Entomological Society of America* 93: 956–965.

504.   Fullard JH, Dawson JW, Jacobs DS (2003) Auditory encoding during the last moment of a moth's life. *Journal of Experimental Biology* 206: 281–294.

505.   Fullard JH, Ratcliffe JM, Soutar AR (2004) Extinction of the acoustic startle response in moths endemic to a bat-free habitat. *Journal of Evolutionary Biology* 17: 856–861.

506.   Fuller RC, Houle D, Travis J (2005) Sensory bias as an explanation for the evolution of mate preferences. *American Naturalist* 166: 437–446.

507.   Funk DH, Tallamy DW (2000) Courtship role reversal and deceptive signals in the long-tailed dance fly, *Rhamphomyia longicauda*. *Animal Behaviour* 59: 411–421.

508.   Furlow FB (1997) Human neonatal cry quality as an honest signal of fitness. *Evolution and Human Behavior* 18: 175–194.

509.   Fusani L, Gahr M, Hutchison JB (2001) Aromatase inhibition reduces specifically one display of the ring dove courtship behavior. *General and Comparative Endocrinology* 122: 23–30.

510.   Gallup AC, White DD, Gallup GG (2007) Handgrip strength predicts sexual behavior, body morphology, and aggression in male college students. *Evolution and Human Behavior* 28: 423–429.

511.   Gamboa GJ (1978) Intraspecific defense: advantage of social cooperation among paper wasp foundresses. *Science* 199: 1463–1465.

512.   Gamboa GJ (2004) Kin recognition in eusocial wasps. *Annales Zoologici Fennici* 41: 789–808.

513.   Gangestad SW, Simpson JA (2000) Trade-offs, the allocation of reproductive effort, and the evolutionary psychology of human mating. *Behavioral and Brain Sciences* 23: 624–644.

514.   Gangestad SW, Simpson JA, Cousins AJ, Garver-Apgar CE, Christensen PN (2004) Women's preferences for male behavioral displays change across the menstrual cycle. *Psychological Science* 15: 203–207.

515.   Gangestad SW, Thornhill R (2008) Human oestrus. *Proceedings of the Royal Society of London, Series B* 275: 991–1000.

516.   Garamszegi LZ, Eens M (2004) Brain space for a learned task: strong intraspecific evidence for neural correlates of singing behavior in songbirds. *Brain Research Reviews* 44: 187–193.

517.   Garamszegi LZ (2005) Bird songs and parasites. *Behavioral Ecology and Sociobiology* 59: 169–180.

518.   Garcia J, Ervin FR (1968) Gustatory-visceral and telereceptor-cutaneous conditioning: Adaptation in internal and external milieus. *Communications in Behavioral Biology (A)* 1: 389–415.

519.   Garcia J, Hankins WG, Rusiniak KW (1974) Behavioral regulation of the milieu interne in man and rat. *Science* 185: 824–831.

520.   Garstang M (2004) Long-distance, low-frequency elephant communication. *Journal of Comparative Physiology A* 190: 791–805.

521.   Garver-Apgar CE, Gangestad SW, Thornhill R, Miller RD, Olp JJ (2006) Major histocompatibility complex alleles, sexual responsivity, and unfaithfulness in romantic couples. *Psychological Science* 17: 830–835.

522. Gaskett AC, Herberstein ME (2008) Orchid sexual deceit provokes pollinator ejaculation. *American Naturalist* 171: E206–E212.

523. Gaulin SJC, FitzGerald RW (1986) Sex differences in spatial ability: An evolutionary hypothesis and test. *American Naturalist* 127: 74–88.

524. Gaulin SJC, FitzGerald RW (1989) Sexual selection for spatial-learning ability. *Animal Behaviour* 37: 322–331.

525. Gaulin SJC, Boster JS (1990) Dowry as female competition. *American Anthropologist* 92: 994–1005.

526. Gaulin SJC, McBurney DH (2003) *Psychology, An Evolutionary Approach*, 2nd edn. Prentice Hall, Upper Saddle River, NJ.

527. Geary DC (1998) *Male, Female: The Evolution of Human Sex Differences*. American Psychological Association, Washington, DC.

528. Gentner TQ, Hulse SH (2000) European starling preference and choice for variation in conspecific male song. *Animal Behaviour* 59: 443–458.

529. Gesquiere LR, Wango EO, Alberts SC, Altmann J (2007) Mechanisms of sexual selection: Sexual swellings and estrogen concentrations as fertility indicators and cues for male consort decisions in wild baboons. *Hormones and Behavior* 51: 114–125.

530. Getting PA (1983) Mechanisms of pattern generation underlying swimming in *Tritonia*. II. Network reconstruction. *Journal of Neurophysiology* 49: 1017–1035.

531. Getting PA (1989) A network oscillator underlying swimming in *Tritonia*. In: Jacklet JW (ed) *Neuronal and Cellular Oscillators*. Dekker, New York.

532. Getz LL, Carter CS (1996) Prairie-vole partnerships. *American Scientist* 84: 56–62.

533. Getz LL, McGuire B, Carter CS (2003) Social behavior, reproduction and demography of the prairie vole, *Microtus ochrogaster*. *Ethology, Ecology and Evolution* 15: 105–118.

534. Ghalambor CK, Martin TE (2001) Fecundity-survival trade-offs and parental risk-taking in birds. *Science* 292: 494–497.

535. Gibson RM (1996) A re-evaluation of hotspot settlement in lekking sage grouse. *Animal Behaviour* 52: 993–1005.

536. Gibson RM, Aspbury AS, McDaniel LL (2002) Active formation of mixed-species grouse leks: a role for predation in lek evolution? *Proceedings of the Royal Society of London, Series B* 269: 2503–2507.

537. Gil D, Leboucher G, Lacroix A, Cue R, Kreutzer M (2004) Female canaries produce eggs with greater amounts of testosterone when exposed to preferred male song. *Hormones and Behavior* 45: 64–70.

538. Gil D, Naguib MKR, Rutstein A, Gahr M (2006) Early condition, song learning, and the volume of song brain nuclei in the zebra finch (*Taeniopygia guttata*). *Journal of Neurobiology* 66: 1602–1612.

539. Gilardi JD, Duffey SS, Munn CA, Tell LA (1999) Biochemical functions of geophagy in parrots: Detoxification of dietary toxins and cytoprotective effects. *Journal of Chemical Ecology* 25: 897–922.

540. Gill FB, Wolf LL (1975) Economics of feeding territoriality in the golden-winged sunbird. *Ecology* 56: 333–345.

541. Gill FB, Wolf LL (1978) Comparative foraging efficiencies of some montane sunbirds in Kenya. *Condor* 80: 391–400.

542. Gillespie DOS, Russell AF, Lummaa V (2008) When fecundity does not equal fitness: evidence of an offspring quantity versus quality trade-off in pre-industrial humans. *Proceedings of the Royal Society of London, Series B* 275: 713–722.

543. Gintis H (2008) Behavior: Punishment and cooperation. *Science* 318: 1345–1346.

544. Giron D, Ross KG, Strand MR (2007) Presence of soldier larvae determines the outcome of competition in a polyembryonic wasp. *Journal of Evolutionary Biology* 20: 165–172.

545. Glickman SE, Frank LG, Licht P, Yalckinkaya T, Siiteri PK, Davidson J (1993) Sexual differentiation of the female spotted hyena: One of nature's experiments. *Annals of the New York Academy of Sciences* 662: 135–159.

546. Glickman SE, Cunha GR, Drea CM, Conley AJ, Place NJ (2006) Mammalian sexual differentiation: lessons from the spotted hyena. *Trends in Endocrinology and Metabolism* 17: 349–356.

547. Gobbini MI, Haxby JV (2006) Neural systems for recognition of familiar faces. *Neuropsychologia* 45: 32–41.

548. Gobes SMH, Bolhuis JJ (2007) Birdsong memory: A neural dissociation between song recognition and production. *Current Biology* 17: 789–793.

549. Gomendio M, Garcia-Gonzalez F, Reguera P, Rivero A (2008) Male egg carrying in *Phyllomorpha laciniata* is favoured by natural not sexual selection. *Animal Behaviour* 75: 763–770.

550. Gonzalez-Voyer A, Székely T, Drummond H (2007) Why do some siblings attack each other? Comparative analysis of aggression in avian broods. *Evolution* 61: 1946–1955.

551. Goodall J (1988) *In the Shadow of Man*. Houghton Mifflin, Boston.

552. Goodisman MAD, Kovacs JL, Hoffman EA (2007) The significance of multiple mating in the social wasp *Vespula maculifrons*. *Evolution* 61: 2260–2267.

553. Göth A, Evans CS (2004) Social responses without early experience: Australian brush-turkey chicks use specific visual cues to aggregate with conspecifics. *Journal of Experimental Biology* 207: 2199–2208.

554. Gottschall JA, Gottschall TA (2003) Are per-incident rape-pregnancy rates higher than per-incident consensual pregnancy rates? *Human Nature* 14: 1–20.

555. Gotzek D, Ross KG (2007) Genetic regulation of colony social organization in fire ants: an integrative overview. *Quarterly Review of Biology* 82: 201–226.

556. Gould JL (1982) Why do honey bees have dialects? *Behavioral Ecology and Sociobiology* 10: 53–56.

557. Gould SJ, Lewontin RC (1979) The spandrels of San Marco and the Panglossian paradigm: A critique of the adaptationist programme. *Proceedings of the Royal Society of London, Series B* 205: 581–598.

558. Gould SJ (1981) Hyena myths and realities. *Natural History* 90: 16–24.

559. Gould SJ (1986) Evolution and the triumph of homology, or why history matters. *American Scientist* 74: 60–69.

560. Gould SJ (1996) The diet of worms and the defenestration of Prague. *Natural History* 105: 18–24ff.

561. Goymann W, East ML, Hofer H (2001) Androgens and the role of female 'hyperaggressiveness' in spotted hyenas (*Crocuta crocuta*). *Hormones and Behavior* 39: 83–92.

562. Grammar K, Fink B, Møller AP, Thornhill R (2003) Darwinian aesthetics: sexual selection and the biology of beauty. *Biological Reviews* 78: 385–407.

563. Granados-Fuentes D, Tseng A, Herzong ED (2006) A circadian clock in the olfactory bulb controls olfactory responsivity. *Journal of Neuroscience* 26: 12219–12225.

564. Grant BS, Owen DF, Clarke CA (1996) Parallel rise and fall of melanic peppered moths in America and Britain. *Journal of Heredity* 87: 351–357.

565. Grant BS (1999) Fine tuning the peppered moth paradigm. *Evolution* 53: 980–984.

566. Grant BS, Wiseman LL (2002) Recent history of melanism in American peppered moths. *Journal of Heredity* 93: 86–90.

567. Grant PR, Grant BR (2002) Unpredictable evolution in a 30–year study of Darwin's finches. *Science* 296: 707–711.

568. Graves JA, Whiten A (1980) Adoption of strange chicks by herring gulls, *Larus argentatus*. *Zeitschrift für Tierpsychologie* 54: 267–278.

569. Gray EM (1997) Do red-winged blackbirds benefit genetically from seeking copulations with extra-pair males? *Animal Behaviour* 53: 605–623.

570. Gray EM (1997) Female red-winged blackbirds accrue material benefits from copulating with extra-pair males. *Animal Behaviour* 53: 625–639.

571. Gray JR, Thompson PM (2004) Neurobiology of intelligence: Science and ethics. *Nature Reviews Neuroscience* 5: 471–482.

572. Greene E (1987) Individuals in an osprey colony discriminate between high and low quality information. *Nature* 329: 239–241.

573. Greene E, Orsak LT, Whitman DW (1987) A tephritid fly mimics the territorial displays of its jumping spider predators. *Science* 236: 310–312.

574. Greene E, Lyon BE, Muehter VR, Ratcliffe L, Oliver SJ, Boag PT (2000) Disruptive sexual selection for plumage colouration in a passerine bird. *Nature* 407: 1000–1003.

575. Greenspan RJ (2004) E Pluribus Unum, Ex Uno Plura: Quantitative and single-gene perspectives on the study of behavior. *Annual Review of Neuroscience* 27: 79–105.

576. Greenwood PJ (1980) Mating systems, philopatry, and dispersal in birds and mammals. *Animal Behaviour* 28: 1140–1162.

577. Grémillet D, Pichegru L, Kuntz G, Woakes AG, Wilkinson S, Crawford RJM, Ryan PG (2008) A junk-food hypothesis for gannets feeding on fishery waste. *Proceedings of the Royal Society of London, Series B* 275: 1149–1156.

578. Grether GF (2000) Carotenoid limitation and mate preference evolution: A test of the indicator hypothesis in guppies (*Poecilia reticulata*). *Evolution* 54: 1712–1714.

579. Grether GF, Kasahara S, Kolluru GR, Cooper EL (2004) Sex-specific effects of carotenoid intake on the immunological response to allografts in guppies (*Poecilia reticulata*). *Proceedings of the Royal Society of London, Series B* 271: 45–49.

580. Griesser M, Ekman J (2005) Nepotistic mobbing behaviour in the Siberian jay, *Perisoreus infaustus*. *Animal Behaviour* 60: 345–352.

581. Griffin DR (1958) *Listening in the Dark*. Yale University Press, New Haven, CT.

582. Griffith SC, Owens IPF, Thuman KA (2002) Extra pair paternity in birds: a review of interspecific variation and adaptive function. *Molecular Ecology* 11: 2195–2212.

583. Griffith SC (2007) The evolution of infidelity in socially monogamous passerines: Neglected components of direct and indirect selection. *American Naturalist* 169: 274–281.

584. Grim T (2007) Experimental evidence for chick discrimination without recognition in a brood parasite host. *Proceedings of the Royal Society of London, Series B* 274: 373–381.

585. Griskevicius V, Tybur JM, Sundie JM, Cialdini RB, Miller GF, Kenrick DT (2007) Blatant benevolence and conspicuous consumption: When romantic motives elicit costly displays. *Journal of Personality and Social Psychology* 93: 85–102.

586. Gross MR, MacMillan AM (1981) Predation and the evolution of colonial nesting in bluegill sunfish (*Lepomis macrochirus*). *Behavioral Ecology and Sociobiology* 8: 163–174.

587. Gross MR (1982) Sneakers, satellites, and parentals: Polymorphic mating strategies in North American sunfishes. *Zeitschrift für Tierpsychologie* 60: 1–26.

588. Gross MR, Sargent RC (1985) The evolution of male and female parental care in fishes. *American Zoologist* 25: 807–822.

589. Gross MR (1991) Evolution of alternative reproductive strategies: frequency-dependent sexual selection in male bluegill sunfish. *Philosophical Transactions of the Royal Society of London, Series B* 332: 59–66.

590. Gross MR (1996) Alternative reproductive strategies and tactics: Diversity within species. *Trends in Ecology and Evolution* 11: 92–98.

591. Grozinger CM, Sharabash NM, Whitfield CW, Robinson GE (2003) Pheromone-mediated gene expression in the honey bee brain. *Proceedings of the National Academy of Sciences USA* 100 (Suppl. 2): 14519–14525.

592. Grunt JA, Young WC (1953) Consistency of sexual behavior patterns in individual male guinea pigs following castration and androgen therapy. *Journal of Comparative Physiology and Psychology* 46: 138–144.

593. Gubernick DJ, Teferi T (2000) Adaptive significance of male parental care in a monogamous mammal. *Proceedings of the Royal Society of London, Series B* 267: 147–150.

594. Gurney ME, Konishi M (1980) Hormone-induced sexual differentiation of brain and behavior in zebra finches. *Science* 208: 1380–1383.

595. Guzmán-Novoa E, Prieto-Merlos D, Uribe-Rubio JL, Hunt GJ (2003) Relative reliability of four field assays to test defensive behaviour of honey bees (*Apis mellifera*). *Journal of Apicultural Research* 42: 42–46.

596. Gwinner E, Dittami J (1990) Endogenous reproductive rhythms in a tropical bird. *Science* 249: 906–908.

597. Gwinner E (1996) Circannual clocks in avian reproduction and migration. *Ibis* 138: 47–63.

598. Gwynne DT (1981) Sexual difference theory: Mormon crickets show role reversal in mate choice. *Science* 213: 779–780.

599. Gwynne DT (1983) Beetles on the bottle. *Journal of the Australian Entomological Society* 23: 79.

600. Gwynne DT, Bussiere LF, Ivy TM (2007) Female ornaments hinder escape from spider webs in a role-reversed swarming dance fly. *Animal Behaviour* 73: 1077–1082.

601. Hack MA (1998) The energetics of male mating strategies in field crickets. *Journal of Insect Behavior* 11: 853–868.

602. Hadfield JD, Burgess MD, Lord A, Phillimore AB, Clegg SM, Owens IPF (2006) Direct versus indirect sexual selection: genetic basis of colour, size and recruitment in a wild bird. *Proceedings of the Royal Society of London, Series B* 273: 1347–1353.

603. Haesler S, Rochefort C, Georgi B, Licznerski P, Osten P, Scharff C (2007) Incomplete and inaccurate vocal imitation after knockdown of FoxP2 in songbird basal ganglia nucleus area X. *PLoS Biology* 5: 1–13.

604. Hahn TP (1995) Integration of photoperiodic and food cues to time changes in reproductive physiology by an opportunistic breeder, the red crossbill, *Loxia curvirostra* (Aves: Carduelinae). *Journal of Experimental Zoology* 272: 213–226.

605. Hahn TP, Wingfield JC, Mullen R, Deviche PJ (1995) Endocrine bases of spatial and temporal opportunism in arctic-breeding birds. *American Zoologist* 35: 259–273.

606. Hahn TP (1998) Reproductive seasonality in an opportunistic breeder, the red crossbill, *Loxia curvirostra*. *Ecology* 79: 2365–2375.

607. Hahn TP, Pereyra ME, Sharbaugh SM, Bentley GE (2004) Physiological responses to photoperiod in three cardueline finch species. *General and Comparative Endocrinology* 137: 99–108.

608. Halgren E, Dale AM, Sereno MI, Tootell RBH, Marinkovic K, Rosen BR (1999) Location of human face-selective cortex with respect to retinotopic areas. *Human Brain Mapping* 7: 29–37.

609. Hamilton WD (1964) The genetical theory of social behaviour, I, II. *Journal of Theoretical Biology* 7: 1–52.

610. Hamilton WD (1971) Geometry for the selfish herd. *Journal of Theoretical Biology* 31: 295–311.

611. Hamilton WD, Zuk M (1982) Heritable true fitness and bright birds: A role for parasites? *Science* 218: 384–387.

612. Hamilton WD (1995) *The Narrow Roads of Gene Land. Vol. 1. Evolution of Social Behaviour*. W.H. Freeman, Oxford.

613. Hamilton WD (1996) Foreword. In: Turillazzi S, West-Eberhard MJ (eds) *Natural History and the Evolution of Paper Wasps*. Oxford University Press, Oxford, pp v–vi.

614. Hammock EAD, Young LJ (2004) Functional microsatellite polymorphism associated with divergent social structure in vole species. *Molecular Biology and Evolution* 21: 1057–1063.

615. Hamner WM (1964) Circadian control of photoperiodism in the house finch demonstrated by interrupted-night experiments. *Nature* 203: 1400–1401.

616. Hanby JP, Bygott JD (1987) Emigration of subadult lions. *Animal Behaviour* 35: 161–169.

617. Hanifin CT, Brodie ED, Jr, Brodie ED, III (2008) Phenotypic mismatches reveal escape from arms-race coevolution. *PLoS Biology* 6: doi:10.1371/journal.pbio.0060060.

618. Harbison H, Nelson DA, Hahn TP (1999) Long-term persistence of song dialects in the mountain white-crowned sparrow. *Condor* 101: 133–148.

619. Harcourt AH, Harvey PH, Larson SG, Short RV (1981) Testis weight, body weight and breeding system in primates. *Nature* 293: 55–57.

620. Hardouin LA, Reby D, Bavoux C, Burneleau G, Bretagnolle L (2007) Communication of male quality in owl hoots. *American Naturalist* 169: 552–562.

621. Hardy CL, Van Vugt M (2006) Nice guys finish first: The competitive altruism hypothesis. *Personality and Social Psychology Bulletin* 32: 1402–1413.

622. Hare B, Brown M, Williamson C, Tomasello M (2002) The domestication of social cognition in dogs. *Science* 298: 1634–1636.

623. Harlow HF, Harlow MK (1962) Social deprivation in monkeys. *Scientific American* 207 (Nov.): 136–146.

624. Harlow HF, Harlow MK, Suomi SJ (1971) From thought to therapy: Lessons from a primate laboratory. *American Scientist* 59: 538–549.

625. Hartung J (1982) Polygyny and inheritance of wealth. *Current Anthropology* 23: 1–12.

626. Hartung J (1995) Love thy neighbor: The evolution of in-group morality. *Skeptic* 3: 86–99.

627. Harvey PH, Pagel MD (1991) *The Comparative Method in Evolutionary Biology.* Oxford University Press, London.

628. Haselton MG (2003) The sexual overperception bias: Evidence of a systematic bias in men from a survey of naturally occurring events. *Journal of Research in Personality* 37: 34–47.

629. Haskell DG (1999) The effect of predation on begging-call evolution in nestling wood warblers. *Animal Behaviour* 57: 893–901.

630. Hasselquist D (1998) Polygyny in great reed warblers: A long-term study of factors contributing to fitness. *Ecology* 79: 2376–2350.

631. Hatchwell BJ, Komdeur J (2000) Ecological constraints, life history traits and the evolution of cooperative breeding. *Animal Behaviour* 59: 1079–1086.

632. Hauber ME, Kilner RM (2007) Coevolution, communication, and host chick mimicry in parasitic finches: who mimics whom? *Behavioral Ecology and Sociobiology* 61: 497–503.

633. Haugen TO, Winfield IJ, Vøllestad LA, Fletcher JM, James JB, Stenseth NC (2006) The ideal free pike: 50 years of fitness-maximizing dispersal in Windermere. *Proceedings of the Royal Society of London, Series B* 273: 2917–2924.

634. Hauser MD, Chen MK, Chen F, Chuang E, Chuang E (2003) Give unto others: genetically unrelated cotton-top tamarin monkeys preferentially give food to those who altruistically give food back. *Proceedings of the Royal Society of London, Series B* 270: 2363–2370.

635. Hauser MD (2006) *Moral Minds: How Nature Designed Our Universal Sense of Right and Wrong.* Ecco Press, New York.

636. Hausfater G (1975) Dominance and reproduction in baboons (*Papio cynocephalus*): A quantitative analysis. *Contributions in Primatology* 7: 1–150.

637. Haydock J, Koenig WD (2002) Reproductive skew in the polygynandrous acorn woodpecker. *Proceedings of the National Academy of Sciences USA* 99: 7178–7183.

638. Hayes LD (2000) To nest communally or not to nest communally: A review of rodent communal nesting and nursing. *Animal Behaviour* 59: 677–688.

639. Healy SD, Rowe C (2007) A critique of comparative studies of brain size. *Proceedings of the Royal Society of London, Series B* 274: 453–464.

640. Heath KM, Hadley C (1998) Dichotomous male reproductive strategies in a polygynous human society: Mating versus parental effort. *Current Anthropology* 39: 369–374.

641. Hedwig B (2000) Control of cricket stridulation by a command neuron: efficacy depends on the behavioral state. *Journal of Neurophysiology* 83: 712–722.

642. Heeb P, Schwander T, Faoro S (2003) Nestling detectability affects parental feeding preferences in a cavity-nesting bird. *Animal Behaviour* 66: 637–642.

643. Heinrich B (1979) *Bumblebee Economics.* Harvard University Press, Cambridge, MA.

644. Heinrich B (1984) *In a Patch of Fireweed.* Harvard University Press, Cambridge, MA.

645. Heinrich B (1988) Winter foraging at carcasses by three sympatric corvids, with emphasis on recruitment by the raven, *Corvus corax. Behavioral Ecology and Sociobiology* 23: 141–156.

646. Heinrich B (1989) *Ravens in Winter.* Summit Books, New York.

647. Heinrich B (2004) *The Geese of Beaver Bog.* HarperCollins Publishers, New York.

648. Heinze J, Hölldobler B, Peeters C (1994) Conflict and cooperation in ant societies. *Naturwissenschaften* 81: 489–497.

649. Heistermann M, Ziegler T, van Schaik CP, Launhardt K, Winkler P, Hodges JK (2001) Loss of oestrus, concealed ovulation and paternity confusion in free-ranging Hanuman langurs. *Proceedings of the Royal Society of London, Series B* 268: 2445–2451.

650. Henrich J (2006) Cooperation, punishment, and the evolution of human institutions. *Science* 312: 60–61.

651. Herberstein ME, Craig CL, Coddington JA, Elgar MA (2000) The functional significance of silk decorations of orb-web spiders: A critical review of the empirical evidence. *Biological Reviews* 75: 649–669.

652. Hibbitts TJ, Whiting MJ, Stuart-Fox DM (2007) Shouting the odds: vocalization signals status in a lizard. *Behavioral Ecology and Sociobiology* 61: 1169–1176.

653. Hidalgo-Garcia S (2006) The carotenoid-based plumage coloration of adult Blue Tits *Cyanistes caeruleus* correlates with the health status of their brood. *Ibis* 148: 727–734.

654. Hill GE (2002) *A Red Bird in a Brown Bag: The Function and Evolution of Colorful Plumage in the House Finch.* Oxford University Press, Oxford.

655. Hitchcock CL, Sherry DF (1990) Long-term memory for cache sites in the black-capped chickadee. *Animal Behaviour* 40: 701–712.

656. Hofer H, East ML (2003) Behavioral processes and costs of co-existence in female spotted hyenas: a life history perspective. *Evolutionary Ecology* 17: 315–331.

657. Hofer H, East ML (2008) Siblicide in Serengeti spotted hyenas: a long-term study of maternal input and cub survival. *Behavioral Ecology and Sociobiology* 62: 341–351.

658. Hogg JT (1984) Mating in bighorn sheep: Multiple creative male strategies. *Science* 225: 526–529.

659. Höglund J, Lundberg A (1987) Sexual selection in a monomorphic lek-breeding bird: Correlates of male mating success in the great snipe *Gallinago media. Behavioral Ecology and Sociobiology* 21: 211–216.

660. Höglund J, Alatalo RV (1995) *Leks.* Princeton University Press, Princeton, NJ.

661. Högstedt G (1983) Adaptation unto death: Function of fear screams. *American Naturalist* 121: 562–570.

662. Hohoff C, Franzen K, Sachser N (2003) Female choice in a promiscuous wild guinea pig, the yellow-toothed cavy (*Galea musteloides*). *Behavioral Ecology and Sociobiology* 53: 341–349.

663. Holden C, Mace R (1997) Phylogenetic analysis of the evolution of lactose digestion in adults. *Human Biology* 69: 605–628.

664. Holden CJ, Sear R, Mace R (2003) Matriliny as daughter-biased investment. *Evolution and Human Behavior* 24: 99–112.

665. Holekamp KE (1984) Natal dispersal in Belding's ground squirrels (*Spermophilus beldingi*). *Behavioral Ecology and Sociobiology* 16: 21–30.

666. Holekamp KE, Sherman PW (1989) Why male ground squirrels disperse. *American Scientist* 77: 232–239.

667. Holland B, Rice WR (1998) Chase-away sexual selection: Antagonistic seduction versus resistance. *Evolution* 52: 1–7.

668. Holland B, Rice WR (1999) Experimental removal of sexual selection reverses intersexual antagonistic coevolution and removes a reproductive load. *Proceedings of the National Academy of Sciences USA* 96: 5083–5088.

669. Hölldobler B, Wilson EO (1990) *The Ants*. Harvard University Press, Cambridge, MA.

670. Hölldobler B, Wilson EO (1994) *Journey to the Ants*. Harvard University Press, Cambridge, MA.

671. Holley AJF (1984) Adoption, parent-chick recognition, and maladaptation in the herring gull *Larus argentatus*. *Zeitschrift für Tierpsychologie* 64: 9–14.

672. Holloway CC, Clayton DF (2001) Estrogen synthesis in the male brain triggers development of the avian song control pathway in vitro. *Nature Neuroscience* 4: 170–175.

673. Holmes WG, Sherman PW (1982) The ontogeny of kin recognition in two species of ground squirrels. *American Zoologist* 22: 491–517.

674. Holmes WG, Sherman PW (1983) Kin recognition in animals. *American Scientist* 71: 46–55.

675. Holmes WG (1986) Identification of paternal half-siblings by captive Belding's ground squirrels. *Animal Behaviour* 34: 321–327.

676. Holmes WG (2004) The early history of Hamiltonian-based research on kin recognition. *Annales Zoologici Fennici* 41: 691–711.

677. Hoogland JL, Sherman PW (1976) Advantages and disadvantages of bank swallow (*Riparia riparia*) coloniality. *Ecological Monographs* 46: 33–58.

678. Hoogland JL (1998) Why do female Gunnison's prairie dogs copulate with more than one male? *Animal Behaviour* 55: 351–359.

679. Hoover JP, Reetz MJ (2006) Brood parasitism increases provisioning rate, and reduces offspring recruitment and adult return rates, in a cowbird host. *Oecologia* 149: 165–173.

680. Hoover JP, Robinson SK (2007) Retaliatory mafia behavior by a parasitic cowbird favors host acceptance of parasitic eggs. *Proceedings of the National Academy of Sciences USA* 104: 4479–4483.

681. Hopkins CD (1998) Design features for electric communication. *Journal of Experimental Biology* 202: 1217–1228.

682. Hori M (1993) Frequency-dependent natural selection in the handedness of scale-eating cichlid fish. *Science* 260: 216–219.

683. Hosken DJ, Stockley P (2004) Sexual selection and genital evolution. *Trends in Ecology and Evolution* 19: 87–93.

684. Hotker H (2000) Intraspecific variation in size and density of avocet colonies: Effects of nest-distances on hatching and breeding success. *Journal of Avian Biology* 31: 387–398.

685. Houston DC, Gilardi JD, Hall AJ (2001) Soil consumption by elephants might help to minimize the toxic effects of plant secondary compounds in forest browse. *Mammal Review* 31: 249–254.

686. Howlett RJ, Majerus MEN (1987) The understanding of industrial melanism in the peppered moth (*Biston betularia*) (Lepidoptera: Geometridae). *Biological Journal of the Linnean Society* 30: 31–44.

687. Hrdy SB (1977) *The Langurs of Abu*. Harvard University Press, Cambridge, MA.

688. Hrdy SB (1999) *Mother Nature: A History of Mothers, Infants, and Natural Selection*. Pantheon, New York.

689. Hristov I, Conner WE (2005) Sound strategy: acoustic aposematism in the bat-tiger moth arms race. *Naturwissenschaften* 92: 164–169.

690. Huang Z-Y, Robinson GE (1992) Honeybee colony integration: Worker-worker interactions mediate hormonally regulated plasticity in division of labor. *Proceedings of the National Academy of Sciences USA* 89: 11726–11729.

691. Hughes WOH, Oldroyd BP, Beekman M, Ratnieks FLW (2008) Ancestral monogamy shows kin selection is key to evolution of sociality. *Science* 320: 1213–1216.

692. Huk T, Winkel W (2006) Polygyny and its fitness consequences for primary and secondary female pied flycatchers. *Proceedings of the Royal Society of London, Series B* 273: 1681–1688.

693. Hume DK, Montgomerie RD (2001) Facial attractiveness signals different aspects of "quality" in women and men. *Evolution and Human Behavior* 22: 93–112.

694. Hunt J, Simmons LW (2002) Confidence of paternity and paternal care: covariation revealed through the experimental manipulation of the mating system in the beetle *Onthophagus taurus*. *Journal of Evolutionary Biology* 15: 784–795.

695. Hunt J, Brooks R, Jennions MD, Smith MJ, Bentsen CL, Bussière LF (2004) High-quality male field crickets invest heavily in sexual display but die young. *Nature* 432: 1024–1027.

696. Hunt S, Bennett ATD, Cuthill IC, Griffiths R (1998) Blue tits are ultraviolet tits. *Proceedings of the Royal Society of London, Series B* 265: 451–455.

697. Hunt S, Cuthill IC, Bennett ATD, Griffiths R (1999) Preferences for ultraviolet partners in the blue tit. *Animal Behaviour* 58: 809–815.

698. Hunt S, Kilner RM, Langmore NE, Bennett ATD (2003) Conspicuous, ultraviolet-rich mouth colours in begging chicks. *Proceedings of the Royal Society of London, Series B* 270: S25–S28.

699. Hunter ML, Krebs JR (1979) Geographic variation in the song of the great tit (*Parus major*) in relation to ecological factors. *Journal of Animal Ecology* 48: 759–785.

700. Hurst LD, Peck JR (1996) Recent advances in understanding of the evolution and maintenance of sex. *Trends in Ecology and Evolution* 11: 46–52.

701. Husak JF, Fox SF, Lovern MB, Van den Bussche RA (2006) Faster lizards sire more offspring: sexual selection on whole-animal performance. *Evolution* 60: 2122–2130.

702. Husak JF, Irschick DJ, Meyers JJ, Lailvaux SP, Moore IT (2007) Hormones, sexual signals, and performance of green anole lizards (*Anolis carolinensis*). *Hormones and Behavior* 52: 360–367.

703. Huxley TH (1910) *Lectures and Lay Sermons*. E. P. Dutton, New York.

704. Hyman J, Hughes M, Searcy WA, Nowicki S (2004) Individual variation in the strength of territory defense in male song sparrows: Correlates of age, territory tenure, and neighbor aggressiveness. *Behaviour* 141: 15–27.

705. Ijichi N, Shibao H, Miura T, Matsumoto T, Fukatsu T (2004) Soldier differentiation during embryogenesis of a social aphid, *Pseudoregma bambucicola*. *Entomological Science* 7: 141–153.

706. Irons W (1979) Cultural and biological success. In: Chagnon NA, Irons W (eds) *Evolutionary Biology and Human Social Behavior: An Anthropological Perspective*. Duxbury Press, North Scituate, MA.

707. Itô Y (1989) The evolutionary biology of sterile soldiers in aphids. *Trends in Ecology and Evolution* 4: 69–73.

708. Jacobs LF, Gaulin SJC, Sherry DF, Hoffman GE (1990) Evolution of spatial cognition: Sex-specific patterns of

spatial behavior predict hippocampal size. *Proceedings of the National Academy of Sciences USA* 87: 6349–6352.

709.  Jamieson IG (1989) Behavioral heterochrony and the evolution of birds' helping at the nest: An unselected consequence of communal breeding? *American Naturalist* 133: 394–406.

710.  Jamieson IG (1991) The unselected hypothesis for the evolution of helping behavior: Too much or too little emphasis on natural selection? *American Naturalist* 138: 271–282.

711.  Jamieson IG (1997) Testing reproductive skew models in a communally breeding bird, the pukeko, *Porphyrio porphyrio*. *Proceedings of the Royal Society of London, Series B* 264: 335–340.

712.  Jarvis ED, Ribeiro S, da Silva ML, Ventura D, Vielliard J, Mello CV (2000) Behaviourally driven gene expression reveals song nuclei in hummingbird brain. *Nature* 406: 628–632.

713.  Jarvis JUM, Bennett NC (1993) Eusociality has evolved independently in two genera of bathygerid mole-rats—but occurs in no other subterranean mammal. *Behavioral Ecology and Sociobiology* 33: 253–260.

714.  Jasiénska G, Ziomkiewicz A, Ellison PT, Lipson SF, Thune I (2004) Large breasts and narrow waists indicate high reproductive potential in women. *Proceedings of the Royal Society of London, Series B* 271: 1213–1217.

715.  Jasiénska G, Lipson SF, Ellison PT, Thune I, Ziomkiewicz A (2006) Symmetrical women have higher potential fertility. *Evolution and Human Behavior* 27: 390–400.

716.  Jaycox ER, Parise SG (1980) Homesite selection by Italian honey bee swarms, *Apis mellifera ligustica* (Hymenoptera: Apidae). *Journal of the Kansas Entomological Society* 53: 171–178.

717.  Jaycox ER, Parise SG (1981) Homesite selection by swarms of black-bodied honey bees, *Apis mellifera caucasia* and *A. m. carnica*. *Journal of the Kansas Entomological Society* 54: 697–703.

718.  Jeffreys AJ, Wilson V, Thein SL (1985) Hypervariable "minisatellite" regions in human DNA. *Nature* 314: 67–73.

719.  Jennions MD, Backwell PRY (1996) Residency and size affect fight duration and outcome in the fiddler crab *Uca annulipes*. *Biological Journal of the Linnean Society* 57: 293–306.

720.  Jennions MD, Petrie M (2000) Why do females mate multiply? A review of the genetic benefits. *Biological Reviews* 75: 21–64.

721.  Jenssen TA, Lovern MB, Congdon JD (2001) Field-testing the protandry-based mating system for the lizard, *Anolis carolinensis*: does the model organism have the right model? *Behavioral Ecology and Sociobiology* 50: 162–171.

722.  Jeon J, Buss DM (2007) Altruism toward cousins. *Proceedings of the Royal Society of London, Series B* 274: 1181–1187.

723.  Jesseau SA, Holmes WG, Lee TM (2008) Mother-offspring recognition in communally nesting degus, *Octodon degus*. *Animal Behaviour* 75: 573–582.

724.  Jiguet F, Bretagnolle V (2006) Manipulating lek size and composition using decoys: An experimental investigation of lek evolution models. *American Naturalist* 168: 758–768.

725.  Jiménez JA, Hughes KA, Alaks G, Graham L, Lacy RC (1994) An experimental study of inbreeding depression in a natural habitat. *Science* 266: 271–273.

726.  Jin H, Clayton DF (1997) Localized changes in immediate-early gene regulation during sensory and motor learning in zebra finches. *Neuron* 19: 1049–1059.

727.  Jinks RN, Markley TL, Taylor EE, Perovich G, Dittel AI, Epifanio CE, Cronin TW (2002) Adaptive visual metamorphosis in a deep-sea hydrothermal vent crab. *Nature* 420: 68–70.

728.  Johansson J, Turesson H, Persson A (2004) Active selection for large guppies, *Poecilia reticulata*, by the pike cichlid, *Crenicichla saxatilis*. *Oikos* 105: 595–605.

729.  Johns T (1990) *With Bitter Herbs They Shall Eat It: Chemical Ecology and the Origins of Human Diet and Medicine.* University of Arizona Press, Tucson, AZ.

730.  Johnsen A, Andersson S, Ornberg J, Lifjeld JT (1998) Ultraviolet plumage ornamentation affects social mate choice and sperm competition in bluethroats (Aves: *Luscinia s. svecica*): A field experiment. *Proceedings of the Royal Society of London, Series B* 265: 1313–1318.

731.  Johnsen A, Andersen V, Sunding C, Lifjeld JT (2000) Female bluethroats enhance offspring immunocompetence through extra-pair copulations. *Nature* 406: 296–299.

732.  Johnson CH, Hasting JW (1986) The elusive mechanisms of the circadian clock. *American Scientist* 74: 29–37.

733.  Johnsson JI, Sundström F (2007) Social transfer of predation risk information reduces food locating ability in European minnows (*Phoxinus phoxinus*). *Ethology* 113: 166–173.

734.  Johnstone RA (2000) Models of reproductive skew: A review and synthesis. *Ethology* 106: 5–26.

735.  Jones BC, DeBruine LM, Perrett DI, Little AC, Feinberg DR, Smith MJL (2008) Effects of menstrual cycle phase on face preferences. *Archives of Sexual Behavior* 37: 78–84.

736.  Jones CM, Healy SD (2006) Differences in cue use and spatial memory in men and women. *Proceedings of the Royal Society of London, Series B* 273: 2241–2247.

737.  Jones IL, Hunter FM (1998) Heterospecific mating preferences for a feather ornament in least auklets. *Behavioral Ecology* 9: 187–192.

738.  Jones JS, Wynne-Edwards KE (2000) Paternal hamsters mechanically assist the delivery, consume amniotic fluid and placenta, remove fetal membranes, and provide parental care during the birth process. *Hormones and Behavior* 37: 116–125.

739.  Jones OD (1999) Sex, culture, and the biology of rape: Toward explanation and prevention. *California Law Review* 87: 827–942.

740.  Jouventin P, Barbraud C, Rubin M (1995) Adoption in the emperor penguin, *Aptenodytes forsteri*. *Animal Behaviour* 50: 1023–1029.

741.  Jukema J, Piersma T (2005) Permanent female mimics in a lekking shorebird. *Biology Letters* 2: 161–164.

742.  Just J (1988) Siphonoecetinae (Corophiidae). 6: A survey of phylogeny, distribution, and biology. *Crustaceana*, Supplement 13: 193–208.

743.  Kalko EKV (1995) Insect pursuit, prey capture and echolocation in pipistrelle bats (Microchiroptera). *Animal Behaviour* 50: 861–880.

744.  Kalmijn AJ (1982) Electric and magnetic field detection in elasmobranch fishes. *Science* 218: 916–918.

745.  Kanazawa S (2003) Can evolutionary psychology explain reproductive behavior in the contemporary United States? *Sociological Quarterly* 44: 291–302.

746.  Kannisto V, Christensen K, Vaupel JW (1997) No increased mortality in later life for cohorts born during famine. *American Journal of Epidemiology* 145: 987–994.

747.  Kanwisher N, McDermott J, Chun MM (1997) The fusiform face area: A module in human extrastriate cortex specialized for face perception. *Journal of Neuroscience* 17: 4302–4311.

748.  Kanwisher N, Downing P, Epstein R, Kourtzi Z (2001) Functional neuroimaging of human visual recognition. In: Cabeza R, Kingstone A (eds) *The Handbook of Functional Neuroimaging.* MIT Press, Cambridge, MA.

749.  Kaplan H, Hill K (1985) Hunting ability and reproductive success among male Ache foragers: Preliminary results. *Current Anthropology* 26: 131–133.

750.  Kasumovic MM, Andrade MCB (2006) Male development tracks rapidly shifting sexual versus natural selection pressures. *Current Biology* 16: R242–R243.

751.  Keeton WT (1969) Orientation by pigeons: Is the sun necessary? *Science* 165: 922–928.

752. Kell CA, von Kriegsterin K, Rosler R, Kleinschmidt A, Laufs H (2005) The sensory cortical representation of the human penis: revisiting somatotopy in the male homunculus. *Journal of Neuroscience* 25: 5984–5987.

753. Keller L, Ross KG (1998) Selfish genes: A green beard in the red fire ant. *Nature* 394: 573–575.

754. Keller LF, Grant PR, Grant BR, Petren K (2001) Heritability of morphological traits in Darwin's Finches: misidentified paternity and maternal effects. *Heredity* 87: 325–336.

755. Kemp DJ (2002) Sexual selection constrained by life history in a butterfly. *Proceedings of the Royal Society of London, Series B* 269: 1341–1345.

756. Kemp DJ, Wiklund C (2003) Residency effects in animal contests. *Proceedings of the Royal Society of London, Series B* 271: 1707–1711.

757. Kemp DJ (2007) Female butterflies prefer males bearing bright iridiscent ornamentation. *Proceedings of the Royal Society of London, Series B* 274: 1043–1047.

758. Kemp DJ (2008) Female mating biases for bright ultraviolet iridescence in the butterfly *Eurema hecabe* (Pieridae). *Behavioral Ecology* 19: 1–8.

759. Kempenaers B, Verheyen GR, van der Broeck M, Burke T, van Broeckhoven C, Dhondt AA (1992) Extra-pair paternity results from female preference for high quality males in the blue tit. *Nature* 357: 494–496.

760. Kempenaers B, Verheyen GR, Dhondt AA (1997) Extrapair paternity in the blue tit (*Parus caeruleus*): Female choice, male characteristics, and offspring quality. *Behavioral Ecology* 8: 481–492.

761. Kenrick DT, Sadalla EK, Groth G, Trost MR (1990) Evolution, traits, and the stages of human courtship: Qualifying the parental investment model. *Journal of Personality* 58: 97–116.

762. Kenrick DT, Keefe RC (1992) Age preferences in mates reflect sex differences in reproductive strategies. *Behavioral and Brain Sciences* 15: 75–133.

763. Kenrick DT (1995) Evolutionary theory versus the confederacy of dunces. *Psychological Inquiry* 6: 56–61.

764. Kenrick DT, Keefe RC, Gabrielidis C, Cornelius JS (1996) Adolescents' age preferences for dating partners: Support for an evolutionary model of life-history strategies. *Child Development* 67: 1499–1511.

765. Kent DS, Simpson JA (1992) Eusociality in the beetle *Australoplatypus incompertus* (Coleoptera: Curculionidae). *Naturwissenschaften* 79: 86–87.

766. Kerverne EB (1997) An evaluation of what the mouse knockout experiments are telling us about mammalian behaviour. *Bioessays* 19: 1091–1098.

767. Kessel EL (1955) Mating activities of balloon flies. *Systematic Zoology* 4: 97–104.

768. Ketterson ED, Nolan V, Jr. (1999) Adaptation, exaptation, and constraint: A hormonal perspective. *American Naturalist* 154 Supplement: S4–S25.

769. Kettlewell HBD (1955) Selection experiments on industrial melanism in the Lepidoptera. *Heredity* 9: 323–343.

770. Kilner RM, Noble DG, Davies NB (1999) Signals of need in parent-offspring communication and their exploitation by the common cuckoo. *Nature* 397: 667–672.

771. Kilner RM, Madden JR, Hauber ME (2004) Brood parasitic cowbird nestlings use host young to procure resources. *Science* 305: 877–879.

772. Kimball RT, Braun EL, Ligon JD, Lucchini V, Randi E (2001) A molecular phylogeny of the peacock-pheasants (Galliformes: *Polyplectron* spp.) indicates loss and reduction of ornamental traits and display behaviours. *Biological Journal of the Linnean Society* 73: 187–198.

773. Kimchi T, Xu J, Dulac C (2007) A functional circuit underlying male sexual behaviour in the female mouse brain. *Nature* 448: 1009–1015.

774. King AP, West MJ (1983) Epigenesis of cowbird song—a joint endeavour of males and females. *Nature* 305: 704–706.

775. Kirchner WH, Grasser A (1998) The significance of odor cues and dance language information for the food search behavior of honeybees (Hymenoptera: Apidae). *Journal of Insect Behavior* 11: 169–178.

776. Kirkpatrick M (1982) Sexual selection and the evolution of female choice. *Evolution* 36: 1–12.

777. Kirn JR, DeVoogd TJ (1989) The genesis and death of vocal control neurons during sexual differentiation in the zebra finch. *Journal of Neuroscience* 9: 3176–3187.

778. Kitcher P (1985) *Vaulting Ambition*. MIT Press, Cambridge, MA.

779. Kiyota M, Insley SJ, Lance SL (2008) Effectiveness of territorial polygyny and alternative mating strategies in northern fur seals, *Callorhinus ursinus*. *Behavioral Ecology and Sociobiology* 62: 739–746.

780. Klein SL (2000) The effects of hormones on sex differences in infection: From genes to behavior. *Neuroscience and Biobehavioral Reviews* 24: 627–638.

781. Klump GM, Kretzschmar E, Curio E (1986) The hearing of an avian predator and its avian prey. *Behavioral Ecology and Sociobiology* 18: 317–324.

782. Knox TT, Scott MP (2006) Size, operational sex ratio, and mate-guarding success of the carrion beetle, *Necrophila americana*. *Behavioral Ecology* 17: 88–96.

783. Knudsen B, Evans RM (1986) Parent-young recognition in herring gulls (*Larus argentatus*). *Animal Behaviour* 34: 77–80.

784. Kodric-Brown A, Brown JH (1984) Truth in advertising: The kinds of traits favored by sexual selection. *American Naturalist* 124: 309–323.

785. Kodric-Brown A (1993) Female choice of multiple male criteria in guppies: Interacting effects of dominance, coloration and courtship. *Behavioral Ecology and Sociobiology* 32: 415–420.

786. Koenig WD (1981) Coalitions of male lions: Making the best of a bad job? *Nature* 293: 413–414.

787. Koenig WD, Mumme RL (1987) *Population Ecology of the Cooperatively Breeding Acorn Woodpecker*. Princeton University Press, Princeton, NJ.

788. Koenig WD, Mumme RL, Stanback MT, Pitelka FA (1995) Patterns and consequences of egg destruction among joint-nesting acorn woodpeckers. *Animal Behaviour* 50: 607–621.

789. Koetz AH, Westcott DA, Congdon BC (2007) Spatial pattern of song element sharing and its implications for song learning in the chowchilla, *Orthonyx spaldingii*. *Animal Behaviour* 74: 1019–1028.

790. Kokko H, Jennions MD, Brooks DR (2006) Unifying and testing models of sexual selection. *Annual Review of Ecology, Evolution and Systematics* 37: 43–46.

791. Kölliker M (2007) Benefits and costs of earwig (*Forficula auricularia*) family life. *Behavioral Ecology and Sociobiology* 61: 1489–1497.

792. Komdeur J (1992) Importance of habitat saturation and territory quality for evolution of cooperative breeding in the Seychelles warbler. *Nature* 358: 493–495.

793. Komdeur J (1992) Influence of territory quality and habitat saturation on dispersal options in the Seychelles warbler: An experimental test of the habitat saturation hypothesis for cooperative breeding. *Acta XX Congressus Internationalis Ornithologici* 20: 1325–1332.

794. Komdeur J, Kraaijeveld-Smit F, Kraaijeveld K, Edelaar P (1999) Explicit experimental evidence for the role of mate guarding in minimizing loss of paternity in the Seychelles warbler. *Proceedings of the Royal Society of London, Series B* 266: 2075–2081.

795. Komdeur J (2001) Mate guarding in the Seychelles warbler is energetically costly and adjusted to paternity risk. *Proceedings of the Royal Society of London, Series B* 268: 2103–2111.

796. Komdeur J, Burke T, Richardson DS (2007) Explicit experimental evidence for the effectiveness of proximity as mate-guarding behaviour in reducing extra-pair fertilization in the Seychelles warbler. *Molecular Ecology* 16: 3679–3688.

797. Komers PE, Brotherton PNM (1997) Female space use is the best predictor of monogamy in mammals. *Proceedings of the Royal Society of London, Series B* 264: 1261–1270.

798. Konishi M (1965) The role of auditory feedback in the control of vocalization in the white-crowned sparrow. *Zeitschrift für Tierpsychologie* 22: 770–783.

799. Konishi M (1985) Birdsong: From behavior to neurons. *Annual Review of Neuroscience* 8: 125–170.

800. Kramer MG, Marden JH (1997) Almost airborne. *Nature* 385: 403–404.

801. Krams I, Krama T, Iguane K, Mand R (2007) Long-lasting mobbing of the pied flycatcher increases the risk of nest predation. *Behavioral Ecology* 18: 1082–1084.

802. Krams I, Krama T, Igaune K, Mand R (2008) Experimental evidence of reciprocal altruism in the pied flycatcher. *Behavioral Ecology and Sociobiology* 62: 599–605.

803. Krebs JR (1971) Territory and breeding density in the great tit, *Parus major* L. *Ecology* 52: 2–22.

804. Krebs JR (1982) Territorial defence in the great tit (*Parus major*): Do residents always win? *Behavioral Ecology and Sociobiology* 11: 185–194.

805. Krebs JR, Kacelnik A (1991) Decision-making. In: Krebs JR, Davies NB (eds) *Behavioural Ecology: An Evolutionary Approach*. Blackwell Scientific Publications, Oxford.

806. Krishnamani R, Mahaney WC (2000) Geophagy among primates: Adaptive significance and ecological consequences. *Animal Behaviour* 59: 899–915.

807. Krohmer RW (2004) The male red-sided garter snake (*Thamnophis sirtalis parietalis*): Reproductive pattern and behavior. *ILAR Journal* 45: 65–74.

808. Kroodsma DE, Canady RA (1985) Differences in repertoire size, singing behavior, and associated neuroanatomy among marsh wren populations have a genetic basis. *Auk* 102: 439–446.

809. Kroodsma DE, Konishi M (1991) A suboscine bird (Eastern Phoebe, *Sayornis phoebe*) develops normal song without auditory-feedback. *Animal Behaviour* 42: 477–487.

810. Kroodsma DE, Liu WC, Goodwin E, Bedell PA (1999) The ecology of song improvisation as illustrated by North American sedge wrens. *Auk* 116: 373–386.

811. Kroodsma DE, Sánchez J, Stemple DW, Goodwin E, da Silva ML, Vielliard JME (1999) Sedentary life style of Neotropical sedge wrens promotes song imitation. *Animal Behaviour* 57: 855–863.

812. Kroodsma DE (2005) *The Singing Life of Birds: The Art and Science of Listening to Birdsong*. Houghton Mifflin, New York.

813. Krüger O, Davies NB (2002) The evolution of cuckoo parasitism: a comparative analysis. *Proceedings of the Royal Society of London, Series B* 269: 375–381.

814. Kruuk H (1964) Predators and anti-predator behaviour of the black-headed gull *Larus ridibundus*. *Behaviour Supplements* 11: 1–129.

815. Kruuk H (1972) *The Spotted Hyena*. University of Chicago Press, Chicago.

816. Kruuk H (2004) *Niko's Nature: The Life of Niko Tinbergen and His Science of Animal Behavior*. Oxford University Press, Oxford.

817. Kuhn TS (1996) *The Structure of Scientific Revolutions*, 3rd edn. University of Chicago Press, Chicago, IL.

818. Kukalová-Peck J (1978) Origin and evolution of insect wings and their relation to metamorphosis, as documented by the fossil record. *Journal of Morphology* 158: 53–126.

819. Kusmierski R, Borgia G, Crozier RH, Chan BHY (1993) Molecular information on bowerbird phylogeny and the evolution of exaggerated male characters. *Journal of Evolutionary Biology* 6: 737–752.

820. Kusmierski R, Borgia G, Uy A, Crozier RH (1997) Labile evolution of display traits in bowerbirds indicates reduced effects of phylogenetic constraint. *Proceedings of the Royal Society of London, Series B* 264: 307–313.

821. Kutsukake M, Shibao H, Nikoh N, Morioka M, Tamura T, Hoshino T, Ohgiya S, Fukatsu T (2004) Venomous protease of aphid soldier for colony defense. *Proceedings of the National Academy of Sciences USA* 101: 11338–11343.

822. Kvarnemo C, Moore GI, Jones AG (2007) Sexually selected females in the monogamous Western Australian seahorse. *Proceedings of the Royal Society of London, Series B* 274: 521–525.

823. Kvist A, Lindstrom A, Green M, Piersma T, Visser GH (2001) Carrying large fuel loads during sustained bird flight is cheaper than expected. *Nature* 413: 730–732.

824. LaBas N, Hockman LR (2005) An invasion of cheats: the evolution of worthless nuptial gifts. *Current Biology* 15: 64–67.

825. Lacey EA, Sherman PW (1991) Social organization of naked mole-rat colonies: Evidence for divisions of labor. In: Sherman PW, Jarvis JUM, Alexander RD (eds) *The Biology of the Naked Mole-Rat*. Princeton University Press, Princeton, NJ.

826. Lacey EA, Wieczorek JR (2001) Territoriality and male reproductive success in arctic ground squirrels. *Behavioral Ecology* 12: 626–631.

827. Lack D (1968) *Ecological Adaptations for Breeding in Birds*. Methuen, London.

828. Laeng B, Mathisen R, Johnsen JA (2007) Why do blue-eyed men prefer women with the same eye color? *Behavioral Ecology and Sociobiology* 61: 371–384.

829. Lafferty KD, Goodman D, Sandoval CP (2006) Restoration of breeding by snowy plovers following protection from disturbance. *Biodiversity and Conservation* 15: 2217–2230.

830. Laiolo P, Tella JL, Carrete M, Serrano D, López G (2004) Distress calls may honestly signal bird quality to predators. *Proceedings of the Royal Society of London, Series B* 271: S513–S515.

831. Laist DW (1987) Overview of the biological effects of lost and discarded plastic debris in the marine environment. *Marine Pollution Bulletin* 18: 319–326.

832. Lalumiére ML, Chalmers LJ, Quinsey VL, Seto MC (1996) A test of the mate deprivation hypothesis of social coercion. *Ethology and Sociobiology* 17: 299–318.

833. Lande R (1981) Models of speciation by sexual selection of polygenic traits. *Proceedings of the National Academy of Sciences USA* 78: 3721–3725.

834. Lang AB, Kalko EKV, Romer H, Bockholdt C, Dechmann DKN (2006) Activity levels of bats and katydids in relation to the lunar cycle. *Oecologia* 146: 659–666.

835. Langer P, Hogendoorn K, Keller L (2004) Tug-of-war over reproduction in a social bee. *Nature* 428: 844–847.

836. Langmore NE, Cockrem JF, Candy EJ (2002) Competition for male reproductive investment elevates testosterone levels in female dunnocks, *Prunella modularis*. *Proceedings of the Royal Society of London, Series B* 269: 2473–2478.

837. Langmore NE, Hunt S, Kilner RM (2003) Escalation of a coevolutionary arms race through host rejection of brood parasitic young. *Nature* 422: 157–160.

838. Langmore NE, Kilner RM (2007) Breeding site and host selection by Horsfield's bronze-cuckoos, *Chalcites basalis*. *Animal Behaviour* 74: 995–1004.

839. Lank DB, Oring LW, Maxson SJ (1985) Mate and nutrient limitation of egg-laying in a polyandrous shorebird. *Ecology* 66: 1513–1524.

840. Lank DB, Smith CM, Hanotte O, Burke T, Cooke F (1995) Genetic polymorphism for alternative mating behaviour in lekking male ruff *Philomachus pugnax*. *Nature* 378: 59–62.

841. Lanyon SM (1992) Interspecific brood parasitism in blackbirds (Icterinae): A phylogenetic perspective. *Science* 255: 77–79.

842. Lappin AK, Brandt Y, Husak JF, Macedonia JM, Kemp DJ (2006) Gaping displays reveal and amplify a mechanically based index of weapon performance. *American Naturalist* 168: 100–113.

843. Lassek WD, Gaulin SJC (2008) Waist-hip ratio and cognitive ability: is glueteofemoral fat a privileged store of neurodevelopmental resources? *Evolution and Human Behavior* 29: 26–34.

844. Latta SC, Brown C (1999) Autumn stopover ecology of the blackpoll warbler (*Dendroica striata*) in thorn scrub forest of the Dominican Republic. *Canadian Journal of Zoology* 77: 1147–1156.

845. Lauder GV, Leroi AM, Rose MR (1993) Adaptations and history. *Trends in Ecology and Evolution* 8: 294–297.

846. Leal M (1999) Honest signalling during prey-predator interactions in the lizard *Anolis cristatellus*. *Animal Behaviour* 58: 521–526.

847. Lebigre C, Alatalo RV, Siitari H, Parri S (2007) Restrictive mating by females on black grouse leks. *Molecular Ecology* 16: 4380–4389.

848. Lecomte N, Kuntz G, Lambert N, Gendner JP, Handrich Y, Le Maho Y, Bost CA (2006) Alloparental feeding in the king penguin *Animal Behaviour* 71: 457–462.

849. Leech SM, Leonard ML (1997) Begging and the risk of predation in nestling birds. *Behavioral Ecology* 8: 644–646.

850. Leitner S, Nicholson J, Leisler B, DeVoogd TJ, Catchpole CK (2002) Song and the song control pathway in the brain can develop independently of exposure to song in the sedge warbler. *Proceedings of the Royal Society of London, Series B* 269: 2519–2524.

851. Lema SC, Nevitt GA (2004) Variation in vasotocin immunoreactivity in the brain of recently isolated populations of a death valley pupfish, *Cyprinodon nevadensis*. *General and Comparative Endocrinology* 135: 300–309.

852. Lemaster MP, Mason RT (2001) Evidence for a female sex pheromone mediating male trailing behavior in the red-sided garter snake, *Thamnophis sirtalis parietalis*. *Chemoecology* 11: 149–152.

853. Lemon WC, Barth RH (1992) The effects of feeding rate on reproductive success in the zebra finch, *Taeniopyga guttata*. *Animal Behaviour* 44: 851–857.

854. Lendrem DW (1986) *Modelling in Behavioural Ecology: An Introductory Text*. Croom Helm, London.

855. Leonard ML, Horn AG, Porter J (2003) Does begging affect growth in nestling tree swallows, *Tachycineta bicolor*? *Behavioral Ecology and Sociobiology* 54: 573–577.

856. Leoncini I, Le Conte Y, Costagliola G, Plettner E, Toth AL, Wang M, Huang Z, Bécard J-M, Crauser D, Slessor KN, Robinson GE (2004) Regulation of behavioral maturation by a primer pheromone produced by adult worker honey bees. *Proceedings of the National Academy of Sciences USA* 101: 17559–17564.

857. Lesch KP (1998) Serotonin transporter and psychiatric disorders: Listening to the gene. *Neuroscientist* 4: 25–34.

858. Levey DJ, Stiles FG (1992) Evolutionary precursors of long-distance migration: Resource availability and movement patterns in Neotropical landbirds. *American Naturalist* 140: 447–476.

859. Levine JD (2004) Sharing time on the fly. *Current Opinion in Cell Biology* 16: 1–7.

860. Li NP, Bailey JM, Kenrick DT, Linsenmeier JAW (2002) The necessities and luxuries of mate preferences: Testing the tradeoffs. *Journal of Personality and Social Psychology* 82: 947–955.

861. Lichtenstein G, Sealy SG (1998) Nesting competition, rather than supernormal stimulus, explains the success of parasitic brown-headed cowbird chicks in yellow warbler nests. *Proceedings of the Royal Society of London, Series B* 265: 249–254.

862. Liebig J, Peeters C, Hölldobler B (1999) Worker policing limits the number of reproductives in a ponerine ant. *Proceedings of the Royal Society of London, Series B* 266: 1865–1870.

863. Ligon JD (1999) *The Evolution of Avian Mating Systems*. Oxford University Press, New York.

864. Lill A (1974) Sexual behavior of the lek-forming white-bearded manakin (*Manacus manacus trinitatis* Hartert). *Zeitschrift für Tierpsychologie* 36: 1–36.

865. Lim MM, Murphy AZ, Young LJ (2004) Ventral striatopallidal oxytocin and vasopressin V1a receptors in the monogamous prairie vole (*Microtus ochrogaster*). *Journal of Comparative Neurology* 468: 555–570.

866. Lim MM, Wang X, Olazábal DE, Ren X, Terwilliger EF, Young LJ (2004) Enhanced partner preference in a promiscuous species by manipulating the expression of a single gene. *Nature* 429: 754–757.

867. Lima SL, Dill LM (1990) Behavioral decisions made under the risk of predation: A review and prospectus. *Canadian Journal of Zoology* 68: 619–640.

868. Lin CP, Danforth BN, Wood TK (2004) Molecular phylogenetics and evolution of maternal care in Membracine treehoppers. *Systematic Biology* 53: 400–421.

869. Lincoln GA, Guinness F, Short RV (1972) The way in which testosterone controls the social and sexual behavior of the red deer stag (*Cervus elaphus*). *Hormones and Behavior* 3: 375–396.

870. Lind J, Fransson T, Jakobsson S, Kullberg C (1999) Reduced take-off ability in robins (*Erithacus rubecula*) due to migratory fuel load. *Behavioral Ecology and Sociobiology* 46: 65–70.

871. Lindauer M (1961) *Communication among Social Bees*. Harvard University Press, Cambridge, MA.

872. Little AC, Burt DM, Penton-Voak IS, Perrett DI (2001) Self-perceived attractiveness influences human female preferences for sexual dimorphism and symmetry in male faces. *Proceedings of the Royal Society of London, Series B* 268: 39–44.

873. Little AC, Jones BC, Burriss RP (2007) Preferences for masculinity in male bodies change across the menstrual cycle. *Hormones and Behavior* 51: 633–639.

874. Little AC, Jones BC, Burt DM (2007) Preferences for symmetry in faces change across the menstrual cycle. *Biological Psychology* 76: 209–216.

875. Lloyd JE (1965) Aggressive mimicry in *Photuris*: Firefly *femmes fatales*. *Science* 149: 653–654.

876. Lloyd JE (1975) Aggressive mimicry in *Photuris* fireflies: Signal repertoires by *femmes fatales*. *Science* 197: 452–453.

877. Lloyd JE (1980) Insect behavioral ecology: Coming of age in bionomics or compleat biologists have revolutions too. *Florida Entomologist* 63: 1–4.

878. Lockard RB, Owings DH (1974) Seasonal variation in moonlight avoidance by bannertail kangaroo rats. *Journal of Mammalogy* 55: 189–193.

879. Lockard RB (1978) Seasonal change in the activity pattern of *Dipodomys spectabilis*. *Journal of Mammalogy* 59: 563–568.

880. Lockhart DJ, Barlow C (2001) Expressing what's on your mind: DNA arrays and the brain. *Nature Reviews Neuroscience* 2: 63–68.

881. Loesche P, Stoddard PK, Higgins BJ, Beecher MD (1991) Signature versus perceptual adaptations for individual vocal recognition in swallows. *Behaviour* 118: 15–25.

882. Loher W (1972) Circadian control of stridulation in the cricket *Teleogryllus commodus* Walker. *Journal of Comparative Physiology* 79: 173–190.

883. Loher W (1979) Circadian rhythmicity of locomotor behavior and oviposition in female *Teleogryllus commodus*. *Behavioral Ecology and Sociobiology* 5: 383–390.

884. Lohmann KJ, Lohmann CMF, Ehrhart LM, Bagley DA, Swing T (2004) Geomagnetic map used in sea-turtle navigation. *Nature* 428: 909–910.

885. Lore R, Flannelly K (1977) Rat societies. *Scientific American* 236 (May): 106–116.

886. Lorenz KZ (1952) *King Solomon's Ring*. Crowell, New York.

887. Lotem A (1993) Learning to recognize nestlings is maladaptive for cuckoo *Cuculus canorus* hosts. *Nature* 362: 743–745.

888. Lotem A, Nakamura H, Zahavi A (1995) Constraints on egg discrimination and cuckoo-host co-evolution. *Animal Behaviour* 49: 1185–1209.

889. Lougheed LW, Anderson DJ (1999) Parent blue-footed boobies suppress siblicidal behavior of offspring. *Behavioral Ecology and Sociobiology* 45: 11–18.

890. Low BS, Heinen JT (1993) Population, resources, and environment: Implications of human behavioral ecology for conservation. *Population and Environment* 15: 7–41.

891. Loyau A, Saint Jalme M, Cagniant C, Sorci G (2005) Multiple sexual advertisements honestly reflect health status in peacocks (*Pavo cristatus*). *Behavioral Ecology and Sociobiology* 58: 552–557.

892. Loyau A, Saint Jalme M, Mauget R, Sorci G (2007) Male sexual attractiveness affects the investment of maternal resources into the eggs in peafowl (*Pavo cristatus*). *Behavioral Ecology and Sociobiology* 61: 1043–1052.

893. Loyau A, Saint Jalme M, Sorci G (2007) Non-defendable resources affect peafowl lek organization: A male removal experiment. *Behavioural Processes* 74: 64–70.

894. Lummaa V, Vuorisalo T, Barr RG, Lehtonen L (1998) Why cry? Adaptive significance of intensive crying in human infants. *Evolution and Human Behavior* 19: 193–202.

895. Lummaa V (2003) Early developmental conditions and reproductive success in humans: Downstream effects of prenatal famine, birthweight, and timing of birth. *American Journal of Human Biology* 15: 370–379.

896. Lundberg P (1985) Dominance behavior, body-weight and fat variations, and partial migration in European blackbirds *Turdus merula*. *Behavioral Ecology and Sociobiology* 17: 185–189.

897. Lundberg P (1988) The evolution of partial migration in birds. *Trends in Ecology and Evolution* 3: 172–176.

898. Lung O, Tram U, Finnerty CM, Eipper-Mains MA, Kalb JM, Wolfner MF (2002) The *Drosophila melanogaster* seminal fluid protein Acp62F is a protease inhibitor that is toxic upon ectopic expression. *Genetics* 160: 211–224.

899. Luschi P, Hays GC, Del Seppia C, Marsh R, Papi F (1998) The navigational feats of green sea turtles migrating from Ascension Island investigated by satellite telemetry. *Proceedings of the Royal Society of London, Series B* 265: 2279–2284.

900. Lutz W, Testa MR, Penn DJ (2006) Population density is a key factor in declining human fertility. *Population and Environment* 28: 69–81.

901. Lycett JE, Dunbar RIM (1999) Abortion rates reflect the optimization of parental investment strategies. *Proceedings of the Royal Society of London, Series B* 266: 2355–2358.

902. Lynch CB (1980) Response to divergent selection for nesting behavior in *Mus musculus*. *Genetics* 96: 757–765.

903. Lyon BE, Montgomerie RD, Hamilton LD (1987) Male parental care and monogamy in snow buntings. *Behavioral Ecology and Sociobiology* 20: 377–382.

904. Lyon BE (1993) Conspecific brood parasitism as a flexible female reproductive tactic in American coots. *Animal Behaviour* 46: 911–928.

905. Lyon BE, Eadie JM, Hamilton LD (1994) Parental choice selects for ornamental plumage in American coot chicks. *Nature* 371: 240–243.

906. Lyon BE (2003) Ecological and social constraints on conspecific brood parasitism by nesting female American coots (*Fulica americana*). *Journal of Animal Ecology* 72: 47–60.

907. Lyon BE (2007) Mechanism of egg recognition in defenses against conspecific brood parasitism: American coots (*Fulica americana*) know their own eggs. *Behavioral Ecology and Sociobiology* 61: 455–463.

908. MacDonald IF, Kempster B, Zanette L, MacDougall-Shackleton SA (2006) Early nutritional stress impairs development of a song-control brain region in both male and female juvenile song sparrows (*Melospiza melodia*) at the onset of song learning. *Proceedings of the Royal Society of London, Series B* 273: 2559–2564.

909. MacDougall-Shackleton EA, Derryberry EP, Hahn TP (2002) Nonlocal male mountain white-crowned sparrows have lower paternity and higher parasite loads than males singing local dialect. *Behavioral Ecology* 13: 682–689.

910. MacDougall-Shackleton SA, Sherry DF, Clark AP, Pinkus R, Hernandez AM (2003) Photoperiodic regulation of food storing and hippocampus volume in black-capped chickadees, *Poecile atricapillus*. *Animal Behaviour* 65: 805–812.

911. Mace R (1998) The coevolution of human fertility and wealth inheritance strategies. *Philosophical Transactions of the Royal Society of London, Series B* 353: 389–397.

912. Macrae CN, Alnwick KA, Milne AB, Schloerscheidt AM (2002) Person perception across the menstrual cycle: Hormonal influences on social-cognitive functioning. *Psychological Science* 13: 532–536.

913. Madden JR (2001) Sex, bowers and brains. *Proceedings of the Royal Society of London, Series B* 268: 833–838.

914. Madden JR (2003) Bower decorations are good predictors of mating success in the spotted bowerbird. *Behavioral Ecology and Sociobiology* 53: 269–277.

915. Madden JR (2003) Male spotted bowerbirds preferentially choose, arrange and proffer objects that are good predictors of mating success. *Behavioral Ecology and Sociobiology* 53: 263–268.

916. Maguire EA, Burgess N, Donnett JG, Frackowiak RSJ, Frith CD, O'Keefe J (1998) Knowing where and getting there: A human navigation network. *Science* 280: 921–934.

917. Maguire EA, Gadian DG, Johnsrude IS, Good CD, Ashburner J, Frackowiak RSJ, Frith CD (2000) Navigation-related structural change in the hippocampi of taxi drivers. *Proceedings of the National Academy of Sciences USA* 97: 4398–4403.

918. Maguire EA, Wollett K, Spiers HJ (2006) London taxi drivers and bus drivers: A structural MRI and neuropsychological analysis. *Hippocampus* 16: 1091–1101.

919. Maguire EA, Spiers HJ (2007) The neuroscience of remote spatial memory: a tale of two cities. *Neuroscience* 149: 7–27.

920. Mak GK, Enwere EK, Gregg C, Pakarainen T, Poutanen M, Huhtaniemi I, Weiss S (2007) Male pheromone-stimulated neurogenesis in the adult female brain: possible role in mating behavior. *Nature Neuroscience* 10: 1003–1011.

921. Mallach TJ, Leberg PL (1999) Use of dredged material substrates by nesting terns and black skimmers. *Journal of Wildlife Management* 63: 137–146.

922. Mangel M, Clark C (1988) *Dynamic Modelling in Behavioral Ecology*. Princeton University Press, Princeton, NJ.

923. Mant J, Brandli C, Vereecken NJ, Schulz CM, Francke W, Schiestl FP (2005) Cuticular hydrocarbons as sex pheromone of the bee *Colletes cunicularius* and the key to its mimicry by the sexually deceptive orchid, *Ophrys exaltata*. *Journal of Chemical Ecology* 31: 1765–1787.

924. Marasco PD, Catania KC (2007) Response properties of primary afferents supplying Eimer's organ. *Journal of Experimental Biology* 210: 765–780.

925. Marden JH, Waage JK (1990) Escalated damselfly territorial contests and energetic wars of attrition. *Animal Behaviour* 39: 954–959.

926. Marden JH, Kramer MG (1994) Surface-skimming stoneflies: A possible intermediate stage in insect flight evolution. *Science* 266: 427–430.

927. Marden JH (1995) How insects learned to fly. *The Sciences* 35: 26–30.

928. Marden JH, Kramer MG (1995) Locomotor performance of insects with rudimentary wings. *Nature* 377: 332–334.

929. Marden JH, O'Donnell BC, Thomas MA, Bye JY (2000) Surface-skimming stoneflies and mayflies: The taxonomic and mechanical diversity of two-dimensional aerodynamic locomotion. *Physiological and Biochemical Zoology* 73: 751–764.

930. Marden JH, Thomas MA (2003) Rowing locomotion by a stonefly that possesses the ancestral pterygote condition of co-occurring wings and abdominal gills. *Biological Journal of the Linnean Society* 79: 341–349.

931. Maret TJ, Collins JP (1994) Individual responses to population size structure: The role of size variation in controlling expression of a trophic polyphenism. *Oecologia* 100: 279–285.

932. Margulis SW, Saltzman W, Abbott DH (1995) Behavioural and hormonal changes in female naked mole-rats (*Heterocephalus glaber*) following removal of the breeding female from a colony. *Hormones and Behavior* 29: 227–247.

933. Margulis SW, Altmann J (1997) Behavioural risk factors in the reproduction of inbred and outbred oldfield mice. *Animal Behaviour* 54: 397–408.

934. Marler CA, Moore MC (1989) Time and energy costs of aggression in testosterone-implanted free-living male mountain spiny lizards (*Sceloporus jarrovi*). *Physiological Zoology* 62: 1334–1350.

935. Marler CA, Moore MC (1991) Supplementary feeding compensates for testosterone-induced costs of aggression in male mountain spiny lizards, *Sceloporus jarrovi*. *Animal Behaviour* 42: 209–219.

936. Marler CA, Walsberg G, White ML, Moore MC (1995) Increased energy-expenditure due to increased territorial defense in male lizards after phenotypic manipulation. *Behavioral Ecology and Sociobiology* 37: 225–231.

937. Marler P (1955) Characteristics of some animal calls. *Nature* 176: 6–8.

938. Marler P, Tamura M (1964) Culturally transmitted patterns of vocal behavior in sparrows. *Science* 146: 1483–1486.

939. Marler P (1970) Birdsong and speech development: Could there be parallels? *American Scientist* 58: 669–673.

940. Marra PP, Holmes RT (2001) Consequences of dominance-mediated habitat segregation in American Redstarts during the nonbreeding season. *Auk* 118: 92–104.

941. Marshall RC, Buchanan KL, Catchpole CK (2003) Sexual selection and individual genetic diversity in a songbird. *Proceedings of the Royal Society of London, Series B* 270: S248–S250.

942. Martín J, López P (2000) Chemoreception, symmetry and mate choice in lizards. *Proceedings of the Royal Society of London, Series B* 267: 1265–1269.

943. Martin TE, Schwabl H (2008) Variation in maternal effects and embryonic development rates among passerine species. *Philosophical Transactions of the Royal Society of London, Series B* 363: 1663–1674.

944. Maschwitz U, Maschwitz E (1974) Platzende Arbeiterinnen: Eine neue Art der Feindabwehr bei sozialen Hautflüglern. *Oecologia* 14: 289–294.

945. Massaro DW, Stork DG (1998) Speech recognition and sensory integration. *American Scientist* 86: 236–244.

946. Massaro M, Chardine JW, Jones IL (2001) Relationships between black-legged kittiwake nest site characteristics and susceptibility to predation by large gulls. *Condor* 103: 793–801.

947. Mateo JM, Holmes WG (1997) Development of alarm-call responses in Belding's ground squirrels: The role of dams. *Animal Behaviour* 54: 509–524.

948. Mateo JM, Johnston RE (2000) Kin recognition and the "armpit effect": Evidence of self-reference phenotype matching. *Proceedings of the Royal Society of London, Series B* 267: 695–700.

949. Mateo JM (2002) Kin-recognition abilities and nepotism as a function of sociality. *Proceedings of the Royal Society of London, Series B* 269: 721–727.

950. Mateo JM (2006) The nature and representation of individual recognition odours in Belding's ground squirrels. *Animal Behaviour* 71: 141–154.

951. Mather MH, Roitberg BD (1987) A sheep in wolf's clothing: Tephritid flies mimic spider predators. *Science* 236: 308–310.

952. Matsumoto-Oda A, Hamai M, Hayaki H, Hosaka K, Hunt KD, Kasuya E, Kawanaka K, Mitani JC, Takasaki H, Takahata Y (2007) Estrus cycle asynchrony in wild female chimpanzees, *Pan troglodytes schweinfurthii*. *Behavioral Ecology and Sociobiology* 61: 661–668.

953. Matthews LH (1939) Reproduction in the spotted hyena *Crocuta crocuta* (Erxleben). *Philosophical Transactions of the Royal Society of London, Series B* 230: 1–78.

954. Mattila HR, Seeley TD (2007) Genetic diversity in honey bee colonies enhances productivity and fitness. *Science* 317: 362–364.

955. May M (1991) Aerial defense tactics of flying insects. *American Scientist* 79: 316–329.

956. Maynard Smith J (1974) The theory of games and the evolution of animal conflicts. *Journal of Theoretical Biology* 47: 209–221.

957. Mayr E (1961) Cause and effect in biology. *Science* 134: 1501–1506.

958. Mayr E (1963) *Animal Species and Evolution.* Harvard University Press, Cambridge, MA.

959. McCandliss BD, Cohen L, Dehaene S (2003) The visual word form area: expertise for reading in the fusiform gyrus. *Trends in Cognitive Sciences* 7: 293–299.

960. McCracken GF, Bradbury JW (1981) Social organization and kinship in the polygynous bat *Phyllostomus hastatus*. *Behavioral Ecology and Sociobiology* 8: 11–34.

961. McCracken GF (1984) Communal nursing in Mexican free-tailed bat maternity colonies. *Science* 223: 1090–1091.

962. McCracken GF, Gustin MK (1991) Nursing behavior in Mexican free-tailed bat maternity colonies. *Ethology* 89: 305–321.

963. McDonald DB, Potts WK (1994) Cooperative display and relatedness among males in a lek-mating bird. *Science* 266: 1030–1032.

964. McDonald DB (2007) Predicting fate from early connectivity in a social network. *Proceedings of the National Academy of Sciences USA* 104: 10910–10914.

965. McGlothlin JW, Jawor JM, Ketterson ED (2007) Natural variation in a testosterone-mediated trade-off between mating effort and parental effort. *American Naturalist* 170: 864–875.

966. McGraw KJ, Hill GE (2000) Differential effects of endoparasitism on the expression of carotenoid- and melanin-based ornamental coloration. *Proceedings of the Royal Society of London, Series B* 267: 1525–1531.

967. McGraw KJ, Ardia DR (2003) Carotenoids, immunocompetence, and the information content of sexual colors: an experimental test. *American Naturalist* 162: 704–712.

968. McGuire B, Bemis WE (2007) Parental care. In: Wolff JO, Sherman PW (eds) *Rodent Societies: An Ecological and Evolutionary Perspective.* University of Chicago Press, Chicago.

969. McKibbin WF, Shackelford TK, Goetz AT, Starratt VG (2008) Why do men rape? An evolutionary psychological perspective. *Review of General Psychology* 12: 86–97.

970. McNair DB, Massiah EB, Frost MD (2002) Ground-based autumn migration of Blackpoll Warblers at Harrison Point, Barbados. *Caribbean Journal of Science* 38: 239–248.

971. Medvin MB, Stoddard PK, Beecher MD (1993) Signals for parent-offspring recognition: A comparative analysis of the begging calls of cliff swallows and barn swallows. *Animal Behaviour* 45: 841–850.

972. Meire PM, Ervynck A (1986) Are oystercatchers (*Haemoptopus ostralegus*) selecting the most profitable mussels (*Mytilus edulis*)? *Animal Behaviour* 34: 1427–1435.

973. Mello CV, Ribeiro S (1998) ZENK protein regulation by song in the brain of songbirds. *Journal of Comparative Neurology* 383: 426–438.

974. Melo L, Mendes AR, Monteriro da Cruz MAO (2003) Infanticide and cannibalism in wild common marmosets. *Folia Primatologica* 74: 48–50.

975. Mendonca MT, Daniels D, Faro C, Crews D (2003) Differential effects of courtship and mating on receptivity and brain metabolism in female red-sided garter snakes (*Thamnophis sirtalis parietalis*). *Behavioral Neuroscience* 117: 144–149.

976. Mery F, Belay AT, So AKC, Sokolowski MB, Kawecki TJ (2007) Natural polymorphism affecting learning and memory in *Drosophila*. *Proceedings of the National Academy of Sciences USA* 104: 13051–13055.

977. Mesoudi A, Danielson P (2008) Ethics, evolution and culture. *Theory in BioSciences* 127: 229–240.

978. Meyer A, Morrisey JM, Schartl M (1994) Recurrent origin of a sexually selected trait in *Xiphophorus* fishes inferred from a molecular phylogeny. *Nature* 368: 539–542.

979. Michener CD (1974) *The Social Behavior of the Bees: A Comparative Study*. Harvard University Press, Cambridge, MA.

980. Michl G, Török J, Griffith SC, Sheldon BC (2002) Experimental analysis of sperm competition mechanisms in a wild bird population. *Proceedings of the National Academy of Sciences USA* 99: 5466–5470.

981. Milinski M, Semmann D, Krambeck HJ (2002) Donors to charity gain in both indirect reciprocity and political reputation. *Proceedings of the Royal Society of London, Series B* 269: 881–883.

982. Milius S (2004) Where'd I put that? *Science News* 165: 103–105.

983. Miller DE, Emlen JT, Jr. (1975) Individual chick recognition and family integrity in the ring-billed gull. *Behaviour* 52: 124–144.

984. Miller GF, Tybur J, Jordan B (2008) Ovulatory cycle effects on tip earnings by lap-dancers: Economic evidence for human estrus? *Evolution and Human Behavior* 28: 375–381.

985. Miller JA (2007) Repeated evolution of male sacrifice behavior in spiders correlated with genital mutilation. *Evolution* 61: 1301–1315.

986. Miller LA, Surlykke A (2001) How some insects detect and avoid being eaten by bats: Tactics and countertactics of prey and predator. *BioScience* 51: 570–581.

987. Mills MGL (1990) *Kalahari Hyaenas: Comparative Behavioural Ecology of Two Species*. Unwin Hyman, London.

988. Milton K (2004) Ferment in the family tree: Does a frugivorous dietary heritage influence contemporary patterns of human ethanol use? *Integrative and Comparative Biology* 44: 304–314.

989. Mitani JC, Watts DP (2005) Correlates of territorial boundary patrol behaviour in wild chimpanzees. *Animal Behaviour* 70: 1079–1086.

990. Mitchell DP, Dunn PO, Whittingham LA, Freeman-Gallant CR (2007) Attractive males provide less parental care in two populations of the common yellowthroat. *Animal Behaviour* 73: 165–170.

991. Mock DW (1984) Siblicidal aggression and resource monopolization in birds. *Science* 225: 731–733.

992. Mock DW, Ploger BJ (1987) Parental manipulation of optimal hatch asynchrony in cattle egrets: An experimental study. *Animal Behaviour* 35: 150–160.

993. Mock DW, Drummond H, Stinson CH (1990) Avian siblicide. *American Scientist* 78: 438–449.

994. Mock DW (2004) *More Than Kin and Less Than Kind: The Evolution of Family Conflict*. Harvard University Press, Cambridge, MA.

995. Mock DW, Schwagmeyer PL, Parker GA (2005) Male house sparrows deliver more food to experimentally subsidized offspring. *Animal Behaviour* 70: 225–236.

996. Moffat SD, Hampson E, Hatzipantelis M (1998) Navigation in a "virtual" maze: Sex differences and correlation with psychometric measures of spatial ability in humans. *Human Behavior and Evolution* 19: 73–87.

997. Moiseff A, Pollack GS, Hoy RR (1978) Steering responses of flying crickets to sound and ultrasound: Mate attraction and predator avoidance. *Proceedings of the National Academy of Sciences USA* 75: 4052–4056.

998. Möller A, Pavlick B, Hile AG, Balda RP (2001) Clark's nutcrackers *Nucifraga columbiana* remember the size of their cached seeds. *Ethology* 107: 451–461.

999. Møller AP (1988) Female choice selects for male sexual tail ornaments in the monogamous swallow. *Nature* 332: 640–642.

1000. Møller AP (1992) Female swallow preference for symmetrical male sexual ornaments. *Nature* 357: 238–240.

1001. Møller AP, Swaddle JP (1997) *Asymmetry, Developmental Stability, and Evolution*. Oxford University Press, New York.

1002. Møller AP (1999) Asymmetry as a predictor of growth, fecundity and survival. *Ecology Letters* 2: 149–156.

1003. Money J, Ehrhardt AA (1972) *Man and Woman, Boy and Girl*. Johns Hopkins University Press, Baltimore, MD.

1004. Monnin T, Ratnieks FLW, Jones GR, Beard R (2002) Pretender punishment induced by chemical signalling in a queenless ant. *Nature* 419: 61–65.

1005. Mooney R, Hoese W, Nowicki S (2001) Auditory representation of the vocal repertoire in a songbird with multiple song types. *Proceedings of the National Academy of Sciences USA* 98: 12778–12783.

1006. Moore J, Ali R (1984) Are dispersal and inbreeding avoidance related? *Animal Behaviour* 32: 94–112.

1007. Moore MC, Kranz B (1983) Evidence for androgen independence of male mounting behavior in white-crowned sparrows (*Zonotrichia leucophrys gambelii*). *Hormones and Behavior* 17: 414–423.

1008. Moreno J, Veiga JP, Cordero PJ, Mínguez E (1999) Effects of paternal care on reproductive success in the polygynous spotless starling *Sturnus unicolor*. *Behavioral Ecology and Sociobiology* 47: 47–53.

1009. Morley R, Lucas A (1997) Nutrition and cognitive development. *British Medical Bulletin* 53: 123–124.

1010. Morrison CD, Berthoud H-R (2007) Neurobiology of nutrition and obesity. *Nutrition Reviews* 65: 517–534.

1011. Morrison RIG, Davidson NC, Wilson JR (2007) Survival of the fattest: body stores on migration and survival in red knots *Calidris canutus islandica*. *Journal of Avian Biology* 38: 479–487.

1012. Morrow EH, Arnqvist G (2003) Costly traumatic insemination and a female counter-adaptation in bed bugs. *Proceedings of the Royal Society of London, Series B* 270: 2377–2381.

1013. Morrow PA, Bellas TE, Eisner T (1976) Eucalyptus oils in defensive oral discharge of Australian sawfly larvae (Hymenoptera: Pergidae). *Oecologia* 24: 193–206.

1014. Moss C (1988) *Elephant Memories*. William Morrow, New York.

1015. Mouritsen H, Frost BJ (2002) Virtual migration in tethered flying monarch butterflies reveals their orientation mechanisms. *Proceedings of the National Academy of Sciences USA* 99: 10162–10166.

1016. Mueller UG (1991) Haplodiploidy and the evolution of facultative sex ratios in a primitively eusocial bee. *Science* 254: 442–444.

1017. Müller CA, Manser MB (2007) "Nasty neighbours" rather than "dear enemies" in a social carnivore. *Proceedings of the Royal Society of London, Series B* 274: 959–965.

1018. Muller MN, Wrangham R (2002) Sexual mimicry in hyenas. *Quarterly Review of Biology* 77: 3–16.

1019. Muller MN, Thompson ME, Wrangham RW (2006) Male chimpanzees prefer mating with old females. *Current Biology* 16: 2234–2238.

1020. Muller MN, Kahlenberg SM, Thompson ME, Wrangham RW (2007) Male coercion and the costs of promiscuous mating for female chimpanzees. *Proceedings of the Royal Society of London, Series B* 274: 1009–1014.

1021. Mumme RL, Koenig WD, Pitelka FA (1983) Reproductive competition in the communal acorn woodpecker: Sisters destroy each other's eggs. *Nature* 306: 583–584.

1022. Mumme RL, Koenig WD, Ratnieks FLW (1989) Helping behaviour, reproductive value, and the future component of indirect fitness. *Animal Behaviour* 38: 331–343.

1023. Mumme RL (1992) Do helpers increase reproductive success? An experimental analysis in the Florida scrub jay. *Behavioral Ecology and Sociobiology* 31: 319–328.

1024. Murdock GP (1967) *Ethnographic Atlas*. Pittsburgh University Press, Pittsburgh, PA.

1025. Murphy CG (2003) The cause of correlations between nightly numbers of male and female barking treefrogs (*Hyla gratiosa*) attending choruses. *Behavioral Ecology* 14: 274–281.

1026. Musiega DE, Kazadi SN, Fukuyama K (2006) A framework for predicting and visualizing the East African wildebeeste migration-route in variable climatic conditions using geographic information system and remote sensing. *Ecological Research* 21: 530–543.

1027. Myerscough MR (2003) Dancing for a decision: a matrix model for nest-site choice by honeybees. *Proceedings of the Royal Society of London, Series B* 270: 577–582.

1028. Nagarajan R, Lea SEG, Goss-Custard JD (2002) Reevaluation of patterns of mussel (*Mytilus edulis*) selection by European Oystercatchers (*Haematopus ostralegus*). *Canadian Journal of Zoology* 80: 846–853.

1029. Nash DR, Als TD, Maile R, Jones GR, Boomsma JJ (2008) A mosaic of chemical coevolution in a large blue butterfly. *Science* 319: 88–90.

1030. Neal JK, Wade J (2007) Courtship and copulation in the adult male green anole: Effects of season, hormone and female contact on reproductive behavior and morphology. *Behavioural Brain Research* 177: 177–185.

1031. Nealen PM, Perkel DJ (2000) Sexual dimorphism in the song system of the Carolina wren *Thryothorus ludovicianus*. *Journal of Comparative Neurology* 418: 346–360.

1032. Neff BD (2003) Decisions about parental care in response to perceived paternity. *Nature* 422: 716–719.

1033. Nelson DA (1999) Ecological influences on vocal development in the white-crowned sparrow. *Animal Behaviour* 58: 21–36.

1034. Nelson DA (2000) Song overproduction, selective attrition and song dialects in the white-crowned sparrow. *Animal Behaviour* 60: 887–898.

1035. Nelson DA, Hallberg KI, Soha JA (2004) Cultural evolution of Puget Sound white-crowned sparrow song dialects. *Ethology* 110: 879–908.

1036. Nelson DA, Soha JA (2004) Male and female white-crowned sparrows respond differently to geographic variation in song. *Behaviour* 141: 53–69.

1037. Nelson RJ (2005) *An Introduction to Behavioral Endocrinology*, 3rd edn. Sinauer Associates, Sunderland, MA.

1038. Nesse RM (2005) Maladaptation and natural selection. *Quarterly Review of Biology* 80: 62–70.

1039. Nesse RM (2007) Runaway social selection for displays of partner value and altruism. *Biological Theory* 2: 143–155.

1040. Neudorf DL, Sealy SG (2002) Distress calls of birds in a neotropical cloud forest. *Biotropica* 34: 118–126.

1041. Newcomer SD, Zeh JA, Zeh DW (1999) Genetic benefits enhance the reproductive success of polyandrous females. *Proceedings of the National Academy of Sciences USA* 96: 10236–10241.

1042. Newman EA, Hairline PH (1982) The infrared "vision" of snakes. *Scientific American* 20 (Mar.): 116–127.

1043. Newton PN (1986) Infanticide in an undisturbed forest population of hanuman langurs, *Presbytis entellus*. *Animal Behaviour* 34: 785–789.

1044. Nicholls JA, Goldizen AW (2006) Habitat type and density influence vocal signal design in satin bowerbirds. *Journal of Animal Ecology* 75: 549–558.

1045. Nieh JC (1998) The role of a scent beacon in the communication of food location by the stingless bee, *Melipona panamica*. *Behavioral Ecology and Sociobiology* 43: 47–58.

1046. Nieh JC (1999) Stingless-bee communication. *American Scientist* 87: 428–435.

1047. Nieh JC, Contrera FAL, Rangel J, Imperatriz-Fonseca VL (2003) Effect of food location and quality on recruitment sounds and success in two stingless bees, *Melipona mandacaia* and *Melipona bicolor*. *Behavioral Ecology and Sociobiology* 55: 87–94.

1048. Nieh JC (2004) Recruitment communication in stingless bees (Hymenoptera, Apidae, Meliponini). *Apidologie* 35: 159–182.

1049. Nieh JC, Barreto LS, Contrera FAL, Imperatriz-Fonseca VL (2004) Olfactory eavesdropping by a competitively foraging stingless bee, *Trigona spinipes*. *Proceedings of the Royal Society of London, Series B* 271: 1633–1640.

1050. Nijhout HF (2003) Development and evolution of adaptive polyphenisms. *Evolution & Development* 5: 9–18.

1051. Noë R, Sluijter AA (1990) Reproductive tactics of male savanna baboons. *Behaviour* 113: 117–170.

1052. Nolen TG, Hoy RR (1984) Phonotaxis in flying crickets: Neural correlates. *Science* 226: 992–994.

1053. Nonacs P (1986) Ant reproductive strategies and sex allocation theory. *Quarterly Review of Biology* 61: 1–21.

1054. Nonacs P, Reeve HK (1995) The ecology of cooperation in wasps: Causes and consequences of alternative reproductive decisions. *Ecology* 76: 953–967.

1055. Nordby JC, Campbell SE, Beecher MD (1999) Ecological correlates of song learning in song sparrows. *Behavioral Ecology* 10: 287–297.

1056. Norris DR, Marra PP, Kyser TK, Sherry TW, Ratcliffe LM (2004) Tropical winter habitat limits reproductive success on the temperate breeding grounds in a migratory bird. *Proceedings of the Royal Society of London, Series B* 271: 59–64.

1057. Nottebohm F, Arnold AP (1976) Sexual dimorphism in vocal control areas of songbird brain. *Science* 194: 211–213.

1058. Nowicki S, Peters S, Podos J (1998) Song learning, early nutrition, and sexual selection in birds. *American Zoologist* 38: 179–190.

1059. Nowicki S, Searcy WA, Hughes M (1998) The territory defense function of song in song sparrows: A test with the speaker occupation design. *Behaviour* 135: 615–628.

1060. Nowicki S, Hasselquist D, Bensch S, Peters S (2000) Nestling growth and song repertoire size in great reed warblers: Evidence for song learning as an indicator mechanism in mate choice. *Proceedings of the Royal Society of London, Series B* 267: 2419–2424.

1061. Nowicki S, Searcy WA, Peters S (2002) Brain development, song learning and mate choice in birds: a review and experimental test of the "nutritional stress hypothesis." *Journal of Comparative Physiology A* 188: 1003–1114.

1062. Nowicki S, Searcy WA, Peters S (2002) Quality of song learning affects female response to male bird song. *Proceedings of the Royal Society of London, Series B* 269: 1949–1954.

1063. Nunn CL, Gittleman JL, Antonovics J (2000) Promiscuity and the primate immune system. *Science* 290: 1168–1170.

1064. O'Connell-Rodwell CE (2007) Keeping an "Ear" to the ground: Seismic communication in elephants. *Physiology* 22: 287–294.

1065. O'Donnell RP, Shine R, Mason RT (2004) Seasonal anorexia in the male red-sided garter snake, *Thamnophis sirtalis parietalis. Behavioral Ecology and Sociobiology* 56: 413–419.

1066. O'Neill KM (1983) Territoriality, body size, and spacing in males of the bee wolf *Philanthus basilaris* (Hymenoptera; Sphecidae). *Behaviour* 86: 295–321.

1067. Oldroyd BP, Fewell JH (2007) Genetic diversity promotes homeostasis in insect colonies. *Trends in Ecology and Evolution* 22: 408–413.

1068. Olendorf R, Getty T, Scribner K, Robinson SK (2004) Male red-winged blackbirds distrust unreliable and sexually attractive neighbours. *Proceedings of the Royal Society of London, Series B* 271: 1033–1038.

1069. Olson DJ, Kamil AC, Balda RP, Nims PJ (1995) Performance of four seed-caching corvid species in operant tests of nonspatial and spatial memory. *Journal of Comparative Psychology* 109: 173–181.

1070. Ophir AG, Wolff JO, Phelps SM (2008) Variation in neural V!aR predicts sexual fidelity and space use among male prairie voles in semi-natural settings. *Proceedings of the National Academy of Sciences USA* 105: 1249–1254.

1071. Orians GH (1962) Natural selection and ecological theory. *American Naturalist* 96: 257–264.

1072. Orians GH (1969) On the evolution of mating systems in birds and mammals. *American Naturalist* 103: 589–603.

1073. Oring LW, Knudson ML (1973) Monogamy and polyandry in the spotted sandpiper. *The Living Bird* 11: 59–73.

1074. Oring LW (1985) Avian polyandry. *Current Ornithology* 3: 309–351.

1075. Oring LW, Colwell MA, Reed JM (1991) Lifetime reproductive success in the spotted sandpiper (*Actitis macularia*): Sex differences and variance components. *Behavioral Ecology and Sociobiology* 28: 425–432.

1076. Oring LW, Fleischer RC, Reed JM, Marsden KE (1992) Cuckoldry through stored sperm in the sequentially polyandrous spotted sandpiper. *Nature* 359: 631–633.

1077. Orrell KS, Jenssen TA (2002) Male mate choice by the lizard *Anolis carolinensis*: a preference for novel females. *Animal Behaviour* 63: 1091–1102.

1078. Osborne KA, Robichon A, Burgess E, Butland S, Shaw RA, Coulthard A, Pereira HS, Greenspan RJ, Sokolowski MB (1997) Natural behavior polymorphism due to a cGMP-dependent protein kinase of *Drosophila*. *Science* 277: 834–836.

1079. Osorno JL, Drummond H (1995) The function of hatching asynchrony in the blue-footed booby. *Behavioral Ecology and Sociobiology* 37: 265–274.

1080. Östlund-Nilsson S, Holmlund M (2003) The artistic three-spined stickleback (*Gasterosteus aculeatus*). *Behavioral Ecology and Sociobiology* 53: 214–220.

1081. Östlund S, Ahnesjö I (1998) Female fifteen-spined sticklebacks prefer better fathers. *Animal Behaviour* 56: 1177–1183.

1082. Ostner J, Heistermann M (2003) Intersexual dominance, masculinized genitals and prenatal steroids: comparative data from lemurid primates. *Naturwissenschaften* 90: 141–144.

1083. Ostner J, Chalise MK, Koeing A, Launhardt K, Nikolet J, Podzuweit D, Borries C (2006) What Hanuman langurs know about female reproductive status. *American Journal of Primatology* 68: 701–712.

1084. Outlaw DC, Voelker G, Mila B, Girman DJ (2003) Evolution of long-distance migration in and historical biogeography of *Catharus* thrushes: A molecular phylogenetic approach. *Auk* 120: 299–310.

1085. Outlaw DC, Voelker G (2006) Phylogenetic tests of hypotheses for the evolution of avian migration: a case study using the Motacillidae. *Auk.* 123: 455–466.

1086. Owens DD, Owens MJ (1996) Social dominance and reproductive patterns in brown hyenas, *Hyaena brunnea*, of the central Kalahari desert. *Animal Behaviour* 51: 535–551.

1087. Owens IPF, Short RV (1995) Hormonal basis of sexual dimorphism in birds: Implications for sexual selection theory. *Trends in Ecology and Evolution* 10: 44–47.

1088. Owings DH, Coss RG (1977) Snake mobbing by California ground squirrels: Adaptive variation and ontogeny. *Behaviour* 62: 50–69.

1089. Packer C (1977) Reciprocal altruism in *Papio anubis*. *Nature* 265: 441–443.

1090. Packer C (1986) The ecology of sociality in felids. In: Rubenstein DI, Wrangham RW (eds) *Ecological Aspects of Social Evolution*. Princeton University Press, Princeton, NJ.

1091. Packer C, Scheel D, Pusey AE (1990) Why lions form groups: Food is not enough. *American Naturalist* 136: 1–19.

1092. Packer C, Gilbert DA, Pusey AE, O'Brien SJ (1991) A molecular genetic analysis of kinship and cooperation in African lions. *Nature* 351: 562–565.

1093. Packer C (1994) *Into Africa*. University of Chicago Press, Chicago.

1094. Page RA, Ryan MJ (2008) The effect of signal complexity on localization performance in bats that localize frog calls. *Animal Behaviour* 76: 761–769.

1095. Page RE, Robinson GE, Fondrk MK, Nasr ME (1995) Effects of worker genotypic diversity on honey-bee colony development and behavior (*Apis mellifera* L.). *Behavioral Ecology and Sociobiology* 36: 387–396.

1096. Page TL (1985) Clocks and circadian rhythms. In: Kerkut GA, Gilbert LI (eds) *Comprehensive Insect Physiology, Biochemistry, and Pharmacology*. Pergamon Press, New York.

1097. Pagnucco K, Zanette L, Clinchy M, Leonard ML (2008) Sheep in wolf's clothing: host nestling vocalizations resemble their cowbird competitor's. *Proceedings of the Royal Society of London, Series B* 275: 1061–1065.

1098. Palmer CT (1991) Human rape: Adaptation or by-product? *Journal of Sex Research* 28: 365–386.

1099. Palombit RA, Seyfarth RM, Cheney DL (1997) The adaptive value of "friendships" to female baboons: Experimental and observational evidence. *Animal Behaviour* 54: 599–614.

1100. Pangle KL, Peacor S, Johannsson OE (2007) Large nonlethal effects of an invasive invertebrate predator on zooplankton population growth rate. *Ecology* 88: 402–412.

1101. Panhuis TM, Wilkinson GS (1999) Exaggerated male eye span influences contest outcome in stalk-eyed flies (Diopsidae). *Behavioral Ecology and Sociobiology* 46: 221–227.

1102. Papaj DR, Messing RH (1998) Asymmetries in physiological state as a possible cause of resident advantage in contests. *Behaviour* 135: 1013–1030.

1103. Parker GA (1970) Sperm competition and its evolutionary consequences in the insects. *Biological Reviews* 45: 526–567.

1104. Parker GA, Baker RR, Smith VGF (1972) The origin and evolution of gamete dimorphism and the male-female phenomenon. *Journal of Theoretical Biology* 36: 529–553.

1105. Parker GA (2006) Sexual conflict over mating and fertilization: an overview. *Philosophical Transactions of the Royal Society of London, Series B* 361: 235–259.

1106. Partecke J, von Haeseler A, Wikelski M (2002) Territory establishment in lekking marine iguanas, *Amblyrhynchus cristatus*: support for the hotshot mechanism. *Behavioral Ecology and Sociobiology* 51: 579–587.

1107. Partecke J, Gwinner E (2007) Increased sedentariness in European blackbirds following urbanization: a consequence of local adaptation? *Ecology* 88: 882–890.

1108. Patricelli GL, Uy JAC, Walsh G, Borgia G (2002) Male displays adjusted to female's response. *Nature* 415: 279–280.

1109. Patricelli GL, Uy JAC, Borgia G (2003) Multiple male traits interact: attractive bower decorations facilitate attractive behavioural displays in satin bowerbirds. *Proceedings of the Royal Society of London, Series B* 270: 2389–2395.

1110. Patricelli GL, Uy JAC, Borgia G (2004) Female signals enhance the efficiency of mate assessment in satin bowerbirds (*Ptilonorhynchus violaceus*). *Behavioral Ecology* 15: 297–304.

1111. Pawłowski B, Dunbar RIM (1999) Impact of market value on human mate choice decisions. *Proceedings of the Royal Society of London, Series B* 266: 281–285.

1112. Peakall R (1990) Responses of male *Zaspilothynnus trilobatus* Turner wasps to females and the sexually deceptive orchid it pollinates. *Functional Ecology* 4: 159–167.

1113. Pelli DG, Farell B, Moore DC (2003) The remarkable inefficiency of word recognition. *Nature* 423: 752–756.

1114. Pengelley ET, Asmundson SJ (1974) Circannual rhythmicity in hibernating animals. In: Pengelley ET (ed) *Circannual Clocks*. Academic Press, New York.

1115. Penn DJ (2003) The evolutionary roots of our environmental problems: Toward a Darwinian ecology. *Quarterly Review of Biology* 78: 275–301.

1116. Penn DJ, Smith KR (2007) Differential fitness costs of reproduction between the sexes. *Proceedings of the National Academy of Sciences USA* 104: 553–558.

1117. Pennisi E (2000) Fruit fly genome yields data and a validation. *Science* 287: 1374.

1118. Pereyra ME, Sharbaugh SM, Hahn TP (2005) Interspecific variation in photo-induced GnRH plasticity among nomadic carduuline finches. *Brain Behavior and Evolution* 66: 35–49.

1119. Perrigo G, Bryant WC, vom Saal FS (1990) A unique neural timing system prevents male mice from harming their own offspring. *Animal Behaviour* 39: 535–539.

1120. Perusse D (1993) Cultural and reproductive success in industrial societies: Testing the relationship at the proximate and ultimate levels. *Behavioral and Brain Sciences* 16: 267–283.

1121. Petit LJ (1991) Adaptive tolerance of cowbird parasitism by prothonotary warblers: A consequence of site limitation? *Animal Behaviour* 41: 425–432.

1122. Petrie M, Halliday T, Sanders C (1991) Peahens prefer peacocks with elaborate trains. *Animal Behaviour* 41: 323–332.

1123. Petrie M (1992) Peacocks with low mating success are more likely to suffer predation. *Animal Behaviour* 44: 585–586.

1124. Petrie M (1994) Improved growth and survival of offspring of peacocks with more elaborate trains. *Nature* 371: 585–586.

1125. Petrie M, Halliday T (1994) Experimental and natural changes in the peacock's (*Pavo cristatus*) train can affect mating success. *Behavioral Ecology and Sociobiology* 35: 213–217.

1126. Petrusková T, Petrusek A, Pavel V, Fuchs R (2007) Territorial meadow pipit males (*Anthus pratensis*; Passeriformes) become more aggressive in female presence. *Naturwissenschaften* 94: 643–650.

1127. Pfaff JA, Zanetter L, MacDougall-Shackleton SA, MacDougall-Shackleton EA (2007) Song repertoire size varies with HVC volume and is indicative of male quality in song sparrows (*Melospiza melodia*). *Proceedings of the Royal Society of London, Series B* 274: 2035–2040.

1128. Pfennig DW, Collins JP (1993) Kinship affects morphogenesis in cannibalistic salamanders. *Nature* 362: 836–838.

1129. Pfennig DW, Sherman PW, Collins JP (1994) Kin recognition and cannibalism in polyphenic salamanders. *Behavioral Ecology* 5: 225–232.

1130. Pfennig DW, Rice AM, Martin RA (2007) Field and experimental evidence for competition's role in phenotypic divergence. *Evolution* 61: 257–271.

1131. Phelps SM, Ophir AG (2009) Monogamous brains and alternative tactics: Neuronal V1aR, space use and sexual infidelity among male prairie voles. In: Dukas R, Ratcliffe JM (eds) *Cognitive Ecology*. University of Chicago Press, Chicago.

1132. Phillips BL, Shine R (2006) An invasive species induced rapid adaptive change in a native predator: cane toads and black snakes in Australia. *Proceedings of the Royal Society of London, Series B* 273: 1545–1550.

1133. Phillips RA, Furness RW, Stewart FM (1998) The influence of territory density on the vulnerability of Arctic skuas *Stercorarius parasiticus* to predation. *Biological Conservation* 86: 21–31.

1134. Picciotto MR (1999) Knock-out mouse models used to study neurobiological systems. *Critical Reviews in Neurobiology* 13: 103–149.

1135. Pierce GJ, Ollason JG (1987) Eight reasons why optimal foraging theory is a complete waste of time. *Oikos* 49: 111–118.

1136. Pierotti R, Murphy EC (1987) Intergenerational conflicts in gulls. *Animal Behaviour* 35: 435–444.

1137. Pietrewicz AT, Kamil AC (1977) Visual detection of cryptic prey by blue jays (*Cyanocitta cristata*). *Science* 195: 580–582.

1138. Pietsch TW, Grobecker DB (1978) The compleat angler: Aggressive mimicry in the antennariid anglerfish. *Science* 201: 369–370.

1139. Pike N (2007) Specialised placement of morphs within the gall of the social aphid *Pemphigus spyrothecae*. *BMC Evolutionary Biology* 7: 18.

1140. Pike N, Whitfield JA, Foster WA (2007) Ecological correlates of sociality in *Pemphigus* aphids, with a partial phylogeny of the genus. *BMC Evolutionary Biology* 7: 185.

1141. Pike TW, Blount JD, Lindstrom J, Metcalfe NB (2007) Dietary carotenoid availability influences a male's ability to provide parental care. *Behavioral Ecology* 18: 1100–1105.

1142. Pillsworth EG, Haselton MG, Buss DM (2004) Ovulatory shifts in female sexual desire. *Journal of Sex Research* 41: 55–65.

1143. Pinker S (1994) *The Language Instinct*. W. Morrow & Co., New York.

1144. Pinker S (2002) *The Blank Slate: The Modern Denial of Human Nature*. Viking, New York.

1145. Pinxten R, de Ridder E, Eens M (2003) Female presence affects male behavior and testosterone levels in the European starling (*Sturnus vulgaris*). *Hormones and Behavior* 44: 103–109.

1146. Piper WH, Evers DC, Meyer MW, Tischler KB, Kaplan JD, Fleischer RC (1997) Genetic monogamy in the common loon (*Gavia immer*). *Behavioral Ecology and Sociobiology* 41: 25–32.

1147. Pischedda A, Chippindale AK (2006) Intralocus sexual conflict diminishes the benefits of sexual selection *PLoS Biology* 4: 2099–2103.

1148. Pitcher T (1979) He who hesitates lives: Is stotting antiambush behavior? *American Naturalist* 113: 453–456.

1149. Pitkow LJ, Sharer CA, Ren XL, Insel TR, Terwilliger EF, Young LJ (2001) Facilitation of affiliation and pair-bond formation by vasopressin receptor gene transfer into the ventral forebrain of a monogamous vole. *Journal of Neuroscience* 21: 7392–7396.

1150. Pitnick S, Garcia-Gonzalez F (2002) Harm to females increases with male body size in *Drosophila melanogaster*. *Proceedings of the Royal Society of London, Series B* 269: 1821–1828.

1151. Pizzari T, Birkhead TR (2000) Female feral fowl eject sperm of subdominant males. *Nature* 405: 787–789.

1152. Place NJ, Glickman SE (2004) Masculinization of female mammals: Lessons from *Nature*. In: Baskin L (ed) *Hypospadias and Genital Development*. Kluwer Academic/Plenum Publishers, New York.

1153. Plaistow S, Siva-Jothy MT (1996) Energetic constraints and male mate securing tactics in the damselfly *Calopteryx splendens xanthosoma* (Charpentier). *Proceedings of the Royal Society of London, Series B* 263: 1233–1238.

1154. Platt JR (1964) Strong inference. *Science* 146: 347–353.

1155. Platzen D, Magrath RD (2004) Parental alarm calls suppress nestling vocalization. *Proceedings of the Royal Society of London, Series B* 271: 1271–1276.

1156. Plomin R, Fulker DW, Corley R, DeFries JC (1997) Nature, nurture, and cognitive development from 1 to 16 years: A parent-offspring adoption study. *Psychological Science* 8: 442–447.

1157. Pluhácek J, Bartos L, Vichová J (2006) Variation in incidence of male infanticide within subspecies of plains zebra (*Equus burchelli*). *Journal of Mammalogy* 87: 35–40.

1158. Polak M, Wolf LL, Starmer WT, Barker JSF (2001) Function of the mating plug in *Drosophila hibisci* Bock. *Behavioral Ecology and Sociobiology* 49: 196–205.

1159. Polak M (ed) (2003) *Developmental Instability: Causes and Consequences*. Oxford University Press, New York.

1160. Porter RH, Tepper VJ, White DM (1981) Experiential influences on the development of huddling preferences and "sibling" recognition in spiny mice. *Developmental Psychobiology* 14: 375–382.

1161. Powell AN, Collier CL (2000) Habitat use and reproductive success of western snowy plovers at new nesting areas created for California least terns. *Journal of Wildlife Management* 64: 24–33.

1162. Powell GVN, Bjork RD (2004) Habitat linkages and the conservation of tropical biodiversity as indicated by seasonal migrations of three-wattled bellbirds. *Conservation Biology* 18: 500–509.

1163. Powzyk JA, Mowry CB (2003) Dietary and feeding differences between sympatric *Propithecus diadema diadema* and *Indri indri*. *International Journal of Primatology* 24: 1143–1162.

1164. Pravosudov VV, Clayton NS (2002) A test of the adaptive specialization hypothesis: Population differences in caching, memory, and the hippocampus in black-capped chickadees (*Poecile atricapilla*). *Behavioral Neuroscience* 116: 515–522.

1165. Pravosudov VV, de Kort SR (2006) Is the western scrub-jay (*Aphelocoma californica*) really an underdog among food-caching corvids when it comes to hippocampal volume and food caching propensity? *Brain, Behavior and Evolution* 67: 1–9.

1166. Preston-Mafham R, Preston-Mafham K (1993) *The Encyclopedia of Land Invertebrate Behaviour*. MIT Press, Cambridge, MA.

1167. Prete FR (1995) Designing behavior: A case study. *Perspectives in Ethology* 11: 255–277.

1168. Pribil S, Searcy WA (2001) Experimental confirmation of the polygyny threshold model for red-winged blackbirds. *Proceedings of the Royal Society of London, Series B* 268: 1643–1646.

1169. Proctor HC (1991) Courtship in the water mite *Neumania papillator*: Males capitalize on female adaptations for predation. *Animal Behaviour* 42: 589–598.

1170. Proctor HC (1992) Sensory exploitation and the evolution of male mating behaviour: A cladistic test. *Animal Behaviour* 44: 745–752.

1171. Pruett-Jones S, Pruett-Jones M (1994) Sexual competition and courtship disruptions: Why do male bowerbirds destroy each other's bowers? *Animal Behaviour* 47: 607–620.

1172. Prum RO (1999) Development and evolutionary origin of feathers. *Journal of Experimental Zoology* 285: 291–306.

1173. Pryke SR, Andersson S (2003) Carotenoid-based epaulettes reveal male competitive ability: experiments with resident and floater red-shouldered widowbirds. *Animal Behaviour* 66: 217–224.

1174. Puce A, Allison T, McCarthy G (1999) Electrophysiological studies of human face perception. III: Effects of top-down processing of face-specific potentials. *Cerebral Cortex* 9: 445–458.

1175. Pulido F, Berthold P, Mohr G, Querner U (2001) Heritability of the timing of autumn migration in a natural bird population. *Proceedings of the Royal Society of London, Series B* 268: 953–959.

1176. Pulido F (2007) The genetics and evolution of avian migration. *BioScience* 57: 165–174.

1177. Purseglove JW, Brown EG, Green CL, Robbins SRJ (1981) *Spices*. Longman, London.

1178. Purves D, Brannon EM, Cabeza R, Huettel SA, LaBar KS, Platt ML, Woldorff M (2007) *Principles of Cognitive Neuroscience*. Sinauer Associates, Sunderland, MA.

1179. Pusey AE, Packer C (1987) The evolution of sex-biased dispersal in lions. *Behaviour* 101: 275–310.

1180. Pusey AE, Packer C (1994) Infanticide in lions. In: Parmigiani S, vom Saal FS (eds) *Infanticide and Parental Care*. Harwood Academic Press, Chur, Switzerland.

1181. Pusey AE, Wolf M (1996) Inbreeding avoidance in animals. *Trends in Ecology and Evolution* 11: 201–206.

1182. Queller DC, Strassmann JE, Hughes CR (1993) Microsatellites and kinship. *Trends in Ecology and Evolution* 8: 285–288.

1183. Queller DC (1996) The measurement and meaning of inclusive fitness. *Animal Behaviour* 51: 229–232.

1184. Queller DC (1997) Why do females care more than males? *Proceedings of the Royal Society of London, Series B* 264: 1555–1557.

1185. Queller DC, Strassmann JE (1998) Kin selection and social insects. *BioScience* 48: 165–175.

1186. Queller DC, Zacchi F, Cervo R, Turillazzi S, Henshaw MT, Santorelli LA, Strassmann JE (2000) Unrelated helpers in a social insect. *Nature* 405: 784–787.

1187. Queller DC (2006) To work or not to work. *Nature* 444: 42–43.

1188. Quillfeldt P (2002) Begging in the absence of sibling competition in Wilson's storm-petrels, *Oceanites oceanicus*. *Animal Behaviour* 64: 579–587.

1189. Quinn JL, Creswell W (2006) Testing domains of danger in the selfish herd: sparrowhawks target widely spaced redshanks in flocks. *Proceedings of the Royal Society of London, Series B* 273: 2521–2526.

1190. Quinn JS, Woolfenden GE, Fitzpatrick JW, White BN (1999) Multi-locus DNA fingerprinting supports genetic monogamy in Florida scrub-jays. *Behavioral Ecology and Sociobiology* 45: 1–10.

1191. Quinn VS, Hews DK (2000) Signals and behavioural responses are not coupled in males: Aggression affected by replacement of an evolutionarily lost colour signal. *Proceedings of the Royal Society of London, Series B* 267: 755–758.

1192. Racey PA, Skinner JD (1979) Endocrine aspects of sexual mimicry in spotted hyenas *Crocuta crocuta*. *Journal of Zoology* 187: 315–326.

1193. Rachlow JL, Berkeley EV, Berger J (1998) Correlates of male mating strategies in white rhinos (*Ceratotherium simum*). *Journal of Mammalogy* 79: 1317–1324.

1194. Ralls K, Brugger K, Ballou J (1979) Inbreeding and juvenile mortality in small populations of ungulates. *Science* 206: 1101–1103.

1195. Ralph MR, Foster RG, Davis FC, Menaker M (1990) Transplanted suprachiasmatic nucleus determines circadian rhythm. *Science* 247: 975–978.

1196. Ramirez MI, Azcarate JG, Luna L (2003) Effects of human activities on monarch butterfly habitat in protected mountain forests, Mexico. *Forestry Chronicle* 79: 242–246.

1197. Rasmussen KM (2001) The "fetal origins" hypothesis": challenges and opportunities for maternal and child nutrition. *Annual Review of Nutrition* 21: 73–95.

1198. Ratcliffe JM, Fenton MB, Galef BG (2003) An exception to the rule: common vampire bats do not learn taste aversions. *Animal Behaviour* 65: 385–389.

1199. Ratnieks FLW, Keller L (1998) Queen control of egg fertilization in the honey bee. *Behavioral Ecology and Sociobiology* 44: 57–62.

1200. Redondo T, Carranza J (1989) Offspring reproductive value and nest defense in the magpie (*Pica pica*). *Behavioral Ecology and Sociobiology* 25: 369–378.

1201. Reed WL, Clark ME, Parker PG, Raouf SA, Arguedas N, Monk DS, Snadjr E, Nolan V, Jr, Ketterson ED (2006) Physiological effects on demography: A long-term experimental study of testosterone's effects on fitness. *American Naturalist* 167: 667–683.

1202. Reeve HK, Westneat DF, Noon WA, Sherman PW, Aquadro CF (1990) DNA "fingerprinting" reveals high levels of inbreeding in colonies of the eusocial naked mole-rat. *Proceedings of the National Academy of Sciences USA* 87: 2496–2500.

1203. Reeve HK, Sherman PW (1993) Adaptation and the goals of evolutionary research. *Quarterly Review of Biology* 68: 1–32.

1204. Reeve HK, Keller L (1997) Reproductive bribing and policing as evolutionary mechanisms for the supression of within-group selfishness. *American Naturalist* 150 Supplement: S42–S58.

1205. Reeve HK (2000) Review of *Unto Others: The Evolution and Psychology of Unselfish Behavior*. *Evolution and Human Behavior* 21: 65–72.

1206. Reeve HK, Starks PT, Peters JM, Nonacs P (2000) Genetic support for the evolutionary theory of reproductive transactions in social wasps. *Proceedings of the Royal Society of London, Series B* 267: 75–79.

1207. Reeve HK, Sherman PW (2001) Optimality and phylogeny. In: Orzack SH, Sober E (eds) *Adaptationism and Optimality*. Cambridge University Press, Cambridge.

1208. Reichard M, Le Comber SC, Smith C (2007) Sneaking from a female perspective. *Animal Behaviour* 74: 679–688.

1209. Reid JM, Monaghan P, Ruxton GD (2002) Males matter: The occurrence and consequences of male incubation in starlings (*Sturnus vulgaris*). *Behavioral Ecology and Sociobiology* 51: 255–261.

1210. Reid JM, Arcese P, Cassidy ALEV, Heibert SM, Smith JNM, Stoddard PK, Marr AB, Keller LK (2005) Fitness correlates of song repertoire size in free-living song sparrows (*Melospiza melodia*). *American Naturalist* 165: 299–310.

1211. Reppert SM, Zhu HS, White RH (2004) Polarized light helps monarch butterflies navigate. *Current Biology* 14: 155–158.

1212. Reyer H-U (1984) Investment and relatedness: A cost/benefit analysis of breeding and helping in the pied kingfisher. *Animal Behaviour* 32: 1163–1178.

1213. Reynolds AM, Smith AD, Reynolds DR, Carreck NL, Osborne JL (2007) Honeybees perform optimal scale-free searching flights when attempting to locate a food source. *Journal of Experimental Biology* 210: 3763–3770.

1214. Reynolds JD, Gross MR (1990) Costs and benefits of female mate choice: Is there a lek paradox? *American Naturalist* 136: 230–243.

1215. Reynolds SM, Dryer K, Bollback J, Uy JAC, Patricelli GL, Robson T, Borgia G, Braun MJ (2007) Behavioral paternity predicts genetic paternity in Satin Bowerbirds (*Ptilonorhynchus violaceus*), a species with a non-resource-based mating system *Auk* 124: 857–867.

1216. Rhoden PK, Foster WA (2002) Soldier behaviour and division of labour in the aphid genus *Pemphigus* (Hemiptera, Aphididae). *Insectes Sociaux* 49: 257–263.

1217. Rhodes G, Proffitt F, Grady JM, Sumich A (1998) Facial symmetry and the perception of beauty. *Psychonomic Bulletin and Review* 5: 659–669.

1218. Rhodes G (2006) The evolutionary psychology of facial beauty. *Annual Review of Psychology* 57: 199–226.

1219. Richard F-J, Tarpy D, Grozinger C (2007) Effects of insemination quantity on honey bee queen physiology. *PLoS ONE* doi:10.1371/journal.pone.0000980.

1220. Richerson PJ, Boyd R (2005) *Not by Genes Alone: How Culture Transformed Human Evolution*. University of Chicago Press, Chicago, IL.

1221. Ridley AR, Child MF, Bell MBV (2007) Interspecific audience effects on the alarm-calling behaviour of a kleptoparasitic bird. *Biology Letters* 3: 589–591.

1222. Ridley AR, Raihani NJ (2007) Variable postfledging care in a cooperative bird: causes and consequences. *Behavioral Ecology* 18: 994–1000.

1223. Riedel S, Hinz A, Schwarz R (2000) Attitude towards blood donation in Germany: Results of a representative survey. *Infusion Therapy and Transfusion Medicine* 27: 196–199.

1224. Rintamäki PT, Alatalo RV, Höglund J, Lundberg A (1995) Male territoriality and female choice on black grouse leks. *Animal Behaviour* 49: 759–767.

1225. Rintamäki PT, Lundberg A, Alatalo RV, Höglund J (1998) Assortative mating and female clutch investment in black grouse. *Animal Behaviour* 56: 1399–1403.

1226. Rintamäki PT, Höglund J, Alatalo RV, Lundberg A (2001) Correlates of male mating success on black grouse (*Tetrao tetrix* L.) leks. *Annales Zoologici Fennici* 38: 99–109.

1227. Robert D, Amoroso J, Hoy RR (1992) The evolutionary convergence of hearing in a parasitoid fly and its cricket host. *Science* 258: 1135–1137.

1228. Roberts SC, Havlicek J, Flegr J, Hruskova M, Little AC, Jones BC, Perrett DI, Petrie M (2004) Female facial attractiveness increases during the fertile phase of the menstrual cycle. *Proceedings of the Royal Society of London, Series B* 271: S270–S272.

1229. Robinson GE (1998) From society to genes with the honey bee. *American Scientist* 86: 456–462.

1230. Robinson GE (2004) Beyond nature and nurture. *Science* 304: 397–399.

1231. Roby JL, Whittenburg KP (2005) The feasibility of intrafamily and in-country adoptions on the Marsh Islands. *Families in Society* 86: 547–557.

1232. Rodd FH, Hughes KA, Grether GF, Baril CT (2002) A possible non-sexual origin of mate preference: are male guppies mimicking fruit? *Proceedings of the Royal Society of London, Series B* 269: 475–481.

1233. Roeder KD, Treat AE (1961) The detection and evasion of bats by moths. *American Scientist* 49: 135–148.

1234. Roeder KD (1963) *Nerve Cells and Insect Behavior*. Harvard University Press, Cambridge, MA.

1235. Roeder KD (1970) Episodes in insect brains. *American Scientist* 58: 378–389.

1236. Roff DA, Fairbairn DJ (2007) The evolution and genetics of migration in insects. *BioScience* 57: 155–164.

1237. Rohwer S, Spaw CD (1988) Evolutionary lag versus bill-size constraints: A comparative study of the acceptance of cowbird eggs by old hosts. *Evolutionary Ecology* 2: 27–36.

1238. Rose M (1998) *Darwin's Spectre*. Princeton University Press, Princeton, NJ.

1239. Rose NA, Deutsch CJ, Le Boeuf BJ (1991) Sexual behavior of male northern elephant seals: III. The mounting of weaned pups. *Behaviour* 119: 171–192.

1240. Rosengaus RB, Maxmen AB, Coates LE, Traniello JFA (1998) Disease resistance: A benefit of sociality in the dampwood termite *Zootermopsis angusticollis* (Isoptera: Termopsidae). *Behavioral Ecology and Sociobiology* 44: 125–134.

1241. Rosenqvist G (1990) Male mate choice and female-female competition for mates in the pipefish *Nerophis ophidion*. *Animal Behaviour* 39: 1110–1116.

1242. Ross KG (1986) Kin selection and the problem of sperm utilization in social insects. *Nature* 323: 798–800.

1243. Rowland WJ (1994) Proximate determinants of stickleback behavior: an evolutionary perspective. In: Bell M, Foster S (eds) *The Evolutionary Biology of the Threespine Stickleback*. Oxford University Press, Oxford.

1244. Rowley I, Chapman G (1986) Cross-fostering, imprinting, and learning in two sympatric species of cockatoos. *Behaviour* 96: 1–16.

1245. Royle NJ, Hartley IR, Parker GA (2002) Begging for control: when are offspring solicitation behaviours honest? *Trends in Ecology and Evolution* 17: 434–440.

1246. Rubenstein DR, Chamberlain CP, Holmes RT, Ayres MP, Waldbauer JR, Graves GR, Tuross NC (2002) Linking breeding and wintering ranges of a migratory songbird using stable isotopes. *Science* 295: 1062–1065.

1247. Rubenstein DR, Hobson KA (2004) From birds to butterflies: animal movement patterns and stable isotopes. *Trends in Ecology and Evolution* 19: 256–263.

1248. Rubenstein DR (2007) Female extrapair mate choice in a cooperative breeder: trading sex for help and increasing offspring heterozygosity. *Proceedings of the Royal Society of London, Series B* 274: 1895–1903.

1249. Rubenstein DR, Lovette IJ (2007) Temporal environmental variability drives the evolution of cooperative breeding in birds. *Current Biology* 17: 1414–1419.

1250. Rudge DW (2006) Myths about moths: a study in contrasts. *Endeavour* 30: 19–23.

1251. Ruegg KC, Smith TB (2002) Not as the crow flies: a historical explanation for circuitous migration in Swainson's thrush (*Catharus ustulatus*). *Proceedings of the Royal Society of London, Series B* 269: 1375–1381.

1252. Ruegg KC, Hijmans RJ, Moritz C (2006) Climate change and the origin of migratory pathways in the Swainson's thrush, *Catharus ustulatus*. *Journal of Biogeography* 33: 1172–1182.

1253. Ruffieux L, Elouard JM, Sartori M (1998) Flightlessness in mayflies and its relevance to hypotheses on the origin of insect flight. *Proceedings of the Royal Society of London, Series B* 265: 2135–2140.

1254. Runcie MJ (2000) Biparental care and obligate monogamy in the rock-haunting possum, *Petropseudes dahli*, from tropical Australia. *Animal Behaviour* 59: 1001–1008.

1255. Rundus AS, Owings DS, Joshi SS, Chinn E, Giannini N (2007) Ground squirrels use an infrared signal to deter rattlesnake predation. *Proceedings of the National Academy of Sciences USA* 104: 14372–14374.

1256. Russell AF, Langmore NE, Cockburn A, Astheimer LB, Kilner RM (2007) Reduced egg investment can conceal helper effects in cooperatively breeding birds. *Science* 317: 941–944.

1257. Russell AF, Young AJ, Spong G, Jordan NR, Clutton-Brock TH (2007) Helpers increase the reproductive potential of offspring in cooperative meerkats. *Proceedings of the Royal Society of London, Series B* 274: 513–520.

1258. Rutowski RL (1998) Mating strategies in butterflies. *Scientific American* 279 (July): 64–69.

1259. Ryan MJ, Tuttle MD, Taft LK (1981) The costs and benefits of frog chorusing behavior. *Behavioral Ecology and Sociobiology* 8: 273–278.

1260. Ryan MJ (1985) *The Túngara Frog*. University of Chicago Press, Chicago.

1261. Ryan MJ, Wagner WE, Jr. (1987) Asymmetries in mating behavior between species: Female swordtails prefer heterospecific males. *Science* 236: 595–597.

1262. Ryan MJ, Fox JH, Wilczynski W, Rand AS (1990) Sexual selection for sensory exploitation in the frog *Physalaemus pustulosus*. *Nature* 343: 66–67.

1263. Ryan MJ, Keddy-Hector A (1992) Directional patterns of female mate choice and the role of sensory biases. *American Naturalist* 139: S4–S35.

1264. Ryan MJ, Rand AS (1999) Phylogenetic influence on mating call preferences in female tungara frogs, *Physalaemus pustulosus*. *Animal Behaviour* 57: 945–956.

1265. Rydale J, Roininen H, Philip KW (2000) Persistence of bat defence reactions in high Arctic moths (Lepidoptera). *Proceedings of the Royal Society of London, Series B* 267: 553–557.

1266. Rydell J, Arlettaz R (1994) Low-frequency echolocation enables the bat *Tadarida teniotis* to feed on tympanate insects. *Proceedings of the Royal Society of London, Series B* 257: 175–178.

1267. Ryner LC, Goodwin SF, Castrillon DH, Anand A, Baker BS, Hall JC, Taylor BJ, Wasserman SA (1996) Control of male sexual behavior and sexual orientation in *Drosophila* by the *fruitless* gene. *Cell* 87: 1079–1089.

1268. Sacks OW (1985) *The Man Who Mistook His Wife for a Hat and Other Clinical Tales*. Summit Books, New York.

1269. Sadowski JA, Moore AJ, Brodie ED, III (1999) The evolution of empty nuptial gifts in a dance fly, *Empis snoddyi* (Diptera: Empididae): Bigger isn't always better. *Behavioral Ecology and Sociobiology* 45: 161–166.

1270. Safran RJ, Adelman JS, McGraw KJ, Hau M (2008) Sexual signal exaggeration affects male physiological state in barn swallows. *Current Biology* 18: R461–R462.

1271. Sahlins M (1976) *The Use and Abuse of Biology*. University of Michigan Press, Ann Arbor.

1272. Saino N, Ninni P, Calza S, Martinelli R, de Bernardi F, Møller AP (2000) Better red than dead: Carotenoid-based mouth coloration reveals infection in barn swallow nestlings. *Proceedings of the Royal Society of London, Series B* 267: 57–61.

1273. Salewski V, Bruderer B (2007) The evolution of bird migration—a synthesis. *Naturwissenschaften* 94: 268–279.

1274. Sánchez F, Korine C, Steeghs M, Laarhoven LJ, Cristescu SM, Harren FJM, Dudley R, Pinshow B (2006) Ethanol and methanol as possible odor cues for Egyptian fruit bats (*Rousettus aegyptiacus*). *Journal of Chemical Ecology* 32: 1289–1300.

1275. Sandberg R, Moore FR (1996) Migratory orientation of red-eyed vireos, *Vireo olivaceus*, in relation to energetic condition and ecological context. *Behavioral Ecology and Sociobiology* 39: 1–10.

1276. Sandercock BK, Jaramillo A (2002) Annual survival rates of wintering sparrows: Assessing demographic consequences of migration. *Auk* 119: 149–165.

1277. Sargent RC (1989) Allopaternal care in the fathead minnow, *Pimephales promelas*: Stepfathers discriminate against their adopted eggs. *Behavioral Ecology and Sociobiology* 25: 379–386.

1278. Sargent TD (1976) *Legion of Night, The Underwing Moths*. University of Massachusetts Press, Amherst, MA.

1279. Sato T (1994) Active accumulation of spawning substrate: A determinant of extreme polygyny in a shell-brooding cichlid fish. *Animal Behaviour* 48: 669–678.

1280. Sax A, Hoi H, Birkhead TR (1998) Copulation rate and sperm use by female bearded tits, *Panurus biarmicus*. *Animal Behaviour* 56: 1199–1294.

1281. Scantlebury M, Speakman JR, Oosthuizen MK, Roper TJ, Bennett NC (2006) Energetics reveals physiologically distinct castes in a eusocial mammal. *Nature* 440: 795–797.

1282. Schaller GB (1964) *The Year of the Gorilla*. University of Chicago Press, Chicago.

1283. Schamel D, Tracy DM, Lank DB (2004) Male mate choice, male availability and egg production as limitations on polyandry in the red-necked phalarope. *Animal Behaviour* 67: 847–853.

1284. Scheel D, Packer C (1991) Group hunting behaviour of lions: A search for cooperation. *Animal Behaviour* 41: 711–722.

1285. Schieb JE, Gangestad SW, Thornhill R (1999) Facial attractiveness, symmetry and cues of good genes. *Proceedings of the Royal Society of London, Series B* 266: 1913–1917.

1286. Schlaepfer MA, Runge MC, Sherman PW (2002) Ecological and evolutionary traps. *Trends in Ecology and Evolution* 17: 474–480.

1287. Schmitt DP, Shackelford TK, Duntley J, Tooke W, Buss DM (2001) The desire for sexual variety as a key to understanding basic human mating strategies. *Personal Relationships* 8: 425–455.

1288. Schneider JM, Lubin Y (1997) Infanticide by males in a spider with suicidal maternal care, *Stegodyphus lineatus*. *Animal Behaviour* 54: 305–312.

1289. Schneider JS, Stone MK, Wynne-Edwards KE, Horton TH, Lydon J, O'Malley B, Levine JE (2003) Progesterone receptors mediate male aggression toward infants. *Proceedings of the National Academy of Sciences USA* 100: 2951–2956.

1290. Schoech SJ (1998) Physiology of helping in Florida scrub-jays. *American Scientist* 86: 70–77.

1291. Schwabl H (1983) Auspragung und Bedeutung des Teilzugverhaltnes einer sudwestdeutschen Population der Amsel *Turdus merula*. *Journal für Ornithologie* 124: 101–116.

1292. Schwabl H, Mock DW, Gieg JA (1997) A hormonal mechanism for parental favouritism. *Nature* 386: 231.

1293. Schwagmeyer PL (1994) Competitive mate searching in the 13–lined ground squirrel: Potential roles of spatial memory? *Ethology* 98: 265–276.

1294. Schwagmeyer PL (1995) Searching today for tomorrow's mates. *Animal Behaviour* 50: 759–767.

1295. Schwander T, Humbert JY, Brent CS, Cahan SH, Chapuis L, Renai E, Keller L (2008) Maternal effect on female caste determination in a social insect. *Current Biology* 18: 265–269.

1296. Schwensow N, Eberle M, Sommer S (2008) Compatibility counts: MHC-associated mate choice in a wild promiscuous primate. *Proceedings of the Royal Society of London, Series B* 275: 555–564.

1297. Sealy SG (1995) Burial of cowbird eggs by parasitized yellow warblers: An empirical and experimental study. *Animal Behaviour* 49: 877–889.

1298. Searcy WA, Nowicki S, Hughes M, Peters S (2002) Geographic song discrimination in relation to dispersal distances in song sparrows. *American Naturalist* 159: 221–230.

1299. Seeley TD (1977) Measurement of nest cavity volume by the honey bee (*Apis mellifera*). *Behavioral Ecology and Sociobiology* 2: 201–227.

1300. Seeley TD (1995) *The Wisdom of the Hive*. Harvard University Press, Cambridge, MA.

1301. Seeley TD, Buhrman SC (1999) Group decision making in swarms of honey bees. *Behavioral Ecology and Sociobiology* 45: 19–32.

1302. Seeley TD, Visscher PK (2003) Choosing a home: how the scouts in a honey bee swarm perceive the completion of their group decision making. *Behavioral Ecology and Sociobiology* 54: 511–520.

1303. Seeley TD, Tarpy DR (2007) Queen promiscuity lowers disease within honeybee colonies. *Proceedings of the Royal Society of London, Series B* 274: 67–72.

1304. Seidelmann K (2006) Open-cell parasitism shapes maternal investment patterns in the Red Mason bee *Osmia rufa*. *Behavioral Ecology* 17: 839–846.

1305. Semel B, Sherman PW (2001) Intraspecific parasitism and nest-site competition in wood ducks. *Animal Behaviour* 61: 787–803.

1306. Semple S, McComb K, Alberts S, Altmann J (2002) Information content of female copulation calls in yellow baboons. *American Journal of Primatology* 56: 43–56.

1307. Senar JC, Figuerola J, Pascual J (2002) Brighter yellow blue tits make better parents. *Proceedings of the Royal Society of London, Series B* 269: 257–261.

1308. Senghas A, Kita S, Özyürek A (2004) Children creating core properties of language: Evidence from an emerging sign language in Nicaragua. *Science* 305: 1779–1782.

1309. Shackelford TK, Buss DM, Weekes-Shackelford VA (2003) Wife killings committed in the context of a lovers triangle. *Basic and Applied Social Psychology* 25: 137–143.

1310. Shavit A, Millstein RL (2008) Group selection is dead! Long live group selection. *BioScience* 58: 574–575.

1311. Sheldon BC (1993) Sexually-transmitted disease in birds: occurrence and evolutionary significance. *Philosophical Transactions of the Royal Society of London, Series B* 339: 491–497.

1312. Shelly TE (2001) Lek size and female visitation in two species of tephritid fruit flies. *Animal Behaviour* 62: 33–40.

1313. Sherman G, Visscher PK (2002) Honeybee colonies achieve fitness through dancing. *Nature* 419: 920–922.

1314. Sherman PW (1977) Nepotism and the evolution of alarm calls. *Science* 197: 1246–1253.

1315. Sherman PW (1981) Kinship, demography and Belding's ground squirrel nepotism. *Behavioral Ecology and Sociobiology* 8: 251–259.

1316. Sherman PW (1988) The levels of analysis. *Animal Behaviour* 36: 616–618.

1317. Sherman PW, Jarvis JUM, Alexander RD (eds) (1991) *The Biology of the Naked Mole-Rat*. Princeton University Press, Princeton, N J.

1318. Sherman PW, Hash GA (2001) Why vegetable dishes are not very spicy. *Evolution and Human Behavior* 22: 147–163.

1319. Sherry DF (1984) Food storage by black-capped chickadees: Memory of the location and contents of caches. *Animal Behaviour* 32: 451–464.

1320. Sherry DF, Forbes MRL, Kjurgel M, Ivy GO (1993) Females have a larger hippocampus than males in the brood-parasitic brown-headed cowbird. *Proceedings of the National Academy of Sciences USA* 90: 7839–7843.

1321. Shine R, Langkilde T, Mason RT (2003) Cryptic forcible insemination: male snakes exploit female physiology, anatomy, and behavior to obtain coercive matings. *American Naturalist* 162: 653–667.

1322. Shorey L (2002) Mating success on white-bearded manakin (*Manacus manacus*) leks: male characteristics and relatedness. *Behavioral Ecology and Sociobiology* 52: 451–457.

1323. Shreeves G, Cant MA, Bolton A, Field J (2003) Insurance-based advantages for subordinate cofoundresses in a temperate paper wasp. *Proceedings of the Royal Society of London, Series B* 270: 1617–1622.

1324. Shuster SM (1989) Male alternative reproductive strategies in a marine isopod crustacean (*Paracerceis sculpta*): the use of genetic markers to measure differences in the fertilization success among alpha, beta, and gamma-males. *Evolution* 43: 1683–1689.

1325. Shuster SM, Wade MJ (1991) Equal mating success among male reproductive strategies in a marine isopod. *Nature* 350: 608–610.

1326. Shuster SM (1992) The reproductive behaviour of alpha, beta, and gamma morphs in *Paracerceis sculpta*: A marine isopod crustacean. *Behaviour* 121: 231–258.

1327. Shuster SM, Sassaman CA (1997) Genetic interaction between male mating strategy and sex ratio in a marine isopod. *Nature* 338: 373–377.

1328. Shuster SM, Wade MJ (2003) *Mating Systems and Strategies*. Princeton University Press, Princeton, N.J.

1329. Silk JB (1980) Adoption and kinship in Oceania. *American Anthropologist* 82: 799–820.

1330. Simmons LW (2001) *Sperm Competition and Its Evolutionary Consequences in the Insects*. Princeton University Press, Princeton, NJ.

1331. Simmons LW, Emlen DJ (2006) Evolutionary trade-off between weapons and testes. *Proceedings of the National Academy of Sciences USA* 103: 16346–16351.

1332. Simmons P, Young D (1999) *Nerve Cells and Animal Behaviour*, 2nd edn. Cambridge University Press, Cambridge.

1333. Simpson SJ, Sword GA, Lorch PD, Couzin ID (2006) Cannibal crickets on a forced march for protein and salt. *Proceedings of the National Academy of Sciences USA* 103: 4152–4156.

1334. Sinervo B, Miles DB, Frankino WA, Klukowski M, DeNardo DF (2000) Testosterone, endurance, and darwinian fitness: Natural and sexual selection on the physiological bases of alternative male behaviors in side-blotched lizards. *Hormones and Behavior* 38: 222–233.

1335. Singer P (1981) *The Expanding Circle: Ethics and Sociobiology.* Farrar, Straus, and Giroux, New York.

1336. Singh D (1993) Adaptive significance of female physical attractiveness: Role of the waist-to-hip ratio. *Journal of Personality and Social Psychology* 65: 293–307.

1337. Singh D, Bronstad PM (2001) Female body odour is a potential cue to ovulation. *Proceedings of the Royal Society of London, Series B* 268: 797–801.

1338. Sisneros JA, Bass AH (2003) Seasonal plasticity of peripheral auditory frequency sensitivity. *Journal of Neuroscience* 23: 1049–1058.

1339. Sisneros JA (2007) Sacculat potentials of the vocal plainfin midshipman fish, *Porichthys notatus. Journal of Comparative Physiology A* 193: 413–424.

1340. Skals N, Anderson P, Kanneworff M, Löfstedt C, Surlykke A (2005) Her odours make him deaf: crossmodal modulation of olfaction and hearing in a male moth. *Journal of Experimental Biology* 208: 595–601.

1341. Skinner BF (1966) Operant behavior. In: Honig W (ed) *Operant Behavior.* Appleton-Century-Crofts, New York.

1342. Slabbekoorn H, den Boer-Visser A (2006) Cities change the songs of birds. *Current Biology* 16: 2326–2331.

1343. Slagsvold T (1998) On the origin and rarity of interspecific nest parasitism in birds. *American Naturalist* 152: 264–272.

1344. Slagsvold T, Hansen BT (2001) Sexual imprinting and the origin of obligate brood parasitism in birds. *American Naturalist* 158: 354–367.

1345. Slagsvold T, Hansen BT, Johannessen LE, Lifjeld JT (2002) Mate choice and imprinting in birds studied by cross-fostering in the wild. *Proceedings of the Royal Society of London, Series B* 269: 1449–1455.

1346. Small TW, Sharp PJ, Deviche P (2007) Environmental regulation of the reproductive system in a flexibly breeding Sonoran Desert bird, the Rufous-winged Sparrow, *Aimophila carpalis. Hormones and Behavior* 51: 483–495.

1347. Smiseth PT, Bu RJ, Eikenaes AK, Amundsen T (2003) Food limitation in asynchronous bluethroat broods: effects on food distribution, nestling begging, and parental provisioning rules. *Behavioral Ecology* 14: 793–801.

1348. Smiseth PT, Lennox L, Moore AJ (2007) Interaction between parental care and sibling competition: Parents enhance offspring growth and exacerbate sibling competition. *Ecology* 88: 3174–3182.

1349. Smith EA, Borgerhoff Mulder M, Hill K (2001) Controversies in the evolutionary social sciences: A guide for the perplexed. *Trends in Ecology and Evolution* 16: 128–135.

1350. Smith HG (1991) Nestling American robins compete with siblings by begging. *Behavioral Ecology and Sociobiology* 29: 307–312.

1351. Smith HG, Montgomerie RD (1991) Sexual selection and the tail ornaments of North American barn swallows. *Behavioral Ecology and Sociobiology* 28: 195–201.

1352. Smith MD, Conway CJ (2007) Use of mammalian manure by nesting burrowing owls: a test of four functional hypotheses. *Animal Behaviour* 73: 65–73.

1353. Smith MS, Kish BJ, Crawford CB (1987) Inheritance of wealth as human kin investment. *Ethology and Sociobiology* 8: 171–182.

1354. Smith RL, Larsen E (1993) Egg attendance and brooding by males of the giant water bug *Lethocerus medius* (Guerin) in the field (Heteroptera, Belostomatidae). *Journal of Insect Behavior* 6: 93–106.

1355. Smith RL (1997) Evolution of paternal care in giant water bugs (Heteroptera: Belostomatidae). In: Choe JC, Crespi BJ (eds) *Social Competition and Cooperation among Insects and Arachnids, II Evolution of Sociality.* Cambridge University Press, Cambridge.

1356. Smith SM (1978) The 'underworld' in a territorial species: Adaptive strategy for floaters. *American Naturalist* 112: 571–582.

1357. Smith TB, Skulason S (1996) Evolutionary significance of resource polymorphisms in fishes, amphibians, and birds. *Annual Review of Ecology and Systematics* 27: 111–133.

1358. Smithseth PT, Ward RJS, Moore AJ (2007) Parents influence asymmetric sibling competition: experimental evidence with partially dependent young. *Ecology* 88: 3174–3182.

1359. Snow DW (1956) Courtship ritual: The dance of the manakins. *Animal Kingdom* 59: 86–91.

1360. Sober E, Wilson DS (1998) *Unto Others: The Evolution and Psychology of Unselfish Behavior.* Harvard University Press, Cambridge, MA.

1361. Sockman KW, Sewall KB, Ball GF, Hahn TP (2005) Economy of mate attraction in the Cassin's finch. *Biology Letters* 1: 34–37.

1362. Sogabe A, Matsumoto K, Yanagisawa Y (2007) Mate change reduces the reproductive rate of males in a monogamous pipefish *Corythoichthys haematopterus*: The benefit of long-term pair bonding. *Ethology* 113: 764–771.

1363. Sogabe A, Yanagisawa Y (2007) Sex-role reversal of a monogamous pipefish without higher potential reproductive rate in females. *Proceedings of the Royal Society of London, Series B* 274: 2959–2963.

1364. Soha JA, Nelson DA, Parker PG (2004) Genetic analysis of song dialect populations in Puget Sound white-crowned sparrows. *Behavioral Ecology* 15: 636–646.

1365. Sokolowski M, Wahlsten D (2001) Gene-environment interaction. In: Chin H, Moldin SO (eds) *Methods in Genomic Neuroscience.* CRC Press, Boca Raton, FL.

1366. Soler M, Soler JJ, Martinez JG, Møller AP (1995) Magpie host manipulation by great spotted cuckoos: Evidence for an avian Mafia? *Evolution* 49: 770–775.

1367. Soma KK, Tramontin AD, Wingfield JC (2000) Oestrogen regulates male aggression in the non-breeding season. *Proceedings of the Royal Society of London, Series B* 267: 1089–1092.

1368. Sommer V (1994) Infanticide among the langurs of Jodhpur: Testing the sexual selection hypothesis with a long-term record. In: Parmigiani S, vom Saal FS (eds) *Infanticide and Parental Care.* Harwood Academic Press, Chur, Switzerland.

1369. Sommer V (1994) Infanticide among free-ranging langurs (*Presbytis entellus*) of Jodhpur (Rajasthan/India): Recent observations and a reconsideration of hypotheses. *Primates* 28: 163–197.

1370. Sordahl TA (2004) Field evidence of predator discrimination abilities in American Avocets and Black-necked Stilts. *Journal of Field Ornithology* 75: 376–386.

1371. Sorenson LG (1994) Forced extra-pair copulation in the white-cheeked pintail: Male tactics and female responses. *Condor* 96: 400–410.

1372. Sorenson MD, Payne RB (2001) A single ancient origin of brood parasitism in African finches: implications for host-parasite coevolution. *Evolution* 55: 2550–2567.

1373. Sosis R (2004) The adaptive value of religious ritual. *American Scientist* 92: 166–172.

1374. Soukup SS, Thompson CF (1998) Social mating system and reproductive success in house wrens. *Behavioral Ecology* 9: 43–48.

1375. Spencer KA, Buchanan KL, Goldsmith AR, Catchpole CK (2003) Song as an honest signal of developmental stress in the zebra finch (*Taeniopygia guttata*). *Hormones and Behavior* 44: 132–139.

1376. Spencer KA, Wimpenny JH, Buchanan KL, Lovell PG, Goldsmith AR, Catchpole CK (2005) Developmental stress affects the attractiveness of male song and female choice in the zebra finch (*Taeniopygia guttata*). *Behavioral Ecology and Sociobiology* 58: 423–428.

1377. Stander PE (1992) Cooperative hunting in lions: The role of the individual. *Behavioral Ecology and Sociobiology* 29: 445–454.

1378. Stapley J, Whiting MJ (2005) Ultraviolet signals fighting ability in a lizard. *Biology Letters* 2: 169–172.

1379. Starks PT, Blackie CA, Seeley TD (2000) Fever in honeybee colonies. *Naturwissenschaften* 87: 229–231.

1380. Starks PT (2001) Alternative reproductive tactics in the paper wasp *Polistes dominulus* with specific focus on the sit-and-wait tactic. *Annales Zoologici Fennici* 38: 189–199.

1381. Starks PT (2002) The adaptive significance of stabilimenta in orb-webs: a hierarchical approach. *Annales Zoologici Fennici* 39: 307–315.

1382. Starks PT (2003) Selection for uniformity: xenophobia and invasion success. *Trends in Ecology and Evolution* 18: 159–162.

1383. Starks PT (2004) Recognition Systems. *Annales Zoologici Fennici* 41: 689–892.

1384. Stein Z, Susser M, Saenger G, Marolla F (1972) Nutrition and mental performance. *Science* 178: 708–713.

1385. Stenmark G, Slagsvold T, Lifjeld JT (1988) Polygyny in the pied flycatcher, *Ficedula hypoleuca*: A test of the deception hypothesis. *Animal Behaviour* 36: 1646–1657.

1386. Sterck EHM, Watts DP, van Schaik CP (1997) The evolution of female social relationships in nonhuman primates. *Behavioral Ecology and Sociobiology* 41: 291–310.

1387. Stern DL, Foster WA (1996) The evolution of soldiers in aphids. *Biological Reviews* 71: 27–80.

1388. Stern DL, Foster WA (1997) The evolution of sociality in aphids: A clone's-eye-view. In: Choe J, Crespi B (eds) *Social Competition and Cooperation in Insects and Arachnids: II Evolution of Sociality*. Princeton University Press, Princeton.

1389. Stevens M, Cuthill IC, Windsor AMM, Walker HJ (2006) Distruptive contrast in animal camouflage. *Proceedings of the Royal Society of London, Series B* 273: 2433–2438.

1390. Stevens M (2007) Predator perception and the interrelation between different forms of protective coloration. *Proceedings of the Royal Society of London, Series B* 274: 1457–1464.

1391. Stoltz JA, Elias DO, Andrade MCB (2008) Females reward courtship by competing males in a cannibalistic spider. *Behavioral Ecology and Sociobiology* 62: 689–697.

1392. Stoutamire WP (1974) Australian terrestrial orchids, thynnid wasps and pseudocopulation. *American Orchid Society Bulletin* 43: 13–18.

1393. Stow A, Briscoe D, Gillings M, Holley M, Smith S, Leys R, Silberbauer T, Turnbull C, Beattie A (2007) Antimicrobial defences increase with sociality in bees. *Biology Letters* 3: 422–424.

1394. Stowers L, Holy TE, Meister M, Dulac C, Koentges G (2002) Loss of sex discrimination and male-male aggression in mice deficient for *TRP2*. *Science* 295: 1493–1500.

1395. Strand CR, Small TW, Deviche P (2007) Plasticity of the Rufous-winged Sparrow, *Aimophila carpalis*, song control regions during the monsoon-associated summer breeding period. *Hormones and Behavior* 52: 401–408.

1396. Strassmann JE, Hughes CR, Queller DC, Turillazzi S, Cervo R, Davis SK, Goodnight KF (1989) Genetic relatedness in primitively eusocial wasps. *Nature* 342: 268–269.

1397. Strassmann JE (2001) The rarity of multiple mating by females in the social Hymenoptera. *Insectes Sociaux* 48: 1–13.

1398. Strum SC (1987) *Almost Human*. W.W. Norton, New York.

1399. Stuart-Fox DM, Moussalli A, Marshall NJ, Owens IPF (2003) Conspicuous males suffer higher predation risk: visual modelling and experimental evidence from lizards. *Animal Behaviour* 66: 541–550.

1400. Stumpner A, Lakes-Harlan R (1996) Auditory interneurons in a hearing fly (*Therobia leonidei*, Ormiini, Tachinidae, Diptera). *Journal of Comparative Physiology A* 178: 227–233.

1401. Stutt AD, Siva-Jothy MT (2001) Traumatic insemination and sexual conflict in the bed bug *Cimex lectularius*. *Proceedings of the National Academy of Sciences USA* 98: 5683–5687.

1402. Sullivan JP, Jassim O, Fahrbach SE, Robinson GE (2000) Juvenile hormone paces behavioral development in the adult worker honey bee. *Hormones and Behavior* 37: 1–14.

1403. Sumner S, Casiraghi M, Foster W, Field J (2002) High reproductive skew in tropical hover wasps. *Proceedings of the Royal Society of London, Series B* 269: 179–186.

1404. Sündstrom L (1994) Sex ratio bias, relatedness asymmetry and queen mating frequency in ants. *Nature* 367: 266–268.

1405. Surlykke A (1984) Hearing in notodontid moths: a tympanic organ with a single auditory neuron. *Journal of Experimental Biology* 113: 323–334.

1406. Surlykke A, Fullard JH (1989) Hearing of the Australian whistling moth, *Hecatesia thyridion*. *Naturwissenschaften* 76: 132–134.

1407. Susser M, Stein Z (1994) Timing in prenatal nutrition: A reprise of the Dutch famine study. *Nutrition Reviews* 52: 84–94.

1408. Svensson BG (1997) Swarming behavior, sexual dimorphism, and female reproductive status in the sex role-reversed dance fly species *Rhamphomyia marginata*. *Journal of Insect Behavior* 10: 783–804.

1409. Svensson GP, Löfstedt C, Skals N (2004) The odour makes the difference: male moths attracted by sex pheromones ignore the threat by predatory bats. *Oikos* 104: 91–97.

1410. Swaddle JP (1999) Limits to length asymmetry detection in starlings: Implications for biological signalling. *Proceedings of the Royal Society of London, Series B* 266: 1299–1303.

1411. Swaddle JP, Reierson GW (2002) Testosterone increases perceived dominance but not attractiveness in human males. *Proceedings of the Royal Society of London, Series B* 269: 2285–2289.

1412. Swaddle JP (2003) Fluctuating asymmetry, animal behavior, and evolution. *Advances in the Study of Behavior* 32: 169–205.

1413. Swaisgood RR, Rowe MP, Owings DH (2003) Antipredator responses of California ground squirrels to rattlesnakes and rattling sounds: the roles of sex, reproductive parity, and offspring age in assessment and decision-making rules. *Behavioral Ecology and Sociobiology* 55: 22–31.

1414. Swan LW (1970) Goose of the Himalayas. *Natural History* 79: 68–75.

1415. Sweeney BW, Vannote RL (1982) Population synchrony in mayflies: A predator satiation hypothesis. *Evolution* 36: 810–821.

1416. Symons D (1979) *The Evolution of Human Sexuality*. Oxford University Press, New York.

1417. Székely T, Thomas GH, Cuthill IC (2006) Sexual conflict, ecology, and breeding systems in shorebirds. *BioScience* 56: 801–808.

1418. Szigeti B, Török J, Hegyi G, Rosivall B, Hargitai R, Szöllõsi E, Michl G (2007) Egg quality and parental ornamentation in the blue tit *Parus caeruleus*. *Journal of Avian Biology* 38: 105–112.

1419. Taborsky M (1994) Sneakers, satellites, and helpers: Parasitic and cooperative behavior in fish reproduction. *Advances in the Study of Behavior* 23: 1–100.

1420. Taborsky M, Grantner A (1998) Behavioural time-energy budgets of cooperatively breeding *Neolamprologus pulcher* (Pisces: Cichlidae). *Animal Behaviour* 56: 1375–1382.

1421. Takahashi M, Arita H, Hiraira-Hasegawa M, Hasegawa T (2008) Peahens do not prefer peacocks with more elaborate trains. *Animal Behaviour* 75: 1209–1219.

1422. Tallamy DW (2001) Evolution of exclusive paternal care in arthropods. *Annual Review of Entomology* 46: 139–165.

1423. Tarpy DR, Nielsen DI (2002) Sampling error, effective paternity, and estimating the genetic structure of honey bee colonies (Hymenoptera: Apidae). *Annals of the Entomological Society of America* 95: 513–528.

1424. Tarpy DR (2003) Genetic diversity within honeybee colonies prevents severe infections and promotes colony growth. *Proceedings of the Royal Society of London, Series B* 270: 99–103.

1425. Taylor ML, Wedell N, Hosken DJ (2007) The heritability of attractiveness. *Current Biology* 17: R959–R960.

1426. Temrin H, Arak A (1989) Polyterritoriality and deception in passerine birds. *Trends in Ecology and Evolution* 4: 106–108.

1427. Thom MD, Hurst JL (2004) Individual recognition by scent. *Annales Zoologici Fennici* 41: 765–787.

1428. Thomas MA, Walsh KA, Wolf MR, McPheron BA, Marden JH (2000) Molecular phylogenetic analysis of evolutionary trends in stonefly wing structure and locomotor behavior. *Proceedings of the National Academy of Sciences USA* 97: 13178–13183.

1429. Thomas ML, Payne-Makrisa CM, Suarez AV, Tsutsui ND, Holway DA (2006) When supercolonies collide: territorial aggression in an invasive and unicolonial social insect. *Molecular Ecology* 15: 4303–4315.

1430. Thompson ME, Jones JH, Pusey AE, Brewer-Marsden S, Goodall J, Marsden D, Matsuzawa T, Nishida T, Reynolds V, Sugiyama Y, Wrangham RW (2007) Aging and fertility patterns in wild chimpanzees provide insights into the evolution of menopause. *Current Biology* 17: 2150–2156.

1431. Thompson ME, Kahlenberg SM, Gilby IC, Wrangham RW (2007) Core area quality is associated with variance in reproductive success among female chimpanzees at Kibale National Park. *Animal Behaviour* 73: 501–512.

1432. Thornhill R (1975) Scorpion-flies as kleptoparasites of web-building spiders. *Nature* 258: 709–711.

1433. Thornhill R (1976) Sexual selection and nuptial feeding behavior in *Bittacus apicalis* (Insecta: Mecoptera). *American Naturalist* 119: 529–548.

1434. Thornhill R (1981) *Panorpa* (Mecoptera: Panorpidae) scorpionflies: Systems for understanding resource-defense polygyny and alternative male reproductive efforts. *Annual Review of Ecology and Systematics* 12: 355–386.

1435. Thornhill R, Alcock J (1983) *The Evolution of Insect Mating Systems*. Harvard University Press, Cambridge, MA.

1436. Thornhill R, Thornhill NW (1983) Human rape: An evolutionary analysis. *Ethology and Sociobiology* 4: 137–173.

1437. Thornhill R, Gangestad SW (1999) Facial attractiveness. *Trends in Cognitive Sciences* 3: 452–460.

1438. Thornhill R, Palmer CT (2000) *A Natural History of Rape: The Biological Bases of Sexual Coercion*. MIT Press, Cambridge, MA.

1439. Thünken T, Bakker TCM, Baldauf SA, Kullmann H (2007) Active inbreeding in a cichlid fish and its adaptive significance. *Current Biology* 17: 225–229.

1440. Tibbetts EA (2002) Visual signals of individual identity in the wasp *Polistes fuscatus*. *Proceedings of the Royal Society of London, Series B* 269: 1423–1428.

1441. Tibbetts EA, Reeve HK (2003) Benefits of foundress associations in the paper wasp *Polistes dominulus*: Increased productivity and survival, but no assurance of fitness returns. *Behavioral Ecology* 14: 510–514.

1442. Tibbetts EA, Dale J (2004) A socially enforced signal of quality in a paper wasp. *Nature* 432: 218–222.

1443. Tibbetts EA (2007) Dispersal decisions and predispersal behavior in *Polistes* paper wasp 'workers.' *Behavioral Ecology and Sociobiology* 61: 1877–1883.

1444. Tinbergen N, Perdeck AC (1950) On the stimulus situations releasing the begging response in the newly hatched herring gull (*Larus argentatus* Pont.). *Behaviour* 3: 1–39.

1445. Tinbergen N (1951) *The Study of Instinct*. Oxford University Press, New York.

1446. Tinbergen N (1958) *Curious Naturalists*. Doubleday, Garden City, NY.

1447. Tinbergen N (1959) Comparative studies of the behaviour of gulls (Laridae): a progress report. *Behaviour* 15: 1–70.

1448. Tinbergen N (1960) *The Herring Gull's World*. Doubleday, Garden City, New York.

1449. Tinbergen N (1963) On the aims and methods of ethology. *Zeitschrift für Tierpsychologie* 20: 410–433.

1450. Tishkoff S, Reed F, Ranciaro A, Voight BF, Babbitt CC, Silverman JS, Powell K, Mortensen HM, Hirbo JB, Osman M, Ibrahim M, Omar S, Lema G, Nyambo TB, Ghori J, Bumpstead S, Pritchard JK, Wray GA, Deloukas P (2007) Convergent evolution of lactase persistence in Africa and Europe. *Nature Genetics* 39: 31–40.

1451. Tobias J (1997) Asymmetric territorial contests in the European robin: The role of settlement costs. *Animal Behaviour* 54: 9–21.

1452. Toh KL, Jones CR, He Y, Eide EJ, Hinz WA, Virshup DM, Ptácek LJ, Fu Y-H (2001) An h*Per*2 phosphorylation site mutation in familial advanced sleep phase syndrome. *Science* 291: 1040–1043.

1453. Toma DP, Bloch G, Moore D, Robinson GE (2000) Changes in *period* mRNA levels in the brain and division of labor in honey bee colonies. *Proceedings of the National Academy of Sciences USA* 97: 6914–6919.

1454. Tomkins JL, Simmons LW (1998) Female choice and manipulations of forceps size and symmetry in the earwig *Forficula auricularia* L. *Animal Behaviour* 56: 347–356.

1455. Tomkins JL, Hazel W (2007) The status of the conditional evolutionarily stable strategy. *Trends in Ecology and Evolution* 22: 522–528.

1456. Toth AL, Robinson GE (2007) Evo-devo and the evolution of social behavior. *Trends in Genetics* 23: 334–341.

1457. Toth AL, Varala K, Newman TC, Miguez FE, Hutchinson SK, Willoughby DA, Simons JF, Egholm M, Hunt JH, Hudson ME, Robinson GE (2007) Wasp gene expression supports an evolutionary link between maternal behavior and eusociality. *Science* 318: 441–444.

1458. Tovée MJ, Maisey DS, Emery JL, Cornelissen PL (1999) Visual cues to female physical attractiveness. *Proceedings of the Royal Society of London, Series B* 266: 211–218.

1459. Townsend JM (1989) Mate selection criteria: A pilot study. *Ethology and Sociobiology* 10: 241–253.

1460. Trainor BC, Marler CA (2002) Testosterone promotes paternal behaviour in a monogamous mammal via conversion to estrogen. *Proceedings of the Royal Society of London, Series B* 269: 823–829.

1461. Trainor BC, Bird IM, Alday NA, Schlinger BA, Marler CA (2003) Variation in aromatase activity in the medial preoptic area and plasma progesterone is associated with the onset of paternal behavior. *Neuroendocrinology* 78: 36–44.

1462. Traniello JFA, Rosengaus RB, Savoie K (2002) The development of immunity in a social insect: Evidence for the group facilitation of disease resistance. *Proceedings of the National Academy of Sciences USA* 99: 6838–6842.

1463. Tregenza T, Simmons LW, Wedell N, Zuk M (2006) Female preference for male courtship song and its role as a signal of immune function and condition. *Animal Behaviour* 72: 809–818.

1464. Trivers RL (1971) The evolution of reciprocal altruism. *Quarterly Review of Biology* 46: 35–57.

1465. Trivers RL (1972) Parental investment and sexual selection. In: Campbell B (ed) *Sexual Selection and the Descent of Man*. Aldine, Chicago.

1466. Trivers RL (1974) Parent-offspring conflict. *American Zoologist* 14: 249–264.

1467. Trivers RL, Hare H (1976) Haplodiploidy and the evolution of the social insects. *Science* 191: 249–263.

1468. Trivers RL (1985) *Social Evolution*. Benjamin Cummings, Menlo Park, CA.

1469. Tsai ML, Dai CF (2003) Cannibalism within mating pairs of the parasitic isopod, *Ichthyoxenus fushanensis*. *Journal of Crustacean Biology* 23: 662–668.

1470. Tschinkel WR, Porter SD (1988) Efficiency of sperm use in queens of the fire ant, *Solenopsis invicta* (Hymenoptera: Formicidae). *Annals of the Entomological Society of America* 81: 777–781.

1471. Tschirren B, Fitze PS, Richner H (2005) Carotenoid-based nestling colouration and parental favouritism in the great tit. *Oecologia* 143: 477–482.

1472. Tsubaki Y, Hooper RE, Siva-Jothy MT (1997) Differences in adult and reproductive lifespan in the two male forms of *Mnais pruinosa costalis* Selys (Odonata: Calopterygidae). *Research in Population Ecology* 39: 149–155.

1473. Turek FW, McMillan JP, Menaker M (1976) Melatonin: effects of the circadian rhythms of sparrows. *Science* 194: 1441–1443.

1474. Tuttle EM, Pruett-Jones S, Webster MS (1996) Cloacal protuberances and extreme sperm production in Australian fairy-wrens. *Proceedings of the Royal Society of London, Series B* 263: 1359–1364.

1475. Tybur JM, Miller GF, Gangestad SW (2007) Testing the controversy—An empirical examination of adaptationists' attitudes toward politics and science. *Human Nature* 18: 313–328.

1476. Tyler WA (1995) The adaptive significance of colonial nesting in a coral-reef fish. *Animal Behaviour* 49: 949–966.

1477. Uetz GW, Smith EI (1999) Asymmetry in a visual signaling character and sexual selection in a wolf spider. *Behavioral Ecology and Sociobiology* 45: 87–94.

1478. Urban MC (2007) Risky prey behavior evolves in risky habitats. *Proceedings of the National Academy of Sciences USA* 104: 14377–14382.

1479. Urquhart FA (1960) *The Monarch Butterfly*. University of Toronto Press, Toronto.

1480. Uy JAC, Borgia G (2000) Sexual selection drives rapid divergence in bowerbird display traits. *Evolution* 54: 273–278.

1481. Uy JAC, Patricelli GL, Borgia G (2001) Complex mate searching in the satin bowerbird *Ptilonorhynchus violaceus*. *American Naturalist* 158: 530–542.

1482. Vahed K (1998) The function of nuptial feeding in insects: Review of empirical studies. *Biological Reviews* 73: 43–78.

1483. Vallet E, Beme I, Kreutzer M (1998) Two-note syllables in canary songs elicit high levels of sexual display. *Animal Behaviour* 55: 291–297.

1484. van Gils JA, Schenk IW, Bos O, Piersma T (2003) Incompletely informed shorebirds that face a digestive constraint maximize net energy gain when exploiting patches. *American Naturalist* 161: 777–793.

1485. van Schaik CP, Kappeler PM (1997) Infanticide risk and the evolution of male-female associations in primates. *Proceedings of the Royal Society of London, Series B* 264: 1687–1694.

1486. van Staaden MJ, Romer H (1998) Evolutionary transition from stretch to hearing organs in ancient grasshoppers. *Nature* 394: 773–778.

1487. Vaughan AE (2003) The association between offender SES and victim-offender relationship in rape offences—revised. *Sexualities, Evolution & Gender* 5: 103–105.

1488. Veiga JP (2003) Infanticide by male house sparrows: gaining time or manipulating females? *Proceedings of the Royal Society of London, Series B* 270: S87–S89.

1489. Velho TAF, Pinaud R, Rodigues PV, Mello CV (2005) Co-induction of activity-dependent genes in songbirds. *European Journal of Neuroscience* 22: 1667–1678.

1490. Vereecken NJ, Mahé G (2007) Larval aggregations of the blister beetle *Stenoria analis* (Schaum) (Coleoptera: Meloidae) sexually deceive patrolling males of their host, the solitary bee *Colletes hederae* Schmidt & Westrich (Hymenoptera: Colletidae). *Annales de la Société Entomologique de France* 43: 493–496.

1491. Vieites DR, Nieto-Román S, Barluenga M, Palanca A, Vences M, Meyer A (2004) Post-mating clutch piracy in an amphibian. *Nature* 431: 305–308.

1492. Vincent ACJ, Sadler LM (1995) Faithful pair bonds in wild seahorses, *Hippocampus whitei*. *Animal Behaviour* 50: 1557–1569.

1493. Vincent ACJ, Marsden AD, Evans KL, Sadler LM (2004) Temporal and spatial opportunities for polygamy in a monogamous seahorse, *Hippocampus whitei*. *Behaviour* 141: 141–156.

1494. Vining DR, Jr. (1986) Social versus reproductive success: The central theoretical problem of human sociobiology. *Behavioral and Brain Sciences* 9: 167–187.

1495. Visscher PK (2003) How self-organization evolves. *Nature* 421: 799–800.

1496. Vitousek MN, Mitchell MA, Woakes AJ, Niemack MD, Wikelski M (2007) High costs of female choice in a lekking lizard. *PLoS ONE* 2: e567. doi:510.1371/journal. pone.0000567.

1497. Vleck CM, Brown JL (1999) Testosterone and social and reproductive behaviour in *Aphelocoma* jays. *Animal Behaviour* 58: 943–951.

1498. von Engelhardt N, Kappeler PM, Heistermann M (2000) Androgen levels and female social dominance in *Lemur catta*. *Proceedings of the Royal Society of London, Series B* 267: 1533–1539.

1499. von Frisch K (1956) *The Dancing Bees*. Harcourt Brace Jovanovich, New York.

1500. von Frisch K (1967) *The Dance Language and Orientation of Bees*. Harvard University Press, Cambridge, MA.

1501. Waage JK (1973) Reproductive behavior and its relation to territoriality in *Calopteryx maculata* (Beauvois) (Odonata:Calopterygidae). *Behaviour* 47: 240–256.

1502. Waage JK (1979) Dual function of the damselfly penis: Sperm removal and transfer. *Science* 203: 916–918.

1503. Waage JK (1997) Parental investment—minding the kids or keeping control? In: Gowaty PA (ed) *Feminism and Evolutionary Biology: Boundaries, Interactions, and Frontiers*. Chapman and Hall, New York.

1504. Wade J (2005) Current research on the behavioral neuroendocrinology of reptiles. *Hormones and Behavior* 48: 451–460.

1505. Wagner RH, Helfenstein F, Danchin E (2004) Female choice of young sperm in a genetically monogamous bird. *Proceedings of the Royal Society of London, Series B* 271: S134–S137.

1506. Walcott C (1972) Bird navigation. *Natural History* 81: 32–43.

1507. Walcott C (1996) Pigeon homing: Observations, experiments and confusions. *Journal of Experimental Biology* 199: 21–27.

1508. Wallraff HG, Chappel J, Guilford T (1999) The roles of the sun and the landscape in pigeon homing. *Journal of Experimental Biology* 202: 2121–2126.

1509. Walter C (2008) Affairs of the lips: why we kiss. *Scientific American* 19 (Jan.): 24–29.

1510. Walther BA, Gosler AG (2001) The effects of food availability and distance to protective cover on the winter foraging behaviour of tits (Aves: *Parus*). *Oecologia* 129: 312–320.

1511. Ward MP, Weatherhead PJ (2005) Sex-specific differences in site fidelity and the cost of dispersal in yellow-headed blackbirds. *Behavioral Ecology and Sociobiology* 59: 108–114.

1512. Ward P, Zahavi A (1973) The importance of certain assemblages of birds as "information-centres" for food finding. *Ibis* 115: 517–534.

1513. Ward PI (2007) Postcopulatory selection in the yellow dung fly *Scathophaga stercoraria* (L.) and the mate-now-choose-later mechanism of cryptic female choice. *Advances in the Study of Behavior* 37: 343–369.

1514. Warner RR (1984) Mating behavior and hermaphroditism in coral reef fishes. *American Scientist* 72: 128–136.

1515. Watson PJ (1998) Multi-male mating and female choice increase offspring growth in the spider *Neriene litigiosa* (Linyphiidae). *Animal Behaviour* 55: 387–403.

1516. Watt PJ, Chapman R (1998) Whirligig beetle aggregations: What are the costs and the benefits? *Behavioral Ecology and Sociobiology* 42: 179–184.

1517. Watts DP, Muller M, Amsler SJ, Mbabazi G, Mitani JC (2006) Lethal group aggression by chimpanzees in Kibale National Park, Uganda. *American Journal of Primatology* 68: 161–180.

1518. Watts HE, Holekamp KE (2007) Hyena societies. *Current Biology* 17: R657–R660.

1519. Watts HE, Holekamp KE (2009) Ecological determinants of survival and reproduction in the spotted hyena. *Journal of Mammalogy.* In press.

1520. Waynforth D, Dunbar RIM (1995) Conditional mate choice strategies in humans—evidence from lonely hearts advertisements. *Behaviour* 132: 755–779.

1521. Weathers WW, Sullivan KA (1989) Juvenile foraging proficiency, parental effort, and avian reproductive success. *Ecological Monographs* 59: 223–246.

1522. Webster MS (1994) Female-defence polygyny in a Neotropical bird, the Montezuma oropendula. *Animal Behaviour* 48: 779–794.

1523. Webster MS, Robinson SK (1999) Courtship disruptions and male mating strategies: Examples from female-defense mating systems. *American Naturalist* 154: 717–729.

1524. Webster MS, Reichart L (2005) Use of microsatellites for parentage and kinship analyses in animals. *Molecular Evolution* 395: 222–238.

1525. Webster MS, Tarvin KA, Tuttle EM, Pruett-Jones S (2007) Promiscuity drives sexual selection in a socially monogamous bird. *Evolution* 61: 2205–2211.

1526. Wedekind C, Milinski M (1996) Human cooperation in the simultaneous and the alternating Prisoner's Dilemma: Pavlov versus Generous Tit-for-Tat. *Proceedings of the National Academy of Sciences USA* 93: 2686–2689.

1527. Wedell N, Tregenza T (1999) Successful fathers sire successful sons. *Evolution* 53: 620–625.

1528. Wehner R, Wehner S (1990) Insect navigation: Use of maps or Ariadne's thread? *Ethology, Ecology and Evolution* 2: 27–48.

1529. Wehner R, Lehrer M, Harvey WR (1996) Navigation: Migration and homing. *Journal of Experimental Biology* 199: 1–261.

1530. Weidinger K (2000) The breeding performance of blackcap *Sylvia atricapilla* in two types of forest habitat. *Ardea* 88: 225–233.

1531. Weimerskirch H, Salamolard M, Sarrazin F, Jouventin P (1993) Foraging strategy of wandering albatrosses through the breeding season: A study using satellite telemetry. *Auk* 110: 325–342.

1532. Weimerskirch H, Martin J, Clerquin Y, Alexandre P, Jiraskova S (2001) Energy saving in flight formation. *Nature* 413: 697–698.

1533. Weiss MR (2003) Good housekeeping: why do shelter-dwelling caterpillars fling their frass? *Ecology Letters* 6: 361–370.

1534. Wells JCK (2003) Parent-offspring conflict theory, signaling of need, and weight gain in early life. *Quarterly Review of Biology* 78: 169–202.

1535. Welty J (1982) *The Life of Birds*, 3rd edn. Saunders College Publishing, Philadelphia, PA.

1536. Wenner AM, Wells P (1990) *Anatomy of a Controversy.* Columbia University Press, New York.

1537. Wenninger EJ, Averill AL (2006) Influence of body and genital morphology on relative male fertilization success in oriental beetle. *Behavioral Ecology* 17: 656–663.

1538. Wenseleers T, Ratnieks FLW (2006) Enforced altruism in insect societies. *Nature* 444: 50.

1539. Wenseleers T, Ratnieks FLW (2006) Comparative analysis of worker reproduction and policing in eusocial Hymenoptera supported relatedness theory. *American Naturalist* 168: E163–E179.

1540. Werner NY, Balshine S, Leach B, Lotem A (2003) Helping opportunities and space segregation in cooperatively breeding cichlids. *Behavioral Ecology* 14: 749–756.

1541. West-Eberhard MJ (1975) The evolution of social behavior by kin selection. *Quarterly Review of Biology* 50: 1–33.

1542. West-Eberhard MJ (1979) Sexual selection, social competition, and evolution. *Proceedings of the American Philosophical Society* 123: 222–234.

1543. West-Eberhard MJ (2003) *Developmental Plasticity and Evolution.* Oxford University Press, New York.

1544. West MJ, King AP (1990) Mozart's starling. *American Scientist* 78: 106–114.

1545. West PM, Packer C (2002) Sexual selection, temperature, and the lion's mane. *Science* 297: 1339–1343.

1546. West SA, Griffiths SW, Gardner A (2008) Social semantics: how useful has group selection been? *Journal of Evolutionary Biology* 21: 374–385.

1547. Westneat DF, Stewart IRK (2003) Extra-pair paternity in birds: Causes, correlates, and conflict. *Annual Review of Ecology and Systematics* 34: 365–396.

1548. Wharton KE, Dyer FC, Huang ZY, Getty T (2007) The honeybee queen influences the regulation of colony drone production. *Behavioral Ecology* 18: 1092–1099.

1549. White PA (2008) Maternal response to neonatal sibling conflict in the spotted hyena, *Crocuta crocuta. Behavioral Ecology and Sociobiology* 62: 353–361.

1550. White SA, Nguyen T, Fernald RD (2002) Social regulation of gonadotropin-releasing hormone. *Journal of Experimental Biology* 205: 2567–2581.

1551. Whiteman EA, Côté IM (2004) Monogamy in marine fishes. *Biological Reviews* 79: 351–375.

1552. Whitfield CW, Cziko A-M, Robinson GE (2004) Gene expression profiles in the brain predict behavior in individual honey bees. *Science* 302: 296–299.

1553. Whitfield CW, Ben-Shahar Y, Brillet C, Leoncini I, Crauser D, LeConte Y, Rodriguez-Zas S, Robinson GE (2006) Genomic dissection of behavioral maturation in the honey bee. *Proceedings of the National Academy of Sciences USA* 103: 16068–16075.

1554. Whitfield DP (1990) Individual feeding specializations of wintering turnstone *Arenaria interpres. Journal of Animal Ecology* 59: 193–211.

1555. Whitham TG (1979) Habitat selection by *Pemphigus* aphids in response to resource limitation and competition. *Ecology* 59: 1164–1176.

1556. Whitham TG (1979) Territorial defense in a gall aphid. *Nature* 279: 324–325.

1557. Whitham TG (1980) The theory of habitat selection examined and extended using *Pemphigus* aphids. *American Naturalist* 115: 449–466.

1558. Whitham TG (1986) Costs and benefits of territoriality: Behavioral and reproductive release by competing aphids. *Ecology* 67: 139–147.

1559. Whiting MJ (1999) When to be neighbourly: Differential agonistic responses in the lizard *Platysaurus broadleyi*. *Behavioral Ecology and Sociobiology* 46: 210–214.

1560. Whittingham LA, Dunn PO, Macgrath RD (1997) Relatedness, polyandry and extra-group paternity in the cooperatively-breeding white-browed scrubwren (*Sericornis frontalis*). *Behavioral Ecology and Sociobiology* 40: 261–270.

1561. Whittingham LA, Dunn PO (1998) Male parental effort and paternity in a variable mating system. *Animal Behaviour* 55: 629–640.

1562. Wickler W (1968) *Mimicry in Plants and Animals*. World University Library, London.

1563. Wickler W, Seibt U (1981) Monogamy in Crustacea and man. *Zeitschrift für Tierpsychologie* 57: 215–234.

1564. Widdig A, Bercovitch FB, Streich WJ, Sauermann U, Nürnberg P, Krawczak M (2004) A longitudinal analysis of reproductive skew in male rhesus macaques. *Proceedings of the Royal Society of London, Series B* 271: 819–826.

1565. Widemo F, Owens IPF (1995) Lek size, male mating skew and the evolution of lekking. *Nature* 373: 148–151.

1566. Wiederman MW, Allgeier ER (1992) Gender differences in mate selection criteria: Sociobiological or socioeconomic explanation? *Ethology and Sociobiology* 13: 115–124.

1567. Wiederman MW, Kendall E (1999) Evolution, sex, and jealousy: Investigation with a sample from Sweden. *Evolution and Human Behavior* 20: 121–128.

1568. Wiens JJ (2001) Widespread loss of sexually selected traits: how the peacock lost its spots. *Trends in Ecology and Evolution* 19: 517–523.

1569. Wiersma P, Verhulst S (2005) Effects of intake rate on energy expenditure, somatic repair and reproduction of zebra finches. *Journal of Experimental Biology* 208: 4091–4098.

1570. Wigby S, Chapman T (2004) Female resistance to male harm evolves in response to manipulation of sexual conflict. *Evolution* 58: 1028–1037.

1571. Wikelski M, Baurle S (1996) Pre-copulatory ejaculation solves time constraints during copulations in marine iguanas. *Proceedings of the Royal Society of London, Series B* 263: 439–444.

1572. Wiklund C, Karlsson B, Leimar O (2001) Sexual conflict and cooperation in butterfly reproduction: a comparative study of polyandry and female fitness. *Proceedings of the Royal Society of London, Series B* 268: 1661–1667.

1573. Wiklund CG, Andersson M (1994) Natural selection of colony size in a passerine bird. *Journal of Animal Ecology* 63: 765–774.

1574. Wilbrecht L, Crionas A, Nottebohm F (2002) Experience affects recruitment of new neurons but not adult neuron number. *Journal of Neuroscience* 22: 825–831.

1575. Wilkinson GS (1984) Reciprocal food sharing in the vampire bat. *Nature* 308: 181–184.

1576. Wilkinson GS, Dodson GN (1997) Function and evolution of antlers and eye stalks in flies. In: Choe JC, Crespi BJ (eds) *The Evolution of Mating Systems in Insects and Arachnids*. Cambridge University Press, Cambridge.

1577. Williams CK, Lutz RS, Applegate RD (2003) Optimal group size and northern bobwhite coveys. *Animal Behaviour* 66: 377–387.

1578. Williams GC (1966) *Adaptation and Natural Selection*. Princeton University Press, Princeton, N J.

1579. Williams GC (1975) *Sex and Evolution*. Princeton University Press, Princeton, NJ.

1580. Williams GC (1996) *The Pony Fish's Glow*. Basic Books, New York.

1581. Williams JM, Oehlert GW, Carlis JV, Pusey AE (2004) Why do male chimpanzees defend a group range? *Animal Behaviour* 68: 523–532.

1582. Williams TC, Williams JM (1978) An oceanic mass migration of land birds. *Scientific American* 239 (Oct.): 166–176.

1583. Willows AOD (1971) Giant brain cells in mollusks. *Scientific American* 224 (Feb.): 68–75.

1584. Wilmer JW, Allen PJ, Pomeroy PP, Twiss SD, Amos W (1999) Where have all the fathers gone? An extensive microsatellite analysis of paternity in the grey seal (*Halichoerus grypus*). *Molecular Ecology* 8: 1417–1429.

1585. Wilson AB, Martin-Smith KM (2007) Genetic monogamy despite social promiscuity in the pot-bellied seahorse (*Hippocampus abdominalis*). *Molecular Ecology* 16: 2345–2352.

1586. Wilson DS, Wilson EO (2007) Rethinking the theoretical foundation of sociobiology. *Quarterly Review of Biology* 82: 327–348.

1587. Wilson EO (1971) *The Insect Societies*. Harvard University Press, Cambridge, MA.

1588. Wilson EO (1975) *Sociobiology, The New Synthesis*. Harvard University Press, Cambridge, MA.

1589. Wilson EO (1976) Academic vigilantism and the political significance of sociobiology. *BioScience* 26: 187–190.

1590. Wilson EO, Holldobler B (2005) Eusociality: Origin and consequences. *Proceedings of the National Academy of Sciences USA* 102: 13367–13371.

1591. Wilson M, Daly M, Gordon S (1998) The evolved psychological apparatus of humans is one source of environmental problems. In: Caro T (ed) *Behavioral Ecology and Conservation Biology*. Oxford University Press, New York, NY.

1592. Wilson MI, Daly M (1997) Life expectancy, economic inequality, homicide, and reproductive timing in Chicago neighborhoods. *British Medical Journal* 314: 1271–1274.

1593. Wilson RS, Angelitta MJ, Jr, James RS, Navas C, Seebacher F (2007) Dishonest signals of strength in male slender crayfish (*Cherax dispar*) during agonistic encounters. *American Naturalist* 170: 284–291.

1594. Wiltschko R, Wiltschko W (2003) Avian navigation: from historical to modern concepts. *Animal Behaviour* 65: 257–272.

1595. Windmill JFC, Jackson JC, Tuck EJ, Robert D (2006) Keeping up with bats: Dynamic auditory tuning in a moth. *Current Biology* 16: 2418–2423.

1596. Windmill JFC, Fullard JH, Robert D (2007) Mechanics of a 'simple' ear: tympanal vibrations in a noctuid moth. *Journal of Experimental Biology* 210: 2637–2548.

1597. Winfree R (1999) Cuckoos, cowbirds and the persistence of brood parasitism. *Trends in Ecology and Evolution* 14: 338–343.

1598. Wingfield JC, Moore MC (1987) Hormonal, social and environmental factors in the reproductive biology of free-living male birds. In: Crews D (ed) *Psychobiology of Reproductive Behavior: An Evolutionary Perspective*. Prentice-Hall, Englewood Cliffs, NJ.

1599. Wingfield JC, Jacobs J, Hillgarth N (1997) Ecological constraints and the evolution of hormone-behavior interrelationships. *Annals of the New York Academy of Sciences* 807: 22–41.

1600. Wingfield JC, Ramenofsky M (1997) Corticosterone and facultative dispersal in response to unpredictable events. *Ardea* 85: 155–166.

1601. Winker K, Pruett CL (2006) Seaonal migration, speciation, and morphological convergence in the genus *Catharus* (Turdidae). *Auk* 123: 1052–1068.

1602. Winking J, Kaplan H, Gurven M, Rucas S (2007) Why do men marry and why do they stray? *Proceedings of the Royal Society of London, Series B* 274: 1643–1649.

1603. Winkler SM, Wade J (1998) Aromatase activity and regulation of sexual behaviors in the green anole lizard. *Physiology & Behavior* 64: 723–731.

1604. Winterer G, Goldman D (2003) Genetics of human prefrontal function. *Brain Research Reviews* 43: 134–163.

1605. Winterhalder B, Smith EA (2000) Analyzing adaptive strategies: Human behavioral ecology at twenty-five. *Evolutionary Anthropology* 9: 51–72.

1606. Wirsing AJ, Heithaus MR, Dill LM (2007) Can you dig it? Use of excavation, a risky foraging tactic, by dugongs is sensitive to predation danger. *Animal Behaviour* 74: 1085–1091.

1607. Wise KK, Conover MR, Knowlton FF (1999) Response of coyotes to avian distress calls: Testing the startle-predator and predator-attraction hypotheses. *Behaviour* 136: 935–949.

1608. Wojcieszek JM, Nicholls JA, Goldizen AW (2007) Stealing behavior and the maintenance of a visual display in the satin bowerbird. *Behavioral Ecology* 18: 689–695.

1609. Wolanski E, Gereta E, Borner M, Mduma S (1999) Water, migration and the Serengeti ecosystem. *American Scientist* 87: 526–533.

1610. Wolf N (1990) *The Beauty Myth*. Chatto & Windus, London.

1611. Wolff JO, Mech SG, Dunlap AS, Hodges KE (2002) Multi-male mating by paired and unpaired female prairie voles (*Microtus ochrogaster*). *Behaviour* 139: 1147–1160.

1612. Wolff JO, Macdonald DW (2004) Promiscuous females protect their offspring. *Trends in Ecology and Evolution* 19: 127–134.

1613. Wolff JO, Sherman PW (eds) (2007) *Rodent Societies, An Ecological & Evolutionary Perspective*. University of Chicago Press, Chicago, IL.

1614. Wong MYL, Munday PL, Buston PM, Jones GP (2008) Monogamy when there is potential for polygyny: tests of multiple hypotheses in a group-living fish. *Behavioral Ecology* 19: 353–361.

1615. Woodroffe R, Vincent A (1994) Mother's little helpers: Patterns of male care in mammals. *Trends in Ecology and Evolution* 9: 294–297.

1616. Woods WA, Hendrickson H, Mason J, Lewis SM (2007) Energy and predation costs of firefly courtship signals. *American Naturalist* 170: 702–708.

1617. Woodward J, Goodstein D (1996) Conduct, misconduct and the structure of science. *American Scientist* 84: 479–490.

1618. Woolfenden GE, Fitzpatrick JW (1984) *The Florida Scrub Jay: Demography of a Cooperative-Breeding Bird*. Princeton University Press, Princeton, NJ.

1619. Woolley SC, Doupe AJ (2008) Social context-induced song variation affects female behavior and gene expression. *PLoS Biology* 6: doi:10.1371/journal.pbio.0060062.

1620. Woyciechowski M, Kabat L, Król E (1994) The function of the mating sign in honey bees, *Apis mellifera* L.: New evidence. *Animal Behaviour* 47: 733–735.

1621. Wrangham RW (1999) Evolution of coalitionary killing. *Yearbook of Physical Anthropology* 42: 1–30.

1622. Wynne-Edwards VC (1962) *Animal Dispersion in Relation to Social Behaviour*. Oliver & Boyd, Edinburgh.

1623. Yack JE, Fullard JH (1990) The mechanoreceptive origin of insect tympanal organs: A comparative study of similar nerves in tympanate and atympanate moths. *Journal of Comparative Neurology* 300: 523–534.

1624. Yack JE (1992) A multiterminal stretch receptor, chordotonal organ, and hair plate at the wing-hinge of *Manduca sexta*: Unravelling the mystery of the noctuid moth ear B cell. *Journal of Comparative Neurology* 324: 500–508.

1625. Yack JE, Fullard JH (2000) Ultrasonic hearing in nocturnal butterflies. *Nature* 403: 265–266.

1626. Yack JE, Kalko JEV, Surlykke A (2007) Neuroethology of ultrasonic hearing in nocturnal butterflies (Hedyloidea). *Journal of Comparative Physiology A* 193: 577–590.

1627. Yager DD, May ML (1990) Ultrasound-triggered, flight-gated evasive maneuvers in the flying praying mantis, *Parasphendale agrionina*. II. Tethered flight. *Journal of Experimental Biology* 152: 41–58.

1628. Yapici N, Kim Y-J, Ribiero C, Dickson BJ (2008) A receptor that mediates the post-mating switch in *Drosophila* reproductive behaviour. *Nature* 451: 33–38.

1629. Yoder JM, Marschall EA, Swanson DA (2004) The cost of dispersal: predation as a function of movement and site familiarity in ruffed grouse. *Behavioral Ecology* 15: 469–476.

1630. Yom-Tov Y, Geffen E (2006) On the origin of brood parasitism in altricial birds. *Behavioral Ecology* 17: 196–205.

1631. Young AJ, Carlson AA, Monfort SL, Russell AF, Bennett NC, Clutton-Brock TH (2006) Stress and the suppression of subordinate reproduction in cooperatively breeding meerkats. *Proceedings of the National Academy of Sciences USA* 103: 12005–12010.

1632. Young LJ, Wang Z (2004) The neurobiology of pair bonding. *Nature Neuroscience* 7: 1048–1054.

1633. Young MW (2000) Marking time for a kingdom. *Science* 288: 451–453.

1634. Zach R (1979) Shell-dropping: Decision-making and optimal foraging in northwestern crows. *Behaviour* 68: 106–117.

1635. Zahavi A (1975) Mate selection—A selection for a handicap. *Journal of Theoretical Biology* 53: 205–214.

1636. Zedrosser A, Støen O-G, Saebø S, Swenson JR (2007) Should I stay or should I go? Natal dispersal in the brown bear. *Animal Behaviour* 74: 369–376.

1637. Zeh JA, Zeh DW (1996) The evolution of polyandry I: Intragenomic conflict and genetic incompatibility. *Proceedings of the Royal Society of London, Series B* 263: 1711–1717.

1638. Zeh JA (1997) Polyandry and enhanced reproductive success in the harlequin beetle-riding pseudoscorpion. *Behavioral Ecology and Sociobiology* 40: 111–118.

1639. Zeh JA, Zeh DW (1997) The evolution of polyandry II: Post-copulatory defenses against genetic incompatibility. *Proceedings of the Royal Society of London, Series B* 264: 69–75.

1640. Zeh JA, Newcomer SD, Zeh DW (1998) Polyandrous females discriminate against previous mates. *Proceedings of the National Academy of Sciences USA* 95: 13273–13736.

1641. Zeh JA, Zeh DW (2003) Toward a new sexual selection paradigm: Polyandry, conflict and incompatibility. *Ethology* 109: 929–950.

1642. Zera AJ, Denno RF (1997) Physiology and ecology of dispersal polymorphism in insects. *Annual Review of Entomology* 42: 207–231.

1643. Zera AJ, Potts J, Kobus K (1998) The physiology of life-history trade-offs: Experimental analysis of a hormonally induced life-history trade-off in *Gryllus assimilis*. *American Naturalist* 152: 7–23.

1644. Zera AJ, Zhao Z, Kaliseck K (2007) Hormones in the field: evolutionary endocrinology of juvenile hormone and ecdysteroids in field populations of the wing-dimorphic cricket *Gryllus firmus*. *Physiological and Biochemical Zoology* 80: 592–606.

1645. Zhu H, Sauman I, Yuan A, Emery-Le M, Emery P, Reppert SM (2008) Cryptochromes define a novel circadian clock mechanism in monarch butterflies that may underlie sun compass navigation. *PLoS Biology* 6: 1–18.

1646. Ziegler HP, Marler P (eds) (2008) *Neuroscience of Birdsong*. Cambridge University Press, New York.

1647. Zucker I (1983) Motivation, biological clocks and temporal organization of behavior. In: Satinoff E, Teitelbaum P (eds) *Handbook of Behavioral Neurobiology: Motivation*. Plenum Press, New York.

1648. Zuk M, Simmons LW, Cupp L (1993) Calling characteristics of parasitized and unparasitized populations of the field cricket *Teleogryllus oceanicus*. *Behavioral Ecology and Sociobiology* 33: 339–343.

1649. Zuk M, Johnsen TS, MacLarty T (1995) Endocrine-immune interactions, ornaments and mate choice in red jungle fowl. *Proceedings of the Royal Society of London, Series B* 260: 205–210.

1650. Zuk M, Kolluru GR (1998) Exploitation of sexual signals by predators and paraistoids. *Quarterly Review of Biology* 73: 415–438.

1651. Zuk M, Rotenberry JT, Tinghitella RM (2006) Silent night: adaptive disappearance of a sexual signal in a parasitized population of field crickets. *Biology Letters* 2: 521–524.

# Illustration Credits

# Index

Hippocampus
  behavioral differences in birds, 72–73
  in brown-headed cowbirds, 102
  navigation in humans, 137–138, *139*
  sex differences in birds, *102*
*Hippocampus whitei*, 381–382
Hirundidae, 194
Högstedt, Goran, 210, 211
Holekamp, Kay, 291
Holland, Brett, 372
Home range, 275
Homeobox genes, 87
Homicide, 531, 542
Homing pigeon, 139–140, 141
Homosexual women, 530
Honest signals
  in disputes, 320–324
  nestlings and, 312–313, 314
Honey bee dances
  adaptive value, 242–243
  evolutionary history, *246*
  nest selection and, 252–253, *254*
  origin and modification, 243–246
  overview, 238–241
Honey bees
  clock genes, 158–159
  communal defense, 198
  development of behavior, 64–67
  *for* gene, 87–88
  habitat selection and hive splitting, 252–253, *254*
  learning, 98–99
  mating system, 379–380
  navigation, 139–140
  obligate altruism, 490–491
  polyandry, 400–401
  response to fungal pathogens, 461
Hoopoe, 265
Hoover, Jeffrey, 441
Hopi people, 237
Hoplophorionini, *423*
Hori, Michio, 229, 230
Hormones
  reproductive behavior and, 172–175
  *See also* Estrogen; Sex hormones; Testosterone
Horned pig, *343*
Horned scarab beetle, 347–348
Horse lubber grasshopper, *203*
Horses, 227
Horseshoe crab
  book gills, 296, *297*
  explosive breeding assemblage, 410
  overharvesting, 222
  satellite males, 348–349
Horsfield's bronze-cuckoo, 442, *443*
*Horvathinia, 429*
Hotshot hypothesis, 412–414
Hotspot hypothesis, 412–414
House mice
  artificial selection, 17–18
  infanticide, 169–170
  odor cues and mating preferences, 168–169
  social environment and behavioral priorities, 168–170
  *See also* Mice
House sparrow, 449
House wren, 380
Household income
  human mate preferences and, 522, 523–524
  *See also* Wealth

Howlett, R. J., 201
*Hox* genes, 87
Hoy, Ronald, 121, 126
Hrdy, Sarah, 20
*5-HTT* gene, 87
Hughes, William, 498
Human behavior
  adaptationist view, 508–518
  adaptive mate preferences (*see* Adaptive mate preferences)
  adaptive parental care, 535–540
  arbitrary culture theory, 513–516
  cultural evolution theory, 517–518
  evolutionary psychology, 541–545
  indirect reciprocity, 508–510
  irrational, 517–518
  marriage, 538–540
  sociobiology controversy, 510–513
Human behavioral ecology, 511
Humans
  adaptive mechanisms of perception, 134–136
  cortical sensory map, *131*
  eating of dirt, 237
  face recognition, 135–136
  facial symmetry and attractiveness, 92, *93*
  intelligence, 86–87
  language acquisition and, 87
  lipreading, 134–135
  navigation, 137–139
  nutritional deprivation and intellectual development, 89–91
  single-gene effects in, 86–87
  spatial learning, 101
  spices and, 236–237
  twin studies on verbal and spatial cognitive abilities, *80*
Hummingbirds, 43–45, *46*
Huxley, T. H., 512
Hyaenidae, 288
*Hydrocyrius, 429*
*Hylopsar*
  *H. cupreocauda, 482*
  *H. purpureiceps, 482*
*Hymenocera picta*, 381
Hymenoptera
  evolution of eusociality in, 498, *499*
  haplodiploid sex determination and eusociality, 491–497
  overview of haplodiploid sex determination, *483*
  relatedness, 491, 492, *493*, 495, 498
*Hypargos niveoguttatus, 439*
*Hypergerus atriceps, 439*
*Hypolimnas bolina, 360*
Hypotheses, testing, 12–13, 22–24
*Hypsoprora, 423*
Hypsoprorini, *423*

Iberian rock lizard, 91–92
*Ichthyoxenus fushanensis*, 375
Ideal free distribution theory, 252
"If–then logic," 12–13
Illegitimate receivers, 317–319
Illegitimate signalers, 324–325, *326*
Immune system
  in polyandrous species, 393–394, 398
  testosterone and, 176
Imprinting, 69–70
Inbreeding, 398
Inbreeding depression, 259–260, 261

Inclusive fitness, *471*
  concept of, 473–474
  defined, 473
  helpers at the nest (*see* Helpers at the nest)
  pied kingfisher studies, 474–476
Income, human mate preferences and, 522, 523–524
Indigo bunting, 408
Indirect fitness, *471*, 473, 514
Indirect reciprocity, 508–510
Indirect selection, 471–473
*Indri indri, 386*
Infanticide
  evolutionary theory, 19–21
  group selection theory, 21–22
  Hanuman langurs, 19–23, 24
  house mice, 169–170
  human, 535
  lions, 23–24
  male monogamy and protective behavior, 385
  testing alternative hypotheses, 22–24
Information center hypothesis, 232
Innate releasing mechanism, 110
Insects
  helpers at the nest, 483–488
  honest signals in disputes, 321–322
  wings, 296–299
Instinct, 110
Instinct theory
  code breaking and, 111–112
  overview, 110
  int-1 sensory interneuron, 121–122
Intelligence
  human, 86–87
  maternal undernutrition and, 89–91
Interneurons, 115
Interspecific brood parasitism
  adaptive responses to, 439–444
  overview, 435–438
Intrasexual selection, 342
*Iridomyrmex, 499*
Irrational behavior, in humans, 517–518
Isolation experiments, 90, *91*
Ivy bee, 111

Jamieson, Ian, 476
Japanese quail, 171–172, 174
Jays
  helpers at the nest, 477, *478*, 479
  spatial learning, 99–100
  *See also specific kinds*
Jealousy, sexual, 531
Jeon, Joonghwan, 514
Jiguet, Frédéric, 412
Jumping spiders, 206
Junco, 177
*Junco hyemalis, 439*
Juvenile hormone
  circadian rhythms in crickets, 156, 157
  flight behavior in winged crickets, 156
  honey bees, 67

Kaas, Jon, 129
Kafue lechwe, *415*
Kamil, Alan, 202
Kangaroo paw flowers, 339
Katydid
  lunar cycle and, 162
  parasitoid fly, 127
  reproductive behavior, 335, 339

**About the Book**

Editor: Graig Donini
Production Editor: Laura Green
Copy Editor: Lou Doucette
Indexer: Grant Hackett
Production Manager: Christopher Small
Book Design and Production: Joanne Delphia
Illustration Program: Elizabeth Morales